Chhorn E. Lim, PhD
Carl D. Webster, PhD
Editors

Tilapia
Biology, Culture, and Nutrition

Pre-publication
REVIEWS,
COMMENTARIES,
EVALUATIONS . . .

"The incredible expansion of worldwide production of tilapia over the past two decades has been fueled by an exploding demand. Scientists and producers have responded by accumulating vast amounts of knowledge to refine and intensify production methods. This book thoroughly reviews all of the critical topics relevant to tilapia aquaculture and presents the latest information attained by scientists and producers worldwide. It is the most complete compilation of information on tilapia aquaculture to date. Each chapter details the history of the topic from the early efforts to the techniques currently utilized by the industry in different regions of the world. This book is a must for everyone with interest in tilapia aquaculture."

Cortney L. Ohs, PhD
Assistant Professor of Aquaculture
University of Florida

"*Tilapia: Biology, Culture, and Nutrition* is, without a doubt, the most comprehensive publication produced to date on the culture of this intriguing and very important group of culture species. The editors, who are well known for their research contributions on tilapia culture, have solicited contributions to this important book from a number of other tilapia experts. Every chapter contains a wealth of information and the chapter on culture of tilapia in saline water is particularly impressive with respect to the amount of material it contains. Research on tilapia in marine culture systems, while having been ongoing for several years, has not, to my knowledge, previously been so comprehensively reviewed. The book could easily have been titled: *Everything You Ever Wanted to Know About Tilapia Culture.* It will undoubtedly be used by students, educators, researchers, and tilapia culturists for many years to come. I certainly plan to keep a copy close at hand."

Robert R. Stickney, PhD
Director, Texas Sea Grant College Program
and Professor of Oceanography
Texas A&M University

More pre-publication
REVIEWS, COMMENTARIES, EVALUATIONS . . .

"*Tilapia: Biology, Culture, and Nutrition* is a splendid mix of pure and applied science on commonly cultured tilapia. It provides an up-to-date review of the scientific literature of tilapia ranging from its biology to culture in the full range of fresh- and saltwater culture systems from extensive to super-intensive. Both the pure scientist and the culturist in the field will find a thorough review of tilapia information. The authors include many of those currently involved in aquaculture research and production, including H. R. Schmittou and his hands-on cage culture development work in China. The information on cage culture will be hard to find elsewhere. Whether you're interested in tilapia hormones, nutrition, culture, diseases, or marketing, this book should complete your collection on tilapia."

William H. Daniels, PhD
Associate Professor,
Auburn University

"This book is a valuable addition to the tilapia culture literature. The text is quite informative . . . nontraditional information is presented. In particular, information on flowing water and cage culture, and BMPs for culture are important additions to tilapia culture knowledge. The segment on the effect of unstable environments and culture ponds on sexual maturity are particularly informative."

Joseph E. Morris, PhD
Associate Professor,
Iowa State University

Food Products Press®
An Imprint of The Haworth Press, Inc.
New York • London • Oxford

NOTES FOR PROFESSIONAL LIBRARIANS AND LIBRARY USERS

This is an original book title published by Food Products Press, an imprint of The Haworth Press, Inc. Unless otherwise noted in specific chapters with attribution, materials in this book have not been previously published elsewhere in any format or language.

CONSERVATION AND PRESERVATION NOTES

All books published by The Haworth Press, Inc. and its imprints are printed on certified pH neutral, acid-free book grade paper. This paper meets the minimum requirements of American National Standard for Information Sciences-Permanence of Paper for Printed Material, ANSI Z39.48-1984.

Tilapia
Biology, Culture, and Nutrition

FOOD PRODUCTS PRESS®
Aquaculture
Carl David Webster, PhD
Senior Editor

Tilapia: Biology, Culture, and Nutrition edited by Chhorn E. Lim and Carl D. Webster

Tilapia
Biology, Culture, and Nutrition

Chhorn E. Lim
Carl D. Webster
Editors

Food Products Press®
An Imprint of The Haworth Press, Inc.
New York • London • Oxford

For more information on this book or to order, visit
http://www.haworthpress.com/store/product.asp?sku=5513

or call 1-800-HAWORTH (800-429-6784) in the United States and Canada
or (607) 722-5857 outside the United States and Canada

or contact orders@HaworthPress.com

Published by

Food Products Press®, an imprint of The Haworth Press, Inc., 10 Alice Street, Binghamton, NY 13904-1580.

PUBLISHER'S NOTE
The development, preparation, and publication of this work has been undertaken with great care. However, the Publisher, employees, editors, and agents of The Haworth Press are not responsible for any errors contained herein or for consequences that may ensue from use of materials or information contained in this work. The Haworth Press is committed to the dissemination of ideas and information according to the highest standards of intellectual freedom and the free exchange of ideas. Statements made and opinions expressed in this publication do not necessarily reflect the views of the Publisher, Directors, management, or staff of The Haworth Press, Inc., or an endorsement by them.

Cover photos courtesy of Brendan G. Lim.

Cover design by Kerry E. Mack.

Library of Congress Cataloging-in-Publication Data

Tilapia : biology, culture, and nutrition / Chhorn E. Lim, Carl D. Webster, editors.
p. cm.
Includes bibliographical references and index.
ISBN-13: 978-1-56022-888-2 (hc. : alk. paper)
ISBN-10: 1-56022-888-1 (hc. : alk. paper)
ISBN-13: 978-1-56022-318-4 (pbk. : alk. paper)
ISBN-10: 1-56022-318-9 (pbk. : alk. paper)
1. Tilapia. I. Lim, Chhorn. II. Webster, Carl D.

SH167.T54T58 2006
639.3'774—dc22

2005022483

This book is dedicated to my wife, Brenda; our children, Chheang Chhun, Lisa, Chhorn Jr., and Brendan; our grandson, Bryant; our brothers, Korn, Daniel, Thong Hor, Muoy Hor, and Trauy; our sisters, Huoy Khim, Chhay Khim, Seang Hay, and Huoy Teang; and our nephews and nieces for their patience and unconditional love.

Chhorn E. Lim

This book is dedicated to my wife, Caroline; my daughters, NancyAnn and Catherine (and Emma) ; my other "children," Lydia, Shyron, KC, Poppins, Michael, (Tillie) and Iggy; my parents (for all of their love, support, and encouragement); my brothers, Tom and Pete; Anthony Fusco (Memorial Middle School, New Jersey); William Musgrave (Monmouth Regional High School, New Jersey); Leland Pollack (Drew University, New Jersey); and Ralph Axtell (Southern Illinois University at Edwardsville, Illinois). My thoughts and love, as always, to my beloved Samwise, Barley, Darwin, and Misty, who are waiting patiently for me.

Carl D. Webster

We both would like to extend a special dedication to Dr. R. T. Lovell, Professor Emeritus, Fisheries and Allied Aquaculture, Auburn University, Alabama.

CONTENTS

ABOUT THE EDITORS

Chhorn Lim has some 30 years of experience in aquaculture nutrition and feed development research and as a research administrator. Currently, he is the Lead Scientist of Nutrition, Immune System Enhancement and Physiology at the Aquatic Animal Health Research Unit of the U.S. Department of Agriculture, Agricultural Research Service, Auburn, Alabama. He also serves as an Affiliate Researcher of the University of Hawaii Institute of Marine Biology and Affiliate Professor and Graduate Faculty Member of the Auburn University Department of Fisheries and Allied Aquacultures, Auburn, Alabama. He has served as an advisor and committee member for several MSc and PhD students and as a mentor for several postdoctoral students and visiting scientists. Dr. Lim has performed numerous long- and short-term consultancies and has received several honors and awards in recognition of his outstanding contributions and achievements. He has been invited as a keynote speaker and lecturer and to organize workshops and conferences. He is a member of several professional organizations, an editorial board member of the *Journal of Applied Aquaculture,* and an associate editor of the *Journal of the World Aquaculture Society.* He has authored or co-authored more than 100 articles in peer-reviewed, technical journals, book chapters, and lay publications and has served as co-editor of four books, one published by AOAC Press, one by CABI Publishing, and two by The Haworth Press.

Carl D. Webster has almost 20 years of experience in aquaculture nutrition and diet development research. Currently, he is Professor and Principal Investigator for Aquaculture at the Aquaculture Research Center, Kentucky State University, where he conducts research into nutrient requirements and practical diet formulations for fish and crustacean species that are currently, or have potential to be, cultured. Dr. Webster has been elected twice to serve as Secretary/Treasurer of the U.S. Chapter of the World Aquaculture Society (now known as the U.S. Aquaculture Society, a Chapter of the World Aquaculture Society), has been elected President of the U.S. Chapter of the World Aquaculture Society, and currently serves as a Past President. He was the Program Chair of the Aquaculture America '99 conference held in Tampa, Florida, and the Aquaculture America '03 conference held in Louisville, Kentucky. He has been, or is, listed in *Manchester Who's Who, American Men and Women of Science,* and *Who's Who in Agriculture Higher Ed-*

doi:10.1300/5513_a

ucation. He is a member of several professional organizations and is editor of the *Journal of Applied Aquaculture*. He has authored or co-authored more than 100 publications in peer-reviewed, technical journals and lay publications, and has served as co-editor of four books.

CONTRIBUTORS

Claude E. Boyd, Department of Fisheries and Allied Aquacultures, Auburn University, Auburn, Alabama.

C. Bauer Duke III, Aquaculture/Fisheries Center, University of Arkansas at Pine Bluff, Pine Bluff, Arkansas.

Carole R. Engle, Aquaculture/Fisheries Center, University of Arkansas at Pine Bluff, Pine Bluff, Arkansas.

Joyce J. Evans, USDA-ARS, Aquatic Animal Health Research Laboratory, Chestertown, Maryland.

Kevin Fitzsimmons, University of Arizona, Tucson, Arizona.

E. Gordon Grau, University of Hawaii, Department of Zoology and Hawaii Institute of Marine Biology, Kaneohe, Hawaii.

Bartholomew W. Green, USDA-ARS, Aquaculture Systems Research Unit, Pine Bluff, Arkansas.

Tetsuya Hirano, University of Hawaii, Department of Zoology and Hawaii Institute of Marine Biology, Kaneohe, Hawaii.

Steven G. Hughes, Cheyney University, Cheyney, Pennsylvania.

Phillip H. Klesius, USDA-ARS, Aquatic Animal Health Research Unit, Auburn, Alabama.

Menghe H. Li, Thad Cochran National Warmwater Aquaculture Center, Mississippi State University, Stoneville, Mississippi.

C. Greg Lutz, Aquaculture Research Station, Louisiana State University, Baton Rouge, Louisiana.

David J. Pasnik, USDA-ARS, Aquatic Animal Health Research Laboratory, Chestertown, Maryland.

Ronald P. Phelps, Department of Fisheries and Allied Aquacultures, Auburn University, Auburn, Alabama.

doi:10.1300/5513_b

Thomas J. Popma, Department of Fisheries and Allied Aquacultures, Auburn University, Auburn, Alabama.

N. Harold Richman III, University of Hawaii, Department of Zoology and Hawaii Institute of Marine Biology, Kaneohe, Hawaii.

Larry G. Riley, University of Hawaii, Department of Zoology and Hawaii Institute of Marine Biology, Kaneohe, Hawaii.

M. Fernanda Rodriguez, National Center for Cool and Cold Water Aquaculture, USDA-ARS, Kearneysville, West Virginia.

H. R. Schmittou, International Center for Aquaculture and Aquatic Environments, Department of Fisheries and Allied Aquacultures, Auburn University, Auburn, Alabama.

Andre Seale, University of Hawaii, Department of Zoology and Hawaii Institute of Marine Biology, Kaneohe, Hawaii.

William L. Shelton, Zoology Department, University of Oklahoma, Norman, Oklahoma.

Brian S. Shepherd, Great Lakes WATER Institute, Great Lakes Aquaculture Center/ARS-USDA, Milwaukee, Wisconsin.

Craig A. Shoemaker, USDA-ARS, Aquatic Animal Health Research Unit, Auburn, Alabama.

Richard W. Soderberg, Mansfield University, Mansfield, Pennsylvania.

Mathilakath M. Vijayan, Department of Biology, University of Waterloo, Waterloo, Canada.

Wade O. Watanabe, The University of North Carolina at Wilmington, Center for Marine Science, Wilmington, North Carolina.

Gregory M. Weber, National Center for Cool and Cold Water Aquaculture, USDA-ARS, Kearneysville, West Virginia.

De-Hai Xu, USDA-ARS, Aquatic Animal Health Research Unit, Auburn, Alabama.

Yang Yi, Asian Institute of Technology, Pathumthani 12120, Thailand.

Preface

Tilapia, because of their enormous adaptability and ability to reproduce under a wide range of physical and environmental conditions, excellent growth rates on a wide variety of natural and prepared diets, resistance to handling and disease-causing agents, and broad consumer appeal as a food fish, are the most successfully cultured fish species worldwide. Although they are endemic to tropical freshwater in Africa, Jordan, and Israel, their distribution has widened following introductions elsewhere in the early part and after the middle of the twentieth century. They are now cultured in virtually all types of production systems; in both fresh and saltwater; and in tropical, subtropical, and temperate climates. Tilapia dominate both small- and large-scale aquaculture in many tropical and subtropical countries, both as a low-priced product for mass consumption as a staple protein source and as a high-value, upscale product for export markets. They are increasingly being seen as the species of choice for intensive aquaculture and are likely to become the most important of all cultured fish in the twenty-first century.

In the past two decades, as a result of technological improvements, tilapia farming has expanded rapidly worldwide at a rate of approximately 12 to 15 percent annually and is predicted to continue to grow steadily for the foreseeable future. During this period, a number of books and conference proceedings dealing with various aspects of tilapia biology, aquaculture, and exploitation have been published. The information contained in these publications has contributed greatly to the successful development and expansion of the tilapia aquaculture industry. In the past few years, however, considerable technological advances have been made, and this book puts together the currently available information on tilapia aquaculture into a single, comprehensive volume.

The book begins with an exhaustive review of tilapia biology. This is followed by chapters on the prospects and potential for global production, physiological aspects of growth, recent directions in genetics, seed production, and hormonal manipulation of sex. The current state of commercial tilapia culture is discussed in three chapters on different production systems: pond production, culture in flowing water, and cage culture. The

doi:10.1300/5513_c

chapter on farming in saline water presents the most comprehensive review of knowledge about all stages of tilapia production, production systems, and socioeconomic impacts. The management of soil and water in ponds and the improvement of effluent quality to minimize impact on the environment are discussed in the following chapter. Four chapters review current knowledge on nutrient requirements, nonnutrient dietary components, feed formulation and processing, and feeding practices. Common parasites and diseases, as well as their prevention and control, and vaccinology against streptococcal disease, are then extensively discussed. The penultimate chapter elaborates the techniques used for harvest, handling, and processing. The book concludes with a comprehensive chapter on marketing and economics.

This book will be invaluable for students, aquaculture scientists, extension specialists, producers, nutritionists, and feed formulators. Although the information contained in this book can never be complete, it is hoped that the book will fulfill its intended purpose of providing state-of-the-art comprehensive information on the various phases of tilapia husbandry, thus contributing to the sustainability and stimulating the development and expansion of the tilapia aquaculture industry.

Acknowledgments

The editors gratefully acknowledge the contributions made by the chapter authors. The preparation of this book has involved the cooperative efforts of many people, to whom we are extremely grateful. A special acknowledgment goes to Ms. Michelle Coyle, who worked patiently and tirelessly to get the book ready. Our gratitude is also extended to our families for their enormous encouragement, patience, and support.

doi:10.1300/5513_d

Chapter 1

Biology

William L. Shelton
Thomas J. Popma

INTRODUCTION

Fossil remains of tilapia that are 18 million years old have been found in Africa (Fryer and Iles 1972). The Nile tilapia is depicted in Egyptian paintings dating to approximately 5,000 years before the present. Regional interest in tilapia was stimulated by its appearance in Java early in the twentieth century, but few would have predicted the expanding importance of this group of cichlids by the beginning of the following century (Balarin and Hatton 1979; Balarin and Haller 1982). In the twentieth century, tilapia were introduced into some 90 countries for aquaculture and fisheries, including through pan-African transplants (Courtenay 1997; Pullin et al. 1997). Although the first introduction outside the African continent was documented early in the twentieth century, most of the remaining transfers occurred after mid-century. The commercial and biological importance of tilapia grew as technical research was published (Table 1.1). Tilapia are now grown commercially in almost 100 countries and have become one of the most important food fishes in the world. Ninety-eight percent of tilapia production is outside the fish's native ranges.

Even though tilapia culture currently makes up only approximately 5 percent of total fish farming, it is second only to the group of carps that account for more than 70 percent of the total. Tilapia are likely to become the most important of all cultured fishes in the twenty-first century: yield has increased by 12 percent per year since 1984 (Fitzsimmons 2000; Shelton 2002a). Worldwide production of tilapia exceeded 2.2 MMT (million metric tons) in 2002; 68 percent was from aquaculture, which is up from 57 percent in 2000 (FAO 2001).

doi:10.1300/5513_01

TABLE 1.1. Summary of tilapia symposium proceedings.

Title	Editor(s)/author(s)	Date	Source
Tilapia, a guide to their biology and culture in Africa	J.D. Balarin and J.P. Hatton	1979	Stirling University, Scotland
The biology and culture of tilapias	R.S.V. Pullin and R.H. Lowe-McConnell	1982	ICLARM Conference Proceedings 7
International symposium on tilapia in aquaculture	L. Fishelson and Z. Yaron	1983	Tel Aviv University, Israel
Second international symposium on tilapia in aquaculture	R.S.V. Pullin, T. Bhukaswan, K. Tonguthai, and J.L. Mclean	1988	ICLARM Conference Proceedings 15
Third international symposium on tilapia in aquaculture	R.S.V. Pullin, J. Lazard, M. Legendre, J.B. Amon Kothias, and D. Pauly	1996	ICLARM Conference Proceedings 41
Fourth international symposium on tilapia in aquaculture, 2 vols.	K. Fitzsimmons	1997	NRAES B 106
Fifth international symposium on tilapia in aquaculture, 2 vols.	K. Fitzsimmons	2000	American Tilapia Association
Tilapia aquaculture in the Americas, 2 vols.	B.A. Costa-Pierce and J.E. Rakocy	1997 2000	WAS
Tilapias: Biology and exploitation	M.C.M. Beveridge and B.J. McAndrew	2000	Kluwer Publishers Fish and Fisheries, Ser. 25

Source: Modified from Shelton 2002.
ICLARM, International Center for Living Resource Management, Manila, Philippines; NRAES, Northeast Regional Agricultural Engineering Service, Coop. Extension, New York; WAS, World Aquaculture Society, Baton Rouge, Louisiana.

TAXONOMIC RELATIONSHIPS

Tilapia are endemic to Africa, Jordan, and Israel, where more than 70 species have been identified (Philippart and Ruwet 1982; Macintosh and

Little 1995; McAndrew 2000). However, relatively few species are commercially important, and even fewer are of aquacultural significance. Commercially important tilapia are currently divided into three major taxonomic groups, according to Trewavas (1982, 1983), based largely on reproductive characteristics. All are nest builders and substrate spawners, except in the following instances.

- *Tilapia* spp. guard the developing eggs and fry in the nests.
- *Oreochromis* spp. females incubate eggs and fry orally.
- *Sarotherodon* spp. males and/or females incubate eggs and fry orally.

The ongoing state of taxonomic flux means that scientific names in various literature sources differ. Fifty years ago, all commercially important tilapia were grouped in the genus *Tilapia* (Lowe 1959), but in the mid-1970s the mouthbreeding species were separated from those that incubated their eggs externally and placed in the genus *Sarotherodon* (Trewavas 1982). Then, in 1983, Trewavas separated the maternal mouthbreeding species of *Sarotherodon* into the genus *Oreochromis,* and the spelling of specific names was changed to conform to the rules of nomenclature. Consequently, for example, the important aquacultural species, the Nile tilapia, now reported as *Oreochromis niloticus,* was called *Sarotherodon niloticus* in the literature of the late 1970s and early 1980s, and prior to that, *Tilapia nilotica.*

Thys (1980) considers that *Tilapia* constitute a monophyletic group and rejects the classification proposed by Trewavas. The North American position was to resist the change and retain a single genus of *Tilapia* (Robins et al. 1991) until 2004, when the three genera position was accepted (Nelson et al. 2004). Most fish culture literature follows the taxonomic classification of Trewavas, and here we use the three genera in deference to the biology of the group. According to Trewavas (1983), the genera *Tilapia, Sarotherodon,* and *Oreochromis* can be distinguished on the basis of reproductive and developmental features, biogeography, and feeding and structural characteristics. The common and scientific names of commercially important species are

1. Congo tilapia, *Tilapia rendalli;*
2. Galilee tilapia, *Sarotherodon galilaeus;*
3. Black-chinned tilapia, *S. melanotheron;*
4. Mozambique tilapia (Robins et al. 1980 recommended this common name [also called Java tilapia]), *Oreochromis mossambicus;*
5. Zanzibar tilapia, *O. urolepis hornorum;*
6. Nile tilapia, silver perch (Jamaica), mojarra plateada (Colombia), red tilapia, or blond tilapia (McAndrew et al. 1988 described color mutants "red" and "blond"), *O. niloticus;*
7. Blue tilapia, *O. aureus;*

8. "Red" tilapia, *Oreochromis* spp. (hybrid usually of *O. mossambucus* or *O. hornorum* × *O. niloticus* or *O. aureus;* progeny breeding does not usually result in breeding true for color); and
9. "White" or "pearl" tilapia, *Oreochromis* spp. (hybrid developed from *O. niloticus* × *O. aureus* crosses; lowered fecundity and poor survival relative to parental stocks)

Primary Commercially Important Culture Species

The suitability of tilapia for various types of culture relates to their ease of propagation; handling tolerance; good growth on natural food and with a variety of supplemental feeds (manufactured and by-products) under various degrees of intensification; tolerance of a wide range of environmental conditions, including resistance to poor water quality and disease; and being perceived as a palatable, marketable, and nutritious product (Teichert-Coddington et al. 1997; Balarin and Haller 1982). The most important tilapia species in aquaculture are *O. niloticus, O. aureus,* and various hybrid combinations of these species with *O. mossambicus.* The Nile tilapia grows faster than the blue tilapia after the first year in Egyptian lakes (Payne and Collinson 1983). All have been extensively transplanted. All are similarly suitable for transplantation, but their intolerance to low temperatures (outside the 20°C winter isotherm) restricts culture to warmer climates or to conditions where warm water is available. The Mozambique tilapia was the first to be dispersed (Balarin and Haller 1982; Philippart and Ruwet 1982; Pullin et al. 1997), but its precocious maturity and proclivity to overpopulate made it a poor emissary for tilapia culture. Nevertheless, *O. mossambicus* still accounts for approximately 10 percent of contemporary global tilapia production (Young and Muir 2000). Further, the characteristics of the Mozambique tilapia stimulated efforts to develop population control (Hickling 1960, 1968; Avault and Shell 1968; Lessent 1968; Clemens and Inslee 1968; Chen 1969). Important milestones for reproduction management in the development of tilapia culture are listed in Table 1.2.

In contrast, the blue tilapia and the Nile tilapia have much better culture characteristics, and the techniques developed to control unwanted reproduction in the Mossambique tilapia (i.e., monosexing through hybridization and steroid-induced sex reversal) have been modified and applied to both (Eckstein and Spira 1965; Jalabert et al. 1974; Guerrero 1975; Pruginin et al. 1975; Shelton et al. 1978; Rothbard et al. 1983). With these technological achievements, progress in commercialization was rapid. The Nile tilapia rose rapidly to the greatest international prominence among tilapia in

TABLE 1.2. Pertinent literature for milestones in the development of tilapia aquaculture.

Sex reversal and sex determination	Hybridization and sex determination	Ploidy manipulation and sex determination	Fry production
Yamamoto 1958	Hickling 1960, 1968		Uchida and King 1962
Eckstein and Spira 1965	Lessent 1968		Maar et al. 1966
Clemens and Inslee 1968	Chen 1969		
Jalabert et al. 1974	Avault and Shell 1968		
Nakamura and Takahashi 1974	Jalabert et al. 1971		
Guerrero 1975	Pruginin et al. 1975		Rothbard and Pruginin 1975
Shelton et al. 1978	Mires 1977		
Tayamen and Shelton 1978	Avtalion and Hammerman 1978		
Hopkins et al. 1979	Balarin and Hatton 1979		Hulata et al. 1981
Shelton et al. 1981	Lovshin 1982		Mires 1982
Rothbard et al. 1983	Avtalion 1982		
Calhoun and Shelton 1983	Wohlfarth and Hulata 1983	Chourrout and Itzkovich 1983	Hughes and Behrends 1983
Shelton et al. 1983		Penman et al. 1987	
Macintosh et al. 1985		Don and Avtalion 1988	Guerrero and Guerrero 1988
Mair et al. 1987	Lester et al. 1989	Pandian and Varadaraj 1988	Rana 1988
Popma and Green 1990	Wohlfarth and Wedekin 1991	Mair, Scott, Penman, Beardmore, and Skibinski 1991; Mair, Scott, Penman, Skibinski, and Beardmore 1991	Scott et al. 1989

TABLE 1.2 *(continued)*

Popma and Green 1990	Wohlfarth and Wedekin 1991	Mair, Scott, Penman, Beardmore, and Skibinski 1991; Mair, Scott, Penman, Skibinski, and Beardmore 1991	Scott et al. 1989
McAndrew 1993	Muller-Belecke and Horstgen-Schwark 1995	Trombka and Avtalion 1993	Pandian and Varadaraj 1990
Phelps and Popma 2000		Mair 1993	Green et al. 1997
Penman and McAndrew 2000		Hussain et al. 1993	Mair et al. 1997
Teichert-Coddington et al. 2000		Myers et al. 1995	Tuan et al. 1999
Smith and Phelps 2001		Shirak et al. 1998	Little and Hulata 2000
Phelps et al. 2001			

Source: Modified from Shelton 2002.

the latter part of the 1990s, accounting for more than 60 percent of world tilapia production, largely because of its fast growth rate, adaptability to various culture conditions, and consumer acceptance (Macintosh and Little 1995; Shelton 2002a).

This chapter discusses only those cichlid species included in the subfamily Tilapiinae that are native to Africa and the vicinity, excluding the haplochromine complex of the Rift Valley lakes and the cichlids of the Indian Subcontinent and Central and South America (Trewavas 1983). Furthermore, the discussion will focus on the economically more important species of *Oreochromis.*

Morphological, Meristic, and Feeding Relationships

Adult tilapia have predominately vegetarian diets, varying from macrophytic to phytoplanktivorous. Related to feeding habits are associated structural features, such as terminal mouth; slender, notched teeth; long gill rakers; and intestines suitable for processing a plant diet (7 to 14 times standard length [SL]). *Tilapia* spp. are generally more herbivorous, and many species feed on macrophytes. They have 7 to 16 gill rakers on the lower part of the first arch in contrast to 13 to 28 for the two other genera. Jaw and pharyngeal dentition of *Tilapia* is coarser than that of *Sarotherodon,* but *Oreochromis* spp. have teeth similar in size to those of *Tilapia. Sarotherodon* spp. rarely eat plants other than phytoplankton. *Tilapia* have slightly lower numbers of vertebrae (26 to 30; mode = 28) than *Sarotherodon* (26 to 31; mode = 29) and *Oreochromis* (27 to 34; mode = 30), and the number of lateral line scales is generally one or two more than the vertebrae count. Belly scales are smaller than the scales on the flank in *Oreochromis,* whereas in *Sarotherodon* the two types are little different.

Distinguishing Anatomical Characteristics

Tilapia belong to the family Cichlidae, which can be distinguished from other families of bony fishes by the presence of an interrupted lateral line running superior along the anterior part of the fish and inferior along the posterior portion, and by the presence of a single nostril (nares) on either side of the snout. Tilapia can generally be separated from other native cichlids in non-African countries by their coloration or the presence of a pharyngeal plate used in the grinding of vegetable matter. The pharyngeal plate of rasping teeth in adult fish feels similar to sandpaper.

The "group" of African cichlids has the following characteristics:

Body: medium high to medium elongate, depth 28 to 60 percent standard length; mouth terminal or nearly so; outline of head generally curved but sometimes concave in adults or with an occipital hump (Figure 1.1).

Fins: Dorsal—XI to XIX/9 to 16; anal—III to VI/7 to 13; pectoral—13–15, acuminate, never rounded; caudal—membranes naked to densely covered with small scales on hind margin, slightly rounded to slightly or moderately emarginated.

Scales: slightly ctenoid, but never feeling rough when touched, rather large on back and flanks, sometimes small on chest and belly, 12 to 20 rows around caudal peduncle; split lateral line separated by two scale rows—upper lateral line extending posterior to below the end of the dorsal fin, lower lateral line running from below the end of the upper lateral line onto the caudal peduncle and caudal fin base.

Teeth: in rows or bands on jaws; all are tricuspid or bicuspid; pharyngeal teeth variable.

Color: dark spot on the upper posterior corner of the operculum and one on the anterior basal corner of the soft dorsal fin (called the "tilapia mark"),

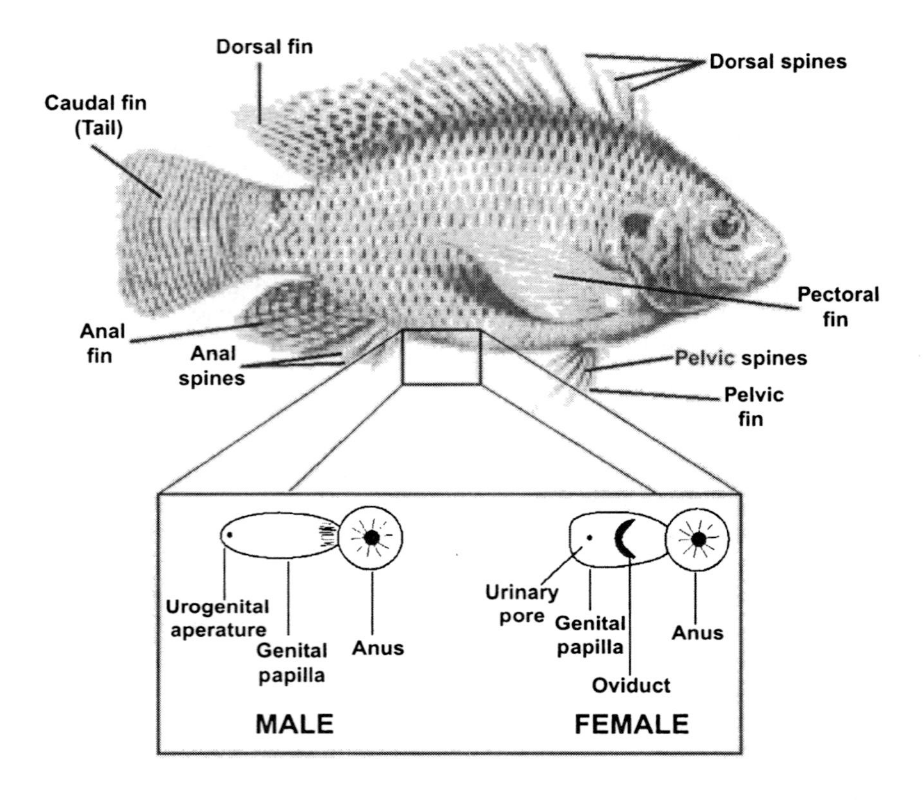

FIGURE 1.1. Generic illustration of the main morphological features of *Oreochromis niloticus. Source:* From Popma and Masser 1999.

well developed in young but often faded in adults; never "egg spots" on anal fin; with or without dark vertical and horizontal bars on sides, the prominence of which is regulated by behavior; never silvery reflecting colors on dorsal fin.

Sexual dimorphic characteristics vary, but color patterns are useful to differentiate some, and urogenital features permit the sexing of adult fish. The papilla of males is slightly larger, more pointed, and has a single terminal urogenital opening, whereas the papilla of the female has a subterminal urinary pore and a horizontal slitlike genital opening midway between the tip and anus (Figure 1.1). In all the *Oreochromis* spp., males grow faster and to a larger size than females. This difference in growth is pronounced after the onset of sexual maturity, and several factors are considered relevant:

1. the importance of territory in reproductive strategy, which may have favored the selection of larger males;
2. the anabolic effects of androgens;
3. the greater energy requirement for egg formation than for testicular development;
4. the lack of feeding during brooding by females.

Distinguishing Characteristics of Representative Species

Mozambique Tilapia (O. mossambicus)

1. *Fins:* Dorsal—XV to XVII/10 to 12 (mode XVI/12); anal—III/9 to 10; pectoral—14 to 15 rays; caudal—scaleless except at base;
2. *Lateral-line scales*—29 to 30;
3. *Vertebrae*—28 to 31 (mode 30);
4. *First lower arch gill rakers*—14 to 20 (mode 18, mean 17.5);
5. Breeding colors of male—Body and vertical fins are black with bright carmine margins on dorsal, caudal, and anal fins, and red pectoral rays; upper parts of the head are black, but lips, cheek, lower parts of operculum and preoperculum, and branchiostegal membrane become white.

The lower parts of the body or the whole body may be mottled with vertical black blotches depending on the position of an individual in the social hierarchy.

Mozambique Tilapia can be readily distinguished from the two other principal culture species, the Nile tilapia and blue tilapia, by the presence of yellow pigmentation in the gular region (most notable when comparing juveniles), an upturned, protruding snout and black color in older males, and a lack of vertical banding on the tail.

Blue Tilapia (O. aureus)

1. Fins: Dorsal—XV to XVI/12 to 16 (mode XVI/13); anal—III/9 to 11; pectoral—15 rays;
2. *Lateral-line scales*—28 to 29;
3. *Vertebrae*—28 to 31 (mode 30);
4. *First lower arch gill rakers*—18 to 26;

5. *Caudal fin*—Never bears regular dark vertical stripes, which are characteristic of *O. niloticus,* but groups of melanophores may randomly align in vertical patterns, sides with 11 to 12 narrow vertical bars; prominence varies relative to stress and social standing;
6. *Breeding colors of male*—Bright metallic blue head; caudal and dorsal margins pink to red.

Nile Tilapia (O. niloticus)

1. *Fins:* Dorsal—XVI to XVIII/13 to 15 (mode XVII/13); anal—III/9 to 11;
2. Vertebrae—30 to 32 (mode 31);
3. *First lower arch gill rakers*—19 to 25;
4. *Caudal fin*—Dark vertical bars;
5. *Breeding colors of male*—Rose-red to pink on head and lower parts of body; dorsal and caudal fins with reddish cast and dark margins.

The Nile tilapia can be distinguished from the blue tilapia by the relatively strong vertical banding in the caudal fin of both sexes and by the gray-pink pigmentation of the gular region.

BIOGEOGRAPHY

The general native range of *Sarotherodon* is West Africa and drainages of the Congo, Nile, and Jordan rivers. The genera *Oreochromis* and *Tilapia* have the same distribution plus drainages south of equatorial Africa. The Mozambique tilapia has been introduced into 64 countries, mostly before the 1960s, whereas *O. aureus* and *O. niloticus* have been introduced into 24 and 57 countries, respectively, primarily after the 1960s. The environmental impact of the various tilapia outside their natural range is discussed by Pullin et al. (1997).

Mozambique Tilapia (O. mossambicus)

Native Range

The characterization of a definitive native range for the Mozambique tilapia is somewhat confused by its extensive transplantation and recent taxonomic disputes. The native range is restricted to the eastward-flowing rivers of southeastern Africa from the lower Zambezi River and its delta and the lower Shire River southwards, including the Sabi, Lundi, Limpopo, Incomati and Pongolo rivers, and to coastal rivers and lagoons southward to Algoa Bay and the Bushmans River (Jubb 1974; Trewavas 1983). This range may appear to contradict earlier reports because of recent taxonomic changes (Trewavas 1966, 1983). Populations exist in the Hunyani and Shangani rivers and the middle Zambezi River drainage, but whether these are native or introduced is unclear. Records from the middle Zambezi usually refer to *O. mortimeri,* but introduction through stocking has complicated interpretation of the natural distribution (Pullin 1988). The range extends to the north of the Zambezi delta along the coast of Mozambique.

In the southern portions of the native range, *O. mossambicus* populations tend increasingly to be restricted to freshwater areas as one approaches the coast; at the southernmost limit the fish occurs more frequently in estuaries (Jubb 1967; Allanson et al. 1971), probably as a result of its being more tolerant of lower temperatures with increasing salinity. Gaigher (1973) showed that Mozambique tilapia prefer pools and avoid swift water in the Limpopo River system. Pienaar (1968) also found that they prefer quiet pools and avoid open tidal areas where current velocity exceeds 10 $cm \cdot s^{-1}$, and similar conditions in estuaries (Whitfield and Blaber 1979).

Introductions

Virtually all of the Mozambique tilapia in Indonesia, Asia, Europe, Central, North, and South America are descended from a few fish found in East Java in 1939. Subsequent additional transfers have usually involved similarly few individuals, resulting in several genetic bottlenecks. The initial discovery of feral Mozambique tilapia in Indonesia is described by Schuster (1952, pp. 94–96, 103–107):

> In July 1939 a Javanese overseer of fisheries in East Java reported having obtained a new species of fish from a small lagoon of the Serang River on the south coast of Java . . . Five specimens, two of which had eggs and fry in their mouths, were in a small homestead-pond in the village Papungan near Kediri. These were later identified as *Tilapia mossambica.* Attempts to keep *T. mossambica* under control failed because its easy spawning habits and quick growth made it very attractive to fish-growers. It soon spread and became adapted to all types of water: ponds, ditches, brooks, tanks, lakes, swamps and marshes and spawned everywhere. It soon reached the coast, where for the first time it was cultivated in 1942 in test tambacks near Surabaya. The author was unable to observe the results of the first tests because he became a prisoner-of-war. When he came to Java after the war, *Tilapia* already had achieved an important place among tamback fish . . . the Japanese Army appropriated *Tilapia,* bringing it to all places.

The origin of these first five tilapia in Java is unclear, but they are presumed to have come from the aquarium trade (Atz 1954; Chen 1966; Trewavas 1968). Similar uncertainties attend various details of Mozambique tilapia introductions to other countries. Balarin and Hatton (1979) list the fish as being present in the Caribbean before 1940; however, all references in the literature put it in that part of the world after 1949, when it was introduced into St. Lucia. The Mozambique tilapia was introduced to Puerto Rico in 1958 by the U.S. Public Health Service as a possible snail predator (Chimits 1955; Furguson 1977). Lobel (1980) discussed the introduction in 1958 on Fanning Atoll in the central Pacific by personnel of the U.S. Bureau of Commercial Fisheries. The fish were released in a saltwater pond on the atoll and in a freshwater pond on nearby Washington Island. Salinity tolerance up to 69 ppt is reported (see section "Salinity"). Mozambique tilapia successfully competed with the native marine species and by 1978 had colonized a 16-km stretch of the atoll coastline.

Tracing the introduction of *O. mossambicus* into various African countries is somewhat confused by the taxonomic history. Many reservoirs and ponds in South Africa were stocked (Meschkat 1967; Balarin and Hatton 1979). Both *O. mossambicus* and *O. mortimeri* were introduced into Rhodesia (Zimbabwe) and Zambia (Jubb 1974). See also Trewavas (1983) for information on introductions.

The earliest record of an introduction into the United States was by the New York Aquarium in 1933 from Portuguese East Africa (Mozambique) (Atz 1954) of fish captured in tidal waters and transported in seawater aboard a steamship. Fish culture research on tilapia in the United States started at Auburn University, Alabama, in 1954: approximately 20 fry were obtained from the Steinhart Aquarium, California (Swingle 1958). The origin of the Steinhart Aquarium specimens is unclear, although its contact with the baitfish industry in California might suggest that the fish came from Hawaii, where *O. mossambicus* had arrived in 1951 (60 fry from Singapore) for testing as baitfish for the skipjack industry (King and Wilson 1957; Uchida and King 1962). For a more complete history of tilapia introductions, see Lee et al. (1980), Courtenay et al. (1984), Fuller et al. (1999), Robins et al. (1991), and Lever (1996).

Blue Tilapia (O. aureus)

Native Range

The blue tilapia is distributed in tropical and subtropical Africa and the Middle East. It occurs in the Senegal and Niger rivers in West Africa but not in the lower reaches, and in Lake Chad and the lower Nile delta from Cairo to the Mediterranean coast but not in the middle and upper river reaches. It is native to the Jordan River and Lake Kinneret (Sea of Galilee) but not coastal drainages of Israel. Throughout most of the range of blue tilapia, *O. niloticus* co-occurs, except in the Jordan River drainage, where all early records of the Nile tilapia actually related to the blue tilapia (Trewavas 1983).

Introductions

The blue tilapia has been transplanted less than either the Mozambique or Nile tilapia. The earliest introduction outside Africa was into the United States in 1957 (Pullin et al. 1997): fish were transferred from Israel to Auburn University, Alabama, for fish culture investigations. However, they were identified as Nile tilapia (McBay 1962; Avault and Shell 1968; Shelton and Smitherman 1984; Hargreaves 2000). Subsequent taxonomic information from Trewavas (1983) clarified the nomenclature. Only ten fish were introduced, and six died; thus, the founder stock consisted of one female and three males (Smitherman 1988). Most of the blue tilapia distributed in the United States originated from this source. Introductions to a total of slightly more than 20 other countries began in the 1960s (Pullin et al. 1997).

Nile Tilapia (O. niloticus)

Native Range

The Nile tilapia is distributed in tropical and subtropical Africa but not into the Jordan Valley of the Middle East; several subspecies are recognized (Trewavas 1983). It is widely distributed in West Africa in the Volta, Gambia, Senegal, and Niger River basins, but absent from Sierra Leone, Liberia, Zaire, and much of Ivory Coast and Cameroon. It is present throughout the Nile River basin. It is native to the Lake Chad basin and to lakes Tanganyika, Albert, Edward, and Kivu (Trewavas 1983; Pullin 1988).

Introductions

From the 1960s onward, the Nile tilapia was exported to approximately 46 countries outside Africa and 11 countries within Africa (Pullin et al. 1997). *Tilapia zillii* was stocked into Lake Victoria in 1954 to boost the fisheries, but the Nile tilapia was also inadvertently introduced (Lowe-McConnell 1982; Trewavas 1983). Then the piscivorous Nile perch *Lates niloticus* was stocked in 1959. These introductions have been cited as the cause of the collapse of the artisanal haplochromine fishery; however, although there is little doubt they had an impact, catch records clearly show a deteriorating condition beginning a decade earlier, most probably because of overfishing (Lowe-McConnell 1987; Acere 1988).

The Nile tilapia was initially introduced into the United States from Pentecoste, Brazil, in 1974, where it was being used in hybridization studies (Lovshin 1982). This stock originated from Ivory Coast and was imported to Brazil in 1971 (Shelton and Smitherman 1984; Pullin et al. 1997). Some 100 fry were transferred to Auburn (Smitherman 1988). In 1982, Auburn University, Alabama, received 200 fry of the Ghana strain from Israel, and 86 Egyptian strain subadults were imported after collection in the vicinity of Cairo. Nile tilapia from these different geographic regions were studied to examine potential genetic effects on growth and in relation to sex determination. Although these transfers involved the same subspecies *(O. niloticus niloticus),* different strains or geographic races have not performed uniformly in hybridization studies (Wohlfarth and Hulata 1989). Color mutants such as "red" and "blond" were isolated from Lake Manzala, Egypt; stocks were imported in 1979 to the University of Stirling, Scotland, and breeding populations have been maintained for scientific studies (McAndrew et al. 1988).

BIONOMICS AND LIFE HISTORY

Reproduction

A feature of tilapia reproduction is the plasticity of initial sexual maturation relative to size and age. Disparate size-related maturity occurs in other species, such as the common carp (Shelton et al. 1995); however, this stunting phenomenon in tilapia delayed their wide acceptance for culture until practical controls were applied. Sexual maturation, even in tilapia of the same genetic stock, is delayed in stable lentic environments, whereas in unstable lotic systems and smaller water bodies, such as in fish-farming ponds, individuals may mature and breed in less than half the time or at less than half the size than those in more stable habitats (Lowe-McConnell 1982; Philippart and Ruwet 1982). The evolutionary significance of this reproductive tactic is enigmatic and has been the object of much research.

General

Tilapia spp. are nesting substrate spawners and brooders; *Sarotherodon* and *Oreochromis* spp. are also nesting substrate spawners but incubate eggs and larvae in their mouth. *Tilapia* eggs adhere to the substrate, whereas the eggs of *Sarotherodon* have much reduced adhesiveness and *Oreochromis* eggs are without stickiness. Breeding is described as a lek (arena) system, where males excavate nests communally and defend territory within a spawning area; there is no pair bonding. Sexual dimorphism is minimal in *Tilapia* and *Sarotherodon,* but *Oreochromis* males have conspicuous breeding colors and grow more rapidly after sexual maturity. Egg care in *Tilapia* is primarily by the female, while the male guards the territory; in *Sarotherodon* one or both parents brood eggs, but only the females of *Oreochromis* incubate them. After hatching, *Tilapia* larvae are transferred to a pit, where secretions from their head glands permit adherence to the substrate; larvae of the other two genera have only vestigial adhesive glands, and the young are sheltered for 7 to 17 days after hatching. During the posthatching period, once they are able to swim effectively, juveniles of the mouthbrooding *Oreochromis* and *Sarotherodon* spp. engage in forays of increasing duration outside of the mouth, returning for refuge when necessary.

Optimal spawning temperature is between 25°C and 30°C for *O. niloticus,* between 20°C and 35°C for *O. aureus* (Green et al. 1997), and between 22°C and 25°C for *O. mossambicus* (Mironova 1977). Blue tilapia (misidentified as Nile tilapia) fish mature at about 10 cm total length (TL) and aged 50 days; they spawn in temperate southern ponds in the United States beginning in April at a water temperature of 23°C; reproductive cycles are at 5 to 8 week intervals (McBay 1962). The average duration of the incubation period and parental care is approximately 20 days.

Reproductive isolation along generic lines appears effective. Wohlfarth and Hulata (1983) report no natural intergeneric spawning between *Oreochromis, Sarotherodon,* and *Tilapia* spp. No evidence for hybridization between sympatric Nile and blue tilapia was evident in Egypt (Payne and Collinson 1983). However, interspecific hybridization among *Oreochromis* spp. under culture conditions, first reported by Hickling (1960), is common, and this mechanism has been used for monosex culture. The diploid chromosome number ($2n$) for the three *Oreochromis* species is 44 (Trewavas 1983), although Wohlfarth and Hulata (1983) cite one source that suggests a diploid set of 40 for the Nile tilapia.

Sexual Dimorphism

Males are larger than females after reaching maturity. Sexual dimorphism and dichromatism are most pronounced during the spawning season. In the Mozambique tilapia, males becomes very dark in color; the upper parts of the head are black, and lips, cheek, lower operculum, preoperculum, and branchiostegal membrane are white. Males also have a pronounced concave aspect to the upper profile of the snout (Jubb 1974). The male papilla is white, which contrasts sharply with the black body color (Fryer and Iles 1972; Trewavas 1983). Breeding males of blue tilapia have a metallic blue head color, and the margins of the caudal and dorsal fins are pink to red. Nile tilapia breeding males are pinkish on the head and lower body; the dorsal and caudal fins have a reddish cast, and the margins are dark. Male and female papillae are dimorphic in terms of relative size, shape, and structure; a male papilla is more pointed that that of a female, is slightly larger, and has a single terminal urogenital opening, whereas the female papilla has a subterminal urinary pore and a horizontal slitlike genital opening (Rothbard and Pruginin 1975). The papilla of both sexes become turgid and erect in profile during courtship and spawning (Shelton 2000). This may occur several hours before actual spawning, perhaps in conjunction with ovulation. Color patterns and behavioral roles are discussed by Neil (1966) and Lanzing and Bower (1974).

Age/Size at Sexual Maturity

The three tilapia most frequently cultured, all of the genus *Oreochromis,* are maternal mouthbrooders. Sexual maturity in tilapias is a function of age, size, and environmental conditions. In general, *O. mossambicus* reaches sexual maturity at a smaller size and younger age than either *O. niloticus* or *O. aureus.*

Maturity in Wild Populations

Size at maturity in natural populations of Mozambique tilapia ranges from 11.0 to 28.5 cm (Fryer and Iles 1972). In Lake Moyua, Nicaragua, where year-round temperatures were over 25°C. Riedel (1965) reported maturity was reached between 12 and 14 cm (90 to 100 g) at between 5 and 6 months of age. Bruton and Allanson (1974) observed that in Lake Sibaya, South Africa, females mature at approximately 8 cm SL in 1 year and males at 10 cm SL in 2 years (SL = 0.071 + 0.83 TL; Hecht 1980). The average size at which 50 percent of the population were mature was 10 cm SL for females and 12 cm SL for males. The smallest actively breeding fish were 8.4 cm SL females and 12.0 cm SL males. Nile tilapia can mature sexually at 6 months of age at approximately 40 g (Macintosh and Little 1995). Tilapia populations in large lakes mature older and larger than the same species grown in culture ponds. For example, *O. niloticus* usually matures at approximately 10 to 12 months and 350 to 500 g in several East African lakes, whereas fish from the same population under conditions of near-maximum growth will reach sexual maturity in farm ponds at an age of 5 to 6 months and size of 150 to 200 g.

Persistent low temperatures, as at high altitudes, may delay maturity and the time taken to initiate reproduction: in tropical Africa at 1,700 m, Nile tilapia mature at 6 to 9 months of age but do not produce fry until 7 to 12 months of age; at 2,000 m, onset of reproduction is delayed by another month or two.

Maturity in Cultured Populations

In contrast to Nile tilapia maturation at more than 150 g in natural lakes, for cultured populations if growth is slow, as in some farm ponds, sexual maturity will be delayed a month or two, but fish may spawn at weights as low as 20 g. Under fast growing conditions in culture ponds, *O. mossambicus* may reach sexual maturity at as young as 3 months of age, at which time they seldom exceed 60 to 100 g. In poorly fertilized ponds, fish may mature weighing as little as 15 g. Vass and Hofstede (1952) reported that in Indonesia, *O. mossambicus* reached maturity in ponds between 8 and 9 cm at 3.5 months in either fresh or brackish water. Mozambique tilapia in Hawaii spawned in culture tanks at 8 to 9 cm SL and 2 to 3 months old (Uchida and King 1962). Peters (1963) found mature fish at approximately 6 cm and 2.7 g. The length (L) to weight (W) relationship for 109 fish, 1.7 to 7.3 cm SL, was

$$\log W = 3.088 \log L - 4.8935$$

Sex Ratios

Reports suggest that sex determination in tilapia is thermolabile during early juvenile development, which can influence sex ratio (Baroiller et al. 1999). Baroiller et al. (1996) and Baras et al. (2000) demonstrated that exposure of Nile and blue tilapia, respectively, to high temperatures (34°C to 35°C) during the first month of exogenous feeding resulted in a sex ratio shift toward males. Several models have been proposed to explain the genetic basis for sex determination in tilapia (Chen 1969; Avtalion and Hammerman 1978; Mair, Scott, Penman, Beardmore, and Skibinski 1991; Mair, Scott, Penman, Skibinski, and Beardmore 1991; Wohlfarth and Wedekind 1991; Devlin and Nagahama 2002); however, although a variety of evidence suggests that sex determination is largely monofactorial in tilapia, a more complete understanding may require further elaboration of modifier genes that may affect the primary pathway.

Population

Most reports of sex ratio in natural populations are based on sampling with fishing gear such as gill nets, but collection with entangling gear can be size selective and, therefore, bias the estimate of sex ratio. De Silva and Chandrasoma (1980) reported a sex ratio of 1.85:1.0 (female:male) for *O. mossambicus* in waters of Sri Lanka, based on gill-net catches. At less than 20 cm in length, females predominated; the female to male ratio decreased with size, and there were no females in the 32 to 35 cm range. The sex ratio in Plover Cove Reservoir, Hong Kong, was approximately 1:1 for fish 8 to 19 cm SL captured in gill nets (Hodgkiss and Man 1978). At lengths less than 8 cm SL, females predominated, but no females were taken above 28 cm SL, and males predominated between 19 and 28 cm SL. Bruton and Boltt (1975) also noted that in Lake Sibaya, South Africa, fewer females were found at larger sizes. Riedel (1965) reported a relatively constant sex ratio of 1:2.3 (female:male) for Lake Moyua, Nicaragua, based on gill-net samples. In general, Fryer and Iles (1972) noted an overall predominance of males in populations of Mozambique tilapia, but populations in the lower Shire River, Malawi, were found to have a female:male sex ratio of 1:1.09 (Willoughby and Tweddle 1978).

Genetic Basis

The sex ratios of young fish can give some insight into the basis of sex determination, but, as mentioned above, this is more complex than is generally recognized. A typical vertebrate pattern based on sex chromosomes was initially proposed as an explanation of the unusual sex ratios in hybrids between different *Oreochromis* spp. (Hickling 1960; Chen 1969). Subsequent hybridization studies suggested that two species have homogametic females (i.e., XX: *O. niloticus* and *O. mossambicus*), and it is proposed that males of *O. aureus* and *O. urolepis hornorum* are homogametic (ZZ). Crosses between a ZZ male and an XX female produce nearly all-male progeny (Wohlfarth and Hulata 1983). As mentioned, this serendipitous finding by Hickling (1960) stimulated many later studies and subsequent applications to reproductive control in fish culture.

The progeny sex ratio from intraspecific spawning of individual pairs should provide a relatively good indication of the genetic basis of sex determination relative to the homogametic/heterogametic model and uncomplicated by hybrid genomes. The progeny sex ratio based on a simple Mendelian sex inheritance should be 1:1 (female:male). Shelton et al. (1978) presented evidence that supported this hypothesis for *O. aureus* and

O. niloticus. Offspring from populations of randomly mating broodstock were used in steroid-induced sex-reversal studies and for the companion control groups. The composite population sex ratio in each control group was statistically 1:1 (female:male) for blue tilapia (653:630 to 1.04:1.00) and for Nile tilapia (177:206 to 1.0:1.1). However, progeny sex ratios from individual pair matings for each of these species included some that were different from 1:1; approximately 20 percent of the progeny sex ratios from 126 blue tilapia pair spawnings and 71 Nile tilapia pair spawnings were statistically different from the expected ratio (Shelton et al. 1983). It is of interest that for the summed total of all pair-spawned progeny groups sexed (8,914 blue tilapia and 5,496 Nile tilapia), 49.4 and 54.7 percent, respectively, were males, which is equivalent to the population sex ratio observed in random matings. These studies were conducted within the normal temperature range for spawning of 24° C to 30° C. Two similar, but independent, studies verified these findings. Tuan et al. (1999) found that from 95 pair spawnings of *O. niloticus*, 50.5 percent of the 7,822 progeny were males, but the sex ratio of progeny from individual pair spawnings varied from approximately 18 to 100 percent males. Phelps et al. (1999) compared progeny sex ratios from pair spawnings of three races of Nile tilapia and found similar numbers of departures from 1:1 in each group.

Courtship, Territoriality, and Spawning

Although some details vary from species to species, reproductive behavior is generally similar within various *Oreochromis* spp. Photoperiod is an important environmental cue that regulates annual cycles, but average water temperatures in excess of 20°C initiate seasonal reproduction. Diurnal light cycles coordinate the onset of daily rhythms; under laboratory conditions, spawning activity begins between 9 and 14 h into the photoperiod (Shelton 2000). Rising temperature seems to be a primary stimulus in nontropical climates, with spawning commencing in April or May in temperate areas; in tropical regions, spawning may continue year-round. The breeding behavior of Nile and blue tilapia begins when water temperature reaches 20°C, and spawning starts at approximately 22°C.

Spawning is more seasonally restricted in the temperate portions of tilapia's range than in the equatorial areas (Payne and Collinson 1983). In the delta region of Egypt, reproduction of Nile tilapia is somewhat more restricted than that of blue tilapia; the latter spawns from May through September, whereas the former begins to spawn in April and may stop earlier in the summer. Nile tilapia incubate eggs and fry for a period of approximately ten days; a short nursing period of one to four days may follow during which a female continues to protect the fry even though they are freely feeding (Macintosh and Little 1995). A complete reproductive cycle lasts from four to six weeks between consecutive batches of mature ova. The interval between spawnings by an individual female is approximately 22 to 44 days. The number of seasonal spawnings can be 6 to 11 in suitably warm climates (Chimits 1955; Turner and Robinson 2000). Lee (1979) demonstrated that removal of fertilized eggs from a female's mouth will shorten the interval of the interovulatory cycle for blue and Nile tilapia to as little as 10 to 11 days. Females incubating eggs can be recognized by the bulging of their gular area and the fact they do not open their mouths as much as usual during respiration.

Oreochromis males are highly polygamous and may spawn with several females in quick succession (Neil 1966; Turner and Robinson 2000). They establish a territory in which they excavate a nest down to a solid substrate with their mouths, ejecting material a short distance away. Males will also establish a spawning site on a solid substrate, such as concrete or the bottom of an aquarium or tank. Although nest excavation is a requisite innate behavior, spawning does not require actual nest construction. The male remains in the nest, defending the area against other males and displaying to attract a female. Males defend their territories by head-to-tail threat displays and by circling during which fins are erect and operculi may be flared; color pattern may change during agonistic behavior to show a lateral barring pattern that is more prominent than usual. Female hierarchy or nip orders can often be discerned from color intensity and barring pattern. The papillae of males and females are turgid and erect during active periods of reproduction (Shelton 2000).

Side nipping is a common aggressive behavior and is not restricted to male-male encounters as agonistic behavior toward females is similar; however, the females' reaction can result in courtship rather than continued aggression. In the confines of aquaria, females that are not receptive and are unable to seek refuge may be subject to such intense courtship that they are stressed and physically injured (side nipping and scale loss). Under these restricted conditions, females may be intensely hierarchical and agonistic. The nesting areas are usually shallower than 1 m (a minimum of 41 cm and maximum of 8.5 m for Mozambique tilapia; Bruton and Boltt 1975). In a study by Bruton and Boltt (1975), the size of nests increased with depth and ranged from 20 cm in diameter (0.6 m depth) to 142 cm (5 m). Studies on Lake Sibaya, South Africa, showed that nests were either directly associated with weed beds or in close proximity to them (Boltt et al. 1969; Jubb 1974; Bruton and Boltt 1975).

Ripe females school in midwater with nonterritorial males and egg-brooding females. When eggs hatch, the brooding females leave the school for more secluded areas. Some prespawning ripe females depart a school and approach a spawning arena where there are nest-guarding males. Under laboratory conditions at 26°C with a regulated light cycle, courtship activity of Nile tilapia begins between 9 and 14 h into the daylight phase; an erect papilla is an indication of readiness to spawn (Shelton 2000). A male tries to attract a female to his nest by leading behavior. The female might circle the nest while the male displays laterally with extended fins and branchiostegals and continued attempts to lead the female to the nest. If she follows into the nest, a series of mutual circling displays may ensue. She may make a pass over the substrate, tilting sideways and vibrating forward, at which time the male responds with a similar behavior. This skimming motion may be repeated several times without egg or sperm deposition. If eggs are laid (two to four eggs per gram of body weight), the male will also deposit milt and the female quickly circles and picks up the eggs. This behavior is repeated several times over a period of 15 to 30 min, after which the female is driven away, carrying the eggs in her buccal cavity (Neil 1966). Nile tilapia (75 to 500 g) deposit from 50 to 2,000 eggs per spawn (Rana 1988). A male will spawn many times per season but with only one female at a time. The number of times a female spawns per season varies but is related to the progressive maturity of intraovarian eggs, which is hormonally controlled in response to egg and larva brooding (Lee 1979).

From four to seven batches of eggs are spawned in one mating (Turner and Robinson 2000). The female takes the eggs into her mouth shortly after they are deposited and after, or while, the male distributes milt over them. There has been some disagreement over whether the eggs are fertilized in the nest and then picked up or fertilized in the female's mouth (Riedel 1965; Neil 1966). Fertilization is not an instantaneous process; it requires the entry of a spermatozoon through the micropyle, activation of the egg, with subsequent cytoplasmic reorganization, and reinitiation of the meiotic events. Thus, effectively, fertilization takes place in the oral cavity. Under aquarium conditions where there are several females, the one highest in the nip order usually spawns before the others. In some cases, a subordinate female gathers a few eggs spawned by another and incubates them (personal observation). After spawning the female is chased away and reenters a school for approximately seven days. After the eggs hatch, she again leaves the school for whatever cover is available in shallow water areas (Neil 1966; Jubb 1967; Turner and Robinson 2000). Eggs hatch in two to three days. Fry remain in the female's mouth until yolk-sac absorption, five to eight days later, by which time the swim bladder has developed, permitting the fry to swim well. Nevertheless, they often seek refuge in the female's mouth at the slightest sign of danger.

Eggs

The eggs of tilapia are surrounded by several membranes, the nomenclature of which is not uniform (chorion, follicular epithelium, zona radiata, zona pellucida, vitelline membrane, etc.). These membranes in various teleosts undergo differential swelling after fertilization in relation to a particular reproductive guild (demersal and adhesive, demersal and nonadhesive, semibuoyant, or pelagic). All species of tilapia have demersal eggs. Fertilized eggs are typically ellipsoid or ovoid and turgid from water infusion ("water hardened") and cytoplasmic reorganization. The eggs of bottom-spawning *Tilapia* have adhesive threads or glutinosus secretions on the outer membrane that attach the egg to a substrate. *Sarotherodon* eggs have remnants of these threads that do not cause adhesiveness; *Oreochromis* eggs lack these structures (Kraft and Peters 1963; Rana 1988). Scanning electron microphotographs of the egg surface for Nile tilapia published by Rana (1988) illustrate these threads and the micropyle. The formation of the micropyle is described by Aravindan and Padmanabham (1972). A detailed account of egg development for Mozambique tilapia is given by Vaas and Hofstede (1952).

Tilapia eggs are smaller (1.1 to 2.0 mm) than those of *Sarotherodon* (2.4 to 3.0 mm) or *Oreochromis* (2.8 to 4.3 mm). Riedel (1965) described the eggs of Mozambique tilapia as yellowish in freshwater and whitish in brackish water, with a diameter of approximately 0.7 mm. Panikkar and Tampi (1954) observed eggs to be yellow in color, pear-shaped, and 2.0 mm by 1.75 mm in size. Chen (1976) states that eggs are elliptical with a long axis of approximately 2.5 mm. Eggs of *O. aureus* and *O. niloticus* are oval and orange-yellow in color; they range in size from 1.94 to 2.95 mm (Payne and Collinson 1983). Generally, egg size of tilapia is species specific, although age or size of the female and nutritional state can influence development (Rana 1988). Older female Nile tilapia produce larger eggs, which in turn results in larger fry. Eggs range in volume from approximately 2.8 to 11.1 mm^3; 1-year-old females produce eggs of 1.9 to 2.8 mg mean weight, and 2-year-old females produce eggs of up to 3.7 mg mean weight (Macintosh and Little 1995).

Fecundity

The number of ripe ovarian eggs in *Tilapia, Sarotherodon,* and *Oreochromis* spp. ranges from 1,000 to 2,500. *Tilapia* spp. have smaller eggs with less yolk than do species in the two other genera. The fecundity of the Mozambique tilapia based on mature eggs ranges from approximately

100 to 1,700 for females of 12 to 30 cm TL. Chimits (1957) reported that an 8 cm Mozambique tilapia produces 100 to 150 eggs per spawning early in maturity and that the number of eggs spawned increases with successive spawnings; a 6-month-old fish can produce more than 1,000 eggs per spawning. Nile tilapia (75 to 500 g body weight) can deposit 50 to 2,000 eggs per spawning.

Fecundity can be expressed as the number of mature ova in the ovary, the number of ovulated eggs, or the number of eggs deposited during spawning. These numbers may differ for fish of equal size; not all mature eggs will be ovulated, and not all ovulated eggs will be deposited during spawning. The number of ripe gonadal ova in *Tilapia, Sarotherodon,* and *Oreochromis* spp. ranges from 1,000 to 2,500. *Tilapia* spp. have smaller eggs with less yolk than do species in the other two genera. The fecundity of the Mozambique tilapia based on mature eggs ranges from approximately 100 to 1,700 for females of 12 to 30 cm TL. Chimits (1957) reported that an 8 cm Mozambique tilapia produces 100 to 150 eggs per spawning early in maturity and that the number increases with successive spawnings; a 6-month-old fish can produce more than 1,000 eggs per spawning. Rana (1988) reported fecundity from a hatchery population of Mozambique tilapia based on egg counts of freshly spawned clutches:

> $F = 33.25\ \mathrm{SL}^{1.10}$ for fish 9 to 20 cm, or $F = 99.78\ W^{0.40}$ for fish 25 to 271 g in Weight.

Rana (1988) characterized fecundity for Nile tilapia based on freshly spawned clutches as:

> $F = 5.31\ \mathrm{SL}^{1.74}$ for fish 7 to 25 cm SL, or $F = 39.41\ W^{0.56}$ for fish 16 to 498 g.

The numbers of unovulated, mature ova in a ripe female for the Nile and blue tilapia sampled from two Egyptian lakes followed the same pattern, and thus were pooled (Payne and Collinson 1983):

> $F = 2.34\ \mathrm{SL}^{2.2}$ or $F = 1.33\ \mathrm{TL}^{2.23}$ for fish 9 to 33 cm SL, where $\mathrm{TL} = 1.33\ \mathrm{SL}^{0.995}$.

Babiker and Ibrahim (1979) reported the number of ovulated oocytes in the ovaries of Nile tilapia from the White Nile area as

> $F = 2.895\ \mathrm{TL}^{2.017}$ for fish 11 to 32 cm TL, or $F = 16.12\ W^{0.83}$ for fish 50 to 600 g.

The relation between offspring numbers and fecundity is critical in terms of overpopulation and seedstock production. Riedel (1965) calculated that

Mozambique tilapia in Lake Moyua, Nicaragua, produced between 2,000 and 5,000 eggs per year in five to six spawnings and that the number of recruits per female per season was 350 to 500. In fry production ponds (0.17 m^2) harvested six times per day, 13,000 to 48,000 young were collected per week. Fry production for culture, including sex-reversal treatments, is discussed in other chapters; it also has been reviewed extensively elsewhere (Green et al. 1997).

Embryonic Development

Incubation time for Nile tilapia is inversely and linearly related to temperature (Rana 1988); time to hatching varies from approximately 2–3 days at 34°C to 8 days h at 17°C, as described by the relationship

$$\text{days} = 12.8 - 0.32 \times \text{temperature (°C)}$$

For hybrid tilapia (*O. niloticus* females × *O. aureus* males), hatching occurs in approximately 50 h at 25°C to 29°C (Rothbard and Pruginin 1975). Mozambique tilapia eggs hatch in approximately 60 h at 28°C (Chen 1976).

Temperature has an important effect on various biological relationships. This effect is particularly applicable to induced ploidy manipulation, where temperature-sensitive developmental events relative to treatment time are critical. Time between mitotic cell divisions during early, synchronous cleavage has been used as an index of temperature-affected developmental rate. This mitotic interval (tau-zero, τ_0) is useful in predicting the occurrence of events at different temperatures. Shelton (1999) described a mitotic interval (τ_0) relationship for Nile tilapia relative to the time to first cleavage *(T)*; at any temperature, the time to the first mitotic cleavage is more than twice the absolute time between consecutive later divisions. The relationship describing time to the first division (T in minutes) at different temperatures (C in degrees Celsius) is

$$T = 9.25 \times 10^4 C^{-2.12}$$

The mitotic interval relationship is

$$\tau_0 = 2.61 \times 10^5 C^{-2.7}$$

Shirak et al. (1998) state that at 28°C, time to first cleavage is 77 min for blue tilapia and the mitotic interval is 28 min. This compares with 79 and 32 min, respectively, for Nile tilapia at the same temperature.

Embryogeny of *O. niloticus* is described by Galman (1980) for development at 26°C to 27°C and by Rana (1988) at 28°C. After a spermatozoon penetrates the micropyle, cellular reorganization ensues, resulting in the formation of a cytoplasmic cap beneath the micropyle on the yolk surface.

Sperm entry stimulates the resumption of chromosomal redistribution during the second meiotic division and the formation of the second polar body. Fertilization is completed with the union between the paternal and maternal chromosome sets. Subsequently, the newly formed zygotic nucleus initiates the first mitotic division (karyokinesis), which is followed by cleavage furrow development (cytokinesis) to form the two-cell zygote. Cleavage is meroblastic as in other teleosts, and cell proliferation forms the blastula. Cells of the blastoderm spread laterally over the yolk in a single-cell layer (epiboly); when epibolic cell growth meets at the vegetal pole, it is termed closure of the blastopore, or the yolk-plug stage. Gastrulation occurs during epiboly and is completed by closure of the blastopore. This is critical in development as the anlagen of the central nervous system is forming.

Fertilized eggs cleave and initiate gastrulation after approximately 12 to 15 h at 26°C to 27°C (Galman 1980) or approximately 10 to 12 h at 28°C (Rana 1988); the blastopore is closed after 24 to 30 h. By 72 h, the three main divisions of the brain, the optic buds, the otic capsule with otolith nuclei, and a few somites have formed, and the tail is free from the yolk surface. Pectoral fin rudiments and a continuous finfold are present approximately 72 h after fertilization in Nile tilapia. Pigmentation, particularly in the eye and on the yolk surface, is present after between 72 and 100 h in normally pigmented fish and in the "red" color mutant of the Nile tilapia but not in the "blond", where, even through swim-up melanophore cells are present, they remain devoid of pigment (Rana 1988; McAndrew et al. 1988; Shelton 2002b). In Mozambique and Nile tilapia, eggs hatch in 3 to 5 days; total time from spawning until the young are no longer taken into the mouth is approximately 10 to 14 days (Panikkar and Tampi 1954; Riedel 1965; Rana 1988).

Larval History

The yolk sac is spherical and relatively large for warmwater fish in Nile and Mozambique tilapia. Eyes are pigmented, and some chromatophores are present on the surface of the yolk at hatching. Yolk absorption and swim bladder development permit buoyancy control at the same time as functional development of the mouth approximately 4 to 5 days after hatching. Feeding begins between 8 and 10 days after hatching. The incubation and brooding period for eggs and larvae of Mozambique tilapia kept in plastic pools was reported to be 20 to 22 days (Bruton and Boltt 1975). Panikkar and Tampi (1954) measured newly hatched fry at 5 mm. The size of Nile and Mozambique tilapia larvae is affected by the size of the egg relative to the amount of yolk stored (Rana 1988). On the second day fry reach approximately 5 to 6 mm, the mouth is formed, gill arches are developing, and the yolk sac is still present. Uchida and King (1962) found that 26 percent of Mozambique fry ranging from 7.8 to 10.3 mm had remnants of the yolk sac. At 8 to 12 days, the larvae are approximately 8 mm, most of the yolk has been absorbed, the swim bladder is newly developed, and the pharyngeal arches and jaw have formed and are functional; the larvae can swim with some buoyancy control, and they begin to forage outside the female's mouth.

These sizes for free-swimming fry under culture are similar to sizes reported from natural populations (Bruton and Boltt 1975). Bogdonova (1970) found that Mozambique fry at 24°C to 26°C began to feed on the seventh day after hatching, even though a considerable amount of yolk remained. Fry are 9 to 10 mm SL when females leave the brooding areas. Fry form large schools in water approximately 1 to 15 cm deep during the day, but they move into deeper water at night. In tank culture, fry have been reported to form tight schools near the surface after departing the females' mouths (Hida et al. 1962). Neil (1966) also found that young formed schools throughout tidal pools in Hawaii.

The main foods of Mozambique fry in Lake Sibaya, South Africa, were diatoms and some amphipods, insects, and harpacticoid copepods (Bruton and Boltt 1975). Juveniles fed on detritus rich in diatoms and periphyton. Vaas and Hofstete (1952) observed that young fed on diatoms, unicellular green algae, small Crustacea, and periphyton. Uchida and King (1962) reported some cannibalism in young under culture conditions. Bogdonova (1970) found that survival of fry until transition to active feeding was 93 percent. Using relative fecundity based on Riedel's (1965) data, survival from fertilized egg (in the mouth) to fry is 36 percent.

Gonadal Differentiation

Sexual development in teleosts is protracted and variable (Francis 1992; Devlin and Nagahama 2002). The gonadal developmental sequence is directed by the mechanism of genetic sex determination but also influenced by environmental factors that affect ontogeny such as temperature, population density, and food (Shelton et al. 1995). Timmermans and Van Winkoop (1993) describe the process of gonadal differentiation using a common carp model. Although various species have different periods of gonadal differentiation based on age and size, the basic chronological process is similar. Time of gonadal sex differentiation is important to the success of steroid-induced sex reversal, and although size is a practical indicator of age, rate of growth can be affected by a number of environmental factors (Shelton et al. 1978, 1995; Shelton 1989). Sex differentiation for Mozambique tilapia occurs when they are between 15 and 30 mm (at 35 to 48 days old) (Clemens and Inslee 1968). Nakamura and Takahashi (1973) reported primordial gonadal development beginning at 8 to 10 days and being complete after approximately 20 days at 20°C. Dutta (1979) examined *O. aureus* gonadal differentiation at two temperatures (21°C and 31°C) and at two stocking densities. Cytological differentiation was delayed at the lower temperature but occurred at similar sizes. Ovaries were differentiating when fish were between 13 and 15 mm TL, starting approximately 24 or 14 days after hatching at 21°C and 31°C, respectively. Avendia-Casauay and Carino (1988) described gonadal differentiation for *O. niloticus* at 25°C to 26°C. Primordial germ cells were present at hatching, but the paired gonadal anlagen were not present until 9 to 10 days after hatching; cytological differentiation began at 30 to 33 days (9 to 12 mm) and was complete after 48 to 56 days.

Management of Reproduction

Management of a biological system requires knowledge of the genetics of sex determination, developmental biology, endocrine physiology, and other biological factors (Shelton 1989). Induced ovulation has been a major development in the culture of species such as the Chinese carps and striped bass; comparably significant for the culture of tilapia has been the technological control of excessive recruitment. Applied management to control unwanted reproduction has included predator stocking, mechanical regulation (hapa and cage culture), hybridization, hormone-induced sex reversal, and ploidy manipulation (gynogenesis, androgenesis, polyploidization), sometimes in combination. Some important reviews provide detailed information on these subjects (Jalabert and Zohar 1982; Guerrero 1982; Lovshin 1982; Baroiller and Jalabert 1989; Wohlfarth and Hulata 1989; Zohar 1989; Dunham 1990; McAndrew 1993; Pandian and Koteeswaran 1998; Pandian and Sheela 1995; Little and Hulata 2000; Phelps and Popma 2000; Penman and McAndrew 2000). Various related components of reproduction management, such as seed production, hormonal manipulation of sex, and genetic applications in tilapia reproduction, are discussed in other chapters in this volume.

ENVIRONMENTAL BIOLOGY

Water Quality Parameters

Tilapia are more tolerant than most commonly cultured fish to various "adverse" environmental factors such as salinity, high ammonia concentrations, low dissolved oxygen, and high water temperature, but they are limited by sensitivity to low water temperature. *Oreochromis* spp. vary in their abilities to function at higher levels of salinities and to tolerate lower temperatures. Juvenile Mozambique tilapia appear to be hardier than adults for some environmental conditions; they can withstand greater variation in temperature and, apparently, lower oxygen concentrations (Caulton and Hill 1973, 1975; Bruton and Boltt 1975; Perez and Maclean 1975; Caulton 1978).

Water Temperature

The general influence of temperature on tilapia is summarized by Balarin and Haller (1982), Chervinski (1982), Philippart and Ruwet (1982), and Wohlfarth and Hulata (1983). The preferred range is between 28°C and 35°C; no reproduction occurs below 20°C and growth is poor between 10°C and 15°C. The upper lethal limits are near 42°C, and lower lethal limits are approximately 8°C to 12°C. Temperature effects can also be reflected in reduced productivity at altitudes up to 1,000 m. Therezien (1968) found that cultured populations in the Malagasy Republic (Madagascar) did not reproduce above 1,000 m. In its native range, *O. mossambicus* is rarely found above 916 m (Gaigher 1973).

Water temperature for optimal tilapia growth is between 29°C and 31°C; generally, tilapia do not grow well below 16°C (Stickney 1986a). When fish are fed to satiation, growth at the preferred temperature is typically three times greater than at 22°C. Maximum feed consumption at 22°C is only 50 to 60 percent as high as at 26°C. Tilapia reportedly tolerate temperatures up to 40°C, but stress-induced disease and mortality are problematic when temperatures exceed 37°C or 38°C (Allanson and Noble 1964; Wohlfarth and Hulata 1983). At the other extreme, handling at lower temperatures can also result in stress-induced trauma, and in mortality at temperatures lower than 17°C or 18°C.

Oreochromis mossambicus is generally tolerant of a wide range of water quality. However, it is a tropical species and has some temperature limitations. Preferred temperature range is between 20°C and 35°C. Reproduction takes place at 25°C to 36°C. Mozambique tilapia exhibit daily and seasonal movements in natural freshwater lakes in South Africa, such as Lake Sabaya. In the colder months, adults remain in deep water offshore until the water begins to warm, at which point they move into littoral areas (Bruton and Boltt 1975). Daily movements are similar, in that there is a general movement into the shallows during the day as these waters warm, then a movement back to deeper water at night.

The inability of tilapia to tolerate low temperatures is a serious constraint for commercial culture in temperate regions. Feeding generally ceases when water temperature falls below 16°C or 17°C. The lethal low temperature for most species is 10°C or 11°C, but *O. aureus* can survive exposure to 8°C or 9°C for several days. When *O. aureus* is hybridized with other *Oreochromis* spp., its cold tolerance appears to be heritable. Cold tolerance is a trait that may be heritable and can thus be modified through selection (Beherends et al. 1990); further, modification of cold tolerance through genetic manipulation would be commercially important if it could be achieved.

The native distribution of Mozambique tilapia appears to be limited chiefly by low temperature toward the southern part of its range and by geographic barriers and low temperature inland. Mozambique tilapia is somewhat south-temperate, and blue tilapia is similarly north-temperate, so both are slightly more temperature tolerant than many species. They are able to tolerate 8°C to 10°C, whereas the more tropical Nile tilapia is killed at temperatures below 12°C. Reproduction for Nile tilapia is inhibited at water temperatures below 20°C, slow in waters of 21°C to 24°C, and most frequent in waters above 25°C. Fry production decreases when average daily water temperature is below 24°C.

Temperatures ranging from 18°C to 35°C can be tolerated without any apparent ill effects. But below 20°C, *O. mossambicus* move to deeper waters and become less active (Bruton and Boltt 1975; Willoughby and Tweddle 1978). Below 15.5°C fish stop feeding (Kelly 1957), and kidney damage has been shown to occur in fish kept below 14°C for 3 days (Allanson 1966; Allanson et al. 1971). Fish transferred from 25°C to 30°C had an average acclimation rate of 1°C per 150 min (Allanson and Noble 1964); however, fish transferred from 25°C to 15°C were not fully acclimated even after 20 days.

Mozambique tilapia tolerate lower temperatures at higher salinities. They can tolerate temperatures of 11°C at salinities of 5 ppt, whereas they cannot survive at this temperature in freshwater. It has been suggested that the ability to withstand lower temperatures in saline waters is correlated with the maintenance of high plasma sodium and chloride concentrations, a relationship that becomes more apparent by examining the native distribution: *O. mossambicus* occurs more frequently in estuaries than in freshwater at its southerly limit in Africa (Allanson et al. 1971; Perez and Maclean 1975).

Salinity

Tilapia tend to be euryhaline and tolerate a wide range of salinities (Stickney 1986a,b; Suresh and Lin 1992). Freshwater species are most used in commercial culture, but all species tolerate salinities equivalent to brackish water with no adverse effects on growth (Green et al. 1997). Mozambique tilapia is the most saline tolerant of the three commonly cultured tilapia; it can survive salinities to 75 ppt, grows well and reproduces at salinities up to 50 ppt (full-strength seawater and above), grows optimally in brackish water, and grows well at 35 to 40 ppt (Balarin and Haller 1982; Wohlfarth and Hulata 1983).

Mozambique tilapia is very tolerant of salinity and in its native range is found in estuaries. Young survive in water with salinities as high as 70 ppt (Potts et al. 1967). Mozambique tilapia grow faster in 50 percent seawater (17.5 ppt) than in either freshwater or seawater (Canagaratnam 1966). Job (1969) reported that at salinities of 12.5 ppt (approximately isotonic for Mossambique tilapia), 80 g fish grew and survived at low levels of dissolved oxygen, probably as a result of a lower osmotic regulatory energetic cost. Osmoregulation-related metabolic rate increased on either side of an optimum range of 8.8 to 17.5 ppt salinity (Bashamohideen and Parvatheswararao 1972). Blue tilapia survive to 45 ppt and grow well up to 13 to 29 ppt, but they do not reproduce above 19 ppt. In the Nile delta, they are found in fresh and brackish water (Trewavas 1983). Blue tilapia can survive direct transfer from freshwater to water at salinities of 20 to 25 ppt. Nile tilapia are the least saline tolerant of the commercially important species but grow well at salinities up to 15 ppt.

Salinity influences spawning of tilapia: *O. mossambicus* can grow and spawn in full-strength seawater, whereas *O. aureus* and *O. niloticus* can reproduce only in salinities up to 10 to 19 ppt (Wohlfarth and Hulata 1983). Nile and blue tilapia produce fry equally well in freshwater and at 5 ppt salinity, but fry production declines at 10 ppt salinity. Nile tilapia grow and reproduce at salinities up to approximately 20 ppt—slightly higher than blue tilapia. They grow throughout the Nile River system but are less abundant in the Nile Delta than blue tilapia, which in turn grow well in brackish water up to 20 ppt salinity and are confined to the delta region (Payne and Collinson 1983).

Reproductive performance of Mozambique tilapia begins to decline in salinities above 10 ppt when compared with performance in freshwater. Red tilapia hybrids from *O. mossambicus* crosses have a reproductive performance in saltwater similar to pure Mozambique tilapia. Ideally, tilapia hatcheries should be located in freshwater or in water with less than 10 ppt salinity. Tilapia eggs are more tolerant of high salinities and can develop over a wider range than fry: in a study by Rana (1988), approximately 50 percent of Nile tilapia eggs incubated at 25 ppt hatched, but within 24 h of hatching, 50 percent of fry died at 19 ppt.

Dissolved Oxygen

Low dissolved oxygen (DO) is usually the primary water quality constraint on growth in intensively managed ponds. Commonly cultured species of tilapia survive routine dawn DO levels of less than 0.5 mg·L^{-1}, which is considerably below the tolerance level for most other cultured fish. Some reports indicate that Mozambique and Nile tilapia survive to 0.1 mg·L^{-1} (Stickney 1986a). Survival in water with low DO (<1 mg·L^{-1}) is due, in part, to their ability to use oxygen at the air-water interface. In addition, tilapia are physiologically adapted to optimize oxygen utilization at low environmental levels. Blood hemoglobin saturates at low dissolved oxygen tension and offloads rapidly at the tissue level much more efficiently than is the case with most other teleosts (Balarin and Haller 1982). Despite this ability to survive acute low DO concentrations, ponds should be managed to generally maintain DO above 2 mg·L^{-1} at dawn because exposure for prolonged periods depresses metabolism and growth, and frequent exposure to low oxygen is stressful and lowers disease resistance (Teichert-Coddington and Green 1993).

In general, *O. mossambicus* is tolerant of poor water quality, but tolerance is affected by temperature and the age of the fish. Kutty (1972) found that Mozambique tilapia maintained a routine respiratory quotient (RQ) of approximately unity in freshwater of 30°C at high ambient oxygen concentrations. However, at low oxygen concentrations (<0.2 mg·L^{-1}) the RQ increased to 8.0.

Oxygen consumption for Mozambique tilapia was equal to

$$Q = 0.407W^{0.73} \text{ at } 25°\text{C}$$
$$Q = 0.262W^{0.74} \text{ at } 20°\text{C}$$

where oxygen consumption is measured in milliliters per hour and wet weight of fish *(W)* is in grams (Mironova 1975, 1976). Nile tilapia between 100 and 300 g used oxygen at 0.5 to 0.7 mg·h^{-1} (Balarin and Haller 1982); the relationship is

$$\text{Oxygen consumption (mg·kg}^{-1}\text{h}^{-1}) = 2.115W^{-0.61}$$

where weight is in grams. Job (1969a,b) demonstrated a rise in metabolism as a function of temperature up to a peak at 30°C. Respiration was independent of dissolved oxygen at temperatures between 15°C and 30°C until the partial pressure of oxygen dropped to 50 mm Hg, which is equivalent to 30 percent saturation (Josman 1971). A decline in metabolic rate also occurs when temperatures are below 20°C.

Hydrogen Ion Concentration (pH)

Tilapia seem to grow best in water that is near neutral or slightly alkaline. Tilapia production is not seriously affected by pH, but the lethal limit for high pH is 11–12, and commonly cultured tilapia species tolerate low pH to approximately 5. Murthy et al. (1981) reported lethal pH limits of 3.7 and 10.3 for 10 g Mozambique tilapia. Fish death is probably the result of lethal levels of free carbon dioxide at lower pH rather than the acidic reaction. Carbon dioxide levels between 50 and 100 $mg \cdot L^{-1}$ cause distress and can be lethal if prolonged (Balarin and Haller 1982). The concentration of gaseous carbon dioxide affects the efficient binding of oxygen on hemoglobin at low oxygen tension. Growth is reduced in acidic waters (possibly because of lower production of natural food organisms). Diurnal variation in pH occurs because of photosynthesis; afternoon levels can reach 10 in unbuffered water, but fluctuations are smaller in waters with high alkalinity (Boyd 1990). Uchida and King (1962) reported carbonate levels ranging from 2.4 to 80.6 $mg \cdot L^{-1}$ and bicarbonate levels from 2.14 to 15.2 $mg \cdot L^{-1}$ in tanks where Mozambique tilapia were raised. Tilapia tolerate total alkalinities of 700 to 3,000 $mg \cdot L^{-1}$.

Ammonia

Ammonia toxicity is closely correlated with pH and, to a lesser extent, water temperature and DO concentration. Low DO increases ammonia toxicity, however, in fish ponds this is largely balanced by decreased toxicity produced by increasing carbon dioxide concentration, which lowers pH. As pH increases above neutral, a greater percentage of total ammonia is converted from the ionic form (NH_4^+) to the toxic un-ionized (NH_3) gaseous form; concentration also increases with increasing temperature (Soderberg 1997). Ammonia is more toxic at higher temperature; the ranges for percentage un-ionized form given below reflect conditions at 24°C to 32°C.

At pH 7, less than 1 percent of the total ammonia is in the toxic un-ionized form; at pH 8, approximately 5 to 9 percent is un-ionized; at pH 9, between 30 and 50 percent; and at pH 10, from 80 to 90 percent is NH_3. Consequently, ammonia toxicity is more problematic in poorly buffered ponds (alkalinity below 30 $mg \cdot L^{-1}$ CaCO3), which frequently experience afternoon pH levels of 9 or even 10. The negative effect of high un-ionized ammonia concentrations in the afternoon, when pH in ponds usually reaches its peak, is greatly reduced if pH falls to near-neutral levels during the evening.

Mass mortality of tilapia occurs within a couple of days of their sudden transfer to water with un-ionized ammonia concentrations greater than 2 mg·L^{-1}. The lethal level of NH_3 is approximately 2.3 mg·L^{-1}; the level of nitrite (NO_2) tolerated is approximately 2.1 mg·L^{-1} (Balarin and Haller 1982). However, approximately 50 percent of fish acclimated to sublethal levels will survive 3 or 4 days at un-ionized ammonia concentrations as high as 3 mg·L^{-1}. Prolonged exposure (several weeks) to un-ionized ammonia concentration greater than 1 mg·L^{-1} causes losses, especially among fry and juveniles in water with low DO. The first mortalities from prolonged exposure begin at concentrations as low as 0.2 mg·L^{-1}. Un-ionized ammonia begins to depress the appetite of tilapia at concentrations as low as 0.08 mg·L^{-1}.

Turbidity

Mozambique tilapia are highly tolerant of turbid conditions; they grow well in waters of the turbid Pongola River, South Africa (Bardach et al. 1972). This tolerance could be expected given the fish's native habitat in coastal rivers of southeast Africa, which commonly carry a heavy silt load. Balarin and Haller (1982) report that tilapia tolerate up to 13,000 mg·L^{-1}; however, higher turbidities reduce light penetration and thus affect primary productivity. Similar to many cichlids, advanced juveniles and adult tilapia are strongly territorial, and vision is important for interactions; higher turbidity can reduce aggressiveness. A consequence of territorial behavior is unequal growth at high densities when limited food is concentrated in few places.

Disease Tolerance

Tilapia are more resistant to viral, bacterial, and parasitic diseases than other commonly cultured fish, but a wide range of epizootics can occur (Roberts and Sommerville 1982; Tonguthai and Chinabut 1997; Plumb 1999). At temperatures greater than 16°C to 18°C and in the absence of severe environmental stress, tilapia rarely become diseased. However, viral, bacterial, and parasitic problems are more likely following stress from low temperature, handling, severe crowding, or poor water quality (Plumb 1997). Uchida and King (1962) reported that fry were most susceptible to infectious diseases and ectoparasitic infestation in the first 4 to 5 weeks of life. Fungal infections, especially from *Saprolegnia,* are particularly common after handling when water temperature is below 20°C. Lymphocystis, a viral disease, has been reported in tilapia. Likewise, a whirling viral disease recently caused heavy losses of tilapia fry in Israel (Avtalion and Schlarobersky 1994).

Flexibacter columnaris, which causes columnaris, is the most common myxobacterial pathogen for tilapia. Skin lesions may be problematic under high-temperature and ammonia stress, and gill infections may also cause heavy losses among fry, especially at low temperatures. The most common bacterial diseases are hemorrhagic septicemias, especially from *Aeromonas hydrophila* and, under hyperintensive culture, *Edwardsiella tarda.* In recent years *Streptococcus iniae* infections have also caused heavy losses of tilapia, primarily in recirculating and intensive flow-through systems following skin abrasions at higher temperatures and salinity. Sick fish swim erratically and have a curved body and lesions in the eyes and internal organs.

"Ich" or "white spot," caused by the protozoan parasite *Ichthyophthirius multifiliis,* can result in serious losses of fry and juveniles in intensive recirculating systems, but problems are much less likely in the tropical areas where most commercial tilapia production occurs because water temperatures are generally warmer than 20°C to 24°C, the optimal temperature range for this disease organism. The protozoan *Trichodina* may also reach debilitating densities on stressed fish at low water temperatures.

Monogenean and digenean helminthic parasites are common infestations of tilapia but are normally of low pathogenicity, with little effect on fish growth. Parasitic crustaceans, such as *Argulus, Ergasilus,* and *Lernaea,* have caused serious losses, but most reports are from Africa and from Israel where tilapia are associated with common carp.

Predation

Natural systems may include piscivorous fish and predatory birds. Bruton (1979) found that Mozambique tilapia formed a significant part of the diet of *Clarias gariepinus* in Lake Sibaya, South Africa. Heavy mortalities have been reported among brooding females from piscivorous birds in the lake (Bruton and Boltt 1975). Some culture ponds are constructed deeper to reduce access by wading birds. Fingerlings of selected lines of red and white tilapia are particularly susceptible to predation by birds because they are more visible from overhead. In addition, the red Nile tilapia appear to aggregate near the water surface. Cormorants can cause heavy losses, especially in nursery ponds. Predatory birds often drop live tilapia into adjacent ponds, a potentially serious problem when genetic purity is critical.

Tilapia also feed on midwater invertebrates, and although they are not generally considered piscivorous, juveniles actively attack larval fish. This cannibalistic feeding behavior is an important consideration for tilapia seed production management strategies. Uchida and King (1962) reported some cannibalism in young under culture conditions.

Feeding

Natural Food Habits

Tilapia ingest a wide variety of natural food organisms, including plankton, succulent green leaves, benthic organisms, aquatic invertebrates, larval fish, detritus, and decomposing organic matter. Generally, tilapia are herbivorous, but there are ontogenetic and species differences (Bardach et al. 1972; Ballarin and Haller 1982; Bowen 1982). Juveniles feed on zooplankton, but adults tend to be more omnivorous and filter feed using long, thin, closely spaced gill rakers. However, describing tilapia feeding as filtration is somewhat misleading (see section "Feeding in culture"). The Mozambique tilapia feeds on vegetation and bottom algae, whereas the Nile tilapia more commonly feeds on phytoplankton and detritus, and the blue tilapia is zooplanktivorous and detritivorous (Wohlfarth and Hulata 1983). In Lake Sibaya, South Africa, adult Mozambique tilapia males move into shallow water in early September, when water temperatures are around 19°C, and feed mainly on detritus, periphyton, and insects.

Fry feed mainly on detritus and neuston, and juveniles feed on detritus and periphyton (Bruton and Boltt 1975). Adults usually feed in deeper water (3 to 5 m), whereas juveniles feed near shore, frequently at less than 30 cm (Bowen 1982).

Mozambique tilapia are less efficient than Nile or blue tilapia at ingesting planktonic algae. Feeding on blue-green algae and other primary producers by tilapia has suggested that they are primarily herbivores, but in fact they are more omnivorous (Stickney 1986a). Consumption of plant tissue does not automatically imply the ability to digest and assimilate the nutrients into fish flesh: most fish species derive no nutrition from plant tissue consumed accidentally in pursuit of other food. Tilapia, however, obtain substantial nutritional benefit from plant material. The commercially important *Oreochromis* spp. digest 30 to 60 percent of the protein in planktonic algae, with blue-green algae being digested more efficiently than green algae, which has a more complex cell wall that resists rupture by the low pH in the stomach. Digestion and assimilation of plant material occur along the length of the long intestine, which is usually at least six times the total length of the fish. Processing filamentous and planktonic algae and higher plants is aided by two mechanisms: physical grinding of plant tissues between two pharyngeal plates of fine teeth and a stomach pH below 2, which ruptures the cell walls of blue-green algae and bacteria (Popma and Masser 1999).

Feeding in Culture

Some species of tilapia, including *Tilapia rendalli* and *T. zillii,* actively feed on fresh leaves of succulent plants, but their slower growth and inability to effectively harvest plankton reduces their value as species for commercial culture. Macrophytes are not considered a preferred food for the commercially important *O. niloticus* or *O. aureus,* which are generally ineffective in eliminating already established stands of emergent weeds but are sufficiently herbivorous to prevent establishment of most emergent and many floating plants in aquacultural ponds. Tilapia are often considered filter-feeders because they can efficiently harvest planktonic organisms from the water column. However, tilapia do not physically filter the water through gill rakers as efficiently as silver and bighead carps (Popma and Masser 1999). Tilapia gills secrete a mucous that entraps planktonic cell (Teichert-Coddington et al. 1997). The plankton-rich bolus is then ingested. This mechanism allows tilapia to harvest microphytoplankton as small as 5 μm in diameter. Tilapia feed on concentrations of planktonic algal cells that float to the surface in late morning or early afternoon, which is sometimes mistakenly interpreted as suffering from oxygen stress.

Growth tests suggest protein requirements for blue and Mossambique tilapia of approximately 35 to 40 percent. In general, approximately 25 percent animal protein in artificial diets can be replaced with plant protein. Various studies on three species suggest differences; thus further refinement through comparative studies is desirable (Balarin and Haller 1982). Moreover, ontogenetic changes occur in protein requirements: smaller fish need more protein. The protein requirement for small Mozambique tilapia is 45 to 50 percent, and for adults it is about 20 to 25 percent. For Nile tilapia the corresponding figures are approximately 35 to 40 percent and to 20 to 25 percent; for blue tilapia, 36 percent and 26 to 36 percent.

Animal manures function as both fertilizer and feed in tilapia ponds. Swine and poultry excrete little carbohydrate that is digestible by tilapia, but approximately half the protein remaining in swine manures (undigested dietary protein, bacteria, and sloughed-off intestinal tissue) is digestible. In ponds with supplemental feeding, natural food organisms typically account for 30 to 50 percent of tilapia growth, whereas in full-fed channel catfish ponds only 5 to 10 percent of fish growth comes from the ingestion of natural food organisms (Teichert-Coddington et al. 1997). The ability of a single tilapia species to utilize so many types of natural food (as well as artificial feeds) makes polyculture with other fish less important than for common carp or Chinese carp culture.

Feeding tilapia do not disturb the pond bottom as aggressively as common carp; they effectively browse, primarily during daylight hours, on live benthic invertebrates and bacteria-laden detritus. In general, tilapia utilize natural food organisms so efficiently that standing crops of fish exceeding 3,000 $kg \cdot ha^{-1}$ can be sustained without supplemental feed in well fertilized ponds. The nutritional value of the natural food supply is important, even for commercial operations with heavy feeding, especially in terms of supplementing the nutritional deficiencies in agricultural by-products and incomplete feeds.

REFERENCES

Acere, T.O. (1988). The controversy over Nile perch, *Lates niloticus,* in Lake Victoria, East Africa. *NAGA* 11(4):3-5.

Allanson, B.R. (1966). A note on histological changes in *Tilapia mossambica* Peters exposed to low temperatures. *Newsletter Limnological Society of South Africa* 7:16-19.

Allanson, B.R., A. Bok, and I.N. Van Wyk (1971). The influence of exposure to low temperatures on *Tilapia mossambica* (Peters) (Cichlidae). II. Changes in serum osmolarity, sodium and chloride ion concentrations. *Journal of Fish Biology* 3:181-185.

Allanson, B.R. and R.G. Noble (1964). The tolerance of *Tilapia mossambica* (Peters) to high temperature. *Transactions of the American Fisheries Society* 93: 323-332.

Alvendia-Casauay, A. and V.S. Carino (1988). Gonadal sex differentiation in *Oreochromis niloticus.* In R.S.V. Pullin, T. Bhukasawan, K. Tonguthai, and J.L. Maclean (Eds.), *Second International Symposium on Tilapia in Aquaculture* (pp. 121-124). ICLARM Conference Proceeding 15, Manila, Philippines.

Aravindan, C.M. and K.G. Padmanabhan (1972). Formation of the micropyle in *Tilapia mossambica* (Peters) and *Stigmatogobious javanicus* (Blkr.). *Acta Zoologica Stockholm* 53(1):45-47.

Atz, J.W. (1957). The peregrinating *Tilapia. Animal Kingdom* 57(5):148-155.

Avault, J.W. and E.W. Shell (1968). Preliminary studies with the hybrid *Tilapia nilotica* × *T. mossambica. FAO Fishies Report* 44(4):237-242.

Avtalion, R.R. (1982). Genetic markers in *Sarotherodon* and their use for sex and species identification. In R.S.V. Pullin and R.H. Lowe-McConnell (Eds.), *The biology and culture of tilapias* (pp. 269-277). ICLARM Conference Proceedings 7, Manila, Philippines.

Avtalion, R.R. and I.S. Hammerman (1978). Sex determination in *Sarotherodon (Tilapia).* I. Introduction to a theory of autosomal influence. *Bamidgeh* 30: 110-115.

Avtalion, R.R. and M. Shlarobersky (1994). A whirling viral disease of tilapia larvae. *Israeli Journal of Aquaculture. Bamidgeh* 46:102-104.

Babiker, M.M. and H. Ibrahim (1979). Studies on the biology of reproduction in the cichlid *Tilapia nilotica* (L.): Gonadal maturation and fecundity. *Journal of Fish Biology* 14:437-448.

Balarin, J.D. and R.D. Haller (1982). The intensive culture of tilapia in tanks, raceways and cages. In J.F. Muir and R.J. Roberts (Eds.), *Recent advances in aquaculture* (pp. 265-355). Croom Helm, London, England.

Balarin, J.D. and J.P. Hatton (1979). *Tilapia, A guide to their biology and culture in Africa.* Stirling, Scotland: University of Stirling.

Baras, E., C. Prignon, G. Gohoungo, and C. Melard (2000). Phenotypic sex differentiation of blue tilapia under constant and fluctuating thermal regimes and its adaptive and evolutionary implications. *Journal of Fish Bio*logy 57:210-223.

Bardach, J.E., J.H. Ryther, and W.O. McLarney (1972). *Aquaculture, the farming and husbandry of freshwater and marine organisms.* New York: Wiley-Interscience.

Baroiller, J.-F., A. Fostier, C. Cauty, X. Rognon, and B. Jalabert (1996). Significant effects of high temperatures on sex-ratio of progenies from *Oreochromis niloticus* with sibling sex-reversed male broodstock. In R.S.V. Pullin, J. Lizard, M. Legendre, J.B. Amon Kothias, and D. Pauly (Eds.), *Proceedings of the 3rd international symposium on tilapia in aquaculture* (pp. 246-256). Manila, Philippines: ICLARM Conference Proceedings.

Baroiller, J,-F., Y. Guigen, and A. Fostier (1999). Endocrine and environmental aspects of sex differentiation in fish. *Cellular and molecular Biology Research* 55: 910-931.

Baroiller, J.-F. and B. Jalabert (1989). Contribution of research in reproductive physiology to the culture of tilapias. *Aquatic Living Resources* 2:105-116.

Bashamohideen, M. and V. Parvatheswararao (1972). Adaptations to osmotic stress in the fresh-water euryhaline teleost *Tilapia mossambica.* IV. Changes in blood glucose, liver glycogen and muscle glycogen levels. *Marine Biology (Berlin)* 16(1):68-74.

Behrends, L.L., J.B. Kingsley, and M.J. Bulls (1990). Cold tolerance in maternal mouth-brooding tilapias: Phenotypic variation among species and hybrids. *Aqua-culture* 85:271-280.

Bogdanova, L.S. (1970). The transition of *Tilapia mossambica* Peters larvae to active feeding. *Journal of Ichthyology* 10(3):241-248.

Boltt, R.E., B.J. Hill, and A.T. Forbes (1969). The benthos of some southern African lakes. Part I: Distribution of aquatic macrophytes and fish in Lake Sibaya. *Transactions of the Royal Society of South Africa* 38(3):241-248.

Bowen, S.H. (1982). Feeding, digestion and growth—qualitative considerations. In R.S.V. Pullin and R.H. Lowe-McConnell (Eds.), *The biology and culture of tilapias* (pp. 141-156). Manila, Philippines: ICLARM Conference Proceedings 7.

Boyd, C.E. (1990). *Water quality in ponds for aquaculture.* Auburn, AL: Auburn University.

Bruton, M.N. (1979). The food and feeding behavior of *Clarias gariepinus* (Pisces: Clariidae) in Lake Sibaya, South Africa, with emphasis on its role as a predator of Cichlidae. *Transactions of the Zoological Society of London* 35:47-114.

Bruton, M.N. and B.R. Allanson (1974). The growth of *Tilapia mossambica* Peters (Pisces: Cichlidae) in Lake Sibaya, South Africa. *Journal of Fish Biology* 6:701-715.

Bruton, M.N. and R.E. Boltt (1975). Aspects of the biology of *Tilapia mossambica* Peters (Pisces: Cichlidae) in a natural freshwater lake (Lake Sibaya, South Africa). *Journal of Fish Biology* 7:423-445.

Calhoun, R.E. and W.L. Shelton (1983). Sex ratios in progeny from mass spawnings of sex-reversed broodstock of *Tilapia nilotica. Aquaculture* 33:365-371.

Canagaratnam, P. (1966). Growth of *Tilapia mossambica* (Peters) at different salinities. *Bulletin of the Fisheries Research Station of Ceylon* 19(1/2):47-50.

Caulton, M.S. (1978). The effect of temperature and mass on routine metabolism in *Sarotherodon (Tilapia) mossambicus* (Peters). *Journal of Fish Biology* 13:195-201.

Caulton, M.S. and B.J. Hill (1973). The ability of *Tilapia mossambica* (Peters) to enter deep water. *Journal of Fish Bio*logy 5:783-788.

Chen, F.Y. (1966) The identity and origin of the Malayan and Zanzibar "strains" of *Tilapia mossambica. Tropical Fish Culture Research Institute of Malacca, Annual Report,* pp. 36-37.

———. (1969). Preliminary studies on the sex-determining mechanism of *Tilapia mossambica* Peters and *T. hornorum* Trewavas. *Verhandlungen Internationale Vereinigung f ür Theortische und Angewandte Limnologie* 17:719-924.

Chen, T.P. (1976). *Aquaculture practices in Taiwan.* London, England: Fishing News Books Ltd.

Chervinsiki, J. (1982). Environmental physiology of tilapias. In R.S.V. Pullin and R.H. Lowe-McConnell (Eds.), *The biology and culture of tilapias* (pp. 119-128). ICLARM Conference Proceedings 7, Manila, Philippines.

Chimits, P. (1955). Tilapia and their culture: A bibliography. *FAO Fisheries Bulletin* 8(1):1-33.

———. (1957). The tilapias and their culture: A second review and bibliography. *FAO Fisheries Bulletin* 10(1):1-24.

Chourrout, D. and J. Itskovich (1983). Three manipulations permitted by artificial insemination of tilapia: Induced gynogenesis, production of all triploid populations and intergeneric hybridization. In L. Fishelson and Z. Yaron (compilers), *International symposium on tilapia in aquaculture* (pp. 246-255). Israel: Tel Aviv University.

Clemens, H.P. and T. Inslee (1968). The production of unisexual broods of *Tilapia mossambica* sex-reversed with methyl testosterone. *Transactions of the American Fisheries Society* 97:18-21.

Courtenay, W.R., Jr. (1997). Tilapias as non-indigenous species in the Americas: Environmental, regulatory and legal issues. In B.A. Costa-Pierce and

J.E. Rakocy (Eds.), *Tilapia aquaculture in Americas* (Vol 1, pp. 18-33). Baton Rouge, Louisiana: World Aquaculture Society.

Courtenay, W.R., Jr., D.A. Hensley, J.N. Taylor, and J.A. McCann (1984). Distribution of exotic fishes in the continental United States. In W.R. Courtenay Jr. and J.R. Stauffer Jr. (Eds.), *Distribution, biology, and management of exotic fishes,* (pp. 41-77). Baltimore, MD: The Johns Hopkins University Press.

de Graaf, G.J., F. Galemoni, and E.A. Huisman (1999). Reproductive biology of pond reared Nile tilapia, *Oreochromis niloticus* L. *Aquaculture Research* 30:25-33.

De Silva, S.S. and J. Chandrasoma (1980). Reproductive biology of *Sarotherodon mossambicus,* and introduced species, in an ancient man-made lake in Sri Lanka. *Environmental Biology and Fisheries* 5:253-259.

Devlin, R.H. and Y. Nagahama (2002) Sex determination and sex differentiation in fish: An overview of genetic, physiological, and environmental influences. *Aquaculture* 208:191-365.

Don, J. and R.R. Avtalion (1988). Comparative study on the induction of triploidy in tilapia using cold- and heat-shock technique. *Journal of Fish Biology* 32:665-672.

Dunham, R.A. (1990). Production and use of monosex or sterile fishes in aquaculture. *CRC Review of the Aquatic Sciences* 2(1):1-17.

Dutta, O.K. (1979). Factors influencing gonadal differentiation in *Tilapia aurea* Steindachner. Doctoral Dissertation, Auburn University, Auburn, Alabama.

Eckstein, B. and M. Spira (1965). Effect of sex hormone on gonadal differentiation in the cichlid, *Tilapia aurea. Biological Bulletin* 129:482-489.

FAO (2004). State of the world fisheries and aquaculture. *Food and Agricultural Organization,* Part 1, Rome, Italy.

Ferguson, F.F. (1977). *The role of biological agents in the control of Schistosome-bearing snails.* Atlanta, GA: U.S. Public Health Service, Centers for Disease Control.

Fishelson, L. and Z. Yaron (compilers) (1983). *International Symposium on Tilapia in Aquaculture,* Israel: Tel Aviv University.

Fitzsimmons, K. (2000). Tilapia: the most important aquaculture species in the 21st century. In K. Fitzsimmons and J.C. Filho (Eds.), *Tilapia aquaculture, proceedings, 5th international symposium on tilapia in aquaculture* (pp. 3-8). Rio de Janeiro, Brazil: Panorama da Aquicultura.

Francis, R.C. (1992) Sexual lability in teleosts: Developmental factors. *Quarterly Review of Biology* 67(1):1-18.

Fryer, G. and T.D. Iles (1972). *The cichlid fishes of the Great Lakes of Africa.* Neptune, NJ: T.F.H. Publications.

Fuller, P.L., L.G. Nico, and J.D. Williams (1999). *Nonindigenous fishes introduced into inland waters of the United States.* Special Publication 27. Bethesda, MD: American Fisheries Society.

Gaigher, I.G. (1973). The habitat preferences of fishes from the Limpopo River system, Transvaal and Mozambique. *Kaedoe* 16:103-116.

Galman, O.R. (1980). Stages in the early development of *Tilapia nilotica. Fisheries Research Journal of the Philippines* 5(1):7-16.

Green, B.W., K.L. Veverica, and M.S. Fitzpatrick (1997). Fry and fingerling production. In H.S. Egna and C.E. Boyd (Eds.), *Dynamics of pond aquaculture* (pp. 215-243). Boca Raton, FL: CRC Press.

Guerrrero, R.D. (1975). Use of androgens for the production of all-male *Tilapia aurea* (Steindachner). *Transactions of the American Fisheries Society* 104:342-348.

———. (1982). Control of tilapia reproduction. In R.S.V. Pullin and R.H. Lowe-McConnell (Eds.), *The biology and culture of tilapia* (pp. 309-316). Manila, Philippines: ICLARM Conference Proceedings 7.

Guerrero, R.D. and L.A. Guerrero (1988). Feasibility of commercial production of sex-reversed Nile tilapia fingerlings in the Philippines. In R.S.V. Pullin, T. Bhukaswan, K. Tonguthai, and J.L. Maclean (Eds.), *Second International Symposium on Tilapia in Aquaculture* (pp. 183-186). Manila, Philippines: ICLARM.

Hargreaves, J.A. (2000). Tilapia culture in the southeast United States. In B.A. Costa-Pierce and J.E. Rakocy (Eds.), *Tilapia aquaculture in the Americas* (Volume 2, pp. 60-81). Baton Rouge, LA: World Aquaculture Society.

Hecht, T.A. (1980). Comparison of the otolith and scale methods of ageing, and the growth of *Sarotherodon mossambicus* (Pisces: Cichlidae) in a Venda impoundment (Southern Africa). *South Africa Journal of Zoology* 15(4):222-228.

Hickling, C.F. (1960). The Malacca *Tilapia* hybrids. *Journal of Genetics* 57:1-10.

———. (1968). Fish hybridization. *FAO Fisheries Report* 44(4):1-11.

Hopkins, K.D., W.L. Shelton, and C.R. Engle (1979). Estrogen sex-reversal of *Tilapia aurea. Aquaculture* 18:263-268.

Hughes, D.G. and L.L. Behrends (1983). Mass production of *Tilapia nilotica* seed in suspended net enclosures. In L. Fishelson and Z. Yaron (compilers), *International symposium on tilapia in aquaculture* (pp. 394-401). Tel Aviv, Israel: Tel Aviv University.

Hulata, G., S. Rothbard, and G. Wohlfarth (1981). Genetic approach to the production of all-male progeny of tilapia. *European Mariculture Society Special Publication* 6:181-190.

Hussain, M.G, D.J. Penman, B.J. McAndrew, and R. Johnstone (1993). Suppression of the first cleavage in the Nile tilapia, *Oreochromis niloticus* L.: A comparison of the relative effectiveness of pressure and heat shocks. *Aquaculture* 111:263-270.

Jalabert, B., P. Kammacher, and P. Lessent (1971). Determinisme du sexe chez les hybrids entre *Tilapia macrochir* et *Tilapia nilotica.* Etude de la sex-ratio dans les recroisements des hybrids de primiere generation par les especes parents. *Annal de Biologie et Animale Biochemistrie et Biophysica* 11:155-165.

Jalabert, B., J. Moreau, P. Planquette, and R. Billard (1974). Determinisme du sexe chez *Tilapia macrochir* et *Tilapia nilotica:* Action de la methyltestosterone dans l'limentation des alevins sur la differenciation sexuelle; proportion des sexes dans la descendance des males "inverses". *Annal de Biologie et Animale Biochemistrie et Biophysica* 14 (4B): 729-739.

Jalabert, B. and Y. Zohar (1982). Reproductive physiology in cichlid fishes, with particular reference to *Tilapia* and *Sarotherodon.* In R.S.V. Pullin and

R.H. Lowe-McConnell (Eds.), *The biology and culture of tilapias* (pp. 129-140). Manila, Philippines: ICLARM Conference Proceedings 7.

Job, S.V. (1969a). The respiratory metabolism of *Tilapia mossambica* (Teleostei) I. The effect of size, temperature and salinity. *Marine Biology (Berlin)* 2(2):121-126.

———. (1969b). The respiratory metabolism of *Tilapia mossambica* (Teleostei) II. The effect of size, temperature, salinity and partial pressure of oxygen. *Marine Biology (Berlin)* 3(3):222-226.

Josman, V. (1971). Some aspects of the effect of temperature on the respiratory and cardiac activities of the cichlid teleost *Tilapia mossambica.* Master's thesis, Rhodes University, Grahamstown, South Africa.

Jubb, R.A. (1967). *Freshwater fishes of Southern Africa.* Cape Town, South Africa: A. A. Balkema.

———. (1974). The distribution of *Tilapia mossambica* (Peters) 1852, and *Tilapia mortimeri* Trewavas 1966, in Rhodesian waters. *Arnoldia (Rhodesia)* 6(25):1-14.

King, J.E. and P.T. Wilson (1957). Studies on tilapia as skipjack bait. U.S. Fish and Wildlife Service, Special Scientific Report 22.

Kraft, A.V. and H.M. Peters. 1963. Vergleichende studien uber die Oogenese in der Gattung *Tilapia* (Cichlidae, Teleostei). *Zeitschrift für Zell forschung* 61:434-485.

Kutty, M.N. (1972). Respiratory qotient and ammonia excretion in *Tilapia mossambica. Marine Biology (Berlin)* 16(2):126-133.

Lanzing, W.J.R. and C.C. Bower (1974). Development of color patterns in relation to behavior in *Tilapia mossambica* (Peters). *Journal of Fish Biology* 6:29-41.

Lee, D.S., C.R. Gilbert, C.H. Hocutt, R. E. Jenkins, D.E. McAllister, and J.R. Stauffer (1980). *Atlas of North American freshwater fishes.* Raleigh: North Carolina State Museum of Natural History.

Lee, J.-C. (1979). Reproduction and hybridization of three cichlid fishes, *Tilapia aurea* (Stindachner), *T. hornorum* trewavas and *T. nilotica* (Linnaeus) in aquaria and in plastic pools. Doctoral dissertation, Auburn University, Auburn, Alabama.

Lessent, P. (1968). Hybridization of the genus *Tilapia* at the fish culture research station at Bouake, Ivory Coast. *FAO Fish Report* 44(4):148-159.

Lester, L.J., K.S. Lawson, T.A. Abella, and M.S. Palada (1989). Estimated heritability of sex ratio and sexual dimorphism in tilapia. *Aquaculture Fisheries and Management* 20:369-380.

Lever, C. (1996). *Naturalized fishes of the world.* New York: Academic Press.

Lim, C. (1989). Practical feeding—tilapias. In T. Lovell (Ed.), *Nutrition and feeding of fish* (pp. 163-183). New York: Van Nostrand Reinholt.

Little, D.C. and G. Hulata (2000). Strategies for tilapia seed production. In M.C.M. Beveridge and B.J. McAndrew (Eds.), *Tilapias: Biology and exploitation* (pp. 267-326). London, England: Kluwer Academic Publisher.

Lobel, P.S. (1980). Invasion by the Mozambique tilapia (*Sarotherodon mossambicus:* Pices; Cichlidae) of a Pacific Atoll marine ecosystem. *Micronesia* 16(2): 349-355.

Lowe (McConnell), R.H. (1959). Breeding behaviour patterns and ecological differences between *Tilapia* species and their significance for evolution within the genus *Tilapia* (Pisces: Cichlidae). *Proceedings of the zoological society of London* 132: 1-30.

Lovshin, L.L. (1982). Tilapia hybridization. In R.S.V. Pullin and R.H. Lowe-McConnell (Eds.), *The biology and culture of tilapias,* (pp. 279-308). Manila, Philippines: ICLARM, Conference Proceedings 7.

Lowe-McConnell, R.H. (1982). Tilapias in fish communities. In R.S.V. Pullin and R.H. Lowe-McConnell (Eds.), *The biology and culture of tilapias* (pp. 83-113). Manila, Philippines: ICLARM Conference Proceedings 7.

———. (1987). *Ecological studies in tropical fish communities.* London, England: Cambridge University Press.

Maar, A., M.A.E. Mortimer, and I. Van Der Lingen (1966). *Fish culture in central east Africa.* Rome, Italy: FAO.

Macintosh, D.J. and D.C. Little (1995). Nile tilapia *(Oreochromis niloticus).* In N.R. Bromage and R.J. Roberts (Eds.), *Broodstock management and egg and larval quality* (pp. 277-320). Oxford, U.K.: Blackwell Science.

Macintosh, D.J., T.J. Varghese, and G.P. Satyanarayana-Rao (1985). Hormonal sex reversal of wild spawned tilapia in India. *Journal of Fish Biology* 26:87-94.

Mair, G.C. (1993). Chromosome-set manipulation in tilapia—Techniques, problems and prospects. *Aquaculture* 111:227-244.

Mair, G.C., J.S. Abucay, D.O.F. Skibinski, T.A. Abella, and J.A. Beardmore (1997). Genetic manipulation of sex ratio for the large-scale production of all-male tilapia, *Oreochromis niloticus. Canadian Journal of Fisheries and Aquatic Sciences* 54:396-404.

Mair, G.C., D.J. Penman, A. Scott, D.O.F. Skibinski, and J.A. Beardmore (1987). Hormonal sex-reversal and the mechanisms of sex determination in *Oreochromis.* In K. Tiews (Ed.), *Selection, hybridization and genetic engineering in aquaculture* (Volume 2, pp. 289-299). Berlin, Germany: Heenemann.

Mair, G.C., A.G. Scott, D.J. Penman, J.A. Beardmore, and D.O.F. Skibinsksi (1991). Sex determination in the genus *Oreochromis.* 1. Sex reversal, gynogenesis and triploidy in *O. niloticus. Theoretical and Applied Genetics* 82:144-152.

Mair, G.C., A.G. Scott, D.J. Penman, D.O.F. Skibinski, and J.A. Beardmore (1991). Sex determination in the genus *Oreochromis.* 2. Sex reversal, gynogenesis and triploidy in *O. aureus* Steindachner. *Theoretical and Applied Genetics* 82:153-160.

McAndrew, B.J. (1993). Sex control in tilapiines. In J.F. Muir and R.J. Roberts (Eds.), *Recent advances in aquaculture, IV* (pp. 87-98). London, England: Blackwell Scientific Publishers.

———. (2000). Evolution, phylogentic relationships, and biogeography. In M.C.M. Beveridge and B.J. McAndrews (Eds.), *Tilapias: Biology and exploitation* (pp. 1-32). London, England: Kluwer Academic Publisher.

McAndrew, B.J., F.R. Roubal, R.J. Roberts, A.M. Bullock, and M. McEwen (1988). The genetics and histology of red, blond and associated color variants in *Oreochromis niloticus. Genetica* 76:127-137.

McBay, L.G. (1962). The biology of *Tilapia nilotica** Linneaus. *Proceedings of the Southeastern Association of the Game and Fish Commissioners* 15(1961):3-13 *(misidentified = blue tilapia).

Meschkat. A. (1967). The status of warm-water fish culture in Africa. *FAO Fisheries Report* 44(2):88-122.

Mires, D. (1977). Theoretical and practical aspects of the production of all-male *Tilapia* hybrids. *Bamidgeh* 29:94-101.

———. (1982). A study of the problems of the mass production of hybrid tilapia fry. In R.S.V. Pullin and R.H. Lowe-McConnell (Eds.), *The biology and culture of tilapias* (pp. 317-329). Manila, Philippines: ICLARM Conference Proceedings 7.

Mironova, N.V. (1975). Oxygen uptake by *Tilapia mossambica* (Peters). *Hydrobiological Journal* 11(2):73-74.

———. (1976). Changes in the energy balance of *Tilapia mossambica* in relation to temperature and ration size. *Journal of Ichthyology* 16(1):120-129.

———. (1977). Energy expenditure on egg production in young *Tilapia mossambica* and the influence of maintenance conditions on their reproductive intensity. *Journal of Ichthyology* 17(4):627-633.

Muller-Belecke, A. and G. Horstgen-Schwark (1995). Sex determination in tilapia *(Oreochromis niloticus),* sex ratios in homozygous gynogenetic progeny and their offspring. *Aquaculture* 137:57-65.

Myers, J.M., D.J. Penman, Y. Basavaraju, S.F. Powell, P. Baoprasertkul, K.J. Rana, N. Bromage, and B.J. McAndrew (1995). Induction of diploid androgenetic and mitotic gynogenetic Nile tilapia (*Oreochromis niloticus* L.). *Theoretical and Applied Genetics* 90:205-210.

Nakamura, M. and H. Takahashi (1973). Gonadal sex differentiation in *Tilapia mossambica,* with special regard to the time of estrogen treatment effective in inducing complete feminization of genetic males. *Bulletin of the Faculty of Fisheries of Hokkaido University* 24(1):1-13.

Neil, E.H. (1966). Observations on the behavior of *Tilapia mossambica* (Pisces, Cichlidae) in Hawaiian ponds. *Copeia* 1966(1):50-56.

Nelson, J.S. and six committee members (2004). Common and scientific names of fishes from the united states, Canada, and Mexico. American Fisheries Society, Special publication 29, Bethesda, Maryland.

Pandian, T.J. and R. Koteeswaran (1998). Ploidy induction and sex control in fish. *Hydrobiologia* 384:167-243.

Pandian, T.J. and S.G. Sheela (1995). Hormonal induction of sex reversal in fish. *Aquaculture* 138:1-22.

Pandian, T.J. and K. Varadaraj (1988). Techniques for producing all-male and all-triploid *Oreochromis mossambicus.* In R.S.V. Pullin, T. Bhukaswan, K. Tonguthai, and J.L. Maclean (Eds.), *Second international symposium on tilapia in aquaculture* (pp. 243-249). Manila, Philippines: ICLARM Conference Proceedings 15.

Pandian, T.J. and K. Varadaraj (1990). The development of monosex female *Orochromis mossambicus* broodstock by integrating gynogenesis with endocrine sex reversal technique. *Journal of Experimental Zoology* 255:88-96.

Panikkar, N.K. and P.R.S. Tampi (1954). On the mouth-breeding cichlid, *Tilapia mossambica* Peters. *Indian Journal of Fisheries* 1 (1/2):217-230.
Payne, A.I. and R.I. Collinson (1983). A comparison of the biological characteristics of *Sarotherodon niloticus* (L) with those of *S. aureus* (Steindachner) and other tilapia of the delta and lower Nile. *Aquaculture* 30:335-351.
Penman, D.J. and B.J. McAndrew (2000). Genetics for management and improvement of cultured tilapias. In M.C.M. Beveridge and B.J. McAndrew (Eds.), *Tilapias: Biology and exploitation* (pp. 227-266). London, England: Kluwer Academic Publisher.
Penman, D.J., M.S. Shah, J.A. Beardmore, and D.O.F. Skinbinski (1987). Sex ratios of gynogenetic and triploid tilapia. In K. Tiews (Ed.), *Selection, hybridization and genetic engineering in aquaculture* (Volume 2, pp. 267-276). Berlin, Germany: Heeneman.
Perez, J.E. and N. Maclean (1975). The haemoglobins of the fish *Sarotherodon mossambicus* (Peters): Functional significance and ontogenetic changes. *Journal of Fish Biology* 9:445-447.
Peters, H.M. (1963). Untersuchungen zum problem des angeborenen Verhaltens. *Die Naturwissie* 50(22):677-686.
Phelps, R.P., J.T. Arndt, and R.L. Warrington (1999). Methods for strain variations in sex ratio inheritance and methods for the contribution from the male and female genome to sex inheritance. *PD/A CRSP Technical Report* 16: 65-67.
Phelps, R.P. and T.J. Popma (2000). Sex reversal of tilapia. In B.A. Costa-Pierce and J.E. Rakocy (Eds.), *Tilapia Aquaculture in the Americas* (Volume 2, pp. 34-59). Baton Rouge, LA: World Aquaculture Society.
Philippart, J-Cl. and J-Cl. Ruwet (1982). Ecology and distribution of tilapias. In R.S.V. Pullin and R.H. Lowe-McConnell (Eds.), *The biology and culture of tilapias,* (pp. 15-59). Manila, Philippines: ICLARM Conference Proceedings 7.
Pienaar, U. De V. (1968). The freshwater fishes of the Kruger National Park. *Koedoe* 11:1-28.
Plumb, J.A. (1997). Infectious diseases of tilapia. In B.A. Costa-Pierce and J.E. Rakocy (Eds.), *Tilapia aquaculture in the Americas* (Volume 1, pp. 212-228). Baton Rouge, LA: World Aquaculture Society.
———. (1999). *Health maintenance and principle microbial diseases of cultured fishes.* Ames, IA: Iowa State University Press.
Popma, T. and M. Masser (1999). Tilapia: Life history and biology. *SRAC Publication No. 283,* Stoneville, Mississippi: Southern Regional Aquaculture Center.
Potts, W.T.W., M.A. Foster, P.P. Rudy, and G.P. Howells (1967). Sodium and water balance in the cichlid Teleost *Tilapia mossambica. Journal of Experimental Biology* 47(3):461-470.
Pullin, R.S.V. (Ed.) (1988). *Tilapia genetic resources for aquaculture.* Manila, Philippines: ICLARM Conference Proceedings 16.
Pullin, R.S.V. and R.H. Lowe-McConnell (Eds.) (1982). *The biology and culture of tilapias.* Manila, Philippines: ICLARM Conference Proceedings 7.
Pullin, R.S.V., M.L. Palomares, C.V. Casal, M.M. Dey, and D. Pauly (1997). Environmental impacts of tilapias. In K. Fitzsimmons (Ed.), *Tilapia aquaculture,*

Proceedings of the 4th International Symposium on Tilapia in Aquaculture (pp. 554-570). Ithaca, NY: N.E. Regional Agricultural Engineering Service.
Rana, K. (1988). Reproductive biology and the hatchery rearing of tilapia eggs and fry. In J.F. Muir and R.J. Roberts (Eds.), *Recent advances in aquaculture* (Volume 3, pp. 343-406). London, England: Croom Helm.
Riedel, D. (1965). Some remarks on the fecundity of *Tilapia* (*T. mossambica,* Peters) and its introduction into Middle Central America (Nicaragua), together with a first contribution towards the limnology of Nicaragua. *Hydrobiologica* 25:357-388.
Roberts, R.J. and C. Sommerville (1982). Diseases of tilapia. In R.S.V. Pullin and R.H. Lowe-McConnell (Eds.), *The biology and culture of tilapias* (pp. 247-263). Manila, Philippines: ICLARM Conference Proceedings 7.
Robins, C.R. and six committee members (1991). *Common and scientific names of fishes from the United States and Canada* (5th edn.) Bethesda, MD: American Fisheries Society Special Publication 20, American Fisheries Society.
Rothbard, S. and Y. Pruginin (1975). Induced spawning and artificial incubation of *Tilapia. Aquaculture* 5:315-321.
Rothbard, S., E. Solnik, S. Shabbath, R. Amado, and I. Grabie (1983). The technology of mass production of hormonally sex-inversed all-male tilapias. In L. Fishelson and Z. Yaron (compilers), *International Symposium on Tilapia in Aquaculture* (pp. 425-434). Tel Aviv, Israel: Tel Aviv University.
Schuster, W. H. (1952). Fish-culture in brackishwater ponds of Java. *Indo-Pacific Fisheries Council Special Publication No. 1.*
Scott, A.G., D.J. Penman, D.J. Beardmore, and D.O.F. Skinbinski (1989). The "YY" supermale in *Oreochromis niloticus* (L) and its potential in aquaculture. *Aquaculture* 78:237-251.
Shelton, W.L. (1989). Management of finfish reproduction for aquaculture. *Critical Reviews in the Aquatic Sciences* 1(3):497-535.
———. (1999). Nile tilapia gamete management for chromosome manipulation. *PD/A CRSP Technical Report* 16:69-77.
———. (2000). Methods for androgenesis techniques applicable to tilapia. *PD/A CRSP Technical Report* 17:51-55.
———. (2002a). Monosex tilapia production through androgenesis. *PD/A CRSP Technical Report* 19:45-51.
———. (2002b). Tilapia culture in the 21st century. In R.D. Guerrero and M.R. Guerrero-del Castillo (Eds.), *Tilapia farming in the 21st century* (pp. 1-20). Los Banos, Laguna, Philippines: Philippine Fisheries Association, Inc.
Shelton, W.L., K.D. Hopkins, and G.L. Jensen (1978). Use of hormones to produce monosex tilapia for aquaculture. In R.O. Smitherman, W.L. Shelton, and J.H. Grover (Eds.), *Symposium on culture of exotic fishes* (pp. 10-33). Auburn, AL, American Fisheries Society, Fish Culture Section.
Shelton, W.L., F.H. Meriwether, K.J. Semmens, and W.E. Calhoun (1983). Progeny sex ratios from inspecific pair spawnings of *Tilapia aurea* and *T. nilotica.* In L. Fishelson and Z. Yaron (compilers), *International Symposium on Tilapia in Aquaculture* (pp. 270-280). Tel Aviv, Israel: Tel Aviv University.

Shelton, W.L., D. Rodrigez-Guerrero, and J. Lopez-Macias (1981). Factors affecting androgen sex reversal of *Tilapia aurea. Aquaculture* 25:59-65.

Shelton, W.L. and R.O. Smitherman (1984). Exotic fishes in warmwater aquaculture. In W.R. Courtenay, Jr. and J.R. Stauffer, Jr. (Eds.), *Distribution, biology, and management of exotic fishes* (pp. 262-301). Baltimore, MD: The Johns Hopkins University Press.

Shelton, W.L., V. Wanniasingham, and A.E. Hiott (1995). Ovarian differentiation in common carp *(Cyprinus carpio)* in relation to growth rate. *Aquaculture* 137:203-211.

Shirak, A., J. Vartin, J. Don, and R.R. Avtalion (1998). Production of viable diploid mitogynogenetic *Oreochromis aureus* using the cold shock and it optimization through definition of cleavage time. *Israeli Journal of Aquacultur—Bamidgeh* 50:140-150.

Smith, E.S. and R.P. Phelps (2001). Impact of feed storage conditions on growth and efficacy of sex reversal of Nile tilapia. *North American Journal of Aquaculture* 63:242-245.

Smitherman, R.O. (1988). The status of wild and cultured genetic resources in the USA. In R.S.V. Pullin (Ed.), *Tilapia genetic resources for aquaculture* (p. 51). Manila, Philippines: ICLARM Conference Proceedings 16.

Soderberg, R.W. (1997). Factors affecting fish growth and production. In H.S. Egna and C.E. Boyd (Eds.), *Dynamics of Pond Aquaculture* (pp. 199-213). Boca Raton, FL: CRC Press.

Stickney, R.R. (1986a). Tilapia. In R.R. Stickney (Ed.), *Culture of nonsalmonid freshwater fishes* (pp. 57-89). Boca Raton, FL: CRC Press.

———. (1986b). Tilapia tolerance of saline waters: a review. *Progressive fish-culturist* 48:161-167.

Suresh, A.V. and K. Lin (1992). Tilapia culture in saline waters: A review. *Aquaculture* 106:201-226.

Swingle, H.S. (1958). Further experiments with *Tilapia mossambica* as a pondfish. *Proceedings of the Southeastern Association of Game and Fish Commissioners* 11:152-154.

Tayamen, M.M. and W.L. Shelton (1978). Inducement of sex reversal in *Sarotherodon niloticus. Aquaculture* 14:349-354.

Teichert-Coddington, D. and B.W. Green (1993). Tilapia yield improvement through maintenance of minimal oxygen concentrations in experimental grow-out ponds in Honduras. *Aquaculture* 118:63-71.

Teichert-Coddington, D.R., B. Manning, J. Eya, and D. Brock (2000). Concentration of 17a-methyltestosterone in hormone-treated feed: Effects of analytical technique, fabrication, and storage temperature. *Journal of the World Aquaculture Society* 31:42-50.

Teichert-Coddington, D.R., T.J. Popma, and L.L. Lovshin (1997). Attributes of tropical pond-cultured fish. In H.S. Egna and C.E. Boyd (Eds.), *Dynamics of pond aquaculture* (pp. 183-213). Boca Raton, FL: CRC Press.

Therezien, Y. (1968). Influence des saisons sur la croissance de populations de *Tilapia* dams les regions d'altitude de Madagascar. *FAO Fisheries Report* 44(4):328-333.

Thys Van Den Audenaerde, D.F.E. (1980). Good advice to hobbyists on questions of nomenclature and taxonomy. *Buntbarsche Bulletin* 81:7-9.

Timmermans, L.P.M. and A. Van Winkoop (1993). Larval development of gonads and germ cells in teleost fish. In B.T. Walther and H.J. Fyhn (Eds.), *Physiological and Biochemical Aspects of Fish Development* (pp. 67-70). Bergen and Norway: University of Bergen.

Tonguthai, K. and S. Chinabut (1997). Diseases of tilapia. In H.S. Egna and C.E. Boyd (Eds.), *Dynamics of Pond Aquaculture* (pp. 263-287). Boca Raton, FL: CRC Press.

Trewavas, E. (1966). A preliminary review of fishes of the genus *Tilapia* in the eastward-flowing rivers of Africa, with proposals of two new specific names. *Reviews in Zoology and Botany of Africa* 74(3-4):394-424.

———. (1968). The name and natural distribution of the "*Tilapia* from Zanzibar" (Pisces, Cichlidae). *FAO Fisheries Report* 44(5):246-254.

———. (1982). Tilapias: taxonomy and speciation, In R.S.V. Pullin and R.H. Lowe-McConnell (Eds.), *The biology and culture of tilapias* (pp. 3-13). Manila, Philippines: ICLARM Conference Proceedings 7.

———. (1983). *Tilapiine fishes of the genera* Sarotherodon, Oreochromis *and* Danakilia. Cornell Ithaca, NY: University Press.

Trombka, D. and R. Avtalion (1993). Sex determination in tilapia—A review. *Israeli Journal of Aquaculture-Bamidgeh* 45:26-37.

Tuan, P.A., G.C. Mair, D.C. Little, and J.A. Beardmore (1999). Sex determination and the feasibility of genetically male tilapia production in the Thai-Chitralada strain of *Oreochromis niloticus* (L). *Aquaculture* 173:257-269.

Turner, G.F. and R.L. Robinson (2000). Reproductive biology, mating systems and parental care. In M.C.M. Beveridge and B.J. McAndrew (Eds.), *Tilapias: Biology and exploitation* (pp. 33-58). London, England: Kluwer Academic Publishers.

Uchida, R.N. and J.E. King (1962). Tank culture of tilapia. *U.S. Fish and Wildlife Service, Fisheries Bulletin* 62:21-52.

Vass, E.F. and A.E. Hofstede (1952). Studies on *Tilapia mossambica* Peters (ikan Mudjair) in Indonesia. *Contributions of the Inland Fisheries Research Station Bogor, Indonesia* 1:1-68.

Whitfield, A.K. and S.J.M. Blaber (1979). The distribution of the freshwater cichlid *Sarotherodon mossambicus* in estuarine systems. *Environmental Biology and Fisheries* 4(1):77-82.

Willoughby, N.G. and D. Tweddle (1978). The ecology of the commercially important species in the Shire Valley fishery, Malawi. In R.L. Welcomme (Ed.), *Symposium on River and Floodplain Fisheries in Africa* (pp. 137-152). CIFA Technical Paper No. 5.

Wohlfarth, G.W. and G. Hulata (1983). Applied genetics of tilapias. *ICLARM Studies and Reviews 6,* Manila, Philippines.

———. (1989). Selective breeding of cultivated fish. In M. Shilo and S. Sarig (Eds.), *Fish culture in warm water systems: Problems and trends* (pp. 22-63). Boca Raton, FL: CRC Press.

Wohlfarth, G. and H. Wedekind (1991). The heredity of sex determination in tilapas. *Aquaculture* 92:143-156.

Yamamoto, T. (1958). Artificial induction of functional sex reversal in genotypic females of the medaka *(Oryzias latipes). Journal of Experimental Zoology* 137:227-260.

Young, J.A. and J.F. Muir (2000). Economics and marketing. In M.C.M. Beveridge and B.J. McAndrew (Eds.), *Tilapias: Biology and exploitation* (pp. 447-487). London, England: Kluwer Academic Publisher.

Zohar, Y. (1989). Fish reproduction: its physiology and artificial manipulation. In M. Shilo and S. Sarig (Eds.), *Fish culture in warm water systems: Problems and trends* (pp. 65-119). Boca Raton, FL: CRC Press.

Chapter 2

Prospect and Potential for Global Production

Kevin Fitzsimmons

INTRODUCTION

During the 1990s, tilapia products became an important commodity in the international seafood trade. Tilapia farming has grown from an industry based on fish introduced around the world by development agencies to feed the rural poor to highly domesticated livestock production with sales now exceeding $2 billion a year. The description of the tilapia as the aquatic chicken becomes more appropriate every day. As in the case of chicken farming, tilapia farming can be successful on any scale, from subsistence farmers with a few essentially feral fish in a pond to multinational corporations rearing highly domesticated fish with farms and processing plants in several countries. Tilapia have been domesticated more quickly and to a greater extent than any other group of fish. They surpassed salmonids in economic importance in 2004 and may eventually equal the carps.

World production of farmed tilapia exceeded 2,002,087 metric tons (mt) in 2004 (Figure 2.1), with China the major producer and consumer. The mainland provinces' production in 2003 was 897,300 mt, and Taiwan produced another 90,000 mt. Other Asian countries produced 440,000 mt. The United States is the world's major importer of tilapia. Its 2005 imports were 126,000 mt, with a value of $374 million, divided between frozen whole fish, frozen fillets, and fresh fillets. These products represent a live weight of 281,000 mt. Adding the 2005 domestic production of 9,000 mt sets the 2005 U.S. consumption of live weight fish at 290,000 mt, or 638 million pounds. Tilapia have already become one of the most important farm-raised fish and have an increasing role in the international seafood trade.

doi:10.1300/5513_02

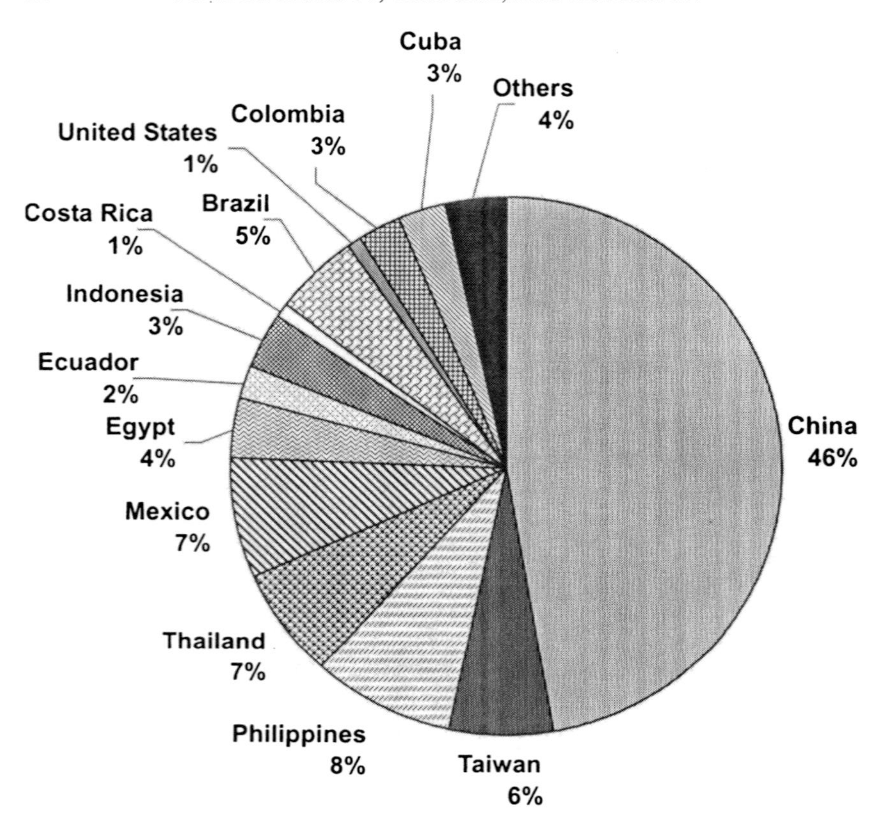

FIGURE 2.1. Origin of tilapia aquaculture production of 1,491,232 mt in 2002.

Humans living where tilapia are native have consumed the fish for centuries. Many common names are found for the fish across Africa, Asia, and the Middle East. In the 1930s, scientists realized the potential of the fish as a food source, efficiently transforming plant materials to fish biomass. Missionaries and others interested in improving the welfare of the rural poor determined that tilapia could be stocked into ponds and lakes as an additional food source. Tilapia could grow with minimal inputs and still make a high-quality contribution to the diet of poor farmers. Subsequently, tilapia were stocked into countries across the tropics and subtropics, often into reservoirs behind newly constructed dams. Tilapia are adept pioneer fish, efficiently utilizing available resources and capitalizing on new and altered ecosystems. Usually, the native fish fauna had not had time to respond to the new lacustrine environment, and officials felt that they were "improving" the fish community. With hindsight, it appears that tilapia have acted alongside other environmental changes to contribute to declines in native fish fauna (Pullin et al. 1997).

In a few countries, notably the cooler subtropical regions, tilapia were introduced to control aquatic vegetation since biological control of nuisance aquatic weeds was considered environmentally preferable to the use of toxic herbicides. In most cases this was an effective solution: tilapia would be stocked into irrigation canals or ponds, consume the aquatic weeds, and then die during cold weather. In some instances, however, the tilapia became established, with breeding populations surviving the cold periods.

The 1980s saw rapid progress in the domestication of tilapia, and in the late 1980s and 1990s large-scale aquaculture operations appeared, beginning the international trade in tilapia products. Global tilapia production and consumption in 2004 trailed behind that of the carps and exceeded that of the salmonids. The rapid improvements in domestication and wider consumption patterns may mean that tilapia may eventually overtake the carps to become the most important farmed fish.

Home Consumption

Home consumption may be the most important use for tilapia from a social standpoint. Millions of small farmers in more than 100 countries supplement their diets with occasional meals of tilapia. The wide distribution of tilapia species in the twentieth century and their adoption by development agencies as the "aquatic chicken" have led to their being available in virtually every tropical and subtropical region. Small-scale culture in ponds and in cages placed in larger bodies of water is encouraged and supported as a method of improving the diet and supplementing household income. The overall economic value is difficult to evaluate, but studies are available for some specific regions (Engle 1997; Neira and Engle 2001; Funez et al. 2001).

Microenterprise

Another important use of tilapia products is in microenterprises developed by small farmers. Some farmers raise extra fish for sale in local markets to supplement household income, and others market value-added product, typically at roadside restaurants. Several countries have seen an expansion in the number of family-owned and -operated roadside stands selling tilapia. The normal operation includes a family-operated pond where the fish are grown. Travelers are seated to have a drink while the proprietor goes to the pond to capture the fish. Alternatively, the fish may be held in a small tank for rapid retrieval. The fish are most frequently scaled, scored on the sides, and then quickly cooked in hot oil. The extremely high temperature cooks the fish quickly, effectively kills most parasites and pathogens, adds flavor, contributes to the caloric value of the meal, and disguises any algae-induced off-flavor. In developing countries, fried fish can help provide adequate dietary protein and calories.

OVERVIEW OF TILAPIA PRODUCED FOR INTERNATIONAL TRADE

In the mid-1980s the only tilapia product traded internationally was whole frozen tilapia grown in and exported from Taiwan. Tremendous increases have since taken place in the number of producing countries exporting tilapia and in the quantity and quality of the processed fish. Frozen fillets from Indonesia, Taiwan, and Jamaica and fresh fillets from Costa Rica, Jamaica, and Colombia opened a floodgate of demand.

The worldwide supply of tilapia grew quickly in the late 1990s (Figure 2.2). The tilapia industry is currently in the middle stage of the market developments previously seen in the salmon and shrimp industries. Commodity prices, which were dependent on wild catches and seasonal availability, are now influenced by the year-round availability and quality of farm-raised product. Rapid expansion of tilapia aquaculture in more than 100 countries

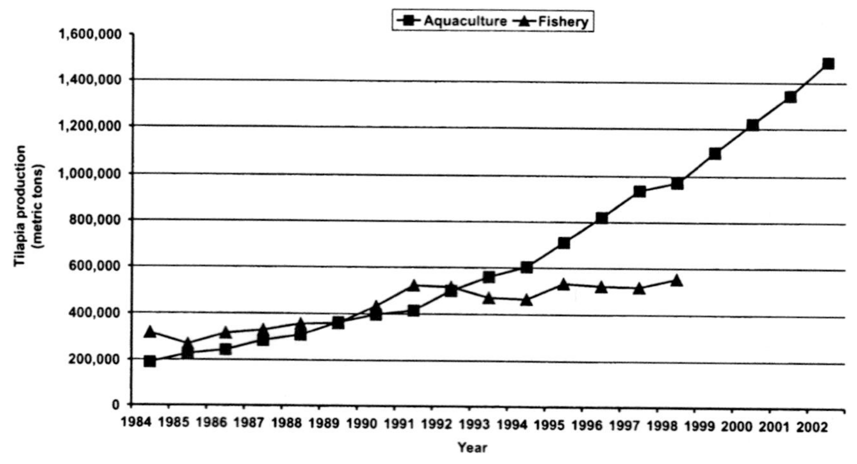

FIGURE 2.2. Production of tilapia from fisheries and aquaculture.

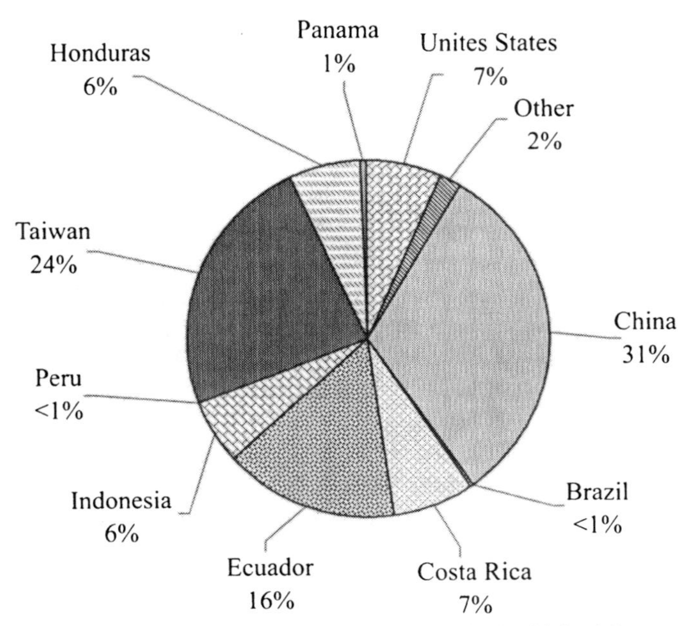

FIGURE 2.3. Sources of tilapia consumed in the United States.

has depressed prices to virtually every grower. Rapid improvements in technology, feeds, genetics, and the experience levels of farmers have allowed producers to reduce costs enough to remain profitable.

By 2002, more than 20 countries were exporting tilapia to the United States (Figure 2.3), and the product forms were many and diverse. Fillets are available in different sizes and packages, with skin on or skin off, deep-skinned, ozone-dipped, carbon monoxide–treated, individually quick-frozen, smoked, in sashimi grade, and as izumi-dai. Whole and gutted tilapia are available. Tilapia skins are sold in frozen, salted, and even deep-fried forms in international trade. Tilapia leather goods, including purses, wallets, belts, vests, and skirts, are now available.

The People's Republic of China has become the single most important producer and supplier of tilapia in the world (Figure 2.1). Chinese production of tilapia grew slowly for several years after Mozambique (Java) tilapia *(Oreochromis mossambicus)* were first introduced in the 1950s. The introduction of blue tilapia *(O. aureus)* and Nile tilapia *(O. niloticus)* in the late 1970s boosted the industry. Production of primarily male fish through the use of hybrid crosses or sex-reversal was a second advance, and the introduction of red-skinned strains has provided another economic impetus to the industry. With annual production of more than 900,000 mt, the mainland provinces produce just under half the world supply. Taiwan contributes another 90,000 mt, a significant portion of which is sold on international markets. In fact, the tilapia industries of Taiwan and the mainland provinces have nearly merged. Technologies, investments, and products flow freely.

The vast majority of tilapia produced in China is consumed locally, and opening additional markets within China is a central challenge to producers. Value-adding, finding new regional markets, product placement, and developing recipe cards have all been suggested as measures to further increase domestic Chinese markets (Fitzsimmons 2001).

Other countries in Southeast Asia are also major producers. Thailand produces more than 100,000 mt per year. Most is consumed domestically, but a significant proportion goes to international trade. The Philippines produces more than 122,227 mt per year, virtually all of which is consumed domestically. Indonesia, producing 169,310 mt, has strong domestic markets and considerable exports of frozen fillets to U.S. and European markets. Vietnam has a relatively small tilapia industry (25,000 mt), but the government has made a commitment to increase production, funding hatcheries and research projects. Malaysia is also rapidly upgrading its industry and hopes to become a significant exporter by 2008.

Tilapia production in the Philippines, Vietnam, and Thailand is certainly underreported in the official statistics. Officials and scientists in these countries provide several related explanations. First, the collection of fisheries statistics is secondary to improvement of physical and technological infrastructure in the aquaculture industry. Second, statistics reported to the United Nations Food and Agriculture Organization (FAO) are frequently several years old. Third, much of the extensive production occurs in remote locations and is either consumed in the producers' households or bartered and so never enters the cash economy, where it could be more easily enumerated (and taxed). Detailed surveys of household food consumption patterns, sales of fish in markets, and sales of feed support these observations.

Ecuador is second to China in the international trade of tilapia products. Shrimp farmers who were devastated by the multiple viral epidemics have converted many of their production facilities to tilapia farming (Figure 2.4). Much of the infrastructure (hatcheries, feedmills, ponds, harvest equipment, processing plants, cold storage, and marketing channels) was easily adapted to tilapia production, allowing Ecuador to quickly become a major producer and the single-most important player in the U.S. market for fresh tilapia fillets.

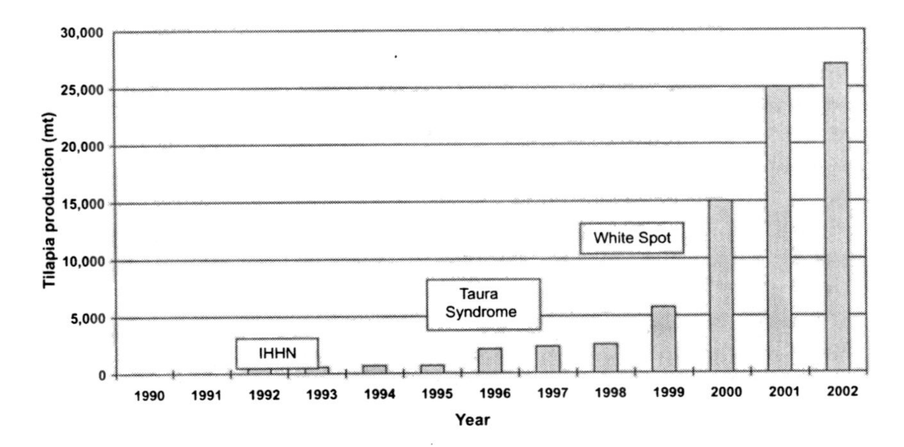

FIGURE 2.4. Ecuador tilapia production generally increases after each major viral shrimp disease outbreak (IHHN, Taura syndrome, and white spot).

Jamaica pioneered the market for fresh tilapia fillets in the United States in the late 1980s. Costa Rica, especially its Aquacorporacion Internacional operation, has overtaken Jamaican producers and further developed the fresh fillet market. The Aquacorp farm and processing plant in Cañas has consistently led the market with new product forms and packaging and in quality, and other producers in Latin America and Asia have improved their operations to match the innovations from Costa Rica. Production in Costa Rica is in a mix of intensive and semi-intensive ponds. New broodstocks are continuously evaluated to improve growth rate and fillet yield. Processing plant has been upgraded several times to increase efficiency and to provide new product forms and packaging. Aquacorporacion Internacional has also been a major investor in market development with its Rain Forest Brand. Other producer companies and countries have now determined that branding as well as generic advertising are needed to compete in this increasingly crowded market.

Zimbabwe (with its major producer, Lake Harvest, based at Lake Kariba) has become a major supplier to the European Union. Fish are grown in cages in Lake Kariba and then processed in a newly constructed plant. Fresh fillets are flown three times a week to Europe. Lake Kariba might be able to accommodate much more production, but political and environmental uncertainties have slowed further development.

REGIONAL PRODUCTION AND MARKETS

Several countries produce large amounts of tilapia but make little contribution to international trade. Egypt is a major producer, but virtually all fish are consumed domestically. Mexico, Colombia, the Philippines, and Cuba are major producers that have little impact on international trade. Mexico and Cuba utilize programs, described as "repopulation aquaculture," in which government hatcheries stock fingerlings into reservoirs, where they are later captured using conventional fishing gear. The FAO does not recognize this as aquaculture as it does not meet the requirement of clear ownership through the life cycle. However, these fish are included in national aquaculture statistics, and, as they are all exotic species and the repopulation efforts are ongoing, they are included in the national statistics presented in this chapter. The uneven quality of these fish and the diverse location of their capture minimize their potential as a product for international trade but provide an obvious benefit to local fishers.

Table 2.1 shows major tilapia products in international trade and the main suppliers.

The Americas

Mexico (110,000 mt, 2004)

Mexico has the largest national tilapia production in the Western Hemisphere (Fitzsimmons 2000). Approximately 40,000 mt are collected byartisanal fishermen from reservoir lakes. Considering that most of these lakes have feral tilapia populations, the efficacy of hatchery programs may be minimal, but all tilapia recovered are included in the Mexican

TABLE 2.1. Major tilapia products in international trade and main suppliers.

Country	Frozen whole fish	Frozen fillets	Fresh fillets	Skins	Leather goods
China	✓	✓			
Ecuador		✓	✓	✓	
Costa Rica			✓	✓	
Indonesia		✓			
Thailand		✓			
Honduras			✓	✓	
Zimbabwe			✓		
Brazil		✓			✓

aquaculture statistics. Many conventional tilapia farms also provide product to municipal markets across the country. Very little product is exported, although a few farms in the northeastern state of Tamaulipas have exported to Texas in the past. Mexico has great potential to be a major producer of tilapia, with its abundant water resources, highly trained biologists, proximity to markets, and North American Free Trade Agreement (NAFTA) trade benefits. One unforeseen NAFTA problem has been imports of Chinese tilapia to Mexico through the United States, where they were relabeled as U.S. products, which are free of tariffs under NAFTA. Mexican officials fear that low-cost Chinese products could adversely affect the Mexican producers.

Brazil (86,400 mt, 2004)

Tilapia were first introduced into Brazil in the 1950s. The fish stocked into newly built reservoirs did contribute to the fishing community's income and diet, but no significant market demand developed. The first significant market was for fee-fishing operations *pesque-pagues* in southeastern Brazil and small-farm production in the arid Northeast. Production in the Northeast followed the general development pattern of introduction for the rural poor. Consumption stayed at a relatively low level and the lack of a vigorous breeding program led to declining growth rates as a result of inbreeding and stunted stocks of fish in ponds owing to overpopulation. Tilapia made a small, but accepted, contribution to the diet of poor farmers.

In the fee-fishing industry of the more prosperous southeastern states, red-skinned varieties of tilapia were most popular with customers, for their color, and with operators, for their visibility in the ponds. People who were impressed with the quality of the red-skin fish from the fee-fishing operations began to ask for the fish through their regular seafood suppliers. Consumers in the São Paulo and Rio de Janeiro regions have increased demand to the level that restaurants frequently carry tilapia among their fish dishes.

Tilapia consumption in Rio de Janeiro received a boost from the Fifth International Symposium on Tilapia in Aquaculture in September 2000. Jomar Carvalho Filho, one of the conference organizers, developed a "gastronomic circuit" in which area restaurants competed over their signature tilapia dishes during the symposium. A front-page newspaper article further publicized the events and the products.

In the Northeast region, several large-scale tilapia farms have been built that use cage operations in reservoirs, flow-through pond farms, and intensive tank-based production systems. Most of these farms raise the Chitralada strain of Nile tilapia, developed in Thailand, as the primary production fish. Brazil will likely compete with China to be the major global tilapia producer. Already Brazil has one of the most diverse tilapia industries, using virtually every production method from low-input to high-tech, computerized recirculating systems. Brazil has the greatest water resources of any tropical country, and virtually the entire nation has adequate-to-ideal environmental conditions for tilapia culture. These environmental conditions, together with a large population, improving infrastructure, strong local demand, and easy access to U.S., European Union, and Latin American markets, indicate that Brazil will be a major force in international tilapia markets.

Another innovative contribution of the industry in Brazil is leather from tilapia skins, converting a waste product into a valuable commodity. Producers have developed a means of curing and dyeing skins that allows them to make a great variety of products. Belts, wallets, purses, and vests were followed by skirts, briefcases, shoes, and even wedding dresses.

Cuba (39,000 mt, 2001)

Cuba has a well-developed industry, with government-supported hatcheries stocking reservoirs and cage culture. Cuban scientists have selectively bred lines of *O. aureus* (Fonticiella and Sonesten 2000) and a genetically modified *O. niloticus* (Martinez et al. 1999), though it is unknown whether the genetically modified fish have been released for general stocking or public consumption. The cessation of the U.S. embargo on Cuban imports would likely lead to a rapid increase of exports of tilapia products to the United States. Cuba has ideal conditions for tilapia production, an established industry, and plenty of U.S. investors who would contribute to preparing products to U.S. specifications.

Colombia (40,000 mt, 2004)

Colombia was among the early exporters of tilapia products. In the late 1980s, Colapia and other large producers provided fresh and frozen fillets to domestic and U.S. markets. During the 1990s, the demand within Colombia grew so fast that exports were eventually dropped in favor of the more lucrative domestic markets. Additional farms have been built in Colombia, but the demand for tilapia products has grown even more. In 2001, Ecuador and Venezuela exported tilapia products to Colombia. With its strong domestic markets, experienced producers, and abundant resources, Colombia can be expected to reemerge as a major exporter once its political and economic instabilities are reduced.

Large cage operations in the state of Huila in southwestern Colombia may bring Colombian tilapia back to international markets. Cages based on the design of Chilean salmon farms have been manufactured in Colombia and stocked into deep reservoirs along with floating feed sheds, sophisticated hatcheries, aerators, and other sophisticated infrastructure.

Costa Rica (17,000 mt, 2004)

Costa Rica has the biggest tilapia farm in the world, in Cañas. Alongside this Aquacorporacion Internacional farm in northwestern Costa Rica, mentioned earlier as a world leader in the development of high-quality tilapia products, several other Costa Rican farms supply local and some international markets. Costa Rica has excellent environmental conditions and a populace intensely interested in maintaining clean water resources in the country. Considering the stable political and economic conditions in Costa Rica, growth of the industry is likely to continue.

Ecuador (35,000 mt, 2004)

Major viral outbreaks in the Ecuadoran shrimp industry have been followed in each case by a rise in tilapia production (Figure 2.4). The growth since the outbreak of white spot disease in 1998 has been the most dramatic. Rapid diversification by shrimp farmers into tilapia production has been stoked by the strong market demand for tilapia products and the apparent benefits of polyculture of tilapia and shrimp. Preliminary studies indicate that tilapia "condition" water by inducing a shift in the microbial populations (bacteria and algae) in the effluent water. Subsequently, shrimp seem to be less susceptible to bacterial and viral infections. Farmers across Ecuador have determined that tilapia farming benefits their overall aquaculture production by improving the survival and sale of shrimp, which have a much higher profit margin than tilapia. The diversification also spreads financial risk, provides additional products to customers, increases the scale of operations (thereby improving cash flows, relations with feed suppliers, and other vendors), and increases employment.

Honduras (15,000 mt, 2004)

Honduras has a long history of tilapia culture by many small pond-based farms across the country (Teichert-Coddington and Green 1997). The export trade has seen rapid expansion in volume since 1997. Large cage operations were begun in Lago Yojoa, near San Pedro Sula, in northeastern Honduras. An intensive raceway-pond system was also built near Lago Yojoa. Both operations produce high-quality products for export to the United States. Rapid growth of the industry and resulting exports are anticipated.

United States (9,000 mt, 2004)

Tilapia production in the United States is mostly devoted to producing live fish for Asian restaurants and groceries. Tilapia farms in most states supply local markets, and several clusters of farms supply regional markets. A concentration of farms in the Coachella Valley of Southern California supplies Los Angles and San Francisco markets. Another group of farms in the Snake River valley of Idaho supply markets in Portland, Seattle, and Vancouver. A small group of farms in western Arizona supply markets in Phoenix and Tucson. All three clusters use low-grade, geothermal waters to achieve year-round growth. The farms in Arizona and California also make use of their locations in low-elevation deserts to further warm their production systems. The farms in the West all grow their fish in round tanks or in-ground raceways. Some are covered with plastic for much of the year to further maintain temperatures, deter predation by birds, and reduce evaporation. In California, much of the water is recycled using settling ponds. In Arizona and Idaho, most of the water is used several times in consecutive growing units (tanks or raceways) before use for irrigation or eventual discharge to a stream.

In the Midwest, and in the eastern and Gulf coastal states, a number of farms supply local markets as well as the large Asian communities in Toronto, New York, Washington, DC, Atlanta, and Houston. These farms use intensive recirculating systems that produce fish year-round. They grow a combination of Nile tilapia, Mozambique tilapia, and red strains. Some deliver fish with their own trucks, others depend on live-haulers who cover huge areas. Hauling live fish from farms in Minnesota or Florida to Toronto and New York is not unusual.

The United States goes through periodic upheavals when gluts of fish are dumped on the market, usually as a result of farms in financial difficulties resorting to sales at any price into established markets rather than carefully developing new markets as earlier farms have done. Attempts to produce fish in the United States for the fillet market have been unable to compete with foreign products.

Production of tilapia in the United States saw a decline in late 2001, mostly related to decreased sales after September 11, 2001. Production levels in 2002 returned to approximately 9,000 mt. Total consumption of tilapia in the United States in 2005 reached a level of 290,000 mt, or 638 million pounds, of live weight fish (Figures 2.5 and 2.6). Statistics released by the National Marine Fisheries Service and the National Fisheries Institute reported U.S. annual per capita consumption increasing from 158 g (0.35 lbs) to 318 g (0.7 lbs) between 2001 and 2004, thus moving tilapia from the tenth most popular seafood item to the sixth most popular. This per capita consumption matches closely the national consumption figure when we calculate the conversion of whole fish to the fillets preferred by most U.S. consumers. The figure corresponds to only one fish per person per year, which leaves tremendous potential in the market.

The mix of tilapia products imported in to the United States has seen a drastic change in recent years (Figure 2.7). Whole frozen tilapia imports have increased, but the imports of fresh and frozen fillets have increased much more rapidly. This is a function of the increased level of processing occurring in the producing countries. Even more notable are the relative values of the products (Figure 2.8), with fresh fillets representing almost 50 percent of the value and less than 25 percent of the volume.

Canada (500 mt, 2002)

The province of Ontario has 5 tilapia farms that supply live fish to the Toronto market. All of these farms are in greenhouses and utilize supplemental

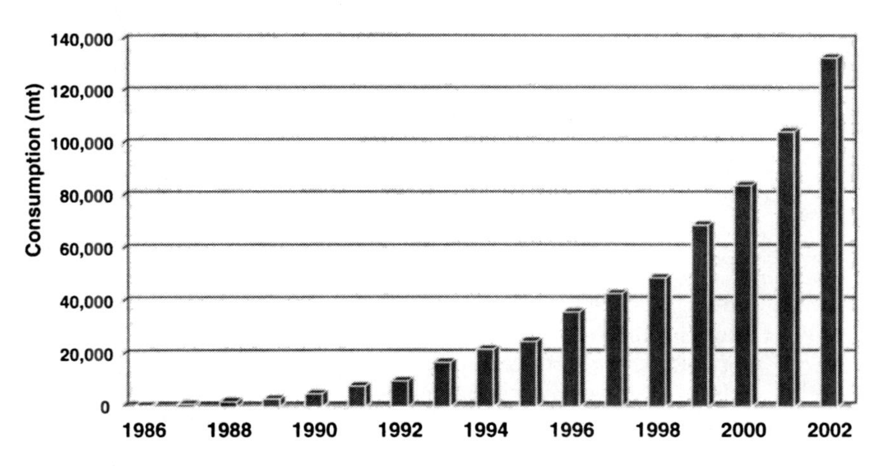

FIGURE 2.5. U.S. consumption of tilapia (live weight equivalent).

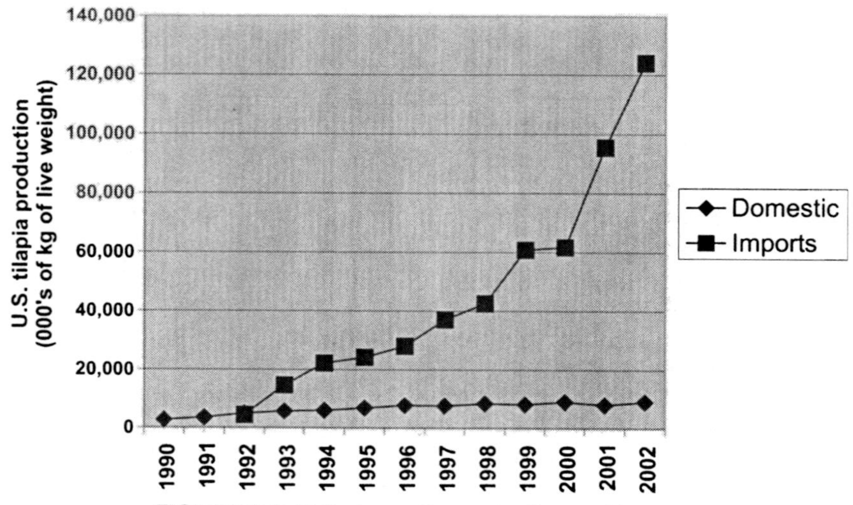

FIGURE 2.6. U.S. domestic production and imports.

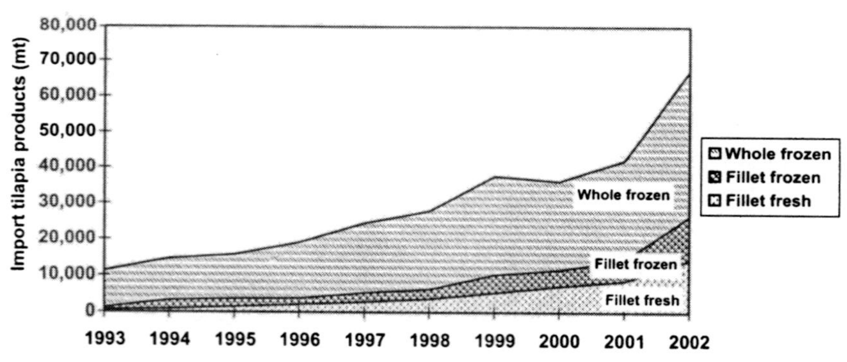

FIGURE 2.7. Volume of tilapia products imported into the United States.

heat to produce fish in the winter months. Some of the farms produce fish only in summer and overwinter breeders and fry. Calgary and Vancouver also have two or three farms each, utilizing greenhouses or well-insulated buildings to maintain warm temperatures. Little growth is expected from these farms as the supply to the live markets from the United States is very strong.

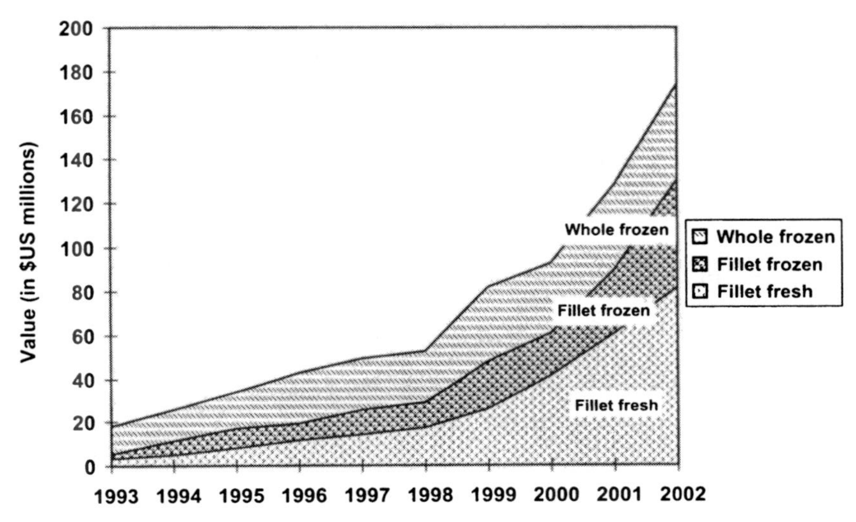

FIGURE 2.8. Value of tilapia products imported into the United States.

Africa

Egypt (220,000, 2004)

Tilapia, native to Egypt, has been part of the local diet for thousands of years. Tilapia held in cages are depicted in ancient paintings. Traditional aquaculture of tilapia is now supplemented with domesticated stocks. Most tilapia production in Egypt comes from small-holder farms consisting of fish grown in conjunction with rice production, small cages placed in irrigation drainage canals, polyculture ponds with carps, and brackish water ponds with mullet. A few large-scale production operations exist, most notable among them the government-supported farm at Maryut. This farm, originally built with international aid monies, provides fingerlings, training, and production ponds and includes a retail seafood outlet.

Among the most important domesticated tilapia strains is the Nile tilapia from Lake Manzala, a fast-growing strain that has been used in several national and international breeding programs. Research and development occurs at the Abbassa National Aquaculture Research Center, which is also the host site for a regional World Fish Center effort. The General Authority for Fish Resources Development (GAFRD) is another important contributor to the tilapia industry in Egypt. Further growth of the industry is predicted based on the Egypt's population size in Egypt and its hosting of the major research location in the region. Production of high-quality fillet products for export will take additional investment.

Eritrea (500 mt, 2002)

The Department of Fisheries of Eritrea has developed a novel full-strength seawater-based tilapia farm in conjunction with a private firm, Seawater Farms, Eritrea. This project grows a red strain of tilapia in effluent from a shrimp farm. The Department of Fisheries also maintains an aquaculture center near the capital, Asmara, which includes a tilapia hatchery.

Kenya (800 mt, 2004)

There are several species of tilapia that are native to Kenya. However, only the people living on the shores of Lake Victoria have traditionally been fishers and consumers of tilapia. Kenya had one of the first commercial tilapia farms, built near Mombasa in the 1970s. Tilapia have also been promoted for to small-scale producers since the 1970s. Production remains a cottage industry, with many farmers growing small numbers of fish. The National Aquaculture and Fisheries Center at Sagana, in central Kenya, is a focal point of research and extension efforts. Moi University, in Eldoret, also runs an active academic and research program. Market demand in Kenya is probably sufficient to absorb all local production for the foreseeable future.

Nigeria (3,948 mt, 2003)

Nigeria also has strong domestic markets and few, if any, tilapia product exports as international trade. In fact, most of the countries of western and central Africa are in a similar situation. Virtually all have significant wild harvests that are supplemented by farmed products.

Ivory Coast (706 mt, 2003)

The Ivory Coast developed a system of production in *acajas*, small ponds with bamboo or wooden posts driven into the sediment. The posts provide substrate for filamentous algae, which in turn are consumed by the tilapia. Posts are removed by hand before pond harvest.

Zimbabwe (5,000 mt, 2004)

Lake Kariba is the focal point for production in Zimbabwe. Lake Harvest Aquaculture Ltd. is the major producer and processor, and Nile tilapia are the primary production fish. Hatchery fry and fingerlings are produced onshore: fry in concrete tanks and fingerlings in ponds. Large modern cages, adapted from the offshore-style salmon cages, are placed in Lake Kariba, where they are serviced by boat.

Asia

China (897,300 mt, 2004)

China has developed its own model for the tilapia industry. Tilapia are grown in almost every province, but the vast majority are grown in the southern provinces of Guangdong, Hainan, Fujian, and Guangxi. Provincial hatcheries produce and sell most of the fry and fingerlings to small farmers. These state-managed hatcheries are meant to operate at a small profit. Farmers grow their fish in small family-operated ponds; production has rapidly intensified and now uses prepared diets. Feeds are produced by many regional feedmills. Farms with electrical connections have begun using feed blowers manufactured in China for export worldwide. Some farms have begun using paddle wheels for aeration, most often powered by diesel or electric engines. Ponds are harvested by either custom harvesters or teams sent from the processing plants. Processing plants are an uneven mix of old and new facilities. The newer plants are often joint ventures with partners from Taiwan, the United States, and Canada. Government-backed loans are available to support new plants.

A processing industry association is actively working to improve standards through the production and processing chain, following Hazard Analysis Critical Control Points (HACCP) techniques and ISO guidelines. As China has improved the consistency and quality of its tilapia products, its pricing structure has had to remain static to compete with products from many other producers. As the cost of living and conducting business in China increases, productivity gains will become more critical to maintain a competitive export trade. Additional growth is expected as production intensifies and additional markets are developed, domestically and in international trade.

Philippines (129,996 mt, 2003)

The Philippines has long been one of the leading research locales for tilapia aquaculture. Tilapia production is common throughout the archipelago and major research institutes are at Iloilo, Los Baños, and the Freshwater Aquaculture Center associated with Central Luzon State University in Muñoz. For many years, the International Center for Living Aquatic Resource Management (ICLARM) was based in Manila. The ICLARM (renamed the World Fish Center) is the aquaculture representative in the Consultative Group on International Agriculture Research (CGIAR), the a group of top-level international agriculture research centers.

Per capita consumption of tilapia in the Philippines, where it is considered a staple part of the diet, may be the highest of any country. Value-added forms are becoming more common as the middle class grows and urban workers look for time-saving preparations in their food purchases. Several processing plants are beginning to provide products for international trade.

Thailand (100,000+ mt, 2003)

Thailand is another major producer and research center for tilapia aquaculture. Virtually all of the production in Thailand occurs in the private sector. A number of very sophisticated hatcheries have developed their own selectively bred strains. Some of the red strains are barely recognizable as tilapia, with altered body forms yielding vastly increased fillets and reduced head and tail, for a high fillet-to-body-weight ratio. The Charoen Pokphand (C.P.) Group has a strain it trade-marked as "Top Tim," aggressively promoted with posters, table tents, and other advertising. The C.P. Group and others have developed supplies and market channels for frozen fillets. Uneven quality has slowed growth, but efforts to improve consistency should improve the market standing of Thai products. Thai farmers are also experimenting with polyculture of tilapia with shrimp and the development of strains of red tilapia that will thrive in higher-salinity waters.

Indonesia (169,310 mt, 2004)

One of the common names for *O. mossambicus* is the Java tilapia. The name was erroneously assigned because one of the early strains was distributed in the 1950s from the island of Java. However, these fish had been introduced in the 1930s from Africa. Tilapia are widely dispersed across Indonesia and now constitute one of the major aquaculture crops. However, the people of Java are still the largest per capita consumers of tilapia in the country. Major cage operations produce fillets for the domestic and international market, and additional farms are planned to further increase exports.

Vietnam (25,000 mt, 2004)

Vietnam has developed a series of government-operated hatcheries-cum-research centers selling high-quality fry and fingerlings to farmers. The goal is to rapidly increase the tilapia aquaculture sector, as they has been achieved with shrimp and basa (*Pangasius* sp.). Vietnam has begun exports of tilapia, but the levels are fairly small compared with production of shrimp and basa. The southern provinces have the best growing conditions, but most of the hatchery facilities have been built in the north and middle provinces, which are more in need of economic development.

Malaysia (22,560 mt, 2003)

There are strong domestic markets for tilapia in Malaysia. A mix of technologies is used, with production scattered across the country. Many new farms are under construction, with several targeting international markets. Aquaculture in Malaysia has received a boost with the transfer of ICLARM, renamed the World Fish Center, to Penang from Manila. Tilapia has been substituted for native fish in several traditional Malaysian recipes and has become one of the staple fish dishes.

Middle East

Israel (6,826 mt, 2003)

Two species of tilapia are native to Israel and have been part of the diet since biblical times. Israel was one of the early centers of domestication of tilapia and continues to be a focal point of research. Researchers in Israel discovered the skewed sex ratios of certain hybrid crosses and developed the primarily male culture systems that have been so important to the industry. New advances include superintensive production systems using greenhouses and integration with vegetable production. Cooperation between Arab and Israeli tilapia scientists has been invaluable to the continued expansion of the industry in the region.

Jordan (650 mt, 2003)

Tilapia comprises the bulk of aquaculture in Jordan, which has 5 major tilapia farms, all in the Jordan River Valley, and 15 small farms in 2004. Virtually all of the farms grow *O. niloticus*. The total production is approximately 1,000 mt/year; 600 mt are consumed in Jordan and 400 mt are exported. The major farms are all in the Jordan River Valley.

Jordan Valley Fish Farm (JVFF) is the biggest farm and a major exporter. To rear its fish JVFF uses an intensive recycling system enclosed in a greenhouse to maintain growth in the winter months. Algae and bacteria are encouraged in the production tanks both as part of the biofiltration system and as an additional food source for the omnivorous tilapia. JVFF uses a depuration system to purge its fish to guard against off-flavors. It also maintains a modern processing plant that produces high-quality products including gutted fish and boneless fillets for domestic and international markets.

The next largest farm is the Al Natoor farm, slightly north along the Jordan Valley. This farm utilizes a mix of raceways and ponds (concrete and soil-lined). A greenhouse is used to cover several raceways, gain additional warmth, and provide protection from predatory and parasite-carrying birds. The raceways ponds have paddle wheels, as does JVFF, to increase dissolved oxygen levels and water mixing.

Most of the farms use well water, because the Jordan River in this area is highly polluted with agricultural wastes. Some of these wells are in geothermal aquifers that provide warm- water for the few times in the year that the weather cools. Some farms utilize their effluents for the irrigation of field crops.

The smaller farms are at higher elevations surrounding Amman. These farms produce tilapia using greenhouses for year-round production and ponds for seasonal growth. Most of the fish from these farms goes directly to Amman or local markets as fresh fish on ice. All of the Jordanian farms feed pelleted diets, some of the feed is produced domestically and some imported from Saudi Arabia.

INDUSTRY PREDICTIONS AND OUTLOOK

1. Further farming intensification will occur in virtually every country.
2. Production will be 75 percent Nile tilapia, 20 percent red strains; blue tilapia and Mossambique tilapia will mostly be used for hybridization.
3. Use of highly domesticated strains will increase rapidly.

4. Production will be 50 percent in semi-intensive or intensive ponds, 25 percent in cages, and 10 percent in intensive recirculating systems, with the balance being "repopulation" and extensive pond operations.
5. Leather goods from skin will become a significant contributor to profitability.
6. Skins will increase in value as a source of gelatin and pharmaceuticals.
7. Processing and value-adding will intensify in producing countries.
8. Polyculture with shrimp will become common in most shrimp farming areas.
9. Production in the United States will increase slowly, intensifying current production.
10. Brazil will challenge China as the world's biggest producer.
11. Worldwide farmed-tilapia production will exceed 2,500,000 mt in 2006 and 3,000,000 mt by 2010, if not sooner.
12. Direct full-time employment in tilapia farming, processing, and marketing will exceed 1 million people worldwide.

Increasing Markets

1. Increasing demand/markets should come about in the producing countries.
2. Opening new markets will be especially critical in China, the European Union, Japan, Korea, Brazil, and the United States.
3. Many techniques can be used to build markets.
4. Many marketing techniques are free or low cost (product placement, samples, live tanks, Web sites).
5. All effective marketing formats require investment (personnel, time, and money).

REFERENCES

El Sayed, A.F.M. (2001). Tilapia culture in Arabia. GlobGlobal Aquaculture Advocate 4 (6): 46.

Engle, C. (1997). Marketing tilapias. In B.A. Costa-Pierce and J.E. Rakocy (Eds.), *Tilapia aquaculture in the Americas* (Volume 1, pp. 244-258). Baton Rouge, LA: World Aquaculture Society.

Fitzsimmons, K. (2000). Tilapia aquaculture in Mexico. In B.A. Costa-Pierce and J.E. Rakocy (Eds.), *Tilapia aquaculture in the Americas* (Volume 2, pp. 171-183). Baton Rouge, LA: World Aquaculture Society.

Fitzsimmons, K. (2001). China and international tilapia markets. In Di Gang (Ed.), *Aquatic Products Processing and Trading Symposium—China, Japan, Korea* (pp. 1-3). China Aquatic Products Processing and Marketing Association, Qingdao, China.

Fonticiella, D.W., and L. Sonesten (2000). Tilapia aquaculture in Cuba. In B.A. Costa-Pierce and J.E. Rakocy (Eds.), *Tilapia aquaculture in the Americas* (Volume 2, pp. 184-203). Baton Rouge, LA: World Aquaculture Society.

Funez, N.O., I. Neira, and C. Engle (2001). Supermarket outlets for tilapia in Honduras: An overview of survey results. In D. Meyer (Ed.), *Memoria: Sesiones de Tilapia* (Volume 6, pp. 82-86). Simposio Centroamericano de Acuacultura, September, Tegucigalpa, Honduras.

Martinez, R., A. Arenal, M.P. Estrada, F. Herrera, V. Huerta, J. Vazquez, T. Sanchez, and J.N.A. de la Fuente (1999). Mendelian transmission, transgene dosage and growth phenotype in trangenic tilapia (*Oreochromis hornorum*) showing ectopic expression of homologous growth hormone. *Aquaculture* 173: 271-283.

Neira, I., and C. Engle (2001). Mercados para tilapia en Nicaragua. Analisis descriptivo: Restaurantes, supermercados, y puestos en mercados municipales. In D. Meyer (Ed.), *Memoria: Sesiones de Tilapia* (Volume 6, pp. 87-91). Simposio Centroamericano de Acuacultura, September, Tegucigalpa, Honduras.

Pullin, R., M. Palomares, C. Casal, M. Dey, and D. Pauly (1997). Environmental impacts of tilapia. In K. Fitzsimmons (Ed.), *Proceedings of the Fourth International Symposium on Tilapia in Aquaculture* (pp. 554-572). NRAES-106. Ithaca, NY.

Teichert-Coddington, D., and B. Green (1997). Experimental and commercial culture of tilapia in Honduras. In B.A. Costa-Pierce and J.E. Rakocy (Eds.), *Tilapia aquaculture in the Americas* (Volume 1, pp. 142-162). Baton Rouge, LA: World Aquaculture Society.

Chapter 3

Control of Growth: Developments and Prospects

Brian S. Shepherd
Gregory M. Weber
Mathilakath M. Vijayan
Andre Seale
Larry G. Riley
M. Fernanda Rodriguez
N. Harold Richman III
Tetsuya Hirano
E. Gordon Grau

INTRODUCTION

Growth and development are governed in fish, as in all animals, through the orderly release of hormones from the neuroendocrine system, which integrates environmental, physiological, and genetic information. Studies on birds and mammals have shown that hormones act to promote growth through specific actions on appetite, digestion, nutrient absorption, and production of muscle and bone (Buttery and Sinnett-Smith 1984; Clancy et al. 1984; Bondi 1989; Garssen and Oldenbroek 1992; Collier and Byatt

Drs. Brian Brazil, Scott Gahr, and Yniv Palti are gratefully acknowledged for their insightful comments and assistance in the preparation of this chapter. The research reported herein and the preparation of this chapter were supported, in part, by grants from the National Research Initiative Competitive Grants Program (NRI/USDA) award number 2002-35206-11629 to B.S.S. and award number 9835206644 to E.G.G. Additional support was obtained through the Edwin W. Pauley and Barbara Pagen-Pauley Foundation and through a grant/cooperative agreement from the National Oceanic and Atmospheric Administration, Project #R/AQ-37 and #R/AQ-62, which is sponsored by the University of Hawaii Sea Grant College Program, SOEST, under Institutional Grant Nos. NA36RG0507 and NA86RG0041 and from the NOAA Office of

doi:10.1300/5513_03

1996; Sillence 2004). Understanding the neuroendocrine system's role in the regulation of growth is therefore essential to understanding how to control growth, and efforts to manipulate growth often involve the intentional manipulation of this system. Even methods to control growth by manipulation of environment, nutrition, and genetics by selection ultimately manipulate the neuroendocrine system and, therefore, would benefit from an understanding of how they affect and involve it. For these reasons, this discussion of the control of growth in tilapia will center on the neuroendocrine regulation of growth.

Although knowledge of the neuroendocrine system has been exploited in the production of mammalian and avian livestock to enhance growth and the efficiency of feed utilization, no systematic application of such endocrine manipulation has taken place in finfish aquaculture, although production would almost certainly benefit greatly from the application of endocrine-based technologies. The development and use of hormone-based technologies in finfish aquaculture are lagging largely because of the lack of basic information on growth regulators and their actions in fish and because of social resistance to such potential strategies as administration of exogenous hormones and transgenesis. A greater understanding of these issues is required to identify and develop the most appropriate technologies for use in aquaculture.

Recent work has focused on clarifying the roles in fish of growth-regulating hormones of the somatotropic axis, which include growth hormone (GH), prolactin (PRL), and the insulin-like growth factors (IGF-I, IGF-II). It is becoming increasingly clear that IGF-I, a GH-dependent anabolic polypeptide, is largely responsible for production of somatic growth in teleost fishes. Thus, the identification of the actions and interactions of these hormones is fundamental to developing and evaluating strategies for altering growth. This becomes even more apparent when one considers that these same hormones are central regulators of osmoregulation (homeostatic mechanisms to regulate salt and water balance), one of the most energy-demanding components of basal metabolism in fishes (Febry and Lutz 1987; Brill et al. 2001) and a process essential to survival. In this regard, many

Sea Grant, Department of Commerce. The views herein are those of the authors and should not be interpreted as necessarily representing the official policies, either expressed or implied, of the U.S. government or any of its subagencies. This chapter is submitted for publication with the understanding that the U.S. government is authorized reproduce and distribute reprints for governmental purposes. Mention of trade names mercial products in this article is solely for the purpose of providing specific in- and does not imply recommendation or endorsement by the U.S. government subagencies. UNIHI-SEAGRANT-BC-96-01.

species of tilapia are euryhaline and can be reared in seawater or fresh water with dramatic differences in expression of these key growth regulators depending on environmental salinity. In fact, the Mozambique tilapia (*Oreochromis mossambicus*) is an important model organism for the study of osmoregulation in vertebrates, and perhaps more is known of the endocrine control of osmoregulation in this species than in most other fish (Nishioka et al. 1988; Grau et al. 1994; McCormick 2001; Manzon 2002). Thus, continued research in this teleost is expected to guide the selection of growth-promoting methods most appropriate to a particular set of rearing conditions and for specific species of tilapia or strains with altered performance traits as a result of selection or transgenesis, so that optimum growth and feed efficiency are achieved without unintended consequences for metabolism or salt and water balance. Because of their centrality to the regulation of growth and osmoregulation, our discussion is based around hormones central to the somatotropic axis, with special attention to their place in environmental physiology.

The androgenic steroid 17α-methyltestosterone (MT) is the only hormonal growth promoter approved for commercial use in tilapia in the United States (under FDA Investigational New Animal Drug exemption, or INAD; Schnick 2003), so its actions are also emphasized in this chapter. This synthetic androgen has been approved for sex inversion of female fry to phenotypic males, which grow faster. This treatment is used in part because of the advantages of having all-male (monosex) populations for tilapia culture: tilapia are prolific breeders, so having mixed sexes in grow-out conditions results in overpopulation of stunted animals or a wide distribution in the size of animals harvested. Although MT treatment has been used only for sex reversal, research suggests that early MT treatment may have growth-promoting effects that extend beyond those attributable to the animals being male alone. Studies undertaken to evaluate the mechanism of its growth-promoting effects have shown that MT acts in part by altering the GH/IGF axis and protein utilization, which is an important component of feed utilization. These findings not only provide insight into the actions of MT but also point to the important implications of sex steroids for growth in tilapia.

Many methods to control growth are based on endocrine manipulations, and although they have been shown effective, they are not without controversy. Consequently, we discuss some of what is known of the endocrine regulation of growth in tilapia, endocrine-based approaches to altering growth performance, and considerations for future research.

RELEVANCE OF NEUROENDOCRINE RESEARCH TO TILAPIA AQUACULTURE

Seafood consumption continues to grow, even though many fisheries decline or approach collapse (NOAA 1988; Harvey 1993; Naylor et al. 2000). These forces have made the United States the second largest importer of seafood products. Approximately 75 percent of the seafood products consumed in the United States are imported, and seafood imports contribute approximately 8 to 10 billion dollars annually to the national trade deficit, making them third only to petroleum and automobiles among commodities (NOAA 2002; Iwamoto 2003; Williams 2004). Increasing demand for seafood has been partly offset by incremental increases in aquaculture production. During the twenty-first century, it is predicted that aquaculture will produce greater than 25 percent of the seafood supply in the United States. This increase is occurring as conventional culture methods near their capacity to realize cost-effective increases (Barge and Phillips 1997; Naylor et al. 2000; Tidwell and Allan 2001; Pauly et al. 2002). New biotechnologies and strategies must be identified and developed if production and efficiency are to be increased (Chen et al. 1996; Bartley 1997a; Halvorson and Quezada 1999; Agresti et al. 2000; Hulata 2001).

Tilapia are a major source of protein in the human diet around the world, and U.S. imports of tilapia over the first half of 2003 were almost 100 million pounds, up 43 percent from the same period in 2002 (USDA/ERS 2003). For instance, in 2003 tilapia imports were 200 to 210 million pounds, and were comparable to the production of veal in the United States in the same year. The overall value of tilapia imports at this time was estimated to be between 240 and 250 million dollars (USDA/ERS 2003).

The rise in aquaculture, including tilapia aquaculture, makes it inevitable that efforts will be directed toward controlling growth, reproduction, and immunity of finfish to enhance production. As in the production of mammalian and avian livestock, finfish culture must develop and exploit knowledge of the neuroendocrine system to improve production (Buttery and Sinnett-Smith 1984; Clancy et al. 1984; Bondi 1989; Garssen and Oldenbroek 1992; Johnsson and Björnsson 1994; Collier and Byatt 1996). Thus far, such endocrine manipulation (via genetic improvement or a combination of other approaches) has not been systematically applied to finfish aquaculture (Mair and Little 1991; Baroiller and D'Cotta 2001; Hulata 2001, 2002; Gabillard et al. 2003; Kohler 2004). Success will depend on developing a thorough understanding of the role that growth-promoting hormones play in other aspects of the biology of commercially important finfish species (Grau et al. 1992) and potential impacts to the environment

(Naylor et al. 2000; Arukwe 2001; Tidwell and Allan 2001; Maclean et al. 2002; Muir and Howard 2002; Pauly et al. 2002; Subasinghe et al. 2003; Westers 2003; Kohler 2004). Studies are clearly needed for characterizing the full spectrum of actions of endogenous growth-regulating hormones and exogenous growth promoters for use in aquaculture species, taking into account the environmental context under which they may be used.

HORMONAL CONTROL OF GROWTH IN TILAPIA

As mentioned, the vertebrate somatotropic axis is formed by the interplay between neuropeptides, GH, PRL, IGF-I and IGF-II, the insulin-like growth factor binding proteins (IGFBPs) and the IGF-receptors (Jones and Clemmons 1995; Duan 1997; Simmen et al. 1998; Butler and Le Roith 2001; Le Roith, Scavo, et al. 2001) (Figure 3.1). The somatotropic axis can be further divided into proximal and distal components. The proximal components include neuropeptides that regulate the release of GH and PRL from the pituitary such as somatostatin (inhibitory for GH), growth hormone–releasing hormone (stimulatory for GH), gonadotropin-releasing hormone (stimulatory for GH and PRL) (Ball 1965; Nagahama et al. 1975; Rivas et al. 1986; Kelley et al. 1988; Nishioka et al. 1988; Peter et al. 1990; Melamed, Eliahu, et al. 1995; Sukumar et al. 1997; Weber et al. 1997; Melamed et al. 1998; Mommsen 1998, 2001; McMahon et al. 2001), and "ghrelin" (stimulatory), a recently discovered hormone secreted by the stomach that is found throughout the central nervous system (Kojima et al. 1999; Galas et al. 2002). Additional proximal components include the GH and PRL receptors, the GH binding protein (truncated receptor), and IGF-I and IGF-II, which have been identified in teleost fishes (Duan 1997; Reinecke and Collet 1998; Sohm et al. 1998; Zhang and Marchant 1999; Lee et al. 2001).

Distal components of the somatotropic axis include the IGF-I (tyrosine kinase), IGF-II (mannose-6-phosphate), insulin (tyrosine kinase), and hybrid receptors and the IGFBPs (Daughaday et al. 1985; Kelley et al. 1996; Chan et al. 1997; Greene and Chen 1998, 1999a,b,c; Maestro et al. 1998; Degger et al. 2000, 2001; Kelley et al. 2001; Mendez, Planas, et al. 2001; Mendez, Smith, et al. 2001; Kelley et al. 2002; Biga et al. 2004). Interestingly, the pattern of GH release, which has been shown to be episodic in vertebrates, is integral to the regulation of the distal components of the somatotropic axis. Moreover, these patterns are sex specific and regulated by neuroendocrine factors, steroids, and the environment (Figures 3.1 and 3.2). For instance, the administration of exogenous GH in a pulsatile rather than a continuous manner is most effective at increasing circulating IGF-I

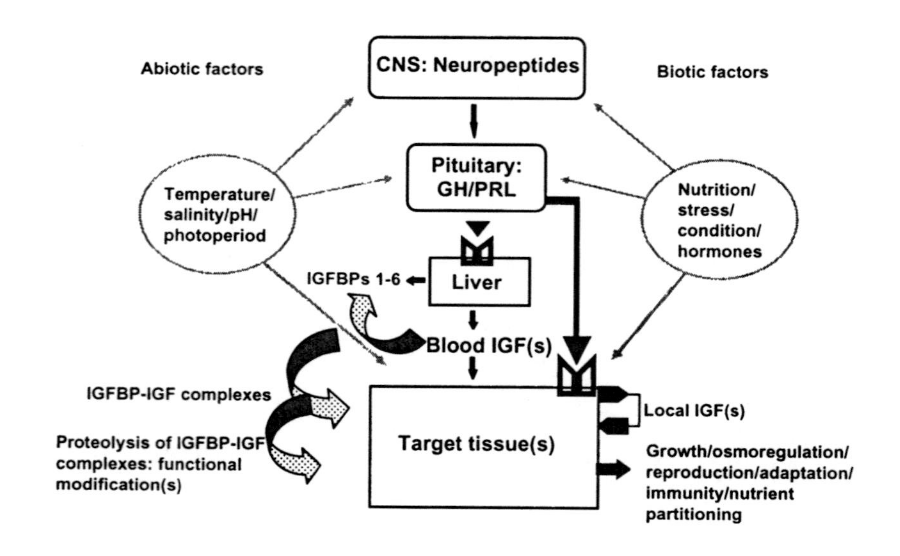

FIGURE 3.1. A schematic representation of the vertebrate somatotropic axis. This axis is regulated, at a number of levels, by a variety of abiotic and biotic factors. A main point of integration is the pituitary, thus affecting the synthesis and release of growth hormone (GH) and prolactin (PRL), among other hormones that are not shown. Other than the pituitary, the function of other endocrine and nonendocrine tissues can be altered by abiotic and biotic factors in such a way as to affect the physiology and endocrinology of the organism. The release of GH and/or PRL can elicit changes in the production of IGFs and the IGFBPs in target tissues that are involved with growth, development, reproduction, immune function, and osmoregulation. Little is known regarding the biology of IGF-IGFBP complexes in teleosts. Another layer of complexity is the function and interaction of the proteins in these polypeptide heteroduplexes, which can be altered by post-translational modifications of the IGFBPs (e.g., glycosylation and phosphorylation), or by proteolysis of the IGFBPs in circulation or upon binding to cognate cell surface receptors.

levels and consequently growth in the rat (Jansson et al. 1982; Clark et al. 1985; Bick et al. 1992). In contrast, continuous infusions of GH are generally more effective at maintaining the distal (some proximal) components of the somatotropic axis (Maiter et al. 1988, 1992; Bick et al. 1990, 1992; Gevers et al. 1996), although growth is still observed. Episodic GH secretory patterns also have been shown to exist in teleosts (carp and trout) (Niu et al. 1993; Zhang et al. 1994); however, the regulation of these patterns and their significance in growth are still unclear.

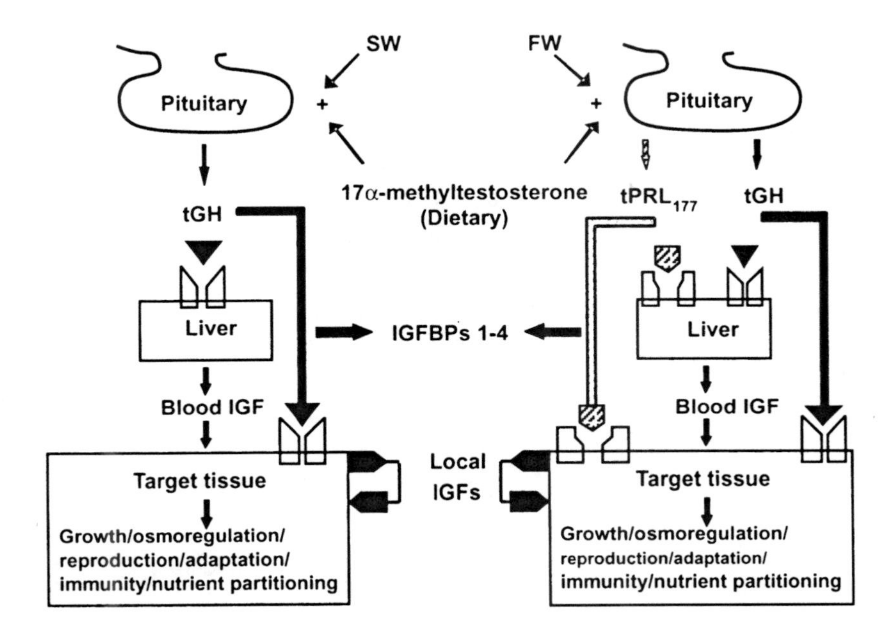

FIGURE 3.2. A schematic representation of the somatotropic axis in the tilapia. Here, the general effects of a specific abiotic factor, environmental salinity, and a biotic factor, dietary administration of the anabolic steroid, 17α-methyltestosterone (MT), are illustrated. In seawater (SW), levels of growth hormone (GH) are elevated, which, in turn, increase levels of IGF-I as well as the higher molecular weight IGFBPs. The net effect is an overall stimulation of the GH-IGF axis, and consequently growth. In SW, levels of the tilapia *(O. mossambicus)* PRLs are significantly lower ($tPRL_{188}$) or undetectable ($tPRL_{177}$); thus it is likely that GH is the only pituitary hormone with somatotropic activity in this environment. The singular effects of MT on the IGFBPs in tilapia have not been independently investigated. In fresh water (FW), levels of the two PRLs are significantly elevated and GH is lower. Consequently, the level of activation of the GH-IGF axis is lower than in SW-adapted tilapia. In the FW example, levels of the two PRLs in the blood and the pituitary (including mRNA) are significantly higher, whereas GH levels are lower than values seen SW-adapted tilapia. However, the addition of MT to the diet further augments levels of GH and the PRLs in FW-adapted tilapia, thus increasing growth. In the FW-adapted animals, GH and PRL_{177} may be acting as growth-promoting hormones (see Figure 3.3); however, growth rate is lower than comparable SW treatments. The lower growth rate in FW-adapted animals is likely due to the increased metabolic demands (O_2 consumption) associated with FW-adaptation. Not shown in this figure are the various long-loop (e.g., IGF-I inhibition of GH release) and short-loop (e.g., GH inhibition of GH secretion) feedback pathways that regulate pituitary function.

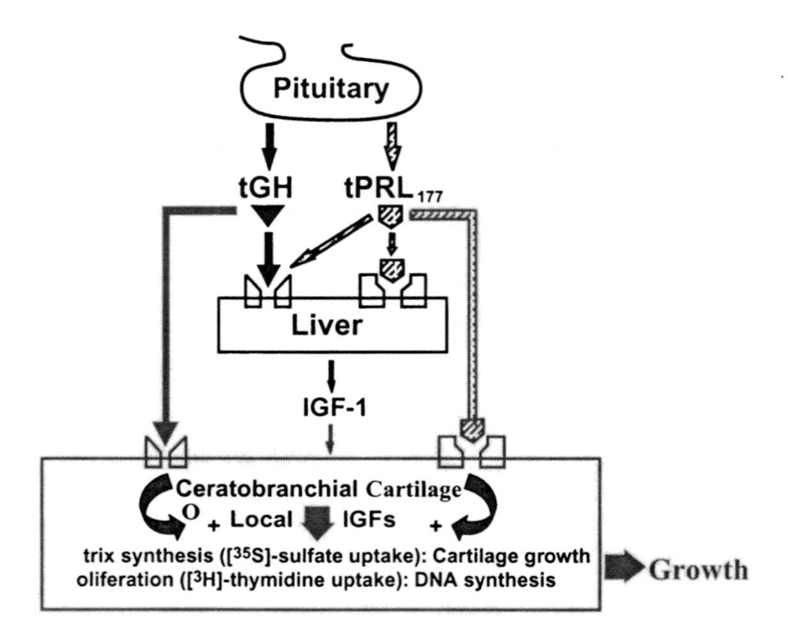

FIGURE 3.3. A schematic representation of the somatotropic axis in the tilapia, and of how GH-dependent, IGF-like activity is measured using the ceratobranchial cartilage assay. This diagram shows our interpretation of how both GH and PRL may regulate growth in the tilapia. Results of early in vivo studies show that the smaller form of tilapia PRL ($tPRL_{177}$) may be part of the somatotropic axis in *O. mossambicus.* GH and $tPRL_{177}$ can stimulate IGF-I mRNA in the gill and liver of hypophysectomized tilapia, whereas $tPRL_{188}$ was without effect. Replacement of either of these native hormones (GH and $tPRL_{177}$) stimulated uptake of $[^{35}S]$-sulfate (cartilage growth) and $[^{3}H]$-thymidine into ceratobranchial cartilage explants in vitro, which is a result of increased levels of GH-dependent plasma IGF-like (IGF-I, IGF-II and Insulin) activity. In this same study, it was shown that the somatotropic actions of $tPRL_{177}$ are mediated by binding to the tilapia GH (tGH) receptor as measured by displacement of $[^{125}I]$-tGH from its receptor, whereas $tPRL_{188}$ did not displace $[^{125}I]$-tGH from its receptor. We propose that in FW, where $tPRL_{177}$ levels are sufficiently high to compete with GH, growth can be mediated by one or both of these hormones. Although the specific effects of androgens on cartilage growth have not been shown, Ng and colleagues (2001) demonstrated that the in vitro treatment of cartilage explants with 17β-estradiol reduces cartilage growth as measured by the ceratobranchial cartilage assay. Again, not shown in this figure are the long-loop and short-loop feedback mechanisms that regulate pituitary function.

PRL has been shown to have growth-related effects in fish, such as actions on lipid metabolism (Meier 1969; Sheridan 1986; O'Connor et al. 1993); however, since PRL has more described actions in vertebrates than any other pituitary hormone (Nicoll 1974, 1981, 1993; Bern 1975, 1983; Clarke and Bern 1980; Bole-Feysot et al. 1998; Goffin et al. 2002), it is difficult to classify the actions as primarily growth regulating in nature. In the tilapia, two forms of PRL have been identified, one containing 177 amino residues ($tPRL_{177}$) and the other containing 188 amino acid residues ($tPRL_{188}$) (Specker, King, Nishioka, et al. 1985; Specker, King, Rivas, et al. 1985). The amino acid sequence identity of the two prolactins is only 69 percent, and each form is encoded by separate genes, suggesting differential actions (Specker, King, Nishioka, et al. 1985; Specker, King, Rivas, et al. 1985; Rentier-Delrue et al. 1989; Specker et al. 1989; Yamaguchi et al. 1991; Rubin and Specker 1992; Rand-Weaver and Kawauchi 1993; Rand-Weaver et al. 1993; Sakamoto et al., 1997; Shepherd, Sakamoto, et al., 1997). Evidence that homologous teleost PRL(s) can stimulate growth in fish is slowly accumulating. Specker and colleagues (1985) reported that the larger of the two PRLs ($tPRL_{188}$) of the tilapia *O. mossambicus*, stimulates growth. In contrast, native $tPRL_{177}$, but not $tPRL_{188}$, raises IGF-I and IGF-II mRNA levels in gill and liver tissue of hypophysectomized tilapia (surgical removal of the pituitary), a novel action that has been attributed to the ability of native tilapia PRL_{177} to bind to tilapia liver GH receptors at concentrations similar to those observed in fresh water-adapted tilapia (Shepherd, Sakamoto, et al., 1997; Figure 3.3). Although PRL has been shown to promote growth or stimulate IGF-like activity in the plasma of tilapia, its potential as a growth promoter in teleosts is less certain.

IGF-I and IGF-II have been identified in all vertebrates examined to date (Chen et al. 1994; Reinecke and Collet 1998). The IGFs are structurally related to the pro-insulin polypeptide and therefore possess insulin-like activity; this is diminished through binding to high-affinity plasma binding proteins called IGFBPs, which have a higher affinity for IGFs than does the insulin receptor. Despite some recent discordant findings (Sjogren et al. 1999; Butler and Le Roith 2001; Le Roith, Bondy, et al. 2001; Le Roith, Scavo, et al. 2001; Yakar et al. 2001), somatic growth is thought to be under the control of bloodborne IGFs produced by the liver as well as IGFs produced locally by the target tissue. The production and release of the IGFs by the liver is strongly dependent on GH, and even though they have been shown to be present in virtually all tissues, they are not always GH dependent, particularly during early development (Jones and Clemmons 1995; Reinecke et al. 1995, 1997, 2000; Reinecke and Collet 1998; Schmid et al. 1999; Butler and Le Roith 2001; Zapf et al. 2002; Caelers et al. 2003). The deficiencies in our understanding of IGF function and regulation in vertebrates are obvious, but the most prominent actions of IGFs include the promotion of tissue/somatic growth, cell motility and mitogenesis, immune function, reproduction, and osmoregulation (Sakamoto et al. 1993; Jones and Clemmons 1995; Zapf et al. 1999; Butler and Le Roith 2001; D'Ercole and Calikoglu 2001; McCormick 2001).

In all vertebrates examined to date, more than 95 percent of circulating IGFs have been shown to be bound to the IGFBPs (Frystyk et al. 1994; Jones and Clemmons 1995; Rajaram et al. 1997; Shimizu et al. 1999). At least six IGFBPs (IGFBP-1 to 6), of varying molecular size, have been characterized in mammals and are regulated by hormones, nutrition, and physiological state of the animal. The IGFBPs are thought to

1. increase the half-life of IGFs in circulation;
2. regulate the interaction of IGFs with membrane receptors;
3. partition the IGFs between the vascular space and tissue compartments;
4. mitigate insulin-like side effects of IGFs; and
5. stimulate mitosis either directly or synergistically with the IGFs (Rosenfeld et al. 1990; Rechler 1993; Jones and Clemmons 1995; Kelley et al. 1996; Rajaram et al. 1997; Duan 2002).

Although the IGF-shuttling roles of IGFBPs are best known, evidence is increasing for IGF-independent and IGF receptor–independent biological actions such as influencing cell adhesion, migration, proliferation, and apoptosis (Rajaram et al. 1997; Yu and Rohan 2000; Duan 2002; Firth and Baxter 2002; Mohan and Baylink 2002). The IGF-independent actions of the IGFBPs are influenced by a variety of posttranslational modifications during synthesis and subsequent modifications via proteolysis in the plasma and at cell surface receptors (Rajaram et al. 1997; Clemmons et al. 1998; Clemmons 2001; Firth and Baxter 2002; Kelley et al. 2002) (Figure 3.1).

In teleost fishes, four binding proteins of various molecular sizes (20 to 29, 31 to 39, 40 to 50, and 76 to 90 kDa) have been identified (Kelley et al. 2001, 2002). With some exceptions (Johnson et al. 2003), the IGFs and IGFBPs are regulated by nutritional and physiological status in teleosts in a manner similar to that reported in mammals (Duan et al. 1994; Moriyama et al. 1994; Perez-Sanchez et al. 1995; Dickhoff et al. 1997; Duan 1997, 1998; Banos et al. 1999; Shimizu et al. 1999, 2000; Company et al. 2001; Kelley et al. 2001; Shimizu, Hara, et al. 2003; Beckman et al. 2004). High molecular weight (76 to 90 kDa) binding protein(s) (IGFBP-4) have been reported in the striped bass (*Morone saxatilis;* Siharath et al. 1995) and channel catfish (*Ictalurus punctatus;* Johnson et al. 2003), although it remains unclear whether these are true IGFBPs since most of the findings are based only on ligand binding assays.

Classification of teleost IGFBPs is based upon comparable molecular weights seen in mammals and comparable responses to endocrine and nutritional manipulation (Kelley et al. 2001). IGFBPs less than 35 kDa in size in teleosts have been demonstrated to be up-regulated in response to stressors such as fasting, hypoxia, handling, and catabolism (a result of acute or chronic stress or cortisol treatment) (Kelley et al. 2001, 2002). In contrast, the 35 to 50 kDa IGFBPs consistently appear to be GH dependent and indicative of anabolic metabolism (Park et al. 2000; Kelley et al. 2002; Kajimura et al. 2003). GH-dependent increases in IGF-I and IGFBP-3 have been demonstrated to exist in the tilapia (Shepherd, Sakamoto, et al. 1997; Park et al. 2000; Cheng et al. 2002), as have effects of catabolic states (stress, starvation, or cortisol treatment) on the various IGFBPs (Park et al. 2000; Kajimura et al. 2002, 2003). Although much information on the occurrence and regulation of the IGFBPs in teleosts has been gained over the past decade, only recently have the teleost homolog to mammalian IGFBPs (−1, −2, and −3) been characterized (Duan et al. 1999; Bauchat et al. 2001; Maures and Duan 2002; Shimizu, Swanson, et al. 2003) and IGF-independent actions been described (Duan et al. 1999; Bauchat et al. 2001; Shimizu, Swanson, et al. 2003).

Growth Hormone, Prolactin, and Insulin-Like Growth Factor-I: Intersection of Growth and Osmoregulation

Growth and osmoregulation are essential to life itself. At first glance, it may not be evident that these two physiological processes are coupled; however, the hormones that regulate vertebrate growth also control osmoregulation (Scanes and Baile 1993; Scanes and Daughaday 1995; Björnsson 1997; Duan 1997; Butler and Le Roith 2001; Le Roith, Bondy, et al. 2001; Le Roith, Scavo, et al. 2001; McCormick 2001; Yakar et al. 2001; Moller 2003). Principal among these regulators are the pituitary hormones GH and PRL. GH promotes somatic growth in most vertebrates (Scanes and Baile 1993; Scanes and Daughaday 1995; Collier and Byatt 1996) and has also been shown to regulate seawater adaptation of euryhaline teleosts such as salmonids, killifish, and tilapia (McCormick 2001). In contrast, PRL is the fresh water osmoregulatory hormone in many teleosts (Hirano 1986; Brown and Brown 1987; McCormick 2001; Manzon 2002) but has also been shown to possess growth-promoting and metabolic actions in a limited number of teleosts (Meier 1969; Clarke et al. 1977; Sheridan 1986; O'Connor et al. 1993; Shepherd, Sakamoto, et al. 1997; Leena et al. 2001). This overlap in hormone function and regulation complicates our understanding of the regulation of growth in teleosts. Consequently, a greater understanding of the integration of these diverse processes by the hormones of the somatotropic axis is required before endocrine manipulation can be used effectively to improve finfish aquaculture.

Growth Hormone and Prolactins in Osmoregulation

Studies in salmonids and other euryhaline teleosts suggest that GH promotes seawater adaptation independent of its somatotropic actions (Sakamoto and Hirano 1993; Björnsson 1997; McCormick 2001). GH treatment has been shown to increase the size and density of gill chloride cells, stimulate gill Na^+,K^+-ATPase activity, increase levels of the cortisol receptor, and reduce the accumulation of plasma electrolytes following seawater entry (McCormick 2001). Evidence for a role of GH in osmoregulation in advanced teleosts, such as tilapia, is only now beginning to emerge. The hypothesis of a possible involvement of GH in seawater osmoregulation in a nonsalmonid teleost stems from the early finding that the in vitro release of GH from the tilapia pituitary is stimulated by physiological increases (hyperosmotic conditions) in medium osmolality (Helms et al. 1987). Subsequent studies have shown that GH treatment facilitates hypo-osmoregulatory mechanisms in the euryhaline tilapia *O. mossambicus* as well as the stenohaline tilapia *O. niloticus* (Xu et al. 1997). Furthermore, Yada and colleagues (1994) found that plasma osmolality and GH levels rise concurrently in the tilapia after transfer from fresh water to seawater. In follow-up studies, the direct effects of homologous GH and PRLs were examined in the tilapia. Here, Sakamoto and colleagues (1997) found that GH treatment lowers plasma osmolality and increases gill Na^+,K^+-ATPase activity in hypophysectomized tilapia transferred to one-fourth from seawater (8 ppt) to full-strength seawater (32 ppt) (Borski et al. 1994; Shepherd, Sakamoto, et al. 1997).

PRL is the most important hormone for the fresh water osmoregulation of teleost fish. It acts on virtually all osmoregulatory tissues to increase ion uptake and reduce water permeability following transfer to freshwater (Clarke and Bern 1980; Hirano 1986; Brown and Brown 1987; Manzon 2002). PRL cells in tilapia respond directly to physiological alterations in extracellular osmolality in both organ and primary cell culture systems (Grau et al. 1994; Seale et al. 2002, 2003a,b). As a result of their direct osmosensitivity and ability to maintain hydromineral balance (via PRL release) in fresh water, PRL cells are considered osmoreceptors; they appear to operate through stretch-activated calcium channels in response to cell swelling (Grau et al. 1994; Seale et al. 2003a,b). GH cells can also respond to changes in extracellular solute concentration in vitro. Whereas

PRL cells respond to reductions in medium osmolality, GH cells have been shown to respond directly to elevations in medium osmolality (Helms et al. 1987; Grau et al. 1994; Seale et al. 2002). Consistent with this notion, the pituitary GH content, the volume of GH-secreting cells, and the activity of GH cells were found to be higher in seawater-adapted tilapia than in fresh water-adapted tilapia (Borski et al. 1994).

Although $tPRL_{177}$ and $tPRL_{188}$ respond similarly and consistently to alterations in medium osmolality, some differences in their actions have been reported. Aside from the somatotropic effect of $tPRL_{177}$, which was discussed in the section on "Hormonal control of growth in tilapia," a study by Sakamoto and colleagues (1997) demonstrated that the larger form of PRL ($tPRL_{188}$) has greater potency than the smaller form ($tPRL_{177}$) in stimulating ion retention. Consequently, it has been established $tPRL_{188}$ in the tilapia, is the "mainline" form and possesses antagonistic actions (lowering gill Na^+,K^+-ATPase) to GH in seawater adaptation (Oshima et al. 1996; Sakamoto et al. 1997; Shepherd, Sakamoto, et al. 1997; Shepherd et al. 1999).

Growth Factors As Intermediaries for the Osmoregulatory and Growth-Promoting Actions of Growth Hormone and Prolactin

The osmoregulatory and somatotropic actions of GH in teleosts have been shown to be mediated by IGF-I and IGF-II (Sakamoto et al. 1993; Chen et al. 1994; McCormick 1995, 2001; Björnsson 1997; Duan 1997, 1998; Mommsen 1998). In teleosts, IGF-I stimulates growth of somatic tissues (McCormick et al. 1992; Negatu and Meier 1995) and inhibits (long-loop feedback) the in vitro release of GH from pituitary tissues in much the same way as it does in mammals (Perez-Sanchez et al. 1992; Blaise et al. 1995; Fruchtman et al. 2000; Kajimura et al. 2002).

Since McCormick and colleagues (1992) demonstrated that in vivo treatment of rainbow trout with bovine IGF-I increases hypo-osmoregulatory function, including growth (McCormick et al. 1991), studies have shown that IGF-I increases gill Na^+,K^+-ATPase in vitro and in vivo (Sakamoto and Hirano 1993; McCormick 2001). Further investigation has shown increased IGF-I gene expression in osmoregulatory tissues, and increased plasma IGF-I levels, in euryhaline teleosts following transfer to seawater or seawater entry during migration (Sakamoto and Hirano 1993; Duan et al. 1995; Sakamoto et al. 1995; Björnsson 1997; Dickhoff et al. 1997; Duan 1997; Beckman et al. 1998; Shepherd, Drennon, et al. 2004). We have observed increases in IGF-I mRNA in gill as well as liver when tilapia were transferred from fresh water to seawater (B. Shepherd, T. Sakamoto, and B. Ron, unpublished observations). IGF-I appears to act specifically to increase the activity of the sodium pump of the gill, Na^+,K^+-ATPase, which mediates the removal of monovalent ions across the gill epithelium in seawater-adapted teleosts (Marshall 1995; McCormick 1995, 2001).

Whether GH and PRL act with or exclusively through IGFs is a matter of debate. The presence of GH and PRL receptors in gill, kidney, and intestine of tilapia suggests that GH and PRL may act directly on osmoregulatory tissues (Fryer 1979a,b; Dauder, Young, and Bern 1990; Dauder, Young, Hass, et al. 1990; Ng et al. 1992; Auperin et al. 1994, 1995; Prunet and Auperin 1994; Sandra et al. 1995, 2000, 2001). Treatment with GH also increases the sensitivity of the sodium pump to bovine IGF-I and produces changes in IGF-I gene expression in gill and kidney that are similar to those observed after seawater entry (Sakamoto and Hirano 1993; McCormick 1995, 2001; Riley, Richman, et al. 2002). Interestingly, treatment of hypophysectomized tilapia with tilapia GH or $tPRL_{177}$ increases IGF-I and IGF-II mRNA levels in gill and liver (B. Shepherd and T. Sakamoto, unpublished observation), further supporting the notion that IGF-I may mediate some of the biological actions of PRL in tilapia (Shepherd, Sakamoto, et al. 1997).

Another mechanism by which salinity may alter the somatotropic axis is at the level of the IGFBPs. We have seen time-course- and salinity-dependent elevations in the higher molecular weight (30 kDa) IGFBPs in rainbow trout gradually transferred to full-strength seawater (Shepherd, Drennan, et al. 2004). This response is consistent with hormonal changes associated with seawater adaptation, wherein GH, IGF-I, and cortisol are transiently elevated (Sakamoto and Hirano 1991; McCormick 2001). In teleosts, cortisol up-regulates low molecular weight IGFBPs, whereas IGF-I and GH up-regulate the higher molecular weight IGFBPs (Kelley et al. 2001, 2002; Kajimura et al. 2003). A similar situation to that seen in the rainbow trout is likely to occur in *O. mossambicus*, which is also euryhaline. In contrast, salinity challenge in the stenohaline *Ictalurus punctatus* resulted in a decrease in the higher molecular weight IGFBPs, despite increases in plasma GH (Drennon et al. 2003; Johnson et al. 2003) and IGF-I levels (B. Shepherd, unpublished data). Differences in the effects of salinity on IGFBPs in the rainbow trout and in the channel catfish highlight species differences and suggest the potential for complex interactions between the effects of various stressors, environmental salinity, cortisol, and IGFBP regulation. An interesting connection has been shown between the affinity of IGF-I and IGF-II for IGFBP-3 and the ionic strength of a physiological solution (Holman and Baxter 1996; Baxter 2001). Interactions were reported to be optimal in solution at physiological ionic strengths and to decrease as ionic strength decreases or becomes hypo-osmotic (Holman and Baxter 1996;

Baxter 2001). Whether a similar relationship exists in teleosts is a matter for further, focused research.

A strong case can thus be made for the mediation of the actions of PRL by growth factors in tilapia. Likewise, good evidence suggests that another factor from the liver, termed "synlactin," may potentiate the actions of PRL (Anderson et al. 1983; Nicoll et al. 1985). Vertebrate synlactin has not been isolated or characterized, but its presence in the blood (bioactivity) is PRL dependent. Synlactin activity can be determined using a standard bioassay that measures the PRL-dependent stimulation of growth in the pigeon-crop mucosa (production of crop-sac milk), a standard bioassay for PRL (Anderson et al. 1983; Nicoll et al. 1985). Synlactin may be a member of the IGF family (Anderson et al. 1983; Nicoll et al. 1990; Nicoll 1993). Indeed, synlactin is a PRL-responsive mitogen that has been identified in the liver of several vertebrates, including the tilapia (Delidow et al. 1986; Nicoll et al. 1990). Increases in synlactin may also account for the stimulation of sulfation activity in the blood and in liver perfusate of rats treated with ovine or rat PRL (Francis and Hill 1975; Hill et al. 1977; Bala et al. 1978). Murphy and colleagues (1988) demonstrated that ovine PRL injections increase levels of hepatic IGF-I mRNA and plasma IGF-I in rats. The use of ovine PRL in rats leaves open the question of possible interactions between a heterologous PRL and rat GH receptors. Leedom and colleagues (2002) determined that the tilapia GH receptor is unable to distinguish the binding of bovine GH or bovine PRL using an established radioreceptor assay, thus leaving open the possibility that the somatotropic actions of heterologous PRL and synlactin activity can be attributed to receptor promiscuity. It remains to be determined whether synlactin activity can be attributed to a novel mitogenic factor that is PRL dependent or to an IGF, IGFBP, or novel IGF-IGFBP complex that results from GH receptor promiscuity.

Androgens and Growth

Sex Reversal

The synthetic steroid MT augments the growth of a variety of fish, including the Mozambique tilapia (Donaldson et al. 1979; Kuwaye et al. 1993; Ron et al. 1995; Sparks et al. 2003). The tilapia, like many teleosts, exhibits a sexually dimorphic pattern of growth, with males growing faster and larger than females. Most studies on androgen manipulation of growth in tilapia have focused on the effects on subsequent growth of early brief applications of MT to fry for 2 to 4 weeks to masculinize genotypic females into faster growing phenotypic males (Yamazaki 1976; Rothbard et al. 1983;

Guerrero and Guerrero 1988; Hiott and Phelps 1993; Lone and Ridha 1993; Varadaraj et al. 1994). However, sex reversal using other steroids via immersion has gained increasing attention in recent years (Contreras-Sanches et al. 2001; Afonso and Leboute 2003; Bart et al. 2003). Unlike dietary administration, immersion is more time intensive, but it may be more environmentally sound as the wastewater containing the steroid can be more readily controlled and disposed of (by carbon filtration), thus minimizing the likelihood of the steroid entering into the environment.

According to Nakamura and colleagues (1998), genetic female tilapia produce estrogens as early as 8 days after hatching. These endogenous estrogens appear to induce ovarian differentiation. In male tilapia, in contrast, androgen synthesis is delayed, with measurable levels occurring only 35 to 40 days after hatching and after differentiation of the testes. It is this "steroid sensitive" phenocritical developmental period when exposure of females to androgenic steroids such as MT or inhibitors of steroid synthesis (Afonso et al. 2001) produces populations with a high frequency of phenotypic males that grow faster than untreated males and females or mixed-sex control animals (Afonso and Leboute 2003).

17α-Methyltestosterone Actions on the Growth of Tilapia Beyond Sex Reversal

Studies have shown that administration of MT for sex reversal can result in males that grow faster than untreated males, suggesting an anabolic action of MT that goes beyond masculinization (Kuwaye et al. 1993). In addition, extended administration of MT through the diet increases the growth of *O. mossambicus* beyond that resulting from short-term treatment, as for sex reversal (Howerton 1988; Howerton et al. 1992; Kuwaye et al. 1993; Ron et al. 1995; Riley, Richman, et al. 2002; Sparks et al. 2003). Repeated studies have shown that exposure to MT leads to growth in *O. mossambicus* that is several times higher than in untreated fish (Howerton 1988; Kuwaye et al. 1993; Ron et al. 1995; Riley, Richman, et al. 2002; Sparks et al. 2003).

The practical application of these findings was tested under full-scale aquaculture pond conditions in southern China. In these studies, *O. mossambicus* were cultivated in fresh water polyculture, as is common practice in China with common carp *(Cyprinus carpio)*, grass carp, (*Ctenopharyngdon idellus*), mud carp *(Cirrhina molitorella)*, silver carp *(Hypophthalmichthys molitrix)*, and bighead carp *(Aristichthys nobilis)*. The fish were reared in ponds approximately one-third of an acre in size. In three separate years, the administration of MT in feed consistently and significantly increased the growth of tilapia by more than 20 percent and of the carp species by 15 to 30 percent. Under actual polyculture conditions, MT also increased the total finfish production of the ponds by approximately 25 percent, with only a 1 to 2 percent increase in the cost of the feed (Kuwaye et al. 1993).

Androgen Actions on the Somatotropic Axis

Androgens have been shown in mammals to augment growth by interacting with endogenous hormones, including members of the GH/PRL family and their associated IGFs. Testosterone, possibly aromatized to estradiol-17β, increases circulating levels of IGF-I (Hagenfeldt et al. 1992; Silva et al. 1992; Hobbs et al. 1993; Keenan et al. 1993; Weissberger and Ho 1993). This effect may result from the stimulation of GH release, which in turn stimulates IGF-I release from the liver (Figure 3.2). Other work in mammals has shown that steroids can also up-regulate hepatic receptors for endogenous growth-regulating hormones, which in itself may partially account for their growth-promoting actions (Rosenfeld and Furlanetto 1985; Simard et al. 1986; Copeland et al. 1990; Coxam et al. 1990; Ulloa-Aguirre et al. 1990; Fernandez et al. 1992; Hasegawa et al. 1992; Hassan et al. 1992; Silva et al. 1992; Borski et al. 1996).

In fish, the specific actions of anabolic steroids on endogenous growth regulators are not as well understood. In tilapia and other teleosts gonadal steroid hormones have been implicated as regulators of GH and other pituitary cell activity (Barry and Grau 1986; Trudeau et al. 1992; Peter and Marchant 1995; Melamed et al. 1998; Ng et al. 2001). For example, work in the tilapia hybrid *O. aureus* × *O. niloticus* has shown that pretreatment with MT increases the subsequent response of pituitary GH cells to growth hormone-releasing factor (GRF) in vitro (Melamed, Eliahu, Levavi-Sivan, et al. 1995; Melamed, Eliahu, Ofir, et al. 1995). In *O. mossambicus*, testosterone and estradiol-17β stimulate release of both pituitary PRLs in vitro and augment gonadotropin-releasing hormone (GnRH)-induced PRL release in vitro (Barry and Grau 1986; Weber et al. 1997). In the tilapia hybrid *O. niloticus* × *O. aureus*, GnRH stimulates GH release (Melamed, Eliahu, Levavi-Sivan, et al. 1995). In tilapia, the effect of androgens is unlikely to be through conversion to estrogens: early exposure of Mozambique tilapia to a synthetic estrogen, ethynylestradiol (EE_2), produces a high proportion of female tilapia that grow more slowly than populations of untreated, mixed-sex controls (Meredith et al. 1999). Similar results of reduced growth were obtained using an in vitro tilapia model to examine the effects of estrogens on cartilage sulfation, a measure of IGF-like bioactivity and cartilage growth (Ng et al. 2001; Figure 3.3). However, in this same study in vivo treatment with estradiol-17β stimulated cartilage growth (uptake of [^{35}S]-labeled sulfate; see Figure 3.3), an effect that may derive from the stimulatory action of estrogen on pituitary PRL and/or GH release in teleosts (Borski et al. 1991; Trudeau et al. 1992; Peter and Marchant 1995; Zou et. al. 1997; Riley, Hirano, and Grau 2002a).

Riley, Hirano, and Grau (2002a) showed that treatment of male and female tilapia with estrogen (E_2) or the nonaromatizable androgen 5α-dihydrotestosterone feminized or masculinized, respectively, the GH/IGF-I axes; however, differential patterns in GH secretion were not shown. E_2-treated males had lower levels of IGF-I and lower growth despite higher levels of plasma GH than untreated controls. In contrast, females treated with 5α-dihydrotestosterone had lower plasma GH levels but elevated plasma IGF-I and liver IGF-I mRNA levels. Although these results suggest that plasma IGF-I levels may be a better indicator for growth than plasma GH, it should be noted that plasma levels of hormones result from the differential between the rates of release and clearance, which were not examined in this study. Furthermore, it is unclear whether gonadal steroids modified the synthesis and release of GH alone or whether either of the PRLs was involved in modification of the somatotropic axis. These findings support the idea that the differences seen in growth rate between male and female tilapia are due, in part, to sex-specific differences in the somatotropic axes.

In *O. mossambicus*, long-term MT treatment and seawater rearing has no discernible effect on plasma GH levels, despite significant elevations in pituitary GH content, GH mRNA, plasma IGF-I levels/activity, and plasma PRL levels (Shepherd, Ron, et al. 1997; Riley, Richman, et al. 2002; Figure 3.2). Specifically, it has been shown that pituitary GH levels are augmented in MT-treated *O. mossambicus* (Shepherd, Ron, et al. 1997; Riley, Richman, et al. 2002). In addition, dietary MT administration increased levels of the two PRLs in the plasma and pituitary of these animals (Shepherd, Ron, et al. 1997). This finding has special importance, since tilapia, $tPRL_{177}$, has subsequently been shown to stimulate the uptake of [^{35}S]- sulfate and [^{3}H]-thymidine by cartilage tissue in vitro, which is an IGF- dependent effect mediated through binding of $tPRL_{177}$ to hepatic GH receptors in tilapia (Shepherd, Sakamoto, et al. 1997; Figure 3.3). This is a reliable bioassay for evaluating the ability of PRL and GH to stimulate growth and cell proliferation as it is a marker of plasma IGF-like bioactivity (Duan and Inui 1990; Duan and Hirano 1992; Duan et al. 1992; Duan 1994; Tsai et al. 1994; Figure 3.3). Indeed, Riley and colleagues (2002) validated earlier findings of Shepherd et al. (1997) by measuring higher levels of plasma IGF-I in tilapia reared in seawater and in those reared in seawater and fed MT above their respective controls.

NONHORMONAL CONTROL OF GROWTH IN TILAPIA

Environmental Salinity

Reports on the effects of environmental salinity on growth of teleosts are increasing and, in a limited fashion, were reviewed by Boeuf and Payan (2001). Tilapia grow significantly faster in seawater (see Chapter 10, "Farming Tilapia in Saline Waters") than in fresh water, and growth is further enhanced when combined with continuous exposure to MT (Kuwaye et al. 1993; Ron et al. 1995; Riley, Richman, et al. 2002; Sparks et al. 2003). Given the potent effect of seawater rearing on growth in *O. mossambicus*, recent studies have been aimed at identifying the specific underlying endocrine mechanism(s) (Figure 3.2). Findings suggest that the enhanced growth in seawater is linked to an activation of the GH/IGF-I axis (Shepherd, Ron, et al. 1997; Riley, Richman, et al. 2002; Figure 3.2). Specifically, it has been shown that pituitary GH levels in Seawater-reared *O. mossambicus* are elevated over those observed in fresh water fish and that MT treatment, regardless of salinity, further augments these levels (Borski et al. 1994; Shepherd, Ron, et al. 1997).

From available data, it appears that optimal growth rate in most euryhaline teleosts occurs at salinities that coincide with lower standard metabolic rates. The effects of salinity and MT on growth are additive: repeated studies have shown that combined exposure to MT and seawater leads to growth several-fold greater than that of untreated fresh water fish (Kuwaye et al. 1993; Ron et al. 1995; Riley, Richman, et al. 2002; Sparks et al. 2003). Of particular interest are the metabolic mechanisms that support enhanced growth in tilapia. Studies have demonstrated that oxygen consumption in seawater-reared tilapia is less than 50 percent of that in fresh water-reared animals, suggesting that the increased growth rate of tilapia in seawater is due, in part, to an increase in the energy available for growth (Ron et al. 1995; Sparks et al. 2003). The combination of prolonged stimulation of the somatotropic axis in seawater-reared tilapia and the ability of GH and IGF-I to attenuate the rise in blood osmolality that accompanies seawater entry (B. Shepherd and T. Sakamoto, unpublished observations) suggest that GH and IGF-I might facilitate seawater adaptation initially and promote growth chronically (Yada et al. 1994; Vijayan et al. 1996; Shepherd, Ron, et al. 1997; Riley, Richman, et al. 2002). Whether activation of the GH/IGF axis at seawater entry causes or results from an increase in metabolic efficiency remains to be determined. In any event, *O. mossambicus* is a natural resident of high-salinity brackish water, and thus it is reasonable to expect that this fish may have an osmoregulatory system that operates more efficiently in seawater, leaving more energy for growth. When this reduced metabolism is accompanied by an activation of the somatotropic axis, the effects on growth are dramatic.

It has been established that this enhanced growth in seawater is linked to an activation of the GH/IGF axis (Shepherd, Ron, et al. 1997; Riley, Richman, et al. 2002). Specifically, seawater rearing is without discernible effect on plasma GH levels, although it produces significant elevations in pituitary GH content, GH mRNA, and plasma IGF-I levels/activity compared with fresh water-reared animals (Borski et al. 1994; Shepherd, Ron, et al. 1997; Riley, Richman, et al. 2002; Figure 3.2). Differences in the growth axis of seawater- and fresh water-reared tilapia are similar to those of MT-treated versus control fish (Riley, Richman, et al. 2002). Riley and colleagues (2002) examined the effects of MT treatment and seawater rearing on plasma levels of GH and IGF-I and on mRNA levels of GH in the pituitary (Figures 3.2 and 3.3). Their results on the effects of MT were discussed in the section "Androgen actions on the somatotropic axis." In many of the studies discussed thus far, we have seen that treatment with MT significantly stimulates growth of the Mozambique tilapia in both fresh water and seawater. Control fish in fresh water were always the smallest. Growth was elevated in seawater control fish and in fresh water fish treated with MT, but the greatest growth was observed in seawater fish treated with MT. In more recent studies, it has been reported that pituitary GH mRNA levels vary in a pattern similar to that of body weight and that plasma IGF-I levels in MT-treated fish in seawater are significantly greater than levels seen in fresh water fish (Riley, Richman, et al. 2002; Riley et al. 2003). These results strongly indicate that the accelerated growth produced by MT treatment and seawater rearing is linked to an activation of the GH/IGF-I axis (Figure 3.2).

Nutrition

Different species can respond quite differently to the same hormone; moreover, the effect of a given hormone within a species may be influenced by environmental factors (e.g. salinity) and diet (McBride and Fagerlund 1973; Degani 1985; Degani and Gallagher 1985; Simone 1990; Kuwaye et al. 1993; Ron et al. 1995; Mancera et al. 2002; Sangiao-Alvarellos et al. 2003; Sparks et al. 2003). Therefore, it is essential to characterize the effects of each hormone and environmental salinity in each species of interest; identify changes in the nutritional requirements induced by hormonal, environmental, and genetic manipulations, and determine the dose of the hormone that elicits the optimal response.

Nutrition is an important component of the response of tilapia to salinity and MT treatment. An important finding from the earlier studies on seawater and MT effects on growth is that the treatment increased growth two- to threefold despite the fact that fish were fed a fixed, limited ration (feed weight to body weight), suggesting an increase in feed efficiency in all cases (Howerton 1988; Howerton et al. 1992; Kuwaye et al. 1993). A later study showed that the effects are even more pronounced if the fish are fed to satiation, with no additional gain for fresh water-reared fish (Ron et al. 1995), suggesting feed was limiting in some earlier studies.

Protein content of the feed also affected growth and endocrine response to salinity and MT treatment in *O. mossambicus* tilapia (Ron 1997; Shepherd, Ron, et al. 1997). High dietary protein increased growth to an appreciable degree only in seawater-reared animals, an effect that was augmented by the addition of MT to the diet. The greatest growth was observed in seawater-reared fish fed an MT-treated 35 percent crude protein diet (versus a 25 percent crude protein diet to satiation three times daily), suggesting the salinity, diet, and MT variables are additive. The only observed alteration in somatotropic axis measures with protein content and MT was a transient increase in both PRLs in the fresh water-fed fish (Shepherd, Ron, et al. 1997). It may be concluded that the additional dietary protein/energy offered by the satiation study (Ron et al. 1995) and the dietary protein study (Ron 1997; Shepherd, Ron, et al. 1997) yields no growth advantage in fresh water-reared tilapia but that additional dietary protein or higher ration, alone or in combination with MT, may offer substantial growth-promoting benefits to seawater-reared tilapia.

Results from these in vivo studies suggest that salinity and MT treatment alter protein and carbohydrate metabolism in the tilapia. Similar conclusions are supported by a number of earlier studies undertaken to examine the effects of MT treatment and environmental salinity on the in vitro uptake of nutrients by the intestine of *O. mossambicus*, which showed that MT treatment in tilapia significantly increases the active (Na^+-dependent) uptake of D-glucose. This increase was due, in part, to elevated glucose transporter density in the brush border epithelium within 30 min of exposure to MT (Reshkin et al. 1989; Hazzard and Ahearn 1990; Ikemoto et al. 1991). Using a modified version (Collie and Stevens 1985) of the everted-sleeve technique (Karasov and Diamond 1983; Karasov et al. 1983) to examine intestinal nutrient transport, Ron (1997) examined amino acid and sugar transport in intact intestinal preparations from tilapia reared in fresh water and seawater. The active uptake of D-glucose was higher in the intestine of fresh water-reared tilapia than in the intestine of seawater-reared tilapia. Conversely, active uptake of the amino acid L-proline was significantly greater in the intestine of seawater-reared tilapia and was further increased by MT treatment (Ikemoto et al. 1991; Ron 1997). In rainbow trout, MT and 11-ketotestosterone also stimulated intestinal uptake of L-leucine (Habibi and Ince 1984). The effect appeared to be specific for androgens among sex steroids since 17β-estradiol was without effect in the rainbow trout (Habibi et al. 1983). GH also has been found to stimulate both the active and passive uptake of L-proline and L-leucine in the intestine of coho salmon and of striped bass hybrids, *M. saxatilis* × *M. chrysops* (Collie and Stevens 1985; Sun and Farmanfarmaian 1992a,b). In both species, GH increased both the density of transport carriers and the total surface area of the intestine.

The effects of growth-promoting hormones on metabolism and nutrition, are complex. The pituitary hormone GH and the synthetic androgen MT have been shown to stimulate food consumption and increase feed conversion efficiency (weight gain per unit of feed consumed) in finfish and other vertebrates (Fagerlund et al. 1979, 1983; Lone and Matty 1980; Nimala and Pandian 1983; Sindhu and Pandian 1984; Degani 1985; Degani and Gallagher 1985; Gill et al. 1985; Varadaraj and Pandian 1988; Ridha and Lone 1990; Foster et al. 1991; Killian and Kohler 1991; Sun and Farmanfarmaian 1992a,b; Johnsson and Björnsson 1994; Collier and Byatt 1996). Total body protein and lipid mobilization (in vivo and in vitro) have been found to be increased in American eel, rainbow trout, channel catfish, goldfish, and other teleosts following either androgen or GH/PRL treatment (De Vlaming et al. 1975; Prack et al. 1980; Degani 1985; Sheridan 1986; Ostrowski and Garling 1987a; Gannam and Lovell 1991; O'Connor et al. 1993; Fauconneau et al. 1996). These hormones also increase the level of protein in specific tissues (brain, kidney, liver, and muscle) of the carp, Indian catfish, chum salmon, and a striped bass hybrid, *M. saxatilis* × *M. chrysops* (Lone and Matty 1980, 1982; Ando et al. 1986; Singh and Prasad 1987; Sun and Farmanfarmaian 1992a,b; Fauconneau et al. 1996). The elevation of tissue protein in hormone-treated animals may be due to a combined increase in the fractional rate of protein synthesis and amino acid availability, possibly via increased intestinal transport (Matty and Cheema 1978; Collie and Stevens 1985; Inui and Ishioka 1985; Matty 1986; Foster et al. 1991; Sun and Farmanfarmaian 1992b; Ron 1997; Sillence 2004). This increase may be secondary to a rise in protein synthesis and, in certain cases at least, to a reduction in the proportion of protein required relative to the total dietary energy necessary to support optimal growth (Fagerlund et al. 1983; Ufodike and Madu 1986; Ostrowski and Garling 1987a,b; Gogoi and Keshavanath 1988).

NOVEL APPROACHES TO GROWTH ENHANCEMENT

The interrelationships between the hormones that regulate growth and development and their range of biological actions in teleosts remain to be fully elucidated. Furthermore, little is known about the integration of environmental and physiological cues that direct the actions of the genes and hormones involved in growth, metabolism, reproduction, and adaptation in fish. Nevertheless, advances have been made towards identifying approaches to alter growth through manipulations of the endocrine system, including selective breeding, administration of exogenous growth promoters, increased GH production via transgenesis, and alterations of the somatotropic axis via environmental or nutritional manipulation. The ensuing discussion is neither an endorsement nor an indictment of these approaches; no one single approach satisfies all interests and concerns about growth promotion in finfish aquaculture. However, there are options from which to choose, with more promising techniques in the wings. A better understanding of the ramifications of each approach is clearly needed to optimize methods for promoting growth and feed efficiency in tilapia in a way that best addresses all concerns. This understanding begins with basic knowledge of growth physiology in these species.

Selective Breeding

Given the high genetic diversity within aquaculture species, tremendous potential exists for an increase in productivity through the application of genetic techniques such as selective breeding (biometrical and molecular methods), chromosomal set manipulation, hybridization, and gynogenesis (Halvorson and Quezada 1999; Hulata 2001). Selective breeding has been the mainstay for manipulating growth since humankind first started to breed livestock. Although it is still based primarily on classical genetics using phenotypical observations of traits of interest, such as growth rate or body size, advances in genomics and proteomics have the potential to increase genetic gain in aquaculture.

Selective breeding programs require a long-term commitment in terms of resources (animal, human, scientific, and financial) to fully utilize the genetic variation that exists within the species of interest (Bartley 1997b) and to avoid the inbreeding suppression of performance traits (Shirak et al. 2002). Proteomics, which, like genomics, can accelarate progress, is a valuable approach in that a broad spectrum of physiologically meaningful differences can be characterized, such as variation in the proteins of the somatotropic axis and the occurrence of novel proteins or hormones associated with variation in the phenotypes of interest. These traits may include improved growth or food conversion efficiency, and such information can be used to make more informed selection decisions. Both genomics and proteomics requires a large investment initially. For proteomics, this initial investment is tied to producing two-dimensional electrophoresis (2DE) maps of protein expression, protein identification using peptide mass fingerprinting or sequencing via mass spectrometry, and bioinformatics resources for interpretation of results. Once such resources are available, the generation of additional 2DE maps for comparing phenotypes is relatively straightforward and inexpensive.

Genomics may benefit or aid selective breeding programs, in a number of ways. Once resources have been developed, including

1. gene arrays and their bioinformatics support resources;
2. comprehensive genetic and physical maps (a genetic map is based on recombination data; a physical map is based on physically localizing genes or markers to chromosome regions);
3. sequence of the genes and chromosomal region of interest; and
4. identification of genetic markers, including type I markers, which are genes for specific proteins such as hormones (a candidate gene approach), and type II markers (among others) such as microsatellites (simple sequence repeats distributed throughout the genome).

These genomic tools can be used to

1. track families;
2. determine the pedigree of an individual organism;
3. characterize strains within a species; and
4. identify loci associated with traits of economic importance.

Most traits of economic importance are quantitatively inherited and typically show a continuous distribution that is attributable to the interaction between the environment and the collective action of many genes, termed quantitative trait loci (QTL) (Smith and Smith 1993; Grandillo and Tanksley 1996; Heyen et al. 1999). A QTL is a locus that may have an unknown primary function, but that affects a particular quantitative trait (Mayo and Franklin 1998). QTLs are identified by linkage analysis, which is the identification of cosegregation between a DNA marker allele (e.g., microsatellites) and the trait of interest (Andersson 2001; Flint and Mott 2001; Doerge 2002). It has been suggested that including marker data in the selection criteria, referred to as marker-assisted selection (MAS), may increase the accuracy of selection (Meuwissen and van Arendonk 1992; Beuzen et al. 2000). The basic theory is that selection for markers with easily detectable phenotypes can simplify the recovery of linked genes of interest that are themselves difficult to resolve (Arus and Gonzalez 1994). The main advantage is that studies can be developed without any prior knowledge about the biochemical or the physiological functions of the QTLs (Elo et al. 1999).

For example, such DNA-based approaches (MAS) can enhance selective breeding programs by characterizing the physiological and genetic bases of differences among desired phenotypes. QTL analysis has led to the identification of candidate genes responsible for improved growth and lean body mass in cattle, swine, and sheep. Specifically, these efforts have mapped increased growth and lean body mass in swine to the *IGF-2* locus (Jeon et al. 1999; Nezer et al. 1999; Andersson 2001), which is the result of a single-base mutation in intron 3 of this gene (Van Laere et al. 2003), a noncoding region to which this QTL maps. Similarly, the double-muscling phenotype, reported in the domestic Belgian blue bull, has been mapped to the *mh* (muscle hypertrophy) locus on chromosome 2 (Charlier et al. 1995), and the candidate gene, myostatin, has been mapped to this locus (Grobet et al. 1997; McPherron and Lee 1997). Specifically, this work has shown that recessive (loss-of-function) mutations in the myostatin gene are responsible for this double-muscling phenotype (McPherron and Lee 1997; Grobet et al. 1998).

It is evident that genomics and proteomics techniques hold potential to enhance traditional genetics approaches to improve aquaculture production. From these examples, it is clear that the more we understand of growth physiology, the better informed screening methods (use of candidate genes) and selection decisions will be in the selective breeding process. Although comprehensive programs are expensive and expansive and take many generations to advance, tilapia are prolific breeders, develop rapidly, and are thus amenable to selective breeding efforts. An extensive selective breeding project, GIFT (Genetic Improvement of Farmed Tilapia; (http://www.worldfishcenter.org/resprg_1.htm), which is aimed at characterizing hybrids and improved strains of the Nile tilapia for distribution in Asia, Africa, and elsewhere for commercialization (Hulata 2001; Subasinghe et al. 2003). Concerted efforts are under way to map the genome in the Nile tilapia (Kocher et al. 1998; Agresti et al. 2000; McConnell et al. 2000; Shirak et al. 2002; Lee et al. 2003; Rutten et al. 2004) and other important teleosts (Hulata 2001; Clark 2003). However, evidence of commercial benefits derived from these efforts is limited (Hulata 2001). Despite the infancy of the incorporation of these techniques into aquaculture breeding programs, studies have demonstrated that such approaches can link physiological (Streelman and Kocher 2002) and developmental (Shirak et al. 2002) phenomena to regulatory elements in the tilapia genome.

Anabolic Steroids

The potential for anabolic steroids to promote growth in tilapia has been shown in numerous studies discussed in this chapter. Clearly, MT administration can cost-effectively improve growth rate and feed efficiency under diverse environmental conditions. Scientific evidence is accumulating to support the contention that MT is cleared quickly from treated animals, so that appropriate depuration procedures will result in the elimination of any potential health risks (Johnstone et al. 1983; Goudie et al. 1986a,b; Rothbard et al. 1990; Curtis et al. 1991; Ron 1997). Unfortunately, the use of anabolic steroids to promote growth is not well accepted by the public, and government approval for the use of steroids beyond sex inversion is less certain. Although a depletion period of 14 days has been proposed as adequate for clearance of MT in rainbow trout prior to human consumption (Vick and Hayton 2001), the potential impacts of MT and its metabolites on the environment are also legitimate concerns (Arukwe 2001). Risks to human health can come not only from consumption of MT-treated fish that have not undergone an adequate depletion period but also through exposure to these compounds after biological clearance or release into the environment. Clearly, more and better information on risks and benefits of anabolic steroid treatment will make appropriate decisions to use or approve the treatments more likely. Since MT appears to have growth-promoting effects beyond simple sex inversion of females to males (Kuwaye et al. 1993), the approval of growth-promoting treatment of these animals with MT, using protocols approved for sex inversion, is being pursued enthusiastically. U.S. Food and Drug Administration approval for this use of MT seems to be very near (Schnick 2003, 2005).

Biotechnology (Transgenics and Recombinant Hormones)

Transgenic growth enhancement is the clearest example of the potential role of biotechnology in aquaculture, and the development and application of transgenic approaches to problems in finfish aquaculture have been reviewed several times over more than a decade (MacLean and Penman 1990; Fletcher and Davies 1991; Houdebine and Chourrout 1991; Gong and Hew 1995; Chen, Vrolijk, et al. 1996; Iyengar et al. 1996; Devlin 1997; Maclean et al. 2002; Zbikowska 2003). Work in salmonids and more recently the mud loach (*Misgurnus mizolepis*) has described spectacular increases (10- to 35-fold) in growth rate of transgenic versus nontransgenic siblings (Devlin et al. 1994, 2001; Nam et al. 2001). Results of GH transgenics in the genus *Oreochromis*, were reviewed by Maclean et al. (2002); however, GH transgenesis yields less dramatic increases in growth rate (1.5- to 2.5-fold) and feed conversion efficiency in this genus (Maclean et al. 2002; Zbikowska 2003). Although most transgenic work has been conducted in tilapia hybrids, which are generally stenohaline, seawater rearing alone in *O. mossambicus* may be as effective at promoting growth as GH transgenesis (Howerton 1988; Watanabe et al. 1988; Howerton et al. 1992; Kuwaye et al. 1993; Ron 1997; Riley, Richman, et al. 2002; Riley et al., 2003).

Transgenic animal models have been, and continue to be, valuable for scientific study, but they have shown that the output of additional GH genes can produce a variety of undesirable outcomes such as glomerulosclerosis, impaired physiology, deformities, and alterations in the synthesis and release of other key hormones in teleost and mammalian models (Kayes 1977a,b; Agellon et al. 1988; Pursell et al. 1989; Dunham et al. 1992; Nagi et al. 1992; Yang et al. 1993; Devlin et al. 1995, 2000, 2001; Chen, Lu, et al. 1996; Jonsson et al. 1996; Farrell et al. 1997; Abrahams and Sutterlin 1999. Moreover, studies on potential adverse effects of GH transgenesis in teleosts are lacking, although some problems such as deformities (Dunham et al. 1992; Mori and Devlin 1999; Devlin et al. 2001) and alterations in behavior have been reported (Johnsson et al. 1996; Abrahams and Sutterlin 1999).

Legitimate concerns surround the exogenous application of GH or GH transgenesis, to finfish growth enhancement. First is the issue of safety to the animal and the environment. Unfortunately, very little information exists regarding the toxicity or impairments that may result from chronic GH application in teleosts. Environmentally, the integration of a transgene or genetic material from animals that have been selectively bred for improved traits, such as better growth, poses a tangible threat to wild fish populations, particularly given the absence of methods to ensure 100 percent infertility in transgenic or selectively improved animals (Reichardt 2000; Devlin et al. 2001; Check 2002; Muir and Howard 2002; Zbikowska 2003). As an example, the escape or accidental release of GH transgenic salmonids, which have been shown to possess superior abilities to wild populations in competing for food and other resources, is a very real concern (Reichardt 2000; Muir and Howard 2002; Zbikowska 2003). Nontransgenic animals selectively bred for superior traits could also compete with wild fish stocks for valuable resources.

In terms of safety to the animal, although evidence for remarkable increases in somatic growth in transgenic teleosts is good, the uncontrolled output of GH from additional GH transgenes may well short-circuit the natural regulatory pathways (e.g., GH and IGF feedback at the tissue-specific level) that control important endocrine axes (Rivas et al. 1986; Maiter et al. 1988, 1992; Dubowsky and Sheridan 1995; Pesek and Sheridan 1995; Yada et al. 1995; Conley et al. 1998; Mori and Devlin 1999; Agustsson and Bjornsson 2000). The use of exogenous GH has similar problems. GH is a large immunogenic polypeptide that is expensive to produce in its pure form (Kayes 1977a,b). However, recent studies have shown that feeding crude (or purified) extracts of the recombinant GH protein or yeast expressing a teleost GH transgene effectively promoting growth in teleosts (Moriyama et al. 1989, 1993; McLean and Donaldson 1990; Down and Donaldson 1991; Kawauchi et al. 1991; McLean et al. 1993, 1999; Ho et al. 1998; Jeh et al. 1998; Ben-Atia et al. 1999; Jin et al. 1999; Leedom et al. 2002).

Finally, given the consumer backlash against the use of genetically modified (GM) plant products (corn, soy, and rice), current public perception regarding the health risks associated with consumption of GM plant products (e.g., the recent widespread, expensive recall of StarLink GM corn; Anonymous 2000), issues regarding risk assessment and ecological safety (invasiveness and crop volunteerism, intra- and interspecific hybridization, effects on non-target organisms, and target organism resistance) and food safety (toxicity and allergenicity; Stewart et al. 2000), and movements to establish guidelines to label products containing GM organisms, it is difficult to predict whether GM finfish will enjoy widespread acceptance in various market sectors, although transgenic tilapia hybrids that carry a homologous GH transgene have been formally approved for human consumption in Cuba (de la Fuente et al. 1999; Maclean et al. 2002). Perhaps the most promising application of transgenic animals (teleosts) is their continued use as research organisms including the development of animal bioreactors for the production of important polypeptides and products for human medicine and biomedical research (Chen, Vrolijk, et al. 1996; Dodd et al. 2000; Dooley and Zon 2000; Goldman et al. 2001; Maclean et al. 2002; Ward and Lieschke 2002; Zbikowska 2003).

Branched-Chain Amino Acids

The use of particular amino acids to enhance growth via pituitary hormone (GH and PRL) release is another way of altering growth in teleosts that has been receiving increasing attention. In humans and other mammals, the amino acid arginine is a potent GH secretogogue (stimulatory) that has been used to test pituitary function (GH release; Merimee et al. 1965, 1967; Harvey 1995; Harvey and Daughaday 1995). Arginine works to elevate plasma GH levels, in part, by altering levels of the two neurohormones somatostatin and growth hormone-releasing hormone (Alba-Roth et al. 1988; Frohman et al. 1990; Jaffe et al. 1996; Hanew and Utsumi 2002).

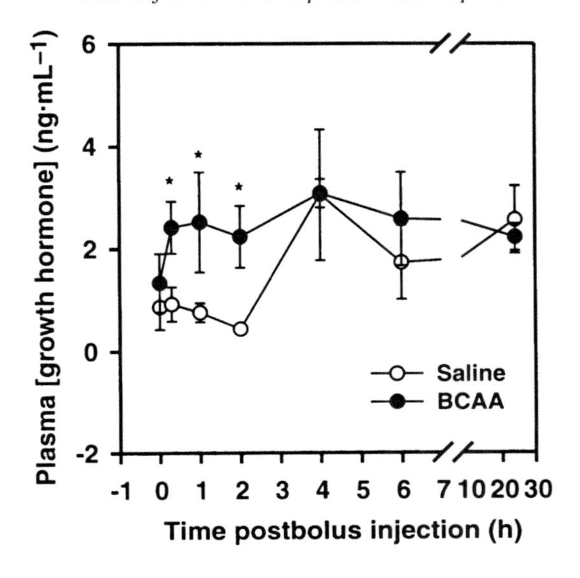

FIGURE 3.4. Effects of branched-chain amino acid (leucine, iso-leucine, and valine in equimolar amounts) infusion (30 μmol·kg^{-1} body weight, dissolved in Tilapia Ringer's) on plasma GH levels in cannulated (~600 g body weight) *O. mossambicus*. Values are in ng·mL^{-1} ± SEM. *$P < 0.05$ compared with saline control at the same time point (Fisher's Protected Least Significant Difference Test).

Other amino acids have also been shown to stimulate GH release, but branched-chain amino acids (BCAAs—leucine, iso-leucine, and valine) have been found to be less effective or ineffective in altering GH release (Chromiak and Antonio 2002). However, in the diet of rainbow trout these amino acids were shown to result in increased growth and lean body mass (Choo et al. 1991; Rodehutscord et al. 1997), phenomena that are characteristic of exogenous GH treatment (Collier and Byatt 1996). This connection encouraged our belief that BCAAs may influence pituitary hormone release in teleosts. Work conducted in *O. mossambicus* (Figures 3.4 and 3.5) demonstrated a rapid and sustained increase in circulating PRL and GH levels following an infusion of BCAAs (equimolar amounts of leucine, isoleucine, and valine) into cannulated tilapia. As a consequence, we are interested in testing diets with varying amounts of the BCAAs and of arginine to see whether the somatotropic axis and thus growth, is stimulated in tilapia and other economically important teleosts.

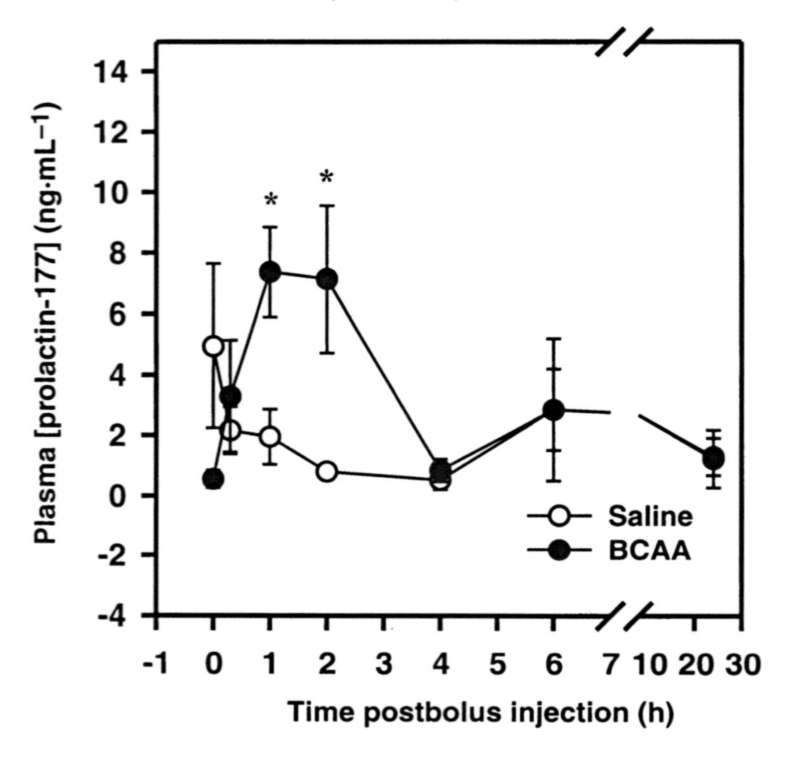

FIGURE 3.5. Effects of branched-chain amino acid (leucine, iso-leucine, and valine in equimolar amounts) infusion (30 μmol·kg^{-1} body weight dissolved in Tilapia Ringer's) on plasma $tPRL_{177}$ levels in cannulated (~600 g body weight) *O. mossambicus*. Values are in ng·mL^{-1} ± SEM. $^*P < 0.05$ compared with saline control at the same time point (Fisher's Protected Least Significant Difference Test). Levels of $tPRL_{188}$ were not determined.

A New Regulator of the Somatotropic Axis

It is evident that current and alternative approaches to enhancing aquaculture production need to be explored and developed further. This requires an understanding of the primary and secondary effects of growth-regulating hormones and the mechanisms that regulate them, for which safe, noninvasive, and economical methods to stimulate growth in finfish aquaculture must be identified. Most desirable would be an approach that safely enhances the natural pulsatile output of pituitary GH.

In the late 1970s and early 1980s a number of small peptides were developed that exhibited a weak but nonspecific stimulatory effect on the mammalian pituitary via binding to an opioid orphan receptor in the central nervous system (Bowers et al. 1984; Bowers 1993, 1998). Additional studies led to the development of Met-enkephalin derivatives that exhibited some ability to directly alter the release of GH from the pituitary in vitro but were inactive in vivo (Momany et al. 1981; Bowers et al. 1991). Subsequent modifications resulted in the development of hexapeptides. One of the earliest hexapeptides, GHRP-2 (KP-102) (D-Ala-D-β-Nal-Ala-Trp-D-Phe-Lys-NH_2), appeared to be the most potent and specific to the somatotrope (Bowers 1993; Hashizumi et al. 1999). Collectively, these peptides are referred to as growth hormone secretogogues (GH secretogogues), or growth hormone-releasing peptides. GHRP-1 through GHRP-6 have been studied extensively for more than 20 years and have been shown to stimulate the release of GH and growth in mammals (Baker et al. 1984; Mericq et al. 1998; Lee et al. 2000), to induce milk production in cattle (Croom et al. 1984), and to induce feeding behavior in rats (Okada et al. 1996). More importantly, GHRPs have been shown to restore growth in humans and rats with diseases of the somatotropic axis (Thorner et al. 1997; Wells et al. 1997; Bowers 1998; Camanni et al. 1998; Mericq et al. 1998), in growth-retarded mammalian models (Wells et al. 1997; Bowers 1998), and in other important animals such as swine and cattle (Roh et al. 1996; Phung et al. 2000). The efficacy of the GHRPs in humans and other mammals, lies in their ability to safely stimulate pulsatile GH release, and consequently growth, with no side effects (Bowers 1993, 1998; Camanni et al. 1998). More recently, nonpeptidyl GHRP mimetics have been developed, tested, and found to be effective at stimulating pulsatile GH release in mammals (Smith et al. 1997, 1999; Bowers 1998; Camanni et al. 1998) and in domestic fowl (Geris et al. 1998, 2001). The GHRPs have been known for some time to bind to a specific G-protein coupled receptor, called the growth hormone-secretogogue receptor (GHS-R), at the levels of the hypothalamus and the pituitary, to control pituitary hormone release (Bowers 1998; Camanni et al. 1998). Only recently, however, has this receptor been cloned and its endogenous ligand, "ghrelin," identified (Howard et al. 1996; Kojima et al. 1999; Palyha et al. 2000).

Shepherd and colleagues (2000) found that intraperitoneal injections of GHRP-2 were effective at elevating plasma GH levels in *O. mossambicus*, demonstrating the presence of this novel GH regulatory axis in teleosts. Subsequent work in tilapia has demonstrated that rat ghrelin and homologous tilapia ghrelin, which are natural ligands (secreted by the stomach) for the GHS-R, are also effective at stimulating both GH and PRL release in vitro (Riley et al. 2002b; Kaiya et al. 2003b). Although the nature, regulation, and function of ghrelin are beyond the scope of this review (see St-Pierre et al. 2003), its cDNAs have been cloned and the native peptides have been shown to stimulate GH and PRL release and appetite in a number of teleost species (Unnippan et al. 2002; Kaiya, Kojima, Hosada, Moriyama, et al. 2003; Kaiya et al. 2003a,b; Parhar et al. 2003). In teleosts, GHRPs (Shepherd et al. 2000) and Ghrelin appear to have a differential effect in that the latter seems to exhibit nonspecific effects on pituitary hormone release (i.e., GH, PRL, and luteinizing hormone release) (Riley, Hirano, and Grau 2002b; Kaiya et al. 2003a,b; Unniappan and Peter 2004). In contrast, Xiao and colleagues (2002) were unable to show a stimulatory effect for two GHRPs (GHRP-6 and hexarelin) on GH release in juvenile grass carp, although human growth hormone–releasing hormone (hGHRH), salmon gonadotropin-releasing hormone (sGnRH-A), and dopamine were effective. The possible taxonomic or developmental reasons for the species-specific response to GHRPs and ghrelin(s) are of great interest, and these differences highlight the following points: (1) the specificity of the GHRPs for GH release, combined with their resistance to proteolysis, makes them excellent candidates as growth promoters (as a feed additive), and (2) no single growth-promoting approach may not be broadly applicable to all species. These points underscore the necessity for further focused research on the endocrine regulation of growth and development in each species of interest.

Although much is known about the in vitro regulation of GH release in some teleosts, overall little is known about the in vitro and in vivo regulation of the GH gene or the control of GH secretory patterns in teleosts. An understanding of the importance of underlying patterns in GH secretion and the involvement of these patterns in the function of GH in teleost physiology is crucial but lacking. An approach that safely enhances endogenous GH secretory patterns in teleosts should yield promising benefits to finfish aquaculture. Finding that endogenous GH levels and growth can be safely enhanced by GHRPs, nonpeptidyl mimetics, or ghrelin, will provide the basis for deciding whether a more general use of such an approach will be advantageous. Other benefits also may be attained by altering/enhancing the natural pulsatility of pituitary hormone release. For example, work in mammals and teleosts supports the role of GH and PRL as immunostimulants (Calduch-Giner et al. 1995; Sakai et al. 1996; Foster et al. 1998; Sakai 1999; Yada et al. 1999, 2002; Harris and Bird 2000). If this is so, an approach that safely stimulates pituitary hormone release could yield many as yet unforeseen benefits to the aquaculture industry.

CONCLUSIONS

Methyltestosterone treatment and environmental salinity (seawater) appear to increase the nutritional requirements of tilapia. Available evidence suggests that the acceleration of growth by MT and salinity is supported not only by increased food consumption but also by improved food and protein utilization and by elevations in endogenous growth-promoting hormones. The hypothesis that the actions of MT and salinity are mediated, in part, by stimulation of the GH/IGF axis finds support from evidence showing that many of the actions of MT on protein metabolism in the whole animal are similar to those of GH and IGF-I. To our knowledge, the actions of androgens and hormones of the somatotropic axis on protein and carbohydrate metabolism have not been well studied in teleosts and warrant greater consideration. Furthermore, research efforts to integrate advanced approaches such as use of growth promoters, selective breeding, and molecular biotechnology into growth enhancement in economically important teleosts, should be encouraged.

It is clear that MT treatment and seawater rearing directly affect the levels of endogenous growth-promoting hormones and intestinal nutrient absorption in tilapia. The influence of anabolic hormones on carbohydrate and protein metabolism suggests that increased availability of these nutrients may be required for the full effects of the growth promoters and salinity to be expressed. Until recently, a diet designed for trout (Purina trout chow™) has been used by us, and others, in growth experiments. Such diets contain a much higher protein and lipid content (and less carbohydrate) than is either required or typically used for tilapia. As the studies discussed here show such a diet is likely to fully support the anabolic actions of MT, but other, less expensive formulations can achieve similar results, especially in fresh water rearing. This is an important issue to fish nutritionists and the aquaculture industry. The high cost and declining availability of fish meal (Naylor et al. 2000), create strong motivation to identify and use alternative, less costly, and sustainable nutrient sources in finfish feeds. However, caution must be exercised when designing feed formulations for growth-accelerated animals since sufficient nutrients may not be supplied by diet formulations in which a higher proportion of calories is supplied by carbohydrates at the expense of protein content or in which the source of protein is inappropriately divided between plant and fishmeal sources. For this reason, the supplementation of feeds with high quality sources of nutrients can be justified only to the degree that the added cost yields a commensurate increase in production efficiency.

REFERENCES

Abrahams, M.V. and A. Sutterlin (1999). The foraging and antipredator behavior of growth-enhanced transgenic Atlantic salmon. *Animal Behavior* 58:933-942.

Afonso, L.O.B. and E.M. Leboute (2003). Sex reversal in Nile tilapia: Is it possible to produce all male stocks through immersion in androgens? *World Aquaculture* 34:16-19.

Afonso, L.O.B., G.I. Wassermann, and R.T. Oliveira (2001). Sex reversal in Nile tilapia (*Oreochromis niloticus*) using a nonsteroidal aromatase inhibitor. *Journal of Experimental Zoology* 290:177-181.

Agellon, L.B., C.J. Emery, J. Jones, S.L. Davies, A.D. Dingle, and T.T. Chen (1988). Growth enhancement by genetically-engineered rainbow trout growth hormone. *Canadian Journal of Fisheries and Aquatic Sciences* 45:146-151.

Agresti, J.J., S. Seki, A. Cnaani, S. Poompuang, E.M. Hallerman, N. Umiel, G. Hulata, G.A.E. Gall, and B. May (2000). Breeding new strains of tilapia: Development of an artificial center of origin and linkage map based upon AFLP and microsatellite loci. *Aquaculture* 185:43-56.

Agustsson, T. and T.B. Bjornsson (2000). Growth hormone inhibits growth hormone secretion from the rainbow trout pituitary in vitro. *Comparative Biochemistry and Physiology Part C* 126:299-303.

Alba-Roth, J., O.A. Muller, J. Schopohl, and K. Werder (1988). Arginine stimulates growth hormone secretion by suppressing endogenous somatostatin secretion. *Journal of Clinical Endocrinology and Metabolism* 67:1186.

Anderson, T.R., J. Rodrigues, and C.S. Nicoll (1983). The synlactin hypothesis: Prolactin's mitogenic action may involve synergism with a somatomedin-like molecule. In E.M. Spencer (Ed.), *Insulin-like growth factors/somatomedins* (pp. 71-78). New York: de Gruyter.

Andersson, L. (2001). Genetic dissection of phenotypic diversity in farm animals. *Natural Review of Genetics* 2:130-138.

Ando, S., R. Yamazaki, and M. Hatano (1986). Effect of 17α-methyltestosterone on muscle composition of chum salmon. *Bulletin of the Japanese Society for Scientific Fisheries* 52:565-571.

Anonymous (2000). FDA releases list of recalled corn products. *USA Today*, New York.

Arukwe, A. (2001). Cellular and molecular responses to endocrine-modulators and the impact on fish reproduction. *Marine Pollution Bulletin* 442:643-655.

Arus, P. and J.M. Gonzalez (1994). Marker-assisted selection. In M.D. Hayward, N.D. Bosemark, and I. Romagusa (Eds.), *Plant breeding principles and prospects* (pp. 314-331). London: Chapman & Hall.

Auperin, B., F. Rentier-Delrue, J. Martial, and P. Prunet (1994). Characterization of a single prolactin (PRL) receptor in tilapia (*Oreochromis niloticus*) which binds both PRLI and PRLII. *Journal of Molecular Endocrinology* 13:241-251.

———. (1995). Regulation of gill prolactin receptors in tilapia (*Oreochromis niloticus*) after a change in salinity or hypophysectomy. *Journal of Endocrinology* 145:213-220.

Baker, P.K., S.D. Conner, M.E. Doscher, L.A. Kraft, and C.A. Ricks (1984). Use of a synthetic growth hormone releasing hexapeptide to increase rate of gain in rats. *Journal of Animal Science* 59:220.
Bala, R.M., H.G. Bohnet, J.N. Carter, and H.G. Friesen (1978). Effect of ovine prolactin on serum somatomedin bioactivity in hypophysectomized female rats. *Canadian Journal of Physiological Pharmacology* 56:984-987.
Ball, J.N. (1965). Effects of autotransplantation of different regions of the pituitary gland on freshwater survival in the teleost *Poecilia latipinna. General and Comparative Endocrinology* 33:v-vi.
Baños, N., J.V. Planas, J. Gutierrez, and I. Navarro (1999). Regulation of plasma insulin-like growth factor-I levels in brown trout (*Salmo trutta*). *Comparative Biochemistry and Physiology Part C* 124:33-40.
Barge, U. and M.J. Phillips (1997). Review of the state world aquaculture: Environment and sustainability. FAO Fisheries Department, FAO Fisheries Circular, 1997, No. 886 FIRI/C886 (Rev. 1), Rome, Italy.
Baroiller, J.F. and H. D'Cotta (2001). Environmental and sex determination in farmed fish. *Comparative Biochemistry and Physiology Part C* 130:399-409.
Barry, T.P. and E.G. Grau (1986). Estradiol-17β and thyrotropin-releasing hormone stimulate prolactin release from the pituitary gland of a teleost fish in vitro. *General and Comparative Endocrinology* 62:306-314.
Bart, A.N., A.R.S.B. Athuada, M.S. Fitzpatrick, and W.M. Contreras-Sanches (2003). Ultrasound enhanced immersion protocols for masculinization of Nile tilapia, *Oreochromis niloticus. Journal of the World Aquaculture Society* 34: 210-216.
Bartley, D. (1997a). *Review of the state of world aquaculture: Biodiversity and genetics*. FAO Fisheries Department, FAO Fisheries Circular, 1997, No. 886 FIRI/C886 (Rev. 1), Rome, Italy.
———. (1997b). Review of the state of world aquaculture: Biodiversity and genetics. FAO Inland Water Resources and Aquaculture Service, Fisheries Resource Division, FAO Fisheries Circular, No. 886, Rev.1, 1997, FAO Fisheries Circular. No. 886. Available online at http://www.fao.org/docrep/003/w7499e/w7499e00.htm, Rome, Italy.
Bauchat, J.R., W.H. Busby, A. Garmong, P. Swanson, J. Moore, M. Lin, and C. Duan (2001). Biochemical and functional analysis of a conserved IGF-binding protein isolated from rainbow trout (*Oncorhynchus mykiss*) hepatoma cells. *Journal of Endocrinology* 170:619-628.
Baxter, R.C. (2001). Inhibition of the insulin-like growth factor (IGF)-IGF-binding protein interaction. *Hormone Research* 55:68-72.
Beckman, B.R., D.A. Larsen, S.L. Moriyama, B. Lee-Pawlak, and W.W. Dickhoff (1998). Insulin-like growth factor-I and environmental modulation of growth during smoltification of spring chinook salmon (*Oncorhynchus tshawytscha*). *General Comparative Endocrinology* 109:325-335.
Beckman, B.R., M. Shimizu, B.A. Gadberry, and K.A. Cooper (2004). Response of the somatotropic axis of juvenile coho salmon to alterations in plane of nutrition

with an analysis of the relationships among growth rate and circulating IGF-I and 41 kDa IGFBP. *General Comparative Endcorinology* 135:334-344.

Ben-Atia, I., M. Fine, A. Tandler, B. Funkenstein, S. Maurice, B. Cavari, and A. Gertler (1999). Preparation of recombinant gilthead seabream (*Sparus aurata*) growth hormone and its use for stimulation of larvae growth by oral administration. *General and Comparative Endocrinology* 113:155-164.

Bern, H.A. (1975). On two possible primary activities of prolactins: Osmoregulatory and developmental. *Verhandlunger der Deutschen Zoologischen Gesellschaft* 1975:40-46.

———. (1983). Functional evolution of prolactin and growth hormone in lower vertebrates. *AmerAmerican Zoologist* 23:663-671.

Beuzen, N.D., M.J. Stear, and K.C. Chang (2000). Molecular markers and their use in animal breeding. *The Veterinarian Journal* 160:42-52.

Bick, T., T. Amit, R.J. Barkey, P. Hertz, M.B.H. Youdim, and Z. Hochberg (1990). The interrelationship of growth hormone (GH), liver membrane GH receptor, serum GH-binding protein activity, and insulin-like growth factor I in the male rat. *Endocrinology* 126:1914-1920.

Bick, T., Z. Hochberg, T. Amit, O.G.P. Isaksson, and O. Jansson (1992). Roles of pulsatility and continuity of growth hormone (GH) administration in the regulation of hepatic GH-receptors, and circulating GH-binding protein and insulin-like growth factor-I. *Endocrinology* 131:423-429.

Biga, P.R., G.T. Schelling, R.W. Hardy, K.D. Cain, K.E. Overturf, and T.L. Ott (2004). The effects of recombinant bovine somatotropin (rbST) on tissue IGF-I, IGF-I receptor, and GH mRNA levels in rainbow trout, *Oncorhynchus mykiss*. *General and Comparative Endocrinology* 135:324-333.

Björnsson, B.T. (1997). The biology of salmon growth hormone: From daylight to dominance. *Fish Physiology and Biochemistry* 17:9-24.

Blaise, O., C. Weil, and P.-Y. Le Bail (1995). Role of IGF-I in the control of GH secretion in rainbow trout (*Oncorhychus mykiss*). Growth Regulation 5:142-150.

Boeuf, G. and P. Payan (2001). How should salinity influence fish growth? *Comparative Biochemistry and Physiology C* 130:411-423.

Bole-Feysot, C., V. Goffin, M. Edery, N. Binart, and P.A. Kelly (1998). Prolactin (PRL) and its receptor: Actions, signal transduction pathways and phenotypes observed in PRL receptor knockout mice. *Endocrine Reviews* 19:225-268.

Bondi, A.A. (1989). *Animal nutrition*. New York: John Wiley & Sons.

Borski, R.J., L.M. Helms, N. H. Richman III, and E.G. Grau (1991). Cortisol rapidly reduces prolactin release and cAMP and $^{45}Ca^{2+}$ accumulation the cichlid fish pituitary in vitro. *Proceedings of the National Academy of Science* 88:2758-2762.

Borski, R.J., J. Yoshikawa, S.S. Madsen, R.S. Nishioka, C. Zabetian, H.A. Bern, and E.G. Grau (1994). Effects of environmental salinity on pituitary growth hormone content and cell activity in the euryhaline tilapia, *Oreochromis mossambicus*. *General and Comparative Endocrinology* 95:483-494.

Bowers, C.Y. (1993). GH releasing peptides-structure and kinetics. *Journal of Pediatric Endocrinology* 6:21-31.

———. (1998). Growth hormone-releasing peptide (GHRP). *Cell and Molecular Life Sciences* 54:1316-1329.

Bowers, C.Y., F.A. Momany, G.A. Reynolds, and A. Hong (1984). On the in vitro and in vivo activity of a new synthetic hexapeptide that acts on the pituitary to specifically release growth hormone. *Endocrinology* 114:1537-1545.

Bowers, C.Y., A.O. Sartor, G.A. Reynolds, and T.M. Badger (1991). On the actions of the growth hormone-releasing hexapeptide, GHRP. *Endocrinology* 128:2027-2035.

Brill, R., Y. Swimmer, C. Taxoboel, K. Cousins, and T. Lowe (2001). Gill and intestinal Na^+-K^+-ATPase activity, and estimated maximal osmoregulatory costs in three high-energy-demand teleosts: Yellowfin tuna (*Thunnus albacares*), skipjack tuna *(Katsuwonus pelamis)*, and dolphin fish *(Coryphaena hippurus)*. Marine Biology 138: 935-944.

Brown, P.S. and S.C. Brown (1987). Osmoregulatory actions of prolactin and other adenohypophysial hormones. In P.K.T. Pang, M.P. Schreibman, and W.H. Sawyer (Eds.), *Vertebrate endocrinology: Fundamentals and biomedical implications* (pp. 45-84). London: Academic Press.

Butler, A.A. and D. Le Roith (2001). Control of growth by the somatotropic axis: Growth hormone and the insulin-like growth factors have related and independent roles. *Annual Review of Physiology* 63:141-164.

Buttery, P.J. and P.A. Sinnett-Smith (1984). The mode of action of anabolic agents with special reference to their effects on protein metabolism—Some speculations. In J.F. Roche and D. O'Callaghan (Eds.), *Manipulation of growth in farm animals* (pp. 211-228). Boston, MA: Martinus Nijhoff Publishers.

Caelers, A., A.C. Schmid, A. Hrusovsky, and M. Reinecke (2003). Insulin-like growth factor II mRNA is expressed in the neurones of the brain of the bony fish *Oreochromis mossambicus, the tilapia. EuroEuropean Journal of Neuroscience* 18: 355-363.

Calduch-Giner, J.A., A. Sitja-Bobadilla, P. Alvarez-Pellitero, and J. Perez-Sanchez (1995). Evidence for a direct action of GH on haemopoietic cells of a marine fish, the gilthead sea bream *(Sparus aurata). Journal of Endocrinology* 146:459-467.

Camanni, F., E. Ghigo, and E. Arvat (1998). Growth hormone-releasing peptides and their analogs. *Frontiers in Neuroendocrinology* 19:47-72.

Chan, S.J., E.M. Plisetskaya, E. Urbinati, Y. Jin, and D.F. Steiner (1997). Expression of multiple insulin and insulin-like growth factor receptor genes in the salmon gill cartilage. *Proceedings of the National Academy of Science* 94: 12446-12451.

Charlier, C., W. Coppieters, F. Farnir, L. Grobet, P. Leroy, C. Michaux, M. Mni, A. Schwers, P. Vanmanshoven, and R. Hanset (1995). The *mh* gene causing double-muscling in cattle maps to bovine chromosome 2. *Mammalian Genome* 6:788-792.

Check, E. (2002). Environmental impact tops list of fears about transgenic animals. *Nature (London)* 418:805.

Chen, T.T., J.K. Lu, M.J. Shamblott, C.M. Cheng, and C.-M. Lin (1996). Transgenic fish: Ideal models for basic research and biotechnological applications. In J.D. Ferraris and S.R. Palumbi (Eds.), *Molecular zoology: Advances, strategies and protocols* (pp. 401-433). New York: Wiley-Liss.

., A. Marsh, M. Shamblott, K.-M. Chan, Y.-L. Tang, C.M. Cheng, and . Yang (1994). Structure and evolution of fish growth hormone and insulin-like growth factor genes. In N.M. Sherwood and C.L. Hew (Eds.), *Fish physiology: Molecular endocrinology of fish* (Volume 13, pp. 179-209). New York: Academic Press.

Chen, T.T., N.H. Vrolijk, J.-K. Lu, C.-M. Lin, R. Reimschuessel, and R.A. Dunham (1996). Transgenic fish and its application in basic and applied research. *Annual Review of Biotechnology* 2:205-236.

Cheng, R., K.M. Chang, and J.L. Wu (2002). Different temporal expressions of tilapia *(Oreochromis mossambicus)* insulin-like growth factor-I and IGF binding protein-3 after growth hormone induction. *Marine Biotechnology* 4:218-225.

Choo, P.-S., T.K. Smith, C.Y. Cho, and H.W. Ferguson (1991). Dietary excess of leucine influence growth and body composition of rainbow trout. *Journal of Nutrition* 121:1932-1939.

Chromiak, J.A. and J. Antonio (2002). Use of amino acids as growth hormone-releasing agents by athletes. *Nutrition* 18:657-661.

Clancy, M.J., J.M. Lester, and J.F. Roche (1984). The effects of anabolic agents on the fibers of the L. dorsi muscle of male cattle. In J.F. Roche and D. O'Callaghan (Eds.), *Manipulation of growth in farm animals* (pp. 100-108). Boston, MA: Martinus Nijhoff Publishers.

Clark, M.S. (2003). Genomics and mapping of Teleostei (bony fish). *Comparative Functional Genomics* 4:182-193.

Clark, R.G., J.-O. Jansson, O. Isaksson, and I.C.A.F. Robinson (1985). Intravenous Growth hormone: Growth responses to patterned infusions in hypophy-sectomized rats. *Journal of Endocrinology* 104:53-61.

Clarke, W.C. and H.A. Bern (1980). Comparative endocrinology of prolactin. In C.H. Li. (Ed.), *Hormonal proteins and peptides*, (Volume 8, pp. 105-197). New York: Academic Press.

Clarke, W.C., S. Walker Farmer, and K.M. Hartwell (1977). Effect of teleost pituitary growth hormone on growth of Tilapia mossambicus and on growth and seawater adaptation of sockeye salmon (*Oncorhynchus nerka*). *General and Comparative Endocrinology* 33:174-178.

Clemmons, D.R. (2001). Use of mutagenesis to probe IGF-binding protein structure/function relationships. *Endocrine Reviews* 22:800-817.

Clemmons, D.R., W.H. Busby, J.B. Clarke, A. Parker, C. Duan, and T.J. Nam (1998). Modifications of insulin-like growth factor binding proteins and their role in controlling IGF actions. *Endocrine Journal* 45:S1-S8.

Collie, N.L. and J.J. Stevens (1985). Hormonal effects on L-proline transport in coho salmon *(Oncorhyncus kisutch)* intestine. *General and Comparative Endocrinology* 59:399-409.

Collier, R.J. and J.C. Byatt (1996). Somatotropin in domestic animals. In A. Altman (Ed.), *Biotechnology in domestic animals* (pp. 483-497). New York: Marcel Dekker, Inc.

Company, R., A. Astola, C. Pendon, M.M. Valdivia, and J. Perez-Sanchez (2001). Somatotropic regulation of fish growth and adiposity: Growth hormone (GH)

and somatolactin (SL) relationship. *Comparative Biochemistry and Physio*
Part C 130:435-445.

Conley, L.K., R.C. Gaillard, A. Guistina, R.S. Brogan, and W.B. Wehrenberg (1998). Effects of repeated doses and continuous infusions of the growth hormone-releasing peptide hexarelin in conscious male rats. *Journal of Endocrinology* 158:367-375.

Contreras-Sanches, W.M., M.S. Fitzpatrick, and C.B. Schreck (2001). Masculinization of tilapia by immersion in trenbolone acetate: Growth performance of trenbolone acetate-immersed tilapia. In K.M. A. Gupta, D. Burke, J. Burright, X. Cummings, and H. Egna (Eds.), *Eighteenth annual technical report. Pond dynamics/aquaculture* (pp. 42-46). CRSP Corvallis, OR: Oregon State University.

Copeland, K.C., M.M. Desouza, and P.C. Gibson (1990). Influence of gonadal steroids on rat pituitary growth hormone secretion. *Research and Experimental Medicine* 190:137-144.

Coxam, V., M.J. Davicco, D. Durand, D. Bauchart, F. Opmeer, and J.P. Barlet (1990). Steroid hormones may modulate hepatic somatomedin C production in newborn calves. *Biology of the Neonate* 58:16-23.

Croom, W.J., E.S. Leonard, P.K. Baker, L.A. Draft, and C.A. Ricks (1984). The effects of synthetic growth hormone releasing hexapeptide BI-679 on serum growth hormone levels and production in lactating dairy cattle. *Journal of Animal Science* 67:109.

Curtis, L.R., F.T. Diren, M.D. Hurley, W.K. Steim, and R.A. Tubb (1991). Disposition and elimination of 17α-methyltestosterone in Nile tilapia *(Oreochromis niloticus). Aquaculture* 99:193-201.

Dauder, S., G. Young, and H.A. Bern (1990). Effect of hypophysectomy, replacement therapy with ovine prolactin, and cortisol and triiodothyronine treatment on prolactin receptors of the tilapia *(Oreochromis mossambicus). General and Comparative Endocrinology* 77:378-385.

Dauder, S., G. Young, L. Hass, and H.A. Bern (1990). Prolactin receptors in liver, kidney, and gill of the tilapia *(Oreochromis mossambicus)*: Characterization and effect of salinity on specific binding of iodinated ovine prolactin. *General and Comparative Endocrinology* 77:368-377.

Daughaday, W.H., M. Kapadia, C.E. Yanow, K. Fabrick, and I.K. Mariz (1985). Insulin-like growth factors I and II in nonmammalian sera. *General and Comparative Endocrinology* 59:316-325.

de la Fuente, J., I. Guillen, R. Martinez, and M.P. Estrada (1999). Growth regulation and enhancement in tilapia: basic research findings and their applications. *Genetic Analysis: Biomolecular Engineering*. 15:85-90.

Degani, G. (1985). The influence of 17α-methyltestosterone on body composition of eels (*Anguilla anquilla* (L.)). *Aquaculture* 50:23-30.

Degani, G. and M.L. Gallagher (1985). Effects of dietary 17α-methyltestosterone and bovine growth hormone on growth and food conversion of slow- and normally-growing American elvers (*Anguilla rostrata*). C). *Canadian Journal of Fisheries and Aquatic Sciences* 42:185-189.

/ ›., H.A. Richards, C. Collet, and Z. Upton (2001). Production, in vitro .acterization, in vivo clearance and tissue localisation of recombinant barramundi *(Lates calcarifer)* insulin-like growth factor II. *General and Comparative Endocrinology* 123:38-50.

Degger, G., Z. Upton, K. Soole, C. Collet, and N. Richardson (2000). Comparison of recombinant barramundi and human insulin-like growth factor (IGF)-I in juvenile barramundi *(Lates calcarifer)*: In vivo metabolic effects, association with circulating IGF-binding proteins and tissue localisation. *General and Comparative Endocrinology* 117:395-403.

Delidow, B.C., N. Herbert, S. Steiny, and C.S. Nicoll (1986). Secretion of prolactin-synergizing activity (synlactin) by the liver of ectothermic vertebrates in vitro. *Journal of Experimental Zoology* 238:147-153.

D'Ercole, A.J. and A.S. Calikoglu (2001). Editorial Review: The case of local versus endocrine IGF-I actions: The jury is still out. *Growth Hormone & IGF Res.* 11:251-265.

De Vlaming, V.L., M. Sage, and R. Tiegs (1975). A diurnal rhythm of pituitary prolactin activity with diurnal effects of mammalian and teleostean prolactin on total body lipid deposition and liver lipid metabolism in teleost fishes. *Journal of Fish Biology* 7:717-726.

Devlin, R.H. (1997). Transgenic salmonids. In L.M. Houdebine (Ed.), *Transgenic animals: Generation and use* (pp. 105-117). Amsterdam, the Netherlands: Harwood Academic Press.

Devlin, R.H., C.A. Biagi, T.Y. Yesaki, D.E. Smailus, and J.C. Byatt (2001). Growth of domesticated transgenic fish. *Nature (London)* 409:781-782.

Devlin, R.H., P. Swanson, W.C. Clarke, E. Plisetskaya, W.W. Dickhoff, S. Moriyama, T.Y. Yesaki, and C.-L. Hew (2000). Seawater adaptability and hormone levels in growth-enhanced transgenic coho salmon, *Oncorhynchus kisutch. Aquaculture* 191:367-385.

Devlin, R.H., T.Y. Yesaki, C.A. Biagl, E.M. Donaldson, P. Swanson, and W.-K. Chan (1994). Extraordinary salmon growth. *Nature* 371:209-210.

Devlin, R.H., T.Y. Yesaki, E.M. Donaldson, and C.-L. Hew (1995). Transmission and phenotypic effects of an antifreez/GH gene construct in coho salmon *(Oncorhynchus kistutch). Aquaculture* 137:161-169.

Dickhoff, W.W., B.R. Beckman, D.A. Larsen, C. Duan, and S. Moriyama (1997). The role of growth hormone in endocrine regulation of salmon smoltification. *Fish Physiology and Biochemistry* 17:231-236.

Dodd, A., P.M. Curtis, L.C. Williams, and D.R. Love (2000). Zebrafish: Bridging the gap between development and disease. *Human Molecular Genetics* 9:2443-2449.

Doerge, R.W. (2002). Mapping and analysis of quantitative trait loci in experimental populations. *Natural Review of Genetics* 3:43-52.

Donaldson, E.M., U.H.M. Fagerlund, D.A. Higgs, and J.R. McBride (1979). Hormonal Enhancement of Growth in Fish. In W.S. Hoar, D.J. Randall, and J.R. Brett (Eds.), *Fish physiology: Bioenergetics and growth* (Volume 8, pp. 456-597). New York: Academic Press, Inc.

Dooley, K. and L.I. Zon (2000). Zebrafish: A model system for the study of human disease. *Current Opinions in Genetic Development* 10:252-256.

Down, N.E. and E.M. Donaldson (1991). Prospects for growth acceleration in salmonids utilizing biologically active proteins and peptides. *Canadian Technical Report on Fisheries and Aquatic Sciences* 1831:141-152.

Drennon, K., B. Small, J.T. Silverstein, S. Moriyama, H. Kawauchi, and B.S. Shepherd (2003). Development of an enzyme-linked immunosorbent assay for the measurement of plasma growth hormone levels in channel catfish *(Ictalurus punctatus)*: Assessment of environmental salinity and GH-secretagogues on plasma GH levels. *General and Comparative Endocrinology* 133:314-312.

Duan, C. (1994). Incorporation of ^{35}S-sulfate into branchial cartilage: A biological model to study hormonal regulation of skeletal growth in fish. In P.W. Hochachka and T.P. Mommsen (Eds.), *Biochemistry and molecular biology of fishes: Analytical techniques*, (Volume 3, pp. 525-533). New York: Elsevier.

———. (1997). The insulin-like growth factor system and its biological actions in fish. *American Zoologist* 37:491-503.

———. (1998). Nutritional and developmental regulation of insulin-like growth factors in fish. *Journal of Nutrition* 128:306S-314S.

———. (2002). Specifying the cellular responses to IGF signals: Roles of IGF-binding proteins. *Journal of Endocrinology* 175:41-54.

Duan, C., J. Ding, Q. Li, W. Tsai, and K. Pozios (1999). Insulin-like growth factor binding protein 2 is a growth inhibitory protein conserved in zebrafish. *Proceedings of the National Academy of Science* 96:15274-15279.

Duan, C., S.J. Duguay, P. Swanson, W.W. Dickhoff, and E.M. Plisetskaya (1994). Tissue specific expression of insulin-like growth factor-I messenger ribonucleic acids in salmonids: Development, hormonal and nutritional regulation. In K.G. Davey, R.E. Peter, and S.S. Tobe (Eds.), *Perspectives in comparative endocrinology* (pp. 365-372). Ottawa: National Research Council of Canada.

Duan, C. and T. Hirano (1992). Effects of insulin-like growth factor-I and insulin on the in vitro uptake of sulfate by eel branchial cartilage: Evidence for the presence of independent hepatic and pancreatic sulphation factors. *Journal of Endocrinology* 133:211-219.

Duan, C. and Y. Inui (1990). Effects of recombinant eel growth hormone on the uptake of [^{35}S]-sulfate by ceratobranchial cartilages of the Japanese eel, *Anguilla japonica. General and Comparative Endocrinology* 79:320-325.

Duan, C., T. Noso, S. Moriyama, H. Kawauchi, and T. Hirano (1992). Eel insulin: Isolation, characterization and stimulatory actions on [^{35}S]-sulfate and [^{3}H]-thymidine uptake in the branchial cartilage of the eel in vitro. *Journal of Endocrinology* 133:221-230.

Duan, C., E. Plisetskaya, and W.W. Dickhoff (1995). Expression of insulin-like growth factor I in normally and abnormally developing coho salmon *(Oncorhynchus mykiss). Endocrinology* 136:446-452.

Dubowsky, S. and M.A. Sheridan (1995). Chronic ovine growth hormone exposure to rainbow trout, *Oncorhynchus mykiss,* reduces plasma insulin concentration, elevates plasma somatostatin-14 concentration, and reduces growth hormone binding capacity. *Experimental Clinical Endocrinology* 14:107-111.

Dunham, R.A., A.C. Bamboux, P.L. Duncan, M. Haya, T.T. Chen, C.-M. Lin, K. Kight, I. Gonzalez-Villasenor, and D.A. Powers (1992). Transfer, expression, and inheritance of salmonid growth hormone genes in channel catfish, IctaIctalurus punctatus, and effects on performance traits. *Molecular Marine Biology and Biotechnology* 1:380-389.

Elo, K.T., J. Vilkki, D.J. de Koning, R.J. Velmala, and A.V. Maki-Tanilla (1999). A quantitative trait locus for live weight maps to bovine chromosome 23. *Mammalian Genome* 10:831-835.

Fagerlund, U.H.M., D.A. Higgs, J.R. McBride, M.D. Plotnikoff, B.S. Dosanjh, and J.R. Markert (1983). Implications of varying dietary protein, lipid and 17α-methyltestosterone content on growth and utilization of protein and energy in juvenile coho salmon (*Oncorhynchus kisutch*). *Aquaculture* 30:109-124.

Fagerlund, U.H.M., J.R. McBride, and E.T. Stone (1979). A test of 17α-methyltestosterone as a growth promoter in a coho salmon hatchery. *American Fisheries Science* 108:476-472.

Farrell, A.P., W. Bennett, and R.H. Devlin (1997). Growth-enhanced transgenic salmon can be inferior swimmers. *Canadian Journal of Zoology* 75:335-337.

Fauconneau, B., M.P. Mady, and P.Y. LeBail (1996). Effect of growth hormone on muscle protein synthesis in rainbow trout *(Oncorhynchus mykiss)* and Atlantic salmon *(Salmo salar). Fish Physiology and Biochemistry* 15:49-56.

Febry, R. and P. Lutz (1987). Energy partitioning in fish: The activity-related cost of osmoregulation in an euryhaline cichlid. *Journal of Experimental Biology* 128:63-85.

Fernandez, G., F. Sanchez-Franco, M.T. De Los Frailes, R.M. Tolon, M.J. Lorenzo, J. Lopez, and L. Cacicedo (1992). Regulation of somatostatin and growth hormone-releasing hormone factor by gonadal steroids in fetal rat hypothalamic cells in culture. *Regulatory Peptides* 42:135-144.

Firth, S.M. and R.C. Baxter (2002). Cellular actions of the insulin-like growth factor binding proteins. *Endocrinology Review* 23:824-854.

Fletcher, G.L. and P.L. Davies (1991). Transgenic fish for aquaculture. *Genetic. Engineering*. 13:331-370.

Flint, J. and R. Mott (2001). Finding the molecular basis of quantitative traits: Successes and pitfalls. *Natural Review of Genetics* 2:437-445.

Foster, A.R., D.F. Houlihan, C. Gray, F. Medale, B. Fauconneau, S.J. Kaushik, and P.-Y. Le Bail (1991). The effects of ovine growth hormone on protein turnover in rainbow trout. *General and Comparative Endocrinology* 82:111-120.

Foster, M., E. Montecino-Rodriguez, R. Clark, and K. Dorshkind (1998). Regulation of B and T cell development by anterior pituitary hormones. *Cellular and Molecular Life Sciences* 54:1076-1082.

Francis, M.J.O. and D.J. Hill (1975). Prolactin-stimulated production of somatomedin by rat liver. *Nature* 255:167-168.

Frohman, K.A., T.R. Downs, I.J. Clarke, and G.B. Thomas (1990). Measurement of growth hormone-releasing hormone and somatostatin in hypothalamic-portal plasma of unanesthetized sheep. *Journal of Clinical Investigation* 86:17-24.

Fruchtman, S., L. Jackson, and R.J. Borski (2000). Insulin-like growth factor I disparately regulates prolactin and growth hormone synthesis and secretion: Studies using the teleost pituitary model. *Endocrinology* 141.

Fryer, J.N. (1979a). Prolactin-binding sites in tilapia *(Sarotherodon mossambicus)* kidney. *General and Comparative Endocrinology* 39:397-403.

———. (1979b). A radioreceptor assay for purified teleost growth hormone. *General and Comparative Endocrinology* 39:123-130.

Frystyk, J., C. Skaerbaek, B. Dinesen, and H. Ørskov (1994). Free insulin-like growth factors (IGF-I and IGF-II) in human serum. *FEBS Letters* 348:185-191.

Gabillard, J.-C., C. Weil, P.-Y. Rescan, I. Navarro, J. Gutierrez, and P.-Y. Le Bail (2003). Environmental temperature increases plasma GH levels independently of nutritional status in rainbow trout *(Oncorhynchus mykiss). General and Comparative Endocrinology* 133:17-26.

Galas, L., N. Chartrel, M. Kojima, K. Kanagawa, and H. Vaudry (2002). Immunohistochemical localization and biochemical characterization of ghrelin in the brain and stomach of the frog *Rana esculenta. Journal of Comparative Neurology* 450:34-44.

Gannam, A.L. and R.T. Lovell (1991). Effects of feeding 17α-methyltestosterone, 11-ketotestosterone, 17β-estradiol, and 3,5,3'-triiodothyronine to channel catfish, *Ictalurus punctatus. Aquaculture* 92:277-288.

Garssen, G.J. and J.K. Oldenbroek (1992). Somatotropins and animal production. In *Biotechnological innovations in animal productivity* (pp. 79--103). Oxford, England: Butterworth/Heinemann.

Geris, K.L., G.J. Hickey, L.R. Berghman, T.J. Visser, E.R. Kuhn, and V.M. Darras (1998). Pituitary and extrapituitary action sites of the novel nonpeptidyl growth hormone (GH) secretagogue L-692,492 in the chicken. *General and Comparative Endocrinology* 111:186-196.

Geris, K.L., G.J. Hickey, A. Vanderghote, E.R. Kuhn, and V.M. Darras (2001). Synthetic growth hormone secretagogues control growth hormone secretion in the chicken at pituitary and hypothalamic levels. *Endocrine* 14:67-72.

Gevers, E., J.M. Wit, and I.C.A.F. Robinson (1996). Growth, growth hormone (GH)-binding protein, and GH receptors are differentially regulated by peak and trough components of the GH secretory pattern in the rat. *Endocrinology* 137: 1013-1018.

Gill, J.A., J.P. Sumpter, E.M. Donaldson, H.M. Dye, L. Souza, T. Berg, J. Wypych, and K. Langley (1985). Recombinant chicken and bovine growth hormones accelerate growth in aquacultured juvenile Pacific salmon *(Onchorhynchus kisutch). Bio/Technology* 3:643-646.

Goffin, V., N. Binart, P. Touraine, and P. Kelly (2002). Prolactin: the new biology of an old hormone. *Annual Review of Physiology* 64:47-67.

Gogoi, D. and P. Keshavanath (1988). Digestibility and growth promoting potential of 17α-methyltestosterone incorporated diets in the Mahseer, *Tor khudree*. In M.M. Joseph (Ed.), *The First Indian Fisheries Forum Proceedings* (pp. 99-102). Mangalore, Karnataka.

Goldman, D., M. Hankin, Z. Li, X. Dai, and J. Ding (2001). Transgenic zebrafish for studying nervous system development and regeneration. *Transgenic Research* 10:21-33.

Gong, Z. and C. Hew (1995). Transgenic fish in aquaculture and developmental biology. *Current Topics in Developmental Biology* 30:177-214.

Goudie, C.A., W.C. Shelton, and N.C. Parker (1986a). Tissue distribution and elimination of radiolabelled methyltestosterone fed to adult blue tilapia. *Aquaculture* 58:227-240.

———. (1986b). Tissue distribution and elimination of radiolabelled methyltestosterone fed to sexually undifferentiated blue tilapia. *Aquaculture* 58:215-226.

Grandillo, S. and S.D. Tanksley (1996). QTL analysis of horticultural traits differentiating the cultivated tomato from the closely related species *Lycopersicon pimpinellifolium. Theoretical and Applied Genetics* 92:935-951.

Grau, E.G., H.A. Bern, T.T. Chen, R. Collier, J. Davidson, E. Donaldson, T. Hirano, A. Kapuscinski, D. Powers, J. Specker, et al. (1992). *Marine resource development: Enhancement of fish growth and development—A proposed national research initiative.* La Jolla, CA: California Sea Grant College Program.

Grau, E.G., N.H. Richmann, III, and R.J. Borski (1994). Osmoreception and a simple endocrine reflex of the prolactin cell of the tilapia *Oreochromis mossambicus.* In K.G. Davey, R.E. Peter, and S.S. Tobe (Eds.), *Perspectives in comparative endocrinology* (pp. 251-256). Ottawa: National Research Council of Canada.

Greene, J.E. and T.T. Chen (1998). Temporal expression pattern of insulin--like growth factor mRNA during embryonic development in a teleost, rainbow trout (*Oncorhynchus mykiss*). *Molecular Marine Biology and Biotechnology* 6:144-151.

———. (1999a). Characterization of teleost insulin receptor family members. I. Developmental expression of insulin receptor messenger RNAs in rainbow trout. *General and Comparative Endocrinology* 115:254-269.

———. (1999b). Characterization of teleost insulin receptor family members. II. Developmental expression of insulin-like growth factor type 1 receptor messenger RNAs in rainbow trout. *General and Comparative Endocrinology* 115:270-281.

———. (1999c). Quantitation of IGF-I and IGF-II and multiple insulin receptor family member messenger RNAs during embryonic development in rainbow trout. *Molecular and Reproductive Development* 54:348-361.

Grobet, L., D. Poncelet, L.J. Royo, B. Brouwers, D. Pirottin, C. Michaux, F. Menissier, M. Zanotti, S. Dunner, and M. Georges (1998). Molecular definition of an allelic series of mutations disrupting the myostatin function and causing double-muscling in cattle. *Mammalian Genome* 9:210-213.

Grobet, L., L.J. Royo Martin, D. Poncelet, D. Pirottin, M. Brouwers, J. Riquet, A. Schoeberlein, S. Dunner, F. Menissier, J. Massabanda, et al. (1997). A deletion in the bovine myostatin gene causes the double-muscling phenotype in cattle. *Nature Genetics* 17:71-74.

Guerrero, R.D. and L.A. Guerrero (1988). Feasibility of commercial production of sex-reversed Nile tilapia fingerlings in the Philippines. In R.S.V. Pullin,

T. Bhukaswan, K. Tonguthai, and J.L. Maclean (Eds.), *The second international symposium on tilapia in aquaculture* (pp. 183-186). Manila: ICLARM.

Habibi, H.R. and B.W. Ince (1984). A study of androgen-stimulated L-leucine transport by the intestine of rainbow trout *(Salmo gairderi Richardson)* in vitro. *Comparative Biochemistry and Physiology* 79A:143-149.

Habibi, H.R., B.W. Ince, and A.J. Matty (1983). Effects of 17α-methyltestosterone and 17β-oestradiol on intestinal transport and absorption of L-(^{14}C)-leucine in vitro in rainbow trout (*Salmo gairdneri*). *Journal of Comparative Physiology* 151.

Hagenfeldt, Y., K. Linde, H.E. Sjoberg, W. Zumkeller, and S. Arvers (1992). Testosterone increases serum 1,25-dihydroxyvitamin D and insulin-like growth factor-I in hypogonadal men. *Journal of Andrology* 15:93-102.

Halvorson, H.O. and F. Quezada (1999). Increasing public involvement in enriching our fish stocks through genetic involvement. *Genetic Analysis: Biomolecular Engeneering* 15:75-84.

Hanew, K. and A. Utsumi (2002). The role of endogenous GHRH in arginine-, insulin-, clonidine- and L-dopa-induced GH release in normal subjects. *European Journal of Endocrinology* 146:197-202.

Harris, J. and D.J. Bird (2000). Modulation of the fish immune system by hormones. *Veternary Immunology and Immunopathology* 77:163-176.

Harvey, D.J. (1993). Outlook for U.S. Aquaculture. In *Agriculture outlook conference, Session 20*. Washington, DC: United States Department of Agriculture.

Harvey, S. (1995). Growth hormone release: Feedback regulation. In S. Harvey, C.G. Scanes, and W.H. Daughaday (Eds.), *Growth hormone* (pp. 163-184). London: CRC Press.

Harvey, S. and W.H. Daughaday (1995). Growth hormone release: Profiles. In S. Harvey, C.G. Scanes, and W.H. Daughaday (Eds.), *Growth hormone* (pp. 193-223). London: CRC Press.

Hasegawa, O., H. Sugihara, S. Minami, and I. Wakabyashi (1992). Masculinization of growth hormone (GH) secretory pattern by dihydrotestosterone is associated with augmentation of hypothalamic somatostatin and GH-releasing hormone mRNA levels in ovariectomized adult rats. *Peptides* 13:475-481.

Hashizumi, T., M. Kawai, K. Ohtsuki, A. Ishii, and M. Numata (1999). Oral administration of peptidergic growth hormone (GH) secretagogue KP-102 stimulates GH release in goats. *Domestic Animal Endocrinology* 16:31-39.

Hassan, H.A., R.A. Merkel, W.J. Enright, and H.A. Tucker (1992). Androgens modulate growth hormone-releasing factor-induced GH release from bovine anterior pituitary cells in static culture. *Domestic Animal Endocrinology* 9:209-208.

Hazzard, C.E. and G.A. Ahearn (1990). Effects of 17α-methyltestosterone on transintestinal transport of D-glucose in an herbivorous fish. *FASEB Journal* 4:729 [Abstract].

Helms, L.M.H., E.G. Grau, S.K. Shimoda, R.S. Nishioka, and H.A. Bern (1987). Studies on the regulation of growth hormone release from the proximal pars distalis of male tilapia, OreoOreochromis mossambicus, in vitro. *General and Comparative Endocrinology* 65:48-55.

Heyen, D.W., J.I. Weller, M. Ron, M. Band, J.E. Beever, E. Feldmesser, Y. Da, G.R. Wiggans, P.M. VanRaden, and H.A. Lewin (1999). A genome scan for QTL influencing milk production and health traits in diary cattle. *Physiological Genomics* 1:165-175.

Hill, D.J., M.J.O. Francis, and R.D.G. Milner (1977). Action of rat prolactin on plasma somatomedin levels in the rat and on somatomedin release from perfused rat liver. *Journal of Endocrinology* 75:137-143.

Hiott, A.E. and R.P. Phelps (1993). Effects of initial age and size on sex reversal of OreoOreochromis mossambicus fry using methyltestosterone. *Aquaculture* 112:301-308.

Hirano, T. (1986). The spectrum of prolactin action in teleosts. In C.L. Ralph (Ed.), *Progress in Clinical and Biological research: Comparative endocrinology: Developments and directions* (Volume 205, pp. 53-74). New York: Alan R. Liss.

Ho, W.K., Z.Q. Meng, H.R. Lin, C.T. Poon, Y.K. Leung, K.T. Yan, N. Dias, A.P. Che, J.-L. Liu, W.M. Zeheng, et al. (1998). Expression of grass carp growth hormone by baculovirus in silkworm larvae. *Biochimica et Biophysica Acta*. 1381: 331-339.

Hobbs, C.J., S.R. Plymate, C.J. Rosen, and R.A. Adler (1993). Testosterone administration increases insulin-like growth factor-I levels in normal men. *Journal of Clinical Metabolism* 77:776-779.

Holman, S.R. and R.C. Baxter (1996). Insulin-like growth factor binding protein 3: Factors affecting binary and ternary complex formation. *Growth Regulation* 6:42-47.

Houdebine, L.M. and D. Chourrout (1991). Trangenesis in fish. *Experentia* (Basel) 47:891-897.

Howard, A.D., S.D. Feighner, and D.F. Cully (1996). A receptor in pituitary and hypothalamus that functions in growth hormone release. *Science* 273:974-977.

Howerton, R.D. (1988). The effects of the synthetic steroid, 17α-methyltestosterone and the thyroid hormone, triodo-L-thyronine on growth of the euryhaline tilapia, *Oreochromis mossambicus*. Master's thesis, Department of Zoology. University of Hawaii, Honolulu, Hawaii.

Howerton, R.D., D.K. Okimoto, and E.G. Grau (1992). The effect of orally administered 17α-methyltestosterone and triiodothyronine on growth and proximate body composition of seawater-adapted tilapia (*Oreochromis mossambicus*). *Aquaculture and Fisheries Management* 23:123-128.

Hulata, G. (2001). Genetic manipulations in aquaculture: A review of stock improvement by classical and modern technologies. *Genetica* 111:155-173.

———. (2002). Aquaculture in the 21st century. *Israeli Journal of Aquaculture* 54:52.

Ikemoto, B.N., C.E. Hazzard, and G.A. Ahearn (1991). Insulin and 17α-methyltestosterone have similar and additive actions on transintestinal D-glucose transport in herbivorous teleosts. *FASEB Journal* 5:A1108.

Inui, Y. and H. Ishioka (1985). In vivo and in vitro effects of growth hormone on the incorporation of [^{14}C]-leucine into protein of liver and muscle of the eel. *General and Comparative Endocrinology* 59:295-300.

Iwamoto, R. (2003). Marine fish enhancement and aquaculture research. Northwest Fisheries Science Center, NWFSC Issue Paper REUT 6203, 2003, HQ ID 283, p. 2.

Iyengar, A., F. Muller, and N. Maclean (1996). Regulation and expression of transgenes in fish—A review. *Transgenic Research* 5:147-166.

Jaffe, C.A., R. DeMott-Friberg, and A.L. Barkan (1996). Endogenous growth hormone (GH)-releasing hormone is required for GH response to pharmacological stimuli. *Journal of Clinical Investigation* 97:934-940.

Jansson, J.-O., A.-W. Kerstin, S. Eden, K.-G. Thorngren, and O. Isaksson (1982). Effect of frequency of growth hormone administration on longitudinal bone growth and body weight in hypophysectomized rats. *Acta Physiologica Scandanavia* 114:261-265.

Jeh, H.S., C.H. Kim, H.K. Lee, and K. Han (1998). Recombinant flounder growth hormone from Escherichia coli: Overexpression, efficient recovery, and growth-promoting effect on juvenile flounder by oral administration. *Journal of Biotechnology* 60:183-193.

Jeon, J.-T., O. Carlborg, A. Tornsten, E. DGiuffra, V. Amarger, P. Chardon, L. Andersson-Eklund, K. Andersson, I. Hansson, K. Lundstrom et al. (1999). A paternally expressed QTL affecting skeletel and cardiac muscle mass in pig maps to the IGF2 locus. *Nature Genetics* 21:157-158.

Jin, M., B. Junjie, L. Xinhui, L. Jainren, J. Qing, and Z. Hongjun (1999). Expression of rainbow trout growth hormone cDNA in Saccharomyces cerevisiae. *Chinese Journal of Biotechnology* 15:219-224.

Johnson, J., J.T. Silverstein, W.R. Wolters, M. Shimizu, W.W. Dickhoff, and B. Shepherd (2003). Disparate regulation of the insulin-like growth factor-binding proteins in a primitive, Ictalurid, teleost *(Ictalurus punctatus). General and Comparative Endocrinology* 134:122-130.

Johnsson, J.I. and B.T. Björnsson (1994). Growth hormone increases growth rate, appetite and dominance in juvenile rainbow trout, *Oncorhynchus mykiss. Animal Behavior* 48:177-186.

Johnsson, J.I., E. Petersson, E. Jonsson, B.T. Bjornsson, and T. Jarvi (1996). Domestication and growth hormone alter antipredator behaviour and growth patterns in juvenile brown trout, *Salmo trutta. Canadian Journal of Fisheries and Aquatic Sciences* 53:1546-1554.

Johnstone, R., D.J. McIntosh, and R.S. Wright (1983). Elimination of orally administered 17α-methyltestosterone by *Oreochromis mossambicus* (tilapia) and *Salmo gairdneri* (rainbow trout) juveniles. *Aquaculture* 35:249-257.

Jones, J.I. and D.R. Clemmons (1995). Insulin-like growth factors and their binding proteins: Biological actions. *Endocrinology Review* 16:3-34.

Jonsson, E., J.I. Johnsson, and T.B. Bjornsson (1996). Growth hormone increases predation exposure of rainbow trout. *Proceedings of the Royal Society of London Bulletin of Biology* 263:647-651.

Kaiya, H., M. Kojima, H. Hosoda, S. Moriyama, A. Takahashi, H. Kawauchi, and K. Kangawa (2003). Peptide purification, complementary deoxyribonucleic acid

(DNA) and genomic DNA cloning, and functional characterization of ghrelin in rainbow trout. *Endocrinology* 144:5215-5226.

Kaiya, H., M. Kojima, H. Hosoda, L.G. Riley, T. Hirano, E.G. Grau, and K. Kangawa (2003a). Amidated fish ghrelin: Purification, cDNA cloning in the Japanese eel and its biological activity. *Journal of Endocrinology* 176:415-423.

———. (2003b). Identification of tilapia ghrelin and its effects on growth hormone and prolactin release in the tilapia, *Oreochromis mossambicus. Comparative Biochemistry and Physiology B Biochemistry and Molecular Biology* 135: 421-429.

———. Dual mode of cortisol action on GH/IGF-I/IGFBPs in the tilapia, *Oreochromis mossambicus. Journal of Endocrinology* 178:91-99.

Kajimura, S., K. Uchida, T. Yada, T. Hirano, K. Aida, and E.G. Grau (2002). Effects of insulin-like growth factors (IGF-I and -II) on growth hormone and prolactin release and gene expression in euryhaline tilapia, *Oreochromis mossambicus. General and Comparative Endocrinology* 127:223-231.

Karasov, W.H. and J.M. Diamond (1983). A simple method for measurement of solute uptake in vitro. *Journal of Comparative Physiology* 152:105-116.

Karasov, W.H., R.S.I. Pond, D.H. Solberg, and J.M. Diamond (1983). Regulation of proline and glucose transport in mouse intestine by dietary substrate levels. *Proceedings of the National Academy of Science* 80:7674-7677.

Kawauchi, H., S. Moriyama, and T. Hirano (1991). Oral administration of recombinant salmon growth hormone to rainbow trout. In B. Fauconneau and F. Takashima (Eds.), *Growth determinants in aquaculture* (pp. 109-120). Third France-Japan Conference in Oceanography. Nantes, France.

Kayes, T. (1977a). Effects of hypophysectomy, beef growth hormone replacement therapy, pituitary autotransplantation, and environmental salinity on growth in the black bullhead (*Ictalurus melas*). *General and Comparative Endocrinology* 33:371-381.

———. (1977b). Effects of temperature on hypophyseal (growth hormone) regulation of length, weight, and allometric growth and total lipid and water concentrations in the black bullhead (*Ictalurus melas*). General and Comparative Endocrinology 33:382-393.

Keenan, B.S., G.E. Richards, S.W. Ponder, J.S. Dallas, M. Nagamani, and E.R. Smith (1993). Androgen-stimulated pubertal growth: The effects of testosterone and dihydrotestosterone on growth hormone and insulin-like growth factor-I in treatment of short stature and delayed puberty. *Journal of Clinical Endocrinology and Metabolism* 76:996-1001.

Kelley, K.M., J.T. Haigwood, M. Perez, and M.M. Galima (2001). Serum insulin-like growth factor binding proteins (IGFBPs) as markers for anabolic stress/ catabolic conditions in fishes. *Comparative Biochemistry and Physiology, Part B* 129:229-236.

Kelley, K.M., R.S. Nishioka, and H.A. Bern (1988). Novel effect of vasoactive intestinal polypeptide and peptide histidine isoleucine: Inhibition of in vitro secretion of prolactin in the tilapia, *Oreochromis mossambicus. General and Comparative Endocrinology* 72:97-106.

Kelley, K.M., Y. Oh, S.E. Gargosky, Z. Gucev, T. Matsumoto, V. Hwa, L. Ng, D.M. Simpson, and R.G. Rosenfeld (1996). Insulin-like growth factor-binding proteins (IGFBPs) and their regulatory dynamics. *International Journal of Biochemistry and Cellular Biology* 28:619-637.

Kelley, K.M., K.E. Schmidt, L. Berg, K. Sak, M.M. Galima, C. Gillespie, L. Balogh, A. Hawayek, J.A. Reyes, and M. Jamison (2002). Comparative endocrinology of the insulin-like growth factor-binding protein. *Journal of Endocrinology* 175:3-18.

Killian, H.S. and C.C. Kohler (1991). Influence of 17α-methyltestosterone and red tilapia under two thermal regimes. *Journal of the World Aquaculture Society* 22:83-94.

Kocher, T.D., W.J. Lee, H. Sobolewska, D.J. Penman, and B. McAndrew (1998). A genetic linkage map of a cichlid fish, the tilapia *(Oreochromis niloticus). Genetics* 148:1225-1232.

Kohler, C.C. (2004). A white paper on the status and needs of tilapia aquaculture in the north central region. North Central Regional Aquaculture Center (NCRAC), White paper, January 14, 2004, p. 10.

Kojima, M., H. Hosoda, Y. Date, M. Nakazato, H. Matsuo, and K. Kangawa (1999). Ghrelin is a growth-hormone-releasing acylated peptide from stomach. *Nature (London)* 402:656-660.

Kuwaye, T.T., D.K. Okimoto, S.K. Shimoda, R.D. Howerton, H.-R. Lin, P.K.T. Pang, and E.G. Grau (1993). Effect of 17α-methyltestosterone on the growth of the euryhaline tilapia, *Oreochromis mossambicus,* in fresh water and in seawater. *Aquaculture* 113:137-152.

Lee, B.Y., D.J. Penman, and T.D. Kocher (2003). Identification of a sex-determining region in the Nile tilapia *(Oreochromis niloticus)* using bulked segregant analysis. *Animal Genetics* 34:379-383.

Lee, H.G., R.A. Vega, L.T. Phung, N. Matsunaga, H. Kuwayama, and H. Hidari (2000). The effect of growth hormone-releasing peptide-2 (KP-102) administration on plasma insulin-like growth factor (IGF-I) and IGF-binding proteins in holstein steers on different planes of nutrition. *Domestic Animal Endocrinology* 18:293-308.

Lee, L.T.O., G. Nong, Y.H. Chan, D.L.Y. Tse, and C.H.K. Cheng (2001). Molecular cloning of a teleost growth hormone receptor and its functional interaction with human growth hormone. *Gene* 270:121-129.

Leedom, T.A., K. Uchida, T. Yada, N.H.I. Richman, J.C. Byatt, R.J. Collier, T. Hirano, and E.G. Grau (2002). Recombinant bovine growth hormone treatment of tilapia: Growth response, metabolic clearance, receptor binding and immunoglobin production. *Aquaculture* 207:359-380.

Leena, S., Shameena, B., and O.V. Oommen (2001) In vivo and in vitro effects of prolactin and growth hormone on lipid metabolism in a teleost, *Anabas testudineus* (Bloch). *Comparative Biochemistry and Physiology B Biochemistry and Molecular Biology* 128:761-766.

Le Roith, D., C. Bondy, S. Yakar, J.-L. Liu, and A. Butler (2001). The somatomedin hypothesis: 2001. *Endocrinology Review* 22:53-74.

Le Roith, D., L. Scavo, and A.A. Butler (2001). What is the role of circulating IGF-I? *Trends in Endocrinology and Metabolism* 12:48-52.

Lone, K.P. and A.J. Matty (1980). The effect of feeding methyltestosterone on the growth and body composition of common carp *(Cyprinus carpio L.). General and Comparative Endocrinology* 40:409-424.

———. (1982). Cellular effects of adrenosterone feeding to juvenile carp, *Cyprinus carpio* L., effect on liver, kidney, brain and muscle protein and nucleic acids. *Journal of Fish Biology* 21:33-45.

Lone, K.P. and M.T. Ridha (1993). Sex reversal and growth of *Oreochromis spilurus* (Gunther) in brackish and seawater by feeding 17α-methyltestosterone. *Aquaculture and Fisheries Management* 24:593-602.

MacLean, N. and D. Penman (1990). The application of gene manipulation to aquaculture. *Aquaculture* 85:1-20.

Maclean, N., M.A. Rahman, F. Sohm, G. Hwang, A. Iyengar, H. Ayad, A. Smith, and H. Farahmand (2002). Transgenic tilapia and the tilapia genome. *Gene* 295:265-277.

Maestro, M.A., E. Mendez, E. Bayraktaroglu, N. Banos, and J. Gutierrez (1998). Appearance of insulin and insulin-like growth factor-I (IGF-I) receptors through-out the ontogeny of brown trout *(Salmo trutta fario). Growth Hormone & IGF Research* 1998:195-204.

Mair, G.C. and D.C. Little (1991). Population control in farmed tilapias, *NAGA, The ICLARM Quarterly*. World Fisheries Center (formerly ICLARM), pp. 8-13.

Maiter, D., L.E. Underwood, M. Maes, M.L. Davenport, and J.M. Ketelslegers (1988). Different effects of intermittent and continuous growth hormone (GH) administration on serum somatomedin-C/insulin-like growth factor I and liver GH receptors in hypophysectomized rats. *Endocrinology* 123:1053-1059.

Maiter, D., J.L. Walker, E. Adam, B. Moats-Staats, N. Mulumba, J.-M. Ketelslegers, and L.E. Underwood (1992). Differential regulation by growth hormone (GH) of insulin-like growth factor-I and GH receptor/binding protein gene expression in rat liver. *Endocrinology* 130:3257-3264.

Mancera, J.-M., R. Laiz Carrion, and M. del Pilar Martin del Rio (2002). Osmoregulatory action of PRL, GH and cortisol in the gilthead seabream (*Sparus aurata L*). General and Comparative Endocrinology 129:95-103.

Manzon, L.A. (2002). The role of prolactin in fish osmoregulation: A review. *General and Comparative Endocrinology* 125:291-310.

Marshall, W.S. (1995). Transport processes in isolated teleost epithelia: opercular epithelium and urinary bladder. In W.S. Hoar, D.J. Randall, and A.P. Farrell (Eds.), *Fish physiology: Cellular and molecular approaches to fish ionic regulation*, (Volume 14, pp. 1-23). San Francisco, CA: Academic Press.

Matty, A.J. (1986). Nutrition, hormones and growth. *Fish Physiology and Biochemistry* 2:1-4.

Matty, A.J. and I.R. Cheema (1978). The effect of some steroid hormones on the growth and protein metabolism of rainbow trout. *Aquaculture* 14:163-178.

Maures, T.J. and C. Duan (2002). Structure, developmental expression and physiological regulation of zebrafish IGF binding-protein-I. *Endocrinology* 143:2722-2731.

Mayo, O. and I.R. Franklin (1998). The place of QTL in the basis of quantitative genetics. I. General considerations. *Proceedings of the 6th World Congress of Genetics and Applied Livestock Production* 26:77-80.

McBride, J.R. and U.H.M. Fagerlund (1973). The use of 17α-methyltestosterone for promoting weight increases in juvenile Pacific salmon. *Journal of the Fisheries Research Board of Canada* 30:1099-1104.

McConnell, S.K.J., C. Beynon, J. Leamon, and D.O.F. Skibinski (2000). Microsatellite marker based genetic linkage maps of *Oreochromis aureus* and *O. niloticus* (Cichlidae): Extensive linkage group segment homologies revealed. *Animal Genetics* 31:214-218.

McCormick, S.D. (1995). Hormonal control of gill Na^{+}, K^{+}-ATPase and chloride cell function. In W.S. Hoar, D.J. Randall, and A.P. Farrell (Eds.), *Fish physiology: Cellular and molecular approaches to fish ionic regulation*, (Volume 14, pp. 285-315. San Francisco, CA: Academic Press.

———. (2001). Endocrine control of osmoregulation in teleost fish. *American Zoologist* 41:781-794.

McCormick, S.D., K.M. Kelley, G. Young, R.S. Nishioka, and H.A. Bern (1992). Stimulation of coho salmon growth by insulin-like growth factor-I. *General and Comparative Endocrinology* 86:398-406.

McCormick, S.D., T. Sakamoto, S. Hasegawa, and T. Hirano (1991). Osmoregulatory actions of insulin-like growth factor-I in rainbow trout *(Oncorhynchus mykiss). Journal of Endocrinology* 130:87-92.

McLean, E. and E.M. Donaldson (1990). Absorption of bioactive proteins by the gastrointestinal tract of fish: A review. *Journal of Aquatic Animal Health* 2:1-11.

McLean, E., E.D. Donaldson, E. Teskeredzic, and L.M. Souza (1993). Growth enhancement following dietary delivery of recombinant porcine somatotropin to diploid and triploid coho salmon *(Oncorhynchus kisutch). Fish Physiology and Biochemistry* 11:363-369.

McLean, E., B. Ronsholdt, C. Sten, and Najamuddin (1999). Gastrointestinal delivery of peptide and protein drugs to aquacultured teleosts. *Aquaculture* 177:231-247.

McMahon, C.D., R.P. Radcliff, K.J. Lookingland, and H.A. Tucker (2001). Neuroregulation of growth hormone secretion in domestic animals. *Domestic Animal Endocrinology* 20:65-87.

McPherron, A.C. and S.-J. Lee (1997). Double muscling in cattle due to mutations in the myostatin gene. *Proceedings of the National Academy of Science* 94: 12457-12461.

Meier, A.H. (1969). Diurnal variations of metabolic responses to prolactin in lower vertebrates. *General and Comparative Endocrinology* 2 (Suppl.) 2:55-62.

Melamed, P., N. Eliahu, B. Levavi-Sivan, M. Ofir, O. Farchi-Pisanty, F. Rentier-Delrue, J. Smal, Y. Zvi, and Z. Naor (1995). Hypothalamic and thyroidal regulation of growth hormone in tilapia. *General and Comparative Endocrinology* 97:13-30.

Melamed, P., N. Eliahu, M. Ofir, B. Levavi-Sivan, J. Smal, F. Rentier-Delrue, and Z. Yaron (1995). The effects of gonadal development and sex steroids on growth hormone secretion in the male tilapia hybrid *(Oreochromis niloticus × O. aureus). Fish Physiology and Biochemistry* 14:267-277.

Melamed, P., H. Rosenfeld, A. Elizur, and Z. Yaron (1998). Endocrine regulation of gonadotropin and growth hormone gene transcription in fish. *Comparative Biochemistry and Physiology* 119C:325-338.

Mendez, E., J.V. Planas, J. Castillo, I. Navarro, and J. Gutierrez (2001). Identification of a type II insulin-like growth factor receptor in fish embryos. *Endocrinology* 142:1090-1097.

Mendez, E., A. Smith, M.L. Figueiredo-Garutti, J.V. Planas, I. Navarro, and J. Gutierrez (2001). Receptors for insulin-like growth factor-I (IGF-I) predominate over insulin receptors in skeletal muscle throughout the life cycle of brown trout, *Salmo trutta. General and Comparative Endocrinology* 122:148-157.

Meredith, H.O., N.H. Richman III, R.J. Collier, A.P. Seale, L.G. Riley, C. Ball, S.K. Shimoda, M.H. Stetson, and E.G. Grau (1999). Pesticide effects on prolactin release from the rostral pars distalis in vitro and their effects on growth in vivo in the tilapia *(Oreochromis mossambicus).* In D.S. Hanshel, M.C. Black, and M.C. Harrasse (Eds.), *Environmental toxicology and risk assessment* (Volume 8) West Conshohocken, PA: American Society for Testing and Materials.

Mericq, V., F. Cassorla, T. Salazar, A. Avila, G. Iniguez, C.Y. Bowers, and G.R. Marriam (1998). Effects of eight months treatment with graded doses of a growth hormone (GH)-releasing peptide in GH-deficient children. *Journal of Clinical Endocrinology and Metabolism* 83:2355-2360.

Merimee, T.J., D.A. Lillicrap, and D. Rabinowitz (1965). Effect of arginine on serum-levels of human growth-hormone. *Lancet* 2:668-670.

Merimee, T.J., D. Rabinowitz, L. Riggs, J.A. Burgess, D.L. Rimoin, and V.A. McKusick (1967). Plasma growth hormone after arginine infusion: Clinical experiences. *New England Journal of Medicine* 276:434-439.

Meuwissen, T.H.E. and J.A.M. van Arendonk (1992). Genetics and breeding: Potential improvements in the rate of genetic gain from marker-assisted selection in diary cattle breeding schemes. *Journal of Dairy Science* 75: 1651-1659.

Mohan, S. and D.J. Baylink (2002). IGF-binding proteins are multifunctional and act via IGF-dependent and -independent mechanisms. *Journal of Endocrinology* 175: 19-31.

Moller, J. (2003). Effects of growth hormone on fluid homeostasis: Clinical and experimental aspects. *Growth Hormone & IGF Research.* 13:55-74.

Momany, F.A., C.A. Bowers, G.A. Reynolds, D. Chang, A. Hong, and A. Newlander (1981). Design, synthesis, and biological activity of peptides which release growth hormone. *Endocrinology* 108:31-39.

Mommsen, T.P. (1998). Growth and metabolism. In D.H. Evans (Ed.), *The physiology of fishes* (pp. 80-85). Boca Raton, FL: CRC Press.

———. (2001). Paradigms in growth in fish. *Comparative Biochemistry and Physiology, Part B* 129:207-219.

Mori, T. and R.H. Devlin (1999). Transgene and host growth hormone gene expression in pituitary and nonpituitary tissues of normal and growth hormone transgenic salmon. *Molecular and Cell Endocrinology* 149:129-139.

Moriyama, S., T. Hirano, and H. Kawauchi (1989). Intestinal uptake of recombinant salmon growth hormone following intragastric or rectal administration to rainbow trout. Presented at the First International Marine Biotechnology Conference. Tokyo, Japan.

Moriyama, S., P. Swanson, M. Nishii, A. Takahashi, H. Kawauchi, W. Dickhoff, and E. Plisetskaya (1994). Development of a homologous radioimmunoassay for coho salmon Insulin-like growth factor I. *General and Comparative Endocrinology* 96:149-161.

Moriyama, S., H. Yamamoto, S. Sugimoto, T. Abe, T. Hirano, and H. Kawauchi (1993). Oral administration of recombinant salmon growth hormone to rainbow trout, *Oncorhynchus mykiss. Aquaculture* 112:99-106.

Muir, W.M. and R.D. Howard (2002). Assessment of possible ecological risks and hazards of transgenic fish with implications for other sexually reproducing organisms. *Transgenic Research* 11:101-114.

Murphy, L.J., K. Tachibana, and H.G. Friesen (1988). Stimulation of hepatic insulin-like growth factor-I gene expression by ovine prolactin: Evidence for intrinsic somatogenic activity in the rat. *Endocrinology* 122:2027-2033.

Nagahama, Y., R.S. Nishioka, H.A. Bern, and R.L. Gunther (1975). Control of prolactin secretion in teleosts, with special reference to *Gillichthys mirabilis* and *Tilapia mossambica. General and Comparative Endocrinology* 25:166-188.

Nagi, J., M.P. Sabour, and B. Benkel (1992). Reproductive impairment in mice with the rat growth hormone transgene. *Journal of Animal Breeding and Genetics* 109:291-300.

Nakamura, M., T. Kobayashi, X. Chang, and Y. Nagahama (1998). Gonadal sex differentiation in teleost fish. *Journal of Experimental Biology* 281:362-372.

Nam, Y.K., J.K. Noh, Y.S. Cho, H.J. Cho, K.-N. Cho, C.G. Kim, and D.S. Kim (2001). Dramatically accelerated growth and exraordinary gigantism of transgenic mud loach, *Misgurnus mizolepis. Transgenic Research* 10:353-362.

Naylor, R.L., R.J. Goldburg, J.H. Primavera, N. Kautsky, M.C.M. Beveridge, J. Clay, C. Folke, J. Lubchenco, H. Mooney, and M. Troell (2000). Effect of aquaculture on world fish supplies. *Nature (London)* 405:1017-1024.

Negatu, Z. and A.H. Meier (1995). In vitro incorporation of [^{14}C]-glycine into muscle protein of gulf killifish *(Fundulus grandis)* in response to insulin-like growth factor-I. *General and Comparative Endocrinology* 98:193-201.

Nezer, C., L. Moreau, B. Brouwers, W. Coppieters, J. Detilleux, R. Hanset, L. Karim, A. Kvasz, P. Leroy, and M. Georges (1999). An imprinted QTL with major effect on muscle mass and fat deposition mpas to the IGF2 locus in pigs. *Nature Genetics* 21:155-156.

Ng, K.P., J.P. Datuin, and H.A. Bern (2001). Effects of estrogens on in vitro and in vivo on cartilage growth in the tilapia (*Oreochromis mossambicus*). *General and Comparative Endocrinology* 121:295-304.

Ng, T.B., Leung, T.C., Cheng, C.H., and Woo, N.Y. (1992). Growth hormone binding sites in tilapia (*Oreochromis mossambicus*) liver. *General and Comparative Endocrinology* 86:111-118.

Nicoll, C.S. (1974). Physiological actions of prolactin. In E. Knobil and W.H. Sawyer (Eds.), *Handbook of physiology: Endocrinology IV* (Volume 2, pp. 253-291). Baltimore, MD: Williams and Wilkins.

———. (1981). Role of prolactin in water and electrolyte balance in vertebrates. In R. Jaffe (Ed.), *Prolactin: Current endocrinology* (pp. 127-166). New York: Elsevier.

———. (1993). Role of prolactin and placental lactogens in vertebrate growth and development. In M.P. Schreibman, C.G. Scanes, and P.K.T. Pang (Eds.), *The endocrinology of growth, development, and metabolism in vertebrates* (pp. 151-182). New York: Academic Press.

Nicoll, C.S., N.J. Hebert, B.C. Delidow, D.E. English, and S.M. Russell (1990). Prolactin and synlactin: Comparative aspects. In A. Epple, C.G. Scanes, and M.H. Stetson (Eds.), *Progress in comparative endocrinology* (pp. 211-218). New York: Wiley-Liss, Inc.

Nicoll, C.S., N.J. Hebert, and S.M. Russell (1985). Lactogenic hormones stimulate the liver to secrete a factor that acts synergistically with prolactin to promote growth of the pigeon crop-sac mucosal epithelium in vivo. *Endocrinology* 80:641-655.

Nimala, A.R.C. and T.J. Pandian (1983). Effect of steroid injection in food utilization in Channa striatus. *Proceedings of the Indian Academy of Science* 92:221-229.

Nishioka, R.S., K.M. Kelly, and H.A. Bern (1988). Control of prolactin and growth hormone secretion in teleost fishes. *Zoological Science* 5:267-280.

Niu, P.-D., J. Perez-Sanchez, and P.-Y. Le Bail (1993). Development of a protein binding assay for teleost insulin-like growth factor (IGF)-like: Relationships between growth hormone (GH) and IGF-like in the blood of rainbow trout (*Oncorhynchus mykiss*). *Fish Physiology and Biochemistry* 11:381-391.

NOAA (1988). *Aquaculture and capture fisheries: Impacts in U.S. seafood markets*. Washington, DC: National Marine Fisheries Service, U.S. Department of Commerce, 88-272-P.

———. (2002). Imports and exports of fishery products, annual summary 2002. NOAA Fisheries: Fisheries statistics and economics. Available online at http://www.st.nmfs.gov/st1/trade/trade2002.pdf.

O'Connor, P.K., B. Reich, and M.A. Sheridan (1993). Growth hormone stimulates hepatic lipid mobilization in rainbow trout, *Oncorhynchus mykiss*. *Journal of Comparative Physiology* 163B:427-431.

Okada, K., S. Ishii, S. Minami, H. Sugihara, T. Shibasaki, and I. Wakabayashi (1996). Intracerebroventricular administration of the growth hormone-releasing peptide KP-102 increases food intake in free-feeding rats. *Endocrinology* 137: 5155-5157.

Oshima, N., M. Makino, S. Iwamuro, and H.A. Bern (1996). Pigment dispersion by prolactin in cultured xanthophores and erythrophores of some fish species. *Journal of Experimental Zoology* 275:45-52.

Ostrowski, A.C. and D.L. Garling (1987a). Changes in dietary protein to metabolizable energy needs of 17α-methyltestosterone treated rainbow trout. *Journal of the World Aquaculture Society* 18:61-70.

———. (1987b). Effect of 17α-methyltestosterone treatment and withdrawal on growth and dietary protein utilization of juvenile rainbow trout fed practical diets varying in protein level. *Journal of the World Aquaculture Society* 18:71-77.

Palyha, O.C., S.D. Feighner, C.P. Tan, K.K. McKee, D.L. Hreniuk, Y.D. Gao, K.D. Schleim, L. Yang, G.J. Moriello, R. Nargund, et al. (2000). Ligand and activation domain of human orphan growth hormone (GH) secretagogue receptor (GHS-R) conserved from pufferfish to human. *Molecular Endocrinology* 14:160-169.

Parhar, I.S., H. Sato, and Y. Sakuma (2003). Ghrelin gene in cichlid fish is modulated by sex and development. *Biochemical and Biophysical Research Communications* 305:169-175.

Park, R., B.S. Shepherd, R.S. Nishioka, E.G. Grau, and H.A. Bern (2000). Effects of homologous pituitary hormone treatment on serum insulin-like growth factor-binding proteins (IGFBPs) in hypophysectomized tilapia, *Oreochromis mossambicus*, with special reference to a novel 20 kDa IGFBP. *General and Comparative Endocrinology* 117:404-412.

Pauly, D., V. Christensen, S. Guenette, T.J. Pitcher, U.R. Sumaila, C.J. Walters, R. Watson, and D. Zeller (2002). Towards sustainability in world fisheries. *Nature (London)* 418:689-695.

Perez-Sanchez, J., H. Marti-Palanca, and S.J. Kaushik (1995). Ration size and protein intake affect circulating growth hormone concentration, hepatic growth hormone binding and plasma insulin-like growth factor-I immunoreactivity in a marine teleost, the gilt head sea bream (*Sparus aurata*). *Journal of Nutrition* 125:546-552.

Perez-Sanchez, J., C. Weil, and P.-Y. Le Bail (1992). Effects of human insulin-like growth factor-I on release of growth hormone by rainbow trout *(Oncorhynchus mykiss)* pituitary cells. *Journal of Experimental Zoology* 262:287-290.

Pesek, M.J. and M.A. Sheridan (1995). Fasting alters somatostatin binding to liver membranes of rainbow trout *(Oncorhynchus mykiss)*. *Journal of Endocrinology* 150:179-186.

Peter, R.E. and T.A. Marchant (1995). The endocrinology of growth in carp and related species. *Aquaculture* 129:299-321.

Peter, R.E., K.-L. Yu, T.A. Marchant, and P.M. Rosenblum (1990). Direct neural regulation of the teleost adenohypophysis. *Journal of Experimental Zoology* Suppl. 4:84-89.

Phung, L.T., H. Inoue, V. Nou, H.G. Lee, N. Matsunaga, S. Hidaka, H. Kuwama, and H. Hidari (2000). The effects of growth hormone-releasing peptide-2 (GHRP-2) on the release of growth hormone and growth performance in swine. *Domestic Animal Endocrinology* 18:279-291.

Prack, M., M. Antoine, M. Caiati, M. Roskowski, T. Treacy, M.J. Vodicnik, and V.L. De Vlaming (1980). The effects of mammalian prolactin and growth hormone on goldfish (*Carassius auratus*) growth, plasma amino acid levels and liver amino acid uptake. *Comparative Biochemistry and Physiology* 67:307-310.

Prunet, P. and B. Auperin (1994). Prolactin receptors. In N.M. Sherwood and C.L. Hew (Eds.), *Fish physiology: Molecular endocrinology of fish* (Volume 13, pp. 367-391). New York: Academic Press.

Pursell, V.G., C.A. Pinkert, K.F. Miller, D.J. Bolt, R.G. Campbell, R.D. Palmiter, R.L. Brinster, and R.E. Hammer (1989). Genetic engineering of livestock. *Science* 244:1281-1288.

Rajaram, S., D.J. Baylink, and S. Mohan (1997). Insulin-like growth factor-binding proteins in serum and other biological fluids: Regulation and functions. *Endocrinology Review* 18:801-831.

Rand-Weaver, M. and H. Kawauchi (1993). Growth hormone, prolactin and somatolactin: A structural overview. In P.W. Hochachka and T.P. Mommsen (Eds.), *Biochemistry and molecular biology of fishes: Analytical techniques*, (Volume 2, pp. 39-56). New York: Elsevier.

Rand-Weaver, M., H. Kawauchi, and M. Ono (1993). Evolution of the structure of the growth hormone and prolactin family. In M.P. Schreibman, C.G. Scanes, and P.K.T. Pang (Eds.), *The endocrinology of growth, development and metabolism in vertebrates* (pp. 13-42). San Diego, CA: Academic Press.

Rechler, M.M. (1993). Insulin-like growth factor binding proteins. *Vitamins and Hormones* 47:1-114.

Reichardt, T. (2000). Will souped up salmon sink or swim? *Nature (London)* 406:10-12.

Reinecke, M., I. Broger, R. Brun, J. Zapf, and C. Maake (1995). Immunohistochemical localization of Insulin-like growth factor I and II in the endocrine pancreas of birds, reptiles and amphibia. *General and Comparative Endocrinology* 100:385-396.

Reinecke, M. and C. Collet (1998). The phylogeny of the insulin-like growth factors. *International Review of Cytology* 183:1-94.

Reinecke, M., A. Schmid, R. Ermatinger, and D. Loffing-Cueni (1997). Insulin-like growth factor I in the teleost *Oreochromis mossambicus*, the tilapia: Gene sequence, tissue expression, and cellular localization. *Endocrinology* 138:3613-3619.

Reinecke, M., A.C. Schmid, B. Heyberger-Meyer, E.B. Hunziker, and J. Zapf (2000). Effect of growth hormone and insulin-like growth factor I (IGF-I) on the expression of IGF-I messenger ribonucleic acid and peptide in rat tibial growth plate and articular chondrocytes in vivo. *Endocrinology* 141:2847-2853.

Rentier-Delrue, F., D. Swennen, P. Prunet, M. Lion, and J.A. Martial (1989). Tilapia prolactin: molecular cloning of two cDNAs and expression in *Escherichia coli*. DNA 8:261-270.

Reshkin, S.J., M.L. Grover, W.D. Howerton, E.G. Grau, and G.A. Ahearn (1989). Dietary hormone modification of growth, intestinal ATPase activity and glucose transport in the omnivorous teleost, *Oreochromis mossambicus*. *American Journal of Physiology* 256:E610-E618.

Ridha, M. and K.P. Lone (1990). Effect of oral administration of different levels of 17α-methyltestosterone on the sex reversal, growth and food conversion efficiency of the tilapia *Oreochromis spilurus. Aquaculture and Fisheries Management* 21:391-397.

Riley, L.G., T. Hirano, and E.G. Grau (2002a). Disparate effects of gonadal steroid hormones on plasma and liver mRNA levels of insulin-like growth factor-I and vitellogenin in the tilapia, *Oreochromis mossambicus. Fish Physiology and Biochemistry* 26:223-230.

———. (2002b). Rat ghrelin stimulates growth hormone and prolactin release in the tilapia, *Oreochromis mossambicus.* Zoological Science 19:797-800.

———. (2003). Effects of transfer from seawater to fresh water on the growth hormone/insulin-like growth factor-I axis and prolactin in the tilapia, *Oreochromis mossambicus. Comparative Biochemistry and Physiology B Biochemistry and Molecular Biology* 136:647-655.

Riley, L.G., N.H. Richman, T. Hirano, and E.G. Grau (2002). Activation of the growth hormone/insulin-like growth factor axis by treatment with 17α-methyltestosterone and seawater rearing in the tilapia, *Oreochromis mossambicus. General and Comparative Endocrinology* 127:285-292.

Rivas, R.J., R.S. Nishioka, and H.A. Bern (1986). In vitro effects of somatostatin and urotensin II on prolactin and growth hormone secretion in tilapia, *Oreochromis mossambicus. General and Comparative Endocrinology* 63:245-251.

Rodehutscord, M., A. Becker, M. Pack, and E. Pfeffer (1997). Response of rainbow trout *(Oncorhynchus mykiss)* to supplements of individual essential amino acids in a semipurified diet, including an estimate of the maintenance requirement for essential amino acids. *Journal of Nutrition* 126:1166-1175.

Roh, S.G., N. Matsunaga, S. Hidaka, and H. Hidari (1996). Characteristics of growth hormone secretion responsiveness to growth hormone-releasing peptide-2 (GHRP-2 or KP-102) in calves. *Endocrine Journal* 43:291-298.

Ron, B. (1997). Relationships among nutrition, salinity, and the exogenous growth promoter 17α-methyltestosterone, and growth in the euryhaline tilapia, *Oreochromis mossambicus.* Doctoral dissertation, Department of Zoology, University of Hawaii, Honolulu, Hawaii.

Ron, B., S.K. Shimoda, G.K. Iwama, and E.G. Grau (1995). Relationships among ration, salinity, 17α-methyltestosterone and growth in the euryhaline tilapia, *Oreochromis mossambicus. Aquaculture* 135:185-193.

Rosenfeld, R.G., G. Lamson, H. Pham, Y. Oh, C. Conover, D.D. De Leon, S.M. Donovan, I. Ocrant, and L. Guidice (1990). Insulin-like growth factor-binding proteins. *Recent Progress in Hormone Research* 46:99-163.

Rosenfeld, R.L. and R. Furlanetto (1985). Physiologic testosterone or estradiol induction of puberty increases plasma somatomedin-C. *Journal of Pediatrics* 107:415-421.

Rothbard, S., E. Solnik, S. Shabbath, R. Amado, and I. Grabie (1983). The technology of mass production of hormonally sex-inversed all-male tilapias. In L. Fishelson and Z. Yaron (Eds.), *Proceedings: International symposium on tilapia in aquaculture* (pp. 570-579). Tel Aviv, Israel: Tel Aviv University Press.

Rothbard, S., Y. Zohar, N. Zmora, B. Levavi-Sivan, B. Moav, and Z. Yaron (1990). Clearance of 17α-methyltestosterone from muscle of sex-inversed tilapia hybrids treated for growth enhancement with two doses of androgen. *Aquaculture* 89:365-376.

Rubin, D.A. and J.L. Specker (1992). In vitro effects of homologous prolactins on testosterone production by testes of tilapia (*Oreochromis mossambicus*). *General and Comparative Endocrinology* 87:189-196.

Rutten, M.J.M., H. Komen, R.M. Derrenberg, M. Siwek, and H. Bovenhuis (2004). Genetic characterization of four strains of Nile tilapia *(Oreochromis niloticus L.)* using microsatellite markers. *Animal Genetics* 35:93-97.

Sakai, M. (1999). Current research status of fish immunostimulants. *Aquaculture* 172:63-92.

Sakai, M., M. Kobayashi, and H. Kawauchi (1996). In vitro activiation of fish phagocytic cells by growth hormone, prolactin and somatolactin. *Journal of Endocrinology* 151:113-118.

Sakamoto, T. and T. Hirano (1991). Growth hormone receptors in the liver and osmoregulatory organs of rainbow trout: Characterization and dynamics during adaptation to seawater. *Journal of Endocrinology* 130:425-433.

———. (1993). Expression of insulin-like growth factor I gene in osmoregulatory organs during seawater adaptation of the salmonid fish: Possible mode of osmoregulatory action of growth hormone. *Proceedings of the National Academy of Science* 90:1912-1916.

Sakamoto, T., T. Hirano, S. Madsen, R.S. Nishioka, and H.A. Bern (1995). Insulin-like growth factor I gene expression during parr-smolt transformation in coho salmon. *Zoological Science* 12:249-252.

Sakamoto, T., S.D. McCormick, and T. Hirano (1993). Osmoregulatory actions of growth hormone and its mode of action in salmonids: A review. *Fish Physiology and Biochemistry* 11:155-164.

Sakamoto, T., B.S. Shepherd, S.S. Madsen, R.S. Nishioka, K. Siharath, N.H.I. Richman, E.G. Grau, and H.A. Bern (1997). Osmoregulatory actions of growth hormone and prolactin in an advanced teleost. *General and Comparative Endocrinology* 106:95-101.

Sandra, O., P. Le Rouzic, C. Cauty, M. Edery, and P. Prunet (2000). Expression of the prolactin receptor (tiPRL-R) gene in tilapia *Oreochromis niloticus*: Tissue distribution and cellular localization in osmoregulatory organs. *Journal of Molecular Endocrinology* 24:215-224.

Sandra, O., P. Le Rouzic, F. Rentier-Delrue, and P. Prunet (2001). Transfer of tilapia (*Oreochromis niloticus*) to a hyperosmotic environment is associated with sustained expression of prolactin receptor in intestine, gill and kidney. *General and Comparative Endocrinology* 123:295-307.

Sandra, O., F. Sohm, A. de Luze, P. Prunet, M. Edery, and P.A. Kelley (1995). Expression cloning of a cDNA encoding a fish prolactin receptor. *Proceedings of the National Academy of Science* 92:6037-6041.

Sangiao-Alvarellos, S., R. Laiz-Carrion, J.M. Guzman, M.P. Martin del Rio, J.M. Miguez, J.-M. Mancera, and J.L. Soengas (2003). Acclimation of *S. aurata* to various salinities alters energy metabolism of osmoregulatory and nonosmoregulatory organs. *American Journal of Physiology and Regular Integrated Comparative Physiology* 285:R897-R907.

Scanes, C.G. and C. Baile (1993). Manipulation of animal growth. In M.P. Schreibman, C.G. Scanes, and P.K.T. Pang (Eds.), *The endocrinology of growth, development, and metabolism in vertebrates* (pp. 541--558). New York: Academic Press.

Scanes, C.G. and W.H. Daughaday (1995). Growth hormone action: Growth. In S. Harvey, C.G. Scanes, and W.H. Daughaday (Eds.), *Growth hormone* (pp. 351-370). London: CRC Press.

Schmid, A.C., E. Naf, W. Kloas, and M. Reinecke (1999). Insulin-like growth factor-I and -II in the ovary of a bony fish, *Oreochromis mossambicus*, the tilapia: In situ hybridisation, immunohistochemical localisation, Northern blot and cDNA sequences. *Molecular and Cell Endocrinology* 156:141-149.

Schnick, R. (2003). National news: Status of drug approval for 17α-methyl-testosterone. North Central Regional Aquaculture Center (NCRAC). Fall, 2003 Newsletter, p. 1.

Schnick, R. (2005). Tenth annual summary of activity highlights for the National Coordinator for Aquaculture new animal drug applications (May 15, 2004-May 14, 2005). Available at http://ag.anse.Purdue.edu/aguanic/jsa/aquadrugs/index. htm.

Seale, A.P., N.H. Richman, T. Hirano, I. Cooke, and E.G. Grau (2003a). Cell volume increase and extracellular Ca^{2+} are needed for hyposmotically induced prolactin release in tilapia. *American Journal of Physiology* 284:C1280-C1289.

——— (2003b). Evidence that signal transduction for osmoreception is mediated by stretch- activated ion channels in tilapia. *American Journal of Physiology* 284:C1290-C1296.

Seale, A.P., L.A. Riley, T.A. Leedom, S. Kajimura, R.M. Dores, T. Hirano, and E.G. Grau (2002). Effects of environmental osmolality on release of prolactin, growth hormone and ACTH from the tilapia pituitary. *General and Comparative Endocrinology* 128:91-101.

Shepherd, B.S., K. Drennon, J. Johnson, J. W. Nichols, R. C. Playle, T.D. Singer, and M.M. Vijayan (2004). Salinity acclimation affects the somatotropic axis in rainbow trout. *American Journal of Physiology: Regulatory, Integrative and Comparative Physiology* 288: R1385-R1395.

Shepherd, B.S, S. Eckert, I. Parhar, M. Vijayan, I. Wakabayashi, G. Grau, and T. Chen (2000). The hexapeptide KP-102 (D-Ala-D-β-Nal-Ala-Trp-D-Phe-lys-NH_2) stimulates growth hormone release in a cichlid fish (*Oreochromis mossambicus*). *Journal of Endocrinology* 167:R7-R10.

Shepherd, B.S., B. Ron, A. Burch, R. Sparks, N.H. Richman, S.K. Shimoda, M.H. Stetson, C. Lim, and E.G. Grau (1997). Effects of salinity, dietary level of protein and 17α-methyltestosterone on growth hormone (GH) and prolactin ($tPRL_{177}$ and $tPRL_{188}$) levels in the tilapia, *Oreochromis mossambicus. Fish Physiology and Biochemistry* 17:279-288.

Shepherd, B.S., T. Sakamoto, S. Hyodo, C. Ball, R.S. Nishioka, H.A. Bern, and E.G. Grau (1999). Is the basic regulation of pituitary prolactin ($tPRL_{177}$ and $tPRL_{188}$)

secretion and gene expression in the euryhaline tilapia (*Oreochromis mossambicus*) hypothalamic or environmental? *Journal of Endocrinology* 161: 121-129.

Shepherd, B.S., T. Sakamoto, R.S. Nishioka, N.H. Richman, I. Mori, S.S. Madsen, T.T. Chen, T. Hirano, H.A. Bern, and E.G. Grau (1997). Somatotropic actions of the homologous growth hormone (tGH) and prolactin ($tPRL_{177}$) in the euryhaline teleost, *Oreochromis mossambicus. Proceedings of the National Academy of Science* 94:2068-2072.

Sheridan, M.A. (1986). Effects of thyroxin, cortisol, growth hormone, and prolactin on lipid metabolism of coho salmon, *Oncorhynchus kisutch*, during smoltification. *General and Comparative Endocrinology* 64:220-238.

Shimizu, M., A. Hara, and W.W. Dickhoff (2003). Development of an RIA for salmon 41 kDa IGF-binding protein. *Journal of Endocrinology* 178:275-283.

Shimizu, M., P. Swanson, and W.W. Dickhoff (1999). Free and protein-bound insulin-like growth factor-1 (IGF-I) and IGF-binding proteins in plasma of coho salmon, *Oncorhynchus mykiss. General and Comparative Endocrinology* 115: 398-405.

Shimizu, M., P. Swanson, A. Hara, and W.W. Dickhoff (2003). Purification of a 41-kDa insulin-like growth factor binding protein from serum of chinook salmon, *Onchorhynchus tshawytscha. General and Comparative Endocrinology* 132: 103-111.

Shimizu, M., P. Swanson, F. Haruhisa, A. Hara, and W.W. Dickhoff (2000). Comparison of extraction methods and assay validation for salmon insulin-like growth factor using commercially available components. *General and Comparative Endocrinology* 119:26-36.

Shirak, A., Y. Palti, A. Cnaani, G. Korol, G. Hulata, M. Ron, and R.R. Avtalion (2002). Association between loci with deleterious alleles and distorted sex ratios in an inbred line of tilapia *(Oreochromis aureus). Journal of Heredity* 93:270-276.

Siharath, K., R.S. Nishioka, S.S. Madsen, and H.A. Bern (1995). Regulation of IGF-binding proteins by growth hormone in the striped bass, *Morone saxatilis. Molecular Marine Biology and Biotechnology* 4:171-178.

Sillence, M.N. (2004). Technologies for the control of fat and lean deposition in livestock. *Veterinary Journal* 167:242-257.

Silva, M.E.R., R.A. Magalhaes, N.V. Gonfinetti, O.A. Germek, M. Nery, B.L. Wajchenberg, and B. Liberman (1992). Effects of testosterone on growth hormone secretion and somatomedin-C generation in prepubertal growth hormone deficient male patients. *Brazilian Journal of Medical Biology Research* 25: 1117-1126.

Simard, J., J.-F. Hubert, T. Hosseinzadeh, and F. Labrie (1986). Stimulation of growth hormone release and synthesis by estrogens in rat anterior pituitary cells in culture. *Endocrinology* 119:2004-2011.

Simmen, F.A., L. Badinga, M.L. Green, I. Dwak, S. Song, and R.C. Simmen (1998). The porcine insulin-like growth factor system: At the interface of nutrition, growth and reproduction. *Journal of Nutrition* 128: 315S-320S.

Simone, D.H. (1990). The effects of the synthetic steroid 17α-methyltestosterone on the growth and organ morphology of the channel catfish (*Ictalurus punctatus*). *Aquaculture* 84:81-93.

Sindhu, S. and T.J. Pandian (1984). Effect of administering different doses of 17α-methyltestosterone in *Heteropneustes fossilis. Proceedings of the Indian Academy of Science and Animal Science* 93:511-516.

Singh, A.K. and M. Prasad (1987). Effect of bovine growth hormone (bGH) on biochemical constituents responsible for somatic growth and vitellogenesis in hypophysectomized catfish *Heteropneustes fossilis* during preparatory and spawning phases of the reproductive cycle. *Naturalia* 11-12:97-106.

Sjogren, K., J.L. Liu, K. Blad, S. Skrtic, O. Vidal, V. Wallenius, D. Le Roith, J. Tornell, O.G.P. Isaksson, J.-O. Jansson et al. (1999). Liver-derived insulin-like growth factor-I (IGF-I) is the principal source of IGF-I in blood but is not required for postnatal body growth in mice. *Proceedings of the National Academy of Science* 96:7088-7092.

Smith, C. and D.B. Smith (1993). The need for close linkages in marker-assisted selection for economic merit in livestock. *Animal Breeding Abstracts* 61:197-204.

Smith, R.G., O.C. Palyha, S.D. Feighner, C.P. Tan, K.K. McKee, D.L. Hreniuk, L. Yang, G. Morriello, R. Nargund, A.A. Patchett et al. (1999). Growth hormone releasing substances: Types and their receptors. *Hormone Research* 51:1-8.

Smith, R.G., L.H. Van der Ploetg, A.D. Howard, S.D. Feighner, K. Cheng, G.J. Hickey, M.J.J. Wyvratt, M.H. Fisher, R.P. Nargund, and A.A. Patchett (1997). Peptidomimetic regulation of growth hormone secretion. *Endocrine Reviews* 18:621-645.

Sohm, F., I. Manfroid, A. Pezet, F. Rentier-Delrue, M. Rand-Weaver, P.A. Kelly, G. Boeuf, M.-C. Postel-Vinay, A. de Luze, and M. Edery (1998). Identification and modulation of a growth hormone-binding protein in rainbow trout (*Oncorhynchus mykiss*) plasma during seawater adaptation. *General and Comparative Endocrinology* 111:216-224.

Sparks, R., B.S. Shepherd, B. Ron, N.H. Richman, III, L.G. Riley, G. Iwama, T. Hirano, and E.G. Grau (2003). Effects of environmental salinity and 17α-methyltestosterone on growth and oxygen consumption the tilapia, *Oreochromis mossambicus. Comparative Biochemistry and Physiology, Part B* 136:657-665.

Specker, J.L., P.S. Brown, and S.C. Brown (1989). Unequal activities of the two tilapia prolactins in the whole-animal transepithelial potential bioassay using the red eft. *Fish Physiology and Biochemistry* 7:119-124.

Specker, J.L., D.S. King, R.S. Nishioka, K. Shirahata, K. Yamaguchi, and H.A. Bern (1985). Isolation and characterization of two prolactins released in vitro by the pituitary of cichlid fish. *Proceedings of the National Academy of Science* 82:7490-7494.

Specker, J.L., D.S. King, R.J. Rivas, and B.K. Young (1985). Partial characterization of two prolactins from a cichlid fish. In R.M. MacLeod, M.O. Thorner, and U. Scapagnini (Eds.), *Prolactin: Basic and clinical correlates* (pp. 427-435). 1 Padova: Liviana Press.

Stewart, C.N., H.A. Richards, and M.D. Halfhill (2000). Transgenic plants and biosafety: Science, misconceptions and public perceptions. *BioTechniques* 29: 832-843.

St-Pierre, D.H., L. Wang, and Y. Tache (2003). Ghrelin: A novel player in the gut-brain regulation of growth hormone and energy balance. *News in Physiology Science* 18:242-246.

Streelman, J.T. and T.D. Kocher (2002). Microsatellite variation associated with prolactin expression and growth of salt-challenged tilapia. *Physiological Genomics* 9:1-4.

Subasinghe, R.P., D. Curry, S.E. McGladdery, and D. Bartley (2003). *Recent technological innovations in aquaculture*. FAO, Recent Technological Innovations, June 9.

Sukumar, P., A.D. Munro, E.Y.M. Mok, S. Subburaju, and T.J. Lam (1997). Hypothalmic regulation of the pituitary-thyroid axis in the tilapia *Oreochromis mossambicus*. *General and Comparative Endocrinology* 106:73-84.

Sun, L.Z. and A. Farmanfarmaian (1992a). Age-dependent effects of growth hormone on striped bass hybrids. *Comparative Biochemistry and Physiology* 101: 237-248.

———. (1992b). Biphasic action of growth hormone on intestinal amino acid absorption in striped bass hybrids. *Comparative Biochemistry and Physiology* 103:381-390.

Thorner, M.O., I.M. Chapman, B.D. Gaylinn, S.S. Pezzoli, and M.L. Hartman (1997). Growth hormone-releasing hormone and growth hormone-releasing peptide as therapeutic agents to enhance growth hormone secretion in disease aging. *Recent Progress in Hormone Research* 52:215-246.

Tidwell, J.H. and G.L. Allan (2001). Fish as food: Aquaculture's contribution. *EMBO Reports* 2:958-963.

Trudeau, V.L., G.M. Somoza, C.S. Nahorniak, and R.E. Peter (1992). Interactions of estradiol with gonadotropin-releasing hormone and thyrotropin-releasing hormone in the control of growth hormone secretion in the goldfish. *Neuroendocrinology* 56:483-490.

Tsai, P.I., S.S. Madsen, S.D. McCormick, and H.A. Bern (1994). Endocrine control of cartilage growth in coho salmon: GH influence in vivo on the response to IGF-I in vitro. *Zoological Science* 11:299-303.

Ufodike, E.B.C. and C.T. Madu (1986). Effects of methyltestosterone on food utilization and growth of *Sarotherodon niloticus* fry. *Bulletin of the Japanese Society for Scientific Fisheries* 52:1919-1922.

Ulloa-Aguirre, A., R.M. Blizzard, E. Garcia-Rubi, A.D. Rogol, K. Link, C.M. Christie, M.L. Johnson, and J.D. Veldhuis (1990). Testosterone and oxandrolone, a nonaromatizable androgen, specifically amplify the mass and rate of growth hormone (GH) secreted per burst without altering GH secretory burst duration or frequency or the GH half-life. *Journal of Clinical Endocrinology and Metabolism* 71:846-854.

Unniappan, S., X. Lin, L. Cervini, J. Rivier, H. Kaiya, K. Kanagawa, and R.E. Peter (2002). Goldfish ghrelin: Molecular characterization of the complementary deoxyribonucleic acid, partial gene structure and evidence for its stimulatory role in food intake. *Endocrinology* 143:4143-4146.

Unniappan, S. and R.E. Peter (2004). In vitro and in vivo effects of ghrelin on luteinizing hormone and growth hormone release in goldfish. *American Journal of Physiology: Regulatory, Integrative and Comparative Physiology*. 286: R1093-R1101.

USDA/ERS (2003). *Aquaculture situation and outlook report*. Washington DC., October 9.

Van Laere, A.-S., M. Nguyen, M. Braunschweig, C. Nezer, C. Collette, L. Moreau, A.L. Archibald, C.S. Haley, N. Buys, M. Tally et al. (2003). A regulatory mutation in *IGF2* causes a major QTL effect on muscle growth in the pig. *Nature (London)* 425:832-836.

Varadaraj, K., S.S. Kumari, and T.J. Pandian (1994). Comparison of conditions for hormonal sex reversal of mozambique tilapias. *Progressive Fish-Culturist* 56: 81-90.

Varadaraj, K. and T.J. Pandian (1988). Food consumption and growth efficiency of normal and phenotypic male *Oreochromis mossambicus*. In R.S.V. Pullin, T. Bhukaswan, K. Tonguthai, and J.L. Maclean (Eds.), *The second international symposium on tilapia in aquaculture* (pp. 429-432). Manila, Philippines: International Commission for Living and Aquatic Living Resources (ICLARM).

Vick, A.M. and W.L. Hayton (2001). Methyltestosterone pharmacokinetics and oral bioavailability in rainbow trout *(Oncorhynchus mykiss)*. *Aquatic Toxicology* 52:177-188.

Vijayan, M.M., J.D. Morgan, T. Sakamoto, E.G. Grau, and G.K. Iwama (1996). Food-deprivation affects seawater acclimation in tilapia: Hormonal and metabolic changes. *Journal of Experimental Biology* 199:2467-2475.

Ward, A.C. and G.J. Lieschke (2002). The zebrafish as a model system for human disease. *Frontiers in Bioscience* 7:827-833.

Watanabe, W.O., L.J. Ellingson, R.I. Wicklund, and B.L. Olla (1988). The effects of salinity on growth, food consumption and conversion in juvenile, monosex male Florida red tilapia. In R.S.V. Pullin (Ed.), *The second international symposium on tilapia in aquaculture* (pp. 515-523). Bangkok, Thailand: International Commission for Living and Aquatic Living Resources (ICLARM).

Weber, G.M., J.F.F. Powell, M. Park, W.H. Fischer, A.G. Craig, J.E. Rivier, U. Nanakorn, I.S. Parhar, S. Ngamvongchon, E.G. Grau et al. (1997). Evidence that gonadotropin-releasing hormone (GnRH) functions as a prolactin-releasing factor in a teleost fish *(Oreochromis mossambicus)* and primary structures for three native GnRH molecules. *Journal of Endocrinology* 155:121-132.

Weissberger, A.J. and K.K.Y. Ho (1993). Activation of the somatotrophic axis by testosterone in adult males: Evidence for the role of aromatization. *Journal of Clinical Endocrinology and Metabolism* 76:1407-1412.

Wells, T., D.M. Flavell, S.E. Wells, D.F. Carmignac, and I.C.A.F. Robinson (1997). Effects of growth hormone secretogogues in the transgenic growth-retarded (Tgr) rat. *Endocrinology* 138:580-587.

Westers, H. (2003). *A white paper on the status and concerns of aquaculture effluents in the north central region.* North Central Regional Aquaculture Center, White Paper, November 30.

Williams, D. (2004). US trade gap hits record. "News.com.au", March 11. Available online at http://finance.news.com.au/common/printpage/ 0,6093, 8933919,00.htm.

Xiao, D., O.L. Anderson, A.O.L. Wong, and H.-R. Lin (2002). Lack of growth hormone-releasing peptide-6 action on in vin vivo and in vin vitro growth hormone secretion in sexually immature grass carp *(Ctenopharyngodon idellus). Fish Physiology and Biochemistry* 26:315-327.

Xu, B., H. Miao, P. Zhang, and D. Li (1997). Osmoregulatory actions of growth hormone in juvenile tilapia *(Oreochromis niloticus). Fish Physiology and Biochemistry* 17:295-301.

Yada, T., E.G. Grau, and T. Hirano (1995). Suppression of prolactin release in vitro from the rainbow trout pituitary, with special reference to the structural arrangement of the pituitary cells. *Zoological Science* 12:231-238.

Yada, T., T. Hirano, and E.G. Grau (1994). Changes in plasma levels of the two prolactins and growth hormone during adaptation to different salinities in the euryhaline tilapia, *Oreochromis mossambicus. General and Comparative Endocrinology* 93:214-223.

Yada, T., M. Nagae, S. Moriyama, and T. Azuma (1999). Effects of prolactin and growth hormone on plasma immunoglobulin M levels of hypophysectomized rainbow trout, *Oncorhynchus mykiss. General and Comparative Endocrinology* 115:46-52.

Yada, T., K. Uchida, S. Kajimura, T. Azuma, T. Hirano, and E.G. Grau (2002). Immunomodulatory effects of prolactin and growth hormone in the tilapia, *Oreochromis mossambicus. Journal of Endocrinology* 173:483-492.

Yakar, S., J.L. Liu, B. Stannard, A. Butler, D. Accili, B. Sauer, and D. Le Roith (2001). Normal growth and development in the absence of hepatic insulin-like growth factor-I. *Proceedings of the National Academy of Science* 96:7324-7329.

Yamaguchi, K., J.L. Specker, D.S. King, Y. Yokoo, R.S. Nishioka, T. Hirano, and H.A. Bern (1991). Complete amino acid sequence of a pair of fish (tilapia) prolactins, $tPRL_{177}$ and $tPRL_{188}$. *Journal of Biological Chemistry* 263:9113-9121.

Yamazaki, F. (1976). Application of hormones in fish culture. *Journal of the Fisheries Research Board of Canada* 33:948-958.

Yang, C.W., L.J. Striker, C. Pesce, W.Y. Chen, E.P. Peten, S. Elliot, T. Doi, J.J. Kopchick, and G.E. Striker (1993). Glomerulosclerosis and body growth are mediated by different portions of bovine growth hormone: Studies in transgenic mice. *Laboratory Investigations* 68:62-70.

Yu, H. and T. Rohan (2000). Role of the insulin-like growth factor family in cancer development and progression. *Journal of the National Cancer Institute* 92:1472-1489.

Zapf, J., E.R. Froesch, and C. Schmid (1999). Metabolic effects of IGFs. In R.G. Rosenfeld and C.T. Roberts (Eds.), *The IGF System: Molecular biology, physiology and clinical applications 17* (pp. 577-616). Portland, OR: Humana Press.

Zapf, J., M. Gosteli-Peter, G. Weckbecker, E.B. Hunziker, and M. Reinecke (2002). The somatostatin analog octreotide inhibits GH-stimulated, but not IGF-I-stimulated, bone growth in hypophysectomized rats. *Endocrinology* 143:2944-2952.

Zbikowska, H.M. (2003). Fish can be first—Advances in fish transgenesis for commercial application. *Transactions in Research* 12:379-389.

Zhang, W.-M., H.-R. Lin, and R.E. Peter (1994). Episodic growth hormone secretion in grass carp, *Ctenopharyngodon idellus* (C & V). *General and Comparative Endocrinology* 95:337-341.

Zhang, Y. and T.A. Marchant (1999). Identification of serum GH-binding proteins in the goldfish *(Carassius auratus)* and comparison with mammalian GH-binding proteins. *Journal of Endocrinology* 161:255-262.

Zou, J.J., V.J. Trudeau, Z. Cui, J. Brechin, K. Mackenzie, Z. Zhu, D.F. Houlihan, and R.E. Peter (1997). Estradiol stimulates growth hormone production in female goldfish. *General and Comparative Endocrinology* 106:102-112.

Chapter 4

Recent Directions in Genetics

C. Greg Lutz

INTRODUCTION

As the importance of tilapia in global aquaculture has increased over the past several decades, so has the intensity and diversity of investigations into their genetic improvement. Ongoing efforts to improve tilapia stocks throughout the world involve traditional animal breeding and quantitative genetic approaches as well as innovative chromosomal manipulations, physiological alteration of sex determination, gene transfer, molecular discrimination of populations, and gene mapping. This chapter focuses on the foundations of these efforts and their development in recent years.

TRADITIONAL ANIMAL BREEDING APPROACHES

Traditional approaches to genetic improvement focus primarily on the characterization and use of additive and dominance genetic effects. Practical applications include selection (strain or species performance evaluation), crossbreeding, and hybridization. Although this area of research has been discounted in some circles in recent years as less powerful and more time consuming than certain more innovative approaches, it is still the basis for evaluation and application of tilapia genetic improvement in other areas of research, such as production of transgenic lines. Even more important, these approaches are still the only practical means of improving tilapia stocks for producers in most countries.

Another important, albeit less recognized, aspect of traditional animal breeding involves maternal effects and their inheritance. Maternal effects frequently contribute tangibly to phenotypic variation in aquaculture species, including tilapia. Maternal effects can involve characteristics such as egg quality or size and, as is especially important in mouthbrooding tilapia, care of eggs or fry.

doi:10.1300/5513_04

Strain Effects Within Tilapia Species

Inasmuch as aquaculture species represent isolated breeding populations with distinct genetic attributes, the first step of selection is to characterize available strains or lines and subsequently select one or more that excel in certain areas from which to form a base population for improvement. This is of particular significance in tilapia owing to the frequent use of very small founder populations and accelerated inbreeding among isolated production stocks.

Several production-related traits were examined in the various *Oreochromis* spp. many decades ago as tilapia culture became established worldwide (Lowe 1955; Maar et al. 1966; Yashouc and Helevy 1971; Fryer and Iles 1972; Jhingran and Gopalakrishnan 1974; Wohlfarth and Hulata 1981). Temperature tolerance, salinity tolerance, fecundity, feeding habits, and growth have all been evaluated under a variety of conditions, but results have often been somewhat variable among studies evaluating the same species. Such variability may be explained in part by differences among particular strains within species, and this phenomenon certainly appears to exist in tilapia (Eknath et al. 1993).

Genotype-environment interactions can also contribute to conflicting evaluations of specific tilapia strains or species. Genetic attributes that result in superiority in one environment may actually be detrimental under other circumstances. Romana-Eguia and Doyle (1992) illustrated this point in an examination of the performance of three strains of Nile tilapia *(Oreochromis niloticus)* under conditions of alternating adequate and poor nutrition. Eguia and Eguia (1993) reported significant differences in growth among three strains of Nile tilapia under restrictive and nonrestrictive feeding regimes.

Eguia (1996) also reported significant genotype-environment interactions in seed production of four genetically diverse red tilapia strains, and Elghobashy et al. (2000) documented dramatic genotype-environment interactions among four strains of Nile tilapia in Egypt. Although one strain in the study, designated as Maryout, exhibited the highest daily weight gain under pond and cage culture conditions, it performed poorly in mixed rice-fish culture ponds. Manzala strain, which performed comparatively poorly under pond conditions, displayed the highest weight gain in rice-fish culture trials.

Genetic correlations between production-related traits can also complicate performance evaluations in tilapia. Bolivar et al. (1993) demonstrated notable differences in the relationship between female reproductive performance and growth in seven distinct Nile tilapia strains originating in Egypt, Ghana, Senegal, Israel, Singapore, Taiwan, and Thailand. Three female reproductive types were distinguishable: early spawning, late-spawning, and non-spawning. In some strains, the growth of late and nonspawning females was nearly equal to that of males.

Lourdes and Cuvin-Aralar (1994) demonstrated that tilapia strains exhibited distinct differences in heavy metal accumulation, and Balfry et al. (1994) found distinct differences in the natural immune response of two experimental strains infected with *Vibrio parahaemolyticus.* Most recently, Sifa et al. (2002) demonstrated inferior cold tolerance in the GIFT strain of Nile tilapia (see, next section, "Selection") compared with two other commercial strains in China.

Selection

The practice of selection is based on the exploitation of additive genetic effects. The underlying principle is that some significant portion of the variation in observable performance is due to individual genotypes and that some component of these genotypic influences is directly heritable from parent to offspring (Figure 4.1). Additive genetic effects of distinct alleles persist from one generation to the next, regardless of whatever alleles they may pair up with (hence the term *additive*). Thus, the concept of heritability, as used in animal breeding, is merely a description of the relative portion of observed variance within a population that can be attributed to additive genetic effects.

Even when traits are highly heritable, significant improvement through selection requires some measurable amount of genetic variation upon which to exert a directional preference. Highly inbred lines tend to exhibit lower heritabilities as a result of reduced genetic variation, and thus observable differences in individual performances within an inbred population are largely the result of random environmental influences. This is important in tilapia production. Consider a large commercial facility using, say, 40,000 adult fish for broodstock. Although this breeding population might seem to be far too large ever to be impacted by inbreeding, this stock would most likely have been developed from as few as 1,000 fish, which in turn may have been shipped from a facility that has a breeding population of fewer than 50 fish and obtained its original stocks from another, similarly small population. This scenario is widespread in tilapia production throughout Asia and even more so in the Americas.

Many examples of the impacts of inbreeding on genetic variation and heritability in tilapia can be found in the literature. After applying selection to a population of Nile tilapia with little effect, Huang and Liao (1990) determined that mass selection was not an effective means of improving

FIGURE 4.1. Growth variation is typical in tilapia stocks, even within full-sib families. The extent to which this variation reflects genetic differences among individuals will determine the potential for improvement through selection.

growth. However, their population of fish had been produced from two lines of fish that had been maintained in laboratory tanks for 7 years and had previously been propagated from offspring of only 56 individuals introduced from Japan 20 years earlier.

Small numbers of breeding adults may represent even fewer active breeders if behavioral hierarchies come into play, as is the case with tilapia. As a consequence, the phenomenon of genetic drift must also be considered, whereby certain alleles can be lost from the population over time simply by chance. Drift can be considered the result of random sampling of the alleles present at any given locus from generation to generation. Evidence for such impacts is not limited to the Huang and Liao (1990) study. Hulata et al. (1986) also reported no notable growth improvement in Nile tilapia by mass selection. The fish in their study were, however, descended from a sample of offspring obtained from a stock established 7 years earlier with approximately 50 fish introduced to Israel from Ghana.

In fact, selection is usually a viable approach to improving tilapia stocks, at least where sufficient genetic variation exists. Precedent certainly exists for claims of substantial selective gains in stocks of Nile tilapia. From 1988 through 1997, the Genetic Improvement of Farmed Tilapia (GIFT) project was undertaken in the Philippines as a joint effort by several agencies. The project involved the development of a synthetic base population including eight strains of Nile tilapia: four from the Philippines and four imported directly from Africa (Eknath et al. 1993). By including a number of strains within the base population, inbreeding effects within stocks could be reduced, if not eliminated (see section "Hybridization, crossbreeding and inbreeding").

Combined family and within-family selection for growth was applied to the GIFT stock on an annual basis across a number of generations. The average selection response per generation was 13 percent, and the accumulated response resulted in an 85 percent increase in growth (Gjedrem 1999). After 10 years of selection, the GIFT strain was estimated to exhibit approximately double the genetic potential for growth rate as the fastest-growing original strain (Gjoen 2002). Dey et al. (2000) examined the potential economic impacts of adopting the GIFT strain on "average" and "efficient" tilapia farms in five Asian countries. Compared with locally used stocks in Bangladesh, China, the Philippines, Thailand, and Vietnam, the GIFT strain produced 7 to 36 percent savings in breakeven prices above variable costs.

Brzeski and Doyle (1995) examined results of mass selection for size-specific growth rate of tilapia in a commercial fish farm in Indonesia. Selection was applied in two steps over a 6-month grow-out period, and initial stocking was based on size-graded, rather than age-matched, fingerlings. These procedures reflected the practical needs of small-scale non-technical producers in the region. Realized heritability was estimated to be 0.12, which is not extremely high compared with many other studies using aquatic species and was probably biased downward somewhat by the need to select solely on size rather than size at age. Although the selection response was small (a 2.3 percent improvement in weight gain in one generation), it was statistically significant and might be expected to accumulate appreciably over several generations.

One approach to selection that has particular advantages in tilapia breeding is to estimate an individual's genetic worth, wholly or partially, based on the performance of its immediate family. One notable benefit from this method, known as family selection, lies in the reduction of the generation interval when using information from overall family performance. Family selection relies solely on family averages, and in this practice all or some of the members of families with the highest performance are retained for breeding purposes. Family selection is essential for improvement of slaughter traits such as dress-out percentage (Figure 4.2). For traits with low heritabilities, family selection is generally more efficient than mass selection, but it requires the identification of fish in particular families, either

FIGURE 4.2. Body conformation, and resultant dress-out percentage, is often a goal in on-farm selection, but little research has been published on this topic.

through permanent marking methods or physical segregation in tanks, ponds, or cages.

Within-family selection relies on evaluating individuals based on their superiority over their family means, and this method is often appropriate for tilapia breeding programs (Uraiwan and Doyle 1986). Bolivar (1999) reported on within-family selection trials for growth in Nile tilapia. The selection criterion was fairly easy to track: growth during the 16 weeks following hatching. The rationale was the development of a selection strategy that could be applied with relatively limited facilities. Over 12 generations, within-family selection for growth was applied to tilapia grown in tanks. A continuous linear response for weight at 16 weeks was apparent over the entire course of the study.

Calculations based on observed selection response indicated a heritability for weight at 16 weeks of 0.38 in the base population, with a potential genetic gain of roughly 12 percent per generation. Although the selection program was carried out in tanks, selected lines were evaluated for their performance in hapas and ponds to determine whether gains would be of value under commercial conditions. Substantial selection response was apparent in both environments, suggesting that relatively modest facilities can be utilized to produce significant genetic improvements in extensive tilapia production settings.

Bolivar and Newkirk (2000) illustrated another important aspect of selection as applied to tilapia by asking what selected lines should be compared with in order to accurately assess selection response. They used two distinct control lines to evaluate the results of the selection trial described in Bolivar (1999). One line was developed randomly, whereas the other was based on selecting animals closest to the population mean. Although results were similar, discrepancies between the two control lines resulted in different comparative estimates of selection response.

Within-family selection, as discussed above, may be of particular value when families must be segregated in order to track their performance. Accordingly, within-family and family selection are often combined, especially in tilapia. Even under the greatest standardization attainable, differences can still exist among family groups in terms of culture environments, but with within-family selection these differences do not directly bias the evaluation. An added benefit of within-family selection is the ease of selecting members of every family to replace the preceding generation. When every family contributes equally to the breeding population the effective population size theoretically doubles, slowing the accumulation of inbreeding depression considerably.

Growth versus Maturation

The interaction between potential for sustained growth and age and/or size at maturity has long been an issue in tilapia culture (Oldorf et al. 1989). Kronert et al. (1989) demonstrated moderately high heritabilities and substantial genetic variability in traits related to gonadal development in Nile tilapia. Horstgen-Schwark and Langholz (1998) illustrated the application of family and within-family selection in a study focusing on delaying maturity in Nile tilapia. Their initial breeding population consisted of 35 full-sib families of fish. Half of the fish from each family were slaughtered at 136 days of age, and gonadosomatic index (GSI) and a visual assessment of gonadal development for each slaughtered fish were used as selection criteria for males and females, respectively.

Families were ranked separately for each sex, based on these traits. Selection was directional in each trait: for the lowest values, to reflect late maturation of both males and females. In this sense, family selection was being practiced, but based on separate rankings for male and female maturity within families. Families were also ranked by mean weight, and only those at or above the overall average were considered for selection of late-maturing animals, to avoid the selection of families with high proportions of underdeveloped offspring. As a result, families were selected based not only on late maturation of males and/or females but also on growth.

Females from families with the lowest average gonadal development values were spawned with males from families with the lowest average GSI, but full-sib matings were disallowed to minimize complications from inbreeding depression. Additionally, only the heaviest fish within each selected family were used to produce the next generation, in order to reduce the time interval between generations. As a result, only approximately 25 percent of the late-maturing males and 50 percent of the late-maturing females were spawned. Thus within-family selection for growth was combined with the family selection for maturation and growth that had already been carried out.

Two males and four females were selected randomly from each family to serve as control lines for each generation. Compared with unselected controls, selected lines exhibited a reduction of 1.25 standard deviations for GSI in males and gonadal development in females after two generations of selection. These responses were statistically significant. However, body weights did not differ significantly between the selected and control lines.

A possible explanation for the lack of response in growth rate in the Horstgen-Schwark and Langholz (1998) study is found in results presented by Longalong et al. (1999), who began with a random sample of 20 individual Nile tilapia from each of 42 full-sib families, representing 21 half-sib families, from the third generation of selection for growth in the GIFT project. Sampled fish were individually tagged and stocked communally in an earthen pond. Four weeks after swim-up fry were first observed in the pond, fish were inspected individually for sexual maturation.

Broodstock were then selected from four families with high (>75 percent) or low (<20 percent) rates of maturation among females. Later, single-pair stockings of these mature fish were conducted in hapas. Sixteen pairs of breeders from families with a mean frequency of 83 percent early-maturing females and nine pairs from families with 0 percent early-maturing females eventually produced fry. These offspring were tagged at 3 to 5 g individual weight and stocked communally in three replication ponds. At 2, 3, and 4 weeks, respectively, following observation of fry in each pond, sex, body weight, and sexual maturation were recorded for the stocked fish. After statistical adjustments, the frequency of sexually mature female offspring of broodstock from families with high and low rates of early female maturity was 57.0 and 33.6 percent, respectively. This equated to a highly significant response to selection for early maturation.

After restocking and further grow-out, mean body weights were significantly higher in mixed-sex progeny of broodstock from families with high rates of early-maturing females. The authors suggested that this correlated response reflected a positive genetic association between early maturity (as expressed in females) and growth (of either sex) in this species. This might explain why Horstgen-Schwark and Langholz (1998) observed no growth response in their study. Family selection for late maturation and correlated reduction in growth may have negated any potential genetic gains from selection within and among families for increased growth rate.

Hybridization, Crossbreeding, and Inbreeding

Whereas additive genetic effects are the basis for selection because of their persistence from generation to generation regardless of allelic combinations, dominance genetic effects depend entirely on the interaction of the two alleles present at any given locus. The sum of all these interactions affecting a particular trait and the genetic variance attributable to them provide a basis for realizing significant gains in performance in the course of a single generation. This tendency for combinations of alleles to influence observed performance is termed heterosis and is most commonly associated with traits related to fitness (growth, disease resistance, reproductive performance, etc.). It is most easily measured as the difference between offspring performance and the average performance of the two parents or parental lines.

One common term for heterosis is hybrid vigor, but this phenomenon is often expressed both within species (crossbreeding) and across related species (hybridization). Inbreeding depression is thought to involve the same mechanisms as heterosis, but with a resulting reduction in fitness as more and more loci affecting a trait become homozygous. Within any given population, inbreeding can only accumulate: once genetic variation has been lost at any given locus, it cannot be restored (except through random mutation). In practical terms, heterosis can often be viewed simply as the relief of inbreeding depression.

As discussed, inbreeding levels can accumulate rapidly in tilapia production stocks. Accordingly, crossbreeding isolated lines of tilapia often produces noticeable improvements in growth, fecundity, and hardiness. Marengoni et al. (1998) described a diallel crossbreeding experiment involving three geographically distinct lines of Nile tilapia. After 90 days' growth, average body weight for pure lines was approximately 70 g, while specific heterosis for crosses ranged from 5.6 to 7.7 g, with an average heterosis of 7 g, or approximately 10 percent.

Tilapia hybrids have been evaluated for a number of decades (Hickling 1960; Hulata et al. 1982; Majumdar and McAndrew 1983; Shelton et al. 1983; Lahav and Lahav 1990). Two relatively old but very useful reviews of hybridization in tilapia are Lovshin (1982) and Schwartz (1983). Although most hybridization studies have involved two or more *Oreochromis* spp., hybrids with other genera are also often possible (Baroiller et al. 2000). Perhaps the most commercially exploited hybrid involves using blue tilapia *(O. aureus)* males and Nile tilapia females to produce predominantly male fry. Recent work by McConnell et al. (2000) documents the high degree of similarity between linkage groups in these parental species. Siddiqui and Al-Harbi (1995) compared *O. niloticus* × *O. aureus* hybrids and their parental strains in terms of performance and found the hybrids to be superior in both growth rate and survival. Similarly, Cai et al. (2004) demonstrated that the resistance of *O. niloticus* × *O. aureus* hybrids to *Aeromonas sobria* was higher than that of either parental strain.

Cnaani et al. (2000) described the cold tolerance of F_1 and F_2 hybrids of Mozambique tilapia *(O. mossambicus)* and blue tilapia, concluding that

FIGURE 4.3. Efforts to improve cold tolerance through selection have met with limited success in most tilapia varieties, reinforcing the point that this trait appears to be controlled largely by dominance genetic effects.

cold tolerance was largely influenced by dominance genetic effects. Cold tolerance of first-generation hybrids was comparable to that of blue tilapia, whereas that of second-generation fish and pure Mozambique tilapia was inferior. This tendency for dominance genetic control for cold tolerance has also been seen in blue tilapia hybrids with Nile tilapia (Figure 4.3). Results presented by Tayamen et al. (2002) imply that salinity tolerance is also influenced to some degree by dominance effects. Among crosses and pure lines of four *Oreochromis* spp., the most salt-tolerant were hybrids of blue and Mozambique tilapia with *O. spilurus*.

The fact that approximately half the gains from heterosis persist indefinitely to complement available additive genetic variation (which is often increased through the combination of distinct lines) provides an opportunity for substantial genetic improvement in populations derived from crossbred or hybrid parents. Hybrid or crossbred breeding populations also offer the opportunity to combine distinct attributes of the parental stocks. This method has been successfully utilized in production of synthetic lines of tilapia throughout the world (Figures 4.4 and 4.5), many based originally on

FIGURE 4.4. The Mississippi commercial strain, a synthetic line derived from a three-way cross involving blue, Mozambique, and Nile tilapia.

FIGURE 4.5. Florida Red tilapia, another synthetic strain, was developed from hybrids between Mozambique tilapia and *Oreochromis urolepis hornorum*.

hybrids involving red strains of Mozambique tilapia (Lutz 2001). An excellent practical application of this approach was outlined by Munoz-Cordova and Garduno-Lugo (2003), who carried out extensive hybridization to produce new lines of tilapia for commercial use in Mexico.

An example of improved performance in "hybrid" tilapia under commercial production can be seen in data presented by McGinty and Verdegem (1989). Growth rates of Nile tilapia and Taiwanese red tilapia parental lines, first generation (F_1), second-generation ($F_1 \times F_1 = F_2$), and third-generation ($F_2 \times F_2 = F_3$) hybrids were compared at satiation feeding and approximately 85 percent of satiation for a 126 day grow-out period. Gains in fry production achieved by spawning F_1 hybrid parents were not substantially offset by any reduction in the performance of the F_2 progeny.

Maternal effects, perhaps a more important factor in tilapia production than in many other aquatic species, can also exhibit heterosis. Working with two strains of Nile tilapia, Tave et al. (1990) showed that maternal heterosis made a substantial contribution to growth of F_2 and backcross ($F_1 \times$ parental line) crossbred fish. These fish were larger than their F_1 parents as a result of the superior maternal effects of F_1 females resulting from the original cross.

Color Variants in Tilapia

A number of color variants have been documented in tilapia. Red individuals have been isolated in both Mozambique and Nile tilapia, and a number of hybrid-based red varieties have subsequently been developed (Lovshin 2000; Figure 4.6). Other unusual varieties, such as golden-, khaki-, or pearl-colored lines, have also been developed and characterized (Lutz 2001; Figure 4.7). A literature review suggests that whereas coloration can be under fairly straightforward genetic control in pure species, inheritance becomes much more complex in hybrids and hybrid-based synthetic strains (Lutz 2001). One notable exception was reported by Majumdar et al. (1997) for a line of Taiwanese red tilapia, in which pink coloration exhibited simple dominance over black coloration in crosses and backcrosses with

FIGURE 4.6. Many varieties of red tilapia have been developed from hybrids involving the red color variant of Mozambique tilapia, but unlike that parental line, most are characterized by black mottling.

FIGURE 4.7. In North America and Europe, interest in white tilapia varieties has increased in recent years. White varieties have been documented in both Nile and Mozambique tilapia.

Mozambique tilapia. Similar patterns of inheritance are apparent for the dominant red coloration of the red strain of Nile tilapia in crosses with both blue tilapia and other strains of Nile tilapia.

ARTIFICIAL SPAWNING: IMPLICATIONS FOR GENETIC IMPROVEMENT

Life history and behavioral characteristics of tilapia can often bias the outcome of genetic research and breeding programs. Considering the typical behavior of tilapia, it is not difficult to grasp the concept of excessive reproductive contributions by dominant male and female broodstock at the expense of genetic variation within captive breeding populations. The results of these impacts over a number of generations include accelerated inbreeding and an associated loss of fitness. Even with deliberate breeding strategies, a founding stock of 50 to 60 unrelated tilapia can quickly result in an inbred population due to social hierarchies within breeding groups. Male aggressiveness during courtship and spawning can also complicate efforts to execute directed mating and, in the absence of artificial spawning, can lead to the loss of superior or novel female genotypes.

Although it is not yet commercially feasible, artificial spawning of tilapia can eliminate many of the genetic complications resulting from their social and reproductive aggression. It allows the execution of complex mating designs involving full-and half-sib groups and facilitates chromosome set manipulations and hybridization studies. Female tilapia of most *Oreochromis* spp., hybrids, and hybrid-based synthetic strains will ovulate on a regular basis when held in aquaria, even in the absence of males (Owusu-Frimpong and Hargreaves 1999; Figure 4.8). Additionally, hand-stripping of eggs and milt is relatively easy in most tilapia species (Cheuk et al. 2002; Poleo et al. 2002; Figure 4.9).

Regular reproductive cycles can be established in aquaria-held tilapia under a variety of conditions (C.G. Lutz, unpublished data), but best results with Nile tilapia are usually attained under a natural or 12:12 h photoperiod, at temperatures above 22°C, and with a diet containing at least 40 percent crude protein (El-Naggar et al. 2000). Females must be closely observed, however, to determine when ovulation has commenced and eggs are suitable for stripping. A potential problem with artificially spawning tilapia in this way involves asynchrony among stripped eggs, which under natural spawning conditions might vary by as much as 45 to 60 min in timing of ovulation and fertilization (Herbst et al. 2002).

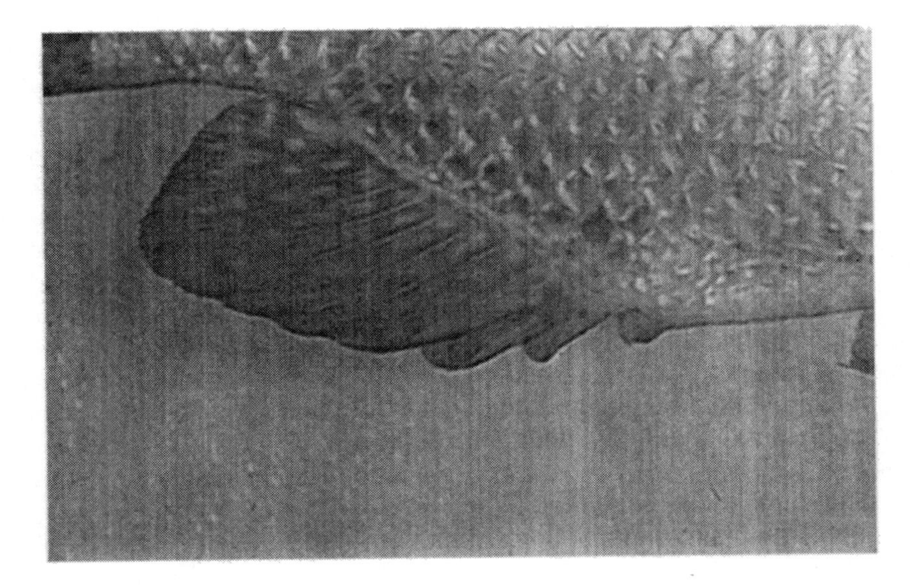

FIGURE 4.8. Female tilapia normally exhibit an extended genital papilla for several hours prior to ovulation and oviposition.

FIGURE 4.9. Artificial spawning of tilapia, as with most teleosts, requires eggs to be stripped within a limited period of time after ovulation. Oviposition normally occurs between 8 and 11 h after sunrise.

Producers and researchers generally have little trouble encouraging tilapia stocked in groups in tanks or ponds to spawn, but this spawning usually occurs on a random, asynchronous schedule. Often, more synchronized spawning is needed to allow for genetic research or improvement. Roderick et al. (1996) reported that Leutirizing Hormone Releasing Hormone analogue (LHRHa; at 30 mg per 100 g body weight) and fertirelin acetate (at 1.0 to 22.5 mg per 100 g body weight) injections increased the number of spawns in *O. niloticus* over a period of 8 days, but spawning patterns were still asynchronous.

Although control over ovulation is still somewhat imprecise in tilapia, when eggs can be stripped reliably, the ability to store sperm on a short-term basis allows for the execution of most common mating designs where more than one female must be mated to the same male (or vice versa). Many common extenders are suitable for storing tilapia sperm, although specific requirements for saltwater-tolerant species and strains (such as Mozambique tilapia, their hybrids, and red varieties based on such hybrids) have not been well defined. Muiruri (1988) and Harvey and Kelley (1988) developed extenders for tilapia milt consisting of easily obtained components.

CHROMOSOMAL MANIPULATIONS

A broad category of related technical approaches involves the manipulation of entire sets of chromosomes. It includes gynogenesis (all-maternal inheritance), androgenesis (all-paternal inheritance), and ploidy manipulations (establishment of additional chromosome sets) (Lutz 2001). The rationale includes the commercial-scale production of sterile stocks (which may be required in the future for commercial culture of genetically modified tilapia), the elucidation of genetic determination of phenotypic sex, and the production of inbred or even cloned lines for crossbreeding studies or other research. Mair (1993) provided a valuable summary of the techniques involved in chromosome manipulation in tilapia and ways to verify their success.

Androgenesis and Gynogenesis

The key to both androgenesis and gynogenesis lies in eliminating the genetic contribution of one parent or the other and subsequently interrupting the normal course of events in a newly fertilized or activated egg in such a way as to restore a $2n$ state, allowing normal development to proceed. Gynogenesis typically involves the use of radiation or ultraviolet light to destroy the genetic material within sperm while still keeping them sufficiently viable to allow motility, egg penetration, and activation. Alternatively, sperm from entirely different species can sometimes be utilized to trigger egg activation and development. Unlike the case of higher vertebrates, eggs of many fish species will begin to divide with only a single chromosome set once they are "activated" by contact with sperm. A number of cell divisions may occur before these haploid larvae eventually cease to develop.

Androgenesis relies on duplicating a single set of chromosomes of paternal origin by interrupting the first cell division of a previously irradiated egg. As a result, androgens are always homozygous for every gene they possess. Once the single chromosome set has completed its first replication, physiological shocks such as temperature or pressure can be applied to prevent the first cell division, resulting in a single, undivided cell with two identical chromosome sets. Whereas pressure and heat shocks in tilapia have been assumed to interfere with spindle formation, and therefore the separation of chromosome sets, Shirak et al. (1998) suggested that cold shocks disrupt the formation of the contractile ring prior to spindle separation.

Two types of gynogenesis can be induced. The first, mitotic gynogenesis, involves the same mechanisms as androgenesis and results in duplication of a single (maternal) set of chromosomes. Mitotic gynogens, therefore, are also completely homozygous. The second approach meiotic gynogenesis, involves the retention within the egg of two maternal chromosome sets that are not identical. As a result, meiotic gynogens possess some degree of heterozygosity. In most fish, including tilapia, a surplus chromosome set left over from egg formation is still present within the egg prior to fertilization. This chromosome set is contained in what is referred to as the second polar body. Once an egg has been fertilized or activated with inert sperm, mechanisms are set in motion within the egg to expel or break down the second polar body. Application of physiological shocks at the precise moment following activation can prevent this process (Figure 4.10). In meiotic gynogenesis, the chromosome sets from the second polar body and the egg nucleus combine to function as a normal $2n$ genome.

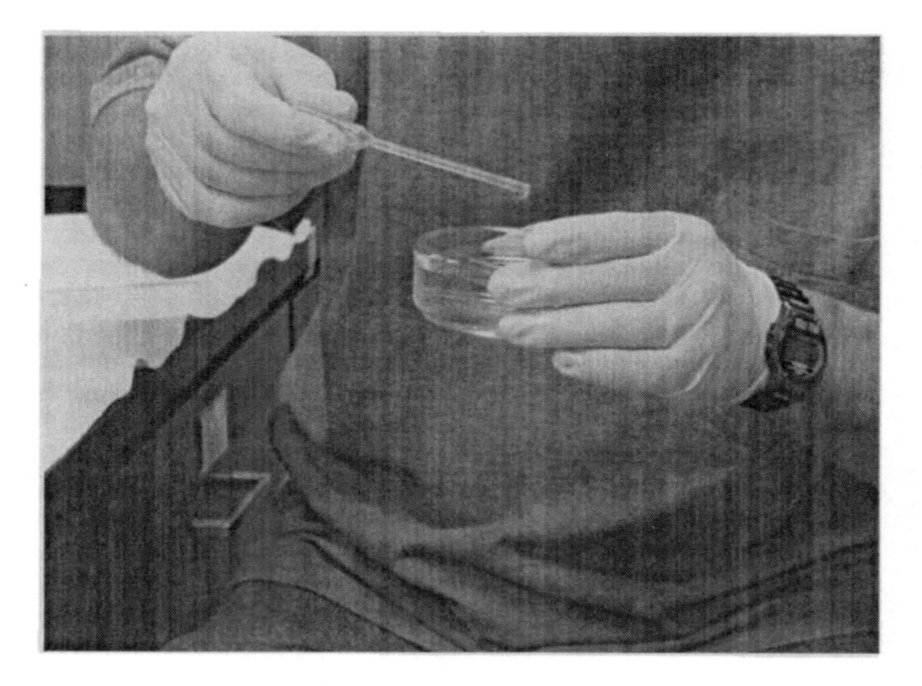

FIGURE 4.10. Androgenesis, gynogenesis, and polyploidy induction in tilapia all require precise control over the timing of egg activation, necessitating strip-spawning and in vitro fertilization techniques.

Hussain et al. (1993) used homozygosity and lack of paternal inheritance as a means of optimizing milt irradiation and pressure and heat shocks for gynogenesis in Nile tilapia. Optimal results for milt irradiation were 300 to 310 $\mu W \cdot cm^{-2}$ for 2 min at 4°C; and optimum pressure shock was 630 $kg \cdot cm^{-2}$ for 2 min at 28°C, applied 40 to 50 min after fertilization; and optimal heat shock was 41°C for 3.5 min 27.5 to 30.0 min after fertilization.

A number of other results have been published describing androgenesis in tilapia. Myers et al. (1995) reported on induction of androgenesis and gynogenesis in Nile tilapia. They destroyed the chromosomes of ovulated tilapia eggs using 5 to 8 min exposure to ultraviolet irradiation at 90 $J \cdot m^{-2} \cdot min^{-1}$ and subsequently fertilized the eggs in vitro with viable sperm. Gynogen and androgen production was achieved using heat shocks (42.5°C) for 3 to 4 min anywhere from 22.5 to 30.0 min after fertilization.

The intensity, timing, and duration of optimal physiological shocks described by Myers et al. (1995) were in good agreement with those of Hussain et al. (1993), and by describing the timing of the first chromosome replication and cell division in Nile tilapia, both studies have provided a starting point for a number of other researchers. Marengoni and Onoue (1998) described the use of Nile tilapia eggs in combination with blue tilapia sperm for the production of androgens, and Campos-Ramos et al. (2003) described gynogenesis protocols for Mozambique tilapia.

Studies involving gynogenesis and androgenesis may appear to have little practical application in tilapia production, but they can shed light on genetic mechanisms for sex determination (see section "Sex determination, manipulation, and control") and on the influence of environmental and genetic factors on production traits. Hussain et al. (1995) measured weight, length, GSI, reproductive response, and various fin ray and scale counts in meiotic gynogens, mitotic gynogens, and normal diploids produced from the same full-sib family of Nile tilapia. For most traits, mean values decreased from diploids to meiotic gynogens to mitotic gynogens, whereas phenotypic variation increased. The authors suggested this trend might reflect the reduction in physiological adaptability often associated with inbreeding depression, and similar trends are often seen over time in isolated tilapia stocks.

Muller-Belecke and Horstgen-Schwark (2002) and Thompson et al. (2002) reported on efforts to develop and evaluate clonal lines of Nile tilapia produced through gynogenesis. The generation of large numbers of identical fish, whether completely homozygous or heterozygous (from crossing two clonal lines), also holds potential for a better understanding of genetic control over important production traits.

Polyploidy

The production of polyploid tilapia, that is, fish with more than the normal two sets of chromosomes, relies on using the same types of shocks as gynogenesis or androgenesis to retain the second polar body or prevent first cleavage in newly fertilized eggs. The only difference is that chromosomes within eggs and sperm are not destroyed but allowed to combine normally. The ultimate goal is large-scale production of triploid ($3n$) tilapia, which are sterile because of their inability to produce normal gametes as a result of possessing an odd number of chromosome sets.

The use of temperature- or pressure-shock-induced triploidy to attain sterile tilapia stocks remains problematic owing to the labor and facilities required to artificially produce such triploids one spawn at a time (Figure 4.11). Advances in artificial spawning, however, may eventually allow profitable production of triploid tilapia. Although interploid triploids, through crossing tetraploid ($4n$) and normal ($2n$) broodstock, have been produced in several aquatic species, efforts to date to produce viable tetraploid tilapia have met with very limited success. One exceptional success was reported by

FIGURE 4.11. Artificial incubation of tilapia eggs is a necessity in studies involving chromosomal manipulations.

Don and Avtalion (1988), who confirmed the production of viable tetraploid blue tilapia through karyological examination and the use of fluorescence-activated cell sorter analysis of erythrocytes.

Bramick et al. (1995) reported on pond trials with heat-shock-induced triploid Nile tilapia. At the end of the grow-out period male triploids averaged 66 percent greater body weight than their diploid counterparts and triploid females exhibited 95 percent greater body weight on average than diploid females. Kim et al. (1991) produced high numbers of triploid Nile tilapia (83.3 percent) using a 14°C cold shock for 60 min beginning 5 min after fertilization. At 6 months of age, gonadal development was greatly reduced in triploids of both sexes. Jeong et al. (1992) reported optimal induction of triploidy (100 percent) with the use of a cold shock (10.0 ± 0.2°C) 5 min after fertilization.

Working with blue tilapia, Chang et al. (1993) reported distinct differences in gonadal development of heat-shock-induced triploids compared with diploids at 24 weeks of age, although growth differences were not apparent. Martinez-Dias and Celis-Maldonado (1994) reported 100 percent induction of triploidy in Nile tilapia using both heat and cold shocks. However, no growth differences were detected up to 3 months of age, and triploid tilapia exhibited variable survival and a high incidence of deformities. Hussain et al. (1995) reported that triploid Nile tilapia of both sexes exhibited no differences in growth or proximate composition from their diploid counterparts.

Byamungu et al. (2001) induced triploidy in blue tilapia through heat-shocking newly fertilized eggs. Approximately 80 percent of the triploid fish they produced were female, and in tanks and ponds receiving organic manure these fish did not outperform normal mixed-sex *O. aureus.* When diploid and triploid females were compared in tank production, however, triploids outperformed diploids slightly under a normal feeding regime and significantly under conditions of restricted feeding.

Sex Determination, Manipulation, and Control

Some investigations focus on masculinization, feminization, and hybridization, with the ultimate goal of producing all-male tilapia fry on a commercial scale. Reasons for pursuing all-male fry production include not only repression of unwanted recruitment but also enhanced growth. Palada de Vera and Eknath (1993) demonstrated that growth divergence between sexes is apparent early in the life cycle, well before phenotypic differences become apparent. Fundamental sex determination in major tilapia species has long been recognized as following two general patterns: XX females and XY males in Nile tilapia and Mozambique tilapia, and ZZ males and WZ females in blue tilapia.

Genetic and Environmental Influences

Worldwide, culturists have noted that certain crosses between distinct tilapia species often result in all-male offspring. Lahav and Lahav (1990) reported on methodology to produce the all-male Nile × blue hybrid and on its performance on a commercial scale (Figure 4.12). Although Hickling (1960) suggested that this type of monosex production resulted from the mating of homogametic females of one species with homogametic males from another, subsequent hybridization trials produced inconsistent or contradictory results, indicating other mechanisms were involved in sex determination, at least in hybrids.

In an effort to tie together a number of diverse results from tilapia hybridization studies, Avtalion and Hammerman (1978) proposed the additional influence of a single autosomal locus, for which some species were normally homozygous dominant (Mozambique and Nile tilapia) and others were homozygous recessive (blue tilapia and *O. urolepsis hornorum*). Although this theory represented a vast improvement in explaining results from many trials, not to mention a monumental intellectual analysis, some data still did not fit the predicted results.

Increasing discrepancies in subsequent hybridization results (Hulata et al. 1982; Majumdar and McAndrew 1983; Shelton et al. 1983) led many breeders and researchers to conclude that inconsistencies were due to contamination of broodstocks (Lahav and Lahav 1990) resulting from negligence or error within hatcheries or from species translocations and introgression in natural systems. A study by Rognon and Guyomard (2003) based on mitochondrial DNA indicated extensive mixing between *O. aureus* and West African stocks of *O. niloticus,* and a previous analysis of geographically diverse populations of these species across all of Africa suggested very similar conclusions with regard to Nile tilapia in the Senegal, Niger, Volta, and Chad basins (Agnese et al. 1997). Similarly, using isozyme and protein markers Macaranas et al. (1986) and Taniguchi et al. (1985) documented substantial introgression of Mozambique tilapia genes into populations of Nile tilapia throughout the Philippines.

Sex determination in tilapia has proven to involve a more complex genetic system than most early investigators could ever have imagined. Mair, Scott, Penman, Beardmore, and Skibinski (1991) demonstrated female homogamety as the norm in Nile tilapia but also revealed the incidence of occasional female heterogamety. In complementary studies, Mair,

FIGURE 4.12. The most common F_1 hybrid used for tilapia production involves crossing a male blue tilapia *(top)* with a female Nile tilapia *(bottom)*. This cross is widely used in China and Israel, and typically yields predominantly male hybrid offspring *(middle)*.

Scott, Penman, Skibinski, and Beardmore (1991) demonstrated that heterogamety is typical for female blue tilapia, but overall results of sex ratios showed a deficit of male progeny. Results suggested a recessive autosomal gene in blue tilapia that interacted with the major sex determining genes.

Muller-Belecke and Horstgen-Schwark (1995) confirmed the mitotic gynogen (completely homozygous) status of a group of 119 tilapia produced in their laboratory, but approximately 35 percent of these fish were male—seemingly in direct conflict with the assumption that female Nile tilapia are homozygous (XX). These and additional seemingly anomalous results led the authors to propose the occurrence of two or even more minor sex-determining mechanisms that when homozygous and acting in combination can supersede the XX-XY mechanism in *O. niloticus*. Homozygosity increases in inbred stocks, and these findings might be related to the tendency for higher incidences of altered sex determination patterns in comparatively inbred tilapia stocks, as has been frequently noted.

Conceptualized systems such as XX-XY and ZZ-WZ can often provide a framework sufficient for enlightening discussions pertaining to practical sex determination and control. Campos-Ramos et al. (2001) identified putative sex chromosomes in blue tilapia through the delineation of regions of restricted pairing in two distinct chromosomes in heterogametic (female, in this species) individuals. One region corresponded with similar findings reported for Nile tilapia by Carrasco et al. (1999), who found an incompletely paired segment of the largest chromosome pair in heterogametic (XY) males. Further work using gene probes derived from Nile tilapia males and females indicated sequence differences between these putative sex chromosomes (Harvey et al. 2002, 2003). Campos-Ramos et al. (2003) reported findings suggesting that the major sex determining loci in Mozambique tilapia also occurred on the largest chromosome pair, but probably in comparative proximity to the centromere.

Lee et al. (2003) approached the physical relationships between sex-determining loci in Nile tilapia somewhat differently. They utilized a technique referred to as bulk segregant analysis to establish a relationship between ten microsatellite markers and phenotypic sex. In two families, they were able to correctly predict phenotypic sex 95 percent of the time using markers linked to sex-determining alleles, specifically UNH995 and UNH104. Unfortunately, a third family from the same population demonstrated no relationship between phenotypic sex and linkage markers in the region in question.

Nonetheless, phenotypic sex in many aquatic species, including tilapia, must be viewed as involving numerous loci and alleles, some of which respond directly to environmental influences. Guan et al. (2000) presented findings on sex determination in Nile tilapia that could possibly shed light on "inconsistencies" in genetic control over sex in this and related species. A gene known to control somatic sex determination in a number of species, which is normally expressed predominantly in vertebrate testes, was confirmed to exist in two distinct forms, in separate loci, in Nile tilapia. Expression of these two genes was mutually exclusive in tilapia testes and ovaries.

Incubation and posthatching temperatures have been implicated as sex-influencing factors in many species. Lester et al. (1989) were among the first to discuss this interpretation of sex determination in Nile tilapia, and Baroiller et al. (1995) demonstrated the masculinizing effect of elevated temperatures in Nile tilapia and the hybrid-based synthetic strain known as Florida red tilapia. Although the genetic control of sex is quite different in these two tilapia, temperature effects were quite similar. Baroiller et al. (1996) demonstrated that elevated temperatures could be used to directly override genetic mechanisms of sex determination in Nile tilapia.

Desprez and Melard (1998) found that rearing blue tilapia fry at 34°C for 40 days after hatching resulted in high numbers of male fingerlings (97.8 percent). Baras et al. (2000) also documented a strong masculinizing effect of constantly high temperatures (35°C) during the first 28 days of exogenous feeding in blue tilapia. Fluctuating temperatures (27°C to 35°C) and high temperatures both resulted in significantly faster and more uniform growth, as well as higher survival, when compared with a constant 27°C environment.

In a related study, Baras et al. (2001) subsequently reported that although high temperatures during the first 28 days of exogenous feeding could be used to produce high percentages (>90 percent) of male Nile tilapia fry, associated mortality resulted in a production loss of more than 70 percent. They suggested that any advantages derived from mostly male stocks would be offset by the loss of production during the high-temperature treatment. Baroiller, Clota, et al. (2000) documented notable genotype/temperature interactions in sex determination of both Nile and blue tilapia progeny groups, and Abucay et al. (1999) demonstrated that elevated temperatures could result in feminization of YY and XY Nile tilapia genotypes.

Transcription of the aromatase gene appears to be an important link in the synthesis of estrogen, and resulting ovarian differentiation, in tilapia. Kwon, Haghpanah, et al. (2000) and Kwon, McAndrew, et al. (2000) implicated the inhibition of aromatase activity in the disruption of Nile tilapia sex determination at high temperatures. They found that high temperatures increased the percentages of males in genetically female (XX) fry, as well as the percentage of females among YY fry, and an aromatase inhibitor showed a strong masculinizing effect and interfered with feminization of YY fry at higher temperatures. Very similar results with XX, XY, and YY Nile tilapia had been reported by Kobayashi et al. (2000). Afonso et al. (2001) demonstrated that oral administration of a nonsteroidal aromatase inhibitor (75 to 100 mg fadrazoe per kilogram of diet) resulted in complete masculinization of Nile tilapia fry, and Kobayashi et al. (2003) reported similar results when treatment occurred between 7 and 14 days after hatching.

D'Cotta et al. (2000, 2001a) also reported a relationship between aromatase gene expression and sex determination in tilapia. At normal temperatures, higher levels of aromatase gene expression in brain tissue were associated with genetically female populations. Female progeny groups in which high temperatures had caused masculinization exhibited reduced aromatase expression, as did normal (XY) males exposed to the same high temperatures. D'Cotta et al. (2001b) found that high temperatures associated with masculinization also stimulated the expression of other genes that exhibited little or no expression in genetic females at normal temperatures (27°C).

Distinct ovarian and brain aromatase genes have been identified in tilapia, and Harvey et al. (2003) presented evidence to suggest that neither gene is located on the same chromosome pair as the primary sex determining factors. Tsai et al. (2001) found that treatment of Mozambique tilapia fry with 17β-estradiol resulted in a notable reduction in expression of aromatase and estrogen receptor mRNA in brain tissue up to day 10 after hatching, but not subsequently. In contrast, treatment of fry with 17α- methyltestosterone resulted in an increase in brain aromatase and estrogen receptor mRNA, but primarily from days 10 to 20 after hatching. These findings reinforce those reported by Kwon, McAndrew, et al. (2000), who found that Nile tilapia fry were most sensitive to aromatase inhibitors during the first week of treatment, between days 7 and 14 after yolk-sac resorption. Similarly, Wang and Tsai (2000) reported that exposure of Mozambique tilapia fry to low temperatures (20°C) before 10 days of age resulted in high proportions of females, whereas elevated temperatures resulted in higher proportions of males, but this effect was noted primarily after 10 days of age.

Practical Applications

Pandian and Varadaraj (1992) provided a review of previous research focusing on endocrine manipulation of phenotypic sex and chromosome manipulation for producing all-male tilapia fingerlings, with a view toward practical application by unskilled fish farmers in less developed countries. Potts and Phelps (1993) subsequently provided useful background and baseline data concerning formulation and feeding rates for feminizing feed to alter the sexual phenotype of genetically male tilapia fry for subsequent use as broodstock for production of YY males *(O. niloticus)* or all-male fry *(O. aureus)*. Ethynylestradiol treatments resulted in progeny groups with 57.9 to 65.0 percent females and an additional 3.0 to 9.4 percent of fish exhibiting ovotestes. Diethylstilbestrol treatments yielded progenies with 60.3 to 80.0 percent females and 0.7 to 7.3 percent fish exhibiting ovotestes. Highest success rates were attained at 400 $mg \cdot kg^{-1}$ of diet.

Mair et al. (1995) reported that genetically male Nile tilapia sired by YY males produced significantly higher yields than mixed-sex or sex-reversed fish in a variety of production environments, and Mair et al. (1997) outlined protocols for producing YY males. Although other genetic factors affect phenotypic sex in their offspring, YY males are nonetheless capable of siring broods with male percentages (approximately 96 percent on average) comparable to some of the best results from feeding masculinizing diets to fry. Unfortunately, the commercial application of YY technology is highly influenced by the strain(s) of Nile tilapia in question. Tuan et al. (1999) demonstrated that sex ratios from YY males in the Thai Chitralada strain of Nile tilapia were highly variable and not suitable for commercial purposes without the application of selection for compatibility of broodstock to produce acceptable results. Abucay and Mair (2000) focused on selecting female lines for compatibility with YY males (to produce higher percentages of male offspring) and superior growth, but commercial applications are still limited on a global basis.

Melard (1995) outlined methods to produce large numbers of all-male blue tilapia based on sex reversal of normal males to phenotypic females. Since males are homogametic in this species (ZZ), crossing sex-reversed ZZ females with normal males should result in all-male ZZ offspring. Blue tilapia fry were fed 17α-ethynylestradiol at 100 to 200 $mg \cdot kg^{-1}$ of diet for 40 days, resulting in 93 to 98 percent females. Some ZZ females did not produce all-male progeny. Fry from those that did were fed 17α-ethynylestradiol at 150 $mg \cdot kg^{-1}$ of diet to produce a second generation of (ZZ) females. This round of sex-reversal produced highly variable feminization rates (18 to 95 percent), but progeny tests on a sample of these ZZ females indicated 97.5 to 100 percent male offspring. One complicating factor in some sex determination studies with blue tilapia is the existence of WW females in addition to the normal WZ genotype (Avtalion and Don 1990).

TRANSGENIC TILAPIA

Production of transgenic organisms involves the physical introduction of genetic constructs into living cells (preferably shortly before or after fertilization) and the incorporation of these constructs into one or more chromosomes within the receiving cells. Multiple copies of a construct can enter a cell, and one critical problem is the potential for introduced constructs to be incorporated within important gene sequences, altering or disrupting their function and expression. For transgenesis to be truly successful, transferred genes must not disrupt existing genes and must be expressed at the appropriate time(s) during the life of the receiving organism. To achieve this goal, introduced constructs often contain promoter sequences to regulate expression and production of gene products. A number of accounts of successful transgenesis in tilapia have been published (Lutz 2001).

Ber et al. (1992) utilized microinjection of DNA into fertilized eggs and exposure of sperm to DNA prior to in vitro fertilization as means to introduce the human growth hormone releasing hormone (GRH) or simple reporter (chloramphenicol acetyl transferase) genes into Nile tilapia. MacLean (1994) presented evaluations of the ability of β-actin promoter sequences from common carp *(Cyprinus carpio carpio)* or rats *(Rattus norvegicus)* to drive a bacterial *lacZ* gene in transgenic Nile tilapia. Results indicated that the carp promoter was roughly one order of magnitude more efficient than that from the rat. Martinez et al. (1994) reported on microinjection of one-cell tilapia embryos. Transmission of the transgene (tilapia growth hormone cDNA) was demonstrated in F_1 and F_2 generations, and accelerated growth was expressed in the initial microinjected individuals and their F_1 offspring.

De la Fuente et al. (1996) transferred constructs containing tilapia growth hormone DNA into tilapia to produce several transgenic lines, including one with accelerated growth. Alam et al. (1996) evaluated 36 transgenic fish produced by MacLean (1994). Only one male produced fertilized embryos that expressed the transgene, at a rate of approximately 15 percent. Different observed patterns of expression were assumed to result from varying levels of integration and the effects of transgene position within the embryos' chromosomes. Such factors have been documented in other transgenic research with tilapia. Alam and Maclean (1997) compared the efficiency of two promoter sequences for *lacZ* genes in Nile tilapia, and Rahman et al. (1997) demonstrated the use of coinjection of a highly integrated reporter gene to improve incorporation of transgenes in Nile tilapia, improving integration rates by a factor of three.

Rahman et al. (1998) produced transgenic tilapia with a construct that incorporated a sockeye salmon *(Oncorhynchus nerka)* growth hormone gene and promoter. These fish, however, exhibited no improvement in growth. In a follow-up study, Rahman and Maclean (1999) produced three lines of transgenic tilapia using a construct built with a chinook salmon *(O. tshawytscha)* growth hormone and a regulatory sequence from ocean pout

(Macrozoarces americanus). The chinook growth hormone was clearly expressed in all transgenic tilapia produced: the average weight of G_1 and G_2 (first- and second-generation transgenic) fish was approximately 300 percent that of controls.

Martinez et al. (1999) produced transgenic tilapia *(O. urolepis hornorum)* by microinjecting developing eggs with constructs containing tilapia growth hormone and regulatory material derived from a human cytomegalovirus. A transgene dosage effect (homozygous [two copies] versus hemizygous [a single transgene copy]) appeared to be present. Transgene dosage effects were also reported by Rahman, Hwang, et al. (2001) in Nile tilapia. Reporter gene expression was found to be associated with the overall number of transgene copies within each of three transgenic lines.

When growth hormone production is involved, transgenic tilapia appear to exhibit enhanced biological efficiency. Martinez et al. (2000) documented significantly higher food conversion efficiency in transgenic lines described in Martinez et al. (1999). Transgenic fish demonstrated higher growth efficiency, synthesis retention, anabolic stimulation, and protein synthesis than controls. Rahman, Ronyai, et al. (2001) reported on grow-out trials on a line of Nile tilapia transgenic for salmon growth hormone. At 7 months of age, transgenic tilapia averaged 653 g, compared with 260 g for their normal, nontrangenic siblings. At the end of a second, shorter growth trial, fish from the transgenic line were approximately four times larger than their normal siblings. Transgenic tilapia were more than 20 percent more efficient at converting food to flesh. A trial indicated this higher efficiency applied to increased digestibility of dietary protein, dry matter, and energy components.

MOLECULAR DISCRIMINATION/CHARACTERIZATION AND GENE MAPPING

Genetic markers have historically been developed and utilized to differentiate distinct species and/or strains of tilapia. In recent years, however, a number of studies have attempted to utilize markers as tools for breeding programs. In fact, tilapia stocks have been purposely developed from several distinct species to facilitate the identification of markers useful for genetic improvement.

Technical Development

A number of rapidly evolving techniques have been applied in the study of genetic markers in tilapia. Bardakci and Skibinski (1994) used random amplified polymorphic DNA (RAPD) techniques to examine polymorphisms within and among blue and Mozambique tilapia and four strains of Nile tilapia. They noted that species differences were more clearly apparent than those among subspecies. Macaranas et al. (1995) examined 30 protein loci electrophoretically to estimate genetic similarity and differences among four African and four domesticated tilapia strains. Naish et al. (1995) compared multilocus DNA fingerprinting and RAPD for the discrimination and characterization of distinct Nile tilapia strains. RAPD resulted in smaller statistical errors in the estimation of within-strain heterozygosity, whereas multilocus DNA fingerprinting was better able to detect genetic differentiation between strains.

Dinesh et al. (1996) utilized RAPD fingerprinting to examine genetic similarities of Nile, blue, and Mozambique tilapia. Highest between-species similarities were found for Nile and Mozambique tilapia. Lee and Kocher (1996) discussed microsatellite marker isolation and use in Nile tilapia. Appleyard and Mather (2000) reported on the utility of using allozyme and RAPD markers in feral populations of Mozambique tilapia in Australia. Allozyme markers typically exhibited codominant inheritance, whereas RAPD markers followed a model of simple dominance.

Marker Applications in Breeding Programs

In tilapia, as in other organisms, individuals with higher heterozygosity levels might be assumed to display some degree of heterosis, and reduced heterozygosity might be associated with inbreeding depression. Appleyard et al. (2001) examined the relationship between individual heterozygosity levels and growth in a Fijian strain of Nile tilapia. Over three generations, no significant association was apparent between individual heterozygosity (measured across 25 allozyme loci) and growth. However, individuals homozygous for one particular microsatellite locus (UNH146) were larger than heterozygotes. The authors concluded that selecting broodstock solely on heterozygosity levels would be an ineffective means of improving growth in this stock.

Nonetheless, highly heterozygous tilapia are proving to be extremely useful in improving our understanding of tilapia genetics and in combining molecular and quantitative approaches to stock improvement. Lines with ancestry from two or more tilapia species can be used to develop genetic (or "genomic") maps for the species and hybrid lines involved. The high degree of genetic variation in the base population allows for the development of a collection of DNA markers that are present in different forms, and analysis of the inheritance of these markers in turn allows for linkage patterns to be identified. Although blue and Nile tilapia have been used successfully to capture genetic diversity in crosses to allow for genetic mapping, McConnell et al. (2000) demonstrated that a large proportion of the linkage groups being developed for these two species exhibit high degrees of similarity.

In tilapia, as in most organisms, chromosomes are seldom passed down intact from one generation to the next. During the process of gamete formation chromosome pairs align and portions of homologous chromosomes can be exchanged, mixing and matching genetic materials provided by the organism's two parents. Linkage is based on the concept that the closer two points, or loci, are on a chromosome, the less likely they are to become separated by one of these events and the more likely they are to find their way into gametes (and offspring) together. When chromosomes from distinct species are involved, parental associations between two such loci are more readily established. A composite linkage map can be used to represent physical relationships between various loci and to associate alleles at these loci with specific species ancestry. Once markers have been identified and characterized, they can be screened for use in marker assisted selection. Identified markers are evaluated for possible association with actual performance measures. Genes that exhibit strong associations with quantitative traits such as growth, environmental tolerances, or other metric characters are termed quantitative trait loci.

Agresti et al. (2000) reported on progress in developing a tilapia stock based on four distinct species. Initial efforts involved two-way crosses among the four species utilized. These first-generation offspring, designated as 2WCs, were used to produce second-generation 2WCs, crossed with nonparental species to produce three-way crosses (3WCs), or crossed with other 2WCs to produce four-way crosses (4WCs). When 3WCs were attempted, however, some crosses were difficult or impossible to complete. Similar problems were encountered with 4WCs, but initial linkage maps were constructed based on the offspring of one particular 3WC.

Researchers elsewhere are working on mapping the centromeres of Nile tilapia chromosomes, which should eventually allow linkage groups to be associated with specific chromosomes in the various tilapia species. An example of other useful outcomes that can be generated from this type of research was given by Shirak et al. (2002), who examined three microsatellite markers for associations with deleterious genes and sex ratios in blue tilapia. Two markers were highly associated with deleterious effects in females, and a third appeared to interact with the others in relation to increased mortality in males. Similarly, Harvey et al. (2003) demonstrated that the brain and ovarian aromatase genes of the Nile tilapia are present on two different chromosomes. The fact that neither is found on the presumed sex chromosomes implies that these genes are not the primary sex determining genes in this species.

This type of work relies on an understanding of the physical structure of tilapia chromosomes. Minrong and Hongxi (1983) provided some of the first descriptions of the karyotypes of Nile and Mozambique tilapia, and Hussain and McAndrew (1994) presented a modified technique for producing high-quality chromosome spreads from samples of embryonic and soft tissues of tilapia. Cattin and Ferrerira (1989) also provided practical techniques for preparing chromosome spreads for tilapia.

Efforts in tilapia linkage mapping may result in substantial improvements to cultured stocks. Cnaani et al. (2003) reported on success in associating markers in a specific chromosome region in hybrid tilapia with quantitative trait loci for growth and cold tolerance. As researchers make progress in combining these techniques with molecular and quantitative genetics approaches, dramatic improvements of tilapia stocks should take place over the next decade.

REFERENCES

Abucay, J.S. and G.C. Mair (2000). Divergent selection for growth in the development of a female line for the production of improved genetically male tilapia (GMT). In K. Fitzsimmons and J.C. Filho (Eds.), *Tilapia aquaculture in the 21st century: Proceedings from the Fifth International Symposium on Tilapia Aquaculture* (pp. 81-89). Rio de Janeiro, Brazil: Panorama do Aquicultura.

Abucay, J.S., G.C. Mair, D.O.F. Skibinski, and J.A. Beardmore (1999). Environmental sex determination: The effect of temperature and salinity on sex ratio in *Oreochromis niloticus* L. *Aquaculture* 173(1-4):219-234.

Afonso, L.O.B., G.J. Wassermann, and R.T. Oliviera (2001). Sex reversal in Nile tilapia (*Oreochromis niloticus*) using a non-steroidal aromatase inhibitor. *Journal of Experimental Biology* 290(2):177-181.

Agnese, J.F., B. Adepo-Gourene, E.K. Abban, and Y. Fermon (1997). Genetic differentiation among natural populations of the Nile tilapia *Oreochromis niloticus* (Teleostei, Cichlidae). *Heredity* 79(1):88-96.

Agresti, J.J., S. Seki, A. Cnaani, S. Poompuang, E.M. Hallerman, N. Umiel, G. Hulata, G.A.E. Gall, and B. May (2000). Breeding new strains of tilapia: Development of an artificial center of origin and linkage map based on AFLP and microsatellite loci. *Aquaculture* 185(1-2):43-56.

Alam, M.D.S., A. Popplewell, and N. MacLean (1996). Germ line transmission and expression of a lacZ containing transgene in tilapia *(Oreochromis niloticus). Transgenic Research* 5(2):87-95.

Alam, S.M., and Maclean (1997). Comparison of the efficiency of expression of two lacZ gene construct containing a short and a long regulatory sequence of the carp beta-actin gene in tilapia. *Journal of Aquaculture in the Tropics* 12(2):103-112.

Appleyard, A.S., and B.P. Mather (2000). Investigation into the mode of inheritance of allozyme and random amplified polymorphic DNA markers in tilapia *Oreochromis mossambicus* (Peters). *Aquaculture Research* 31(5):435-445.

Appleyard, S., J. Renwick, and P. Mather (2001). Individual heterozygosity levels and relative growth performance in *Oreochromis niloticus* (L.) cultured under Fijian conditions. *Aquaculture Research* 32(4):287-296.

Avtalion, R.R., and J. Don (1990). Sex-determining genes in tilapia: A model of genetic recombination emerging from sex ratio results of three generations of diploid gynogenetic *Oreochromis aureus*. *Journal of Fish Biology* 37(1):167-173.

Avtalion, R.R., and I.S. Hammerman (1978). Sex determination in *Sarotherodon* (Tilapia). I. Introduction to a theory of autosomal influence. *Israeli Journal of Aquaculture/Bamidgeh* 30:110-115.

Balfry, S.K., M. Shariff, T.P.T. Evelyn, and G.K. Iwama (1994). The importance of the natural immune system in disease resistance of fishes. In *International Symposium on Aquatic Animal Health: Program and Abstracts* (pp. 99-110). Davis: University of California School of Veterinary Medicine.

Baras, E., B. Jacobs, and C. Melard (2001). Effect of water temperature on survival, growth and phenotypic sex of mixed (XX-XY) progenies of Nile tilapia *Oreochromis niloticus*. *Aquaculture* 192(2-4):187-199.

Baras, E., C. Prignon, G. Gohoungo, and C. Melard (2000). Phenotypic sex differentiation of blue tilapia under constant and fluctuating thermal regimes and its adaptive and evolutionary implications. *Journal of Fish Biology* 57(1):210-223.

Bardakci, F., and D.O.F. Skibinski. (1994). Application of the RAPD technique in tilapia fish: Species and subspecies identification. *Heredity* 73(2):117-123.

Baroiller, J.F., E. Bezaut, S. Bonnet, F. Clota, M. Derivaz, A. D'Hont, B. Fauconneau, J. Lazard, C. Ozouf-Costaz, X. Rognon, et al. (2000). Production of two reciprocal intergeneric hybrids between *Oreochromis niloticus* and *Sarotherodon melanotheron*. In K. Fitzsimmons and J.C. Filho (Eds.), *Tilapia aquaculture in the 21st century: Proceedings from the Fifth International Symposium on Tilapia Aquaculture* (p. 11). Rio de Janeiro, Brazil: Panorama do Aquicultura.

Baroiller, J.F., F. Clota, and H. D'Cotta (2000). Genetic and environmental sex determination in tilapias: A review. In K. Fitzsimmons and J.C. Filho (Eds.), *Tilapia aquaculture in the 21st century: Proceedings from the Fifth International Symposium on Tilapia Aquaculture* (p. 81). Rio de Janeiro, Brazil: Panorama do Aquicultura.

Baroiller, J.F., F. Clota, and E. Geraz (1995). Temperature sex determination in two tilapia, *Oreochromis niloticus* and the red tilapia (red Florida strain): Effect of high or low temperature. In F.W. Goetz and F. Thomas (Eds.), *Proceedings of the Fifth International Symposium on the Reproductive Physiology of Fish* (pp. 158-160). The University of Texas at Austin.

Baroiller, J.F., A. Fostier, C. Cauty, X. Rognon, and B. Jalabert (1996). Effects of high rearing temperatures on the sex ratio of progeny from sex reversed males of *Oreochromis niloticus*. In R.S.V. Pullin, J. Lazard, M. Legendre, J.B. Amon Kottias, and D. Pauly (Eds.), *The Third International Symposium on Tilapia in Aquaculture* (pp. 246-256). Manila, Philippines: International Center for Living Aquatic Resources Management.

Ber, R., R. Avtalion, B. Timan, and V. Daniel (1992). DNA transfer into the genome of tilapia. In B. Moav, V. Hilge, and H. Rosenthal (Eds.), *Progress in aquaculture research, 1992* (pp. 155-165). European Aquaculture Society Special Publication No. 17.

Bolivar, R. (1999). Estimation of response to within-family selection for growth in Nile tilapia. *Dissertation Abstracts International Part B: Science and Engineering* 60(3):934.

Bolivar, R.B., A.E. Eknath, N.L. Bolivar, and T.A. Abella (1993). Growth and reproduction of individually tagged Nile tilapia *(Oreochromis niloticus)* of different strains. *Aquaculture* 111:159-169.

Bolivar, R.B. and G.F. Newkirk (2000). Response to selection for body weight of Nile tilapia *(Oreochromis niloticus)* in different culture environments. In K. Fitzsimmons and J.C. Filho (Eds.), *Tilapia aquaculture in the 21st century: Proceedings from the Fifth International Symposium on Tilapia Aquaculture* (pp. 12-23). Rio de Janeiro, Brazil: Panorama do Aquicultura.

Bramick, U., B. Puckhaber, H.-J. Langholz, and G. Horstgen-Schwark (1995). Testing of triploid tilapia *(Oreochromis niloticus)* under tropical pond conditions. *Aquaculture* 137(1-4):343-353.

Brzeski, V.J., and R.W. Doyle (1995). A test of an on-farm selection procedure for tilapia growth in Indonesia. *Aquaculture* 137(1-4):219-230.

Byamungu, N., V.M. Darras, and E.R. Kuhn (2001). Growth of heat-shock induced triploids of blue tilapia, *Oreochromis aureus,* reared in tanks and ponds in Eastern Congo: Feeding regimes and compensatory growth response of triploid females. *Aquaculture* 198(1-2):109-122.

Cai, W., S. Li, and J. Ma (2004). Disease resistance of Nile tilapia *(Oreochromis niloticus),* blue tilapia *(Oreochromis aureus)* and their hybrid (female Nile tilapia × male blue tilapia) to *Aeromonas sobria. Aquaculture* 229(1-4):79-87.

Campos-Ramos, R., S. Harvey, J. Mazabanda, L. Carrasco, D. Griffin, B. McAndrew, N. Bromage, and D. Penman (2001). Identification of putative sex chromosomes in the blue tilapia, *Oreochromis aureus,* through synaptonemal complex and FISH analysis. *Genetics* 111(1-3):143-153.

Campos-Ramos, R., S.C. Harvey, B.J. McAndrew, and D.J. Penman (2003). An investigation of sex determination in the Mozambique tilapia, *Oreochromis mossambicus,* using synaptonemal complex analysis, FISH, sex reversal and gynogenesis. *Aquaculture* 221(1-4):125-140.

Carrasco, L.A.P., D.J. Penman, and N. Bromage (1999). Evidence for the presence of sex chromosomes in the Nile tilapia *(Oreochromis niloticus)* from synaptonemal complex analysis of XX, XY and YY genotypes. *Aquaculture* 173(1-4):207-218.

Cattin, P.M., and J.T. Ferreira (1989). A rapid non-sacrificial chromosome preparation technique for freshwater teleosts. *South African Journal of Zoology* 24(1): 76-78.

Chang, S.-L., C.-F. Chang, and I.-C. Liao (1993). Comparative study on the growth and gonadal development of diploid and triploid tilapia, *Oreochromis aureus. Journal of Taiwan Fisheries Research* 1(1):43-49.

Cheuk, G.W., G.A. Poleo, E.C. Herbst, and T.R. Tiersch (2002). Intracytoplasmic sperm injection in Nile tilapia. In *Book of abstracts, aquaculture America 2002* (p. 66). Baton Rouge, LA: World Aquaculture Society.

Cnaani, A., G.A.E. Gall, and G. Hulata (2000). Cold tolerance of tilapia species and hybrids. *Aquaculture International* 8(4):289-298.

Cnaani, A., E.M. Hallerman, M. Ron, J.I. Weller, M. Indelman, Y. Kashi, G.A. Gall, and G. Hulata (2003). Detection of a chromosomal region with two quantitative trait loci, affecting cold tolerance and fish size, in an F2 tilapia hybrid. *Aquaculture* 223(1-4):117-128.

D'Cotta, H., A. Fostier, Y. Guiguen, M. Govoroun, and J.-F. Baroiller (2001a). Aromatase plays a key role during normal and temperature-induced sex differentiation of tilapia *Oreochromis niloticus. Molecular Reproduction and Development* 59(3):265-276.

———. (2001b). Search for genes involved in the temperature-induced gonadal sex differentiation in the tilapia, *Oreochromis niloticus. Journal of Experimental Biology* 290(6):574-585.

D'Cotta, H., Y. Guiguen, M. Govoroun, O. McMeel, and J.F. Baroiller (2000). Aromatase gene expression in temperature-induced gonadal sex differentiation of tilapia *Oreochromis niloticus.* In B. Norberg, O.S. Kjesbu, G.L. Taranger, E. Andersson, and S.O. Stefansson (Eds.), *Reproductive biology of fish, Bergen, 1999* (pp. 244-246). Department of Fisheries and Marine Biology, University of Bergen, Norway.

de la Fuente, J., O. Hernandez, R. Martinez, I. Guillan, M.P. Estrada, and R. Lleonart (1996). Generation, characterization and risk assessment of transgenic tilapia with accelerated growth. *Biotecnologia Aplicada* 13(3):221-230.

Desprez, D., and C. Melard (1998). Effect of ambient water temperature on sex determination in the blue tilapia *Oreochromis aureus. Aquaculture* 162(1-2):79-84.

Dey, M.M., A.E. Eknath, L. Sifa, M.G. Hussain, T.M. Thien, N. Van Hao, S. Aypa, and N. Pongthana (2000). Performance and nature of genetically improved farmed tilapia: A bioeconomic analysis. *Aquaculture Economics and Management* 4(1-2):83-106.

Dinesh, K.R., T.M. Lim, W.K. Chan, and V.P.E. Phang (1996). Genetic variation inferred from RAPD fingerprinting in three species of tilapia. *Aquaculture International* 4(1):19-30.

Don, J., and R.R. Avtalion (1988). Production of viable tetraploid tilapia using the cold shock technique. *Bamidgeh* 40(1):17-21.

Eguia, M.R.R. (1996). Reproductive performance of four red tilapia strains in different seed production systems. *Bamidgeh* 48(1):10-18.

Eguia, M.R.R., and R.V. Eguia (1993). Growth response of three *Oreochromis niloticus* strains to feed restriction. *Bamidgeh*, 45(1):8-17.

Eknath, A.E., M.M. Tayamen, M.S. Palada-de Vera, J.C. Danting, R.A. Reyes, E.E. Dionisio, J.B. Capili, H.L. Bolivar, T.A. Abella, A.V. Circa, et al. (1993). Genetic improvement of farmed tilapias: The growth performance of eight strains of *Oreochromis niloticus* tested in different farm environments. *Aquaculture* 111:171-188.

Elghobashy, H.A., A. Rahman, A. El Gamal, and A.M. Khater (2000). Growth evaluation of four local strains of Nile tilapia *(Oreochromis niloticus)* under different farming conditions in Egypt. In K. Fitzsimmons and J.C. Filho (Eds.), *Tilapia aquaculture in the 21st century: Proceedings from the Fifth International Symposium on Tilapia Aquaculture* (pp. 346-351). Rio de Janeiro, Brazil: Panorama do Aquicultura.

El-Naggar, G.O., M.A. El-Nady, M.G. Kamar, and A.I. Al-Kobaby (2000). Effect of photoperiod, dietary protein and temperature on reproduction in Nile tilapia. In K. Fitzsimmons, and J.C. Filho (Eds.), *Tilapia aquaculture in the 21st century: Proceedings from the Fifth International Symposium on Tilapia Aquaculture* (pp. 352-358). Rio de Janeiro, Brazil: Panorama do Aquicultura.

Fryer, G., and T.D. Iles (1972). *The cichlid fishes of the Great Lakes of Africa—Their biology and evolution.* Edinburgh: Oliver and Boyd.

Gjedrem, T. (1999). Aquaculture needs genetically improved animals. *Global Aquaculture Advocate* 2(6):69-70.

Gjoen, H.M. (2002). GIFT program continues: Distribution of fast-growing tilapia to expand. *Global Aquaculture Advocate* 4(6):44.

Guan, G., T. Kobayashi, and Y. Nagahama (2000). Sexually dimorphic expression of two types of DM (Doublesex/Mab-3)-domain genes in a teleost fish, the tilapia *(Oreochromis niloticus). Biochemical and Biophysical Research Communications* 272(3):662-666.

Harvey, B.J., and R.N. Kelley (1988). Practical methods for chilled and frozen storage of tilapia spermatozoa. In R.S.V. Pullin, T. Bhukaswan, K. Tonguthai, and J.L. Maclean (Eds.), *The Second International Symposium on Tilapia in Aquaculture* (pp. 179-189). Manila, Philippines: ICLARM, International Center for Living Aquatic Resources Management.

Harvey, S.C., C. Boonphakdee, R. Campos-Ramos, M.T. Ezaz, D.R. Griffin, N.R. Bromage, and P. Penman (2003). Analysis of repetitive DNA sequences in the sex chromosomes of *Oreochromis niloticus. Cytogenetic and Genome Research* 101(3-4):314-319.

Harvey, S.C., J.Y. Kwon, and D.J. Penman (2003). Physical mapping of the brain and ovarian aromatase genes in the Nile tilapia, *Oreochromis niloticus,* by fluorescence in situ hybridization. *Animal Genetics* 34(1):62-64.

Harvey, S.C., J. Masabanda, L.A.P. Carrasco, N.R. Bromage, D.J. Penman, and D.K. Griffin (2002). Molecular-cytogenetic analysis reveals sequence differences between the sex chromosomes of *Oreochromis niloticus:* Evidence for an early stage of sex-chromosome differentiation. *Cytogenetic and Genome Research* 97(1-2):76-80.

Herbst, E.C., G.A. Poleo, C.G. Lutz, and T.R. Tiersch (2002). Is variable induction of polyploidy in Nile tilapia caused by asynchrony of zygotic development? *Abstracts, 2002 Joint Meeting of the Mississippi and Louisiana Chapters,* American Fisheries Society.

Hickling, C.F. (1960). The Malacca tilapia hybrids. *Journal of Genetics* 57:1-10.

Horstgen-Schwark, G., and H.J. Lanholz (1998). Prospects of selecting for late maturity in tilapia *(Oreochromis niloticus).* III. A selection experiment under laboratory conditions. *Aquaculture* 167(1-2):123-133.

Hussain, M.G., and B.J. McAndrew (1994). An improved technique for chromosome karyotyping from embryonic and soft tissues of tilapia and salmonids. *Asian Fisheries Science* 7(2-3):187-190.

Hussain, M.G., B.J. McAndrew, and D.J. Penman (1995). Phenotypic variation in meiotic and mitotic gynogenetic diploids of the Nile tilapia, *Oreochromis niloticus* (L.). *Aquaculture Research* 26(3):205-212.

Hussain, M.G., D.J. Penman, B.J. McAndrew, and R. Johnstone (1993). Suppression of first cleavage in the Nile tilapia, *Oreochromis niloticus* L. A comparison of the relative effectiveness of pressure and heat shocks. *Aquaculture* 111:263-270.

Huang, C.-M., and I.-C. Liao (1990). Response to mass selection for growth rate in *Oreochromis niloticus. Aquaculture* 85:199-205.

Hulata, G., G.W. Wohlfarth, and A. Halevy (1986). Mass selection for growth rate in the Nile tilapia *(Oreochromis niloticus). Aquaculture* 57:177-184.

Hulata, G., G. Wohlfarth, and S. Rothbard (1982). Progeny-testing selection of tilapia broodstocks producing all-male hybrid progenies—Preliminary results. In N.P. Wilkins and E.M. Gosling (Eds.), *Genetics in aquaculture* (pp. 263-268). Amsterdam: Elsevier.

Jeong, W.G., J.S. Hue, and E.O. Kim (1992). Induction of triploidy, gonadal development and growth in the Nile tilapia, *Oreochromis niloticus. Bulletin of the National Fisheries Research and Development Agency (Korea)* 46:161-171.

Jhingran, V. G., and V. Gopalakrishnan (1974). *A catalogue of cultivated aquatic organisms.* FAO Fish. Tech. Pap 130. F.A.O., Rome.

Kim, D.S., G.C. Choi, and I.-S. Park (1991). Triploidy production of Nile tilapia, *Oreochromis niloticus. Contributions of the Institute of Marine Science, National Fisheries University of Pusan* 23:235-244.

Kobayashi, T., H. Kajiura-Kobayashi, and Y. Nagahama (2000). Gonadal formation and differentiation in a teleost, tilapia *Oreochromis niloticus.* In B. Worberg, O.S. Kjesbu, G.L. Taranger, E. Andersson, and S.O. Stefansson (Eds.), *Reproductive physiology of fish* (p. 259). Department of Fisheries and Marine Biology, University of Bergen, Norway.

———. (2003). Induction of XY sex reversal by estrogen involves altered gene expression in a teleost, tilapia. *Cytogenetics and Genome Research* 101(3-4): 289-294.

Kronert, U., G. Hoerstgen-Schwark, and H.-J. Langholz (1989). Prospects of selecting for late maturity in tilapia *(Oreochromis niloticus)* 1. Family studies under laboratory conditions. *Aquaculture* 77(2-3):113-121.

Kwon, J.-Y., V. Haghpanah, L.M. Kogson-Hurtado, B.J. McAndrew, and D.J. Penman (2000). Masculinization of genetic female Nile tilapia *(Oreochromis niloticus)* by dietary administration of an aromatase inhibitor during sexual differentiation. *Journal of Experimental Biology* 287(1):46-53.

Kwon, J.Y., B.J. McAndrew, and D.J. Penman (2000). Inhibition of aromatase activity suppresses high-temperature feminisation of genetic male Nile tilapia, *Oreochromis niloticus.* In B. Norberg, O.S. Kjesbu, G.L. Taranger, E. Andersson and S.O. Stefansson (Eds.), *Reproductive Biology of Fish, Bergen, 1999*

(p. 268). Department of Fisheries and Marine Biology, University of Bergen, Norway.

Lahav, M., and E. Lahav (1990). The development of all-male tilapia hybrids in Nir David. *Bamidgeh* 42(2):58-61.

Lee, B., D.J. Penman, and T.D. Kocher (2003). Identification of a sex-determining region in Nile tilapia *(Oreochromis niloticus)* using bulked segregant analysis. *Animal Genetics* 34(5):379-383.

Lee, W.-J., and T.D. Kocher (1996). Microsatellite DNA markers for genetic mapping in *Oreochromis niloticus. Journal of Fish Biology* 49(1):169-171.

Lester, L.J., K.S. Lawson, T.A. Abella, and M.S. Palada (1989). Estimated heritability of sex ratio and sexual dimorphism in tilapia. *Aquaculture and Fisheries Management* 20:369-380.

Longalong, F.M., A.E. Eknath, and H.B. Bentsen (1999). Response to bi-directional selection for frequency of early maturing females in Nile tilapia *(Oreochromis niloticus). Aquaculture* 178(1-2):13-25.

Lourdes, M., and A. Cuvin-Aralar (1994). Survival and heavy metal accumulation of two *Oreochromis niloticus* (L.) strains exposed to mixtures of zinc, cadmium and mercury. *The Science of the Total Environment 148(1):31-38.*

Lovshin, L.L. (1982). Tilapia hybridization. In R.S.V. Pullin and R.H. Lowe (Eds.), *The biology and culture of tilapias* (pp. 279-308). Manila, Philippines: International Center for Living Aquatic Resource Management.

———. (2000). Criteria for selecting Nile tilapia and red tilapia for culture. In K. Fitzsimmons and J.C. Filho (Eds.), *Tilapia aquaculture in the 21st century: Proceedings from the Fifth International Symposium on Tilapia Aquaculture* (pp. 49-57). Rio de Janeiro, Brazil: Panorama do Aquicultura.

Lowe, R.H. (1955). The fecundity of *Tilapia* species. *East African Agricultural Journal* 21:45-52.

Lutz, C.G. (2001). *Practical genetics for aquaculture.* Oxford: Fishing News Books, Blackwell Science, Ltd.

Maar, A., M.A.E. Mortimer, and I. Van der Lingen (1966). *Fish culture in central east Africa.* Rome: FAO.

Macaranas, J.M., L.Q. Agustin, M.C.A. Ablan, M.J.R. Pante, A.A. Eknath, and R.S.V. Pullin (1995). Genetic improvement of farmed tilapias: Biochemical characterization of strain differences in Nile tilapia. *Aquaculture International* 3(1):43-54.

Macaranas, J.M., N. Taniguchi, M.J.R. Pante, J.B. Capili, and R.S.V. Pullin (1986). Electrophoretic evidence for extensive hybrid gene introgression into commercial *Oreochromis niloticus* (L.) stocks in the Philippines. *Aquaculture and Fisheries Management* 17(4):249-258.

MacLean, N. (1994). Lac Z reporter gene expression and transmission in transgenic tilapia *Oreochromis niloticus.* In *3RD International Marine Biotechnology Conference: Program* (p. 70). Tromsoe, Norway: Tromsoe University.

Mair, G.C. (1993). Chromosome-set manipulation in tilapia—Techniques, problems and prospects. *Aquaculture* 111:227-244.

Mair, G.C., J.S. Abucay, J.A. Beardmore, and D.O.F. Skibinski (1995). Growth performance trials of genetically male tilapia (GMT) derived from YY-males in

Oreochromis niloticus L.: On station comparisons with mixed sex and sex reversed male populations. *Aquaculture* 137(1-4):313-323.

Mair, G.C., J.S. Abucay, D.O.F. Skibinski, T.A. Abella, and J.A. Beardmore (1997). Genetic manipulation of sex-ratio for the large-scale production of all-male tilapia, *Oreochromis niloticus. Canadian Journal of Fisheries and Aquatic Science* 54(2):396-404.

Mair, G.C., A.G. Scott, D.J. Penman, J.A. Beardmore, and D.O.F. Skibinski (1991). Sex determination in the genus *Oreochromis*: 1. Sex reversal, gynogenesis and triploidy in *O. niloticus* (L.). *Theoretical and Applied Genetics* 82(2):144-152.

Mair, G.C., A.G. Scott, D.J. Penman, D.O.F. Skibinski, and J. Beardmore (1991). Sex determination in the genus *Oreochromis:* 2. Sex reversal, hybridisation, gynogenesis and triploidy in *O. aureus* Steindachner. *Theoretical and Applied Genetics* 82(2):153-160.

Majumdar, K.C., and B.J. McAndrew (1983). Sex ratios from interspecific crosses within the tilapia. In L. Fishelson and Z. Yaron (Eds.), *Proceedings of the International Symposium on Tilapia in Aquaculture* (pp. 261-269). Tel Aviv, Israel: Tel Aviv University.

Majumdar, K.C., K. Nasaruddin, and K. Ravinder (1997). Pink body color shows single gene inheritance. *Aquaculture Research* 28(8):581-589.

Marengoni, N.G., and Y. Onoue (1998). Ultraviolet-induced androgenesis in Nile tilapia, *Oreochromis niloticus* (L.), and hybrid Nile × blue tilapia, *O. aureus* (Steindachner). *Aquaculture Research* 29(5):359-366.

Marengoni, N.G., Y. Onoue, and T. Oyama (1998). Offspring growth in a diallel crossbreeding with three strains of Nile tilapia *Oreochromis niloticus. Journal of the World Aquaculture Society* 29(1):114-119.

Martinez, R., A. Arenal, M.P. Estrada, F. Herrera, V. Huerta, J. Vazquez, T. Sanchez, and J. de la Fuente (1999). Mendelian transmission, transgene dosage and growth phenotype in transgenic tilapia *(Oreochromis hornorum)* showing ectopic expression of homologous growth hormone. *Aquaculture* 173(1-4):271-283.

Martinez, R., M.P. Estrada, O. Hernandez, E. Cabrera, D. Garcia del Barco, R. Lleonart, J. Berlanga, and R. Pimentel (1994). Towards growth manipulation in tilapia (*Oreochromis sp.*): Generation of transgenic tilapia with chimeric constructs containing the tilapia growth hormone cDNA. In *3rd International Marine Biotechnology Conference: Program* (p. 70). Tromsoe University, Tromsoe, Norway.

Martinez, R., J. Juncal, C. Zaldivar, A. Arenal, I. Guillen, V. Morera, O. Carrillo, M. Estrada, A. Morales, and M.P. Estrada (2000). Growth efficiency in transgenic tilapia (*Oreochromis* sp.) carrying a single copy of an homologous cDNA growth hormone. *Biochemical and Biophysical Research Communications* 267(1):466-472.

Martinez-Diaz, H.A., and C.A. Celis-Maldonado (1994). Analisis de viabilidad y crecimiento haste el levante de triploides y diploides de tilapia nilotica (*Oreochromis niloticus*, Linne). *Biologia Cientifica Inpa* 2:33-45.

McConnell, J.S.K., C. Beynon, J. Leamon, and F.D.O. Skibinski (2000). Microsatellite marker based genetic linkage maps of *Oreochromis aureus* and *O. niloticus* (Cichlidae): Extensive linkage group segment homologies revealed. *Animal Genetics* 31(3):214-218.

McGinty, A.S., and M.C. Verdegem (1989). Growth of *Tilapia nilotica,* Taiwanese red tilapia, and their F_1, F_2 and F_3 hybrids. *Journal of the World Aquaculture Society* 20(1):56A.

Melard, C. (1995). Production of a high percentage of male offspring with 17 alpha-ethynylestradiol sex-reversed *Oreochromis aureus.* 1. Estrogen sex-reversal and production of F_2 pseudofemales. *Aquaculture* 130(1):25-34.

Minrong, C., and C. Hongxi (1983). A comparative study of karyotypes in three tilapia fishes. *Acta Genetica Sinica* 10(1):56-62.

Muiruri, R.M. (1988). *Chilled and cryogenic preservation of tilapia spermatozoa.* Master's thesis, Institute of Aquaculture, University of Stirling.

Muller-Belecke, A. and G. Horstgen-Schwark (1995). Sex determination in tilapia *(Oreochromis niloticus)*—Sex ratios in homozygous gynogenetic progeny and their offspring. *Aquaculture* 137(1-4):57-65.

———. (2002). Performance testing of clonal Nile tilapia lines. *Global Aquaculture Advocate* 5(1):32-33.

Munoz-Cordova, G., and M. Garduno-Lugo (2003). Mejoramiento Genetico en Tilapia. Facultad de Medicina Veterinaria y Zootecnia de la Universidad Nacional Autonoma de Mexico, SIGOLFO.

Myers, J.M., D.J. Penman, Y. Basavaraju, S.F. Powell, P. Baoprasertkul, K.J. Rana, N. Bromage, and B.J. McAndrew (1995). Induction of diploid androgenetic and mitotic gynogenetic Nile tilapia (*Oreochromis niloticus* L.). *Theoretical and Applied Genetics* 90(2):205-210.

Naish, K.-A., M. Warren, F. Bardakci, D.O.F. Skibinski, G.R. Carvalho, and G.C. Mair (1995). Multilocus DNA fingerprinting and RAPD reveal similar genetic relationships between strains of *Oreochromis niloticus* (Pisces:Cichlidae). *Molecular Ecology* 4(2):271-274.

Oldorf, W., U. Kronert, J. Balarin, R. Haller, G. Hoerstgen-Schwark, and H.-J. Langholz (1989). Prospects of selecting for late maturity in tilapia *(Oreochromis niloticus).* 2. Strain comparisons under laboratory and field conditions. *Aquaculture* 77(2-3):123-133.

Owusu-Frimpong, M., and J.A. Hargreaves (1999). Visual of olfactory contact with males is not necessary for natural spawning of female blue tilapia *(Oreochromis aureus).* In *Book of abstracts, aquaculture America '99* (p. 62). Baton Rouge, LA: World Aquaculture Society.

Palada-de Vera, M.S., and A.E. Eknath (1993). Predictability of individual growth rates in tilapia. *Aquaculture* 111:147-158.

Pandian, T.J., and K. Varadaraj (1992). Technique to produce 100% male tilapia. *Pakistani Seafood Digest* 5(10-11):3-6.

Poleo, G.A., G.W. Cheuk, and T.R. Tiersch (2002). Microinjection: An alternative way to fertilize fish eggs. In *Book of abstracts, aquaculture America 2002* (p. 267). Baton Rouge, LA: World Aquaculture Society.

Potts, A.C., and R.P. Phelps (1993). Effectiveness of diethylstilbestrol and ethynylestradiol in the production of female Nile tilapia *(Oreochromis niloticus)* and the effect on fish morphology. In M. Carrillo, L. Dahle, J. Morales, P. Sorgeloos, N. Svennevig, and J. Wyban (eds.), *From discovery to commercialization, Oostende, 1993* (p. 255). Special Publication of the European Aquaculture Society No. 19.

Rahman, M.A., G.-L. Hwang, S.A. Razak, F. Sohm, and N. Maclean (2001). Copy number related transgene expression and mosaic somatic expression in hemizygous and homozygous transgenic tilapia *(Oreochromis niloticus)*. *Transgenic Research* 9(6):417-427.

Rahman, M.A., A. Iyengar, and N. Maclean (1997). Co-injection strategy improves integration efficiency of a growth hormone gene construct, resulting in lines of transgenic tilapia *(Oreochromis niloticus)* expressing an exogenous growth hormone gene. *Transgenic Research* 6(6):369-378.

Rahman, M.A., and N. Maclean (1999). Growth performance of transgenic tilapia containing an exogenous piscine growth hormone gene. *Aquaculture* 173(1-4):333-346.

Rahman, M.A., R. Mak, H. Ayad, A. Smith, and N. Maclean (1998). Expression of a novel piscine growth hormone gene results in growth enhancement in transgenic tilapia *(Oreochromis niloticus)*. *Transgenic Research* 7(5):357-369.

Rahman, M.A., A. Ronyai, B.Z. Engidaw, K. Jauncey, G. Hwang, A. Smith, E. Roderick, D. Penman, L. Varadi, and N. Maclean (2001). Growth and nutritional trials on transgenic Nile tilapia containing an exogenous fish growth hormone. *Journal of Fish Biology* 59(1):62-78.

Roderick, E.E., L.P. Santiago, M.A. Garcia, and C.G. Mair (1996). Induced spawning in *Oreochromis niloticus* L. In R.S.V. Pullin, J. Lazard, M. Legendre, J.B. Amon Kottias and D. Pauly (Eds.), *The Third International Symposium on Tilapia in Aquaculture* (p. 548). Makati City, Philippines: ICLARM.

Rognon, X., and R. Guyomard (2003). Large extent of mitochondrial DNA transfer from *Oreochromis aureus* to *O. niloticus* in West Africa. *Molecular Ecology* 12(2):435-445.

Romana-Eguia, M.R.R., and R.W. Doyle (1992). Genotype-environment interaction in the response of three strains of Nile tilapia to poor nutrition. *Aquaculture* 108(1-2):1-12.

Schwartz, F.J. (1983). "Tilapia" hybrids: Problems, value, use and world literature. In L. Fishelson, and Z. Yaron (Eds.), *Proceedings of the International Symposium on Tilapia in Aquaculture, Nazareth, Israel* (pp. 611-622). Tel Aviv, Israel: Tel Aviv University.

Shelton, W.L., F.H. Meriwether, K.J. Semmens, and W.E. Calhoun (1983). Progeny sex ratios from interspecific pair spawnings of *Oreochromis (Tilapia) aureus* and *O. niloticus*. In L. Fishelson, and Z. Yaron (Eds.), *Proceedings of the International Symposium on Tilapia in Aquaculture* (pp. 270-280). Tel Aviv, Israel: Tel Aviv University.

Shirak, A., Y. Palti, A. Cnaani, A. Korol, G. Hulata, M. Ron, and R.R. Avtalion (2002). Association between loci with deleterious alleles and distorted sex ratios in an inbred line of tilapia *(Oreochromis aureus)*. *Journal of Heredity* 93(4):270-276.

Shirak, A., J. Vartin, J. Don, and R.R. Avtalion (1998). Production of viable diploid mitogynogenetic *Oreochromis aureus* using the cold shock and its optimization through definition of cleavage time. *Bamidgeh* 50(3):140-150.

Siddiqui, A.Q., and A.H. Al-Harbi (1995). Evaluation of three species of tilapia red tilapia and a hybrid tilapia as culture species in Saudi Arabia. *Aquaculture* 138:145-157.

Sifa, L., L. Chenhong, M. Dey, F. Gagalac, and R. Dunham (2002). Cold tolerance of three strains of Nile tilapia, *Oreochromis niloticus,* in China. *Aquaculture* 213(1-4):123-129.

Taniguchi, N., J.M. Macaranas, and R.S.V. Pullin (1985). Introgressive hybridization in cultured tilapia stocks in the Philippines. *Bulletin of the Japanese Society of Scientific Fisheries* 51(8):1219-1224.

Tave, D., R.O. Smitherman, V. Jayaprakas, and D.L. Kuhlers (1990). Estimates of additive genetic effects, maternal genetic effects, individual heterosis, maternal heterosis, and egg cytoplasmic effects for growth in *Tilapia nilotica. Journal of the World Aquaculture Society* 21(4):263-270.

Tayaman, M.M., R.A. Reyes, M.J. Danting, A.M. Mendoza, Marquez, E.B., Salguet, A.C., R.C. Gonzales, T.A. Abella, and E.M. Vera-Cruz (2002). Tilapia broodstock development for saline waters in the Philippines. *Naga* 25(1):32-36.

Thompson, K.D., D.J. Penman, and M.R.I. Sarder (2002). Scotland study evaluates immune responses, disease resistance in clonal lines of Nile tilapia. *Global Aquaculture Advocate* 5(1):34-35.

Tsai, C.-L., L.-H. Wang, and L.-S. Fang (2001). Estradiol and para-Chlorophenylalanine downregulate the espression of brain aromatase and estrogen receptor alpha mRNA during the critical period of feminization in tilapia *(Oreochromis mossambicus). Neuroendocrinology* 74(5):325-334.

Tuan, P.A., G.C. Mair, D.C. Little, and J.A. Beardmore (1999). Sex determination and the feasibility of genetically male tilapia production in the Thai-Chitralada strain of *Oreochromis niloticus* (L.). *Aquaculture* 173(3-4):257-269.

Uraiwan, S., and R.W. Doyle (1986). Replicate variance and the choice of selection procedure for tilapia *(Oreochromis niloticus)* stock improvement in Thailand. *Aquaculture* 48:143-157.

Wang, L.-H., and C.-L. Tsai (2000). Effects of temperature on the deformity and sex differentiation of tilapia, *Oreochromis mossambicus. Journal of Experimental Biology* 286(5):534-537.

Wohlfarth, G.W., and G. Hulata (1981). *Applied genetics of tilapias.* ICLARM Studies and Reviews 6. International Center for Living Aquatic Resources Management. Manila.

Yashouv, A., and A. Helevy (1971). Studies on comparative growth of *Tilapia aurea* (Steindachner) and *T. vulcani* (Trewavas) in experimental ponds at Dor. *Fisheries and Fish Breeding in Israel* 6:7-21.

Chapter 5

Fingerling Production Systems

Bartholomew W. Green

INTRODUCTION

Fingerling production remains a significant bottleneck to the continued expansion of tilapia aquaculture throughout the world. Throughout Asia, the Americas, and Africa, tilapia aquaculture continues to expand and represents an important source of fish to domestic and export markets.The development of effective techniques for mass production of monosex (male) tilapia fingerlings, specifically sex-reversal technology, has been an important factor in the rapid expansion of tilapia aquaculture during the past decade. Transfer of fingerling production technologies from government hatcheries and universities to the private sector also contributed significantly to the consistent availability of fingerlings for grow-out.

The Nile tilapia *(Oreochromis niloticus),* Mozambique tilapia *(O. mossambicus),* blue tilapia *(O. aureus),* and red tilapia (*Oreochromis* sp.) are the principal cultured tilapias. Currently, the Nile tilapia is the most commonly cultured tilapia. Red tilapia originate from hybrid lines of fish of diverse origin. The Florida red tilapia, which has been introduced to Latin America, is a hybrid strain descended from the original *O. urolepis hornorum* (F) × *O. mossambicus* (M) cross (Watanabe et al. 1988). Red tilapia from Taiwan, which result from the red *O. mossambicus* (F) × *O. niloticus* (M) cross (Liao and Chang 1983), also have been introduced into Latin America. Introductions into Latin America of red tilapia from the University of Stirling, Scotland, and from Israel also have been reported, but they have not been documented in the scientific literature. These red tilapia subsequently have been crossed with *O. niloticus* and *O. aureus* to develop hybrid lines derived

The author thanks Drs. Carole Engle, Peter Perschbacher, and Jack Lee Wiles for providing comments to improve this manuscript.

doi:10.1300/5513_05

from *O. mossambicus, O. urolepis hornorum, O. niloticus,* and *O. aureus* (Carberry and Hanley 1997; Espejo 1997). Unfortunately, the domestic distributions, international transfers, and breeding history of red tilapia have gone unreported in the scientific literature.

PRODUCTION METHODOLOGIES

Tilapia grow-out ponds are stocked with either mixed-sex or monosex (male) fingerlings. Mixed-sex tilapia fingerlings continue to be stocked in ponds that receive little to no fertilizer or supplemental feed input. Intensified tilapia aquaculture, where substantial applications of fertilizer and/or feed are made, depends primarily on stocking of monosex, especially sex reversed, fingerlings into grow-out ponds. Successful sex reversal requires newly hatched fry (9 to 11 mm total length) that are presumed to have sexually undifferentiated gonads. Sex-reversal technology is discussed in detail in Chapter 6 "Hormone manipulation of sex."

Tilapia fingerlings can be produced in hapas, concrete or fiberglass tanks, and earthen ponds. The life stage to be produced—fertilized eggs, sac fry, or swim-up fry—determines the type of production system used (Table 5.1). A hapa is a net cage constructed from fine (usually 1.6 mm) mesh netting suspended in a body of water, such as a lake, pond, or tank. Mixed age/size fingerlings are produced most often in earthen ponds. Production of uniform age and size of fry for sex-reversal can be accomplished in all production systems, but fertilized eggs and sac fry are best produced in hapas or tanks. Incubation facilities are required if fertilized eggs are collected. Holding facilities in the hatchery are necessary to nurse sac fry to swim-up fry stage. Operation of a hatchery requires a greater initial investment, higher operating costs, and trained technicians. The production of fertilized eggs and sac fry requires more intensive management than the harvesting of swim-up fry from earthen ponds, but daily seed production is higher and total output can be greater.

Earthen Ponds

Production of mixed age/size fingerlings is most often accomplished in earthen ponds stocked with lower numbers of broodfish than food-fish ponds. Reproduction ponds are stocked with 150 to 2,000 kg·ha^{-1} fish biomass, equivalent to 3,000 to 10,000 broodfish per hectare (Broussard et al. 1983; Mires 1983; Little 1989; Green et al. 1994; Hulata 1997). Stocking ra-
ˑ᷉s are one to four female broodfish for each male broodfish. Formulated
ˋ (20 to 40 percent crude protein) are offered to broodfish at 1 to 2 percent

TABLE 5.1. Ranges of broodfish stocking number and biomass, female:male ratio, seed harvest management, and productivity for tilapia seed production in earthen ponds, hapas, or tanks.

Production system	Broodfish			Harvest management	Daily productivity	
	Number	Biomass	Female: Male		Seed/m^{-2}	Seed/g^{-1} female
Mixed age/sex fingerlings—earthen ponds	3,000-10,000 ha^{-1}	150-2,000 kg·ha^{-1}	1-4:1	Partial harvests initiated after 21-28 days, then repeated at 7 day intervals	1.5-4.5	—
Fry for sex reversal—partial harvests—earthen ponds	3,000-40,000 ha^{-1}	2,000-12,000 kg·ha^{-1}	1-4:1	Partial harvests initiated after 10 days then repeated at 1-3 day intervals	2.5-14	0.01-0.16
Fry for sex reversal—complete harvests—earthen ponds	3,000-11,000 ha^{-1}	2,000-6,000 kg·ha^{-1}	1-4:1	Pond drain harvested after 17-20 days depending on water temperature	2.9-17	0.05-0.31
Eggs/sac-fry—hapas	1.0-12.5 m^2	0.3-1.5 kg·m^{-2}	1-4:1	Total harvest of seed after 5-21 days	35-86	0.20-0.96
Eggs/sac-fry—tanks	3-10 m^2	0.5-2.0 kg·m^{-2}	1-4:1	Total harvest of seed after 5-14 days	65-303	0.08-0.89
Eggs—hapas	2-5 m^2	0.6-1.5 kg·m^{-2}	1:1	Total harvest of seed after 5-7 days	65-274	0.20-0.96

Note: The term *tilapia seed* encompasses eggs, sac-fry, and free-swimming fry.

FIGURE 5.1. Mixed age/sex fingerlings are harvested from reproduction ponds at one to two-week intervals using a seine with a small mesh (approximately 6.35 mm^2).

TABLE 5.2. Production (mean ± SD) of hybrid tilapia *(O. niloticus × O. urolepis hornorum)* fingerlings in 0.05 ha earthen ponds stocked with 5,000 broodfish per hectare (four females to one male) in Honduras.

Harvest frequency (days)	Cumulative production (Number per 0.05 ha)	Daily production (Number $m^{-2} \cdot day^{-1}$)	Mean weight (g per fingerling)	Duration (days)
7	104,100 ± 17,644	2.2	0.9 ± 0.1	95
25	49,000 ± 9,037	0.9	1.5 ± 0.2	114

Note: Ponds were subjected to partial harvests at 7 or 25 day intervals beginning 25 or 34 days, respectively, after stocking.

of fish biomass, 5 to 7 days·$week^{-1}$, most commonly as floating extruded feed (Green and Teichert-Coddington 1993; Hulata 1997; Dan and Little 2000; Green and Engle 2000; Little et al. 2000). Partial harvests of reproduction ponds are conducted at weekly to monthly intervals using a seine with 1.0 to 6.2 mm^2 mesh (Figure 5.1). Greater frequency of partial harvests generally

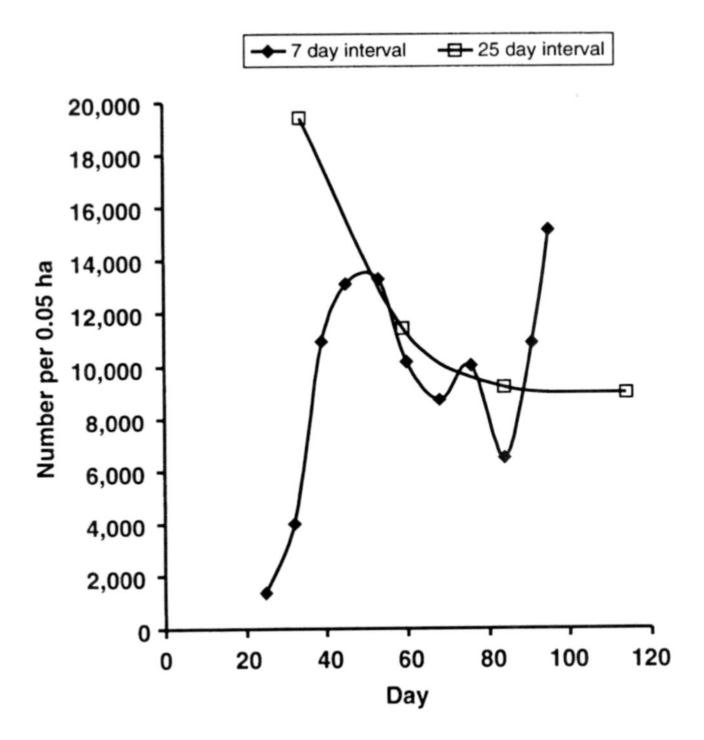

FIGURE 5.2. Mean number of tilapia (*O. niloticus* × *O. urolepis hornorum*) fry captured by seine net during partial harvests of 0.05 ha reproduction ponds conducted at 7- or 25-day intervals in Honduras.

increases fingerling productivity: more frequent partial harvests serve to maintain a more uniform population size structure, which reduces the occurrence of cannibalism of smaller fry by larger fingerlings. Table 5.2 shows that greater numbers of hybrid tilapia *(O. niloticus* × *O. urolepis hornorum)* fingerlings were harvested with weekly partial harvests than with monthly partial harvests. The total number of fingerlings captured during subsequent monthly partial harvests decreased following the initial partial harvest (Figure 5.2). In a study by Holl (1983), the greatest production of 1 to 2 g Nile tilapia fingerlings in Ivory Coast occurred with fortnightly harvests of reproduction ponds. Partial harvests of Nile tilapia reproduction ponds at 30 day intervals in the Philippines yielded 0.26 fingerlings per square meter per day (Broussard et al. 1983), compared with 3.0, 4.0, or 4.5 fingerlings per square meter per day for partial harvests at 27-, 14-, or 7-day intervals, respectively,

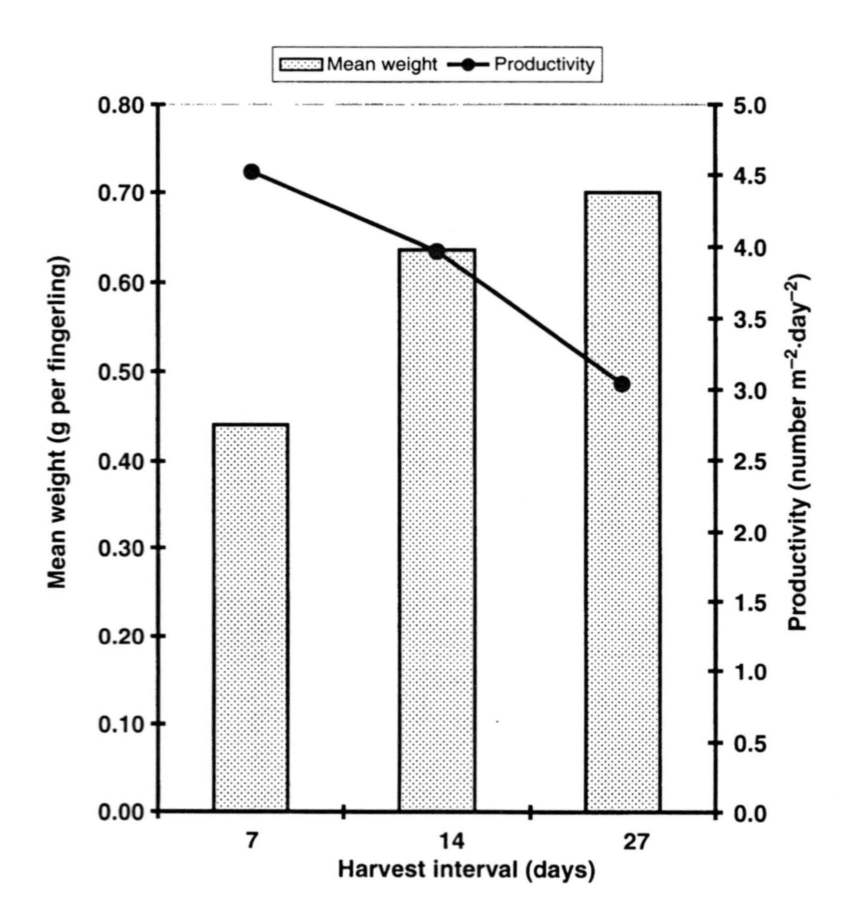

FIGURE 5.3. Mean fingerling weight and productivity of *O. niloticus* × *O. urolepis hornorum* in 0.05 ha earthen reproduction ponds in Honduras subjected to partial harvests with a seine net at 7-, 14-, or 27-day intervals.

in Honduras (Green and Alvarenga, unpublished data; Figure 5.3). Mixed age/size fingerlings harvested from reproduction ponds are stocked directly into grow-out ponds or are nursed to a size, usually 20 to 25 g that permits manual separation ("sexing") of males and females based on external genital morphology, whereupon the nursed male fingerlings are stocked into grow-out ponds.

FIGURE 5.4. Tilapia swim-up fry are collected along the edge of the reproduction pond using a short, fine-mesh (approximately 1.0 mm^2) seine. Fry are harvested one to three times per week.

Extraction of fry for sex reversal can be achieved by frequent collection of swim-up fry along pond edges with dip nets or short seine net (Guerrero 1986; Little 1989; Verdegem and McGinty 1989; Hulata 1997; Figure 5.4) or by complete drain harvest every 16 to 21 days (Rothbard et al. 1983; Popma and Green 1990; Green and Teichert-Coddington 1993; Hulata 1997). When they are released by their mother, swim-up fry prefer to school in the warm, shallow waters near the pond's edge (Philippart and Ruwet 1982; Macintosh and Little 1995; Coward and Bromage 2000), where they can be captured easily with handheld dip nets or small fine-mesh seines. This technique is referred to as "edge seining." In a study by Verdegem and McGinty (1989), daily harvesting of Nile tilapia reproduction ponds by edge seining averaged 2.2 fry per square meter, of which more than 90 percent were of suitable size for sex reversal. Reproduction ponds should be stocked with well-conditioned broodfish; fry should be harvested one to three times per week by seining once or twice around the pond perimeter, and fingerlings that elude capture should be harvested every 2 weeks using a 6.4-mm mesh seine (Verdegem and McGinty 1989). In more intensively managed reproduction ponds, fry are collected from two to six times daily using a 0.06 to 0.25 m^2 frame dip net made from mosquito netting (Guerrero 1986; Little 1989). Frequent harvesting (six times a day) of fry from a 1,740 m^2 depth reproduction pond resulted in 60 percent more fry than harvesting three times per day (Little 1989). Production cycles vary from 30 to 40 days (Guerrero 1986) to almost 120 days (Little 1989).

In Honduras, edge seining for fry is the most common technique used on intensively managed tilapia farms. Two to four female broodfish are stocked per male broodfish into earthen reproduction ponds (average size 900 m^2, 1 m average depth). Partial harvests of fry occur at 1- to 2-day intervals starting 10 to 12 days after stocking and continue for 35 to 50 days, at which time the pond is drained. Harvested fry generally are graded through a 3.2-mm mesh grader before sex reversal. Daily fry yield is reported by farmers to range from 2.4 to 4.9 fry per m^2.

One limitation to harvesting fry by edge seining is that a significant number of fry escape capture and grow to prey upon subsequent generations of fry. Ponds must be drained between cycles to eliminate recruits that have eluded capture, which can represent up to 50 to 60 percent of the net yield even with frequent partial harvests (Little 1989; Verdegem and McGinty 1989). Tilapia fry production can be affected significantly by cannibalism, particularly where swim-up or advanced fry are harvested from earthen ponds. High levels of cannibalism of tilapia fry are possible during the first 10 to 30 days following the start of exogenous feeding, which occurs 6 to 7 days after hatching (Macintosh and De Silva 1984). The degree of cannibalism is inversely proportional to food availability (Macintosh and De Silva 1984; Pantastico et al. 1988) and directly proportional to the size differential within the fry population (Pantastico et al. 1988). More frequent partial harvests likely act to reduce cannibalism by narrowing the fry population size differential and increasing food availability.

Frequent complete harvests by draining of reproduction ponds further improve the efficiency of fry harvesting (Rothbard et al. 1983; Popma and Green 1990; Hulata, 1997), even though in ponds subjected to complete drain harvest up to 30 percent of total fry production is estimated to remain unharvested as fry are trapped in puddles or else eggs or sac-fry are expelled prematurely from the female's mouth (Rothbard et al. 1983). Any puddles that remain after draining should be treated with a fish toxicant before the pond is refilled for subsequent use to eliminate fish that escape harvest (Mires 1983; Popma and Green 1990). To help minimize cannibalism in each subsequent spawning cycle, the mouths of female broodfish should be

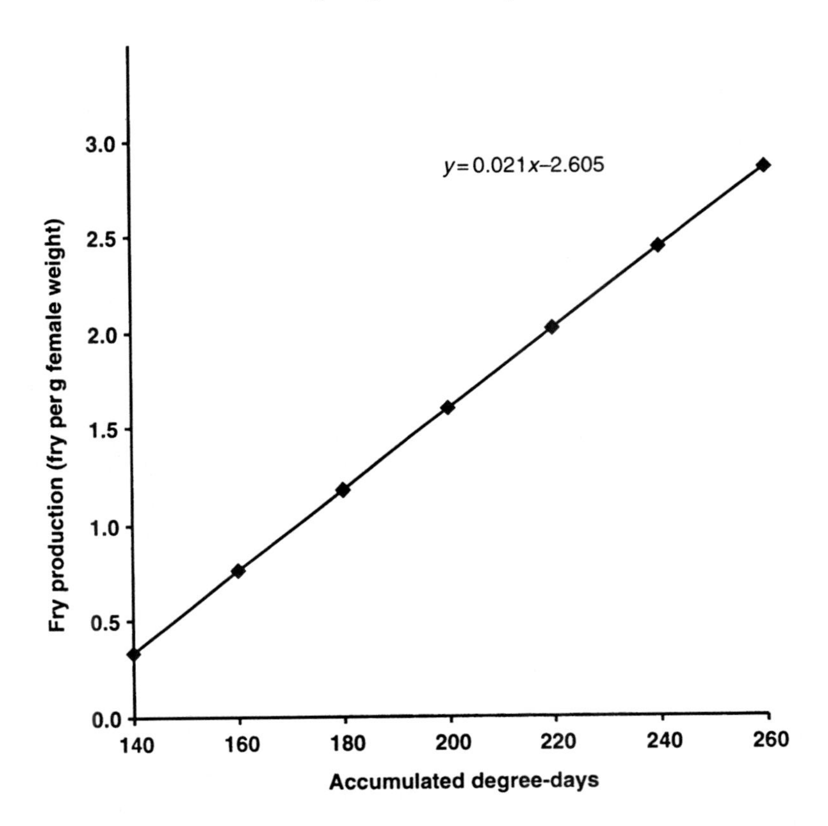

FIGURE 5.5. Regression line showing the relationship between Nile tilapia fry productivity (fry per g female broodfish) and accumulated degree-days above the 15°C threshold temperature in reproduction ponds subjected to complete drain harvests in Honduras.

inspected for eggs, and any eggs found must be removed before these fish are restocked into spawning ponds (Popma and Green 1990).

At the Gan-Shmuel Fish Farm in Israel, reproduction ponds are drained completely every 17 to 19 days, which is the sum of the assumed durations of the different steps in the tilapia reproductive process in ponds (Rothbard et al. 1983). Popma and Green (1990) based their recommendations for the duration of fry production cycles on water temperature. Degree-days, also known as the rule of thermal summation, are used to quantify the effects of temperature on biological processes (Belehradek 1930; Higley et al. 1986;

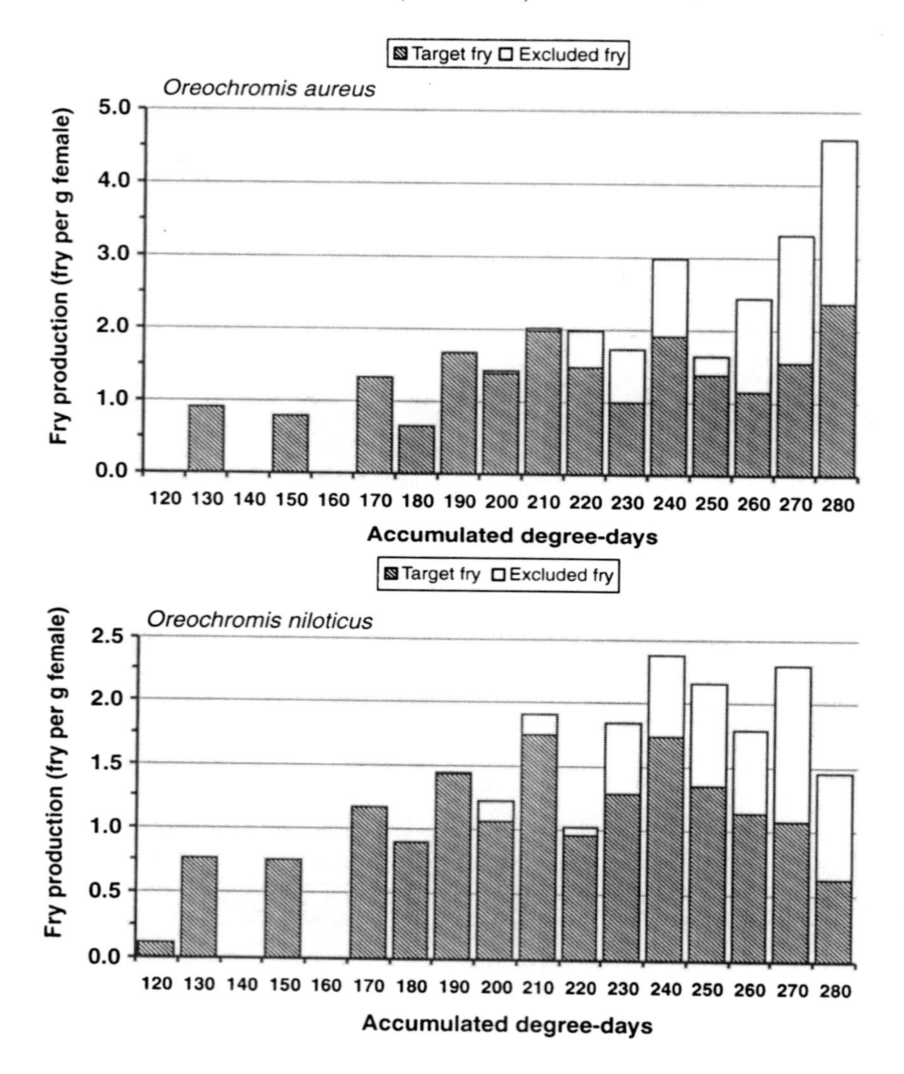

FIGURE 5.6. Mean production of blue tilapia *(top)* and Nile tilapia *(bottom)* fry in response to accumulated degree-days in reproduction ponds in Egypt subjected to complete drain harvest. Target fry are fry of total length 9 to 11 mm that are suitable for sex reversal; excluded fry are greater than 11 mm total length.

Regier et al. 1990), and the calculation was applied by Green and Teichert-Coddington (1993) to tilapia fry production.

Degree-days for tilapia reproduction are calculated as the mean daily water temperature minus the threshold temperature of 15°C, summed over the production period. The threshold temperature, also known as "biological zero," is the temperature at which development ceases because of cold; it is not necessarily 0°C (Belehradek 1930). Tilapia physical activity and feeding cease at water temperatures between 16°C and 20°C (Chervinski 1982); mean lower lethal water temperatures from 10°C to 14°C have been reported for three strains of Nile tilapia (Khater and Smitherman 1988). Thus, 15°C represents a "biological zero" for tilapia reproduction.

The results of research in Honduras showed that although drain harvest of Nile tilapia reproduction ponds did not yield fry at fewer than 140 degree-days, female broodfish were incubating eggs at the lowest degree-day value (119.6) measured (Green and Teichert-Coddington 1993). A positive linear relationship was observed between mean total Nile tilapia fry production and accumulated degree-days (Figure 5.5). Optimal production of Nile tilapia fry suitable for sex reversal occurred at between 195 and 220 cumulative degree-days (Green and Teichert-Coddington 1993). At fewer than 195 degree-days, fry production still was increasing, but fry harvest was suboptimal. At more than 220 degree-days, a linear increase was observed in the percentage of the fry population too large (=13 mm total length) for sex reversal. Similar results were obtained for blue and Nile tilapia fry production for the subtropical limited reproduction season environment in northern Egypt (Green et al. 1995; Figure 5.6). Complete drain-harvesting of ponds to obtain fry suitable for sex reversal should, therefore, generally occur 14 to 20 days after stocking, with the shorter duration corresponding to the period of warmer water temperatures.

Hapas

Hapas (2 to 120 m^2 each) suspended in tanks, ponds, lakes or coastal waters allow for easy management of broodfish (Figure 5.7). Hapas are particularly suited for tilapia reproduction because they require small initial investment and are easy to harvest, and fry mortality during harvest generally is low. However, water quality can deteriorate within hapas if the fine mesh becomes clogged with periphyton (Lovshin and Ibrahim 1988; Little et al. 2000). In a study by Little et al. (2000) costocking of periphyton-consuming fish in the hapa with tilapia failed to improve water quality and led to a significant reduction in tilapia fry productivity. The most effective practical method of controlling biofouling is to exchange hapas at 5 to 10 day

FIGURE 5.7. Tilapia eggs and sac-fry are harvested from hapas at 5 to 10 day intervals. Female broodfish are conditioned at high density in the smaller hapas adjacent to the larger hapa used for seed production. Duplicate or triplicate groups of female broodfish are rotated between the conditioning and reproduction hapas. Note the conditioning hapa being allowed to sun-dry to reduce biofouling.

intervals with hapas that have been cleaned and air-dried (Little and Hulata 2000). The use of hapas for tilapia reproduction also requires incubation facilities for harvested eggs.

The broodfish stocking rate in hapas averages almost 4 fish per square meter (range 1 to 12.5) (Guerrero and Garcia 1983; Lovshin and Ibrahim 1988; Little 1989; Behrends et al. 1993). In a study by Bautista et al. (1988) mean daily production of Nile tilapia seed was similar for stocking rates of 4, 7, and 10 females per square meter. Anywhere from one to ten female broodfish are stocked per male broodfish, the most common broodfish sex ratios are one to four females per male (Guerrero and Garcia 1983; Bautista et al. 1988; Lovshin and Ibrahim 1988; Little 1989; Behrends et al. 1993). The sex ratio of tilapia broodfish stocked into hapas does not appear to affect daily seed production; stocking three, four, five, seven, or ten Nile tilapia female broodfish per male yielded similar daily seed production (Guerrero and Garcia 1983; Bautista et al. 1988). Broodfish are fed daily with a 30 to 40 percent protein floating extruded feed offered at 1 to 2 percent of broodfish biomass (Bhujel 2000; Dan and Little 2000; Little et al. 2000). Feeding broodfish more than once or twice daily does not appear to improve fry production (Bhujel 2000).

Tilapia seed harvested from intensively managed hapas ranges from newly fertilized eggs to sac-fry to actively feeding swim-up fry (Behrends et al. 1993). Management of the reproduction cycle of tilapia in hapas is varied to obtain the desired developmental stage. Hapas are harvested at intervals of between from 1 and 32 days. Swim-up fry younger than 1 week old are harvested from less intensively harvested, less productive systems. Older fry are more difficult to capture with dip nets because they congregate near the bottom of the hapa. Daily yields of tilapia seed, comprised of fertilized eggs through swim-up fry, vary from 2 to 86 seeds per square meter, but generally range from 35 to 86 seeds per square meter for harvest intervals of 5 to 21 days (Guerrero and Garcia 1983; Lovshin and Ibrahim 1988; Little 1989; Behrends et al. 1993). Daily tilapia seed yields can be increased further by harvesting only fertilized eggs. When eggs are collected from the mouths of brooding females every 5 to 7 days, daily productivity increases to 65 to 250 seeds per square meter (Little et al. 1993).

Single-net hapas generally are used for tilapia reproduction, although a double-net hapa has been used in the Philippines and Indonesia, and is recommended for lacustrine aquaculture (Guerrero and Garcia 1983; Costa-Pierce and Hadikusumah 1995). The double-net hapa system involves a smaller, 30 mm mesh hapa suspended inside of a larger, 2 mm mesh hapa. Broodfish are confined to the inner hapa, whereas swim-up fry are able to pass through to the outer hapa. Daily seed production ranges from 6.8 to 12.2 seeds per square meter (Guerrero and Garcia 1983; Costa-Pierce and

Hadikusumah 1995). However, costs are higher for the double hapa system (Little and Hulata 2000).

Tanks

Tanks constructed from fiberglass, concrete, or wood (fitted with a plastic liner) that range from 4 to 100 m^2 and up to 1.5 m deep are used for tilapia seed production (Figure 5.8). As with hapas, tilapia seed that ranges from fertilized eggs to swim-up fry can be produced in tanks. Broodfish stocking rates vary from 3 to 10 fish per square meter, and broodfish sex ratios vary from one female to one male up to ten females to one male

FIGURE 5.8. Tilapia can be reproduced in concrete or fiberglass tanks. Tanks can be in-ground (as shown here) or aboveground.

(Guerrero and Guerrero 1985; Bautista et al. 1988; Little et al. 1993). However, the most common sex ratios used are one to four female broodfish per male. Stocking 4, 7, or 10 females per square meter or increasing the broodfish sex ratio from four to ten females per male does not significantly increase daily production of Nile tilapia seed (Bautista et al. 1988). Broodfish are fed floating extruded feed (30 to 45 percent crude protein) at 1 to 2 percent of fish biomass once or twice daily (Bhujel 2000; Green and Engle 2000; Ridha and Cruz 2000). Daily seed productivity in tanks increases as earlier life stages are harvested, from 8 to 16 seeds per square meter where swim-up fry are harvested (Guerrero and Guerrero 1985), to 42 to 84 seeds per square meter where sac-fry and swim-up fry are harvested (Little 1989), to 239 seeds per square meter where eggs and sac-fry are harvested (Little 1989).

INTERSPAWNING INTERVAL

The removal of eggs and sac-fry from mouth-brooding females shortens the interspawning interval and increases the spawning frequency (Verdegem and McGinty 1987; Little 1989). The removal of eggs and sac-fry at 5 to 10 day intervals significantly increases seed production compared with natural egg incubation (Little et al. 1993) and compared with egg and sac-fry removal at 2 to 4 day intervals (Verdegem and McGinty 1987). The interspawning interval also is shortened through broodfish exchange. For example, each time eggs and sac-fry are collected from mouth-brooding females, either spent females or all females are exchanged for broodfish that have been conditioned for 10 days (Little et al. 1993). During conditioning, only female broodfish are stocked, at high density, and they are fed a high-quality formulated feed for 5 to 10 days (Little 1989). Broodfish exchange and conditioning also increases spawning synchronicity (Lovshin and Ibrahim 1988; Little 1989), possibly through disruption of established social hierarchies and maintenance of elevated concentrations of the gonadal steroids 17β-estradiol and testosterone (Little et al. 1993, 2000).

Daily Nile tilapia seed production for a 10-day harvest interval increases from 31 seeds per kilogram female weight where only swim-up fry are removed to 106 seeds per kilogram female weight where eggs are removed from mouth-brooding females, to 160 seeds per kilogram female weight where seeds are removed and spawned females are exchanged, to 274 seeds per kilogram female weight where seeds are removed and all females are exchanged (Little et al. 1993). Seed harvested at 10 day intervals from all treatments included swim-up fry and fertilized eggs. When the harvest interval was reduced to 5 days, daily productivity (seeds per kilogram female weight) for all treatments, except those in which only swim-up fry were harvested, increased by 17 to 163 percent, and this increase was composed only of fertilized eggs (Little et al. 1993). Significantly greater numbers and eggs per square meter were harvested when all females were exchanged with conditioned females at 3.5 or 7 day intervals compared with no female exchange (Little et al. 2000). Conditioning periods of 10 to 21 days have been reported for Nile tilapia (Lovshin and Ibrahim 1988; Little 1989), but Little (1989) reported lower fecundity (seeds per kilogram female weight per day) for a 20 day compared with a 10 day conditioning period. Conditioning females for 7 days and exchanging females every 7 days may be more economical than a 10 day conditioning period and 5 day harvest interval (Little et al. 2000).

INCUBATION

Eggs harvested from broodfish in hapas or tanks must be incubated in hatchery facilities. Tilapia eggs are negatively buoyant and, in the absence of a current to suspend them in the water column, sink quickly and clump (Rana 1986; Macintosh and Little 1995). A variety of incubation systems have been used successfully for tilapia eggs. When up-welling Zuger-type hatching jars have been used for tilapia eggs, incubation success has been variable because of incubation-induced mechanical injury to eggs and subsequent secondary infection (Rothbard and Hulata 1980; Rana 1986, 1988). Down-welling round-bottom incubators result in 17 to 22 percent greater hatchability and in improved overall survival (85 percent) from egg to 10-day-old swim-up fry compared with overall survival (60 percent) for conical up-welling incubators (Rana 1986, 1988). Down-welling round-bottom incubators come in a variety of sizes and are relatively simple to construct from readily available materials: 2 and 3 L soft drink bottles make good incubators, as do 20 L plastic potable water bottles and MacDonald jars (Little 1989; Macintosh and Little 1995). Egg loading rates range from 650 to 1,350 eggs per liter in small incubators up to 4,000 eggs per liter in large incubators (Little 1989; Macintosh and Little 1995). Water flow rates are 1 $L \cdot min^{-1}$ in small incubators and 1 $L \cdot s^{-1}$ per 10,000 eggs in large incubators (Little 1989; Macintosh and Little 1995). Approximately 500,000 eggs (2.2 kg) can be stocked in down-welling round-bottom incubators (30 cm high, 15 to 20 cm diameter) supplied with 8 to 10 $L \cdot min^{-1}$ of slow sand-filtered water (D. Little, personal communication).

Sac-fry harvested from broodfish also must be held in the hatchery until yolk sac absorption is complete. Sac-fry range in size from 5 to 6 mm total length and can pass through the openings in the 1.6 mm mesh normally used for sex-reversal hapas. Although sac-fry can be reared in fine-mesh hapas in ponds, biofouling of the hapa mesh can lead to water quality problems. Therefore, it is recommended that sac-fry be reared in the hatchery until they attain a total length of 6 to 7 mm. In Thailand, sac-fry are reared in 22.9 × 30.5 cm cake pans suspended in concrete tanks (D. Little, personal communication). A number of 1.25 cm diameter holes for water exchange are drilled along the sides of the pan and covered with mosquito netting. Sac-fry are stocked into pans at 15,000 per pan and each pan is supplied with a water flow of 3 $L{\cdot}min^{-1}$ (D. Little, personal communication).

Maintenance of good water quality is critical to successful hatchery operation. A source of clean, good quality water is necessary for hatcheries using single-pass water, or a properly functioning biofilter is required to maintain low total ammonia nitrogen and nitrite concentrations where water is recirculated (Piper et al. 1982). Slow sand filtration or ultraviolet radiation can be used to reduce bacterial loads in water for incubators (Piper et al. 1982). Concentrations of dissolved oxygen should be maintained as close to 100 percent saturation as possible. Because of their relatively impermeable chorion, the hatchability of tilapia eggs is unaffected by 101 $mg{\cdot}L^{-1}$ un-ionized ammonia or 140 $mg{\cdot}L^{-1}$ total nitrate (Rana 1988). However, at 28°C, for 7- to 10-day-old tilapia fry have the LC_{50} for un-ionized ammonia (or lethal concentration that results in 50 percent mortality) is 0.4 $mg{\cdot}L^{-1}$ (Rana 1988).

FRY REARING

Tilapia swim-up fry must be reared to 1 to 2 g average weight before being stocked into nursery ponds. Generally, fry rearing occurs simultaneously with sex reversal. As with reproduction, tilapia fry can be reared in hapas (2 to 5 m^2 surface area) suspended in fertile earthen ponds (Popma and Green 1990; Green and Teichert-Coddington 1991; Macintosh and Little 1995; Carberry and Hanley 1997), in hapas suspended in concrete tanks (Argue and Phelps, 1996), stocked free in tanks (5 to 25 m^2 surface area) (Rothbard et al. 1983; Carberry and Hanley 1997), or stocked free in earthen ponds (300 to 1,500 m^2) (Phelps et al. 1995; Berman 1997). Stocking rates vary from 2,000 to 16,000 fry per square meter in hapas suspended in either ponds or tanks, from 150 to 750 fry per square meter free in tanks, from 75 to 260 fry per cubic meter free in ponds, from 6,000 to 12,000 fry per cubic meter in recirculating tank systems, and from 8,000 to 18,000 fry per square meter in tanks with continuous water exchange.

The fry-rearing phase generally lasts 28 to 30 days. Fry growth is a function of water temperature, stocking rate, feeding rate, and quality of feed. Growth of fry is exponential, and growth functions have been estimated from specific growth rate data from various sources. Data on Nile tilapia growth from a controlled laboratory study reported by Dambo and Rana (1992) are considered to be very close to maximum growth because of ideal water temperature (30°C), high quality feed (54 percent crude protein), and high feeding rate (fish fed to satiation three to four times daily). Fry were stocked at the equivalent of 2,000 to 20,000 per square meter in 2 L aquaria in a warm-water recirculating system. After the 26-day study, fish averaged 1.0 g each for the highest stocking rate to 1.7g each for the lowest stocking rate (Dambo and Rana 1992; Figure 5.9). In another laboratory study, hybrid tilapia (broodfish were a mix of *O. mossambicus* and *O. urolepis hornorum*) were stocked at rates equivalent to 20,000 to 200,000 per square meter and grown for 56 days under conditions of optimal water quality (Gall and Bakar 1999). The stocking rate did not significantly affect mean final weights, which ranged from 2.25 to 2.62 g per fish (Gall and Bakar 1999). However, the authors reported that mean dissolved oxygen concentration in rearing units decreased from 6.7 to 3.2 $mg \cdot L^{-1}$ as stocking rate increased.

Tilapia farmers occasionally report a final weight of 1.0 g per fish upon completion of the fry-rearing phase. More often fry attain a mean final weight of up to 0.5 g per fish after 30 days of fry rearing (Figure 5.10). Nile tilapia fry grew to mean weights of 0.1 to 0.3 g per fish after 30 days of fry rearing in hapas in ponds stocked at 3,000 to 10,000 fry per square meter (Popma and Green 1990; Green and Teichert-Coddington 1991). Differences in environmental conditions and rearing unit management (stocking rate, feed quality, etc.) are responsible for the variation in growth during the fry-rearing phase.

Maximum feed consumption by tilapia fry varies positively with temperature and declines with increasing fish size throughout the 28 day rearing phase at each specific temperature (J. Newman, personal communication; Table 5.3). Recommended feeding rates for tilapia fry at 28°C to 30°C are 38 percent of biomass per day for the first week; 21 percent of biomass per day for the second week; 15 percent of biomass per day for the third week; and 10 percent of biomass per day for the fourth week (J. Newman, personal communication). Feed protein content for intensive fry rearing generally varies from 25 to 45 percent crude protein, but fry have been reared successfully using a 20 percent crude protein feed (Popma and Green 1990). The daily ration should be divided into at least four meals evenly spaced throughout daylight hours (Popma and Green 1990).

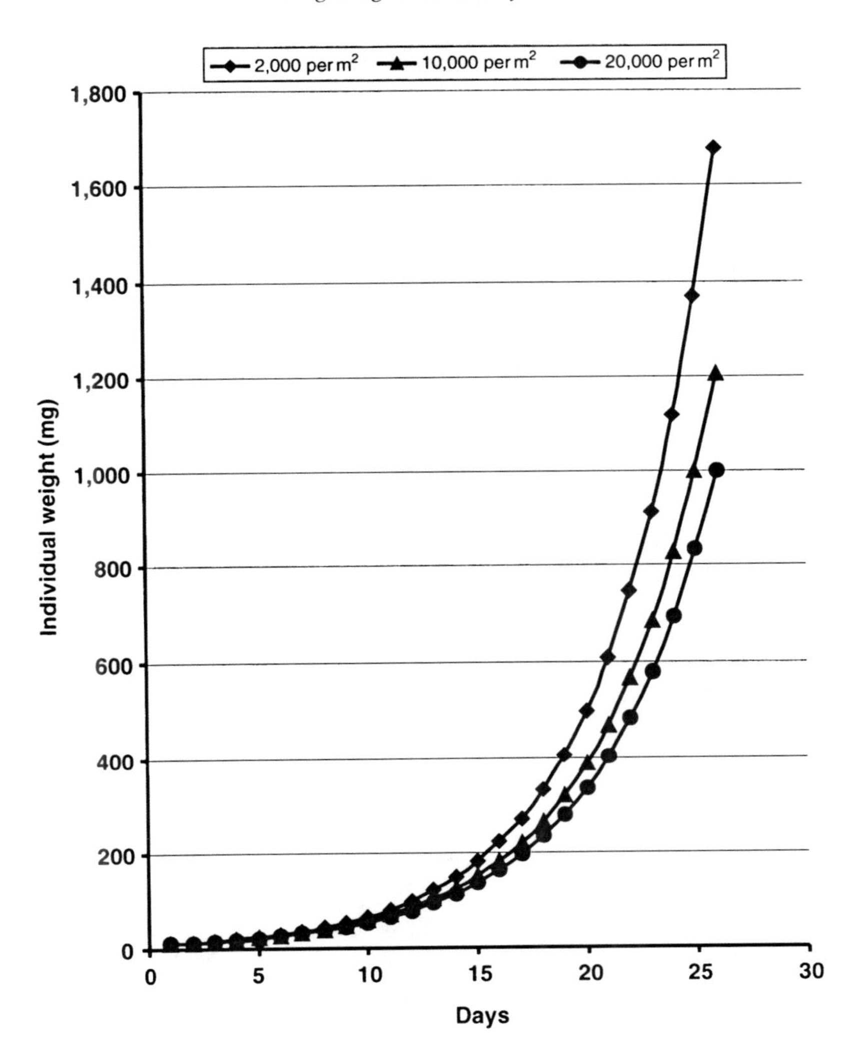

FIGURE 5.9. Growth of Nile tilapia fry at three stocking rates under ideal culture conditions. Equations for growth curves are $wt_t = 8.5849e^{0.203t}$ (2,000 fry·m^{-2}); $wt_t = 8.7408e^{0.190t}$ (10,000 fry·m^{-2}); and $wt_t = 8.8022e^{0.182t}$ (20,000 fry·m^{-2}), where wt = individual weight (mg fry) and t = time (days). Not shown are growth curves for stocking rates of 5,000·m^{-2} ($wt_t = 8.7233e^{0.191t}$) and 15,000·m^{-2} ($wt_t = 8.8022e^{(0.182t)}$). *Source:* Figure redrawn from data in Dambo and Rana (1992).

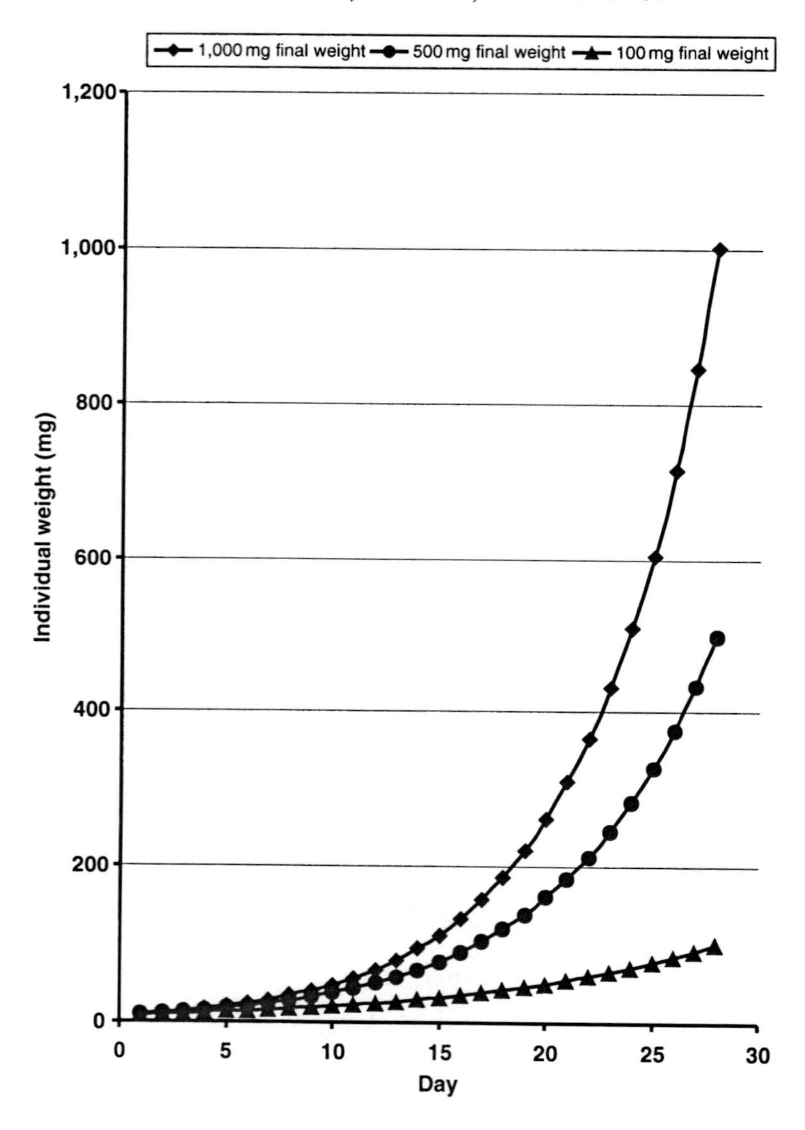

FIGURE 5.10. Range of Nile tilapia fry growth during the fry-rearing phase under farm conditions. Variations in environmental conditions and rearing unit management are responsible for variation in growth curves. Equations for growth curves are: $wt_t = 8.9174e^{0.169t}$ (1,000 mg final weight), $wt_t = 9.1523e^{0.143t}$ (500 mg final weight), $wt_t = 8.5165e^{0.088t}$ (100 mg final weight), where wt = individual weight (mg per fry) and t = time (days).

TABLE 5.3. Daily feeding rates for tilapia fry as a percentage of tilapia biomass during the four-week fry-rearing phase at two water temperature regimes.

	Daily feeding rate	
Week	**21°C ± 1°C**	**29°C ± 1°C**
1	17-20	36-40
2	12-14	20-23
3	9	15
4	7	10

HATCHERY-PHASE FISH HEALTH MANAGEMENT

Although tilapia are generally considered to be hardy fish, they can be infected by a variety of pathogenic organisms at any point during the production process. Commonly reported problems during the hatchery phase of tilapia production include protozoan parasites such as *Tricodina* sp., *Epistylis* sp., *Ichthyobodo* sp., and occasionally *Ichthyophthirius multifiliis* (Carberry and Hanley 1997; Little et al. 1997). Outbreaks of protozoan parasites are more common at low water temperatures (<25°C). Protozoan parasites affect the skin and the gills of tilapia, and heavy infestations cause abundant secretions of mucous. Often, fish heavily infested with protozoan parasites are susceptible to secondary bacterial infections. Treatments for protozoan parasites include two to three applications at 3-day intervals each of formalin at 25 to 50 $mL{\cdot}m^{-3}$ as an indefinite treatment, formalin at 200 $mL{\cdot}m^{-3}$ as a 7 min dip, or salt at 3 $g{\cdot}L^{-1}$ as an indefinite treatment or at 10 to 15 $g{\cdot}L^{-1}$ as a 20 min bath (Conroy and Herman 1970; Tonguthai and Chinabut 1997). In Thailand, it has been reported that newly hatched tilapia fry tolerate indefinite formalin treatments up to a maximum concentration of 15 $mL{\cdot}m^{-3}$, but at this level formalin is ineffective against *Tricodina* sp. (D. Little, personal communication). A more effective treatment is to dip fry into a concentrated formalin bath (200 $mL{\cdot}m^{-3}$) for no more than 7 min (D. Little, personal communication) or to use salt (3 $g{\cdot}L^{-1}$) as an indefinite treatment (Little et al. 1997).

Bacterial infections are caused by *Aeromonas hydrophila, Flexibacter columnaris, Pseudomonas* sp., and *Streptococcus* sp. Bacterial infections are best treated through the use of medicated feed. The recommended treatment for *Aeromonas* sp. and *Pseudomonas* sp. infections is terramycin incorporated into the feed at 4.4 g active drug per kilogram of feed. Fish should be fed medicated feed ad libitum for 10 to 14 days (J. Plumb, personal communication). Outbreaks of *Streptococcus* sp. can be treated with terramycin or erythromycin incorporated into the feed at 4.4 g active drug per kilogram of feed and fed to satiation for 10 to 14 days (J. Plumb, personal communication). Treatment of *F. columnaris* is with potassium permanganate at 2 to 4 $g \cdot m^{-3}$ as an indefinite treatment or 10 $g \cdot m^{-3}$ as a 1-h bath. See Chapter 16 "Parasites and diseases" for a thorough discussion.

FINGERLING NURSERY REARING

Tilapia have attained an average weight of 0.1 to 1.0 g per fish at the end of the fry-rearing phase or at partial harvest of mixed age/size fingerlings from earthen reproduction ponds. Small tilapia fingerlings generally are reared to an advanced fingerling size (10 to 100 g per fish) in nursery ponds before being stocked into grow-out ponds. Direct stocking of these small fingerlings into ponds for grow-out is an inefficient use of pond facilities because of the low fish biomasses during the early stage of production. Nursery rearing can involve more than one phase. Where multiple phases are used, the fish population is thinned periodically to reduce the standing stock and thus maintain rapid fingerling growth. Management of nursery ponds is varied and often farm specific.

Fingerling growth during the nursery phase is affected by food availability and quality. Fingerling standing crop varies with nursery pond management. Snow et al. (1983) reported standing crops of 511 to 605 $kg \cdot ha^{-1}$ for chemically fertilized ponds, 2,873 $kg \cdot ha^{-1}$ for ponds with supplemental feeding only, and 5,450 $kg \cdot ha^{-1}$ in nursery ponds managed with fertilization and supplemental feeding. Growth of fingerlings to a standing crop of 800 to 1,200 $kg \cdot ha^{-1}$ can be attained with the natural food that results from organic fertilization, but supplemental feed is needed to maintain fast growth of fingerlings to higher standing crops (Green 1992). Fingerling standing crops can vary from 7,000 to 14,000 $kg \cdot ha^{-1}$ in intensively managed ponds (Sarig and Marek 1974; Perez 1995; Green and Engle 2000; Figure 5.11). Supplemental feeds offered during nursing vary from 24 to 54 percent protein (Sarig and Marek 1974; Snow et al. 1983; Dambo and Rana 1992). Low-protein diets often are used to supplement natural food in nursery ponds with low standing crop. With a high standing crop or when natural food is limited, higher-protein diets were used.

Nursery pond stocking rates vary from 5 to 90 fingerlings per square meter (Sarig and Marek 1974; Broussard et al. 1983; Perez 1995; Green and Engle 2000). Ponds are harvested after 21 to 97 days, at which time fingerling average weight ranges from 10 to 100 g per fish (Figure 5.12).

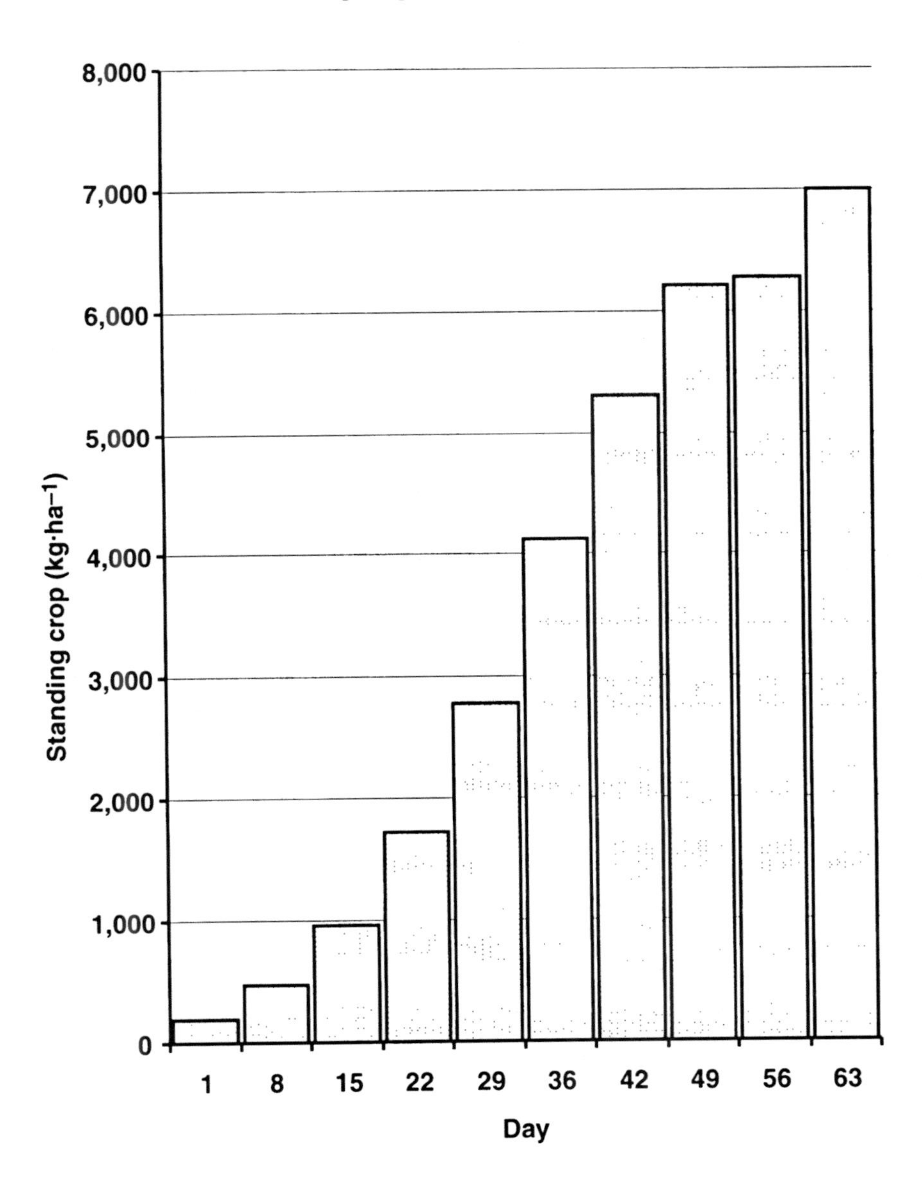

FIGURE 5.11. Increase in Nile tilapia standing crop during 63 days of nursery rearing in a brackish water pond on an Ecuadorian tilapia farm. Fingerlings are stocked at 34 fish·m^{-2}.

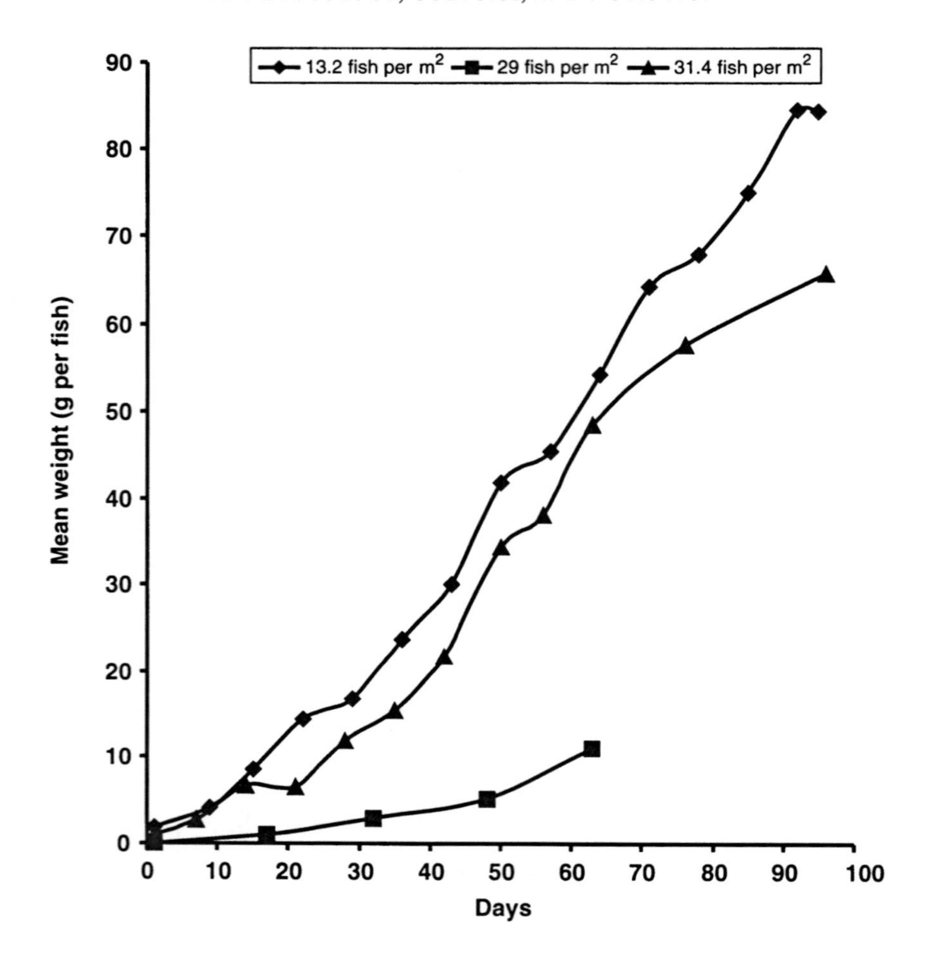

FIGURE 5.12. Growth of Nile tilapia fingerlings during nursery rearing on tilapia farms in Honduras. Fish are stocked at 13.2 to 31.4 fish·m^{-2}. *Source:* Redrawn from data in Green and Engle (2000).

Survival during the nursery phase varies from 60 to 95 percent, averaging 75 percent. Tilapia fingerlings grow between 0.05 and 0.73 g per day during the nursery phase (Figure 5.13). In Costa Rica, small fingerlings (approximately 1 g) from sex-reversal treatments are stocked at 110 fish per square meter for growth to 20 g, followed by a reduction in stocking rate to 28 fingerlings per square meter for growth from 20 to 50 g (Perez 1995). In

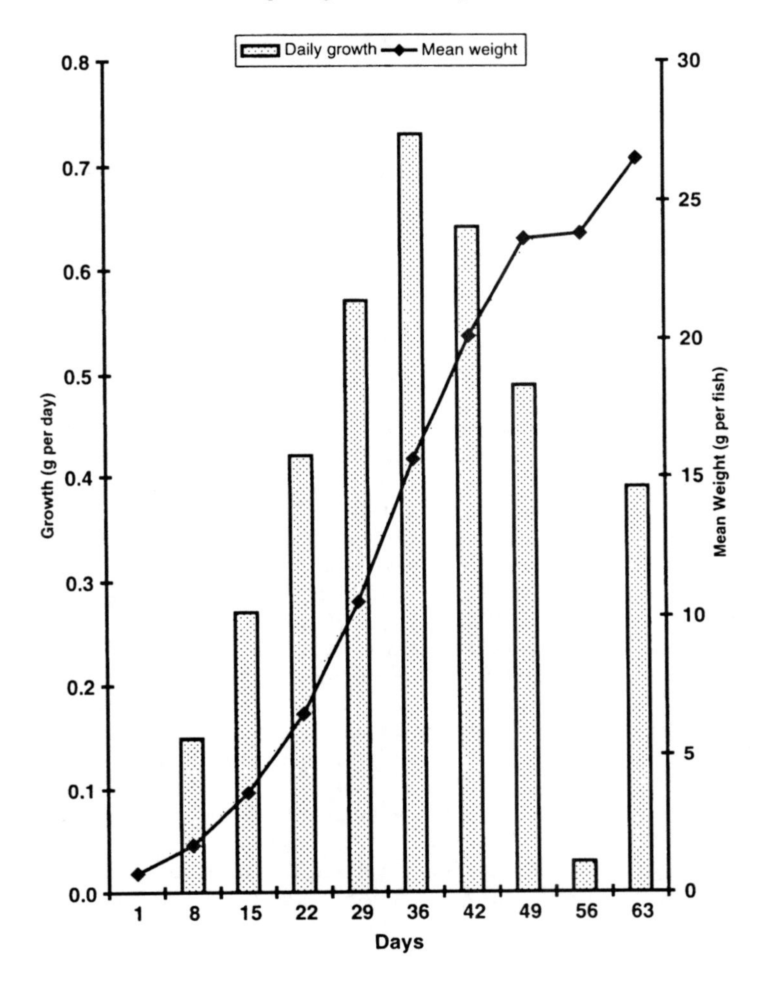

FIGURE 5.13. Mean daily growth and increase in mean weight of Nile tilapia at weekly intervals during 63 days of nursery rearing in a brackish water pond in Ecuador. Fingerling tilapia are stocked at 34 fish·m^{-2}.

Honduras, fingerlings are reared in earthen ponds (mean size 0.27 ha) from 0.5 to 80 g per fish in an average of 67 days (Green and Engle 2000). Fingerlings are stocked into ponds at a mean of 25.6 fish per square meter and survival averages 71 percent. Fish are fed a 32 percent protein commercially formulated extruded ration offered either as a declining percentage of biomass (8 to 15 percent·day^{-1} decreasing to 4 to 9 percent·day^{-1}) or at a constant rate (5 to 10 percent·day^{-1}). The standing crop at harvest ranges from 900 to 11,680 kg·ha^{-1} and averages nearly 5,200 kg·ha^{-1}.

Fingerlings that complete the nursery phase are transferred to grow-out ponds or cages for one to three phases of grow-out to market size. Specifics of grow-out systems in ponds are presented in Chapter 7 "Pond production."

REFERENCES

Argue, B. J., and R. P. Phelps (1996). Evaluation of techniques for producing hormone sex-reversed *Oreochromis niloticus* fry. *Journal of Aquaculture in the Tropics* 11:153-159.

Bautista, A. M., M. H. Carlos, and A. I. San Antonio (1988). Hatchery production of *Oreochromis niloticus* L. at different sex ratios and stocking densities. *Aquaculture* 73: 85-95.

Behrends, L. L., J. B. Kingsley, and A. H. Price III (1993). Hatchery production of blue tilapia, *Oreochromis aureus* (Steindachner), in small suspended net hapas. *Aquaculture and Fisheries Management* 24: 237-243.

Belehradek, J. (1930). Temperature coefficients in biology. *Biological Review of the Cambridge Philosophical Society* 5: 30-58.

Berman, Y. (1997). Producción intensiva de tilapia en agua fluyente. In D. E. Alston, B. W. Green, and H. C. Clifford (Eds.), *IV Symposium on Aquaculture in Central America: Focusing on Shrimp and Tilapia* (pp. 59-63). April 22-24. Tegucigalpa, Honduras: Asociacion Nacional de Acuacultores de Honduras and the Latin American Chapter of the World Aquaculture Society.

Broussard, M. C., R. Reyes, and F. Raguindin (1983). Evaluation of hatchery management schemes for large scale production of *Oreochromis niloticus* fingerlings in Central Luzon, Philippines. In L. Fishelson, and Z. Yaron (compilers), *Proceedings: International Symposium on Tilapia in Aquaculture* (pp. 414-424). Nazareth, Israel, May 8-13. Tel Aviv, Israel: Tel Aviv University.

Bhujel, R. C. (2000). A review of strategies for the management of Nile tilapia *(Oreochromis niloticus)* broodfish in seed production systems, especially hapa-based systems. *Aquaculture* 181: 37-59.

Carberry, J., and F. Hanley (1997). Commercial intensive tilapia culture in Jamaica. In D. E. Alston, B. W. Green, and H. C. Clifford (Eds.), *IV Symposium on Aquaculture in Central America: Focusing on shrimp and tilapia* (pp. 64-67). April 22-24. Tegucigalpa, Honduras: Asociacion Nacional de Acuacultores de Honduras and the Latin American Chapter of the World Aquaculture Society.

Chervinski, J. (1982). Environmental physiology of tilapias. In R. S. V. Pullin, and R. H. Lowe-McConnell (Eds.), *The biology and culture of tilapias* (pp. 119-128). Manila, Philippines: International Center for Living Aquatic Resources Management.

Conroy, D. A., and R. L. Herman (1970). *Textbook of fish diseases.* Neptune City, NJ: T. F. H. Publications.

Costa-Pierce, B. A., and H. Hadikusumah (1995). Production management of double-net tilapia *Oreochromis* spp. hatcheries in a eutrophic tropical reservoir. *Journal of the World Aquaculture Society* 26: 453-459.

Coward, K., and N. R. Bromage (2000). Reproductive physiology of female tilapia broodstock. *Reviews in Fish Biology and Fisheries* 10: 1-25.

Dambo, W. B., and K. J. Rana (1992). Effect of stocking density on growth and survival of *Oreochromis niloticus* (L.) fry in the hatchery. *Aquaculture and Fisheries Management*, 23: 71-80.

Dan, N.C., and D.C. Little (2000). Overwintering performance of Nile tilapia *Oreochromis niloticus* (L.) broodfish and seed at ambient temperatures in northern Vietnam. *Aquaculture Research* 31:485-493.

Espejo, C. (1997). La piscicultura en Colombia: Tecnología de punta en el Departamento del Valle del Cauca. In D. E. Alston, B. W. Green, and H. C. Clifford (Eds.), *IV Symposium on Aquaculture in Central America: Focusing on shrimp and tilapia* (pp. 78-84). April 22-24. Tegucigalpa, Honduras: Asociacion Nacional de Acuacultores de Honduras and the Latin American Chapter of the World Aquaculture Society.

Gall, G. A. E., and Y. Bakar (1999). Stocking density and tank size in the design of breed improvement programs for body size of tilapia. *Aquaculture* 173:197-205.

Green, B. W. (1992). Substitution of organic matter for pelleted feed in tilapia production. *Aquaculture* 101:213-222.

Green, B. W., and C. R. Engle (2000). Commercial tilapia aquaculture in Honduras. In B. A. Costa-Pierce, and J. E. Rakocy (Eds.), *Tilapia aquaculture in the Americas* (Volume 2, pp. 151-170). Baton Rouge, LA: The World Aquaculture Society.

Green, B. W., E. H. Rizkalla, and A. R. El Gamal (1995). Mass production of Nile *(Oreochromis niloticus)* and blue tilapia fry. In H.S. Egna, J. Bowman, B. Goetze, and N. Weidner (Eds.), *Twelfth Annual Administrative Report, Collaborative Research Support Program* (1994) (pp. 192-195). Corvallis: Oregon State University.

Green, B. W., and D. R. Teichert-Coddington (1991). Effects of fry stocking rate, hormone treatment, and temperature on the production of sex-reversed *Oreochromis niloticus*. In H. S. Egna, J. Bowman, and M. McNamara (Eds.), *Eighth Annual Administrative Report, Collaborative Research Support Program* (pp. 22-25). (1990). Corvallis: Oregon State University.

——— (1993). Production of *Oreochromis niloticus* fry for hormonal sex reversal in relation to water temperature. *Journal of Applied Ichthyology* 9:230-236.

Green, B. W., D. R. Teichert-Coddington, and T. R. Hanson (1994). Development of semi-intensive aquaculture technologies in Honduras. International Center for Aquaculture and Aquatic Environments Research and Development Series No. 39, Auburn University, Alabama.

Guerrero, R.D., III. (1986). Production of Nile tilapia fry and fingerlings in earthen ponds at Pila, Laguna, Philippines. In J.L. Maclean, L.B. Dizon and L.V. Hosillos (Eds.), *The First Asian Fisheries Forum* (pp. 49-52). Manila, Philippines: Asian Fisheries Society.

Guerrero, R.D., III, and A.M. Garcia (1983). Studies on the fry production of *Sarotherodon niloticus* in a lake-based hatchery. In L. Fishelson and Z. Yaron (compilers), *Proceedings: International Symposium on Tilapia in Aquaculture* (pp. 386-393). Nazareth, Israel, May 8-13. Tel Aviv, Israel: Tel Aviv University.

Guerrero, R.D., III, and L.A. Guerrero (1985). Further observations on the fry production of *Oreochromis niloticus* in concrete tanks. *Aquaculture* 47: 257-261.

Higley, L. G., L. P. Pedigo, and K. R. Ostlie (1986). DEGDAY: A program for calculating degree-days, and assumptions behind the degree-day approach. *Environmental Entomology* 15: 999-1016.

Holl, M. (1983). Production d'alvins de *Tilapia nilotica* en Station Domaniale. Project PNUD/FAO/IVC/77/003. Developpement de la pisciculture en Eaux continentales en Côte d'Ivoire. Document Technique No. 10.

Hulata, G. (1997). Large-scale tilapia fry production in Israel. *The Israeli Journal of Aquaculture-Bamidgeh* 49: 174-179.

Khater, A.A., and R.O. Smitherman (1988). Cold tolerance and growth of three strains of *Oreochromis niloticus*. In R.S.V. Pullin, T. Bhukaswan, K. Tonguthai, and J. L. Maclean (Eds.), *The Second International Symposium on Tilapia in Aquaculture* (pp. 215-218). ICLARM Conference Proceedings 15. Department of Fisheries, Bangkok, Thailand, and International Center for Living Aquatic Resources Management, Manila, Philippines.

Liao, I. C. and S. L. Chang (1983). Studies on the feasibility of red tilapia culture in saline waters. In L. Fishelson and Z. Yaron (compilers), *Proceedings: International Symposium on Tilapia in Aquaculture* (pp. 524-533). Nazareth, Israel, May 8-13. Tel Aviv, Israel: Tel Aviv University.

Little, D.C. (1989). An evaluation of strategies for production of Nile tilapia (*Oreochromis niloticus* L.) fry suitable for hormonal treatment. Doctoral dissertation, University of Stirling, Stirling, Scotland.

Little, D.C., K. Coward, R.C. Bhujel, T.A. Pham, and N.R. Bromage (2000). Effect of broodfish exchange strategy on the spawning performance and sex steroid hormone levels of *Oreochromis niloticus* broodfish in hapas. *Aquaculture* 186: 77-88.

Little, D.C., and G. Hulata (2000). Strategies for tilapia seed production. In M.C.M. Beveridge, and B.J. McAndrew (Eds.), *Tilapias: Biology and exploitation* (pp. 267-326). United Kingdom: Kluwer Academic Publishers.

Little, D.C., D.J. Macintosh, and P. Edwards (1993). Improving spawning synchrony in the Nile tilapia, *Oreochromis niloticus* (L.). *Aquaculture and Fisheries Management* 24: 399-405.

Little, D.C., W.A. Turner, and R.C. Bhujel (1997). Commercialization of a hatchery process to produce MT-treated Nile tilapia in Thailand. In D.E. Alston, B.W. Green, and H.C. Clifford (Eds.), *IV Symposium on Aquaculture in Central America: Focusing on shrimp and tilapia* (pp. 108-118). April 22-24. Tegucigalpa, Honduras: Asociacion Nacional de Acuacultores de Honduras and the Latin American Chapter of the World Aquaculture Society.

Lovshin, L.L., and H.H. Ibrahim (1988). Effects of broodstock exchange on *Oreochromis niloticus* egg and fry production in net enclosures. In R.S.V. Pullin, T. Bhukaswan, K. Tonguthai and J. L. Maclean (eds.), *The Second International Symposium on Tilapia in Aquaculture* (pp. 231-236). ICLARM Conference

Proceedings 15. Department of Fisheries, Bangkok, Thailand, and International Center for Living Aquatic Resources Management, Manila, Philippines.

Macintosh, D.J., and S.S. De Silva (1984). The influence of stocking density and food ration on fry survival and growth in *Oreochromis mossambicus* and *O. niloticus* female × *O. aureus* male hybrids reared in a closed circulated system. *Aquaculture* 41: 345-358.

Macintosh, D.J., and D.C. Little (1995). Nile tilapia *(Oreochromis niloticus).* In N. R. Bromage and R.J. Roberts (Eds.), *Broodstock management and egg and larval quality* (pp. 277-320). Oxford, England: Blackwell Science Ltd.

Mires, D. (1983). Current techniques for the mass production of tilapia hybrids as practiced at Ein Hamifratz fish hatchery. *Bamidgeh,* 35: 3-8.

Pantastico, J.B., M.M.A. Dangilan, and R.V. Eguia (1988). Cannibalism among different sizes of tilapia *(Oreochromis niloticus)* fry/fingerlings and the effect of natural food. In R.S.V. Pullin, T. Bhukaswan, K. Tonguthai, and J.L. Maclean (Eds.), *The Second International Symposium on Tilapia in Aquaculture* (pp. 465-468). ICLARM Conference Proceedings 15. Department of Fisheries, Bangkok, Thailand, and International Center for Living Aquatic Resources Management, Manila, Philippines.

Perez, H. (1995). Cultivo de tilapia en Centroamerica. First National Meeting on Tilapia Culture in Nicaragua, PRADEPESCA, Managua, Nicaragua.

Phelps, R.P., G. Conterras Salazar, V. Abe, and B. Argue (1995). Sex reversal and nursery growth of Nile tilapia, *Oreochromis niloticus* (L.), free-swimming in earthen ponds. *Aquaculture Research* 26: 293-295.

Philippart, J.-Cl., and J.-Cl. Ruwet (1982). Ecology and distribution of tilapias. In R.S.V. Pullin, and R.H. Lowe-McConnell (Eds.), *The biology and culture of tilapias. ICLARM Conference Proceedings 7* (pp. 15-59). Manila, Philippines: International Center for Living Aquatic Resources Management.

Piper, R.G., I.B. McElwain, L.E. Orme, J.P. McCraren, L.G. Fowler, and J.R. Leonard (1982). *Fish hatchery management.* Washington, DC: United States Department of the Interior, Fish and Wildlife Service.

Popma, T.J., and B.W. Green (1990). *Sex reversal of tilapia in earthen ponds.* International Center for Aquaculture and Aquatic Environments Research and Development Series No. 35, Auburn University, Alabama.

Rana, K.J. (1986). An evaluation of two types of containers for the artificial incubation of *Oreochromis* eggs. *Aquaculture and Fisheries Management* 17: 139-145.

———. (1988). Reproductive biology and hatchery rearing of tilapia eggs and fry. In J.F. Muir and R.J. Roberts (Eds.), *Recent advances in aquaculture* (Volume 3, pp. 343-407). London, England: Croom Helm Ltd.

Regier, H.A., J.A. Holmes, and D. Pauly (1990). Influence of temperature changes on aquatic ecosystems: An interpretation of empirical data. *Transactions of the American Fisheries Society* 119: 374-389.

Ridha, M.T., and E.M. Cruz (2000). Effect of light intensity and photoperiod on Nile tilapia *Oreochromis niloticus* L. seed production. *Aquaculture Research* 31: 609-617.

Rothbard, S., and G. Hulata (1980). Closed-system incubator for cichlid eggs. *The Progressive Fish-Culturist* 42:203-204.

Rothbard, S., E. Solnik, S. Shabbath, R. Amado, and I. Grabie (1983). The technology of mass production of hormonally sex-inversed all-male tilapias. In L. Fishelson and Z. Yaron (compilers), *Proceedings: International Symposium on Tilapia in Aquaculture, Nazareth, Israel* (pp. 425-434). May 8-13. Tel Aviv, Israel: Tel Aviv University.

Sarig, S., and M. Marek (1974). Results of intensive and semi-intensive fish breeding techniques in Israel 1971-1973. *Bamidgeh* 26:28-48.

Snow, J.R., J.M. Berrios-Hernandez, and H.Y. Ye (1983). A modular system for producing tilapia seed using simple facilities. In L. Fishelson and Z. Yaron (compilers), *Proceedings: International Symposium on Tilapia in Aquaculture, Nazareth, Israel* (pp. 402-413). May 8-13. Tel Aviv, Israel: Tel Aviv University.

Tonguthai, K., and S. Chinabut (1997). Diseases of tilapia. In H. Egna and C.E. Boyd (Eds.), *Dynamics of pond aquaculture* (pp. 263-287). Boca Raton, FL: Lewis Publishers.

Verdegem, M.C., and A.S. McGinty (1987). Effects of frequency of egg and fry removal on spawning by *Tilapia nilotica* in hapas. *The Progressive Fish-Culturist* 49:129-131.

———. (1989). Evaluation of edge seining for harvesting *Oreochromis niloticus* fry from spawning ponds. *Aquaculture* 80:195-200.

Watanabe, W.O., K.E. French, L.J. Ellingson, R.I Wicklund, and B.L. Olla (1988). Further investigations on the effects of salinity on growth in Florida red tilapia: Evidence for the influence of behavior. In R.S.V. Pullin, T. Bhukaswan, K. Tonguthai, and J.L. Maclean (Eds.), *The Second International Symposium on Tilapia in Aquaculture* (pp. 525-530). ICLARM Conference Proceedings 15, 623 p. Department of Fisheries, Bangkok, Thailand, and International Center for Living Aquatic Resources Management, Manila, Philippines.

Chapter 6

Hormone Manipulation of Sex

Ronald P. Phelps

INTRODUCTION

Tilapia are a paradox in terms of reproduction. The relative fecundity of the *Oreochromis* species is low, at 6,000 to 13,000 eggs per kilogram per spawn (Siraj et al. 1983). This is compensated for by the high survival of fry as a result of the large size at hatching, the large yolk reserve, and the mouth-brooding maternal care given until the fry are 10 mm or larger. The low fecundity is also compensated for by the frequent spawning of these asynchronous species, whereby the low fecundity per spawn may be parlayed into a yearly quantity of eggs per kilogram equal to that of many groups of synchronously spawning species.

Ideally, a fish species used in aquaculture will not reproduce in the culture environment before reaching market size. From this perspective, tilapia present some challenges to the fish culturist. Most species of tilapia will reach maturity under favorable conditions within 6 to 8 months of hatching at a size often less than 100 g . They will continue to reproduce, with offspring competing with the initial stock for food, resulting in stunted growth and unmarketable fish. Swingle (1960) found that in 169- to 196-day culture cycles of mixed-sex Mozambique tilapia *(Oreochromis mossambicus)*, production exceeded 3,000 $kg \cdot ha^{-1}$, but more than 90 percent of the harvest was composed of fish weighing less than 100 g. Verani et al. (1983) produced 4,944 $kg \cdot ha^{-1}$ of mixed-sex Nile tilapia *(O. niloticus)* in 11 months, but the average weight harvested was less than 100 g. Rakocy (1980) stocked mixed-sex 23 g blue tilapia *(O. aureus)* fingerlings into unaerated static-water ponds, where they were fed intensively for 141 days. The production averaged 14,170 $kg \cdot ha^{-1}$ with an FCR (food conversion ratio) of 1.08, but 61 percent of the production was from reproduction. The initial

doi:10.1300/5513_06

stock of fish averaged 236 g at harvest. High yields and efficient nutrient utilization are meaningless unless a significant portion of the production is marketable. Tilapia have numerous advantages as an aquaculture species (Teichert-Coddington et al. 1997), but the ability to reproduce in the production setting has resulted in the need to develop various techniques to control unwanted reproduction, including stock manipulation (Swingle 1960), polyculture with predatory fish (Lovshin 1975), and monosex culture (Shell 1968).

The use of a predator does not prevent reproduction but can prevent recruitment. Tilapia yields are often low, reduced by the slower-growing females and restricted nutrient inputs. Often, because the predator has lower tolerance to poor water quality, nutrient inputs are curtailed in order to maintain adequate water quality for the predator species.

All-male (monosex) culture of tilapia is preferred because of males' faster growth (Guerrero 1975; Shelton et al. 1978). Several techniques have been adopted: manual sexing (Guerrero 1982), hybridization (Hickling 1960), genetic manipulation (Pandian and Varadaraj 1988), and sex reversal through sex hormone administration (Yamamato 1951; Clemens and Inslee 1968; Shelton et al. 1978; Guerrero 1982) are the most commonly used techniques to obtain monosex populations.

Monosex culture by either manually or mechanically selecting males results in half of the potential fish seed being rejected. Popma et al. (1984) developed an efficient system to produce hand-sexed fingerlings, but 30 percent of a farm's acreage had to be devoted to broodfish and fingerling production to support the remaining 70 percent in foodfish production. Fingerlings were grown to 20 to 30 g for sexing, producing the equivalent of 4,000 $kg \cdot ha^{-1} \cdot year^{-1}$ of females, most of which were discarded.

Selective crosses of a homogametic male of one tilapia species with a homogametic female of another have resulted in all-male hybrids (Lovshin 1982). First reported by Hickling in 1960, this became a common approach to producing males, but it was largely replaced by the mid-1980s owing to difficulties in maintaining two pure lines of broodfish, keeping them separate, and making available the amount of space required. In addition, apparent autosomal influences affected sex ratios, often resulting in populations that were less than 100 percent males, even if pure lines were maintained.

The use of hormones to alter the sex ratios of fish was first demonstrated in species other than tilapia. Berkowitz (1937) evaluated 17β-estradiol as a means of altering the sex of the guppy *(Poecilia reticulata)* and produced intersex fish. Yamamato (1951), working with medaka *(Oryzai latipes),* concluded that sex hormones, in addition to modifying secondary sex characteristics, also affect the gonads. Androgen induced masculinization, and estrogen resulted in feminization. He produced 100 percent medaka females with an estrogen in 1951 and a nearly all-male population with an androgen in 1954 (Yamamoto 1953, 1958). The technique has been successfully implemented to alter the sex ratio of rainbow trout *(Oncorhynchus mykiss)*, goldfish (*Carassius auratus;* Yamazaki 1976), grass carp (*Ctenopharyngodon idella;* Stanley 1976), and tilapia (Clemens and Inslee 1968; Nakamura and Takahashi 1973).

Although commonly referred to as sex reversal, hormone treatment does not alter the genotype of the fish but directs the expression of the phenotype. A treated population of fish may be phenotypically monosex but genetically will have remained as determined at the moment of fertilization. As a result of hormone treatment it is possible to have phenotypical male fish that are genetically female or phenotypical female fish that are genetically male. For convenience, the altering of the phenotype by administration of sex hormones is referred to as sex reversal.

Administration of an androgen (17α-methyltestosterone) is considered to be the most effective and economically feasible method for obtaining all-male tilapia populations (Guerrero and Guerrero 1988). Following the techniques outlined by Popma and Green (1990), less than 15 percent of a farm's acreage would need to be devoted to broodstock and fingerling production to support the remaining acreage of foodfish.

Such efficiency and simplicity have resulted in hormone sex reversal becoming the commercial procedure of choice to produce male tilapia fingerlings and have been a significant factor in the rapid growth of the tilapia industry.

CHEMICALS USED TO DIRECT GONADAL DIFFERENTIATION

Steroids are a group of sterols with several unique properties affecting animal growth and development. Steroids are called androgens if they are able to induce male characteristics and estrogens if they induce female characteristics. Androgens have two physiological actions: (1) androgenic activity, promoting the development of male sex characteristics, and (2) anabolic activity, stimulating protein biosynthesis. Androgens can be classified into two groups: androstene derivatives, which have both androgenic and anabolic properties, and 19-nor-androstene derivatives, which have anabolic properties but only weakly androgenic ones (Camerino and Sciaky 1975). From a sex-reversal perspective, androstene derivatives are of more value because of their potential to direct the sexual development of fish into males. In evaluating a steroid for sex reversal by oral administration, three main criteria should be considered: metabolic half-life, androgenic or estrogenic strength, and solubility in water.

Testosterone is the principal androgen secreted by the testis and the main androgenic steroid in the plasma of human males (Murad and Haynes 1985). It is often used as the standard to evaluate the androgenic properties of a steroid. It is ineffective when given orally and has a short life when given by injection because it is rapidly metabolized by the liver. Synthetic androgens are preferred over natural ones because some can be administered orally and withstand catabolism in the gut. The chemical structure, bonds, and attached groups determine the effectiveness (Brueggemeier 1986). Introduction of a 3-ketone function or a 3α-OH group or reduction of the 4,5-double bond enhances androgenic activity. Alkylation of the 17α-position or the 1α-position allows for oral activity.

Natural and synthetic compounds are considered estrogenic if they are able to activate estrogen receptors and regulate gene transcription. There is no clear relationship of chemical structure to potency such as is seen with androgens. Piferrer (2001) lists three natural estrogens and nine synthetic estrogenic substances that have been used with fish. Estrone and 17β-estradiol are natural steroidal estrogens found in the ovary of tilapia (Katz et al. 1971). The synthetic estrogens most commonly used for sex reversal are the nonsteroidal estrogens ethynylestradiol (EE_2) and diethylstilbestrol (DES). Synthetic estrogens are more potent than natural estrogens when given orally because of their stability in the digestive tract and the liver (White et al. 1973). DES is more potent and was once used as a growth promoter in livestock, until it was banned by the U.S. Food and Drug Administration in 1979. Both are carcinogens (Herbst et al. 1971), and alternative chemicals should be considered. At a proper dose and treatment duration, 17β-estradiol can be used in place of DES. In a study by Al-Albani (1997) Bluegill *(Lepomis macrochirus)* fry given DES at 200 $mg \cdot kg^{-1}$ of diet had a mortality more than double that of nontreated fish, but 17β-estradiol at 100 $mg \cdot kg^{-1}$ gave an all-female population, with a mortality similar to the nontreated fish.

In addition to androgens and estrogens, several other chemicals that hinder normal gonadal development have been tested with tilapia.

Masculinization

Androgens

Evaluation of natural androgens has produced variable results. Eckstein and Spira (1965) used a 1 $mg \cdot L^{-1}$ testosterone bath with blue tilapia with little effect. Al-Daham (1970) obtained 89 percent males using 0.4 $mg \cdot L^{-1}$ of testosterone. Adrenosterone at 5 $mg \cdot L^{-1}$ produced an "all-male" population in which 55 percent of the fish had no gonadal development (Katz et al. 1976). When adrenosterone was given in the diet at 30 and 50 $mg \cdot kg^{-1}$ of diet only 74 and 81 percent males, respectively, were obtained (Guerrero and Guerrero 1997). Nakamura (1981) masculinized Mozambique tilapia by feeding 11-ketotestosterone at 200 $mg \cdot kg^{-1}$ of diet for 19 days. Desprez et al. (2003) obtained 99.1 percent males by feeding 11β-hydroxyandrostenedione at 50 $mg \cdot kg^{-1}$ of diet to Florida red tilapia for 28 days. Haylor and Pascual (1991) prepared a diet for young *O. niloticus* that contained 57 percent ram testis. In 27 fish examined, 23 were males and 4 intersex. Phelps et al. (1996) obtained a 65 percent male population using a diet of which half was freeze-dried bull testes.

A number of synthetic androgens, applied either as a bath or as a feed additive, have altered the sex ratio of tilapia. Clemens and Inslee (1968) produced all-male populations of Mozambique tilapia by incorporating 17α-methyltestosterone into the diet at 10 to 40 $mg \cdot kg^{-1}$. Methyltestosterone (MT) has since become the most commonly used synthetic androgen to alter the sex ratio of fish. It has proven to be effective in a number of species of tilapia and under various management scenarios. Other synthetic androgens that have been incorporated into the diet of tilapia for sex-reversal are listed in Table 6.1.

No single androgen can be considered the best for tilapia sex reversal. The range of dose rates and treatment protocols makes it difficult to compare the efficacy of compounds. One approach is to evaluate these androgens based on the local costs to produce a given number of fish. Phelps et al. (1992) discussed how fluoxymesterone was more expensive than MT had a lower but effective dose, thus compensating for the cost difference. Selection of an androgen to be incorporated into the feed should be based on local availability, costs, government regulations, and human and environmental safety.

Nonsteroidal Compounds

Another approach to producing male populations is the use of nonsteroidal compounds that interfere with steroid binding or metabolism. In the sequence of events associated with gonadal differentiation, endogenous androgens are aromatized into estrogen by an aromatase enzyme. It is possible to block this action by the addition of an aromatase inhibitor. Guiguen et al. (1999) were able to change the sex ratio of an all-female *O. niloticus* population to 77.8 percent male by feeding 11-day-old fry the aromatase

TABLE 6.1. Other synthetic androgens that have been incorporated into the diet of tilapia for sex reversal.

Androgen	Tilapia sp.	Dose rate ($mg \cdot kg^{-1}$)	Efficacy	Reference
1-Dehydrotestosterone	*O. aureus*	15	69% male	Guerrero (1975)
		30	59% male	
		60	44% male	
Ethynyltestosterone	*O. aureus*	15	85% male	Guerrero (1975)
		30	98% male	
		60	100% male	
Luoxymesterone	*O. niloticus*	1	87.3% male	Phelps et al. (1992)
		5	100% male	
		25	100% male	
Mestanolone	*O. niloticus*	5	99.5% male	Soto (1992)
		10	97.0% male	
		20	99.0% male	
Mibolerone 1993	*O. mossambicus*	1.5	84% male	Guerrero and Guerrero (1993)
		1.75	88.0% male	
		2.0	94.0% male	
19-Norethisterone acetate	*O. mossambicus*	1	52% male	Varadaraj (1990)
Trenbolone acetate	*O. aureus*	25	98.3% male	Galvez et al. (1996)
		50	99.3% male	
		100	99.0% male	

inhibitor 1,4,6-androstatriene-3-17-dione at 150 $mg \cdot kg^{-1}$ diet for 30 days. Kwon et al. (2000) fed a genetically female population of 7-day-old *O. niloticus* fry with the aromatase inhibitor fadrozole at 200 to 500 $mg \cdot kg^{-1}$ diet for 30 days and obtained 92.5 to 96.0 percent males. Afonso et al. (2001) were able to produce a 100 percent male *O. niloticus* population by feeding fadrozole at 75 or 100 $mg \cdot kg^{-1}$ diet for 30 days starting with 9-day-old fry.

Steroid hormones bind with receptor sites in the cytoplasm or nucleus of a cell to regulate the transcription of DNA. Blocking of estrogen binding sites is another approach to the production of males. Tamoxifen, an antiestrogen thought to compete for estrogen binding sites, has been evaluated as a feed additive to direct the sex ratio of trout and tilapia. In a study by Himes and Watts (1995), an all-male population was obtained when tilapia fry were fed at 100 $mg \cdot kg^{-1}$ diet, but at 15 and 50 $mg \cdot kg^{-1}$ diet, the percentage of males was 77 and 86 percent, respectively. When rainbow trout fry and tilapia were fed tamoxifen at 20 at 25 $mg \cdot kg^{-1}$ diet, respectively, Guguen et al. (1999) found no effect on sex ratio.

Feminization

Female tilapia are not preferred for culture because of their early maturity and reduced growth rate once they begin to reproduce. However, alteration of the phenotype through the feminization of genetically heterogametic male Nile tilapia offers the possibility of all-male tilapia through a YY breeding program. Similarly interesting to producers is the feminization of homogametic male *O. aureus* into functional females for mating with normal male *O. aureus* to produce all-male offspring. Piferrer (2001) reviewed the use of estrogens in fish and found that approximately 45 different fish species have been treated with 17β-estradiol, 16 with DES, and 12 with EE_2.

Sex reversal to females can be more difficult than sex reversal to males. As gonads begin to differentiate into ovarian tissues before testicular tissues, treatment must begin early and an effective dose must be given until differentiation is complete. Piferrer (2001) discussed how the period of maximum sensitivity to estrogens is earlier than that to androgens. In several of their earlier trials with estrogens, Hopkins (1977) and Jensen and Shelton (1979) had little success directing the sex of tilapia. Hopkins (1977) used 17β-estradiol at 220 $mg \cdot kg^{-1}$ of diet to obtain a population of 59 percent females, 2 percent males, and 39 percent "atypical." Jensen and Shelton (1979) added the natural estrogens estriol, estrone, and 17β-estradiol to the diet of 8 to 11 mm *O. aureus* in an attempt to produce all-female populations. They were unable to skew the sex ratio of treated fish but noted atypical males at the higher treatment rates.

The effectiveness of DES and EE_2 in feminization may be dependent on the species of tilapia and the management conditions. Hopkins et al. (1979) fed DES at 100 $mg \cdot kg^{-1}$ diet to *O. aureus* fry for 5 weeks and produced 64 percent females. Rosentein and Hulata (1993) obtained 98 and 100 percent females in two sets of *O. aureus* fed DES at 100 $mg \cdot kg^{-1}$ diet for 30 days. EE_2 was used at 100 $mg \cdot kg^{-1}$ diet with *O. aureus* for 40 days by Melard (1995) to obtain a 94 percent female population.

Potts and Phelps (1995) fed *O. niloticus* fry diets containing up to 400 mg DES or 200 mg EE_2 per kilogram for 28 days. At 400 mg DES they were able to obtain 92 percent of the population with female-shaped papilla and 80 percent with ovaries. EE_2 at 100 $mg \cdot kg^{-1}$ was not as effective, producing a 65 percent female population. Scott et al. (1989) fed two sets of genetically all-male *O. niloticus* fry DES at 100 $mg \cdot kg^{-1}$ diet and obtained 52 percent females in one set and 84 percent in the other. Ridha and Lone (1995) treated *O. spilurus* swim-up fry for 42 days with EE_2 at 100 $mg \cdot kg^{-1}$ diet and obtained a 92.2 percent female population.

Studies with *O. mossambicus* suggest that they might be easier to feminize with DES or EE_2. Varadaraj (1989) found that all-female populations were obtained by feeding DES at 100 $mg \cdot kg^{-1}$ for 11 days. Basavaraja et al. (1990) fed 5-day-old fry DES at 50 $mg \cdot kg^{-1}$ diet for 30 days and obtained an all-female population. Rosenstein and Hulata (1993) obtained 100 percent females by feeding DES at 100 $mg \cdot kg^{-1}$ or EE_2 at 75 $mg \cdot kg^{-1}$ for 14 days, but the results were not consistent across trials.

Sterilants

Another approach to controlling tilapia reproduction is sterilization. Androgens and estrogens have been reported to cause gonadal damage, rendering fish sterile. Eckstein and Spira (1965) found that treatment with stilbestrol diphosphate at 50 $\mu g \cdot L^{-1}$ resulted in gonadal destruction in *O. aureus* treated for 2 weeks. Adrenosterone at 5 $mg \cdot L^{-1}$ resulted in 55 percent of fish with no gonadal development (Katz et al. 1976). Okoko (1996) found that high doses of MT reduced the gonadosomatic index of "male" *O. niloticus.*

Al-Daham (1970) was able to limit reproduction by treating *O. aureus* fingerlings with the chemosterilants metepa and tretamine at 20 and 0.8 ppm, respectively. These chemicals did not produce monosex populations but reduced gonadosomatic indices in both males and females. Dazdzie (1975) found that methallibure prevented secondary sexual characteristics and controlled tilapia spawning behavior.

MODE OF HORMONE ACTION— OR WHY SEX REVERSAL WORKS

Not all the details of how sexual differentiation is controlled in tilapia have been worked out. Tilapia eggs take 4 to 6 days to hatch depending on water temperature. At hatching, larvae are not well developed, and the gonads have not formed. D'Cotta et al. (2001) monitored the development of *O. niloticus* fry. They found that by 9 days postfertili- zation (PF), yolk-sac absorption is near completion. The fish were 8.94 mm long and a pair of tubular undifferentiated gonads could be found along the dorsal peritoneal wall of the body cavity. At 9 days PF, gonads contained an average of 48 germ cells per gonad. In approximately half of the fish, around 21 days PF, the germ cells began to multiply and had increased 2.7-fold between days 25 and 35 PF. Fish averaged 18 $\pm$ 1.8 and 26.1 $\pm$ 1.1 mm, respectively, at those two times. Between days 28 and 35 PF, germ cells underwent meiosis and a groove was observed along the lateral side of the gonads that later developed into the ovarian cavity. The other half of the fish showed a slower increase in germ cells from days 9 to 35 PF. After day 35 PF germ and somatic cells were actively proliferating. By day 50 PF, when the fish averaged 40.2 $\pm$ 3.7 mm, the gonads had taken on the lobular configuration typical of testes. By 90 days PF, active spermatogenesis was taking place in the testes and previtellogenic oocytes were appearing in the ovaries.

The onset of ovarian differentiation is considered to be the occurrence of meiotic germ cells in the gonad; the formation of the ovarian cavity is a reliable histological criterion for identifying the gonad as an ovary (Nakamura et al. 1998). The appearance of ovocoel and testocoel, indications of sex differentiation into female and male, takes place 16 to 20 days after hatching in *O. mossambicus* (Nakamura and Takahashi 1985) and perhaps as late as 30 to 33 days after hatching in *O. niloticus* (Alvendia-Casauay and Carino 1988). However, biochemically, the onset of differentiation has begun earlier. Yamamoto (1969) demonstrated that steroids could control gonadal differentiation. However, when and from where endogenous steroids that control sexual differentiation appear is unknown. Histologically, in tilapia steroid-producing cells are not evident before ovarian or testicular differentiation (Nakamura and Nagahma 1985, 1989). Bogart (1987) proposed that sex determination is based on the ratio of androgens to estrogens. Baroiller et al. (1999) refined this measure to the ratio of 11-oxgenated androgens (biologically active androgens in fish) to estrogens, which is controlled by the natural metabolism of androgens into estrogens.

As discussed, one mechanism for the metabolism of androgens into estrogens is through aromatization by aromatase enzymes. Cytochrome P450 aromatase is responsible for the aromatization of androstenedione to estrone and of testosterone into 17β-estradiol (Jeyasuria and Pace 1998). D'Cotta et al. (2001) suggested that the level of aromatase directs the gonadal differentiation of *O. niloticus*. Aromatase activity in fish has been detected before the start of gonadal differentiation. In rainbow trout, P450 aromatase activity was found approximately 3 weeks before histological sex differentiation (Lui et al. 2000). In tilapia, aromatase gene expression occurred around the time of hatching (3 to 4 days PF) and a dimorphism in activity appeared from 11 to 27 days PF (Kwon et al. 2001). This dimorphism in activity precedes histological dimorphism of the gonad (days 28 to 35 PF; D'Cotta et al. 2001) and the appearance of steroid-producing cells in the gonad (27 to 30 days PF; Nakamura and Nagahama 1985, 1989). This suggests that a biochemical sexual dimorphism occurs much earlier than is evident morphologically. These differences in timing of dimorphism must be considered later when application of exogenous hormones is discussed.

Aromatase activity is also higher in females than in males during the 27 to 43 day PF period of gonadal differentiation (Kwon et al. 2001), suggesting its role in not only stimulating differentiation but also maintaining it. A high level of aromatase activity favors the conversion of androgens into estrogens, shifting the androgen: estrogen ratio to favor sex differentiation to female. Aromatase activity in tilapia can be suppressed by an aromatase inhibitor, resulting in populations where the sex ratio is highly skewed toward male (Guiguen et al. 1999; Kwon et al. 2000; Afonso et al. 2001). Kitano et al. (2000) found that in the Japanese flounder *(Paralichthys olivaceus)* P450 aromatase mRNA was highly expressed in nontreated females. However, when the androgen MT was fed to larvae, P450 aromatase mRNA was barely expressed in the resultant 100 percent phenotypically male population. The authors concluded that the mechanism by which MT induces genetic females to develop as phenotypic males is suppression of the P450 aromatase gene expression. This also may be the mechanism of directing the sexual differentiation of tilapia to males when androgens are given.

FISH SPECIES EVALUATED

Hormonal sex reversal has been demonstrated in a wide range of families of fish, including Andantidae, Atherinidae, Bothidae, Centrarchidae, Cichlidae, Cyclopteridae, Cyprinidae, Cyprinodontidae, Ictaluridae, Percidae, Poecilidae, Polyodontidae, Salmonidae, and Serranidae. It has been particularly effective in cichlids because gonadal differentiation takes place early in the life history. Among the species of tilapia included in studies to chemically alter the sex ratio are those in Table 6.2. Mouth-brooding species have been successfully sex reversed when hormone treatment

TABLE 6.2. Species of tilapia studied for sex reversal.

Species	Hormone	Mode of treatment	Reference
Oreochromics mossambicus	MT	Feed additive	Clemens and Inslee (1968)
O. niloticus	MT	Feed additive	Tayamen and Shelton (1978)
Tilapia aurea	ET	Feed additive	Guerrero (1975)
O. honorum	ET	Feed additive	Obi and Shelton (1983)
O. spilurus	MT	Feed additive	Lone and Ridha (1993)
T. henedeloti	Testosterone propionate	Immersion	Hackmann (1971)
T. macrochir	MT	Feed additive	Jalabert et al. (1974)
T. zilli	ET	Feed additive	Guerrero (1976)

ET = Ethynyltestosterone; MT = Methyltestosterone.

begins within a few days of hatching. Less success has been achieved with a substrate spawner red belly tilapia *(Tilapia zilli),* where timing and duration of treatment may have been the problem.

Of the mouth-brooders, *O. mossambicus* was among the first to be successfully sex reversed. Clemens and Inslee (1968) obtained 100 percent *O. mossambicus* male populations with MT at 10, 20, 30, 40, and 50 $mg{\cdot}kg^{-1}$ diet by feeding fish daily for 59 days at 6 percent body weight. Guerrero (1975) treated *O. aureus* with MT at 15, 30, and 60 $mg{\cdot}kg^{-1}$ diet at 4 percent body weight and obtained 84, 98, and 85 percent males, respectively. Tayamen and Shelton (1978) fed α−MT at 30 and 60 $mg{\cdot}kg^{-1}$ diet to *O. niloticus* and obtained 99, 100, and 100 percent males in 25, 35, and 59 days, respectively, for the lower dose and 100 percent males for all three durations for the higher dose. Obi and Shelton (1983) fed *O. honorum* with 17 α-ET for 21 and 28 days and obtained 94 and 97 percent males, respectively, with 60 $mg{\cdot}kg^{-1}$ diet and 100 percent males with 30 mg for both treatment durations. Buddle (1984) obtained 95 to 100 percent *O. niloticus* × *O. aureus* hybrid males in tanks and 95 to 99 percent males in cages when fish were fed a diet containing MT at 60 $mg{\cdot}kg^{-1}$ diet for 25 to 28 days. Berger and Rothbard (1987) obtained 97.3 percent males plus 3.3 percent intersex and 99.7 percent males plus

1.3 percent intersex when they treated hybrids (*O. niloticus* × O. mossambicus) for 28 days with 17 α-ET at 60 ppm and 120 $mg \cdot kg^{-1}$ diet, respectively.

PROTOCOLS FOR SEX REVERSAL OF TILAPIA WITH HORMONE-TREATED FEED

Treatment Setting

During sex reversal, all fry must be held in a setting wherein they can be guaranteed to receive a daily intake of hormone from a time before gonadal differentiation has begun until it is complete. Early investigations into sex reversal using hormone-treated feed were conducted in aquaria or troughs receiving clear water (Clemens and Inslee 1968; Guerrero 1975; Tayamen and Shelton 1978). Tanks with flowing water have been successfully used to produce commercial quantities of sex-reversed fry (Rothbard et al. 1983; Guerrero and Guerrero 1988). Indoor tanks often are not as suitable as outdoor tanks because they result in greater mortality. Popma (1987) reported average survival of 40 percent when *O. niloticus* were sex reversed at a density of 3,200 to 4,500 per cubic meter in indoor tanks with a 5 to 7 $percent \cdot h^{-1}$ exchange rate. Vera Cruz and Mair (1994) obtained more than 70 percent survival in outdoor tanks stocked at 5,000 fish per cubic meter with an exchange rate of 100 $percent \cdot day^{-1}$. This difference in survival between indoor and outdoor tanks is common on commercial farms as well.

Initial concerns that tilapia should consume no natural food during hormone treatment proved to be unfounded. Buddle (1984) compared the use of indoor tanks with clear water, outdoor tanks, and hapas in static-water ponds as treatment units for tilapia sex reversal. He obtained 96 to 98 percent males from those held in hapas or treated in indoor or outdoor tanks. Chambers (1984), working with *O. niloticus,* obtained 98.5 percent males and 95 percent survival using hapas placed in a fertile earthen pond or fertile static-water outdoor tanks.

When hapas are used to hold fry for sex reversal, they are stocked at densities of 3,000 to 5,000 fish per square meter (Popma and Green 1990) or 12 fry per liter (MacIntosh and Little 1995). The size of the hapas and the number needed should be proportionate to the quantity of fry available on a given day. Hapas with a water surface area of 2 to 5 m^2 and a water depth of 50 to 60 cm are convenient for management. The mesh size should be no larger than 1.6 mm, but this small mesh will foul during the treatment period, and attention should be paid to preventing the hapas from becoming fouled to the point where dissolved oxygen becomes low. To help ensure overall water quality remains high, 100 to 200 m^2 of pond area should be allowed for 10 to 15 m^2 of hapas.

It has been possible to sex reverse fry stocked free into static- or flowing-water tanks or earthen ponds. Phelps and Cerezo (1992) stocked *O. niloticus* fry into 20 m^2 outdoor concrete tanks at 150 per square meter, fed a 60 $mg \cdot kg^{-1}$ MT-treated feed for 28 days, and obtained a 98.3 percent male population that averaged 1.86 g at the end of the treatment period. Stocking fry directly into earthen ponds has also been effective. Phelps et al. (1995) obtained more than 96 percent males when *O. niloticus* fry were stocked at 200 to 260 per square meter into 215-m^2 earthen ponds and fed 60 $mg \cdot kg^{-1}$ MT-treated feed for 28 days. In a second trial, fish were stocked at only 75 per square meter and the percentage of males was 91.3. Many producers in Colombia successfully sex reverse red tilapia in shallow 15- to 30-m^2 outdoor tanks, stocking fry at 1,000 to 2,000 per square meter and exchanging water 4 to 7 times daily (Popma and Phelps 1998).

Stocking of Fry

Fry are most commonly stocked at densities of 3,000 to 4,000 per square meter of hapa and at higher densities in flowing-water tanks. Vera Cruz and Mair (1994) compared stocking densities of 1,000, 3,000, and 5,000 per square meter of hapa using *O. niloticus* and found best sex reversal at 3,000 and 5,000 per square meter but lower survival at 5,000 per square meter. High densities help insure an active feeding response, so that all fish are consuming feed. Pandian and Vardaraj (1987) observed that fry can establish a feeding order hierarchy, resulting in small fish not consuming adequate quantities of hormone-treated feed for successful sex reversal.

Fry are first graded, if necessary, and counted for stocking. An efficient method is to enumerate the fry by visual comparison. As a standard, fry are added individually into a bucket or pan to give a uniform distribution throughout the container. A second bucket or pan of the same color and size with water to the same depth is prepared and fry are added until the fish density appears the same. Commonly, a 18.7-L bucket is filled with 5 to 10 cm of water and a standard prepared using 1,000 fry per bucket. Care should be given to keeping healthy fish in the standard container for visual estimation and replacing them if they become stressed. If another lot of fish to be counted might be of a different size, then a new standard should be prepared. It is important to try to avoid aquatic insects or plant material being mixed in with the fish being estimated.

Fry can also be enumerated efficiently by weight on a balance accurate at 0.1 g. A known number of fish is weighed in water; then a larger quantity of fish is weighed in water and the number extrapolated. Care should be taken not to transfer additional water to the weighing container when the fish are weighed. Sex-reversible fry should average 10 to 30 mg at the start of treatment.

Feed Preparation

A highly palatable and nutritious feed is needed to obtain an active feed response and effective sex reversal. Commercial diets for young fish are suitable. They are generally more than 40 percent protein, complete in vitamins and minerals, and have fish oil added to increase palatability. Effective diets can be prepared using rice bran or finely ground poultry or hog diets and increasing the protein percentage by adding fish meal. The feed ingredients should be reground, mixed, and passed through a 0.6-mm mesh screen before use. Vitamins and minerals can be added using premixes for other livestock, especially if fry have limited access to natural food.

The feed particle size should be the equivalent of a No. 00 or 0 crumble of a commercial feed (0.42 to 0.59 mm) for the first week of feeding. A No. 1 crumble (0.59 to 0.84 mm) may be given in the second or third week of feeding.

Synthetic steroids are not water soluble and are added to the diet by dissolving an appropriate quantity in alcohol or fish or vegetable oils to prepare a stock solution. Androgens, such as MT, dissolve readily in ethanol, and a stock solution using 95 to 100 percent pure ethanol can be prepared at a concentration of 6 $g \cdot L^{-1}$. Ten milliliters of stock solution added to a carrier and mixed with 1 kg of diet would be adequate to obtain MT at 60 $mg \cdot kg^{-1}$ diet.

Lower strength ethanol or isopropyl alcohol or vegetable oil may be used as a carrier. However, weaker alcohols will add greater quantities of water to the feed, and this additional moisture must be allowed to evaporate off or the feed may become moldy. The prepared feed should contain no more than 10 percent moisture. Excess oil can contribute to rancidity and to hormone loss if oil floats off the feed when it is fed to fish.

The quantity of carrier depends on the type of carrier and the mode of application. When small quantities of feed are prepared, it is convenient to use 200 mL of an alcohol carrier plus the required quantity the stock solution per kilogram of feed. The solution is poured over the feed and thoroughly mixed until all the feed is moist. The hormone can also be applied using a smaller volume of carrier solution by spraying the solution over the feed. Both alcohol and oil have been used as carriers for spray application (McAndrews and Majumdar 1989; Killian and Kohler 1991; Galvez et al. 1995).

When small quantities of feed are prepared, the feed is spread out as a thin layer, sprayed, mixed, and sprayed again. Large quantities of feed can be sprayed in a mixer, allowing enough mixing time to insure all the feed is exposed to the solution. The moist feed is air-dried out of direct sunlight or stirred in the mixer until dry. The dried feed is stored under dark, dry conditions until use. Androgens will break down when exposed to sunlight or high temperatures. The pure hormone, any stock solution, and treated feed should be stored in the dark at room temperature or cooler. Varadaraj et al. (1994) studied the relationship between storage conditions of MT stock solutions and treated feed and the impact on efficacy of sex reversal of *O. mossambicus*. They found that when either the stock solution or feed was exposed to light, the efficacy of treatment was significantly reduced. Smith and Phelps (2001) held a feed containing MT at 60 $mg \cdot kg^{-1}$ in the dark in a freezer, then for various periods in a 4°C refrigerator or at ambient temperature (28 ± 1.5°C), before it was fed to *O. niloticus* fry. They obtained populations of more than 98 percent males using feed stored under any condition tested, including feed refrigerated for 60 days and then held for 26 days at ambient temperature. When first prepared, the feed contained MT at 60.4 $mg \cdot kg^{-1}$; after the feed had been stored in the refrigerator for 60 days and for an additional 26 days at ambient temperature, analysis showed an MT concentration of 54.8 $mg \cdot kg^{-1}$. The feed, which contained 15 percent fat, showed a sight degree of rancidity when it was stored under the harshest conditions, but that did not appear to effect palatability or the effectiveness of the sex-reversal treatment.

Feeding

Young tilapia fry grow rapidly, and depending upon water temperature, consume 20 percent or more of their body weight daily at the start of hormone treatment. Such rapid growth requires the quantity of feed to be adjusted daily, which can be done by preparing a feeding chart based on anticipated growth and making weekly corrections of the assumed growth rate. Given the low average weights involved, it can be difficult to obtain accurate figures in the field. A practical approach is to extrapolate average weight by measuring length and applying a length:weight formula. For sampling, fish are densely crowded together and a random sample of at least 50 fish from each lot to be treated is measured to the nearest millimeter. This length can be used in the following formula:

$$\text{Weight (in grams) of 1,000 fry} = 0.02\ L^3$$

where L is the mean total length in millimeters.

Feeding tables can be based on known mean lengths and anticipated increases in length. Growth estimates are best based on the results from previous sets of fish. A general guide for anticipated growth per day for various ranges of fry size would be for 8 to 12 mm fry, 0.25 to 0.5 mm·day^{-1}; 12 to 17 mm fry, 0.50 to 0.75 mm·day^{-1}; and 17 to 25 mm fry, 0.75 to 1.25 mm·day^{-1}.

Table 6.3 is an extract from a feeding table where the length was known on day 1 at stocking and on day 8 based on a sample measured to the nearest millimeter. During treatment, the fish should be resampled weekly to determine the mean length and recalculate the growth rate. The next week's growth is projected based on the previous week's growth rate.

Feeding can be by hand or automatic feeder. The fish should be fed three or more times per day for best growth. Bocek et al. (1991) found that sex

TABLE 6.3. Sample feeding table for 1,000 tilapia fry, assuming no mortality, fed at 20 percent body weight during week 1 and 15 percent during week 2.

Week	Day	Daily growth (mm·day^{-1})	Fish length (mm)	Weight (g per 1,000 fish)	Feed rate (percentage body wt/day)	Daily diet (g·day^{-1})
1	1	sample	11.0	26.6	20.0	5.3
1	2	0.3	11.3	28.9	20.0	5.8
1	3	0.3	11.6	31.2	20.0	6.2
1	4	0.3	11.9	33.7	20.0	6.7
1	5	0.3	12.2	36.3	20.0	7.2
1	6	0.3	12.5	39.1	20.0	7.8
1	7	0.3	12.8	41.9	20.0	8.4
2	8	sample	12.9	42.9	15.0	6.4
2	9	0.7	13.6	50.3	15.0	7.5
2	10	0.7	14.3	58.5	15.0	8.8
2	11	0.7	15.0	67.5	15.0	10.1
2	12	0.7	15.7	77.4	15.0	11.6

Source: From Popma and Green 1990.

Note: Fish are sampled weekly to determine an accurate length.

reversal could be achieved when fish were fed twice daily, 5 days per week, but best growth was obtained from daily feeding, four times per day. When automatic feeders are used, the daily ration should be divided into four to five portions, so that large quantities will be released at each feeding. When small quantities of feed are released uniformly throughout the day, the larger tilapia dominate the area around the feeder and consume most of the feed, resulting in considerable size variation and often in poor sex reversal.

SEX REVERSAL OF TILAPIA THROUGH IMMERSION (BATH) TREATMENTS

A window appears to occur in the sequence of events leading to gonadal differentiation where a limited exposure once or twice to an exogenous chemical can skew the sex ratio of the treated population. This was best demonstrated in salmon, where MT given one time at concentrations of 400 $\mu g \cdot L^{-1}$ for 2 h before 3 days posthatch fry resulted in a 100 percent male population (Piferrer et al. 1993).

In tilapia, the results have been variable. Several androgens have been added to the holding water using a variety of protocols. Al-Daham (1970) treated *O. aureus* fry for 12 days with testosterone at 0.4 $mg \cdot L^{-1}$ and produced 89 percent males, but a 6 day treatment was ineffective. Eckstein and Spira (1965) treated 4- to 5-week-old *O. aureus* for 5 to 6 weeks with MT with variable results. Torrans et al. (1988) used mestanolone at 0.6 and 1 $mg \cdot L^{-1}$ to treat *O. aureus* fry for 5 weeks and obtained populations that averaged 82 percent male and 18 percent intersex fish. Gale et al. (1995) treated *O. niloticus* fry at 10 and 13 days PF with MT at either 100 or 500 $\mu g \cdot L^{-1}$ for 3 h; they could significantly skew the sex ratio at 100 $\mu g \cdot L^{-1}$ but not at 500 $\mu g \cdot L^{-1}$. They were able to produce 100 percent males in one trial and 94 percent in another using mestanolone at 500 $\mu g \cdot L^{-1}$ following the same protocol. Contreras-Sanchez et al. (1997) obtained more than 90 percent males by exposing *O. niloticus* fry to two 2-h baths of 500 $\mu g \cdot L^{-1}$ trenbolone acetate. Gale et al. (1999) achieved variable results in three trials where *O. niloticus* fry were treated at 10 and 13 days PF for 3 h in 500 $\mu g \cdot L^{-1}$ baths of either MT or 17α-methyldihydrotestosterone (MDHT). The percentage of males ranged from 52 to 87 for the MT treatments and 83 to 100 for the MDHT treatments.

Bath treatments with estrogens have also given mixed results in feminizing tilapia. Rosenstein and Hulata (1992) treated *O. mossambicus* with 17β-estradiol over a range of concentrations and durations with no effect on the sex ratios. Gilling et al. (1996), in a series of trials, found that up to 88 percent female populations could be produced when *O. niloticus* sac fry were treated with EE_2 at 100 $\mu g \cdot L^{-1}$ for 12 or more days where the holding water was replaced every 2 days and additional EE_2 added. They obtained a 91 percent female population using DES at 100 $\mu g \cdot L^{-1}$ for 30 days.

Toxicity is an issue in estrogen bath treatments. Eckstein and Spira (1965) reported high mortality of *O. aureus* fry given stilbestrol diphosphate baths at 400 to 1,000 $\mu g \cdot L^{-1}$. Gilling et al. (1996) had mortalities greater than 50 percent when EE was given at 250 $\mu g \cdot L^{-1}$, DES at 500 $\mu g \cdot L^{-1}$, and EE_2 at 200 $\mu g \cdot L^{-1}$.

EVALUATION OF TREATMENT EFFICACY

Gonadal Examination

Treatment efficacy should be evaluated based on a detailed examination of the gonads of a representative sample of fish. Tilapia can be sexed with reasonable certainty based on the appearance of the genital papilla if they have not been hormone treated, but for hormone-treated fish, the nature of the gonad does not always correspond with the shape of the papilla. Phelps et al. (1993) followed an MT treatment protocol that gave only a partial sex reversal. They carefully examined the papillae of 270 *O. niloticus* for the opening of an oviduct; 82 percent were classified as male based on the appearance of the papilla. Examination of the gonads of the fish with "male" papilla revealed that only 60 percent had gonads with well-developed testicular tissue; 29 percent were intersex, and 11 percent had normal ovaries. Likewise, 12 percent of the fish with female papilla had gonads comprised of only testicular tissue.

Sampling error from an inadequate sample size can contribute to a misinterpretation of efficacy. In commercial-scale production, a minimum of 300 fish should be collected randomly at the end of the hormone treatment period after crowding the fish together. These fish should be grown out to 5 cm or more with a feed that is not hormone-treated in water free of hormone and preserved in 10 percent formalin. A Sample of 100 or more fish representative of the population length-frequency distribution should be selected for gonadal examination. For Nile tilapia that were 11 to more than 17 mm at the end of the treatment period, Popma (1987) found the smaller fish to be more frequently female. In contrast, Hiott (1990) found that when fish were grown to 4 to 11 cm following MT treatment, females were more common among the larger fish.

For gonadal examination, dissecting equipment is needed, along with a microscope, slides, and a stain. Guerrero and Shelton (1974) described a gonadal squash technique using an acetocarmine stain. Other effective stains include fast green and hemotoxilin. Fish should be preserved a minimum of 10 days in formalin before gonadal examination. The gonadal tissue of fish preserved for less than 10 days remains elastic and often breaks when removed. Fish are dissected by making a cut from near the anus to below the base of the pectoral fin. The entire gonad, located on the dorsal portion of the peritoneal lining, should be carefully removed from posterior forward. For efficient use of supplies, four to five sets of gonads are placed on a microscope slide and each is stained with a drop of dye. Another slide is placed on top and the gonads are gently rolled or squashed. When larger fish are examined, an ovary with readily apparent eggs may be obvious in the body cavity, but the gonad may also contain testicular tissue and should be examined microscopically. Thick gonads may need to be sliced longitudinally before they can be examined properly. The entire length should be examined to see whether it contains only one type of gonadal tissue.

In a gonadal squash of an ovary, eggs of various sizes will be evident throughout the gonad (Figure 6.1). It should be possible to focus up and down on an egg to see the nucleus. Testicular tissue is not so obvious. Lobes of the testis will be apparent, but other structures are not as distinct (Figure 6.2). Connective tissue, oviduct, or sperm duct may also appear on the slide.

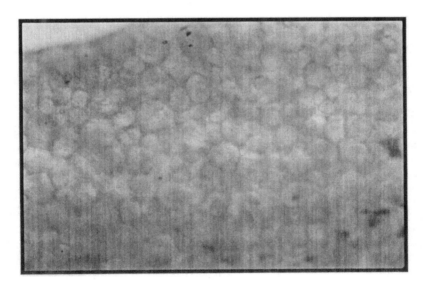

FIGURE 6.1. Tilapia ovary as it appears in a gonadal squash, with individual eggs having an evident nucleus.

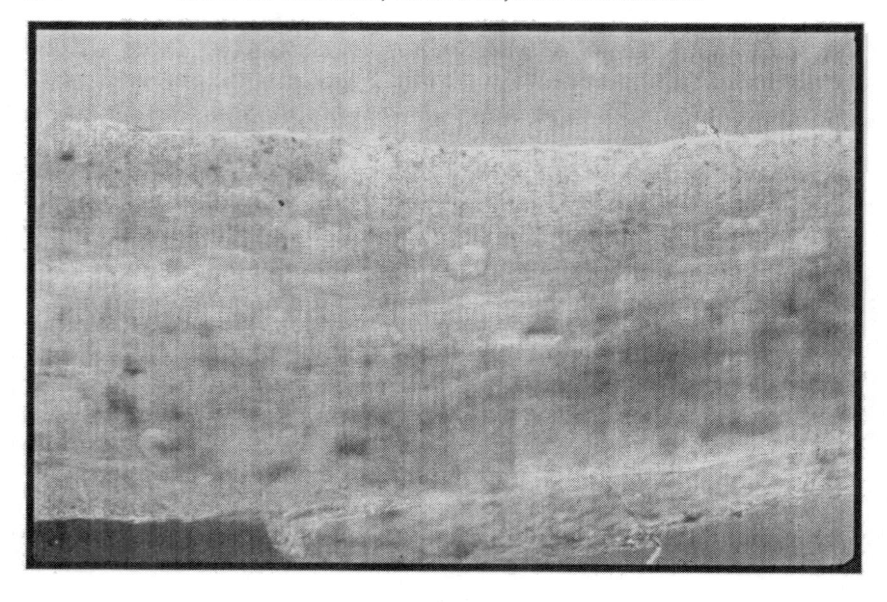

FIGURE 6.2. Testicular tissue of tilapia as it appears in a gonadal squash.

Significance of Intersex Fish

Gonads may be found to contain both ovarian and testicular tissue (ovotestes). Such intersex fish are found at a low frequency in normal tilapia not treated with hormones (Clark and Grier 1985; Okoko 1996); most commonly, small portions of ovary are interspersed within the testis. Intersex hormone-treated fish contain a variety of combinations of ovarian and testicular tissues; they, vary in terms of the percentage of the gonad that is of a given tissue type and where such tissues are located (Figure 6.3).

The reproductive viability of intersex fish is difficult to evaluate. The intersex condition is known only after the fact, when the fish has been killed and the gonad removed. On occasions, bloated hormone-treated adults are found with the anterior portion of the gonad putrefied and no oviduct evident. Clark and Grier (1985) reproduced several apparently male *O. aureus* and found that three were intersex fish with nonvitellogenic oocytes (25 to 75 mm) within the testicular tissue. Nakamura (1975) found that oocyctes in the gonads of intersex *O. mossambicus* had degenerated by 25 to 40 days after androgen treatment.

FIGURE 6.3. Intersex gonads may have sections of ovarian tissue or isolated eggs interspersed within testicular tissue.

Acceptable Frequencies of Males and Management Strategies

The minimum acceptable percentage of males after sex reversal depends on the culture technique and acceptable market size. When androgen treatments are effective, the percentage of females should be less than 5. If a small fish is acceptable in the market, less than 95 percent males may be allowable. Lovshin, Tave et al. (1990) found that even in a mixed-sex culture of *O. niloticus,* young 5 g fish stocked at 10,000 per hectare reached 140 g in 104 days (males = 150 g, females = 126 g) with only minimal reproduction. The impact of a few females can be more significant if a heavier market weight is required. Lovshin, Da Silvar et al. (1990) found that even 2.5 percent females in a tilapia population depressed growth within 4 months when no attempt was made to control recruitment. Reproduction was 58 percent of the weight harvested after 9 months of culture in ponds when females were 2.5 percent of the initial stock. Recruitment is most serious after the offspring from the few females begin to spawn, typically 5 to 7 months after initial stocking of 20 to 30 g "sex-reversed" fish.

In outdoor ponds, production of fish large enough to yield 5 to 7 oz fillets requires a two-phase grow-out or the introduction of predators to control recruitment. Several species of predators, including guapote tigre *(Cichlasoma managuense),* largemouth bass *(Micropterus salmodies),* and tucunare *(Cichla ocellaris),* have been used effectively to control tilapia reproduction in "all-male" tilapia ponds (Dunseth and Bayne 1978; Pike 1983; McGinty 1983, 1985; Green and Teichert-Coddington 1994). Although tilapia can reproduce in recirculating systems, predators are seldom used to control recruitment. Females can be removed to some extent by grading. Pruginen and Shell (1962) were able to grade *O. niloticus* weighing 13 to 17.9 g and separate males with more than 88 percent effectiveness.

FACTORS AFFECTING SEX REVERSAL

Size and Age

The recommendation of Shelton et al. (1978) to begin hormone treatment using 9 to 11 mm fry is still valid, although larger fish have on occasion been successfully sex reversed. Argue (1992) sex reversed *O. niloticus* beginning with fry less than 19 days old but longer than 14 mm and obtained 91 percent males and 4 percent intersex. Hiott and Phelps (1993) obtained 87.3 percent males from treating 12 to 13 mm *O. niloticus* that were less than 10 days old but only 66.7 percent males when they treated 12 to 13 mm fry that were 10 days or older. Nakamura and Iwahashi (1982) obtained 98 percent males when they treated 12.7 mm *O. niloticus,* and 70.2 percent males when fry were held for 10 additional days and measured 15.9 mm at the start of treatment.

Duta (1979) reported that at 31°C, *O. aureus* gonadal tissue had differentiated to ovary at 14 days of age and fry were 15 to 17 mm total length, whereas at 21°C differentiation took place at an age of 24 to 27 days and the fry were 14 to 17 mm long. The relationship between growth rate, temperature, gonadal differentiation, and successful sex reversal is still not clear, but as discussed by Nakamura and Takahashi (1973) hormone treatment must begin before the onset of gonadal differentiation and continue until it is complete—timings that depend on production conditions. Popma and Green (1990) suggested that after a 25 to 28 day hormone treatment, grading may be necessary to remove any fish 13 mm or less that may have not completed gonadal differentiation. Hiott and Phelps (1993) and Argue and Phelps (1996) found that females were more common among the largest of the treated fish.

Fish age and size variation may affect survival. Ellis et al. (1993) studied recently hatched fry, older fry, and a set composed of 50 percent from each size group. They found that after 30 days, the smallest fry were more uniform at harvest. The mixed-size group not only was more variable at harvest but had 38 percent lower survival.

The greatest efficacy of sex reversal and high survival are achieved by beginning hormone treatment with the youngest, smallest fry possible, uniform in size at less than 14 mm.

Treatment Duration

The oral administration of MT-treated feed (30 to 60 $mg \cdot kg^{-1}$ feed) to tilapia fry for 21 to 28 days has yielded populations of no more than 5 percent females under a variety of environmental conditions (Table 6.4). The duration of treatment must be adequate to allow all fish to complete gonadal differentiation. Mbarereehe (1992) found that with *O. niloticus* given MT at 60 $mg \cdot kg^{-1}$ diet, at water temperatures between 18°C and 22°C, a 40 day treatment period resulted in 95 percent males, but a 20 day treatment gave only 69 percent males. Bocek et al. (1991) produced 98 percent male *O. niloticus* by feeding MT at 60 $mg \cdot kg^{-1}$ for 30 days at 21°C to 23°C. At the end of treatment the fish averaged 14.9 mm. Hiott and Phelps (1993), feeding MT at 60 $mg \cdot kg^{-1}$ diet to *O. niloticus* for 28 days at 27°C, found that the effectiveness of hormone sex reversal was correlated with the number of days fry received hormone before reaching 18 mm. They started with fish shorter than 11 mm, which grew rapidly during treatment. Fish received 14 days of hormone feeding before reaching 18 mm and were 95.7 percent males. When they started with 12 to 13 mm fish under the same conditions, fish were larger than 18 mm in 9 days and were only 87.3 percent males. Owusu-Frimpong and Nijjhar (1981) treated 13.5 mm *O. niloticus* fry with MT for 42 days. These fish had reached only 18 to 19 mm after 14 days and were successfully sex reversed. Pandian and Varadaraj (1988) treated 10-day-old *O. mossambicus* with MT for 11 days and obtained 100 percent males; in another trial they treated 13-day-old fry for 13 days and obtained 69 percent males. Duration of treatment should be related to initial size and growth conditions. As a general rule, fish should receive at least 14 days of hormone treatment before reaching 18 mm. If growth is slower, the duration of treatment should be extended until all fish reach this size or a total treatment period of 28 days is exceeded. If growth is too fast, it may be necessary to reduce the feeding frequency or slightly lower the feeding rate to reduce the growth rate.

TABLE 6.4. Effect of various dosages (in $mg \cdot kg^{-1}$ of diet) and treatment periods of 17α-methyltestosterone (MT) on percentage sex inversion and survival of tilapia (*Oreochromis* spp.).

Dose	Days	Percentage males	Percentage survival	Reference
Control	18	56.0	85.0	Guerrero (1975)
MT-15		84.0	88.0	
MT-30		98.0	82.0	
MT-60		85.0	88.0	
Control	19	59.5	70.0	Nakamura (1975)
MT-50		100	80.0	
MT-1,000		61.0	95.0	
MT-1,000	44	0.0	58.0	
Control	59	54.0	97.7	Tayamen and Shelton (1978)
MT-30		100	95.5	
MT-60		100	98.5	
Control	22	50.4	71.0	McGeachin et al. (1987)
MT-60		99.0	70.0	
MT-90		98.0	62.0	
MT-120		96.0	58.0	
Control	19	54.0	96.0	Varadaraj and Pandian (1989) (Feeding ad libitum)
MT-2		54.0	96.0	
MT-5		100	96.0	
MT-10		100	96.0	
MT-20		100	96.0	
MT-30		100	94.0	
MT-40		100	90.0	
Control	28	54.7	94.3	Phelps et al. (1992)
MT-60		97.8	93.1	
Control	28	51.3	85.0	Green and Teichert-Coddington (1994)
MT-60		96.8	88.9	

The most appropriate timing for immersion treatments is not clear. Extended treatment periods of 12 days or more have been effective, whereas short exposures once or twice from 11 to 14 days after hatching have yielded mixed results.

Dose Rate

The effective dose for sex reversal depends on the quantity of feed consumed daily. Varadaraj and Pandian (1989) reported 100 percent masculinization when *O. mossambicus* were fed to satiation a feed containing MT at 5 $mg \cdot kg^{-1}$. When the feeding rate was controlled, using a diet containing the same MT concentration, Pandian and Varadaraj (1987) obtained 60 percent males using a 10 percent body weight per day rate, 80 percent males at 20 percent body weight per day, and 100 percent at 30 percent body weight per day. They concluded that to obtain a 100 percent male population a minimum rate of MT of 1.5 μg $MTday^{-1} \cdot g^{-1}$ of fish needed to be consumed.

Rodriguez-Guerrero (1979), feeding 17α-ET to *O. aureus,* found that an effective dose corresponded to a mean daily hormone intake over 21 days of 0.46 $\mu g \cdot day^{-1}$ per fish. Okoko (1996) fed *O. niloticus* for 28 days at 15 percent body weight per day diets containing MT at 3.75, 7.5, 15, 30, 60, 120, 240, 480, 600, or 1,200 $mg \cdot kg^{-1}$. At 3.75 mg, 80 percent males and 19 percent intersex were produced; at 7.5 mg, the results were 91.7 percent males and 8.3 percent intersex; and at 15 mg, 98.3 percent males and 1.75 percent intersex were produced. It was calculated that daily MT intakes of 0.52 to 2.85 $\mu g \cdot g^{-1}$ of fish gave more than 95 percent male populations.

Excessive androgen intake can reduce treatment efficacy. Nakamura (1975) obtained 100 percent males when feeding MT to *O. mossambicus* at 50 $mg \cdot kg^{-1}$ of diet, but at 1,000 $mg \cdot kg^{-1}$, the percentage of males was 61.4. Okoko (1996) found that doses of MT of 240, 480, 600, or 1,200 $mg \cdot kg^{-1}$ of diet at 15 percent body weight per day resulted in no increase in the percentage of males produced (Figure 6.4).

Exact daily hormone intakes are hard to achieve. Knowing the exact weight and number of fry on a given day is difficult, and appetite may vary from day to day. In a clear-water environment with no natural food available, a diet containing 15 to 30 $mg \cdot kg^{-1}$ may be effective, whereas in an outdoor setting with natural food available an optimum dose may be between 30 and 60 $mg \cdot kg^{-1}$ of diet fed at 15 percent body weight or more.

Environmental Parameters

Environmental factors such as temperature and water quality can impact growth and the treatment duration. Temperature alone can also skew sex ratios. Baroiller et al. (1995) skewed the sex ratio of *O. niloticus* to a greater frequency of males by holding fry at 36°C during gonadal differentiation. Desprez and Melard (1998) found similar results with *O. aureus,* where the

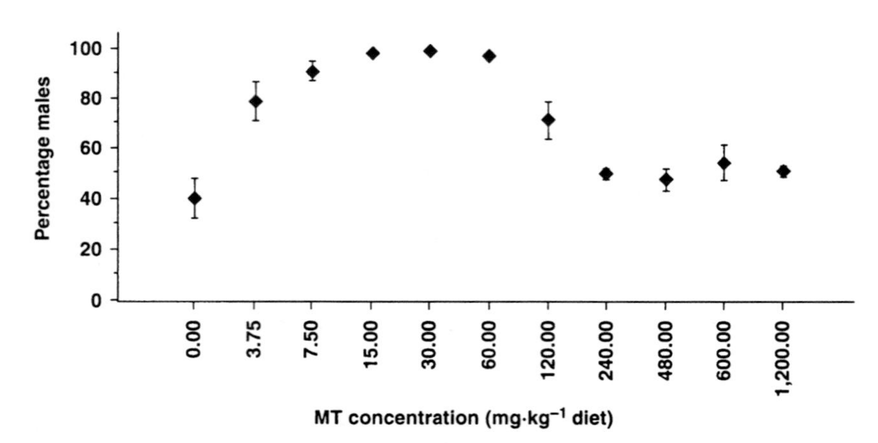

FIGURE 6.4. Percentage of males produced when *O. niloticus* fry were fed a diet containing 0, 3.75, 7.5, 15, 30, 60, 120, 240, 480, 600, or 1,200 mg·kg^{-1} MT for 28 days. Data from Okoko (1996).

sex ratio was altered to 97.8 percent male by holding fry not treated with hormones at 34°C. Using *O. mossambica,* Varadaraj et al. (1994) found that the sex ratio of untreated fry was not altered when they were grown at 22°C, 25°C, 27°C, 33°C, or 38°C, but the effectiveness of a 10 mg·kg^{-1} MT treatment was affected. At 32°C, intersex fish were 35 percent of the population, and at 38°C females were 73 percent of the population and the remaining 27 percent were intersex.

Optimum holding temperature for fry during the sex-reversal period is between 26°C and 28°C. Temperatures below 24°C significantly reduce growth and may result in some fish not having completed gonadal differentiation during the treatment. Lower temperatures also favor more disease problems. Spring water or well water is often too cool for direct use and should be held initially in a reservoir. Even a slight change in temperature can affect growth and survival.

Most sex reversal is carried out in freshwater over a range of alkalinities and hardnesses. The literature does not suggest that efficacy is affected by alkalinity or hardness. Several species of tilapia, such as *O. mossambicus* and *O. spilurus,* can reproduce in brackish water or full-strength seawater, and the fry can be sex reversed under such conditions (Wantanabe et al. 1993; Ridha and Lone 1995). Water quality conditions that allow good growth and survival of fry are appropriate for sex reversal. Dissolved oxygen (DO) concentrations should remain above 4 mg·L^{-1} to ensure a strong feeding response. The fry will tolerate lower levels but are stressed and more susceptible to diseases. Fouling of hapas and the resultant restricted water exchange can contribute to low DO conditions.

No effect of pH on the sex ratio of tilapia has been demonstrated; however, in another genus of cichlids *(Apistogramma)* low pH has skewed the sex ratio to male. Romer and Beisenherz (1996) evaluated several species of *Apistogramma* and found that when eggs and fry were held at 26°C and at a pH of 4.5, the sex ratio was skewed to male, whereas at pH 6.5 the ratio was skewed to female.

Genetic Differences

Variation in efficacy may be associated with differences in susceptibility among families of fish. Contreras et al. (1999) found that the effectiveness of trenbolone acetate immersions in altering sex ratio varied across experiments and thought this might be related to different degrees of sensitivity among broods. The ability of temperature to alter sex ratio has been shown to vary among spawns within the same strain of *O. niloticus* (Mair et al. 1990). Baroiller et al. (1999) found that high temperature (36°C) could alter *O. niloticus* sex ratios; however, the response varied among families. When families of XX genotype fry were held at 36°C, the percentage of males obtained ranged from 5 to 98. At Auburn University, selected families of fry were treated with MT at 60 $mg \cdot kg^{-1}$ of diet for 28 days. The resultant populations were 84 to 100 percent male, whereas, the untreated populations were between 18 and 100 percent male. However, there was no correlation between the sex ratio of the untreated fish of a family and the efficacy of MT treatment (Phelps, unpublished data).

ANABOLIC CONSIDERATIONS

Androgens have both an androgenic and an anabolic effect. The anabolic effect resulting from the use of androgens to sex reverse tilapia is difficult to identify. Several studies that discuss improved growth of sex-reversed fish relative to nontreated fish are comparing the growth of nearly all-male populations with that of a mixed-sex population after hormone treatment; however, the presence of females reduces the growth rate because of the slower growth of females or their reproduction (Guerrero 1975; Hanson et al. 1983; Macintosh et al. 1985; McAndrew and Majumdar 1989). Likewise, comparisons of genetically all-male tilapia and males obtained by sex reversal are often complicated by a few females and their reproduction in one or more of the treatments. Hanson et al. (1983) compared growth of a genetically mixed-sex population sex reversed to male, a genetically female population sex reversed to male, and genetic "males" selected by examination of the genital papilla. Both sets of sex reversed fish (99.4 and 100 percent male, respectively) grew larger than the hand-sexed fish, but the hand-sexed "males" contained 6.2 percent females, which may have depressed the mean final weight. Mair et al. (1995) found that genetically male tilapia (99.1 percent males) and sex-reversed "males" (71.6 percent males) reached similar average weights in 168 days, but reproduction represented 10.9 percent of the total biomass in ponds with sex-reversed males. Tuan et al. (1998), in one trial, found no difference in growth of genetically "male" (96.2 percent male) and sex-reversed "male" (99.3 percent male) populations. In a second trial, sex-reversed "males" (100 percent male) grew larger than the genetic "males," but the genetic "male" population was only 82.6 percent males. Green and Teichert-Coddington (1994) evaluated growth and survival of MT-treated and untreated Nile tilapia fry during the treatment, nursery, and grow-out phases under commercial, semi-intensive conditions in earthen ponds. During treatment, fry were held in hapas suspended in 0.2 ha ponds and were fed either a 0 or 60 $mg \cdot kg^{-1}$ MT diet for 28 days. No significant difference in growth was found during any phase.

Any improved growth of androgen-treated tilapia is related more to the superior growth of males than to the more classical anabolic response of enhanced protein synthesis and increase in muscle mass.

FOOD SAFETY AND ENVIRONMENTAL CONSIDERATIONS

Sex reversal of tilapia should consider food safety and environmental issues associated with the use of steroids. It is the obligation of the producer to insure that the public receives the highest quality fish possible, produced using techniques that have minimal adverse effects on the environment. In the United States, the Food and Drug Administration (FDA) has approved only a few drugs for use in aquaculture. Efforts are now under way to obtain approval for the most commonly used steroid, MT. Numerous scientific studies of MT are available to support the approval process. Other androgens used for sex reversal have a smaller data base and will require considerably more investigation to provide the data needed for the drug approval process.

The short treatment duration and rapid metabolism of MT help insure that tilapia are free of MT before fish reach the consumer. Using Growth data, Dambo and Rana (1992) show that fish reaching 0.5 g at the end of the treatment period consume 29.9 to 33.2 μg MT during the treatment period. Ingested MT is rapidly metabolized and excreted. Curtis et al. (1991) fed tilapia fry for 30 days with a feed containing radioactively labeled MT, and only a trace of MT could be found 10 days after the end of treatment. Goudie et al. (1986) found that when radioactively labeled MT was fed to tilapia, the head and viscera contained more than 90 percent of the MT, and 21 days after treatment less than 1 percent remained. Johnstone et al. (1983) found more than 95 percent of the radioactively labeled MT in the viscera, and no radioactivity could be found 50 h after treatment. This rapid metabolism and excretion of MT by fish treated early in life, combined with the short treatment duration and extended period needed to produce fish of marketable size, ensures a safe consumer product.

No detailed studies on the environmental fate of androgens are available, but under certain conditions they can produce secondary effects. Contreras-Sanchez et al. (2001b), using static-water, indoor aquaria, found that MT could be detected in the water column soon after tilapia were fed MT-treated feed. MT is susceptible to breakdown when exposed to light or high temperatures (American Hospital Formulary Service 1997). Both fungi and bacteria can metabolize exogenous steroids. Many different steroid metabolic reactions, including metabolism of MT, are possible in bacteria (Schubert et al. 1972; Jankov 1977), as well as metabolism of steroids to carbon dioxide and water (Sandor and Mehdi 1979).

In an outdoor pond where fry are treated in hapas, the combination of light, temperature, and microbial degradation should result in the rapid breakdown of MT. In a pilot study without fish, a single 40 $mg \cdot L^{-1}$ dose of testosterone was added to one tank of a two-tank recirculating system (Budworth and Senger 1993). Dilution within the recirculating system resulted in the initial testosterone concentration (17.4 $\mu g \cdot L^{-1}$) being decreased to 10.6 $\mu g \cdot L^{-1}$. The concentration in water, measured by double-antibody radioimmunoassay, peaked in both tanks about 2 h after application and then decreased exponentially to near 0 $\mu g \cdot L^{-1}$ testosterone 18 h after application. Phelps et al. (2000) studied MT given to tilapia fry in outdoor hapas. They found MT levels in water from within the treatment hapa to be in general similar to pretreatment levels in pond water. Likewise, levels in the soil were similar before, during, and after treatment. Contreras-Sanchez et al. (2001a) found similar MT concentrations in water and soil before and after MT treatment for tilapia held in hapas in an outdoor pond.

Nontarget fish may be unaffected when held in the vicinity of fish being fed MT in outdoor tanks or ponds. In an outdoor tank, Phelps et al. (1992) used small hapas spaced approximately 30 cm apart in a 20 m^2 static-water tank to hold fish fed either an MT- or fluroxymesterone-treated feed or an untreated feed. The treatments were randomly assigned within the tank, and there was no evidence of hormone leaching affecting the sex ratio of untreated fish. Soto (1992) used a similar physical setup with eighteen 0.12 m^2

hapas distributed in a 20 m^2 tank to evaluate the androgenic potential of mestanolone to sex reverse *O. niloticus.* No evidence of untreated fish having a skewed sex ratio was reported, even though such fish were surrounded by hormone-treated sets of fish.

Indoors, MT may be more persistent. Abucay et al. (1997) found that reusing water that had held tilapia fry during a 25 day MT treatment could alter sex ratios. When a second group of fish were stocked into such water and given untreated feed, the sex ratio was skewed. They also found that when "all female" fry were stocked into an enclosure inside an aquarium and MT-treated feed was placed on the bottom of the aquarium, where the fish had no access to it, the sex ratio became skewed to males. Contreras-Sanchez et al. (2001b) fed three successive sets of tilapia fry MT-treated feed in static-water aquaria containing soil and found detectable levels of MT in the soil after the third cycle. A fourth and fifth cycle were conducted in the same aquaria without changing the soil or water but with the fry given untreated feed. During the fourth and fifth cycles, an apparent MT level of 1,265 to 3,193 $pg \cdot g^{-1}$ of soil was present, but there was no skewing of the sex ratio of the untreated fish.

In recirculating systems where MT has been given daily, MT may remain in the water column long enough to influence sex ratios. All-female gynogenetic 27- to 40-day-old common carp *(Cyprinus carpio)* were held in a recirculating system (Gomelsky et al. 1994). Fish in one tank were fed an androgen-treated diet, and fish in another tank within the same system were fed an androgen-free diet. Four separate experiments were conducted with different ages and weights of fish, where fish (MT-fed group) in one tank were fed a formulated diet containing MT at 100 $mg \cdot kg^{-1}$ for 40 days and fish (water-exposed group) in the other tank of the same system were fed an androgen-free diet. A control population of the same fish fed the androgen-free diet was grown in an earthen pond. Direct or indirect androgen treatment, either through oral administration (MT-fed group) or water exposure (water-exposed group), induced sex inversion in all experiments. The frequency of males in MT-fed and water-exposed fish populations within the same recirculating system ranged from 46.7 to 96.6 percent, depending on the experiment, whereas control fish from earthen ponds were entirely female. Within each experiment, the percentages of inverted males among the MT-fed and water-exposed fish populations did not differ significantly. Differences in response among the four experiments were attributed to differences in initial age and individual weight of fish used.

These results indicate that unmetabolized MT and metabolites of MT can accumulate in the water of recirculating systems or perhaps static water not exposed to direct sunlight. The degree of accumulation appears to depend on the frequency and dose of MT administered to the target fish.

CONCLUSIONS

Production of male tilapia using androgens is very effective. It does not require that a portion of the production be discarded as in manual selection or that two separate stocks of fish be maintained as in hybridization. Currently available seed production techniques are adaptable to producing appropriately sized fry for sex reversal. The relative ease and predictability of tilapia sex reversal have been major factors in the rapid growth of the industry.

Although a variety of hormones have been used for sex reversal, MT is the most commonly used androgen. Dose rate and treatment durations vary depending on environmental conditions, fish species, and life stage. Tilapia fry less than 14 mm long should be treated for at least 14 days before reaching 18 mm. If growth is slower, the duration of treatment should be extended until all fish reach this size (18 mm) or to at least 28 days. An MT dose rate of 30 to 60 $mg \cdot kg^{-1}$ of diet fed at an initial rate of 20 percent body weight per day three or four times daily should result in successful treatment. Estimates of the efficacy of treatment should be based on gonadal examinations.

As aquaculture supplies an increasing portion of the world's fisheries products, tilapia culture will play a more important role. Sex reversal will remain the industry standard for reproduction control in tilapia.

REFERENCES

Abucay, J.S., G.C. Mair, D.O.F. Skibinski, and J.A. Beardmore (1997). The occurrence of incidental sex reversal in *Oreochromis niloticus* L. In K. Fitzsimmons (Ed.), *Tilapia aquaculture: Proceedings from the Fourth International Symposium on Tilapia in Aquaculture* (pp. 729–738). Ithaca, NY: Northeastern Regional Agricultural Engineering Service Cooperative Extension.

Afonso, L.O.B., G.J. Wassermann, and R.T. De Oliveira (2001). Sex reversal in Nile tilapia *(Oreochromis niloticus)* using a nonsteroidal armomatase inhibitor. *Journal of Experimental Zoology* 290:177-181.

Al-Ablani, S. (1997). Use of synthetic steroids to produce monosex populations of selected species of sunfish (family: Centrachidae). Doctoral dissertation, Auburn University, Auburn, Alabama.

Al-Daham, N.K. (1970). The use of chemosterilants, sex hormones, radiation, and hybridization for controlling reproduction in tilapia species. Doctoral dissertation, Auburn University, Auburn, Alabama.

Alvendia-Casauay, A., and V.S. Carino (1988). Gonadal sex differentiation in *Oreochromis niloticus*. In R.S.V. Pullin, T. Bhukaswan, K. Tonguthai, and J.L. Maclean (Eds.), *The Second International Symposium on tilapia in*

Aquaculture. ICLARM Conference Proceedings (pp. 121-124). Department of Fisheries, Bangkok, Thailand, and International Center for Living Aquatic resources Management, Manilla, Philippines.

American Hospital Formulary Service (AHFS) (1997). 97 Drug Information. In G.K. McEvoy (Ed.), American Hospital Formulary Service, Washington, DC.

Argue, B.J. (1992). An evaluation of several techniques for production of Nile tilapia *(Oreochromis niloticus)* sex-reversed fry. Master's thesis, Auburn University, Auburn, Alabama.

Argue, B.J., and R.P. Phelps (1996). Evaluation of techniques for producing hormone sex-reversed *Oreochromis niloticus* fry. *Journal of Aquaculture in the Tropics* 11:153-159.

Baroiller, J.-F., Y. Guiguen, and A. Fostier (1999). Endocrine and environmental aspects of sex differentation in fish. *Cellular and Molecular Life Scienes* 55:910-931.

Basavaraja, N., M.C. Nadeesha, T.J. Varghese, P. Keshavanath, and G.K. Srikant (1990). Induction of sex reversal in *Oreochromis mossambicus* by diethylstilbestrol. *Journal of Applied Ichthylogy* 6:46-50.

Berger, A., and S. Rothbard (1987). Androgen induced sex-reversal of red tilapia fry stocked in cages within ponds. *Bamidgeh* 39:49-57.

Berkowitz, P. (1937). Effect of oestrogenic substances in *Lebistes reticlatus* (guppy). *Proceedings of the Society for Experimental Medicine* 36:416-418.

Bocek, A.J., R.P. Phelps, and T.J. Popma (1991). Effect of feeding frequency on sex reversal and growth of Nile tilapia, *Oreochromis niloticus. Journal of Applied Aquaculture* 1:97-103.

Bogart, M.H. (1987). Sex determination: A hypothesis based on steroid ratios *Journal of Theoretical Biology* 128:349-357.

Brueggemeier, R.W. (1986). Androgens, anabolics and antiandrogens. In M. Verderame (Ed.), *Handbook of hormones, vitamins, and radiopaques* (pp. 1-49). Boca Raton, FL: CRC Press, Inc.

Buddle, C.R. (1984). Androgen-induced sex-reversal of *Oreochromis* (Trewavas) hybrid fry stocked into cages standing in an earthen pond. *Aquaculture* 40:233-239.

Budworth, P.R., and P.L. Senger (1993). Fish-to-fish testosterone transfer in a recirculating-water system. *The Progressive Fish-Culturist* 55:250-254.

Camerino, B., and R. Sciaky (1975). Structure and effects of anabolic steroids. *Pharmacology and Therapeutics* 1B:233-275.

Chambers, S.A. (1984). Sex reversal of Nile tilapia in the presence of natural food. Master's thesis, Auburn University, Auburn, Alabama.

Clark, B., and H.J. Grier (1985). Testis-ova in spawning blue tilapia, *Oreochromis aureus. Gulf Research Reports* 8:69-70.

Clemens, H.P., and T. Insee. (1968). The production of unisexual brood of *Tilapia mossambica* sex reversed with methyltestosterone. *Transactions of the American Fisheries Society* 97:18-21.

Contreras-Sanchez, W.M., M.S. Fitzpatrick, R.H. Milston, and C.B. Schreck. (1997). Masculinization of Nile tilapia *(Oreochromis niloticus)* by single immersion in 17α-methyldihydrotestosterone and trenbolone acetate. In K. Fitzsimmons (Ed.), *Tilapia aquaculture: Proceedings from the Fourth International Symposium on Tilapia in Aquaculture* (pp. 783-790). Ithaca, NY: Northeastern Regional Agricultural Engineering Service Cooperative Extension.

———. (1999). Masculinization of Nile tilapia with steroids: Alternate treatments and environmental effects. Sixth International Symposium on the Reproductive Physiology of Fish. University of Bergen, Bergen, Norway.

Contreras-Sanchez, W.M., M.S. Fitzpatrick, and C.B. Schreck (2001a). Fate of methyltestosterone in the pond environment: Detection of MT in pond soil from a CRSP site. In A. Gupta, K. McElwee, D. Burke, J. Burright, X. Cummings, and H. Egna (Eds.), *Eighteenth Annual Technical Report* (pp. 79-82). Corvallis, OR: Pond Dynamics/Aquaculture CRSP, Oregon State University.

———. (2001b). Fate of methyltestosterone in the pond environment: Impact of MT-contaminated soil on tilapia sex determination. In A. Gupta, K. McElwee, D. Burke, J. Burright, X. Cummings, and H. Egna (Eds.), *Eighteenth Annual Technical Report* (pp. 83-86). Corvallis, OR: Pond Dynamics/Aquaculture CRSP. Oregon State University.

Curtis, L.R., F.T. Diren, M.D. Hurley, W.K. Seim, and R.A. Tubb (1991). Disposition and elimination of 17-α-methyltestosterone in Nile tilapia *(Oreochromis niloticus). Aquaculture* 99:193-201.

Dadzie, S. (1975). A preliminary report on the use of methallibure in tilapia culture. *African Journal of Hydrobiology and Fisheries* 4:127-140.

Dambo, W.B., and K.J. Rana (1992). Effect of stocking density on growth and survival of *Oreochromis niloticus* (L.) fry in the hatchery. *Aquaculture and Fisheries Management* 23:71-80.

D'Cotta, H., A. Fostier, Y. Guiguen, M. Govoroun, and J.-F. Baroiller (2001). Aromatase plays a key role during normal and temperature induced sex differentiation of tilapia *Oreochromis niloticus. Molecular Reproduction and Development* 59:265-276.

Desprez, D., E. Geraz, M. C. Hoareau, C. Melard, P. Bosc, and J.-F. Baroiller (2003). Production of a high percentage of male offspring with a natural androgen, 11β-hydroxyandrostenedione (11β OHA4), in Florida red tilapia. *Aquaculture* 216: 55-65.

Desprez, D., and C. Melard (1998). Effect of ambient water temperature on sex determinism in the blue tilapia *Oreochromis aureus. Aquaculture* 162:79-84.

Dunseth, D.R., and D.R. Bayne (1978). Recruitment control and production of *Tilapia aurea* (Steindachner) with the predator *Cichlasoma managuense* (Gunther). *Aquaculture* 14:383-390.

Dutta, O.K. (1979). Factors influencing gonadal sex differentiation in *Tilapia aurea* (Steindachner). Doctoral dissertation, Auburn University, Auburn, Alabama.

Eckstein, B., and M. Spira (1965). Effect of sex hormones on gonadal differentiation in a cichlid, *Tilapia aurea. Biological Bulletin* 129:482-489.

Ellis, S.C., W.O. Watanabe, and W.D. Head (1993). Effect of initial age variation on production of Florida red tilapia fry under intensive, brackishwater tank culture. *Aquaculture and Fisheries Management* 24:465-471.

Gale, W.L., M.S. Fitzpatrick, M. Lucero, W.M. Contreas-Sanchez, and C.B. Schreck (1999). Masculization of Nile tilapia *(Oreochromis niloticus)* by immersion in androgens. *Aquaculture* 178:349-357.

Gale, W.I., M.S. Fitzpatrick, and C.B. Schreck (1995). Immersion of Nile tilapia *(Oreochromis niloticus)* in 17α-methyltestosterone and mestanolone for the production of all-male populations. In F. Goetz and P. Thomas (Eds.), *Proceedings of the Fifth International Symposium on the Reproductive Physiology of Fish* (p. 117). Austin: The University of Texas at Austin Printing Department.

Galvez, J.I., P.M. Mazik, R.P. Phelps, and D.R. Mulvaney (1995). Masculinization of channel catfish *Ictalurus punctatus* by oral administration of trenbolone acetate. *Journal of the World Aquaculture Society* 26:378-383.

Galvez, J.I., J.R. Morrison, and R.P. Phelps (1996). Efficacy of trenbolone acetate in sex inversion of the blue tilapia *Oreochromis aureus. Journal of the World Aquaculture Society* 27:483-486.

Gilling, C.J., D.O.F. Skibinski, and J.A. Beardmore (1996). Sex reversal of tilapia fry by immersion in water containing oestrogens. In R.S.V. Pullin, J. Lazard, M. Legendre, J.B. Amon-Kottias, and D. Pauly (Eds.), *The Third Symposium on Tilapia in Aquaculture* (pp. 314-319). Manilla, Philippines: ICLARM Conference Proceedings.

Gomelsky, B., N.B. Cherfas, Y. Peretz, N. Ben-Dom, and G. Hulata (1994). Hormonal sex inversion in the common carp (*Cyprinus carpio* L.). *Aquaculture* 126:265-270.

Goudie, C.A., W.L. Shelton, and N.C. Parker (1986). Tissue distribution and elimination of radiolabelled methyltestosterone fed to sexually undifferentiated blue tilapia. *Aquaculture* 58:215-226.

Green, B.W., and D.R. Teichert-Coddington (1993). Production of *Oreochromis niloticus* fry for hormonal sex reversal in relation to water temperature. *Journal of Applied Icthyology* 9:230-236.

———. (1994). Growth of control and androgen-treated Nile tilapia, *Oreochromis niloticus* (L.), during treatment, nursery and grow-out phases in tropical fish ponds. *Aquaculture and Fisheries Management* 25:613-621.

Guerrero, R.D., III (1975). Use of androgens for the production of all-male *Tilapia aurea* (Steindachner). *Transactions of the American Fisheries Society* 104:342-348.

———. (1976). *Tilapia mossambica* and *T. zilli* treated with ethynyltestosterone for sex reversal. *Kalikasan, Philippine Journal of Biology* 5:187-192.

———. (1982). Control of tilapia reproduction. In R.S.V. Pullin, and R.H. Lowe-McConnell (Eds.), *The biology and culture of tilapias* (pp. 309-316). Manila, Philippines: ICLARM Conference Proceedings 7, International Center for Living Resources Management.

Guerrero, R.D., III, and L.A. Guerrero (1988). Feasibility of commercial production of Nile tilapia fingerlings in Philippines. In R.S.V. Pullin, T. Bhukaswan, K. Tonguthai, and J.L. Maclean (Eds.), *The Second International Symposium on Tilapia in Aquaculture* (pp. 183-186). ICLARM Conference Proceedings 15. Department of Fisheries, Bangkok, Thailand, and International Center for Living Aquatic Resources Management, Manilla, Philippines.

———. (1993). Effect of oral treatment of mibolerone on sex reversal of *Oreochromis mossambicus*. Asian Fisheries Science 6:347-350.

———. (1997). Effects of androstenedione and methyltestosterone on *Oreochromis niloticus* fry treated for sex reversal in outdoor net enclosures. In K. Fitzsimmons (Ed.), *Tilapia aquaculture: Proceedings from the Fourth International Symposium on Tilapia in Aquaculture* (pp. 772-777). Ithaca, NY: Northeastern Regional Agricultural Engineering Service Cooperative Extension.

Guerrero, R.D., and W.I. Shelton (1974). An aceto-carmine squash method for sexing juvenile fishes. *The Progressive Fish-Culturist* 36:56.

Guiguen, Y., J.-F. Baroiller, M.J. Ricordel, K. Iseki, O.M. McMeel, S.A.M. Martin, and A. Fostier (1999). Involvement of estrogens in the process of sex differentiation in two fish species: The rainbow trout *(Oncorhynchus mykiss)* and a tilapia *(Oreochromis niloticus). Molecular Reproduction and Development* 54:154-162.

Hackmann, E. (1971). Paradoxe gonaden differenzierung nach behandlung mit Androgenen bei verschiedenen Cichliden (Teleostei). Doctor's dissertation abst., Johannes Gutenberg University.

Hanson, T.R., R.O. Smitherman, W.L. Shelton, and R.A. Dunham (1983). Growth comparisons of monosex tilapia produced by separation of sexes, hybridization, and sex reversal. In L. Fishelson, and Z. Yaron (Eds.), *International symposium on tilapia in aquaculture* (pp. 570-579). Tel Aviv, Israel: Tel Aviv University.

Haylor, G.S., and A.B. Pascual (1991). Effect of using ram testis in a fry diet for *Oreochromis niloticus* (L.) on growth, survival and resultant phenotypic sex ratio. *Aquaculture and Fisheries Management* 22:265-268.

Herbst, A.L., H. Ulfelder, and D.C. Poskanzer (1971). Adenocarcinoma of the vagina: Association of material stilbestrol therapy with tumor appearance in young women. *New England Journal of Medicine* 284:878-881.

Hickling, C.F. (1960). The Malacca tilapia hybrids. *Journal of Genetics* 57:1-10.

Hines, G.A., and S.A. Watts (1995). Non-steroidal chemical sex manipulation of tilapia. *Journal of the World Aquaculture Society* 26(1):98-102.

Hiott, A.E. (1990). Affects of initial age and size on the sex reversal of the Nile tilapia *Oreochromis niloticus* fry using methyltestosterone. Master's thesis, Auburn University, Auburn, Alabama.

Hiott, A.E., and R.P. Phelps (1993). Effects of initial age and size on sex reversal of *Oreochromis niloticus* fry using methyltestosterone. *Aquaculture* 112:301-308.

Hopkins, K.D. (1977). Sex reversal of genotypic male *Sarotherodon aureus* (Cichlidae). Master's thesis, Auburn University, Auburn, Alabama.

Hopkins, K.D., W.L. Shelton, and C.R. Engle (1979). Estrogen sex-reversal of *Tilapia aurea. Aquaculture* 18:263-268.

Jalabert, B., J. Moeau, P. Planquette, and R. Billard (1974). Determinisme du sexe chez *Tilapia macrochir* et *Tilapia nilotica:* Action de la methyltestosterone dans l'alimentation des alevins sur la differenciation sexuell; Proportion dessexes dans la descendance des males "inverses." *Annales de Biologie Animale, Biochimie, Biophysique* 14:720-739.

Jankov, R.M. (1977). Microbial transformation of steroids. V. Aromatization of the ring A of androstane steroids by *Mycobacterium phlei. Glasnik Hemijskog Drushtva* 42(9-10):655-668 (English abstract).

Jensen G.L., and W.L. Shelton. (1979). Effects of estrogens on *Tilapia aurea:* Implications for production of monosex genetic male tilapia. *Aquaculture* 16:233-242.

Jeyasuria, P., and A.R. Pace (1998). Embryonic brain-gonad axis in temperature-dependent sex determination of reptiles: A role for P450 (Cyp19). *Journal of Experimental Zoology* 281:428-449.

Johnstone, R., D.J. Macintosh, and R.S. Wright (1983). Elimination of orally administered 17 α-methyltestosterone by *Oreochromis mossambicus* (Tilapia) and *Salmo gairdneri* (Rainbow trout) juveniles. *Aquaculture* 35:249-257.

Katz, Y., M. Abraham, and B. Eckstein (1976). Effects of adrenosterone on gonadal and body growth in *Tilapia nilotica* (Teleostei, Cichlidae). *General and Comparative Endocrinology* 29:414-418.

Katz, Y., B. Eckstein, R. Ikan, and R. Gottleieb (1971). Esterone and estradiol-17β in the ovaries of *Tilapia aurea* (Teleostei, Cichlidae). *Comparative Biochemistry and Physiology* 40B:1005-1010.

Killian, H.S., and C.C. Kohler (1991). Influence of 17α-methyltestosterone on red tilapia under two thermal regimes. *Journal of the World Aquaculture Society* 22:83-94.

Kitano, T., K. Takamune, Y. Nagahama, and S-I. Abe (2000). Aromatase inhibitor and 17α-methyltestosterone cause sex reversal from genetic females to phenotypic males and suppression of P450 aromatase gene expression in Japanese flounder *(Paralichththys olivaceus). Molecular Reproduction and Development* 56:1-5.

Kwon, J.Y., V. Haghpanan, L.M. Kogson-Hurtado, B.J. McAndrew, and D.J. Penman (2000). Masculinization of genetic female Nile tilapia *(Oreochromis niloticus)* by dietary administration of an aromatase inhibitor during sexual differentiation. *Journal of Experimental Zoology* 287:46-53.

Kwon, J.Y., B.J. McAndrew, and D.J. Penman (2001). Cloning of brain aromatase gene and expression of brain and ovarian aromatase genes during sexual differentiation in genetic male and female Nile tilapia *Oreochromis niloticus. Molecular Reproduction and Development* 59:359-370.

Lone, K.P., and M.T. Ridha (1993). Sex reversal and growth of *Oreochromis spilurus* (Gunther) in brackish and sea water by feeding 17α-methyltestosterone. *Aquaculture and Fisheries Management* 24:593-602.

Lovshin, L.L. (1975). Progress report on fisheries management in Northeast Brazil. Research and Development Series No. 9. Auburn University, Auburn, AL: International Center for Aquaculture.

———. (1982). Tilapia hybridization. In R.S.V. Pullin, and R.H. Lowe-McConnell (Eds.), *The biology and culture of tilapias* (pp. 279-308). Manila, Philippines: ICLARM Conference Proceedings 7. International Center for Living Resources Management.

Lovshin, L.L., A.B. Da Silva, A. Carneiro-Sobrinho, and F.R. Melo (1990). Effects of *Oreochromis niloticus* females on the growth and yield of male hybrids (*O. niloticus* female × *O. hornorum* male) cultured in earthen ponds. *Aquaculture* 88:55-60.

Lovshin, L.L., D. Tave, and A.O. Lieutaud (1990). Growth and yield of mixed-sex young-of-the-year *Oreochromis niloticus* raised at two densities in earthen ponds in Alabama, U.S.A. *Aquaculture* 89:21-26.

Lui, S., M. Govoroun, H. D'Cotta, M.-J. Ricordel, J.J. Lareyre, O.M. McMeel, T. Smith, Y. Nagahama, and Y. Guiguen (2000). Expression of cytochrome P450 11β (11β-hydroxylase) gene during gonadal sex differentiation and spermatogenesis in rainbow trout, *Oncorhynchus mykiss*. *Journal of Steroid Biochemistry and Molecular Biology* 75:291-298.

Macintosh, D.C., and D.C. Little (1995). Nile tilapia *(Oreochromis niloticus)*. In N.R. Bromage, and R.J. Roberts (Eds.), *Broodstock management and egg and larval quality* (pp. 277-320). London, England: Blackwell Science Ltd.

Macintosh, D.J., T.J. Varghese, and G.P. Satyanarayana Rao (1985). Hormonal sex reversal of wild-spawned tilapia in India. *Journal of Fish Biology* 26:87-94.

Mair, G.C., J.S. Abucay, J.A. Beardmore, and D.O.F. Skibinski (1995). Growth performance trials of genetically male tilapia (GMT) derived from YY-males in *Oreochromis niloticus* L.: On station comparisons with mixed sex and sex reversed male populations. *Aquaculture* 137:313-322.

Mair, G.C., J.A. Beardmore, and D.O.F. Skibinski (1990). Experimental evidence for environmental sex determination in *Oreochromis* species. In Hirano, R., and I. Hanya (Eds.), *Proceedings of the 2nd Asian Fisheries Forum* (pp. 555-558). Manilla, Philippines: Asian Fisheries Society.

Mbarereehe, F. (1992). Contribution a l'etude de l'influence de la temperature et la duree de traitement sur la production des alevins monosexes du *Tilapia nilotica*. Memoire Presente en vue de L'obtention du Diplome d'Ingenier Technicien. Institut Superieur d'Agriculture et d'Elevage de Busogo, Ruhengeri, Rwanda.

McAndrews, B.J., and K.C. Majumdar (1989). Growth studies on juvenile tilapia using pure species, hormone-treated and nine interspecific hybrids. *Aquaculture and Fisheries Management* 20:35-47.

McGeachin, R.B., E.H. Robinson, and W.H. Neil (1987). Effect of feeding high levels of androgens on the sex ratio of *Oreochromis aureus*. *Aquaculture* 61:317-321.

McGinty, A.S. (1983). Population dynamics of peacock bass, *Cicla ocellaris* and *Tilapia nilotica* in fertilized ponds. In L. Fishelson, and Z. Yaron (Eds.), *International Symposium on Tilapia in Aquaculture* (pp. 86-94). Tel Aviv, Israel: Tel Aviv University.

———. (1985). Effects of predation by largemouth bass in fish production ponds stocked with *Tilapia nilotica*. *Aquaculture* 46:269-274.

Melard, C. (1995). Production of a high percentage of male offspring with 17α-ethynylestradiol sex-reversed *Oreochromis aureus*. I. Estrogen sex-reversal and production of F2 pseudofemales. *Aquaculture* 130:25-34.

Murad, F., and R.C. Haynes Jr (1985). Androgens. In A.G. Gilman, S. Goodman, and A. Gilman (Eds.), *The pharmacological basis of therapeutics* (pp. 1440-1457). New York: McMillan Publishing Co.

Nakamura, M. (1975). Dosage-dependent changes in the effect of oral administration of methyl-testosterone on gonadal sex differentiation in *Tilapia mossambica. Bulletin of the Faculty of Fisheries, Hokkaido University* 26:99-108.

———. (1981). Effects of 11-ketotestosterone on gonadal sex differentiation in *Tilapia mossambica. Bulletin of the Faculty of Fisheries, Hokkaido University* 47:1323-1327.

Nakamura M., and M. Iwahashi (1982). Studies on the practical masculinization in *Tilapia nilotica* by oral administration of androgen. *Bulletin of the Japanese Society for Scientific Fisheries* 48:763-769.

Nakamura, M., T. Kobayashi, X.-T. Chang, and Y. Nagahama (1998). Gonadal sex differentation in teleost fish. *Journal of Experimental Zoology* 281:362-372.

Nakamura, M., and Y. Nagahama (1985). Steroid producing cells during ovarian differentiation of the tilapia *Sarotherodon niloticus. Fish Physiology and Biochemistry* 7:211-219.

———. (1989). Differentiation and development of Leydig cells, and changes of testosterone levels during testicular differentiation in tilapia *Oreochromis niloticus. Fish Physiology and Biochemistry* 7:211-219.

Nakamura, M., and H. Takahashi (1973). Gonadal sex differentiation in *Tilapia mossambica* with special regard to the time of estrogen treatment effective in inducing feminization of genetic fishes. *Bulletin of the Faculty of Fisheries, Hokkaido University* 24:1-13.

———. (1985). Sex control in cultured tilapia *(Tilapia mossambica)* and salmon *(Oncorhynchus masou).* In B. Lofts, and W.N. Holmes (Eds.), *Current trends in comparative endocrinology* (pp. 1255-1260). Honk Kong: Hong Kong University Press.

Obi, A., and W.L. Shelton (1983). Androgen and estrogen sex reversal in *Tilapia hornorum.* In L. Fishelson, and Z. Yaron (Eds.), *International Symposium on Tilapia in Aquaculture, Proceedings* (pp. 165-173). Tel Aviv, Israel: Tel Aviv University.

Okoko, M. (1996). Effect of 17α-Methyltestosterone concentrations on the sex ratio, and gonadal development of Nile tilapia *Oreochromis niloticus.* Master's thesis, Auburn University, Auburn, Alabama.

Owusu-Frimpong, M., and B. Nijjhar (1981). Induced sex reversal in *Tilapia nilotica* (Cichlidae) with methyltestosterone. *Hydrobiologia* 78:157-160.

Pandian, T.J., and K. Varadaja (1987). Techniques to regulate sex ratio and breeding in tilapia. *Current Science* 56:337-343.

———. (1988). Techniques for producing all-male and all-triploid *Oreochromis mossambicus.* In R.S.V. Pullin, T. Bhukaswan, K. Tonguthai, and J. L. Maclean (Eds.), *The Second International Symposium on Tilapia in Aquaculture* (pp. 243-249). ICLARM Conference Proceedings 15, Department of Fisheries, Bangkok, Thailand, and International Center for Living Aquatic Resources Management, Manilla, Philippines.

Phelps, R.P., E. Arana, and B. Argue (1993). Relationship between the external morphology and gonads of androgen-treated *Oreochromis niloticus. Journal of Applied Aquaculture* 2:103-108.

Phelps, R.P., and G. Cerezo (1992). The effect of confinement in hapas on sex reversal and growth of *Oreochromis niloticus. Journal of Applied Aquaculture* 1:73-81.

Phelps, R. P., W. Cole, and T. Katz (1992). Effect of fluoxymesterone on sex ratio and growth of Nile tilapia *Oreochromis niloticus* (L.). *Aquaculture and Fisheries Management* 23:405-410.

Phelps, R.P., G. Conterras Salazar, V. Abe, and B.J. Argue (1995). Sex reversal and nursery growth of Nile tilapia, *Oreochromis niloticus* (L.) free-swimming in earthen ponds. *Aquaculture Research* 26:293-295.

Phelps, R.P., M.S. Fitzpatrick, W.M. Contreras-Sanchez, R.L. Warrington, and J.T. Arndt (2000). Detection of MT in pond water after treatment with MT food. In K. McElwee, D. Burke, M. Niles, X. Cummings, and H. Egna (Eds.), *Pond Dynamics/Aquaculture Collaborative Research Support Program. Seventeenth Annual Technical Report* (pp. 57-59). Corvallis: Pond Dynamics/Aquaculture CRSP, Oregon State University.

Phelps, R.P., L.L. Lovshin, and B.W. Green (1996). Sex reversal of tilapia: 17α-methyltestosterone dose rate by environment and efficacy of bull testes. In D. Burke, B. Goetze, D. Clair, and H. Egna (Eds.), *Pond Dynamics/Aquaculture Collaborative Research Support Program. Fourteenth Annual Technical Report* (pp. 89-91). Corvallis: Oregon State University.

Piferrer, F. (2001). Endrocrine sex control strategies for the feminization of teleost fish. *Aquaculture* 197:229-281.

Piferrer, F., I.J. Baker, and E.D. Donaldson (1993). Effects of natural, synthetic, aromatizable, and nonaromatizable androgens in inducing male sex differentiation in genotypic female Chinook salmon *(Oncorhynchus tshawytscha). General and Comparative Endrocrinology* 91:59-65.

Piferrer, F., S. Zanuy, M. Carllo, I.I. Solar, R.H. Delvin, and E.M. Donaldson (1994). Brief treatment with an aromatizase inhibitor during sex differentiation cause chromosomally female salmon to develop as normal, functional males. *Journal of Experimental Zoology* 270:255-262.

Pike, T. (1983). *Oreochromis mossambicus* in Natal: Natural distribution and growth studies. In L. Fishelson, and Z. Yaron (Eds.), *International Symposium on Tilapia in Aquaculture* (pp. 39-47). Tel Aviv, Israel: Tel Aviv University.

Popma, T.J. (1987). *Freshwater fishculture development project, ESPOL, Guayaquil, Ecuador: Final technical report.* Auburn, AL: Department of Fisheries and Allied Aquacultures, Auburn University.

Popma, T.J., and B.W. Green (1990). *Aquaculture production manual: Sex reversal of tilapia in earthen ponds,* Research and Development Series No. 35. Auburn, AL: International Center for Aquaculture, Alabama Agricultural Experiment Station, Auburn University.

Popma, T.J., and R.P. Phelps (1998). Status report to commercial tilapia producers in monosex fingerling production techniques. Latin American Chapter World Aquaculture Society Aquaculture Brazil '98. Recife, Brazil.

Popma, T.J., F.E. Ross, B.L. Nerrie, and J.R. Bowman (1984). *The development of commercial farming of tilapia in Jamaica 1979-1983,* Research and Development

Series No. 31. Auburn, AL: International Center for Aquaculture, Alabama Agricultural Experiment Station, Auburn University.

Potts, A.D., and R.P. Phelps (1995). Use of diethylstilbestrol and ethynylestradiol to feminize Nile tilapia *Oreochromis niloticus* (L.) in an outdoor environment. *Journal of Applied Icthylogy* 7:147-154.

Pruginin, Y., and E.W. Shell (1962). Separation of the sexes of *Tilapia nilotica* with a mechanical grader. *The Progressive Fish-Culturist* 24:37-40.

Rakocy, J.E. (1980). Evaluation of a closed recirculating system for tilapia culture. Doctoral dissertation, Auburn University, Auburn, Alabama.

Ridha, M.T., and K.P. Lone (1995). Preliminary studies on feminization and growth of *Oreochromis spilurus* (Gunther) by oral administration of 17α-ethynloestradiol in sea water. *Aquaculture Research* 26:479-482.

Rodriguez-Guerrero, D. (1979). Factors influencing the androgen sex reversal of *Tilapia aurea.* Master's thesis, Auburn University, Auburn, Alabama.

Romer U., and W. Beisenherz (1996). Environmental determination of sex in *Apistogramma* (Cichlidae). *Journal of Fish Biology* 48:714-725.

Rosenstein, S., and G. Hulata (1992). Sex reversal in the genus *Oreochromis:* 1. Immersion of eggs and embryos in oestrogen solutions is ineffective. *Aquaculture and Fisheries Management* 23:669-678.

———. (1994). Sex reversal in the genus *Oreochromis:* Optimization of feminization protocol. *Aquaculture and Fisheries Management* 25:329-339.

Rothbard S., E. Solink, S. Shabbath, R. Amado, and I. Grabie (1983). In L. Fishelson and Z. Yaron (Eds.), *International Symposium on Tilapia in Aquaculture* (pp. 425-432). Tel Aviv, Israel: Tel Aviv University.

Sandor, T., and A. Z. Mehdi (1979). Steroids and evolution. In E.J.W. Barrington (Ed.), *Hormones and evolution* (pp. 1-72). New York: Academic Press.

Schubert, K., J. Schlegel, H. Groh, G. Rose, and C. Hoerhold (1972). Metabolism of steroid drugs. VIII. Structure-metabolism relations in the microbial hydrogenation of various substituted testosterone derivates. *Endokrinologie* 59(1):99-114 (English abstract).

Scott, A.G., D.J. Penman, J.A. Beardmore, and D.O.F. Skibinski (1989). The "YY" supermale in *Oreochromis niloticus* (L.) and its potential in aquaculture. *Aquaculture* 78:237-251.

Shell, E.W. (1968). Mono-sex culture of male *Tilapia nilotica* Linnnaeus in ponds stocked at three rates. In *Proceedings FAO World Symposium on Warm-Water Pond Fish Culture,* FAO Fisheries Report 44 (pp. 353-356). Rome, Italy.

Shelton, W.L., K.D. Hopkins, and G.L. Jensen (1978). Use of hormones to produce monosex tilapia. In R. O. Smitherman, W.L. Shelton, and J.L. Grover (Eds.), *Culture of Exotic Fishes Symposium Proceedings* (pp. 10-33). Auburn, AL: Fish Culture Section, American Fisheries Society.

Shelton, W.L., D. Rodriguez-Guerrero, and J. Lopez-Macias (1981). Factors affecting sex reversal of *Tilapia aurea. Aquaculture* 25:59-65.

Shepperd, V.D. (1984). Androgen sex inversion and subsequent growth of red tilapia and Nile tilapia. Master's thesis, Auburn University, Auburn, Alabama.

Siraj, S.S., R.O. Smitherman, S. Castillo-Gallusser, and R.A. Dunham (1983). Reproductive traits for three year classes of *Tilapia nilotica* and maternal effects on their progeny. In L. Fishelson, and Z. Yaron (Eds.), *International Symposium on Tilapia in Aquaculture* (pp. 210-218). Tel Aviv, Israel: Tel Aviv University.

Smith, E.S., and R.P. Phelps (2001). Impact of feed storage conditions on growth and efficacy of sex reversal of *Oreochromis niloticus. North American Journal of Aquaculture* 63:242-245.

Soto, P. (1992). Effect of mestanolone (17α-methylandrostan-17β OL-3-one) on sex ratio and growth of Nile tilapia *(Oreochromis niloticus).* Master's thesis, Auburn University, Auburn, Alabama.

Stanley, J.G. (1976). Female homogamety in grass carp *(Ctenopharyngodon idella)* determined by gynogenesis. *Journal of Fisheries Research Board of Canada* 33:1372-1374.

Swingle, H.S. (1960). Comparative evaluation of two tilapia as pondfishes in Alabama. *Transactions American Fisheries Society* 89:142-148.

Tayamen, M.M., and W.L. Shelton (1978). Inducement of sex reversal in *Sarotherodon niloticus* (Linnaeus). *Aquaculture* 14:349-354.

Teichert-Coddington, D.R., T.P. Popma, and L.L. Lovshin (1997). Attributes of tropical pond-cultured fish. In H.S. Egna, and C.E. Boyd (Eds.), *Dynamics of pond aquaculture* (pp. 183-198). Boca Raton, FL: CRC Press.

Torrans, L., F. Meriweather, and F. Lowell (1988). Sex-reversal of *Oreochromis aureus* by immersion in mibolerone, a synthetic steriod. *Journal of the World Aquaculture Society* 19:97-102.

Tuan, P.A., D.C. Little, and G.C. Mair (1998). Genotypic effects on comparative growth performance of all-male tilapia *Oreochromis niloticus* (L.). *Aquaculture* 159:293-302.

Varadaraj, K. (1989). Feminization of *Oreochromis mossambicus* by the administration of diethylstilbestrol. *Aquaculture* 80: 337-341.

———. (1990). Production of monosex male *Oreochromis mossambicus* (Peters) by administering 19-norethisterone acetate. *Aquaculture and Fisheries Management* 21:133-135.

Varadaraj, K., and T.J. Pandian (1989). Monosex male broods of *Oreochromis mossambicus* produced through artificial sex reversal with 17α-methyl-4 androsten-17β-ol-3-one. *Current Trends in Life Science* 15:169-173.

Varadaraj, K., S. Sindhu Kumari, and T.J. Pandian (1994). Comparison of conditions for hormonal sex reversal of mozambique tilapias. *The Progressive Fish-Culturist* 56:81-90.

Vera Cruz, E.M., and G.C. Mair (1994). Conditions for effective androgen sex reversal in *Oreochromis niloticus* (L.). *Aquaculture* 112:137-248.

Verani, J.R., M. Marins, A.B. Da Sivia, and A.C. Sobrinho (1983). Population control in intensive fish culture associating *Oreochromis (Sarotherodon) niloticus* with the natural predator *Cicha ocellaris:* Quantitive analysis. In L. Fishelson and Z. Yaron (Eds.), *International Symposium on Tilapia in Aquaculture* (pp. 580-587). Tel Aviv, Israel: Tel Aviv University.

Wantanabe, W.O., K.W. Mueller, W.D. Head, and S.C. Ellis (1993). Sex reversal of Florida red tilapia in brackish water tanks under different treatment durations of 17α-ethynyltestosterone administered in feed. *Journal of Applied Aquaculture* 2:29-42.

White, A., P. Hundler, and E.L. Smith (1973). *Principles of biochemistry, Fifth edition.* New York: McGraw-Hill.

Yamamato, T.-O. (1951) Artificial induction of sex-reversal in genotypic females of the medake *(Oryzias latipes). Journal of Experimental Zoology* 123:571-594.

———. (1953). Artificial induction of sex-reversal in genotypic males of the medaka *(Oryzias latipes). Journal of Experimental Zoology* 123:571-594.

———. (1958). Artificial induction of functional sex-reversal in genotypic females of the medaka *(Oryzias latipes). Journal of Experimental Zoology* 137:227-262.

———. (1969). Sex differiation. In W.S. Hoar, and D.J. Randall (Eds.), *Fish physiology* (Volume 3, pp. 117-177). New York: Academic Press.

Yamazaki, F. (1976). Application of hormones in fish culture. *Journal of the Fisheries Research Board of Canada* 33:948-958.

Chapter 7

Pond Production

Bartholomew W. Green
C. Bauer Duke III

INTRODUCTION

Tilapia culture began in the first half of the twentieth century at the subsistence level, encouraged by colonials on the African continent. By the 1970s, Nile tilapia, *Oreochromis niloticus*, was considered the best growing of the tilapia being cultured and was used by the United Nations, nongovernmental organizations, and governmental aid agencies in the worldwide war on poverty. Tilapia are resistant to poor water quality and disease, able to convert a wide range of foodstuffs into protein, and have a high-quality flesh. These attributes made tilapia attractive to fish farmers at the commercial level. Now tilapia is raised in a wide variety of settings, and far more are raised for sale than are raised by developing-country families. What was once considered the poor man's fish has become a commodity of great value.

Production strategies evolve continuously, driven by available resources, environment, and exposure to new ideas and markets. Although the choice of production strategy falls to the farmer, it is difficult for the farmer to know all the production options available. However, this is changing with globalization, as technical consultants transfer technologies around the

---

The authors thank the farm owners and managers for sharing data about their farms. Drs. P. Perschbacher, W. Shelton, and N. Stone are thanked for their comments on the manuscript.

doi:10.1300/5513_07

world. Fish farmers and scientists meet annually for technology exchange, and the Internet further increases access to information.

How does the farmer weigh options and decide upon a production strategy? The decision involves analyzing technical, economic, and marketing components in the development of a sound business plan. Economic and marketing considerations are discussed in Chapter 19, "Marketing and Economics." Among the technical considerations are the type, size, and conformation of production unit to be used. Will it be a rectangular pond, rounded-corner rectangular pond, circular pond, or raceway? Who else is using the design and is it profitable? Is the design new research technology not yet operated on a large scale? Is the design based on incomplete understanding of copied technology? "Turnkey" technology is possibly the most dangerous area for the farmer because the term suggests ease of use and no emergencies. Emergencies will always arise: fish will die; problems will occur. The key is to minimize risk situations and turn a profit. It is important to understand also that high production does not necessarily equal profit.

Tilapia grow-out ponds are stocked with either mixed-sex or monosex (male) fingerlings. Mixed-sex tilapia fingerlings continue to be stocked in ponds that receive little to no fertilizer or supplemental feed input. Intensified tilapia aquaculture, involving substantial applications of fertilizer, feed, or both, depends primarily on stocking monosex, especially sex-reversed, fingerlings into grow-out ponds. This chapter addresses production systems for raising tilapia in ponds using data from sex-reversed, all-male populations. (Sex reversal of tilapia is discussed in Chapter 6, "Hormone Manipulation of Sex").

SPECIES SELECTION

The four most commonly cultured tilapia are the Nile tilapia *(Oreochromis niloticus)*, Mozambique tilapia *(O. mossambicus)*, blue tilapia *(O. aureus)*, and red tilapia (*Oreochromis* sp.). Specifics about the biology and fingerling production of these tilapia are presented in Chapters 1 "Biology" and 5 "Fingerling Production Systems." The Nile tilapia is the most commonly cultured tilapia worldwide (Josupeit 2001; Watanabe et al. 2002). In Asia, Nile tilapia comprises 85 percent of tilapia production, except in Indonesia, where Mozambique tilapia comprises 67 percent of production (Dey 2001). Mozambique tilapia is used extensively in South African aquaculture (van der Merwe et al. 2001).

Nile tilapia has often been recommended over Mozambique tilapia for culture purposes because of its perceived faster growth (Pullin 1983, 1985, 1988). However, there are few or no data from trials that compare growth across tilapia species under standardized conditions. The farmer may assume production is the cornerstone of a profitable fish farm and therefore the fastest-growing species should be cultured; however, other factors are also involved in species selection.

Phenotypic color should be considered in tilapia species selection. Some consumers will accept only tilapia with the grayish wild-type coloration. In Louisiana consumers prefer silvery or pearl-colored fish (Lutz 2001). In other places consumers accept only light-colored or red tilapia. Consumers in Los Angeles, California, will not accept red or calico tilapia.

Tilapia are exotic species in the United States, Latin America, and Asia, but they have been introduced widely, intentionally and unintentionally, in these areas. Even within Africa, tilapia have been introduced into waters beyond their native ranges (De Vos et al. 1990; Ogutu-Ohwayo 1990), with positive and negative impacts on receiving waters (De Vos et al. 1990; Ogutu-Ohwayo 1990; Nelson and Eldredge 1991).

Governmental regulatory agencies also can affect species selection. For example, the state of California prohibits the culture of the Nile tilapia. Although *Tilapia zillii* is a permitted culture species in California, it grows poorly. Mozambique tilapia and Zanzibar tilapia *(O. urolepis hornorum)* are both acceptable culture species in California, but the Mozambique tilapia is the most prevalently grown species. Examination of more than 25 years of records from Southern California tilapia farms shows that the Mozambique tilapia present in California may grow as fast as Nile tilapia (Costa-Pierce 1997, 2001). Additional research is necessary to compare growth and production characteristics of different tilapia species under standard conditions.

Hybrid tilapia also are reared in production ponds, but only in a handful of countries. The *O. niloticus* (female) × *O. aureus* (male) hybrid is used in China and Israel (Hulata 1997; Xiao et al. 2002). Many farmers do not use hybrid tilapia because their production requires specialized facilities and management to ensure that parent lines remain pure. In addition, a readily available supply of sex-reversed tilapia fingerlings has obviated the need for tilapia hybridization as a source of monosex fingerlings.

Pond water salinity also must be considered. Tilapia are cultured in brackish-water ponds in many parts of the world, and the commercially important tilapia continue to grow normally at salinity levels as high as 15 to 25 ppt (Wohlfarth and Hulata 1981; Philippart and Ruwet 1982; Balarin 1988). Both Nile and blue tilapia adapt to and grow well in brackish water up to 20 ppt salinity, and Mozambique and the red tilapia appear to grow well in full-strength seawater (34 ppt). Green (1997) discussed brackish-water tilapia fingerling production and acclimation.

EARTHEN VERSUS LINED PONDS

Earthen ponds account for most of the pond area worldwide. Relatively easy and inexpensive to construct, they have been the mainstays of aquaculture for more than 4,000 years. Given appropriate soils, good construction, and good management, earthen ponds can be used for hundreds of years. Ponds also may be used for other purposes. Culture ponds are rotated with fish, grazing animals, rice, and row crops, but such usage is limited (Müller 1978). Occasionally, conditions are such that an earthen pond is not feasible and the pond must be lined to ensure success.

Soil structure, maintenance, production technology, and available materials, alone or in concert, are among the reasons a pond is lined. Frequently, soil structure, mainly the inability to hold water, is the primary factor leading to pond lining. Lining a pond may be unnecessary in areas with abundant sources of water or rainfall. However, liners are a fiscal necessity for ponds in areas with limited or expensive water supply. Sometimes a farm can be located in an area with excellent water but poor soils. Such areas often have inexpensive land, and the cost of liners is not prohibitive in terms of the overall development of the project.

Liners take different forms, including plastics, masonry, and cement. Plastic (e.g., high-density polyethylene and polyvinyl chloride) liners are very popular, especially where entry into ponds is infrequent. The thickness of plastic liners is variable. Plastic liners are less expensive than cement but are prone to puncture, have a shorter lifespan, deteriorate from ultraviolet radiation, and are more difficult to repair. Plastic liners fabricated using a layering process sometimes suffer from layer separation, with bubbles of gas covering several square feet developing between layers.

Ponds can be lined with cement applied in a form, slurry, or watered mix. The form method, although more expensive because of the greater quantity of cement used, is more durable and results in a smoother surface. Using cement slurry is quicker and easier, but likely will require more applications over time. Once the pond is dug, cement slurry is applied, dumping and brooming it into place. Alternatively, dry cement mix can be distributed evenly over the pond shell area and then watered extensively to create a cement liner on the pond. After a few production cycles, slurry-applied concrete may need patching. The slurry and wetting-of-dry-mix methods are advisable only in very stable soils, such as those with high gravel content.

Ponds also can be constructed as large tanks, either above ground or in ground. The sidewalls are constructed of cinder block or brick and covered with a sealing layer of cement. The bottom is then constructed using forms or slurry or by wetting a prelaid dry mix. Cement can also be blown onto the sides and bottoms as "shotcrete."

Wetting a dry-mix layer is cheaper than using formed cement but will require repair sooner. Cement slurry is best for these repairs. Ponds constructed this way may need repair in one to five years, but the delay in capital investment can help the farm establish cash flow because there is less debt.

Today, lined ponds are found in Africa, Asia, Europe, and the Americas. In most cases, these lined ponds are on farms run by large corporations and government agencies. Lined ponds are extremely useful for high-intensity tilapia culture because high rates of water exchange, rapid water movement, and easy solids removal are possible without the turbidity and sedimentation problems associated with earthen ponds. Whereas an earthen tilapia pond may have a carrying capacity of 20,000 $kg \cdot ha^{-1}$, the carrying capacity of a lined tilapia pond can be 140,000 $kg \cdot ha^{-1}$. Despite this, the majority of tilapia produced worldwide will continue to be grown in earthen ponds for the foreseeable future. More complete discussions of pond construction can be found in Wheaton (1977), Avault (1996), and USDA-NRCS (1997).

PRODUCTION STRATEGIES

Successful culture of tilapia is possible using a wide variety of management strategies in ponds filled with fresh to saline water. This chapter addresses pond culture only, not culture in recirculating aquaculture systems or cages. (see Chapter 9 "Cage Culture"). Pond management strategy progresses from fertilization only, to fertilization plus feeds (supplemental or complete), to complete feeds only, to feeds plus aeration and/or water exchange, as production intensity increases from extensive production to hyperintensive production.

Fertilization

The omnivorous feeding habits of the commonly cultured tilapia allow them to be reared successfully in ponds that receive only fertilizer inputs. Because tilapia consume phytoplankton directly, higher fertilization rates are required than would be used for fish, such as the common carp, that do not consume plants directly. Pond fertilization regimes based on organic or chemical fertilization have been used successfully to produce tilapia. In general, fertilized ponds are stocked with 10,000 to 20,000 tilapia per hectare and are managed without water exchange or mechanical aeration. Pond water losses to seepage and evaporation are replaced periodically.

Animal manures have been used successfully as organic fertilizer for tilapia production ponds, but their use may be limited by availability. Small-scale producers integrate farming of livestock (e.g., pigs, chickens, goats, ducks, geese, or cattle) with aquaculture, either by locating an animal enclosure adjacent to the fishpond or by enclosing the animals for the night elsewhere on the farm (Pullin and Shehadeh 1980; Lovshin et al. 1986). Fresh dairy cow manure, layer chicken litter, and chemical fertilizer were compared as fertilizer sources for Nile tilapia production in earthen ponds by Green et al. (1989). The cow manure averaged 21.3 percent dry matter, and the chicken litter averaged 83.3 percent dry matter. Chicken litter was composed of pine sawdust, chicken manure, feathers, and waste feed and had remained in the layer house for 12 to 18 months. Urea and triple superphosphate were used as chemical fertilizers. Ponds were fertilized weekly with similar quantities of nitrogen and phosphorus, on a dry matter basis, from the three fertilizer sources. Tilapia yield after 150 days was significantly greater in ponds fertilized with chicken litter than in ponds fertilized with cow manure or chemical fertilizer (see Table 7.1).

Chicken litter, which has a higher dry matter content, is a more practical organic fertilizer source for tilapia ponds because it can be handled and transported in bulk with relative ease. In order to determine the optimal chicken litter fertilization rate for tilapia production, a study was conducted in Honduras and Panama, where the use of chicken litter was profitable (Green et al. 1990). Ponds were stocked with 10,000 tilapia per hectare and fertilized weekly with 125, 250, 500, or 1,000 kg·ha^{-1} chicken litter (dry matter basis). A 141 to 150-day production cycle was conducted during the rainy and dry seasons in each country. Layer chicken litter, which was used

TABLE 7.1. Summary of mean production (± SD) of all-male Nile tilapia (10,000 ha^{-1}) after 150 days in ponds that received nutrients from one of three fertilizers.

Nutrient source	Fish weight (g per fish)	Gross yield (kg·ha^{-1} per 150 days)	Net yield (kg·ha^{-1} per 150 days)
Layer chicken litter	203.9 ± 16.1*	2,075 ± 177*	1,759 ± 176*
Dairy cow manure	172.1 ± 7.6	1,626 ± 77	1,295 ± 90
Chemical fertilizer	150.4 ± 17.9	1,513 ± 211	1,194 ± 219

Source: After Green et al. 1989.
*Mean is significantly different ($P > 0.05$).

in Honduras, was composed as described in the previous paragraph and averaged 88.9 percent dry matter. Broiler chicken litter, used in Panama, was composed of rice hulls, manure, feathers, and waste feed, had been in the broiler house six months, and averaged 89.8 percent dry matter. Tilapia gross yield increased significantly with increased chicken litter applications in both countries (see Table 7.2). No seasonal differences in tilapia yields were observed in Honduras, but in Panama the dry season yields were greater.

Chemical fertilizers also have been used as the sole nutrient source for tilapia production ponds. Some would argue that chemical fertilizers are more appropriate as a fertilizer source because of their high content of available nutrients, broader availability, ease of transport and use, and relatively low cost. Nitrogen fertilizer appears to be necessary for tilapia production in the tropics. The optimal nitrogen fertilization rate was determined for ponds in Honduras (Green et al. 1999), Kenya (Veverica, Bowman et al. 2001), Thailand (Lin et al. 1999, 2000), and the Philippines (Brown et al. 2001). Pond trials were conducted during the rainy/warm and dry/cool seasons in Honduras, Kenya, and Thailand, and during the dry/cool season in the Philippines. All ponds were stocked with sex-reversed (all-male) tilapia at 1,000 $kg \cdot ha^{-1}$. Ponds were fertilized weekly with nitrogen (as urea or diammonium phosphate) at 0, 10, 20, or 30 $kg \cdot ha^{-1}$. Ponds also were fertilized with phosphorus (8 $kg \cdot ha^{-1}.week^{-1}$), and total alkalinity was 75 $mg \cdot L^{-1}$ as $CaCO_3$ or greater. Ponds were harvested when fish growth ceased. Experiment duration ranged from 90 days in the Philippines to 147 days in Kenya. The response of gross tilapia production to nitrogen input rate was variable across countries (see Table 7.3). The optimal nitrogen fertilization rate was 10 $kg \cdot ha^{-1} \cdot week^{-1}$ in the Philippines, 10 to 20 $kg \cdot ha^{-1} \cdot week^{-1}$ in Kenya, 20 $kg \cdot ha^{-1} \cdot week^{-1}$ in Honduras, and 30 $kg \cdot ha^{-1} \cdot week^{-1}$ in Thailand (Green et al. 1999; Lin et al. 1999, 2000; Brown et al. 2001, Veverica, Bowman et al. 2001). Differences among countries were attributed to differences in pond water quality and local environmental conditions. Cooler pond water temperatures likely were responsible for the generally lower tilapia yield observed during the dry/cool season in each country.

Phosphorus also is a required nutrient to stimulate primary productivity. However, there has been little research conducted to determine the optimal phosphorus fertilization rate in tilapia ponds in the tropics. A study was conducted in Kenya to determine the optimal phosphorus fertilization rate in ponds receiving the optimal nitrogen fertilization rate (20 $kg \cdot ha^{-1} \cdot week^{-1}$) (Veverica, Mirera et al. 2001). Ponds were stocked with sex-reversed Nile tilapia fingerlings (456 $kg \cdot ha^{-1}$; mean weight 16 g) plus *Clarias gariepinus* (1,500 ha^{-1}; mean weight 5 g). Ponds were fertilized weekly with phosphorus (as monoammonium phosphate) at 0, 2, 5, or 8 and nitrogen (as urea) at

TABLE 7.2. Means of all-male Nile tilapia gross yield, net yield, and individual weight in ponds in Honduras and Panama fertilized weekly with chicken litter at four rates on a dry matter basis.

		Honduras		Panama	
Chicken litter ($kg \cdot ha^{-1} \cdot week^{-1}$)	Variable	Rainy season	Dry season	Rainy season	Dry season
125	Gross yield ($kg \cdot ha^{-1}$)	1,179	1,145	827	1,230
	Net yield ($kg \cdot ha^{-1}$)	915	781	584	852
	Individual weight (g)	114.6	116.7	69.8	87.0
250	Gross yield ($kg \cdot ha^{-1}$)	1,649	1,426	1,290	1,844
	Net yield ($kg \cdot ha^{-1}$)	1,381	1,050	1,053	1,483
	Individual weight (g)	154.8	142.9	85.9	114.2
500	Gross yield ($kg \cdot ha^{-1}$)	1,890	1,915	1,708	2,228
	Net yield ($kg \cdot ha^{-1}$)	1,643	1,543	1,472	1,863
	Individual weight (g)	177.1	202.6	113.8	136.4
1,000	Gross yield ($kg \cdot ha^{-1}$)	2,324	2,333	2,729	2,984
	Net yield ($kg \cdot ha^{-1}$)	2,046	1,964	2,495	2,621

TABLE 7.3. Mean gross tilapia yield (kg·ha^{-1}) in ponds fertilized weekly with 0, 10, 20, or 30 kg·ha^{-1} N during the rainy/warm and/or cool/dry season in Honduras, Kenya, the Philippines, and Thailand.

Country	Duration (days)	Nitrogen application rate			
		0	10	20	30
Rainy/warm season					
Honduras	121	1,128	1,891	2,490	1,914
Kenya	133	1,119	1,672	1,720	1,520
Thailand	91	1,221	2,173	1,935	2,410
Dry/cool season					
Honduras	107	1,501	1,611	1,729	1,360
Kenya	133	1,015	2,602	2,953	2,510
Philippines	90	1,138	2,642	2,456	2,671
Thailand	91	818	1,628	1,755	1,938

Note: Ponds were stocked with all-male Nile tilapia (1,000 kg·ha^{-1}). Phosphorus was added weekly to all ponds at 8 kg·ha^{-1}.

20 kg·ha^{-1} (N). Total alkalinity was 75 mg·L^{-1} as $CaCO_3$ or greater. Gross tilapia yield averaged 1,649, 1,839, 2,040, and 2,095 kg·ha^{-1} for the 0, 2, 5, and 8 kg·ha^{-1}·week^{-1} (P) treatments, respectively. The optimal phosphorus fertilization rate for tilapia production was 5 kg·ha^{-1}·week^{-1} (P) when combined with 20 kg·ha^{-1}·week^{-1} (N).

Feeding

A formulated feed can increase tilapia yield. At semi-intensive management levels, formulated feed can be used in combination with fertilization or alone depending upon the feed quality and fish standing stock in the pond. As pond management intensifies through increased stocking rates and the availability of mechanical aeration or water exchange, feeding rate and feed quality must increase in order to sustain rapid fish growth. Because the high standing stock of tilapia in intensively managed ponds derives all its nutrition for growth from formulated feeds, ponds are not fertilized. The nutritional requirements for the different life stages of tilapia are discussed in Chapter 12, "Nutrient Requirements."

Production ponds typically are stocked with tilapia that weigh from 25 to 100 g each. The initial fish standing stock, which depends on the number and size of fish stocked, can be low relative to the standing stock at harvest. Natural pond productivity can be sufficient to sustain rapid tilapia growth during an initial period of the grow-out phase. Thus, efficient feed use could be achieved if feed were offered once the natural productivity could no longer support rapid tilapia growth.

Green (1992) replaced formulated feed by fertilization with chicken litter during the first 60 days of tilapia culture with no significant impact on yield or production economics. Ponds were stocked with sex-reversed Nile tilapia (20,000 ha^{-1}; average weight 18.6 g). The pond management strategies tested were feed only (23 percent crude protein), layer chicken litter only (1,000 $kg \cdot ha^{-1} \cdot week^{-1}$, dry matter basis) for the first 60 days followed by feed only (3 percent of fish biomass daily), or layer chicken litter (500 $kg \cdot ha^{-1} \cdot week^{-1}$, dry matter basis) plus feed (1.5 percent of fish biomass daily). No mechanical aeration or water exchange was used during the 151-day trial. Mean tilapia yields did not differ significantly among treatments and were 4,470, 4,522, and 4,021 $kg \cdot ha^{-1}$ for the feed-only, chicken litter then feed, and chicken litter plus feed treatments, respectively. Fish growth in the chicken litter then feed treatment had slowed by day 61, which indicated that the critical standing crop (Hepher and Pruginin 1981) had been exceeded, that is, that natural pond productivity was insufficient to maintain rapid tilapia growth (Green 1992; Green et al. 2002). Because natural food limitation likely occurred between 40 and 60 days after stocking, tilapia growth should be monitored closely during this period to ensure rapid growth is maintained. Rapid growth resumed once formulated feed was offered.

Individual fish weight has been used as an indicator of when to offer formulated feed in fertilized ponds. Diana et al. (1996) found that delaying provision of formulated feed in fertilized ponds until fish weigh 100 to 150 g each may improve input utilization efficiency. In this study, sex-reversed Nile tilapia (average weight 15 g) were stocked into ponds at 30,000 ha^{-1}. All ponds were fertilized weekly with urea (60 $kg \cdot ha^{-1}$) and triple superphosphate (34 $kg \cdot ha^{-1}$). Formulated feed (30 percent crude protein) was offered daily at 50 percent of the ad libitum rate once fish reached individual weights of 50, 100, 150, 200, or 250 g. Ponds were harvested when fish weighed 500 to 600 g each. No mechanical aeration or water exchange was used. Feed was first offered after 38, 80, 153, 178, or 234 days, respectively, for the 50, 100, 150, 200, or 250 g treatments. Tilapia growth rate in all treatments during the fertilizer-only stage averaged 1.17 $g \cdot day^{-1}$, which was significantly less than the 3.10 $g \cdot day^{-1}$ average growth rate during the feeding stage (see Table 7.4). The critical standing crop was reached before 38 days, as shown by the fact that tilapia growth rate increased significantly once formulated feed was offered. Delaying initiation of feeding until fish weighed 50 to 100 g each did not affect growth, final size, yield, or duration of grow-out compared with the other treatments. However, harvest occurred about 30 days later in ponds where formulated feed was offered once fish

TABLE 7.4. Means of final individual weight, growth rates during the fertilization and feeding stages, yield, and feed conversion ratio for all-male Nile tilapia (30,000·ha^{-1}) reared in fertilized ponds.

Treatment (g)	Duration (days)	Final weight (g per fish)	Growth rate during Fertilizer only (g per day)	Growth rate during Feeding (g per day)	Yield (kg·ha^{-1})	Feed conversion ratio
50	236	593	1.44	2.78	15,396	1.14
100	236	596	1.15	3.29	15,372	0.93
150	265	534	1.21	3.48	14,132	0.93
200	305	627	1.17	3.27	15,920	1.02
250	328	488	1.03	2.76	12,952	0.87

Source: After Diana et al. 1996.
Note: A 30 percent crude protein formulated feed was offered to fish when the average individual weight was 50, 100, 150, 200, or 250 g.

weighed 150 g each. A thorough economic evaluation is needed to determine the most profitable strategy.

Feed utilization efficiency also can be improved by reducing the feeding rate in fertilized ponds—by as much as 50 percent of satiation without reducing tilapia growth and yield (Diana et al. 1994). Sex-reversed Nile tilapia (average weight 10.1 g) were stocked into ponds at 30,000 ha^{-1}. All ponds were fertilized weekly with urea (60 kg·ha^{-1}) and triple superphosphate (35 kg·ha^{-1}). Mechanical aeration or water exchange was not used. Treatments tested in this 155-day study were

1. fertilization only;
2. satiation feeding only;
3. 75 percent of satiation feeding plus fertilization;
4. 50 percent of satiation feeding plus fertilization;
5. 25 percent of satiation feeding plus fertilization.

Tilapia growth and yield were similar in ponds with up to a 50 percent reduction in satiation feeding rate. Thus, tilapia continued to derive substantial nutrition for growth from natural pond productivity. In these ponds growth rate averaged 2.5 g·day^{-1} and yield averaged 9,813 kg·ha^{-1}.

Dissolved Oxygen

Tilapia are able to tolerate low dissolved oxygen (DO) concentration in pond water. The length of this tolerance depends on a variety of factors, including water temperature, fish size and biomass, and biochemical oxygen demand of pond water. Primary production by phytoplankton is the primary source and sink of dissolved oxygen in fertilized ponds (Boyd 1990). Often phytoplankton in fertilized ponds produce a surplus of oxygen, and diurnal DO concentrations remain at acceptable levels. However, dense blooms of phytoplankton, which can develop in both fertilized and fed ponds, can reduce to critical levels or exhaust DO in the water column during the night. A sudden massive die-off of the phytoplankton also can deplete pond DO. Low DO concentrations also occur in intensively managed ponds that receive high inputs of feed daily because wastes from feed stimulate bacterial and phytoplankton productivity. Prolonged exposure to low DO concentration can stress fish and reduce growth or cause mortality.

Mechanical aeration is one method used to increase pond DO concentration and tilapia yield. Teichert-Coddington and Green (1993) found that although maintenance of a minimum pond DO concentration at 10 percent saturation increased tilapia yield compared with no aeration, no additional increase in yield was observed when DO concentration was maintained at 30 percent saturation. Earthen ponds (0.1 ha) used in this 148-day study were stocked with sex-reversed Nile tilapia (20,000 ha^{-1}; average weight 24 g), *Cichlasoma maculiacauda* (290 ha^{-1}, average weight 12 g) and *C. managuense* (500 ha^{-1}; average weight 1.4 g). Aeration was supplied at 3.7 $kW \cdot ha^{-1}$. Ponds were fertilized weekly during the first two months with chicken litter (1,000 $kg \cdot ha^{-1}$, dry matter basis). From day 61, fertilization was suspended and fish were offered a formulated ration (20 percent crude protein) at 3 percent of fish biomass, 6 $days \cdot week^{-1}$. Tilapia yield averaged 4,133 and 4,269 $kg \cdot ha^{-1}$ for the 10 and 30 percent saturation treatments, respectively. These yields were similar and significantly greater than the 3,404 $kg \cdot ha^{-1}$ average yield from unaerated control ponds. In order to maintain pond DO concentrations at 10 and 30 percent saturation, aerators were operated nightly for an average of 4.8 and 9.0 h, respectively. Maintenance of a minimum DO concentration permitted more uniform growth of tilapia, as shown by a higher standard deviation for mean individual weight for the unaerated control treatment (±26.9 g) compared with the 10 percent (±4.7 g) and 30 percent (±6.8 g) saturation treatments.

Pond Management

Critical standing crop and carrying capacity are important concepts in aquaculture (Hepher 1978). For any given pond management strategy, fish will grow at a maximum rate until food or some other environmental factor becomes limiting, causing growth to deviate from its maximum rate. The level at which growth deviates from the maximum is referred to as the critical standing crop. Fish continue to grow once the critical standing crop is exceeded, albeit at a decreasing rate, until growth ceases and conditions in the ponds are sufficient only to maintain the fish population without growth. This point is referred to as the carrying capacity. If the factor that limits growth is removed, fish once again can grow at a maximum rate until another factor limits growth, causing the growth rate to decline until the population reaches a new carrying capacity (see Figure 7.1). Below the critical standing crop, absolute fish growth rate (g·day^{-1}) increases linearly as fish individual weight increases. Above the critical standing crop, yield (kg·ha^{-1}·day^{-1}) continues to increase as long as growth rate decreases more slowly than the increase in fish weight. Thus, maximum yield (kg·ha^{-1}·day^{-1}) occurs between the critical standing crop and the carrying capacity. Total production (kg·ha^{-1}) continues to increase until the carrying capacity is attained; however, the yield equals zero at the carrying capacity. Harvesting ponds near the time of reaching maximum yield generally

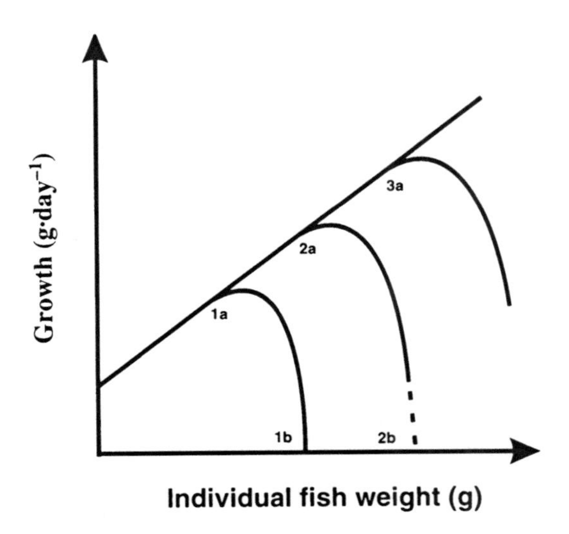

FIGURE 7.1. Generalized relationship between fish growth and individual fish weight. Fish will grow at a maximum rate under a given management system until some factor limits growth (points 1a, 2a, 3a, referred to as the critical standing crop). Growth rate will decline until growth ceases once the fish population reaches the carrying capacity (points 1b, 2b). If the constraint to growth is removed, fish resume maximum growth until the next factor limits growth, and so on. *Source:* After Hepher 1978.

results in the greatest profit. The economics of aquaculture is discussed in Chapter 19, "Marketing and Economics."

Semi-intensively managed production ponds generally are stocked with 10,000 to 30,000 sex-reversed tilapia per hectare and managed with fertilization and/or feeding. Generally fish are produced for local markets. The production cycle involves only one phase; ponds are stocked with 10 to 50 g fish harvested from nursery ponds, and final size is from 150 to 500 g per fish. Ponds are not equipped with mechanical aerators, nor is water exchange practiced. Tractor-powered emergency aeration equipment often is lacking on these farms.

Two or more production phases are used as management intensifies. Two grow-out phases are most common, but some farms use three. Stocking rate, and thus the initial biomass, are manipulated so that fast growth rates and high yields are obtained throughout much of grow-out phases I and II (see Figures 7.2 and 7.3). Once the fish biomass in grow-out phase I (or phases I and II in a three-phase system) is between the critical standing crop and the carrying capacity, either the pond is completely harvested and the fish are restocked at lower rates for continued growth in other ponds or the fish standing stock in the pond is partially harvested, with the harvested fish

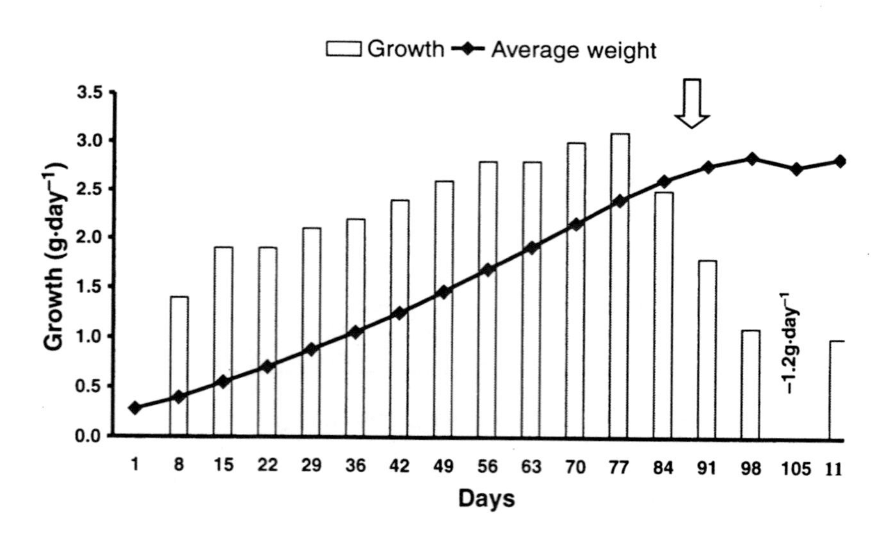

FIGURE 7.2. Growth rate and mean individual weight of sex-reversed red tilapia (*Oreochromis* sp.) stocked at 4 fish per square meter in brackish water ponds during grow-out phase I. The critical standing crop was attained around day 77. The arrow indicates a potential harvest date.

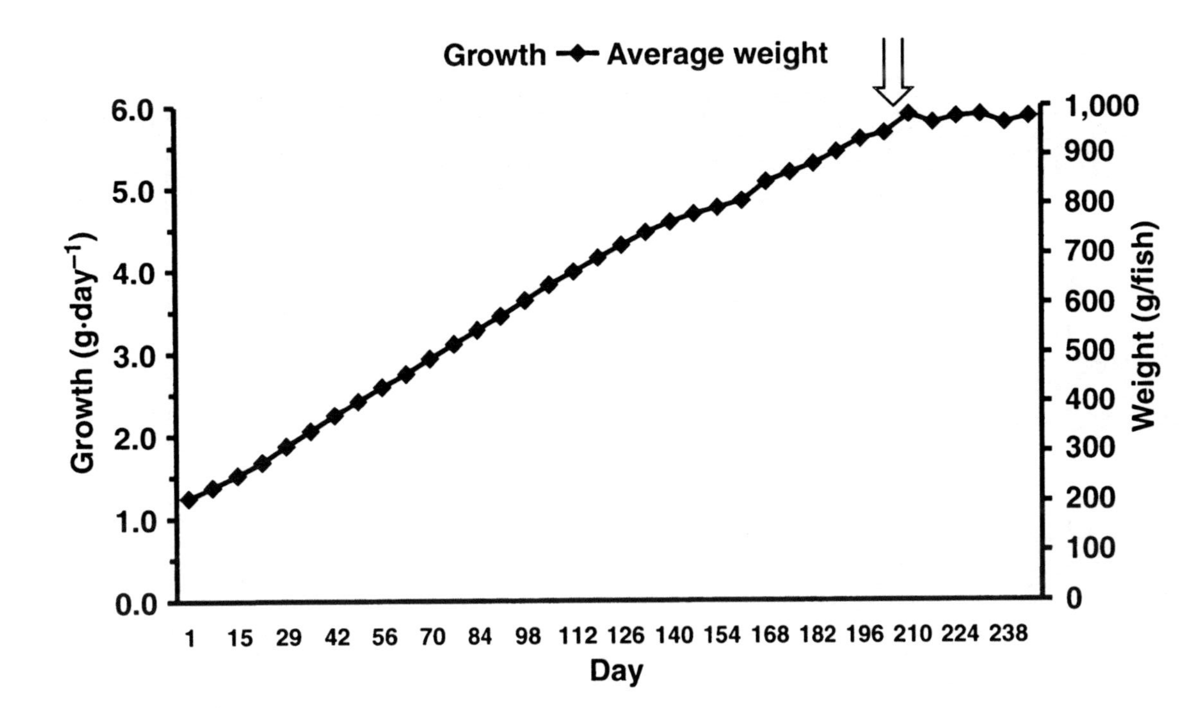

FIGURE 7.3. Growth rate and mean individual weight of sex-reversed red tilapia (*Oreochromis* sp.) stocked at 1 fish per square meter in brackish water ponds during grow-out phase II. The critical standing crop was attained around day 210. The arrow indicates a potential harvest date.

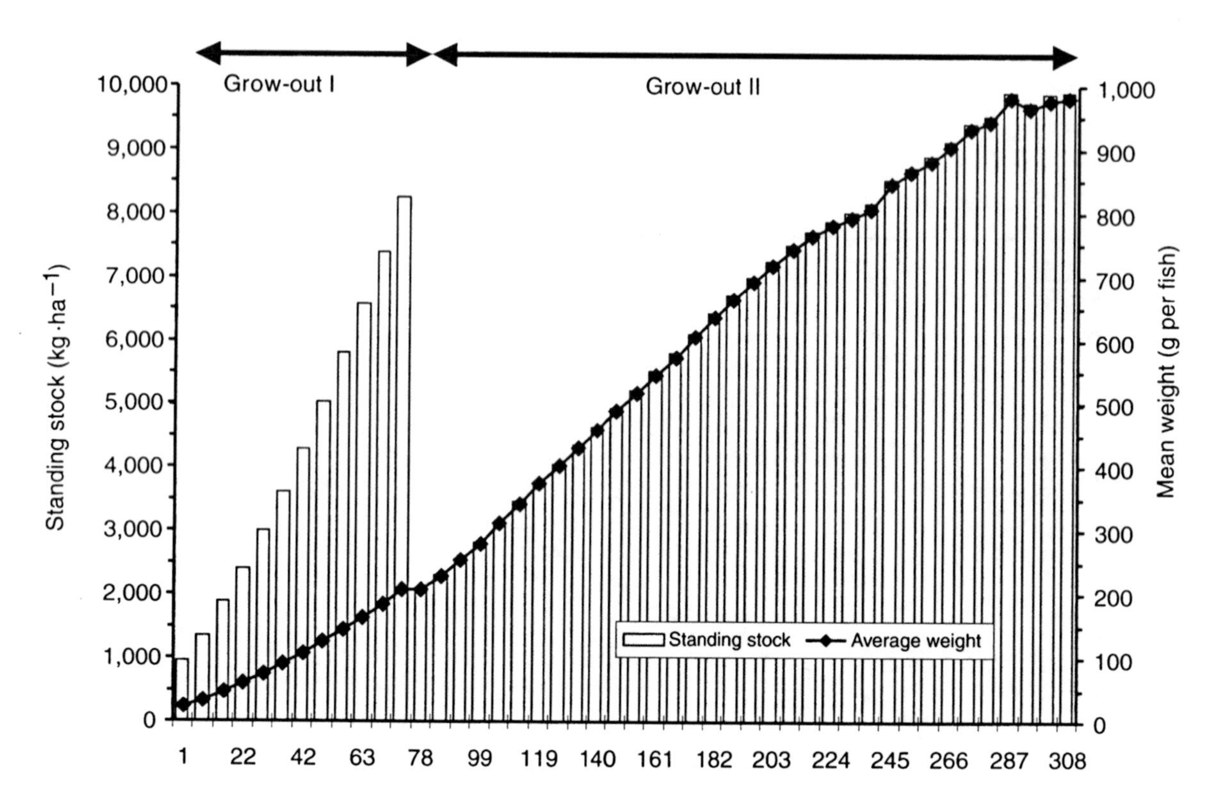

FIGURE 7.4. Change in standing stock (kg·ha^{-1}) and mean individual weight (g per fish) of sex-reversed red tilapia (*Oreochromis* sp.) during grow-out phases I and II in brackish water ponds.

being restocked in another pond for further growth. In the final grow-out phase, fish are harvested and sent to market when the biomass is between the critical standing crop and carrying capacity. Instead of growing by 200 to 400 g each during a production phase, the fish grow 100 to 200 g. Fish are handled more frequently where ponds are managed for smaller growth increments. Figure 7.4 shows the change in tilapia biomass during two-phase grow-out.

A model was developed to compare one versus two production phases, based on a California Mozambique tilapia farm using single-batch culture. The culture system and growth data from this farm were used in the model as the basis for the single grow-out phase culture system. There are 20 ponds (2 nursery, 18 grow-out) of 0.1 ha on the model farm. In the single grow-out phase culture model, fish were grown to an individual weight of 2 g in a hatchery, to 35 to 95 g in nursery ponds, and to approximately 500 g or more in grow-out I ponds (see Table 7.5). In the two grow-out phase culture model, fish were grown to an individual weight of 2 g in a hatchery, to 35 to 80 g in nursery ponds, to 190 to 295 g in grow-out I ponds, and to approximately 500 g or more in grow-out II ponds (see Table 7.6). Data in Tables 7.5 and 7.6 were for the model farm at full production capacity, which is achieved 40 weeks after start-up. Individual weights shown in Tables 7.5 and 7.6 correspond to the beginning of the two-week period. In both models, the largest fish were harvested selectively from nursery ponds for transfer into grow-out I ponds. Thus, with the exception of the final harvest of the nursery pond, the average weight of fish stocked into grow-out I ponds was higher than the nursery pond population average weight. In the two grow-out phase culture model, fish were transferred from grow-out I ponds to grow-out II ponds at the end of the two-week period. Therefore, the higher mean individual weight seen in the grow-out II ponds at stocking reflects the previous two weeks' growth. Fish were graded into two size groups during the transfer from grow-out I to grow-out II ponds. Both models assumed that all fish survived all interpond transfers.

The same pond management strategy was used for both models. The farm used only floating extruded feeds. Fish were fed a 40 percent crude protein feed for the first four to six weeks of culture in nursery ponds, followed by a 36 percent crude protein feed for the remainder of the nursery period and all grow-out phases. Fish were fed three times daily, at 900, 1,200, and 1,500 h. Daily feeding rate was 5 percent of tilapia biomass at the start of the nursery phase and decreased progressively to 1 to 1.5 percent of biomass by the end of grow-out. All ponds were equipped with two 3.73 kW electric paddlewheel aerators that operated continuously. Thirty percent of the pond volume was exchanged daily.

TABLE 7.5. Fish stocking rate (number per square meter), average weight (g per fish), and biomass (kg·/0.1 ha) for individual 0.1 ha ponds on a model tilapia farm in southern California raising sex-reversed Mozambique tilapia *(O. mossambicus)* using one grow-out phase over a 54-week period.

	Nursery						Grow-out I															
	Pond 1			Pond 2			Pond 3		Pond 4		Pond 5		Pond 6		Pond 7		Pond 8		Pond 9			
Week	No.	g	kg	No.	g	kg	g	kg	g	kg	g	kg	g	kg	g	kg	g	kg	g	kg		
0	80	84	6,720	80	84	6,720	249	9,960	249	9,960	226	9,040	226	9,040	204	8,160	204	8,160	163	6,520		
2	40	89	3,560	40	89	3,560	295	11,800	295	11,800	272	10,880	272	10,880	250	10,000	250	10,000	191	7,640		
4	360	2	720	360	2	720	341	13,640	341	13,640	318	12,720	318	12,720	296	11,840	296	11,840	237	9,480		
6	360	3	1,080	360	3	1,080	375	15,000	375	15,000	358	14,320	358	14,320	342	13,680	342	13,680	283	11,320		
8	360	4	1,440	360	4	1,440	408	16,320	408	16,320	392	15,680	392	15,680	376	15,040	376	15,040	330	13,200		
10	360	6	2,160	360	6	2,160	442	17,680	442	17,680	425	17,000	425	17,000	409	16,360	409	16,360	363	14,520		
12	360	15	5,400	360	15	5,400	475	19,000	475	19,000	459	18,360	459	18,360	443	17,720	443	17,720	397	15,880		
14	360	23	8,280	360	23	8,280	509	20,360	509	20,360	493	19,720	493	19,720	476	19,040	476	19,040	430	17,200		
16	320	32	10,240	320	32	10,240	37	1,480	37	1,480	526	21,040	526	21,040	510	20,400	510	20,400	464	18,560		
18	280	41	11,480	280	41	11,480	46	1,840	46	1,840	46	1,840	46	1,840	544	21,760	544	21,760	498	19,920		
20	240	49	11,760	240	49	11,760	54	2,160	54	2,160	55	2,200	55	2,200	54	2,160	54	2,160	531	21,240		
22	200	54	10,800	200	54	10,800	63	2,520	63	2,520	63	2,520	63	2,520	63	2,520	63	2,520	59	2,360		

24	160	59	9,440	160	59	9,440	71	2,840	71	2,840	72	2,880	72	2,880	71	2,840	71	2,840	68	2,720
26	120	64	7,680	120	64	7,680	90	3,600	90	3,600	100	4,000	100	4,000	90	3,600	90	3,600	76	3,040
28	80	69	5,520	80	69	5,520	118	4,720	118	4,720	128	5,120	128	5,120	118	4,720	118	4,720	104	4,160
30	40	74	2,960	40	74	2,960	146	5,840	146	5,840	156	6,240	156	6,240	146	5,840	146	5,840	132	5,280
32	360	2	720	360	2	720	183	7,320	183	7,320	202	8,080	202	8,080	183	7,320	183	7,320	160	6,400
34	360	3	1,080	360	3	1,080	229	9,160	229	9,160	248	9,920	248	9,920	229	9,160	229	9,160	206	8,240
36	360	4	1,440	360	4	1,440	275	11,000	275	11,000	294	11,760	294	11,760	275	11,000	275	11,000	252	10,080
38	360	6	2,160	360	6	2,160	321	12,840	321	12,840	340	13,600	340	13,600	321	12,840	321	12,840	298	11,920
40	360	15	5,400	360	15	5,400	360	14,400	360	14,400	373	14,920	373	14,920	360	14,400	360	14,400	344	13,760
42	360	23	8,280	360	23	8,280	394	15,760	394	15,760	407	16,280	407	16,280	394	15,760	394	15,760	378	15,120
44	360	32	11,520	360	32	11,520	428	17,120	428	17,120	440	17,600	440	17,600	428	17,120	428	17,120	411	16,440
46	360	41	14,760	360	41	14,760	461	18,440	461	18,440	474	18,960	474	18,960	461	18,440	461	18,440	445	17,800
48	320	49	15,680	320	49	15,680	54	2,160	54	2,160	508	20,320	508	20,320	495	19,800	495	19,800	479	19,160
50	280	54	15,120	280	54	15,120	63	2,520	63	2,520	59	2,360	59	2,360	528	21,120	528	21,120	513	20,520
52	240	59	14,160	240	59	14,160	71	2,840	71	2,840	68	2,720	68	2,720	64	2,560	64	2,560	546	21,840
54	200	64	12,800	200	64	12,800	90	3,600	90	3,600	76	3,040	76	3,040	73	2,920	73	2,920	69	2,760

TABLE 7.5 *(continued)*

	Grow-out I																					
	Pond 10		Pond 11		Pond 12		Pond 13		Pond 14		Pond 15		Pond 16		Pond 17		Pond 18		Pond 19		Pond 20	
Week	g	kg	g	kg	g	kg	g	kg	g	kg	g	kg	g	kg	g	kg	g	kg	g	kg	g	kg
0	163	6,520	140	5,600	140	5,600	112	4,480	112	4,480	89	3,560	89	3,560	552	22,080	552	22,080	529	21,160	529	21,160
2	191	7,640	176	7,040	176	7,040	140	5,600	140	5,600	117	4,680	117	4,680	89	3,560	89	3,560	562	22,480	562	22,480
4	237	9,480	222	8,880	222	8,880	176	7,040	176	7,040	146	5,840	146	5,840	117	4,680	117	4,680	94	3,760	94	3,760
6	283	11,320	268	10,720	268	10,720	222	8,880	222	8,880	183	7,320	183	7,320	146	5,840	146	5,840	122	4,880	122	4,880
8	330	13,200	315	12,600	315	12,600	268	10,720	268	10,720	229	9,160	229	9,160	183	7,320	183	7,320	150	6,000	150	6,000
10	363	14,520	350	14,000	350	14,000	315	12,600	315	12,600	275	11,000	275	11,000	229	9,160	229	9,160	196	7,840	196	7,840
12	397	15,880	384	15,360	384	15,360	350	14,000	350	14,000	321	12,840	321	12,840	275	11,000	275	11,000	242	9,680	242	9,680
14	430	17,200	417	16,680	417	16,680	384	15,360	384	15,360	360	14,400	360	14,400	321	12,840	321	12,840	288	11,520	288	11,520
16	464	18,560	451	18,040	451	18,040	417	16,680	417	16,680	394	15,760	394	15,760	360	14,400	360	14,400	334	13,360	334	13,360
18	498	19,920	484	19,360	484	19,360	451	18,040	451	18,040	428	17,120	428	17,120	394	15,760	394	15,760	368	14,720	368	14,720
20	531	21,240	518	20,720	518	20,720	484	19,360	484	19,360	461	18,440	461	18,440	428	17,120	428	17,120	401	16,040	401	16,040
22	59	2,360	552	22,080	552	22,080	518	20,720	518	20,720	495	19,800	495	19,800	461	18,440	461	18,440	435	17,400	435	17,400
24	68	2,720	64	2,560	64	2,560	552	22,080	552	22,080	529	21,160	529	21,160	495	19,800	495	19,800	468	18,720	468	18,720
26	76	3,040	73	2,920	73	2,920	69	2,760	69	2,760	562	22,480	562	22,480	529	21,160	529	21,160	502	20,080	502	20,080

28	104	4,160	101	4,040	101	4,040	87	3,480	87	3,480	74	2,960	74	2,960	562	22,480	562	22,480	536	21,440	536	21,440
30	132	5,280	129	5,160	129	5,160	115	4,600	115	4,600	102	4,080	102	4,080	79	3,160	79	3,160	569	22,760	569	22,760
32	160	6,400	157	6,280	157	6,280	143	5,720	143	5,720	130	5,200	130	5,200	107	4,280	107	4,280	84	3,360	84	3,360
34	206	8,240	203	8,120	203	8,120	180	7,200	180	7,200	158	6,320	158	6,320	135	5,400	135	5,400	112	4,480	112	4,480
36	252	10,080	249	9,960	249	9,960	226	9,040	226	9,040	204	8,160	204	8,160	163	6,520	163	6,520	140	5,600	140	5,600
38	298	11,920	295	11,800	295	11,800	272	10,880	272	10,880	250	10,000	250	10,000	191	7,640	191	7,640	176	7,040	176	7,040
40	344	13,760	341	13,640	341	13,640	318	12,720	318	12,720	296	11,840	296	11,840	237	9,480	237	9,480	222	8,880	222	8,880
42	378	15,120	375	15,000	375	15,000	358	14,320	358	14,320	342	13,680	342	13,680	283	11,320	283	11,320	268	10,720	268	10,720
44	411	16,440	408	16,320	408	16,320	392	15,680	392	15,680	376	15,040	376	15,040	330	13,200	330	13,200	315	12,600	315	12,600
46	445	17,800	442	17,680	442	17,680	425	17,000	425	17,000	409	16,360	409	16,360	363	14,520	363	14,520	350	14,000	350	14,000
48	479	19,160	475	19,000	475	19,000	459	18,360	459	18,360	443	17,720	443	17,720	397	15,880	397	15,880	384	15,360	384	15,360
50	513	20,520	509	20,360	509	20,360	493	19,720	493	19,720	476	19,040	476	19,040	430	17,200	430	17,200	417	16,680	417	16,680
52	546	21,840	542	21,680	542	21,680	526	21,040	526	21,040	510	20,400	510	20,400	464	18,560	464	18,560	451	18,040	451	18,040
54	69	2,760	558	22,320	576	23,040	543	21,720	543	21,720	544	21,760	544	21,760	498	19,920	498	19,920	484	19,360	484	19,360

Note: Grow-out I ponds are stocked at 40 fish per square meter. Fish stocked in nursery ponds were grown to 2 g individual weight in the hatchery. Nursery to grow-out I pond transfer is shown by the shaded boxes on the same line.

TABLE 7.6. Fish stocking rate (number per square meter), average weight (g per fish), and biomass (kg·0.1 ha) for individual 0.1 ha ponds on a model tilapia farm in southern California raising sex-reversed Mozambique tilapia *(O. mossambicus)* using two grow-out phases over a 54-week period.

	Nursery						Grow-out I												Grow-out II	
	Pond 1			Pond 2			Pond 3		Pond 4		Pond 5		Pond 6		Pond 7		Pond 8		Pond 9	
Week	No.	g	kg	No.	g	kg	g	kg	g	kg	g	kg	g	kg	g	kg	g	kg	g	kg
0	160	59	9,440	320	3	960	127	10,160	127	10,160	71	5,680	101	8,080	87	6,960	64	5,120	455	18,200
2	160	64	10,240	320	4	1,280	155	12,400	155	12,400	99	7,920	129	10,320	115	9,200	73	5,840	488	19,520
4	160	69	11,040	320	6	1,920	201	16,080	201	16,080	127	10,160	157	12,560	143	11,440	101	8,080	522	20,880
6	80	74	5,920	320	15	4,800	79	6,320	247	19,760	155	12,400	203	16,240	180	14,400	129	10,320	237	9,480
8	320	79	25,280	320	23	7,360	107	8,560	79	6,320	201	16,080	249	19,920	226	18,080	157	12,560	283	11,320
10	320	3	960	240	32	7,680	135	10,800	107	8,560	35	2,800	272	21,760	249	19,920	203	16,240	329	13,160
12	320	4	1,280	160	41	6,560	172	13,760	135	10,800	44	3,520	46	3,680	272	21,760	249	19,920	369	14,760
14	320	6	1,920	80	49	3,920	218	17,440	172	13,760	52	4,160	54	4,320	54	4,320	272	21,760	402	16,080
16	320	15	4,800	320	54	17,280	241	19,280	241	19,280	61	4,880	63	5,040	63	5,040	54	4,320	436	17,440
18	320	23	7,360	320	3	960	264	21,120	264	21,120	69	5,520	71	5,680	71	5,680	63	5,040	470	18,800
20	320	32	10,240	320	4	1,280	272	21,760	272	21,760	88	7,040	99	7,920	99	7,920	71	5,680	503	20,120
22	240	41	9,840	320	6	1,920	46	3,680	272	21,760	116	9,280	127	10,160	127	10,160	99	7,920	262	10,480

24	160	49	7,840	320	15	4,800	54	4,320	54	4,320	144	11,520	155	12,400	155	12,400	127	10,160	308	12,320
26	80	54	4,320	320	23	7,360	63	5,040	63	5,040	190	15,200	201	16,080	59	4,720	155	12,400	354	14,160
28	320	59	18,880	320	32	10,240	71	5,680	71	5,680	236	18,880	59	4,720	68	5,440	201	16,080	388	15,520
30	320	3	960	240	41	9,840	99	7,920	99	7,920	46	3,680	68	5,440	86	6,880	247	19,760	421	16,840
32	320	4	1,280	160	49	7,840	127	10,160	127	10,160	54	4,320	86	6,880	132	10,560	54	4,320	455	18,200
34	320	6	1,920	160	54	8,640	155	12,400	155	12,400	63	5,040	132	10,560	178	14,240	63	5,040	488	19,520
36	320	15	4,800	160	59	9,440	201	16,080	201	16,080	71	5,680	178	14,240	224	17,920	71	5,680	522	20,880
38	320	23	7,360	80	64	5,120	69	5,520	247	19,760	99	7,920	224	17,920	247	19,760	99	7,920	237	9,480
40	320	32	10,240	320	2	640	88	7,040	69	5,520	127	10,160	247	19,760	270	21,600	127	10,160	283	11,320
42	240	41	9,840	320	3	960	116	9,280	88	7,040	155	12,400	270	21,600	46	3,680	155	12,400	329	13,160
44	160	49	7,840	320	4	1,280	144	11,520	116	9,280	201	16,080	54	4,320	54	4,320	201	16,080	369	14,760
46	80	54	4,320	320	6	1,920	190	15,200	144	11,520	247	19,760	63	5,040	63	5,040	59	4,720	402	16,080
48	320	59	18,880	320	15	4,800	236	18,880	190	15,200	59	4,720	71	5,680	71	5,680	68	5,440	436	17,440
50	320	3	960	320	23	7,360	259	20,720	236	18,880	68	5,440	99	7,920	99	7,920	86	6,880	470	18,800
52	320	4	1,280	320	32	10,240	272	21,760	259	20,720	86	6,880	127	10,160	127	10,160	132	10,560	503	20,120
54	320	6	*1,920*	240	41	9,840	46	3,680	272	21,760	132	10,560	155	12,400	155	12,400	178	14,240	262	10,480

TABLE 7.6 *(continued)*

	Grow-out II																					
	Pond 10		Pond 11		Pond 12		Pond 13		Pond 14		Pond 15		Pond 16		Pond 17		Pond 18		Pond 19		Pond 20	
Week	g	kg	g	kg	g	kg	g	kg	g	kg	g	kg	g	kg	g	kg	g	kg	g	kg	g	kg
0	469	18,760	421	16,840	435	17,400	352	14,080	366	14,640	329	13,160	349	13,960	287	11,480	307	12,280	264	10,560	284	11,360
2	502	20,080	455	18,200	469	18,760	386	15,440	399	15,960	369	14,760	383	15,320	333	13,320	353	14,120	312	12,480	332	13,280
4	536	21,440	488	19,520	502	20,080	419	16,760	433	17,320	402	16,080	416	16,640	373	14,920	387	15,480	358	14,320	372	14,880
6	257	10,280	522	20,880	536	21,440	453	18,120	467	18,680	436	17,440	450	18,000	406	16,240	420	16,800	392	15,680	405	16,200
8	303	12,120	260	10,400	280	11,200	486	19,440	500	20,000	470	18,800	483	19,320	440	17,600	454	18,160	425	17,000	439	17,560
10	349	13,960	287	11,480	307	12,280	237	9,480	257	10,280	503	20,120	517	20,680	474	18,960	487	19,480	459	18,360	472	18,880
12	383	15,320	333	13,320	353	14,120	283	11,320	303	12,120	262	10,480	282	11,280	507	20,280	521	20,840	492	19,680	506	20,240
14	416	16,640	373	14,920	387	15,480	329	13,160	349	13,960	308	12,320	328	13,120	262	10,480	282	11,280	526	21,040	540	21,600
16	450	18,000	406	16,240	420	16,800	363	14,520	383	15,320	354	14,160	368	14,720	308	12,320	328	13,120	262	10,480	282	11,280
18	483	19,320	440	17,600	454	18,160	396	15,840	416	16,640	388	15,520	401	16,040	354	14,160	368	14,720	308	12,320	328	13,120
20	517	20,680	474	18,960	487	19,480	430	17,200	450	18,000	421	16,840	435	17,400	388	15,520	401	16,040	354	14,160	368	14,720
22	282	11,280	507	20,280	521	20,840	463	18,520	483	19,320	455	18,200	469	18,760	421	16,840	435	17,400	388	15,520	401	16,040
24	328	13,120	262	10,480	282	11,280	497	19,880	517	20,680	488	19,520	502	20,080	455	18,200	469	18,760	421	16,840	435	17,400
26	368	14,720	308	12,320	328	13,120	191	7,640	211	8,440	522	20,880	536	21,440	488	19,520	502	20,080	455	18,200	469	18,760

28	401	16,040	354	14,160	368	14,720	237	9,480	247	9,880	226	9,040	246	9,840	522	20,880	536	21,440	488	19,520	502	20,080
30	435	17,400	388	15,520	401	16,040	283	11,320	293	11,720	272	10,880	292	11,680	272	10,880	292	11,680	522	20,880	536	21,440
32	469	18,760	421	16,840	435	17,400	329	13,160	339	13,560	318	12,720	338	13,520	318	12,720	338	13,520	260	10,400	280	11,200
34	502	20,080	455	18,200	469	18,760	369	14,760	379	15,160	359	14,360	378	15,120	359	14,360	378	15,120	287	11,480	307	12,280
36	536	21,440	488	19,520	502	20,080	402	16,080	412	16,480	391	15,640	411	16,440	391	15,640	411	16,440	333	13,320	353	14,120
38	257	10,280	522	20,880	536	21,440	436	17,440	446	17,840	425	17,000	445	17,800	425	17,000	445	17,800	373	14,920	387	15,480
40	303	12,120	260	10,400	280	11,200	469	18,760	479	19,160	459	18,360	479	19,160	459	18,360	479	19,160	406	16,240	420	16,800
42	349	13,960	287	11,480	307	12,280	260	10,400	280	11,200	492	19,680	512	20,480	492	19,680	512	20,480	440	17,600	454	18,160
44	383	15,320	333	13,320	353	14,120	287	11,480	307	12,280	260	10,400	280	11,200	526	21,040	546	21,840	474	18,960	487	19,480
46	416	16,640	373	14,920	387	15,480	333	13,320	353	14,120	287	11,480	307	12,280	237	9,480	257	10,280	507	20,280	521	20,840
48	450	18,000	406	16,240	420	16,800	373	14,920	387	15,480	333	13,320	353	14,120	283	11,320	303	12,120	260	10,400	280	11,200
50	483	19,320	440	17,600	454	18,160	406	16,240	420	16,800	373	14,920	387	15,480	329	13,160	349	13,960	287	11,480	307	12,280
52	517	20,680	474	18,960	487	19,480	440	17,600	454	18,160	406	16,240	420	16,800	369	14,760	383	15,320	333	13,320	353	14,120
54	282	11,280	507	20,280	521	20,840	474	18,960	487	19,480	440	17,600	454	18,160	402	16,080	416	16,640	373	14,920	387	15,480

Note: Fish stocked in nursery ponds were grown to 2 g individual weight in the hatchery. Nursery to grow-out I pond transfer is shown by the shaded boxes on the same line. Grow-out I to grow-out II pond transfer is shown by underlined numbers on adjacent lines. Grow-out I and II ponds are stocked at 80 and 40 fish per square meter, respectively.

Once the full production level is achieved, the model shows consistent predicted annual production. A total of 645,120 kg of tilapia were harvested annually from ponds managed according to the single grow-out phase culture model. Annual tilapia production for the farm totaled 785,600 kg for ponds managed according to the two grow-out phase model. The 22 percent greater annual production in the two grow-out phase model resulted from the 12 grow-out II ponds being harvested at least three times annually. In the single grow-out phase model, the 18 grow-out I ponds were harvested on average twice annually. Fish were harvested for market on 30 occasions in the single grow-out phase model and on 38 occasions in the two grow-out phase model. Tilapia production was similar for the two models and averaged 21,622 and 20,668 kg per 0.1 ha^{-1} for the single and two grow-out phase models, respectively. The total tilapia biomass on the farm at any time ranged from 109,000 to 336,000 kg and from 228,000 to 307,000 kg for the single and two grow-out phase models, respectively (see Figure 7.5). Generally, the total tilapia biomass within any production phase was less variable for the two grow-out phase model than for the single grow-out phase model (see Figure 7.6). Thus, a more consistent supply of fish for the market was produced using the two grow-out phase culture model. The increased productivity obtained with the two grow-out phase culture model requires more fish handling, with the associated stress, greater water requirements, and more labor. An economic analysis, in addition to a general feasibility analysis, is required to decide whether a farm should adopt a new production strategy.

Production parameters in intensively managed ponds vary by grow-out phase (see Table 7.7). Only sex-reversed (male) tilapias are stocked in production ponds. Floating extruded feed (30 to 35 percent crude protein) is offered to fish daily. Feed conversion ratio ranges from 1.7 to 2.5 kg feed per kilogram of fish. Fish growth in ponds is rapid. Mozambique tilapia stocked in ponds at 38 to 40 fish per square meter on a California tilapia farm grew an average of 2.9 $g \cdot day^{-1}$ (see Figure 7.7). Nile tilapia in grow-out phase I ponds stocked at 123 fish per square meter grew an average of 2.9 $g \cdot day^{-1}$, and in grow-out phase II ponds stocked at 57 fish per square meter grew an average of 3.9 $g \cdot day^{-1}$ (see Figure 7.8). At harvest, individual fish range in size from 500 to 1,000 g. Generally fish harvested from intensively managed ponds outside the United States are processed, and either fresh or frozen fillets are exported. Fish produced in the United States generally are sold whole.

Two pond management strategies, single- and multiple-batch culture, address the utilization of pond productive capacity differently. The fish biomass is low during the early stages of each production phase where ponds are stocked with a single cohort of fish, that is, single-batch culture

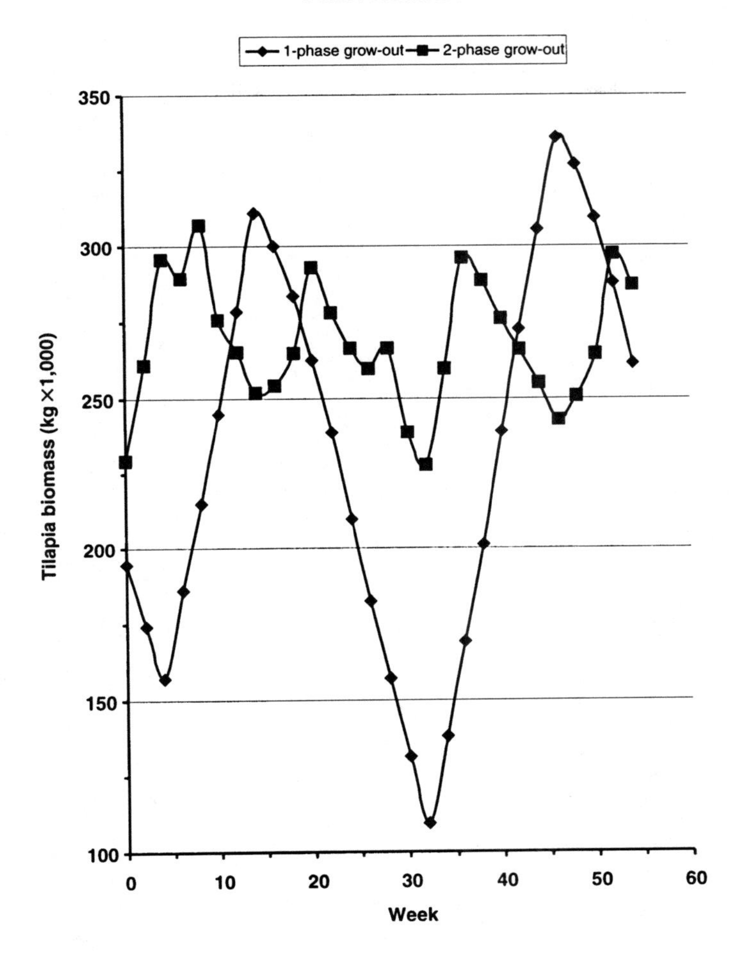

FIGURE 7.5. Total tilapia biomass over a 54-week period on a model farm in southern California that raises sex-reversed Mozambique tilapia *(O. mossambicus)* in twenty 0.1 ha ponds using single or two grow-out phase culture. See text for a description of both management systems.

(see Figure 7.4). In single-batch culture, which is common, ponds are stocked with uniform-size fish that are allowed to grow until the fish biomass is between the critical standing crop and the carrying capacity. However, the productive capacity of the pond, in terms of daily yield, is underutilized during the first two to three months of culture. Multiple-batch culture, where two or more distinct cohorts of fish are stocked, more fully utilizes the productive

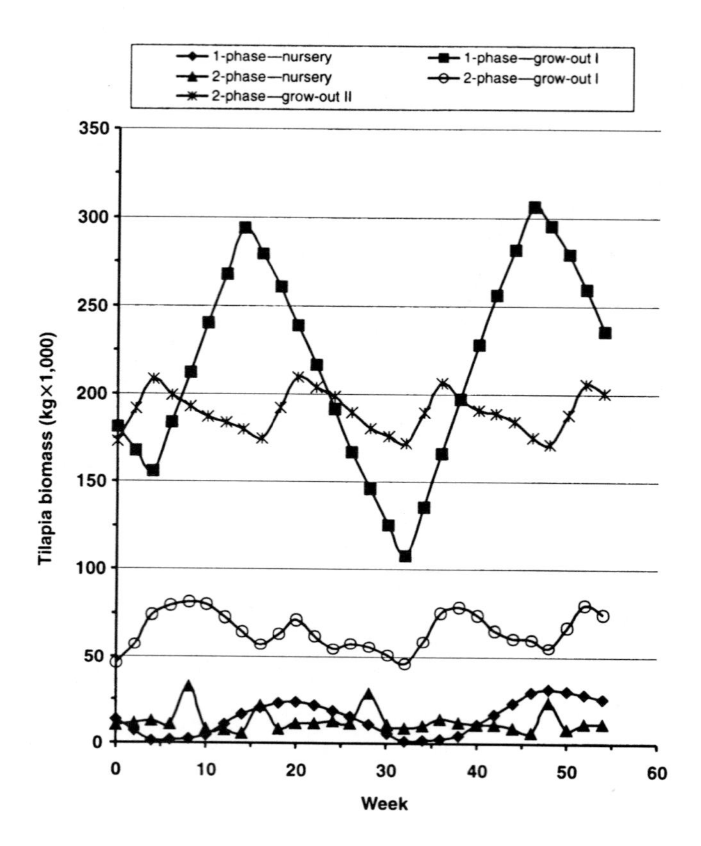

FIGURE 7.6. Total tilapia biomass within each production phase over a 54-week period on a model farm in southern California that raises sex-reversed Mozambique tilapia *(O. mossambicus)* in twenty 0.1 ha ponds using single or two grow-out phase culture. See text for a description of both management systems.

capacity of the pond at all times. Ponds managed for multiple-batch culture are periodically partially harvested to remove the larger fish for additional growth in the next production phase or for market. Following the partial harvest, the pond is restocked (understocked) with smaller fish; the number of fish understocked generally equals the number harvested plus a percentage

TABLE 7.7. Summary of production parameters, as ranges, for grow-out phases I, II, and III on intensively managed tilapia farms, stocking sex-reversed tilapia, in select countries.

Phase	Stocking rate (fish per m^2)	Final Size (g per fish)	Duration (days)	Growth ($g \cdot day^{-1}$)	Survival (%)	Standing stock ($kg \cdot ha^{-1}$)	Aeration ($kW \cdot ha^{-1}$)	Water exchange ($\% \cdot day^{-1}$)
Grow-out I								
Honduras	3-12	100-250	60-90	0.9-1.7	77-90	3,542-12,960	5-7.5	1-20
Jamaica	11-17	150-250	98-100	0.8-2.0	85	16,500	1.5-3.7	75-100
Grow-out II								
Honduras	2-8	225-650	60-240	1.8-2.4	85-97	6,600-24,360	5-7.5	1-20
Jamaica	8.4-10	550-650	130	2.3-3.9	90-92	42,000-59,150	11.2-14.9	150-200
Grow-out III								
Honduras	1.8-4	630-650	90-135	2.2-3.7	90	9,954-23,400	5-7.5	1-20

Source: After Carberry and Hanley 1997; Green and Engle 2000.

(usually less than 25 percent of the total). The partial harvests and restockings occur on a regular basis in multiple-batch culture.

Single-batch culture is a convenient pond management strategy for intensive production systems because it allows the farmer to more closely track the performance of the fish population and control tilapia reproduction and growth. Despite advances in producing populations of male tilapia through sex reversal, the process generally is only 95 to 99.9 percent effective. Reproduction will occur if just one female fish is present in the pond. Unchecked, tilapia reproduction will result in a substantial biomass of mixed-sex tilapia offspring that will reach sexual maturity and reproduce, further exacerbating the problem, especially in multiple-batch culture. Although semi-intensively managed tilapia production ponds have been costocked with a piscivorous fish to prey upon fingerlings produced during grow-out, this approach is impractical for intensively managed ponds. Consequently, many producers utilize some form of single-batch culture so that ponds are harvested once or twice annually, thereby allowing the removal of offspring as well as stunted fish.

Grading

Fish growth is better when fish cultured together are similar in size. Cannibalism is prevented in the hatchery if uniformly sized fish comprise the

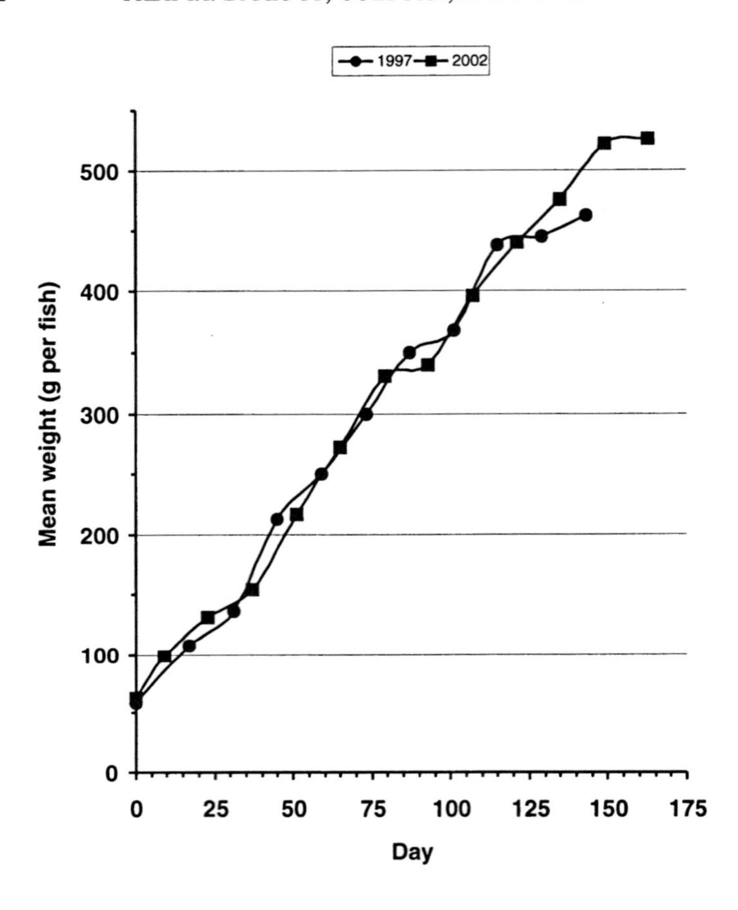

FIGURE 7.7. Growth of sex-reversed Mozambique tilapia *(O. mossambicus)* stocked at 39 to 40 fish per square meter in 0.1 ha ponds on a tilapia farm in southern California during 1997 and 2002. Tilapia are grown in one grow-out phase in single-batch culture.

population. However, there always will be a percentage, albeit small, of the fish population that will grow more slowly during every phase of culture. In addition to concerns about slower growth rates, fish size is very important in marketing harvested fish. Live market vendors and processors generally have strict size tolerances for fish they will accept or purchase from producers. To ensure they produce a consistent, uniformly sized product for

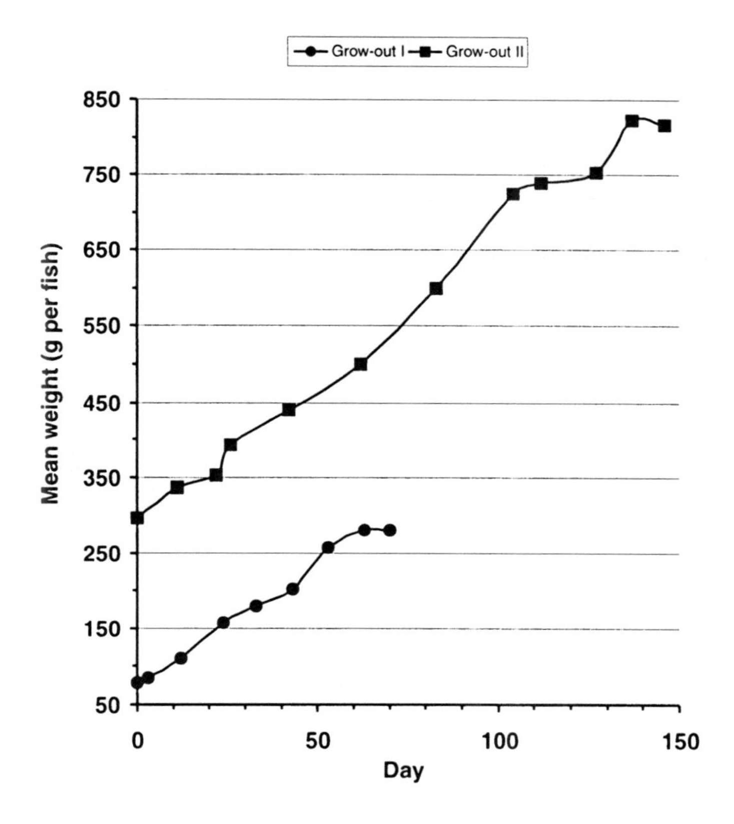

FIGURE 7.8. Growth of sex-reversed Nile tilapia *(O. niloticus)* during grow-out phases I and II on an intensively managed farm in Honduras. Ponds are stocked with 123 fish per square meter and 57 fish per square meter during grow-out phases I and II, respectively. Grow-out phase I lasts 70 days and grow-out phase II lasts 146 days. Final standing stock is 34 $kg{\cdot}m^{-3}$ and 46 $kg{\cdot}m^{-3}$ for grow-out phases I and II, respectively. Water exchange ranges from 380 to 640 $m^3{\cdot}h^{-1}$.

market, many farmers use some grading method each time fish are transferred between ponds and before transport to market.

When fish are graded between production phases, slower-growing fish may be grouped with similar fish from other ponds for additional growth or be discarded or sold as low-value product. Grading fish between production phases also provides an opportunity to eliminate any females from the population, thereby minimizing the potential impact of reproduction during grow-out phases. The decision as to the minimum size of fish to be retained by the grader comes from experience with fish performance during each phase of production.

As live fish vendors easily detect size variation, fish destined for this market should be graded to within 25 g of market weight. Although fish bound for a processor need not be monitored as closely for size variation, the processor still will have tolerances that must be respected. Fish must be large enough to allow processors to obtain the required size classes of fillets. Postharvest handling of tilapia is discussed in Chapter 18, "Harvest, Handling, and Processing."

There are a variety of graders, many of which have been developed empirically by farmers. Some are connected to the harvesting mechanism and fish are graded as they are caught. Other systems must be loaded using a dip net. Floating boxes, crowding systems in circular or raceway form, and in-line graders with dewatering systems are all possibilities. Commercially manufactured graders also are available. All graders use a system of bars spaced to retain a particular size of fish while letting the smaller fish escape.

PVC pipe usually is a poor choice for grader bars. Unless the length is less than 2 feet or there are frequent crosswise stabilizers, PVC pipe warps quickly and it becomes difficult to maintain the desired bar spacing, which results in poor size segregation. Aluminum pipe is a better choice because it is lightweight and holds its form. In the end, the grader must segregate the fish quickly, as gently as possible, and with minimal size variation in the graded fish population.

Off-Flavor

Because tilapia can absorb algal- and microbial-induced off-flavors from the pond water during grow-out, a quality assurance program needs to be in place on farms to ensure that only clean-tasting fish are harvested for market. Avoiding off-flavor fish is equally important for domestic and export markets and is key to establishing and maintaining markets. Many farms, particularly the more intensively managed farms, will purge tilapia before sending them to market. Harvested fish are transferred to large tanks that range from 7 to 500 m^2 in size, are supplied continuously with high-quality, clean water, and may or may not be equipped with aerators. One to two complete water exchanges per hour are conducted during the purging process, which lasts from 15 to 48 h up to 5 to 7 days. Fish are loaded into purging tanks at 5 to 110 $kg \cdot m^{-3}$. Normally fish are taken off feed 24 to 48 h before harvest and are not fed during purging. Tilapia lose weight during purging, from about 2 to 4 percent for short-duration up to 5 to 10 percent for the long-duration purging. Survival during purging is approximately

98 percent. Further discussion of postharvest handling is presented in Chapter 18, "Harvest, Handling, and Processing."

REFERENCES

Avault, J. (1996). *Fundamentals of aquaculture.* Baton Rouge, LA: AVA Publishing Company.

Balarin, J.D. (1988). Development planning for tilapia farming in Africa. In R.S.V. Pullin, T. Bhukaswan, K. Tonguthai, and J.L. Maclean (Eds.), *The Second International Symposium on Tilapia in Aquaculture* (pp. 531–538). ICLARM Conference Proceedings 15. Bangkok, Thailand: Department of Fisheries and Manila, Philippines: International Center for Living Aquatic Resources Management.

Boyd, C.E. (1990). *Water quality in ponds for aquaculture.* Auburn, AL: Alabama Agriculture Experiment Station.

Brown, C.L., R.B. Bolivar, E.B.T. Jimenez, and J.P. Szyper (2001). Global experiment: Optimization of nitrogen fertilization rate in freshwater tilapia production ponds (cool season trial). In A. Gupta, K. McElwee, D. Burke, J. Burright, X. Cummings, and H. Egna (Eds.), *Eighteenth annual technical report pond dynamics/aquaculture CRSP* (pp. 23-26). Corvallis, OR: Oregon State University.

Carberry, J., and F. Hanley (1997). Commercial intensive tilapia culture in Jamaica. In D.E. Alston, B.W. Green, and H.C. Clifford (Eds.), IV SIV Symposium on Aquaculture in Central America: Focusing on Shrimp and Tilapia, 22-24 April 1997, Tegucigalpa, Honduras (pp. 64-67). Baton Rouge, LA: Asociacion Nacional de Acuacultores de Honduras and the Latin American Chapter of the World Aquaculture Society.

Costa-Pierce, B. (1997). Tilapias of the Salton Sea, a marine lake in California. In *Tilapia aquaculture: Proceedings from the Fourth International Symposium on Tilapia Aquaculture* (pp. 584-590). Ithaca, NY: Northeast Regional Agricultural Engineering Service.

———. (2001). Doubling global tilapia production in the 21st century: Species selection, markets and farmer-scientist collaboration. *Aquaculture 2001: Book of abstracts* (p. 143). Lake Buena Vista, Florida. Baton Rouge, LA: World Aquaculture Society.

De Vos, L., J. Snocks, and D.T. van der Audenaerde (1990). The effects of *Tilapia* introductions in Lake Luhondo, Rwanda. *Environmental Biology of Fishes* 27:303-308.

Dey, M.M. (2001). Tilapia production in south Asia and the far east. In *Tilapia: Production, marketing and technological developments: Proceedings of the Tilapia 2001 International Technical and Trade Conference of Tilapia, 28-30 May 2001* (pp. 17-26). Kuala Lumpur, Malaysia.

Diana, J.S., C.K. Lin, and K. Jaiyen (1994). Supplemental feeding of tilapia in fertilized ponds. *Journal of the World Aquaculture Society* 25:497-506.

Diana, J.S., C.K. Lin, and Y. Yi (1996). Timing of supplemental feeding for tilapia production. *Journal of the World Aquaculture Society* 27:410-419.

Green, B.W. (1992). Substitution of organic manure for pelleted feed in tilapia production. *Aquaculture* 101:213-222.

———. (1997). Inclusion of tilapia as a diversification strategy for penaeid shrimp culture. In D.E. Alston, B.W. Green, and H.C. Clifford (Eds.), *IV Symposium on Aquaculture in Central America: Focusing on Shrimp and Tilapia, 22-24 April 1997, Tegucigalpa, Honduras* (pp. 84-93). Baton Rouge, LA: Asociacion Nacional de Acuacultores de Honduras and the Latin American Chapter of the World Aquaculture Society.

Green, B.W., Z. El Nagdy, and H. Hebicha (2002). Evaluation of Nile tilapia pond management strategies in Egypt. *Aquaculture Research* 33:1037-1048.

Green, B.W., and C.R. Engle (2000). Commercial tilapia aquaculture in Honduras. In B.A. Costa-Pierce and J.E. Rakocy (Eds.), *Tilapia aquaculture in the Americas* (Volume 2, pp. 151-170). Baton Rouge, LA: World Aquaculture Society.

Green, B.W., R.P. Phelps, and H.R. Alvarenga (1989). The effect of manures and chemical fertilizers on the production of *Oreochromis niloticus in earthen ponds. Aquaculture* 76:37-42.

Green, B.W., D.R. Teichert-Coddington, C.E. Boyd, N. Claros, and C. Cardona (1999). Global experiment: Optimization of nitrogen fertilization rate in freshwater tilapia production ponds. In K. McElwee, D. Burke, M. Niles, and H. Egna (Eds.), *Sixteenth annual technical report, pond dynamics/aquaculture* CRSP (pp. 27-37). Corvallis, OR: Oregon State University.

Green, B.W., D.R. Teichert-Coddington, and R.P. Phelps (1990). Response of tilapia yield and economics to varying rates of organic fertilization and season in two Central American countries. *Aquaculture* 90:279-290.

Hepher, B. (1978). Ecological aspects of warmwater fishpond management. In S.D. Gerking (Ed.), *Ecology of freshwater fish production* (pp. 447-468). New York: John Wiley.

Hepher, B., and Y. Pruginin (1981). *Commercial fish farming: With special reference to fish culture in Israel.* New York: John Wiley.

Hulata, G. (1997). Large-scale tilapia fry production in Israel. *Israeli Journal of Aquaculture-Bamidgeh* 49:174-179.

Josupeit, H. (2001). Tilapia supply. In *Tilapia: Production, marketing and technological developments: Proceedings of the Tilapia 2001 International Technical and Trade Conference of Tilapia*, 28-30 May 2001 (pp. 4-6). Kuala Lumpur, Malaysia.

Lin, C.K., Y. Yi, R.B. Shivappa, M.A.K. Chowdhury, and J.S. Diana (1999). Global experiment: Optimization of nitrogen fertilization rate in freshwater tilapia production ponds. In K. McElwee, D. Burke, M. Niles, and H. Egna (Eds.), *Sixteenth annual technical report, pond dynamics/aquaculture CRSP* (pp. 49-56). Corvallis, OR: Oregon State University.

Lin, C.K., Y. Yi, H. Tung, and J.S. Diana (2000). Global experiment: Optimization of nitrogen fertilization rate in freshwater tilapia production ponds during cool season. In K. McElwee, D. Burke, M. Niles, X. Cummings, and H. Egna (Eds.), *Seventeenth Aannual technical report, pond dynamics/aquaculture CRSP* (pp. 43-48). Corvallis, OR: Oregon State University.

Lovshin, L.L., N.B. Schwartz, V.G. de Castillo, C.R. Engle, and U.L. Hatch (1986). Cooperatively managed rural Panamanian fish ponds: The integrated approach. International Center for Aquaculture Research and Development Series No. 35, Alabama Agricultural Experiment Station, Auburn, Alabama.

Lutz, C.G. (2001). Field evaluation of hybrid/crossbred varieties of tilapia in greenhouse based recirculating systems. In *Aquaculture 2001: Book of abstracts* (p. 394). Baton Rouge, LA: World Aquaculture Society.

Müller, F. (1978). The aquaculture rotation. *Aquacultura Hungarica* (Szarvas) 1:73-79.

Nelson, S.G., and L.G. Eldredge (1991). Distribution and status of introduced cichlid fishes of the genera *Oreochromis* and *Tilapia* in the islands of the South Pacific and Micronesia. *Asian Fisheries Society* 4:11-22.

Ogutu-Ohwayo, R. (1990). The decline of the native fishes of lakes Victoria and Kyoga (East Africa) and the impact of introduced species, especially the Nile perch, *Lates niloticus*, and the Nile tilapia, *Oreochromis niloticus. Environmental Biology of Fishes* 27:81-96.

Philippart, J.C., and J.C. Ruwet (1982). Ecology and distribution of tilapias. In R.S.V. Pullin and R.H. Lowe-McConnell (Eds.), *The biology and culture of tilapias* (pp. 15-59). ICLARM Conference Proceedings 7. Manilla, Philippines: International Center for Living Aquatic Resources Management.

Pullin, R.S.V. (1983). Choice of tilapia species for aquaculture. In L. Fishelson and Z. Yaron (Eds.), *International Symposium on Tilapia in Aquaculture* (pp. 64-76). Tel Aviv, Israel: Tel Aviv University.

———. (1985). Tilapias: Everyman's fish. *Biologist* 32:84-88.

———. (1988). *Tilapia genetic resources for aquaculture.* Manila, Philippines: International Center for Living Aquatic Resources Management.

Pullin, R.S.V., and Z. Shehadeh (1980). *Integrated agriculture-aquaculture farming systems.* ICLARM Conference Proceedings 4. Manila, Philippines: International Center for Living Aquatic Resources Management, and Laguna, Philippines: Southeast Asian Center for Graduate Study and Research in Agriculture, College, Los Banos.

Teichert-Coddington, D.R., and B.W. Green (1993). Tilapia yield improvement through maintenance of minimal oxygen concentrations in experimental grow-out ponds in Honduras. *Aquaculture* 118:63-71.

USDA-NRCS (1997). *Ponds—Planning, design, construction.* Washington, DC: U.S. Department of Agriculture Natural Resources Conservation Service Handbook Number 590.

Van der Merwe, J.P., J.J. Swart, and P.R. King (2001). Requirements for the successful establishment of aquaculture small farmers in the Western Cape Province of South Africa. In *Aquaculture 2001: Book of abstracts* (p. 657). Baton Rouge, LA: World Aquaculture Society.

Veverica, K.L., J. Bowman, and T. Popma (2001). Global experiment: Optimization of nitrogen fertilization rate in freshwater tilapia production ponds. In A. Gupta, K. McElwee, D. Burke, J. Burright, X. Cummings, and H. Egna (Eds.), *Eighteenth annual technical report, pond dynamics/aquaculture CRSP* (pp. 13-22). Corvallis, OR: Oregon State University.

Veverica, K.L., D. Mirera, and G. Matolla (2001). Optimization of phosphorus fertilization rate in freshwater tilapia production ponds Kenya. In *Aquaculture 2001: Book of abstracts* (p. 663). Baton Rouge, LA: World Aquaculture Society.

Watanabe, W.O., T.M. Losordo, K. Fitzsimmons, and F. Hanley (2002). Tilapia production systems in the Americas: Technological advances, trends, and challenges. *Reviews in Fisgheries Science* 10:465-498.

Wheaton, F.W. (1977). *Aquacultural engineering*. Malabar, FL: Robert E. Krieger.

Wohlfarth, G.W., and G.I. Hulata (1981). *Applied genetics of tilapia*. ICLARM Studies and Reviews 6. Manila, Philippines: International Center for Living Aquatic Resources Management.

Xiao, X., Z. Huang, and M. Liu (2002). The tilapia cultivation in Guandong. In *World Aquaculture 2002: Book of abstracts* (p. 838). Baton Rouge, LA: World Aquaculture Society.

Chapter 8

Culture in Flowing Water

Richard W. Soderberg

INTRODUCTION

Tilapia are usually cultured in static-water ponds at levels of intensity ranging from dependence on natural fertility to utilization of complete dry diets. When fish growth depends upon natural foods produced within the pond, which may be augmented by addition of fertilizers, production is limited by inadequate levels of one or more dietary components. Use of a complete dry diet eliminates these deficiencies and allows for greater fish production. The maximum level of production in fed ponds is determined by the capacities of the ecological mechanisms within the ponds to maintain suitable water quality to ensure a reasonable probability for satisfactory fish growth and health. Low dissolved oxygen (DO) levels, caused primarily by plankton respiration, usually determine the maximum levels of feeding and, hence, fish production. With aeration, maximum production levels are determined by the accumulation of nitrogenous metabolites, the most important of which are ammonia from protein metabolism and nitrite from the nitrification of fish-excreted ammonia.

Further intensification of fish production requires that water be flushed through the ponds to replenish DO and replace used, metabolite-laden water with fresh. The water flow rate eventually becomes too great for the significant production of natural food. Thus, complete diets must be used. The technology of flowing-water aquaculture, with application to tilapia, is the subject of this chapter.

Flowing-water fish culture, in its simplest form, involves channeling the water supply through fish rearing units and subsequent discharge to a receiving stream. Adequate volumes of water of suitable temperature for tilapia culture may be provided by geothermal springs or wells or warm-water

doi:10.1300/5513_08

discharges from industrial processes. In tropical locations, ground water and surface waters may be suitable for flowing-water culture of tilapia.

The fish production from such a system is determined by the DO content of the water, the rate at which fish consume DO, and the minimum DO concentration that the fish can tolerate. The production potentials of flowing-water supplies can usually be increased substantially with aeration. Serial water reuse employing water reconditioning by mechanical aeration or injection of liquid oxygen is common in salmonid culture and appropriate for tilapia culture where large volumes of warm water are available. The limit to which water can be reconditioned by aeration or oxygenation is determined by the accumulation of ammonia, which is toxic in its un-ionized form. For this reason, intensification of flowing-water aquaculture beyond what can be accomplished by replenishment of DO requires ammonia removal.

Technology for ammonia removal, which is accomplished in aquaculture by biological filtration, allows complete, or nearly complete, water recirculation. Thus, the amount of new water added to the production system is negligible, so that tilapia culture may be indoors in temperate climates or be located conveniently for markets or take place where no suitable water supplies are available.

FIGURE 8.1. Flowing-water aquaculture technology applied to tilapia in Kenya.

FLOW-THROUGH CULTURE SYSTEMS

The culture of salmonids in flowing water, which began in the mid-1800s, has been quantified by Haskell (1955, 1959; Haskell et al. 1960), refined by others (Westers 1970; Buss and Miller 1971; Piper 1975), and summarized by Soderberg (1995). Application of this technology to tilapia has been reported by Ray (1978), Lauenstein (1978), Balarin and Haller (1982), and Hargreaves (2000) (see Figure 8.1).

Growth

Haskell (1959) showed that trout growth, in units of length, was a linear function of temperature. Soderberg (1990, 1992) demonstrated that this relationship held for other species, including tilapia. The linear growth equation for blue tilapia *(Oreochromis aureus)* is

$$\Delta L = -0.853 + 0.048T,$$

where ΔL is daily growth in millimeters and T is temperature in degrees Celsius (Soderberg 1990).

The daily growth rate is used to project growth over time and to precisely calculate feed rations. Losordo (1997) noted that Soderberg's (1990) equation underestimated the growth potential of Nile tilapia *(O. niloticus)* and its hybrids, the tilapia usually chosen for culture because of their rapid growth rates (Rakocy 1989), which are traditionally reported in grams per day. Tilapia growth rates using complete diets in static-water ponds have ranged from 1.59 (Fram and Pagan-Font 1978) to 2.80 $g \cdot day^{-1}$ (Collis and Smitherman 1978). These values have limited application to flowing-water fish culture because temperatures and initial fish weights are not specified and the environmental factors that affect fish growth are different (Soderberg 1997).

Muir et al. (2000) reported that a typical growth period of 240 days for recirculating systems resulted in a weight gain from 5 to 350 g (1.44 $g \cdot day^{-1}$). Lauenstein (1978), growing *O. aureus* in geothermal well water at 24°C to 26°C, reported that fish gained 0.4 $g \cdot day^{-1}$ from 1-2 to 25 g, 1.0 $g \cdot day^{-1}$ from 25 to 150 g, and 1.5 $g \cdot day^{-1}$ from 150 to 300-400 g. Rakocy (1989) gives estimated growth rates for *O. niloticus* in 28°C to 30°C water ranging from 0.5 $g \cdot day^{-1}$ for fish from 0.5 to 20 g to 2.9 $g \cdot day^{-1}$ for growth from 250 to 450 g (see Table 8.1). When these growth data are converted to millimeters per day by converting weight to length using the equation

$$W = 0.0000233L^3,$$

TABLE 8.1. Growth rates of *Oreochromis niloticus* in (g·day^{-1}), converted to mm/day using the equation $W = 0.0000233L^3$.

Growth stanza (g)	Growth period (days)	Growth Rate	
		(g·day^{-1})	(mm·day^{-1})
5-20	30	0.5	1.17
20-50	30	1.0	1.13
50-100	30	1.5	1.12
100-250	50	2.5	1.16
250-450	70	2.9	0.68

Source: Soderberg 1995; growth data from Rakocy 1989.

where W = weight in grams and L = length in millimeters (Soderberg 1990), values of ΔL ranging from 0.68 to 1.17 mm.day^{-1} are obtained (Table 8.1). The relatively constant value for ΔL which averages 1.145 mm. day^{-1} for fish weights up to 250 g, demonstrates the linear nature of fish growth first shown by Haskell (1959). The deviation from linear growth for fish approaching 450 g may have caused a lower growth rate in the final culture stanza that is attributable to endogenous influences. Few studies on tilapia growth report data for fish weighing more than 350 g. Most reports on static-water tilapia culture show linear growth up to a size of approximately 250 g (Anderson and Smitherman 1978; Fram and Pagan-Font 1978; Teichert-Coddington and Green 1993; Green and Teichert-Coddington 1994).

When the growth rate data provided by Muir et al. (2000) are converted to length increments, the resulting ΔL value is 0.778 mm.day^{-1} for fish grown from 5 to 350 g. Losordo (1997) reported that *O. niloticus* and *O. niloticus* × *O. aureus* hybrids could be expected to grow 1.27 mm.day^{-1} at 28°C to 30°C. From these growth data, it can be concluded that *O. niloticus* and its hybrids can be expected to grow at least 1.1 mm·day^{-1} to a size of at least 250 g in flowing water at temperatures of 28°C to 30°C. More research should be conducted on the temperature-related growth of tilapia on dry diets because an accurate prediction of the ΔL value is necessary to calculate the most efficient feeding rate.

Feeding

Haskell (1959) introduced a feeding equation for fish in flowing water:

$$F = (3 \times C \times \Delta L/L) \times 100$$

where F = percentage ratio of body weight to feed per day; ΔL = expected daily growth; C = food conversion; and L = fish length on the day of feeding. The values for L and ΔL must be in the same unit of measure. Note that this equation calculates the daily feed ration needed for a given growth increment. The ration changes every day because L increases by the increment of ΔL every day.

Minimum DO Levels

It is well known that tilapia can tolerate hypoxic, and even anoxic, conditions for short periods and thus are better suited than other species to the hypereutrophic conditions that may exist in static-water aquaculture systems. Fry (1957) described how fish respond to hypoxia. The rate of oxygen consumption is constant and at a maximum at DO tensions above an incipient limiting level. Below this level, oxygen consumption is dependent upon environmental oxygen tension and the result is decreased oxygen delivery to the tissues. Incipient limiting levels of oxygen generally average 73 mmHg (3.55 $mg \cdot L^{-1}$ at 28°C) for warm-water fish and 90 mmHg (6.17 $mg \cdot L^{-1}$ at 10°C) for salmonids (Davis 1975). Ross and Ross (1983) determined the incipient limiting DO tension (pC, or critical point) to be 60 mmHg (2.92 $mg \cdot L^{-1}$ at 28°C) for *O. niloticus*. Becker and Fishelson (1986) reported values of 26 to 31 mmHg (28.5 mmHg = 1.39 $mg \cdot L^{-1}$ at 28°C) for the incipient limiting level of DO for *O. niloticus*.

Field observations of DO concentrations below which tilapia growth is reduced range from 2.3 (Coche 1977) to 3.0 $mg \cdot L^{-1}$ (Melard and Philppart 1980). Thus the value of 60 mmHg appears to be a more reasonable value for assignment of DO minima for flowing-water culture of tilapia. Instructions for converting DO concentrations to tension units are provided by Soderberg (1995). Representative examples, using a minimum DO tension criterion of 60 mmHg, are presented in Table 8.2.

Loading, Carrying Capacity, and Rearing Density

It is conventional in flowing-water fish culture to refer to the weight of fish per unit of flow as the loading rate and the weight of fish per unit of rearing space as density. Carrying capacity is the maximum permissible loading rate because it is the fish load that results in an effluent DO concentration at an assigned minimum level, shown in Table 8.2 for tilapia. Loading rate describes the water requirements of fish in terms of DO. Rearing density describes their spatial requirements. Buss et al. (1970) and Piper (1975) first described the independence of these factors.

TABLE 8.2. Solubility of dissolved oxygen (DO) at 1 atm of pressure and DO concentrations corresponding to a tension of 60 mmHg at various temperatures.

Temperature (°C)	Solubility ($mg \cdot L^{-1}$)	Concentration at 60 mmHg ($mg \cdot L^{-1}$)
20	8.84	3.33
21	8.68	3.24
22	8.53	3.21
23	8.38	3.16
24	8.25	3.11
25	8.11	3.06
26	7.99	3.01
27	7.86	2.96
28	7.75	2.92
29	7.64	2.88
30	7.53	2.84
31	7.43	2.80
32	7.33	2.76

The loading rate can be estimated from oxygen consumption as

$$\text{Lr} = (O_a - O_b)(0.00654)/F,$$

where Lr = kilograms of fish per liter of water per minute; O_a = influent DO in milligrams per liter; O_b = effluent DO in milligrams per liter; and F = feeding rate in kilograms of food per kilogram of fish per day (Soderberg 1995). When O_b is the minimum allowable DO, the expression represents the carrying capacity. This formula is based on oxygen consumption measurements for trout fed a diet containing 2.64 total $kcal \cdot kg^{-1}$. Ross and Ross (1983) presented regression equations for the resting oxygen consumption rates of *O. niloticus* (Table 8.3). These expressions may be used to estimate carrying capacity for tilapia in flowing-water aquaculture. For example, the equation for 30°C is

$$\text{Log OC} = 3.34 - 0.586 \text{ (log wt)},$$

where OC = oxygen consumption rate in milligrams of oxygen per kilogram of fish per hour and wt = fish body weight in grams. The estimated oxygen consumption rate for a 200 g fish at 30°C is thus 98.08 $mg \cdot kg^{-1} h^{-1}$. If the influent water contains 7.53 $mg \cdot L^{-1}$ DO and the minimum allowable effluent DO is 2.84 $mg \cdot L^{-1}$ (Table 8.2), 4.69 $mg \cdot L^{-1}$ of DO are available for

TABLE 8.3. Regression equations for resting oxygen consumption rate in *Oreochromis niloticus*.

Temperature (°C)	Regression equation
20	Log OC = 3.00 – 0.777 (log wt)
25	Log OC = 2.80 – 0.350 (log wt)
30	Log OC = 3.34 – 0.586 (log wt)
35	Log OC = 3.03 – 0.255 (log wt)

Source: Adapted from Ross and Ross 1983.
Note: OC = oxygen consumption rate in milligrams of oxygen per kilogram of fish per hour and wt = fish body weight in grams.

respiration by fish in the rearing unit. This can be converted to loading rate as follows:

$$(4.69\ mg{\cdot}L^{-1})/(98.08\ mg{\cdot}kg^{-1}\ L^{-1})\ (60\ min{\cdot}h^{-1}) = 2.86\ kg{\cdot}L^{-1}\ min^{-1}.$$

Thus, the carrying capacity for 200 g tilapia at 30°C is estimated to be 2.86 $kg{\cdot}L^{-1}$ If the equation of Soderberg (1995) is used, assuming a feeding rate of 1.5 percent body weight per day:

$$\text{Carrying capacity} = (7.53 - 2.84)\ (0.00654)/0.015 = 2.04\ kg{\cdot}L^{-1}{\cdot}min^{-1}.$$

Because it is derived from oxygen consumption of fish in actual flowing-water culture conditions, the more conservative estimate from the procedure in Soderberg (1995) might be more appropriate than that provided by the oxygen consumption curves of Ross and Ross (1983).

Fish of all species investigated can be grown at very high densities without adverse effects attributable to behavioral interaction. Schmittou (1969) grew channel catfish *(Ictalurus punctatus)* in cages at densities up to 190 $kg{\cdot}m^{-3}$. Culture densities of at least 160 $kg{\cdot}m^{-3}$ have been recommended for channel catfish (Ray 1978) and lake trout *(Salvelinus namaycush)* (Soderberg and Krise 1986) in raceways. Hickling (1971) reported that common carp *(Cyprinus carpio)* are grown in Japan at densities as high as 278 $kg{\cdot}m^{-3}$. Hargreaves (2000) reported Nile tilapia densities of 160 to 185 $kg{\cdot}m^{-3}$ in raceways in Arkansas with water detention times of 8.1 min.

Rearing Unit Design

Rearing units for fish culture in flowing water are nearly always either plug flow or circulating flow. Plug-flow units are long, narrow channels, often called linear units or raceways. Properly designed raceways have length

FIGURE 8.2. A raceway velocity of 0.033 $m \cdot S^{-1}$ allows solid waste to settle in the fish-free zone between the screen and the dam; it can be removed through a clean-out drain. Water flow is from left to right. *Source:* Soderberg 1995.

to width ratios of 5:1 to 10:1, water exchange rates of four to ten exchanges per hour, depths of 60 cm or less, and water velocities of 0.033 $m \cdot s^{-1}$ (Soderberg 1995). Fish are restricted to the upper 90 percent of the raceway by a screen, and depth is controlled by a dam board at the downstream end. The fish-free zone between the screen and the dam allows solid waste produced by the fish to settle at the end of the raceway, whence it can be removed through a clean-out drain (Figure 8.2). The required linear velocity of 0.033 $m \cdot s^{-1}$ allows for most of the solid waste from the fish that remains in suspension due to fish activity to settle behind the screen (Westers and Pratt 1977). In order to achieve plug flow (i.e., the water moving through the unit without restriction), water must enter and exit the unit over its entire width (Haskell et al. 1960).

Circulating fish culture units are usually circular with a center drain. Water is admitted from a single location along the circumference and flows in a circular pattern to the drain. Circular rearing units exhibit a water velocity gradient, with velocities at the outside edge five to ten times greater than those in raceways. The gradually decreasing velocity toward the center allows solid waste to be swept toward the drain, making circular units self-cleaning; but there is no mechanism to isolate solids from the liquid bulk. Circular units are usually 0.6 to 1.0 m deep and have diameters from 1 to 15 m. Water exchange rates generally do not exceed two exchanges per hour. The relatively high velocities of circular units force the fish to work harder to maintain position, which theoretically diverts energy from growth. The higher water exchange rate of raceways allow the fish to be grown at higher densities than in circular units. Thus, raceways allow a given level of production to be accomplished in less space.

Raceways would appear to be more appropriate for serial reuse than circular units because solids can be collected at the end of each section rather than accumulating through the series. In fact, raceways are generally built in series of three or more units. Losordo (1997) described a double-drain system for circular fish culture units that allows the separation of solid waste from the general flow of the unit: solids that can settle collect in a bowl at the bottom of the unit and are carried away in a separate stream from the liquid bulk, which exits from the surface of the unit. This design may allow circular units to be used successfully for serial water reuse. Drawings of two types of double-drains are found in Losordo (1997) and Losordo et al. (1999).

SERIAL REUSE SYSTEMS

In serial water reuse, water flows from one unit to the next with aeration applied, as necessary, between units to restore DO. Thus, each unit in the series is capable of producing the carrying capacity of fish. Successful serial reuse requires that cleaning waste be separated from the liquid bulk to prevent excessive turbidity in the downstream units. The extent to which water re-aeration allows reuse depends upon the accumulation of un-ionized ammonia, a toxic product of fish protein metabolism.

Aeration

Aeration can be partially accomplished by the water fall from one serial unit to the next and can be restored to 90 percent of saturation by mechanical or diffuser devices (Soderberg 1995). Modern flowing-water aquaculture applications have abandoned mechanical aeration in favor of the use of liquid oxygen. Pure oxygen can be produced on site but is more commonly delivered by truck and stored in vessels that are purchased or leased.

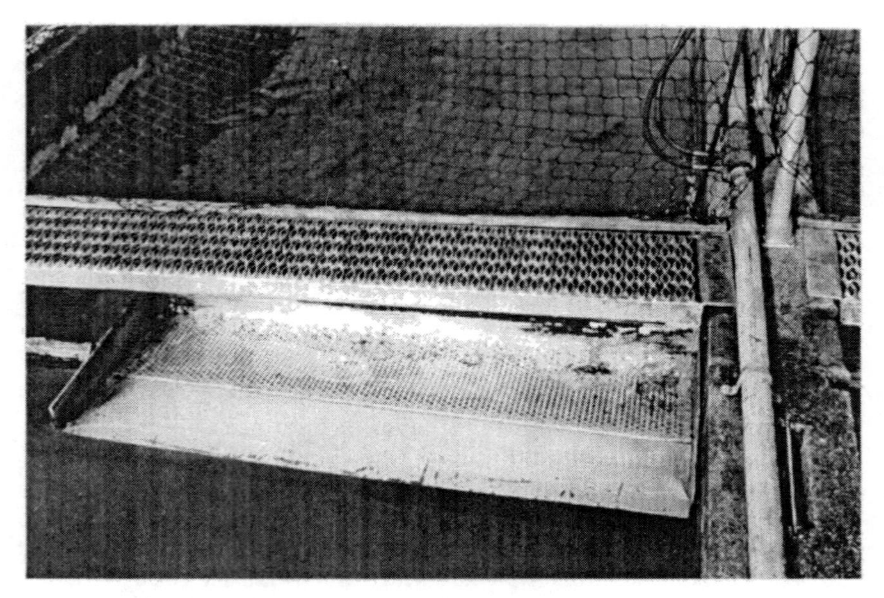

FIGURE 8.3. An LHO (low-head oxygenator) fits conveniently into a raceway.

Speece (1981) described several means by which liquid oxygen can be effectively transferred to water, but either the multistage column system described by Watten (1991) or a downflow-bubble contactor is usually selected for aquaculture applications. The multistage column technology is present in a patented device called a low-head oxygenator (LHO) that accomplishes 90 percent oxygen absorption through a head loss of as little as 0.3 m (Watten 1994). The LHO fits conveniently into a raceway (Figure 8.3). If dual-drain circular tanks are selected for serial reuse, the LHO can be placed in a sump between rearing units, but the best choice in this case is a downflow-bubble contactor. This is a cone-shaped device placed directly in a pipeline, thus eliminating the need for a sump (Figure 8.4). Water enters the top of the cone and oxygen is introduced at the bottom. The cone shape allows for efficient oxygen transfer by reducing the water velocity to less than the rise velocity of the oxygen bubbles (Losordo 1997). Schematic diagrams of both devices can be found in Watten (1991) and Losordo et al. (1999). An important feature of pure oxygen is that DO can easily be maintained at levels above saturation. This allows for carrying capacity to be increased over that possible with water with DO at atmospheric equilibrium. In practice, oxygen flow to the LHO or downflow-bubble contactor is adjusted so that rearing unit effluent DO levels are at or above the

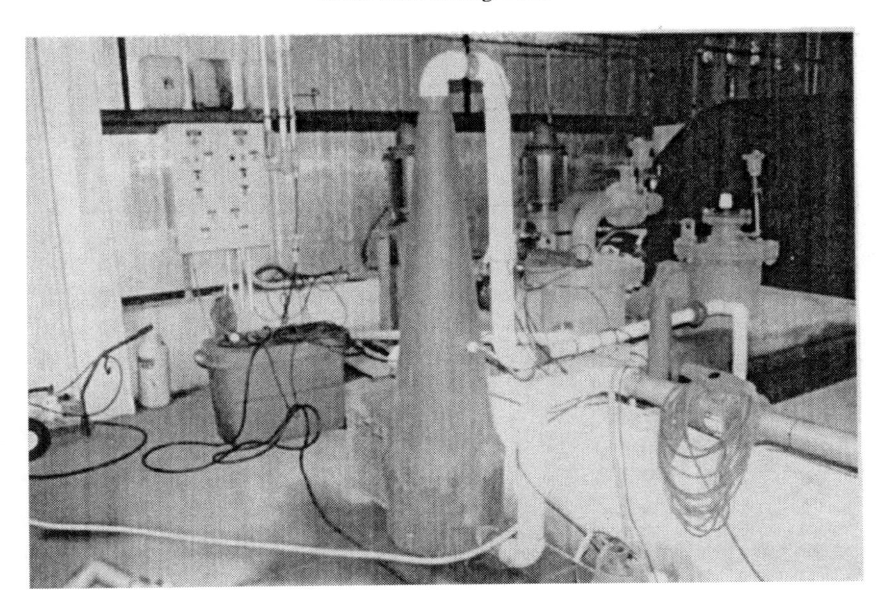

FIGURE 8.4. Downflow-bubble contactor. Water enters the top and oxygen is introduced to the bottom to achieve efficient oxygen transfer.

designated minimum allowable concentration. The current cost of liquid oxygen, including delivery within 100 miles of the air reduction plant, is approximately \$0.75/100 ft^3 (\$0.144 kg^{-1}). Rental of a 1,500 gallon (8,978 kg) on-site storage tank is \$510/month (figures from Air Products, Allentown, PA).

Ammonia

Ammonia is the principal nitrogenous product of fish metabolism. It originates from the deamination of amino acids; thus its production rate can be estimated from the consumption of dietary protein using the expression $A = 56P$, where A = ammonia production rate in grams of ammonia N per kilogram of food and P = decimal fraction of protein in the diet (Soderberg 1995). For example, the ammonia production of 1,000 kg of tilapia fed 30 kg per day of a 36 percent protein diet would be estimated as

$$A = 56 \times 0.36 = 20.16 \text{ g ammonia N per kilogram of feed.}$$

Thus the fish would be expected to generate 604.8 g of ammonia N per day.

TABLE 8.4. Fraction of NH_3 in an ammonia solution.

Temperature	pH				
(°C)	6.5	7.0	7.5	8.0	8.5
20	0.0013	0.0039	0.0124	0.0381	0.1112
21	0.0013	0.0042	0.0133	0.0408	0.1186
22	0.0015	0.0046	0.0143	0.0438	0.1264
23	0.0016	0.0049	0.0153	0.0469	0.1356
24	0.0017	0.0053	0.0164	0.0502	0.1431
25	0.0018	0.0056	0.0176	0.0537	0.1521
26	0.0019	0.0060	0.0189	0.0574	0.1614
27	0.0021	0.0065	0.0202	0.0613	0.1711
28	0.0022	0.0069	0.0216	0.0654	0.1812
29	0.0024	0.0074	0.0232	0.0697	0.1916
30	0.0025	0.0080	0.0248	0.0743	0.2025
31	0.0027	0.0085	0.0265	0.0791	0.2137
32	0.0029	0.0091	0.0283	0.0842	0.2253

Ammonia accumulates in serial reuse aquaculture, and its average daily concentration can be predicted by dividing the total estimated daily production by the total daily water flow. Returning to the example above, if 1,000 kg of tilapia producing 604.8 g of ammonia per day were held in a rearing unit receiving 500 $L \cdot min^{-1}$ of water, the estimated ammonia concentration in the rearing unit effluent is 0.84 $mg \cdot L^{-1}$. If this unit is followed by another producing the same amount of ammonia, the concentration would accumulate to 1.68 $mg \cdot L^{-1}$ at the end of the second unit, and so on.

Aqueous ammonia exists in two molecular forms, NH_3 and NH_4^+, and their combined concentration is referred to as TAN (total ammonia nitrogen). Un-ionized ammonia (NH_3) is toxic to fish, and the equilibrium between the two forms is pH and temperature dependent. Instructions for calculating the un-ionized fraction in a TAN solution are provided by Soderberg (1995), but for most applications Table 8.4 will suffice. Note that for a given temperature, the exposure of fish to NH_3 increases dramatically with pH. For this reason, pH is one of the most important water quality criteria for serial reuse aquaculture, and waters with high pH values may be deemed unsuitable because of ammonia toxicity.

Muir et al. (2000) recommended that NH_3 levels be kept below 1 $mg \cdot L^{-1}$, and ideally below 0.2 $mg \cdot L^{-1}$, for flowing-water tilapia culture, but there are no bioassay data available on the chronic toxicity of NH_3 to tilapia. Ball (1967) reported that the 2-day LC_{50} (concentration resulting in 50 percent mortality) of NH_3 to rainbow trout *(Oncorhynchus mykiss)* is 0.41 $mg \cdot L^{-1}$.

Colt and Tchobanoglous reported that the LC_{50} of NH_3 to channel catfish *(Ictalurus punctatus)* is 2 to 3.1 mgL^{-1}. Rainbow trout and channel catfish exhibit reduced growth at NH_3 concentrations of 0.016 (Larmoyeaux and Piper 1973) and 0.048 $mg{\cdot}L^{-1}$ (Colt and Tchobanoglous 1978), respectively. Thus, 0.039 the application factors between chronic and acute toxicities of NH_3 are 0.039 (0.016/0.41) for rainbow trout and 0.015 to 0.024 for catfish. The 72 h LC_{50} of NH_3 to *O. aurea* has been reported at 2.35 $mg{\cdot}L^{-1}$ (Redner and Stickney 1979). Applying application factors of 0.015 and 0.039 to this value results in projected chronic toxicities of 0.035 to 0.092 $mg{\cdot}L^{-1}$ for tilapia. More research is required to define maximum allowable exposures of tilapia to NH_3 because this is a critical parameter in designing flowing-water aquaculture systems.

Carrying capacity with respect to ammonia accumulation is determined by estimating the fish production resulting in a designated maximum exposure of the fish to NH_3. For example, consider a water supply of 500 L of pH 7 water per minute at 28°C and a feeding rate of 3 percent body weight per day of a 36 percent protein diet. A fish weight of 6,037 kg would result in a maximum ammonia exposure of 0.035 $mg{\cdot}L^{-1}$. The calculation to obtain this value is

$$A = 56 \times 0.36 = \frac{20.16 \text{ g TAN}}{\text{kg food}} \times \frac{0.03\text{k g food}}{\text{kg fish}} = \frac{0.605 \text{ g TAN}}{\text{kg fish}}$$

$$\frac{0.035 \text{ mg}{\cdot}\text{L}^{-1}\text{NH}_3}{0.0069} \text{ (at 28°C, pH7)} = \frac{5.07 \text{ mg TAN}}{\text{L}} \quad \text{(maximum safe exposure)}$$

$$\frac{5.07 \text{ mg TAN}}{\text{L}} \times \frac{500 \text{ L}}{\text{min}} \times \frac{1{,}440 \text{ min}}{\text{day}} \times \frac{\text{kg fish}}{0.605\text{g TAN}} = 6{,}037 \text{ kg per fish}$$

At a pH of 8, the maximum fish load is reduced to 637 kg because of the decreased ionization of ammonia (substitute .0654 from Table 8.4).

CLOSED SYSTEMS

Serial reuse systems can successfully reuse water 10 to 20 times or more, depending upon pH and the accumulation of NH_3. More intensive water reuse requires ammonia removal and more rigorous control of solid waste. Modern technologies for ammonia and solids management permit aquaculture systems to be closed, or nearly closed, with the water requirement limited to evaporation, leakage, and filter backwashing.

FIGURE 8.5. Rotating biological contactor.

Ammonia Removal

Ammonia is removed from water in aquaculture by biological filtration, a process of nitrification by autotrophic bacteria that oxidize ammonium to nitrite (NO_2^-) and then to nitrate (NO_3^-). The biological filter contains a medium with a large surface area that becomes colonized by the nitrification bacteria and through which water from the recirculation stream passes. Many media types and filter configurations have been investigated since the early 1970s, but the state-of-the-art filter is probably an upwelling fluidized bed of fine medium such as sand or plastic beads. The rotating biological contactor (RBC) is also a commonly selected biofilter type for aquaculture.

In fluidized bed filters, water flows up through the medium with sufficient pressure to expand, or fluidize, it. The particles of medium are thus held in motion, so that they cannot make continuous contact with each other and the entire particle surface is available for bacterial colonization (Losordo et al. 1999). The RBC filter is a series of disks on a shaft that rotates slowly in a trough of water. Approximately 40 percent of the disk medium is submerged at any one time (Losordo et al. 1999), but its rotation alternates the disks between the ammonia-rich water and the air. The recycle stream passes through the trough, parallel to the rotating shaft (Figure 8.5). Soderberg (1995) and Losordo et al. (1999) provide schematic diagrams of RBCs.

Biofilters are designed to match the ammonia production of the fish to the nitrification capacity of the filter. The nitrification rate is influenced by temperature, pH, influent ammonia concentration, alkalinity, dissolved oxygen, and organic load. Nitrification bacteria are found in nature in a wide variety of environmental conditions and thus can adapt to temperatures from –5°C to 35°C and pH levels from at least 6 to 9, if changes are not too abrupt (Wheaton et al. 1991). The nitrification rate is directly proportional to temperature (Speece 1973; Soderberg 1995) and influent ammonia concentration (Knowles et al. 1965), but NH_3 is toxic to nitrification bacteria at levels of 0.1 to 1.0 $mg \cdot L^{-1}$ (Anthonisen et al. 1976). The stoichiometric requirement for the oxidation of 1 g of ammonia N is 4.34 g of oxygen, and DO levels of at least 2 $mg \cdot L^{-1}$ are recommended to assure that nitrification is not oxygen limited (Sharma and Albert 1977; Manthe et al. 1985; Wheaten et al. 1991). Nitrification reduces pH by consuming alkalinity and producing hydrogen ions. The conversion of 1 g of ammonia to nitrate requires 7.15 g of alkalinity ($CaCO_3$), and alkalinity values of 75 (Gujer and Boller 1986) to 150 $mg \cdot L^{-1}$ $CaCO_3$ (Allain 1988) must be maintained to ensure optimal operation of a biological filter. The organic load on a biofilter is affected by the efficiency of solids removal and thus is variable in closed aquaculture systems. Organic loading affects nitrification by the colonization of the filter media with heterotrophic bacteria that compete for space with the nitrification flora.

Speece (1973) and Soderberg (1995) provide simple expressions for nitrification rate that result in predictions on the order of 1.5 $g \cdot m^{-2} \cdot day^{-1}$, but Losordo (1997) warns that between 0.25 and 0.55 $g \cdot m^{-2} \cdot day^{-1}$ is a more reasonable range for expected nitrification rates in production-scale biofilters. Malone et al. (1993) reported nitrification rates of 0.280 $g \cdot m^{-2} \cdot day^{-1}$ for an RBC and 0.284 $g \cdot m^{-2} \cdot day^{-1}$ for a fluidized bed in actual aquaculture environments. The actual nitrification rate of a particular biofilter depends upon the efficiency of organic matter control and the allowable ammonia level, which is a function of system pH. For aquaculture applications, 0.28 $g \cdot m^{-2} \cdot day^{-1}$ appears to be a reasonable design criterion.

Originally, biofilters for fish culture contained stone or plastic rings with specific surface areas of 60 to 130 $m^2 \cdot m^{-3}$ (Soderberg 1995). Three millimeter plastic beads have a specific surface area of 1,050 $m^2 \cdot m^{-3}$, whereas the surface area of sand ranges from 2,300 (Malone et al. 1993; Westerman et al. 1996) to 3,937 $m^2 \cdot m^{-3}$ (Owsley 1993). Thus, new-generation biofilters require much less medium and are much smaller than those investigated in the 1970s. The nitrification surface area of an RBC is the total surface area of both sides of all the disks in the unit. Losordo et al. (1999) described an RBC that used blocks of plastic medium with a specific surface area of 200 $m^2 \cdot m^{-3}$ rather than flat or corrugated disks. This medium increased the nitrification capacities of traditional RBC designs, but the authors warned that the filter medium increases in weight as much as tenfold due to the growth of the microbial flora, and the support structure of the unit must be designed to accommodate this weight.

Because biological filtration consumes alkalinity and produces acidity, a means of maintaining suitable pH values and restoring alkalinity to the stoichiometric requirement is necessary for proper biofilter function. Sodium bicarbonate ($NaHCO_3$) is the base of choice for this function because of its high solubility and limited danger of excessive pH rise as a result of overdosing. Base is generally metered in the system flow as necessary to maintain desired water chemistry. Bisogni and Timmons (1991) present a diagram for estimating the amount of base required to maintain alkalinity, and thus pH, based on losses resulting from nitrification. According to their estimates, approximately 0.11 kg of $NaHCO_3$ would be required per kilogram of food fed per day.

Solids Removal

In serial reuse aquaculture, sedimentation provided by raceways or double-drain tanks is sufficient to keep water reasonably free of turbidity,

FIGURE 8.6. Drum filter in use in a closed aquaculture system.

but the accumulation of recirculating solid waste in closed systems must be controlled using mechanical filters. Solids removal has been the greatest technical obstacle to the success of closed recirculating aquaculture systems, but the recent availability and widespread use of drum filters seems to have solved this problem.

Drum filters (Figure 8.6), available from several manufacturers, operate by passing water through a fine-mesh microscreen, usually 40 μm. The screen is in the form of a cylinder. Water passes from the inside to the outside of the drum. When pores clog, as determined by a water-level difference across the screen, the drum rotates, usually 180°, and solids are washed from the screen into a waste discharge line. During typical operation, the screen is washed from several hundred to several thousands times per day (Libey 1993; Summerfelt et al. 1996). Compared with other methods of suspended solids removal, drum filters have high hydraulic capacities, low space requirements, and small water loss to waste disposal. Schematic diagrams of drum filters can be found in Losordo (1997) and Losordo et al. (1999).

Closed System Management

The components described above for in-tank settling of solids, oxygenation, biofiltration, maintenance of pH and alkalinity, and suspended solids screening form the backbone of a state-of-the-art closed aquaculture system. Other components that may be necessary for successful fish production include carbon dioxide (CO_2) stripping, water chemistry manipulation to control nitrite toxicity, ozonation, and foam fractionation.

Carbon dioxide can accumulate in aquaculture systems to levels that interfere with respiration. Furthermore, CO_2 decreases pH, increasing the need for water chemistry manipulation. Losordo (1997) recommends that CO_2 levels be kept below 25 $mg{\cdot}L^{-1}$ to ensure fish health. Since the solubility of CO_2 at tilapia growth temperatures is less than 0.5 $mg{\cdot}L^{-1}$, it might seem obvious that the use of liquid oxygen for aeration would prevent CO_2 levels from approaching 25 $mg{\cdot}L^{-1}$, but this is not necessarily true.

Predicting the quantity of CO_2 removed by aeration is difficult because composition of the gas phase varies as a result of its reaction with bicarbonate in water. The problem is complicated by the difficulty in accurately measuring hyper-supersaturated levels. The titration process removes some of the gas before it is measured, and calculation of CO_2 from alkalinity and pH is not reliable because the bicarbonate system would not be in equilibrium at times of extreme CO_2 levels.

Excess CO_2 can be removed from water by contacting it with air or oxygen or by adding of NaOH to convert CO_2 to bicarbonate (Vinci et al. 1996), but the required level of removal is site specific and should be evaluated during system operation.

Nitrite (NO_2^-) is the intermediate product of nitrification and can accumulate to toxic levels in aquaculture systems. Losordo (1997) recommended that NO_2^- -N levels be kept below 5 $mg \cdot L^{-1}$, but channel catfish exposed to this level for 24 h showed a conversion of 77 percent of their hemoglobin to methemoglobin (Tomasso et al. 1979). It is impossible to assign a maximum tolerable concentration for NO_2^- to fish because its toxicity is highly variable depending upon pH, chloride, and calcium levels. Nitrite toxicity can be controlled by increasing the volume of biofilter medium, ozone addition to oxidize NO_2^- to NO_3^-, or chloride addition. Recommendations on effective NO_2^-:Cl ratios range from 3:1 (Schwedler et al. 1985) to 20:1 (Losordo 1997). Because Ca^{2+} protects fish from NO_2 toxicity, $CaCl_2$ is more effective than NaCl for this purpose.

In-tank settling removes solid waste particles greater than 100 μm in size, and the microscreen filters used for aquaculture most often have pore sizes of 40 μm. Smaller particles, including dissolved organic matter and colloids, may become important in aquaculture where all or nearly all of the water is recirculated. These fine particles can be harmful to fish and provide a substrate for heterotrophic bacteria that compete for space with the nitrification fauna of the biofilter, thereby reducing its performance. The requirement for mechanisms to remove this component of the solids load is controversial; when removal is adopted, ozone, foam fractionation, or both are the processes of choice.

Ozone is a powerful oxidizing agent; when added to water, it reduces NO_2^- to NO_3^- and oxidizes dissolved organic matter and colloidal solids. Ozone-treated water enhances fine solids removal by foam fractionation (Losordo 1997) and improves the performance and reduces the backwash frequency of microscreen filters (Summerfelt et al. 1996). Because its unstable nature makes storage impossible, ozone must be generated on site. It is most commonly added along with oxygen during oxygenation, but care must be taken to not expose fish to toxic levels. Wedemeyer et al. (1979) reported gill damage to fish at ozone levels as low as 5 $\mu g \cdot L^{-1}$. Summerfelt et al. (1996) recommended that ozone be added to the oxygen feed line at 3 to 4 percent of the gas flow in the amount of 25 $g \cdot kg^{-1}$ food. Brazil et al. (1996), in contrast, reported that 13 $g \cdot kg^{-1}$ was a sufficient dose for fine solids control. Bullock et al. (1996) found that 25 $g \cdot kg^{-1}$ was safe to fish when added in the oxygen feed line, but 36 to 39 $g \cdot kg^{-1}$ was occasionally toxic.

Foam fractionation works because of the tendency of fine organic particles to adhere to gas bubbles, forming skimmable suds, or foam. Such foam is sometimes produced in the oxygenation process, but foam fractionation is usually accomplished using devices designed for the process that vigorously mix air into the water to be treated in a closed chamber. These units have a convenient mechanism for collection and removal of the foam.

SUMMARY

Flow-through or serial reuse aquaculture systems are particularly appropriate for tilapia, and this technology can be applied in many tropical locations where large volumes of warm flowing water are available. The disadvantages are that costly complete diets must be used and construction costs may be higher than for extensive earthen ponds. Lutz (2000) estimated that the production costs for tilapia in a raceway system in a tropical location would be \$1.64 kg^{-1}. Flow-through and serial reuse technology can be applied in temperate climates using geothermally heated (Zachritz and Rafferty 2000) or industrially heated (Hubert 1980) water. Raceway systems in subtropical locations may be feasible if economical facilities for overwintering the fish can be provided (Hargreaves 2000). Closed recirculating systems can in theory produce tilapia at an estimated cost of \$3.57 kg^{-1} (Lutz 2000) and have the potential for profitable operation by supplying ethnic live fish markets, but these markets are rather limited and easily saturated. American producers are not currently competitive with Asia or Central America for processed tilapia (Costa-Pierce 2000). The future of tilapia aquaculture in temperate, developed locations lies in marketing campaigns to portray fresh, locally produced tilapia as a high quality product worthy of premium prices.

REFERENCES

Allain, P.A. (1988). Ion shifts and pH management in high density shedding systems for blue crabs (*Callinectus sapidus*) and red swamp crayfish *(Procambarus clarkii)*. Master's thesis, Baton Rouge, LA: Louisiana State University.

Anderson, C.E. and R.O. Smitherman (1978). Production of normal male and androgen sex-reversed *Tilapia aurea* and *T. nilotica* fed a commercial catfish diet in ponds. In R.O. Smitherman, W.L. Shelton, and J.H. Grover (Eds.), *Culture of exotic fishes symposium proceedings* (pp. 34-42). Auburn, AL: Fish Culture Section, American Fisheries Society.

Anthonisen, A.C., R.C. Loehr, T.B.S. Prakasam, and E.G. Srinath (1976). Inhibition of nitrification by ammonia and nitrous acid. *Journal Water Pollution Control Federation* 48:835-852.

Balarin, J.D. and R.D. Haller (1982). The intensive culture of tilapia in tanks, raceways and cages. In J.F. Muir and R.J. Roberts (Eds.), *Recent advances in aquaculture* (pp. 266-355). London, United Kingdom: Croom Helm.

Ball, I.R. (1967). The relative susceptibilities of some species of freshwater fish to poisons-I. Ammonia. *Water Research* 1:767-775.

Becker, K. and L. Fishelson (1986). Standard and routine metabolic rate, critical oxygen tension and spontaneous scope for activity of tilapias. In J.L. Maclean, L.B. Dizon, and L.V. Hosillos (Eds.), *Proceedings of the first Asian fisheries Forum* (pp. 623-628). Manila, Philippines: Asian Fisheries Society.

Bisogni, J.J. Jr. and M.B. Timmons (1991). Control of pH in closed cycle aquaculture systems. In *Engineering aspects of intensive aquaculture* (pp. 333-348). Ithaca, NY: Northeast Regional Agricultural Engineering Service, Cornell University.

Brazil, B.L., S.T. Summerfelt, and G.S. Libey (1996). Application of ozone to recirculating aquaculture systems. In G.S. Libey and M.B. Timmons (Eds.), *Successes and failures in commercial recirculating aquaculture* (pp. 373-389). Ithaca, NY: Northeast Regional Agricultural Engineering Service.

Bullock, G.L., S.T. Summerfelt, A.C. Noble, A.L. Weber, M.D. Durant, and J.T. Hankins (1996). Effects of ozone on outbreaks of bacterial gill disease and numbers of heterotrophic bacteria in a trout culture recycle system. In G.S. Libey and M.B. Timmons (Eds.), *Successes and failures in commercial recirculating aqua-culture* (pp. 598-610). Ithaca, NY: Northeast Regional Agricultural Engineering Service.

Buss, K., D.R. Graff, and E.R. Miller (1970). Trout culture in vertical units. Progressive Fish-Culturist 32:187-191.

Buss, K. and E.R. Miller (1971). Considerations for conventional trout hatchery design and construction in Pennsylvania. *Progressive Fish-Culturist* 33:86-94.

Coche, A.G. (1977). Premiers resultats de l'elevage en cages de *Tilapia nilotica* (L.) dans le Lac Kossou, Cote d'Ivoire. *Aquaculture* 10:109-140.

Collis, W.J. and R.O. Smitherman (1978). Production of tilapia hybrids with cattle manure or a commercial diet. In R.O. Smitherman, W.L. Shelton, and J.H. Grover (Eds.), *Culture of exotic fishes symposium proceedings* (pp. 43-54). Auburn, AL: Fish Culture Section, American Fisheries Society.

Colt, J. and G. Tchobanoglous (1978). Chronic exposure of channel catfish, *IctaIctalurus punctatus*, to ammonia: Effects on growth and survival. *Aquaculture* 15:353-372.

Costa-Pierce, B.A. (2000). Challenges facing the expansion of tilapia aquaculture [preface]. In B.A. Costa-Pierce and J.E. Rakocy (Eds.), *Tilapia aquaculture in the Americas* (Volume 2). Baton Rouge, LA: The World Aquaculture Society.

Davis, J.C. (1975). Minimal dissolved oxygen requirements of aquatic life with special emphasis on Canadian species: A review. *Journal of the Fisheries Research Board of Canada* 32:2295-2332.

Fram, M.J. and F.A. Pagan-Font (1978). Monoculture yield trials of an all-male tilapia hybrid (F *Tilapia nilotica* × M *T. hornorum*) in small ponds in Puerto Rico. In R.O. Smitherman, W.L. Shelton, and J.H. Grover (Eds.), *Culture of exotic fishes symposium proceedings* (pp. 55-64). Auburn, AL: Fish Culture Section, American Fisheries Society.

Fry, F.E. (1957). The aquatic respiration of fish. In M.E. Brown (Ed.), *The physiology of fishes* (Volume 1, pp. 1-63). New York: Academic Press.

Green, B.W. and D.R. Teichert-Coddington (1994). Growth of control and androgen-treated Nile tilapia, *Oreochromis niloticus* (L.), during treatment, nursery and grow-out phases in tropical fish ponds. *Aquaculture and Fisheries Management* 25:613-621.

Gujer, W. and M. Boller (1986). Design of a nitrifying tertiary trickling filter based on theoretical concepts. *Water Research* 20:1353-1362.

Hargreaves, J.A. (2000). Tilapia culture in the southeast United States. In B.A. Costa-Pierce and J.E. Rakocy (Eds.), *Tilapia aquaculture in the Americas,* (Volume 2, pp. 60-81). Baton Rouge, LA: The World Aquaculture Society.

Haskell, D.C. (1955). Weight of fish in per cubic foot of water in hatchery troughs and ponds. *Progressive Fish-Culturist* 17:117-118.

———. (1959). Trout growth in hatcheries. *New York Fish and Game Journal* 6:205-237.

Haskell, D.C., R.O. Davies, and J. Reckahn (1960). Factors in hatchery pond design. *New York Fish and Game Journal* 7:113-129.

Hickling, C.F. (1971). *Fish culture.* London, England: Faber and Faber.

Hubert, W.A. (1980). Aquaculture uses. In L.B. Goss (Ed.), *Factors affecting power plant waste heat utilization* (pp. 18-31). Elmsford, NY: Pergamon Press.

Knowles, G., A.L. Downing, and M.J. Barrett (1965). Determination of kinetic constants for nitrifying bacteria in mixed culture with the aid of an electronic computer. *Journal of General Microbiology* 79:263-278.

Larmoyeaux, J.C. and R.G. Piper (1973). Effects of water reuse on rainbow trout in hatcheries. *Progressive Fish-Culturist* 23:26-29.

Lauenstein, P.G. (1978). Intensive culture of tilapia with geothermally heated water. In R.O. Smitherman, W.L. Shelton, and J.H. Grover (Eds.), *Culture of exotic fishes symposium proceedings* (pp. 82-85). Auburn, AL: Fish Culture Section, American Fisheries Society.

Libey, G.S. (1993). Evaluation of a drum filter for removal of solids from a recirculating aquaculture system. In J.-K. Wang (Ed.), *Techniques for modern aquacul-ture* (pp. 529-532). St. Joseph, MI: American Society of Agricultural Engineers.

Losordo, T.M. (1997). Tilapia culture in intensive recirculating systems. In B.A. Costa-Pierce and J.E. Rakocy (Eds.), *Tilapia aquaculture in the Americas* (Volume 1, pp. 185-211). Baton Rouge, LA: The World Aquaculture Society.

Losordo, T.M., M.B. Masser, and J.E. Rakocy (1999). *Recirculating aquaculture tank production systems: A review of component options.* College Station, TX: Southern Regional Aquaculture Center, Publication No. 453.

Lutz, C.G. (2000). Production economics and potential competitive dynamics of commercial tilapia culture in the Americas. In B.A. Costa-Pierce and J.E. Rakocy (Eds.), *Tilapia aquaculture in the Americas* (Volume 2, pp. 119-132). Baton Rouge, LA: The World Aquaculture Society.

Malone, R.F., B.S. Chitta, and D.G. Drennan (1993). Optimizing nitrification in bead filters for warmwater recirculating aquaculture systems. In J.-K. Wang

(Ed.), *Techniques for modern aquaculture* (pp. 315-325). St. Joseph, MI: American Society of Agricultural Engineers.

Manthe, D.P., R.F. Malone, and H. Perry (1985). Water quality fluctuations in response to variable loading in a commercial blue crab shedding system. *Journal of Shellfish Research* 3:175-182.

Melard, C.H. and J.C. Philppart (1980). Pisciculture intensive de Sarotherodon niloticus (L.) dans les effluents thermiques d'une centrale nucleaire en Belgique. EIFAEIFAC Symposium on New Developments in Utilization of Heated Effluents and of Recirculation Systems for Intensive Aquaculture. Stavanger, Norway. EIFAC/80/DOC.E/11

Muir, J., J. van Rijn, and J. Hargreaves (2000). Production in intensive and recycle systems. In M.C.M. Beveridge and B.J. McAndrews (Eds.), *Tilapias: Biology and exploitation* (pp. 405-445). London, England: Kluwer Publishers.

Owsley, D.E. (1993). Status of U.S. Fish and Wildlife Service hatcheries with reuse systems in region 1. In J.-K. Wang (Ed.), *Techniques for modern aquaculture* (pp. 278-283). St. Joseph, MI: American Society of Agricultural Engineers.

Piper, R.G. (1975). *A review of carrying capacity for fish hatchery rearing units.* U.S. Fish and Wildlife Service Fish Cultural Development Center (Bozeman, MT) Information Leaflet No. 1.

Rakocy, J.E. (1989). *Tank culture of Tilapia.* College Station, TX: Southern Regional Aquaculture Center, Publication No. 282.

Ray, L.E. (1978). Production of tilapia in catfish raceways using geothermal water. In R.O. Smitherman, W.L. Shelton, and J.H. Grover (Eds.), *Culture of exotic fishes symposium proceedings* (pp. 86-89). Auburn, AL: Fish Culture Section, American Fisheries Society.

Redner, B.D. and R.R. Stickney (1979). Acclimation to ammonia by *Tilapia aurea.* Transactions of the American Fisheries Society 108:383-388.

Ross, B. and L.G. Ross (1983). The oxygen requirements of *Oreochromis niloticus under adverse conditions.* In L. Fishelson and Z. Yaron (Eds.), *Proceedings of the First International Symposium on Tilapia in Aquaculture* (pp. 134-143). Tel Aviv, Israel: Tel Aviv University.

Schmittou, H.R. (1969). Developments in the culture of channel catfish, *Ictalurus punctatus R., in cages suspended in ponds.* 23rd Annual Conference of the Southeastern Association of Game and Fish Commissioners, Mobile, Alabama.

Schwedler, T.E., C.S. Tucker, and M.H. Beleau (1985). Non-infectious diseases. In C.S. Tucker (Ed.), *Channel catfish culture* (pp. 497-541). New York: Elsevier.

Sharma, B. and R.C. Albert (1977). Nitrification and nitrogen removal. *Water Research* 11:897-925.

Soderberg, R.W. (1990). Temperature effects on the growth of blue tilapia in intensive culture. *Progressive Fish-Culturist* 52:155-157.

———. (1992). Linear fish growth models for intensive aquaculture. *Progressive Fish-Culturist* 54:255-258.

———. (1995). *Flowing water fish culture.* Boca Raton, FL: Lewis Publishers.

———. (1997). Factors affecting fish growth and production. In H.S. Egna and C.E. Boyd (Eds.), *Dynamics of pond aquaculture* (pp. 199-213). Boca Raton, FL: CRC Press LLC.

Soderberg, R.W. and W.F. Krise (1986). Effects of density on growth and survival of lake trout, *Salvelinus namaycush. Progressive Fish-Culturist* 48:30-32.

Speece, R.E. (1973). Trout metabolism characteristics and the rational design of nitrification facilities for water reuse in hatcheries. *Transactions of the American Fisheries Society* 102:323-334.

———. (1981). Management of dissolved oxygen and nitrogen in fish hatchery waters. In L.J. Allen and E.C. Kinney (Eds.), *Proceedings of the bio-Engineering symposium for fish culture* (pp. 53-62). Fish Culture Section Publication 1. Bethesda, MD: American Fisheries Society.

Summerfelt, S.T., J.A. Hankins, A.L. Weber, and M.D. Durant (1996). Effects of ozone on microscreen filtration and water quality in a recirculating rainbow trout culture system. In G.S. Libey and M.B. Timmons (Eds.), *Successes and failures in commercial recirculating aquaculture* (pp. 163-172). Ithaca, NY: Northeast Regional Agricultural Engineering Service. *Tilapia nilotica. Bulletin of the Japanese Society of Scientific Fisheries.*

Teichert-Coddington, D.R. and B.W. Green (1993). Tilapia yield improvement through maintenance of minimal oxygen concentrations in experimental grow-out ponds in Honduras. *Aquaculture* 118:63-71.

Tomasso, J.R., B.A. Simco, and K.B. Davis (1979). Chloride inhibition of nitrite induced methemoglobinemia in channel catfish *Ictalurus punctatus. Journal of the Fisheries Research Board of Canada* 36:1141-1144.

Vinci, B.J., S.T. Summerfelt, M.B. Timmons, and B.J. Watten (1996). Carbon dioxide control in intensive aquaculture: Design tool development. In G.S. Libey and M.B. Timmons (Eds.), *Successes and failures in commercial recirculating aquaculture* (pp. 399-418). Ithaca, NY: Northeast Regional Agricultural Engineering Service.

Watten, B.J. (1991). Application of pure oxygen in raceway culture systems. In *Engineering aspects of intensive aquaculture* (pp. 311-332). Ithaca, NY: Northeast Regional Agricultural Engineering Service, Cornell University.

———. (1994). Aeration and oxygenation. In M.B. Timmons and T.M. Losordo (Eds.), *Aquaculture water reuse systems: Engineering, design and management.* Developments in Aquaculture and Fisheries Sciences 27 (pp. 311-332). Amsterdam, The Netherlands: Elsevier Scientific Publishing Company.

Wedemeyer, G.A., N.C. Nelson, and W.T. Yasutke (1979). Physiological and biochemical aspects of ozone toxicity to rainbow trout *(Salmo gairdneri). Journal of the Fisheries Research Board of Canada* 36:605-614.

Westerman, P.W., T.M. Losordo, and M.L. Wildhaber (1996). Evaluation of various biofilters in an intensive recirculating fish production facility. *Transactions of the American Society of Agricultural Engineers* 39:723-727.

Westers, H. (1970). Carrying capacity of salmonid hatcheries. *Progressive Fish-Culturist* 32:43-46.

Westers, H. and K.M. Pratt (1977). Rational design of hatcheries for intensive salmonid culture, based on metabolic characteristics. *Progressive Fish-Culturist* 39:157-165.

Wheaton, F.W., J.N. Hochheimer, G.E. Kaiser, and M.J. Krones (1991). Principles of biological filtration. In *Engineering aspects of intensive aquaculture* (pp. 1-31). Ithaca, NY: Northeast Regional Agricultural Engineering Service, Cornell University.

Zachritz, W.H., II, and K. Rafferty (2000). Basic considerations in design of geothermal-based tilapia production systems in the United States. In B.A. Costa-Pierce and J.E. Rakocy (Eds.), *Tilapia aquaculture in the Americas* (Volume 2, pp. 82-89). Baton Rouge, LA: The World Aquaculture Society.

Chapter 9

Cage Culture

H. R. Schmittou

INTRODUCTION

Concept and Origin of Cage Fish Culture

Cage culture is the raising of fish in water-suspended containers that are enclosed on all sides and the bottom by mesh material that secures the fish inside while allowing relatively free exchange of water with the surrounding environment. Cages are used for fish culture throughout the world and come in a broad variety of sizes, shapes, and materials, designed for various fish species in a wide variety of culture environments. A few examples include the nylon mesh hapas used for breeding tilapia in Thailand ponds, wooden slat "boxes" for raising carp in Indonesian lakes, and huge nylon netpens for raising salmon in Norwegian fjords.

Although there are no records of when and where cages were first used for culturing fish, cage-type enclosures have been in use for centuries to hold excess captured fish for later disposal. It is logical to assume some fishermen tried to raise caught fish to a larger size. The earliest records of fish culture in cages date to the late nineteenth century; however, only since the early 1980s has cage fish culture been a significant contributor to aquacultural production. Until recently, the application of cage fish culture for commercial purposes has been constrained primarily by the limited availability of fish feeds of required quality.

Currently, cage culture is practiced primarily in inland freshwater lakes, reservoirs, and coastal saltwater bays throughout much of the world,

doi:10.1300/5513_09

although it may be practiced in any environment suitable for fish life. The People's Republic of China (China) is by far the world's principal producer of total fish and tilapia in cages. According to the *China Fishery Statistics Yearbook* for 2004, China produced approximately 1 million metric tons (mmt) of fish in 77 million m^2 of cages suspended in freshwater and saltwater environments. This amount represents about 5 percent of China's total fish farm production of almost 20 mmt. The 77 million m^2 of cages in 2004 is a 33 percent increase over 2003, and cage fish culture in China will likely continue to expand because

1. China's population is increasing by 17 million people annually while natural fish, land, and water resources are decreasing on a per capita basis;
2. the per capita demand for fish is increasing at an estimated 10 percent per year;
3. the completion of the Three Gorges reservoir will create a huge new area for cage production;
4. the total fish culture production will continue to expand at an annual rate of $\geqslant$10 percent whereas capture fishery production has been stable at about 18 mmt since 2000;
5. the decline of capture fishery will cause a major socioeconomic shift in fishery-related employment and business ventures, potentially involving millions of people, when compared to other resource sectors;
6. cage fish culture production will accelerate at a more rapid growth rate than traditional fish cultures.

China is the world's leading producer and exporter of farmed tilapia. Pond and cage production of tilapia in China in 2005 is expected to top 1 mmt and contributes about 5 percent of the country's total farm production.

In China, one of the primary driving forces behind the expected decline in capture fisheries and a steady shift in fish farming from pond to cage culture is the nation's dense and increasing human population.

Management of cage fish production is principally different from management of ponds. Cage type, design, material, size, and other engineering features vary greatly, all of which affect management results. However, only limited scientific research and development has been applied to cage engineering and in determining the fundamental principles governing production. This chapter discusses some of the principles that have been learned through limited research and extensive trial-and-error experiences.

Contribution of Cage Culture to World Fish Production

Because China is the world's most populous nation and the leading producer of cultured and captured fish, it is the fish culture industry to which the remainder of the world may be compared. Every nation's fishery resource is finite and cannot keep pace with the inevitable human population growth. However, aquacultural production is perhaps the most promising short-term, as well as long-term, means of increasing fish supplies thereby supplementing and replacing the components of natural fishery. World cultured fish production is presently increasing at an average annual rate of about 9 percent. It could increase at an even greater rate simply by accelerating the production intensity, increasing the area of conventional farming, and by expanding into the vast natural waters of inland, coastal, and selected offshore marine environments.

Conventional fish culture methods in inland and coastal ponds are becoming more intensive; however, increasing pond area is very capital intensive and highly competitive with other uses. The raising of fishes in cages is an alternative means of fish production. It has increasing technical, ecological, social, and economic advantages over fisheries and conventional farming. Cage fish culture technology is

1. compatible, not competitive for land and water resources, with other fish production systems and complementary to some;
2. applicable to almost all cultured fish species;
3. ideally applicable in open waters with low capture fish yields and in other areas where fishery development is impractical, such as inland reservoirs, large rivers and open waters of coastal estuaries, tidal streams and related environments of marine lagoons, bays, and other waters relatively protected from turbulent waves and periodic storms;
4. technologically simple and ready for practical application in freshwater environments and for limited application in coastal and marine environments;
5. not necessarily capital intensive;
6. economically and technologically available to all sectors of society, including the uneducated, poor, and small farmers; and
7. more adaptable than conventional aquaculture methods in meeting production to market demand.

Compared to conventional pond fish culture, cage culture has numerous technological and economic advantages. Fish stock and feeds are the same for ponds and cages. However, the environments, and consequently the technologies and costs of managing them, are vastly different. Examples for comparison include fish density (6,000 kg ha^{-1} versus 6,000 kg per 24 m^3, cost ($5,000 per hectare versus $500 per 24 m^3) and permanency of structure (fixed pond versus portable cages).

Chapter Focus

This chapter discusses the production of tilapia in general, in cages, and in a model type of cage culture system termed low volume, high density cage fish culture (LVHD), in particular. LVHD is the raising of fish in "low" cage volumes of 1 to $\leqslant$10 m^3 (typically about 4 m^3) at optimum "high" stocking densities of 300 to 500 individuals or expected harvest weights of from 150 to 250 kg fish per m^3 of cage. This technology is basically the same for all tilapia and numerous other fish species. LVHD culture technology is advanced when compared to traditional cage technology commonly practiced in net cages of $\geqslant$100 m^3 at stocking densities of about 40 fish per m^3 and yields of 20 to 25 $kg \cdot m^{-3}$.

LVHD was introduced into China in 1991, and by 1996 the total production of all species had reached 7,000 mt from 50,000 m^3 of LVHD cages, and in 2002, production was estimated at 21,992 mt from 268,200 m^3 of LVHD cages. Total tilapia production in China was 394,000 mt in 1995 and 897,276 mt in 2004, of which an estimated 88,000 mt, or 10 percent, was from cages of unspecified sizes and technology systems. LVHD cage culture is likely to be the future small-scale cage technology of choice in inland waters globally, especially for tilapia species, mainly because

1. the technology directly and indirectly satisfies the universal fundamental governmental goals of producing high-quality food, creating employment, generating farm income, helping to balance trade revenues, and making better use of natural resources;
2. it is relatively inexpensive and simple, therefore, easily adaptable by poor, landless people;
3. it is applicable in most existing water environments and does not require conversion of land into new bodies of water;
4. it is technically and economically applicable at any magnitude of scale;
5. there is less likelihood of off-flavor fish than in ponds; and
6. it will increase fish production and consumption in inland rural areas in many countries where fish supplies are low.

Larger cages, using modifications of the LVHD technology, will likely be the choice of large-scale commercial producers.

PRINCIPLES OF CAGE FISH CULTURE: FUNDAMENTALS

Fish culture is the raising of fish for profit in controlled, unnatural aquatic ecosystems, such as ponds, raceways, and cages. The unnatural aquatic ecosystems are that of the cage and the surrounding environment. Fish culture in cages is primarily influenced, in descending order, by water quality, feed and feeding, and stock quality. The primary focus of fish culture is to control (manage) these factors at optimum conditions for the species being cultured, at densities and in culture environments that do not naturally exist in nature. Using biological, physical, and chemical factors and with a good understanding of the natural laws that govern them, fish culturists create and temporarily maintain unnatural aquacultural ecosystems by controlling water quality, nutrient quality and quantity, and stock structure.

Although fish culturists have the ability to create and maintain unnatural aquacultural ecosystems, they do so only within the boundaries of the same natural laws that govern natural ecosystems. These natural laws are the "ecological principles of aquaculture" that govern all fish culture technologies however advanced and intensive, wherever located, or however applied. Learning these principles and how to apply them are fundamental to practical management of aquaculture ecosystems for profit. This section is based on these ecological principles of aquaculture as they specifically relate to tilapia culture in cages. It discusses the fundamental principles of cage fish culture, in general, and LVHD culture, in particular.

The concept of LVHD cage fish culture as a specific technology, in contrast to traditional high-volume, low-density cage fish culture, was developed through application of the water exchange principles discussed above. The basis of the concept is, of course, the provision and maintenance of higher water quality resulting from more frequent water exchange in low-volume cages compared to traditional cages. Although low volume is the essential component of LVHD cage fish culture technology, other major components that distinguish the technology from traditional cage cultures are higher fish densities, feeding enclosures for floating or sinking feeds, and opaque cage covers. These together with other principle-based technology components are presented in the following sections and are discussed in more detail elsewhere in this chapter.

Water Quality in Cages

Fish are successfully produced in cages at high densities with optimum yields as high as 250 $kg \cdot m^{-3}$ of cage and maximum yields of over 600 $kg \cdot m^{-3}$.

How are such yields possible? The answer lies in the principles of cage fish culture which have established that the fundamental limiting factor to fish health, production performance, survival, and ultimately profit, is adequate water quality. *Water quality* is a general term that represents the chemical, physical, and biological characteristics of the water ecosystem. In cages, the adverse water quality factors that most affect fish, and consequently limit fish production, are low dissolved oxygen (DO) levels and high levels of metabolic wastes. A fish population in a cage collectively consumes oxygen and produces wastes. The higher the collective weight of the fish population, the higher the oxygen consumption and waste production. Therefore, a constant, adequate, nonstressing water quality level inside the cage is fundamental to good fish health and production performance. Adequate cage water quality is dependent on continuous water exchange with the body of water surrounding the cage. Natural and fish-induced water currents are responsible for continuous water exchanges in cages. The water quality inside a fish cage is directly dependent upon the quality of water surrounding the cage and the rate of exchange of cage water with the surrounding water.

Water Quality of the Open Water Surrounding the Cage

Water quality inside a cage is always equal to or lower than the water quality surrounding the cage. The water quality of surrounding waters is directly related to the general level of nutrient enrichment of those waters and may be generally classified as eutrophic (rich in nutrients), mesotrophic (moderately rich in nutrients), or oligotrophic (poor in nutrients). Phytoplankton density, measured by depth of visibility into the water, is a relative measure of nutrient enrichment of that water. In general, the lower the level of enrichment the higher the water quality and cage production potential. However, regardless of the level of enrichment, small- and large-volume cages are equally influenced by the water quality surrounding the cages. Nevertheless, because of other factors, water quality in small cages will always be equal to or better than water quality in large cages with similar stock densities. Many tilapia species are capable of obtaining nutrients from plankton-rich eutrophic water, but the benefits of "free" food are offset by the negative consequences of poor water quality in the eutrophic water compared to plankton-poor oligotrophic water. LVHD technology is applicable in eutrophic environments whereas large-cage technology is not without some additional means of maintaining high DO levels.

Depth of water visibility may be influenced by suspended soil particles ("silt") in addition to, or exclusive of, phytoplankton. Siltation may be ignored if it is a temporary condition lasting for only 3 or 4 days following heavy rains. However, water quality in persistently silty waters is lower than in planktonic waters of equal depths of visibility. In silty waters, the lower the depth of visibility, the lower the production potential and higher the risk of production failure. Persistently silted waters with visibilities of $\leq$ 40 cm are not recommended environments for cage culture.

Rate of Water Exchange Between the Cage and Surrounding Water

Cage Design and Construction

Mesh Size of the Enclosing Material. Water exchange potential between a cage and its surrounding water is directly influenced by the cage material mesh size, i.e., the size and amount of open space in the enclosing materials. The amount of solid space between open spaces is also a critical factor. Water exchange potential increases with open space size but decreases with increasing amount of solid space between openings. Open space for a given mesh size (e.g., 13 mm bar mesh) may range from $\leq$65 to approximately 90 percent. Fish production in cages enclosed with nets of these extremes would hypothetically be different by at least the same magnitude of 35 percent. Another example of this principle is where a 1 m^3, net-enclosed cage with 90 percent water exchange potential is compared to a 1 m^3, wood-enclosed box "cage" with zero percent water exchange potential. Under normal conditions, the net cage, with one or a more complete water exchanges per minute, could produce an optimum yield of about 200 kg fish whereas the box cage, with no water exchanges, could yield only about 0.25 kg fish. Similarly, a 1 m^3, net-enclosed cage in the same water environment with much smaller mesh size and a 50 percent water exchange potential would produce an optimum yield of only about 100 kg of fish.

Cage Shape and Orientation. Cage shape is not a major factor in water exchange potential, but the relationship between cage shape and water exchange is a basic set of principles that should be understood. Most cages are either rectangular or square in shape, but cylindrical cages are also used sometimes. Water flow through the cages is relatively less restricted in rectangular and square LVHD cages compared to cylindrical cages of the same volume. Water flow potential would be uniform across the entire width of a rectangular or a square cage whereas it would be progressively reduced to the left and right of the center of a cylindrical cage. A rectangular cage may have greater or lesser water exchange when compared to a square cage of the same volume depending on the direction of water current relative to cage shape and orientation.

Cage Size (Lateral Surface Area:Volume Ratio, LSA:V). Among the factors that influence water quality in cages, the LSA:V ratio is the only factor that is different for traditional cages and LVHD cages and, consequently, the one on which these distinctly different technologies are based. Traditional fish culture cages are typically about 100 m^3 and some are more than 1,000 m^3. However, fish production performance per volume of cage is much higher and economically more efficient in small cages of about 1 to 4 m^3. For example, optimum fish yields in cages suspended in a lake may be 200 kg fish per m^3 in a 1 m^3 cage, but only about 25 kg fish per m^3 in a 100 m^3 cage. Optimum fish yields in low-volume cages may be higher than 300 kg fish per m^3 of cage, and, under favorable environmental conditions, yields of over 600 $kg{\cdot}m^{-3}$ have been achieved. Such high yields are not possible in traditional, high-volume cages. The reason is simply that under otherwise equal conditions, complete water exchanges are more frequent in smaller cages.

The smaller the cage, the greater the lateral (total cage side) surface area (m^2):volume (m^3) ratio. Increasing LSA:V will increase water exchange potential from natural and fish-induced water currents. In Figure 9.1, the 1 m^3 cage ($1 \times 1 \times 1$ m^3) has an LSA:V of 4:1 whereas the 32 ($4 \times 4 \times 2$ m^3) and 98 m^3 ($7 \times 7 \times 2$ m^3) cages have LSA:V of only 1:1 and 0.57:1, respectively. These lateral surface areas are only 25 percent and 14 percent of the 1 m^3 cage (see Table 9.1). If optimum yield of a 1 m^3 cage is 200 kg fish, then optimum yields of the 32 and 98 m^3 cages would hypothetically be about 50 and 25 kg fish, respectively. (Note, only lateral surface area, not bottom surface area, is considered because water currents normally move laterally, not vertically.)

Water Volume Flow Through the Cage

Water Current Rate. Water exchange rate in a cage is directly proportional to the water flow rate and the linear distance across the cage; therefore, the smaller the cage, the greater the water exchange rate potential. For example, a water flow rate of 1 $m{\cdot}min^{-1}$ will exchange water one time in 1 min in a 1 m wide (1 m^3) cage, but only one time in 7 min in a 7 m wide (98 m^3) cage. At an oxygen consumption rate of 300 $mg{\cdot}kg^{-1}$ of fish per h at 28°C, 1.0 g of O_2 per min would be consumed by 200 kg fish for every linear meter distance the water flows through the cages. Hypothetically, if oxygen concentration was 3.2 $mg{\cdot}L^{-1}$ as it entered the 1 m^3 cage, it would be 2.2 $mg{\cdot}L^{-1}$ after flowing 1 m where it exits the cage. This level of oxygen is

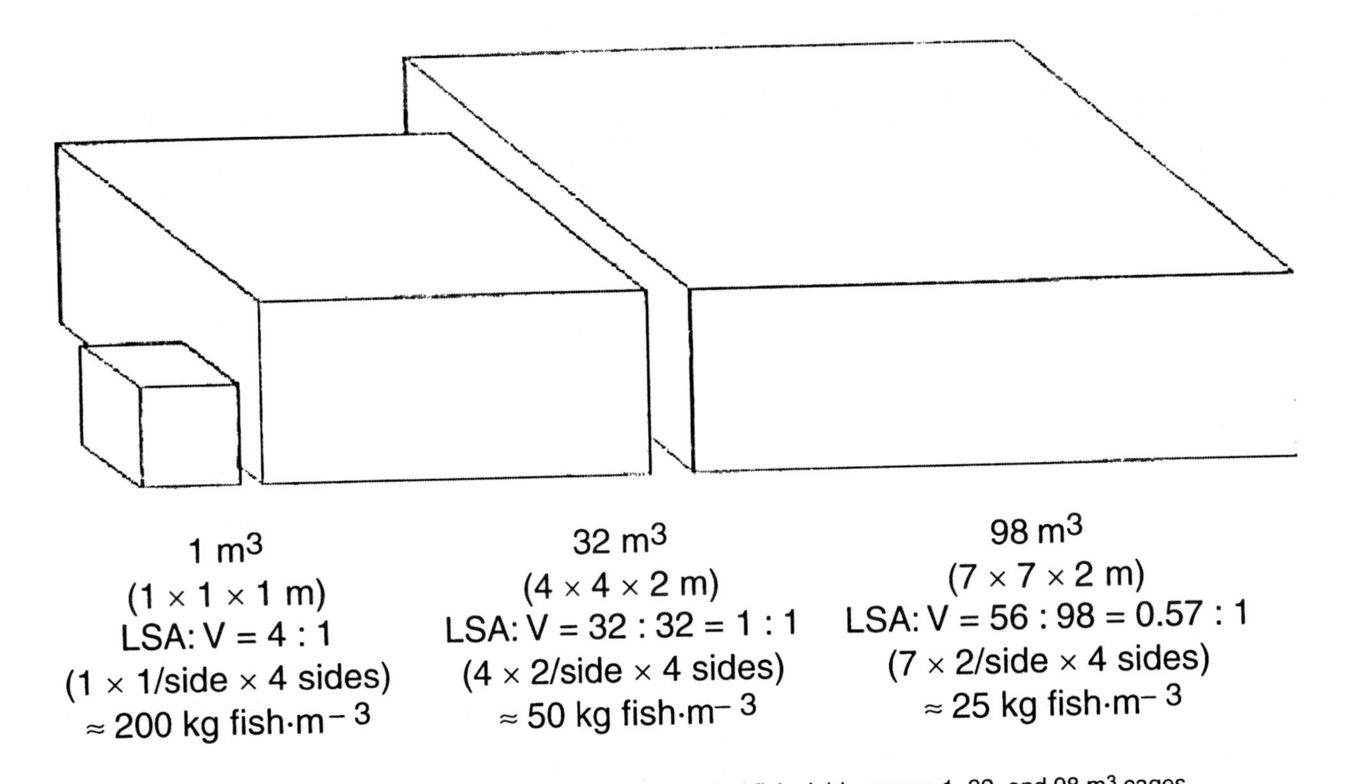

FIGURE 9.1. Comparison of LSA: V ratio and expected fish yields among 1, 32, and 98 m^3 cages.

TABLE 9.1. Comparison of water exchange efficiency potentials of cages of different LSA:V ratios.

Dimensions (m x m x m)	Volume (m^3)	LSA:V (m^2:m^3)	Water exchange efficiency potential (%)
1 × 1 × 1	1	4:1	100
2 × 2 × 1	4	2:1	50
2 × 4 × 1	8	1.5:1	38 (25/50)
4 × 4 × 2	32	1:1	25
7 × 7 × 2	98	0.57:1	14
6 × 11 × 2	132	0.52:1	13 (9/17)
13 × 13 × 2	338	0.31:1	8
11 × 11 × 3	363	0.36:1	9

stressing, for a short duration, but not critical to the health of the fish. However, if oxygen concentration was 3.2 $mg \cdot L^{-1}$ as it entered the 98 m^3 cage, all the fish would be dead from lack of oxygen within the first meter unless they remained evenly distributed, in which case, only the fish beyond 2 m would be dead. Increasing water flow by mechanical means could compensate for volume differences up to the point where fish having to maintain position against the current would increase oxygen demand.

Cage Placement Relative to the General Environment. Choosing a site for situating cages in any water is primarily dictated by two factors: (1) access to the cages for routine management activities and (2) water exchange between the cages, their immediate surrounding water, and water adjacent to the surrounding environment. Optimum water exchange for each cage is at an approximate rate of five complete water changes per minute (range of ≤1 to ≤10 changes). Greater than five water changes per minute may be desirable in cages with near-maximum fish-carrying capacity, high fish metabolism after feeding, high water temperature (>30°C), and/or an oxygen level below saturation. However, there are no benefits to water exchange rates above 6 to 7 times per min. Because oxygen consumption by fish increases exponentially with water flow rate, caged fish should not be cultured in prolonged water currents greater than about 10 $m \cdot min^{-1}$ Fish in low-volume cages suspended in stagnant water will, by their own activity, circulate water through the cage at sufficient rates to prevent respiratory stress, even at oxygen levels as low as 30 percent saturation. However, this would be a temporary and unusual phenomenon.

Cage Positioning Relative to Other Cages. Positioning of cages relative to each other is important for maintaining good water quality within the cages. Two principles are involved: (1) increasing cage density increases fish biomass, resulting in decreased water quality, specifically decreased DO levels in and around the cages and (2) increasing cage density decreases water exchange within the cages. Individual cages must be spaced far enough apart so as not to restrict proper water exchange. Positioning individual cages in a straight line, with at least one cage width between each, is far superior to positioning cages side by side in a chessboard-like arrangement.

Cage Fish Density

Cage fish density should reflect environmental water quality and water exchange factors discussed in the previous sections on cage design and water volume flow. If optimally stocked, water quality inside the cage at the lowest anticipated water quality conditions and the highest standing fish crop should not be negatively affected by fish density.

Fish Nutrition in Cages

Nutritionally complete and water-stable feeds are essential for cage production of tilapia. Because some tilapia have the ability to filter food from their environment, some culturists have incorrectly assumed that nutritionally complete feeds are not necessary for optimal production. The feeds should be pelleted in either a floating (extruded) or a sinking form. A feeding objective is to have 100 percent of the feed contained in the cage until consumed by the fish. However, some feed pellets will either float out, fall through the cage or be washed out of the cages before being consumed unless special techniques are used to contain them. This problem of feed loss increases with decreasing cage volume and increasing fish density. Therefore, it is a more serious problem in LVHD cage cultures than with traditional cage technologies.

A standard method for feeding tilapia in traditional cages is by broadcasting, whereby portions of a designated feeding allowance are manually broadcast over the water surface over a period of several minutes until the full allowance has been fed. While perhaps applicable in traditional cages, this technique is neither practical nor successful in preventing feed loss from LVHD cages. Consequently, feed enclosures, i.e., structures specifically installed in cages to prevent feed loss and achieve the feeding objective, are necessary components of LVHD cages.

Extruded, floating pelleted feeds are preferred for all types of cages. A model floating feed enclosure for LVHD is a box positioned in the top center of the cage that has an open bottom and a removable top cover. The floating feed enclosure should extend 40 cm below and as much as 20 cm above the water surface. The purpose of the feed enclosure is to prevent floating feed loss while allowing the fish free access to the feed. The sides of the feed enclosure are extended below and above the water surface where the feed floats because down currents and splashes created by the feeding fish may pull some of the floating feed down 50 cm or more below the surface and splash some of the feed more than 1 m into the air. A removable cover over the feed enclosure will prevent splash loss. Water surface area inside the feed container should be 20 to 25 percent of the total cage surface area. Down currents and splashes are reduced by minimizing the feed-holding area. It is important to confine floating feed enclosures to the center of the cages and not around the sides. Side positioning, though not as effective in preventing feed loss as center positioning, more importantly, restricts water exchange between the cage and the surrounding environment.

Sinking feeds are less preferable to floating feeds for tilapia cage production, but floating feeds are not yet universally available. Sinking feeds are more difficult to contain in cages than floating feeds, and preventing sinking feed loss requires a different, more complex structure. The difficulties with sinking feed losses are compounded by the inability to observe any losses that would be obvious with floating feeds.

A model sinking feed enclosure for LVHD cages consists of a tube and a "table." The permanently fixed tube (a 10 cm diameter PVC pipe) extends from above the cage surface to a point approximately 15 cm above the center of a screen-covered cage bottom that serves as the feed table. The bottom screen table should cover the entire cage bottom and extend 20 cm up the walls of the lower sides of the cage. Feed is poured into the top of the tube, which passes unmolested to the bottom table before it is accessible to the fish. Both the tube and the screen table are essential. Feed would be lost laterally through the cage sides without the tube and vertically through the cage bottom without the screen table. A fish barrier should be installed in, or at, the bottom of the tube to prevent fish from entering the tube from which they can not escape. Tilapia especially are prone to swimming up unbarred tubes.

Acceptable materials for feed containers vary greatly. The primary criteria for choosing a material are the same as that for the other components of cages, i.e., it must serve its function, it must not be injurious to the fish, and must be durable and reasonably priced.

Cage Covers

Covers may be needed on cages to protect fish from predators and theft. However, opaque covers placed directly on or immediately above cages are a recommended component of LVHD cages. The primary uses for opaque covers are to block sunlight, specifically ultraviolet light, from entering the cage and to restrict fish vision of objects and movement above the cage. Both ultraviolet light and fright stress caused by overhead movement negatively affect fish production performance in cages. Most affected by ultraviolet light are tilapia fry, small, immaturely pigmented fingerlings, and "colored" strains. Reducing the amount of light exposure may also enhance the immune system of fish, further improving production performance. Shading tends to reduce skin pigmentation, which may improve marketability. Also, predaceous birds are not attracted to opaque-covered cages as they are to open top cages. Opaque covers reduce photosynthetic bio-fouling that can restrict water exchange in cages. Opaque covers over fish cages have resulted in 10 percent higher production performance and over 30 percent higher profits compared to production in cages with transparent or no covers. The opaque cover material should cover the entire cage surface, alternatively excluding the floating feed container, at 10 to 15 cm above the water surface. The ideal cage cover material should be lightweight, inexpensive, weather resistant and impervious to light.

FISH STOCK QUALITY AND QUANTITY

Quality and quantity of fish stock are important factors to successful cage culture. Stock quality primarily refers to a fish's genetic potential for cage culture including its species and strain characteristics. Genetic quality factors are among several criteria for choosing a species, strain, or a specific group of fish for raising in cages. Stock quality also refers to the general health, relative size, and other physical and physiological characteristics of a given group of fish being stocked. Stock quantity generally refers to individual number and collective weight of group being stocked.

Criteria for Selecting Fish Stock

Nile, blue, and Java tilapia species (*Oreochromis niloticus*, *O. aureus*, and *O. mossambicus*, respectively), along with their strains and hybrids numbering perhaps upward of more than a hundred, are cultured throughout the world in the tropics and warm temperate regions. All of these varieties are suitable for cage culture, and most of them are already being raised in cages. Some tilapia varieties are better suited than others. Potential for cage culture fundamentally depends on the quality of the fish measured by many criteria, mostly genetic. Considerable research is underway to improve genetic potential of culture species by improving such factors as reproductive potential, environmental tolerance, disease resistance, feed efficiency, and marketability. These, and other genetic-related factors, are being applied to develop identifiable select strains for culture. However, no known efforts are underway to selectively breed or otherwise develop strains of tilapia, or any fish species, specific for cage culture. Perhaps there is no need for a specific cage strain different from one selected for intensive aquacultures in general. One possible criterion for a cage strain might be a greater than normal tolerance for close confinement at high density.

The production performance of fish in cages, or any culture system, will be directly related to the quality of fish stocked. Selection is critically important to culture success and should be done with concern for quality relative to genetic history, general health, individual size, and gender.

Genetic history should be assessed relative to the potential production performance of a specific fish species, strain, or group in cages in a specific environment. The aquacultural ecosystem concept of "fitting the fish to the environment" is demonstrated in this process. The concept is to stock the most appropriate strain, which has been genetically improved through selection, for a specific environment appropriately modified. Each is purposely modified to best match the other. In practice, the cage culturist cannot genetically modify the fish stock but should try to obtain stock already modified. This is accomplished by interviewing the fish supplier and other farmers to determine how well a specific strain or group has actually performed under culture conditions.

Fish for stocking cages should be in good general health and "disease free." Inspection by a certified fish disease specialist is recommended. In lieu of inspection, some key indicators of good health are uniformity of skin color among the group; absence of sores, blotches, spots, frayed fins and deformities; and all the fish vigorously avoiding capture.

The individual size of fish is important only as it relates to uniformity and cage mesh size. Tilapia for stocking should be of relatively uniform size and sufficiently large to escape through the mesh (minimum of 20 g for 13 mm bar mesh). Smaller fish (15 to 50 g) adjust to cage confinement more readily than larger fish. Effects of handling and confinement stress on fish during the first 5 to 10 days after stocking generally increase with increasing size of the fish. Tilapia stocked at a uniform size will tend to grow uniformly. However, with nonuniform stock, the size difference becomes progressively greater during culture. The larger fish in the group will grow normally while the smaller fish will grow at a less than normal rate and may not grow at all during the culture period. The reasons for differential growth are not clear but are assumed to be associated with size intimidation ("bullying") and pheromone repression.

The gender of tilapia stock is especially important. Males are preferred because of their faster growth, larger size, and inability to reproduce. Mixed-sex tilapias are capable of reproducing in cages at low densities, but generally do not at optimum densities in LVHD cages.

Stock Density

Fish stock density refers both to the number or weight of fish per unit volume of cage and to the unit area of water environment. Fish could be stocked so densely that restricted individual or collective space could become limiting to production performance. However, as density increases, both water quality and feed access decrease and limit production performance before restricted space becomes a factor. Therefore, an overstocked fish population is one where its density negatively affects production performance through its effect on water quality and feed access.

Relative to water quality, the primary concern about overstocking is the impact on water quality of the total environment directly, and the cage indirectly. With proper water exchange, water quality inside a cage will not be significantly different from that of the external water environment. Metabolic wastes are directly proportional to stock density. With proper water exchange, the wastes are diluted and disbursed to approximately equal concentrations inside and outside the cage. Therefore, the principal concern for overstocking, relative to effects on water quality, is fish weight per area of total environment, rather than the number or weight per volume of cage.

Stock density per volume of LVHD cage has a direct impact on the potential feed loss from the cage and feed access by the fish. As density increases, feed loss potential increases because of increased fish-induced water turbulence at feeding time. With increasing fish density of over 150 to 200 $kg \cdot m^{-3}$, access to feed at feeding time will become increasingly limited to all fish because of the amount of biomass separating some fish from the feed container. Feed loss and limited access resulting from high fish density can be reduced but not eliminated by feeding more than the fish will consume at one time (>100 percent satiation) and by more frequent feeding. This will reduce feeding frenzy and feed loss, and feed access to all fish will increase with increased feed amounts and exposure time. However, feed efficiency will likely decline.

Feed access may be influenced by fish behavior and physiology as well as by the physical biomass factor. Some caged tilapia populations may have a few individuals (1 to ≤3 percent) that grow either poorly or not at all, although they may maintain their general health and condition, while the other fish grow normally. This phenomenon is thought to be caused by physical aggressive behavior and possibly by pheromone secretions by dominant individuals, resulting in repressed feeding activity among the most subdominant individuals. This phenomenon appears to be exacerbated in poor water quality conditions. Territorial responses, especially aggressive behavior, are observed to be much more prevalent at low stock densities (≤200 fish per m^3). Therefore, feed access may be limited by factors related to both overstocking and understocking.

For stocking purposes, fish density is measured in numbers per cage volume (fish per m^3) or total water environment area (fish per ha). Measures of fish density for expressing standing crop, carrying capacity, and yield are given in weight per cage volume ($kg \cdot m^{-3}$) and total water environment area ($kg \cdot ha^{-1}$). The number of fish to stock in cages is dependent upon total yield and mean weight expected at harvest and may be calculated with the simple equation: $Nc = Wc/w$ where Nc is the number of fish to stock per m^3 of cage, Wc is the expected total fish weight per cubic meter of cage at harvest, w is the desired mean fish weight at harvest.

Expected optimum cage fish yield is usually equal to optimum cage-carrying capacity, and both vary directly with environmental water quality and inversely with cage volume. Expected optimum cage-carrying capacities generally reflect expected water quality conditions based on the nutrient enrichment (fertility) level of the water. Highest optimum carrying capacities and yields would be in low nutrient, oligotrophic water environments. Carrying capacities decline with increased cage volume, but remain relatively constant for cage volumes of 1 through 4 m^3. Fish mortality in cages during culture is usually insignificant (≤4 percent) when healthy fish are stocked and proper management provided. Therefore, the above equation is applicable without compensation for mortality. However, higher mortality (6 to 8 percent) is common for tilapia within a week after stocking at water temperatures either below or above 18°C and 24°C, respectively.

Every water environment is different and the number of fish to collectively stock in cages per area of each water environment must be considered if collective yields are expected to approach carrying capacity of the environment. The number of fish to stock is determined using the same equation as for individual cages except that the expected weight (yield) per water environment area at harvest replaces the expected weight per individual cage volume. Suggested maximum yields of caged tilapia per pond area are 2,750 $kg \cdot ha^{-1}$ without emergency aeration and 6,000 $kg \cdot ha^{-1}$ with emergency aeration. Suggested maximum standing biomass for all caged fish per area of open lakes and reservoirs is 300 $kg \cdot ha^{-1}$ of total water area, 20 mt in a 1 ha designated cage culture area, and 60 mt in a 10 ha designated culture area (see section on "Environmental impact of cage fish culture" in this chapter). Numbers may be increased in large lakes and reservoirs and in open marine environments if water flow is sufficient to assure adequate water exchange.

WATER ENVIRONMENTS AND WATER QUALITY

Aquacultural ecosystems, including those involving cage fish culture, are composed of physical, chemical, and biological factors that interact individually and collectively to influence culture performance. Water is the primary component of all aquacultural ecosystems, therefore, each characteristic of water is a water quality variable. Although all of the impacting variables are important, only those that normally cause fish stress or otherwise limit performance in some way are of concern to the practical aquaculturist. An understanding of the key variables, how they relate to fish production, and how they may be managed to control the aquacultural environment are essential to fish farmers. However, water quality factors for caged tilapia are generally the same as that for other fish culture systems, such as for tilapia raised in open ponds, but for cage application some factors require better understanding and attention to management than for pond systems. Some of the key water quality variables and their management are discussed in the following sections as they generally behave within aquacultural ecosystems and relate to tilapia culture in cages suspended in ponds, reservoirs, and lakes.

Selected Water Quality Factors

Temperature

Temperature impacts cage tilapia culture in two major ways: (1) the temperature of the water where the fish are located and (2) the temperature stratification of the water column in which the cages are located. Water temperature surrounding tilapia in production should ideally be about 26°C to 28°C and within an optimum range of about 23°C to 30°C. Temperatures below 18°C and above 33°C are stressful and should be avoided. Thermally stratified waters may be void of DO in the lower (hypolimnion) stratum where oxygen may be consumed but not produced, and be supersaturated with oxygen in the upper (epilimnion) stratum where photosynthesis is active. Thermally stratified water is suitable for cages only in situations where destratification would not lower cage DO to lethal or stressful levels or the culturist can protect the fish by such means as mechanically aerating water in the cage area or relocating the cages to a DO-safe location.

Light

Light intensity, and perhaps duration, impacts caged tilapia. Daylight, and especially direct sunlight, affects caged fish behavior and production performance apparently by causing stress to the confined fish. Light, specifically in the ultraviolet spectrum, is especially stressful to "colored" (e.g., white and red) tilapia. Light entering the cages should be controlled. Opaque covers over at least some of the cage surface are recommended partly to reduce the amount of direct sunlight entering the cages. The effect of photoperiod on production performance of caged fish is unknown. Compared to caged common carp, *Cyprinus carpio*, at more southerly latitudes with daily photoperiods of about 12 h, common carp cultured at about 45°N latitude in various locations of Heilongjiang, China, have been observed for years to consume more feed and grow proportionately faster with extended photoperiods of up to 14 to 16 h.

Sound

Sound, specifically unnatural and loud sounds, affects fish behavior, and sound-induced fright stress may significantly reduce production performance. Such sounds should be avoided in all aquacultural environments, especially where fish are confined in cages or under handling conditions.

LODOS

Low dissolved oxygen syndrome (LODOS) is the most important water quality variable in aquaculture and is especially critical to caged fish. This requires a much more thorough understanding by culturists than the brief comments here. Oxygen naturally enters and dissolves into standing open waters (ponds, lakes, and reservoirs) primarily through oxygen-releasing photosynthesis (approximately 90 to 95 percent), secondarily by diffusion from the air (most effective when aided by surface agitation), and by incoming water. Oxygen exits, including being consumed within, aquacultural standing waters primarily through plankton respiration, secondarily by fish respiration, respiration of bottom microorganisms, and diffusion. Biochemical oxygen demand (BOD) in fish ponds with intensive feeding varies greatly but may be assumed to be about 0.4 to 0.6 $mg \cdot L^{-1} \cdot h^{-1}$. Oxygen diffuses out of standing open waters only when the surface waters are supersaturated.

LODOS involves a combination of low DO with increased ammonia, increased free carbon dioxide, decreased pH, increased nitrite, increased fish metabolism (e.g., such as for 2 to 3 h after feeding), increased water temperature, abundant gill parasites, and numerous other factors, which when combined can significantly reduce fish production performance. LODOS most commonly occurs at dawn at the lowest DO level in the diurnal DO cycle, but it may occur at any time caused by any number of factors that lower water DO.

Feed

Feed fed to fish indirectly impacts water quality through its waste impact on plankton production. Approximately 80 to 85 percent of the nutrients in pelleted feeds used in aquacultural ecosystems are released into the water as fecal matter or unmetabolized compounds that include phosphorus, ammonia, and carbon dioxide that promote phytoplankton production. Organic matter produced by phytoplankton photosynthesis exceeds by many times the amount of organic matter from fecal wastes. Metabolism of zooplankton, bacteria, and other nonphytoplankton microorganisms may actually be higher than metabolism by fish. Feed wastes increase with feeding rates, and phytoplankton densities increase directly with metabolized feed wastes. As phytoplankton density increases, the depth of photosynthesis decreases and BOD increases. These changes result in critical water quality deterioration manifesting itself in early morning LODOS conditions.

At low feeding levels, the ecosystem is in balance between water quality deterioration resulting from feed wastes and water quality restoration resulting from biological utilization of these same wastes. At high feeding levels and resulting waste loading, the ecosystem balance breaks down primarily because phytoplankton increases proportionately with the metabolic wastes to the point that its contribution to water quality restoration is offset by its impact on water quality deterioration. Ultimately, increase in feeding rate is limited by decrease in water quality. Thus, nutrition, the second limiting factor to cage fish production, gives way to the first limiting factor, water quality, usually in the form of LODOS. In new ponds and early in the production cycle in older ponds, low to moderate quantities of feed wastes may actually improve water quality for up to a few weeks. Feed wastes may improve water quality and wild fish production in oligotrophic lakes and reservoirs as well.

Management techniques to prevent and control water quality deterioration resulting from feed wastes must be based on limiting the feeding amount to a "safe" level relative to methods (environmental modifications) to counter the direct (toxic) and indirect (phytoplankton density and LODOS) effects of the wastes on water quality. Phytoplankton density and scums in ponds may be reduced by filter feeding of tilapia and with algaecides.

Fish Density

Fish density may directly influence two seldom considered water quality factors that may be problems at either low or high density. Fish stock density is a measure of fish numbers or biomass per volume of cage. Aggressive behavior and possibly pheromones are known or suspected density-dependent social interactions that affect caged fish. Aggressive, territorial behavior is an observed problem with tilapia and other species at low densities (e.g., <100 fish per m^3) but not at high densities (e.g., >300 fish per m^3). Pheromone influences on growth and general performance of cultured fish are only vaguely known for ponds and suspected for cages. Pheromone influence may be greater at higher fish densities in cages.

FEEDS AND FEEDING

All caged fish require proteins, lipids, energy, vitamins, and minerals in their diet for growth, reproduction, and other normal physiological functions. These requirements vary somewhat among species and within species relative to the stage of life cycle, sex, reproductive state, and environment. Nutrients for caged tilapia may come from food sources, such as plankton, bacteria, and insects from within the cage and/or from organic matter and processed feeds added to the cage. Foods are defined as natural sources of nutrients produced in the environment whereas feeds are natural and manufactured sources of nutrients produced elsewhere and added to the environment. The majority of cage-cultured tilapia species, including its strains and hybrids, filter suspended plankton, graze mesh-attached organisms, and readily take manufactured feeds. The following refers to nutritionally complete manufactured feeds commonly used in intensive fish culture.

Feeds for caged tilapia must be nutritionally complete and balanced. Although some tilapia, such as the Nile tilapia, may obtain a few essential nutrients by filtering plankton from nutrient-rich waters, they still need a complete diet as if they were being cultured in food-free waters. All of the nutrient requirements for all cultured tilapia are not known at present, but the requirements for Nile tilapia are generally acceptable for other tilapia species.

Only manufactured, dry-pelleted feeds are recommended for caged tilapia. No forms of natural fresh feeds are recommended. The use of raw "trash fish," usually whole fish with low market value or by-products from fish processing, is strongly discouraged. as is the use of natural, single ingredients such as raw or cooked corn and cassava.

Pelleted fish feeds are either compressed or extruded based on their manufacturing process. Although compressed and extruded feeds are formulated from the same ingredients, they are fundamentally different in that in the former the pellets immediately sink when placed in water whereas in the latter they float on the surface. Although extruded feeds are more expensive than compressed feeds of the same formulation, their advantages to caged tilapia allow them to be significantly more economically efficient.

The objective of feeding tilapia in cages is to economically provide proper nutrition for growth and good health while minimizing metabolic waste and ecosystem pollution. The requirements for achieving the objective are providing proper feed quality and quantity, employing an in-cage feed enclosure, and using proper feeding methods. A basic feed quality standard for caged tilapia requires that it be nutritionally complete, water stable for a minimum of 10 min, and freshly manufactured within the past 4 weeks. Only pelleted feeds are recommended and these may be of either sinking or floating type, although the latter is preferred where available.

Daily feeding rates are based on a percentage of the mean weight of fish in a cage (see Table 9.2). Rates may vary according to many factors including species, size, stage in life cycle, water temperature, other water quality variables, feed density, nutritional level, and management system. For any given species, feeding rates are routinely affected by fish size and water temperature. Smaller fish eat proportionately more feed and more often than larger fish. Fish of all sizes eat progressively less and eventually stop taking feed as water temperature decreases or increases beyond their optimum temperature range. The optimum production temperature for tilapia and most warm water fish is approximately 26°C to 28°C, with a range between 23°C and 30°C.

The amount of feed that can be given to caged fish per unit of ecosystem (environment in which the cages are suspended) per day is limited by the effect of feed metabolism on water quality of the ecosystem. As feed quantity is increased, water quality decreases, with low DO usually becoming the first water quality factor to affect fish. Maximum "safe" feeding allowances are generally known for aquaculture systems and are usually based on some minimum DO level expected during a daily cycle (e.g., 2.0 mg $DO \cdot L^{-1}$ at dawn). Because many factors affect water quality, other than simply feed quantity, there is no sure means of accurately predicting the magnitude of DO and water quality deterioration as a result of a measured amount of feed given to the fish. Therefore, maximum "safe" feeding allowances are only guidelines and at times may be "unsafe."

TABLE 9.2. General daily feeding rate and frequency guide for grow-out production of tilapia in cages using a 32 percent protein pelleted feed at 28°C water temperature.

Fish mean weight (g)	Feed rate (%)	Feeding frequency (times per day)
25	4.50	3
50	3.70	3
75	3.40	3
100	3.20	3
150	3.00	2
200	2.80	2
250	2.50	2
300	2.30	2
400	2.00	2
500	1.70	2
600	1.40	2

Note: Calculate feed allowance and adjust for water temperature differences as follows (temperature measured at 50 cm depth):

at 16-19°C, feed at 60 percent of calculated allowance once a day every day;
at 20-24°C, feed at 80 percent of calculated allowance once or twice a day every day;
at 25-31°C, feed at 100 percent of calculated allowance maximum frequency every day;
at 32-33°C, feed at 80 percent calculated allowance once or twice a day;
at ≥34°C, feed only what fish are observed to take—such temperatures should be avoided.

The optimum feeding rates of tilapia in intensive cage cultures usually mean feed allowances of near 100 percent satiation, which is the total amount of feed that fish will consume at a feeding before they stop taking feed. Caged tilapia should require no more than 5 min to consume a feed allowance. Any feed not consumed when fish stop taking feed is in excess of satiation. About 90 percent satiation is usually the target rate for feeding caged tilapia. Moderately lower amounts of ≥80 percent rate will result in better feed conversion but slower growth, whereas higher amounts approaching 100 percent rate will result in poorer feed conversion but faster growth. Satiation is based either on an estimate of standing crop or on observed feeding behavior. In typical single stock or single harvest cage aquacultures, daily amounts of feed per unit area of open water environment increase geometrically from minimum amounts of only about 5 percent of maximum safe rate per day immediately after stocking to 100 percent of maximum safe rate for the last several days before harvest. However, in multiple stock or multiple harvest aquacultures, daily amounts of feed per open water area increase and decrease proportionately with each stock and harvest, but minimum and maximum amounts are usually as high as 35 percent of maximum and never above 100 percent of maximum safe rate per day, respectively.

Daily feed allowance is the quantity of feed given to caged fish (in kg/cage volume/day) and is calculated by multiplying the total standing crop (weight) of fish by the feeding rate (Table 9.2). For example, 146 kg fish·m^{-3} cage standing crop × 2.3 percent feed rate = 3.4 kg feed·m^{-3} cage/day. Note that daily feed allowances increase, but feed rates decrease as fish grow larger.

Feeding schedule refers to the specific time(s) and frequency at which the feed allowance is given to the fish. The feeding frequency for Nile tilapia during optimum growing temperature will vary according to size or stage of the life cycle from up to 12 times per day for newly hatched fry to 3 to 4 times a day for fingerlings, 2 to 3 times per day for grow-out production fish (Table 9.2), and once a day for brood fish. Although multiple daily feedings may increase growth rate, especially for tilapia and fish that do not have stomachs, they may not improve feed conversion efficiency of larger grow-out size fish (>100 g). To demonstrate this point, Nile tilapia were raised from approximately 45 to ≥400 g in LVHD cages suspended in a reservoir with high water quality (Qian Dao, Zhejiang, China). The tilapia were stocked at 400 fish per m^3 in 16 cages. Fish in half the cages were fed twice a day at 9.5-h intervals whereas fish in the other half were fed four times a day at 2.5-h intervals. In 140 feeding days, it was observed that there were no differences in production performances of yield, growth rate, survival, and feed conversion values between tilapia fed daily rations in two or four feedings. Economic differences favored feeding twice a day because of only half the labor requirement.

Tilapia should normally be fed during daylight hours from about 2 h after sunrise to about 2 h before sunset. This rule is not as important in nutrient-poor waters as it is in nutrient-rich waters such as ponds. Never feed caged fish at night in ponds and other nutrient-rich waters.

Methods of feeding caged tilapia are designed to offer the daily feed allowance in a way that it will be 100 percent consumed by the fish. Accomplishing this objective requires water stable feed, optimum feed allowance (≥80 to ≤100 percent satiation), and proper feeding methods, including a proper feed structure for the type of feed used and proper feed distribution techniques. Pelleted feeds must be water stable for at least 10 min before they begin to disintegrate.

Determining the optimum feed allowances for each feeding time per day is a simple concept but sometimes a difficult practice. The concept is to feed the fish at each feeding time carefully measured amounts of feed until the fish stop eating. The amount of feed consumed to that point is considered 100 percent satiation and the maximum end point of the feed allowance optimum range. That same amount should be fed at the same feeding time for the next 7 to 14 days when a new 100 percent satiation optimum feed allowance would be determined. On each successive day during the 7 to 14 days, the set feed amount would become progressively less than the 100 percent satiation and by the 14th day would be near 80 percent satiation or the minimum end point of the feed allowance optimum range. Feed allowance adjustments should be made any time overfeeding or underfeeding is obvious. Although neither is desirable, slight underfeeding is preferable to overfeeding.

When using floating feed, the 100 percent satiation end point is obvious and simple to obtain. However, with sinking feed the satiation end point is neither obvious nor simple to obtain. Feed rate adjustments for sinking feeds may be done based on periodic sampling to determine the average and total weight of fish and then using a feeding table (Table 9.2) to determine the proper feed rate and allowance.

Techniques for feeding caged tilapia vary, but the most basic techniques include the following:

1. extruded feed poured all at once into a floating feed enclosure at the cage surface;
2. compressed feed poured all at once down a tube to a sinking feed enclosure at the cage bottom;
3. compressed (sinking) feed sprinkled moderately into the top center of the cage.

Only the first two techniques are recommended in LVHD cages. The third technique is labor consuming, may favor dominant individuals, and will likely result in feed loss through the cage.

Feed enclosure structures for either sinking or floating feed are essential for feeding fish in LVHD cages. The use of automatic and demand mechanical feeders in cages has received mixed assessments. Some research results show improved growth rate and feed conversion efficiency with mechanical feeders when compared to hand feeding. Other research results have shown the opposite. Mechanical feeders are not recommended, because they reduce management's attention to fish-feeding behavior and general production performance.

HEALTH OF CAGED TILAPIA

Health Concepts Related to Caged Tilapia

Health is defined as the standard or typical condition of the fish with respect to normalcy of body functions and disease at any given time. A healthy fish functions optimally and is free of abnormalities caused by stress and disease. A primary objective of cage culture is to maintain healthy fish populations that are optimally feeding, growing, and normally functioning. The key to achieving this objective is through stress management, preventing and minimizing stress to fish in the culture environment by understanding and managing the various environmental factors that cause stress. One might argue that "fish culturists do not actually culture or raise fish; they actually culture or manage the environment (ecosystem) and the fish raise themselves." Maintaining fish health, therefore, is accomplished by individually and collectively managing the environmental quality factors as near to optimum as is practically possible and essentially within the tolerance level of fish.

Obviously then, good fish health and maintenance of these conditions are primary concerns and critical objectives of cage aquaculture. Fish stress, health, and disease management may be simplistically compared, in human health terms, more to a public health officer concerned with health maintenance and disease prevention rather than to a doctor of internal medicine at a hospital concerned with treating a disease to restore good health.

Health conditions of caged fish populations are the direct outcomes of management. The majority of fish health problems in cage fish populations, especially disease problems, are directly linked to stress on the fish from some environmental factor(s). Direct relationships exist among environmental quality, fish health, and disease. Avoiding stress by maintaining good environmental quality through proper management is essential to the maintenance of a healthy fish population. Thus, emphasis, in order of importance, is on stress (from environmental quality factors), health, and disease. Obviously, not all environmental stressors are under full management control, and some obligate infectious diseases are not predisposed by an environmental stressor. Nevertheless, these and all health problems are subject to some measure of management.

It is an inaccurate, but pervasive, opinion that because of their high density and close confinement, caged fish are more susceptible than fish in other culture systems to disease and general health problems. Perhaps the inaccurate observation is because signs of unhealthy fish and conditions are more visible and readily detected in cages.

"Stress" in caged tilapia is an abnormal physiological condition resulting when collective adaptive responses to an environmental factor(s) are extended to, or approaching, the fish's limit of tolerance for that factor. Cage-culture environments, as all aquacultural ecosystems, are innately unstable, unnatural water environments. In general, the greater the culture intensity the greater the environmental instability. All environmental components—chemical, physical, and biological—are constantly changing. These changes, and the technology procedures involved in raising fish individually and collectively, may stimulate abnormal physiological responses or stress in fish. Stress occurs when an environmental factor (stressor) extends to or beyond the normal optimum range of the fish and disrupts its physiology. Stressors reduce the ability of fish to normally function physiologically and behaviorally. Stressors can be acute or chronic, and their impact on fish is additive and accumulative at least for a short period.

A fundamental management objective of all aquacultures is to avoid and minimize stress on fish. This requires an understanding of stressors and their effects on fish and an ability to recognize fish that are under stress. Obviously, measuring internal physiological responses to stress is not a practical option to a cage culturist. More practical knowledge and means of identifying stressors and stress are required.

The many known types of stressors (i.e., causes of stress) to caged tilapia may be grouped under the four main categories of chemical, physical, biological, and procedural stressors. Some of the most common are briefly discussed below.

1. *Diet* is a stressor when the quantity of food is lacking or limited and the feed provided is deficient in any essential nutrient. Low-quality nutrition and insufficient quantity of feed offered can be direct or indirect stressors causing the fish to be more susceptible to other stressors.

2. *Crowding* of tilapia in cages does not in itself increase stress or the incidence, spread, and severity of disease as is sometimes supposed. Healthy, unstressed commonly cultured fishes, including tilapia, "crowded" at densities of 400 to 700 fish per m^3 and total weights of 150 to $\geq$250 $kg \cdot m^{-3}$ in low-volume cages are unlikely to become diseased. No evidence indicates that "crowded" fish in cages are more susceptible to infectious diseases, or that incidence and severity of disease is greater than for lower culture densities in cages or open water of the same water environment.

Population density in cages may be a biological stressor more often at low density than at high density. In cultural ecosystems, especially at low density, members of some fish species (including tilapia) will try to establish hierarchy of dominant and subdominant individuals of which the latter are chronically stressed. Pheromones are probably involved with the establishment of most and perhaps all of those territorial hierarchies. In culture situations, population-density stressors are from LODOS, metabolic waste build-up, and social interactions such as pheromone-related hierarchies, not from spatial limitations of high densities.

3. *Confinement* is at least temporarily stressing to fish. When first confined in cages, fish generally show signs of stress. It appears that older and larger fish and those least accustomed to confinement are the most affected after stocking into cages.

4. *Light* is a stressor when embryos or fry are exposed to direct sunlight, ultraviolet light, and white light of moderate to high intensity ($\geq$850 Lx = 80 FC), when fish are confined in enclosed facilities (e.g., cages in direct sunlight), and when fish are being subjected to other stressors. This surely must be a major stressor to "colored" (i.e., unnaturally pigmented) strains.

5. *Sound* waves are known to reduce embryonic development of some fish. Fish growth and reproduction are negatively affected by some sounds perhaps because of the fright factor.

6. *LODOS* is usually the most important stressor in aquacultural environments. LODOS is low DO with any combination of likely simultaneous environmental and physiological conditions, such as high CO_2 and decreased water pH, increased blood lactic acid and decreased blood pH, high NH_3 or NO_2^-, high temperature, and numerous other factors. If all other factors are disregarded, reduced DO levels become stressing to fish at about 70 percent saturation for embryos and young fry and at about 60 percent saturation for larger fishes.

7. Temperature is a common but well-understood stressor of tilapia in all aquacultural ecosystems. Temperature is a stressor of all tilapia when it ranges below or above about 15°C and 32°C, respectively, and when it fluctuates rapidly by more than a few degrees (e.g., 3°C to 5°C in less than 1 h), especially if it is an increasing temperature. Tilapia culture is practiced in the temperate regions of the world, although all tilapia species are tropical and, therefore, cold sensitive.

8. *Salinity* is a stressor to some species and strains of tilapia under certain conditions. The Java tilapia is tolerant of, and observably unstressed at, water salinities $\geq$80 ppt whereas Nile tilapia may be stressed at $\leq$7 ppt. If acclimated their hybrid may readily adapt to seawater salinity (about 34 ppt).

9. *Combined* factors may be serious stressors where individually the factors are only mildly stressful or nonstressing. For example, tilapia are generally more susceptible to stress due to cool temperatures when in salt water than in freshwater.

10. *Procedural stressors* are those associated with handling, holding, transporting, and treating cultured fish. Procedural activities include induced spawning, stocking, harvesting, holding fish in tanks, and all other

short-term activities that supplement routine culture. Procedural stressors also include all the chemical, physical, and biological stressors already discussed. A common procedural stressor of tilapia is too rapid temperature and/or salinity acclimation when transferring from one environment to another. Other procedural stressors include (1) the crushing effect of gravity when handling fish out of water, which is particularly stressful to larger fish and groups of fish lifted together en men masse out of water (fish are morphologically and physiologically adapted to the pressurized, buoyant environment of water, but not to the opposite conditions out of water) and (2) the various effects of herbicides, parasiticides, and other chemicals, such as formalin and copper sulfate, which are used to prevent or control pests in aquacultural environments. Anesthetics, such as MS 222, sometimes used when handling and transporting fish, may cause greater stress rather than less stress for which they are intended.

Caged tilapia are subject to all of the above stressors and may be more vulnerable than pond-cultured tilapia to some of these. However, incidence and severity of stress and resulting poor health are more easily recognized and managed in cages, especially in low-volume cages, than in open ponds.

Techniques for preventing and minimizing stress are the principal techniques of any aquacultural technology system. Emphasis is on fish stress because stress predisposes fish to most diseases and affects fish health, thereby decreasing production performance (e.g., fish growth, yield, survival, and feed efficiency). Stress can usually be avoided or minimized by good management practices.

The principal technological components of high-density cage fish culture and their purpose relative to stress management are presented in Table 9.3. By managing the specific chemical, biological, and physical factors discussed earlier that may cause stress, fish health is almost assured but not guaranteed. Diseases do occur in fish that have not been stress mediated.

"Disease" is an abnormal condition of fish wherein body functions are impaired as a consequence of stress (e.g., handling and LODOS), inherent weakness (i.e., genetic and congenital defects), or infection (e.g., *Streptococcus* bacterium). The direct interrelationships between stress, health, and disease are obvious from the above discussion. Stress in fish will proportionally affect fish health and may lead to infectious disease of fish, especially if compounded by two or more stressors. For practical understanding and management, one needs only to consider the cyclic relationship of stress impairing health, resulting in disease, which leads to further stress and further impairment of health.

TABLE 9.3. Principal technological components and their relationship to stress management of tilapia in LVHD cage-culture technology.

Technology component	Stress management component
1. Select a high-quality, relatively stable environment for culture	Water quality variables fluctuate well within tolerance ranges of fish
2. Construct rectangular or square cages of 1 to 8 m^3 volume; enclose with net with ≥13 mm mesh	Water exchanges maintain water quality inside cages within tolerance ranges of fish
3. Place cages in open water, with individual cages spaced apart	See component 2 above
4. Stock selected, healthy fish	Fish have natural resistance to stressors and pathogens
5. Feed proper allowance of nutritionally complete feed of good physical quality in special feeding enclosure	Nutritional requirements for growth and good health are met. Minimum pollution potential wastes are lost to the environment
6. An opaque cover over the cage is needed	This cover will prevent light stress and minimize fright stress

All fish, under certain circumstances, are subject to disease. However, healthy fish, whether in natural or aquacultural environments, have a strong resistance to disease as long as they are not weakened. This does not mean that disease epizootics will not occur in caged fish unless the fish are first predisposed to stress. Such epizootics are not common because obligate disease pathogens are manageable and should be prevented from the fish population at the hatchery, nursery, or elsewhere before stocking in grow-out cages.

Infectious organisms are constant, ubiquitous components of every cage-culture environment, and healthy caged fish will normally harbor some of the potentially pathogenic organisms. However, clinical signs of disease may not occur as long as the fish remain unstressed. The process of infection into disease first requires an abnormal disruption of the fish's physiology, lowering its natural resistance to the invading pathogen. The most common diseases of caged tilapia are caused by bacterial infections. Most of these are predisposed by stress and all are exacerbated by stress. The most common stressors to caged tilapia that lead to disease are, in the order of observed occurrence and severity, handling (prestock seining, holding, transporting, stocking, and post-stock sampling), LODOS, low temperature, high salinity, especially if associated with low temperature, and poor nutrition. Disease epizootics will be uncommon in caged tilapia populations where the above stressors are absent.

Health Management

The maintenance of good fish health is critical to profitable cage fish culture. Slow growth, poor feed efficiency, low yields, increased disease incidence and mortality, and consequently low profitability are the results of poor fish health. Since stress is the fundamental cause of most fish health problems, practical health management is based on the premise, "avoid fish stress and avoid fish health problems," by assuring environmental parameters within optimum tolerance ranges of the fish being cultured. This premise is founded on the management precept "know the fish and know its environment." Sound health management decisions can be made only when there is a clear understanding of what environmental factors are stressors to the specific fish being cultured and at what circumstance and level they become stressing. Good management is the key to avoiding essentially all health problems whether stress-related or not.

Treatment of Diseased Fish in Cages

Fish stock of good genetic quality, handled properly, stocked disease-free in cages in an environment with constant good water quality, and fed properly with nutritionally complete feed are very unlikely to become diseased. However, disease epizootics in caged tilapia do occur, usually caused by bacteria or protozoa. These epizootics usually correct themselves if the fish are in a relatively stress-free environment and receive good nutrition. Nevertheless, sometimes chemotherapeutic treatment with approved chemicals and medicines by the U.S. Food and Drug Administration may be necessary. The treatment may be performed in holding and handling facilities before stocking or in open ponds surrounding the cages, and in cages after stocking. In more serious situations, it may be desirable to transfer fish from cages to holding tanks for treatment.

ENVIRONMENTAL IMPACT OF CAGE FISH CULTURE

Sustainable cage fish culture is dependent on both long-term environmental and economic viability. Environmental viability is dependent on maintaining water quality at or above a minimum standard. In turn, maintenance of a water quality standard is dependent on the collective amount of wastes introduced into the water, resulting from the quantity and quality of feed and other nutrients used to culture the fish, and from all other sources. Cage culture inevitably enriches the surrounding water environment with metabolized feed waste as would organic fertilization. As with fertilization, the amount of feed waste up to a certain threshold is beneficial to the environment because of the increased phytoplankton and other biomass that are

1. important in maintaining good water quality;
2. providing food to some caged fish, including the majority of tilapia; and
3. increasing production of noncaged fish that result from the enrichment.

Beyond this threshold, however, the enriching elements become pollutants.

The threshold is not an absolute amount that can be predicted for any situation, for example, as a road bridge engineer could predict the load limit for a specific support structure. Ecological systems are too complex and dynamic for that degree of accuracy. In addition, the threshold is variable based on limits one wants to set on change of the environment, such as the level of eutrophication. Consequently, thresholds for added enrichment with fish feeds vary with the prevailing trophic state of the environment. For example, the safe feeding threshold for an oligotrophic environment will initially be higher than for a eutrophic environment. Other influencing factors include water surface area, water depth, seasonal fluctuations, amount, seasonality of flow through, and other uses of the environment.

The enriching and polluting substances resulting from cage fish culture are primarily phosphorous and nitrogen contained in the feed. The caged fish species and their numbers and weights are relatively incidental to the noncage environment except when they relate to the amount of feed used to produce them. The quantity of phosphorus and nitrogen in fish feeds vary somewhat with feed quality, but are usually about 12 and 55 kg, respectively, per ton of pelleted feed (Table 9.4). Fish will assimilate some of these nutrients (about 5 and 14 kg, respectively, at FCR = 2.0) leaving the remainder to enter the environment as metabolic waste. These wastes enrich phytoplankton production which stimulates increased biomass at all trophic levels, including the levels occupied by noncaged, wild fishes in the open water. However, increases in biomass result in comparable increases in respiration causing proportional declines in nighttime dissolved oxygen levels. This water quality problem is worsened by a syndrome of chemical, physical, and biological changes that accompany increases in biomass.

The primary enriching-polluting nutrient in freshwater aquacultural environments is phosphorus. Phytoplankton biomass in fish culture environments is usually proportional to the amount of phosphorus, expressed as phosphate (P_2O_5), entering the environment. The quantities of phosphorus in pelleted, including extruded, fish feeds and resulting metabolized feed wastes are about 12 and 7 kg (= 16 kg P_2O_5) per ton of feed, respectively (Table 9.4). Therefore, the threshold for the amount of fish production in

TABLE 9.4. Useful information for assessing environmental impact (eutrophication) of P and N from metabolized feed wastes.

Component	Equivalent
1. Fish dry weight	= 25 percent wet weight
2. Phosphorous (P)	= 0.2 percent of feed (varies from about 0.8 to 2.2 percent)
	= 12 kg ton^{-1} of feed
	= 4.1 percent of fish dry weight
3. At FCR = 2.0 (2.0 kg feed to produce 1.0 kg net wet weight gain of fish)	(a) 1.0 ton feed produces 500 kg fish wet weight
	= 125 kg fish dry weight
	(b) fish assimilate 5 kg P (125 kg fish dry weight × 4.1 percent P)
	(c) therefore, metabolic P waste = 7 kg (12 kg P per ton feed – 5 kg P assimilated)
	(d) 7 kg P × 2.29 (1.0 kg P = 2.29 kg P_2O_5) = 16 kg P_2O_5 waste per ton feed
4. At optimum fertilization in ponds, P_2O_5 = 2.8 kg/week; therefore, 1 ton feed results in enough P_2O_5 (16 kg) waste to optimally fertilize a 1 ha pond for about 6 weeks.	
5. Nitrogen (N)	= 5.5 percent of feed (varies from about 4.5 to 7.0 percent)
	= 55 kg per ton
	= 11.2 percent of fish dry weight
6. At above conditions, the 55 kg in 1 ton feed	(a) is assimilated in fish at 14 kg (125 kg dry fish × 11.2 percent N)
	(b) is released as waste into the environment at 41 kg per ton (55 kg total N–14 kg N in fish)
	(c) results in enough waste N to optimally fertilize a 1 ha pond at 2.8 kg N per week for approximately 15 weeks

cages in a given environment can be generally determined based on the amount of pelleted feed required to produce the fish.

Standard maximum "safe," or sustainable, feeding amounts for Nile or blue tilapia in ponds without aeration mixing is about 45 kg feed per ha per day. This quantity of feed would result in a daily release of about 0.81 kg of P_2O_5 waste per ha per day into the environment causing it to become eutrophic.

A pond in this condition would become very rich in phytoplankton (water visibility index of about 15 cm), thermally and chemically stratified at about 120 cm, anaerobic in the hypolimnion, and DO supersaturated to near 130 percent in the epilimnion in the afternoon and critically unsaturated to as low as about 25 percent at dawn. These would be dangerous conditions for tilapia in a pond, but would be worse for fish in open-water lakes and reservoirs. The conditions would likely be lethal to caged tilapia in these environments.

Similar eutrophic conditions in a natural, open-water environment would greatly reduce native species diversity, destroy species balance, and destabilize the physical, chemical, and biological relationships of the ecosystem. These changes would not be permanent if feeding were discontinued because the original, or very similar, ecosystem would reestablish itself relatively soon after feeding was stopped. However, in the eutrophic state, most human uses of the water would not be greatly affected. Some recreational uses, especially swimming, would be less desirable. The water would also be less desirable for municipal purposes because of the dense phytoplankton. Capture fishing, on the other hand, would likely be measurably improved although fish community composition might change.

At a rate of 15 kg feed (0.27 kg P_2O_5 waste) per ha of natural open waters per day, the waters would be visibly changed, but the amount would probably be an acceptable threshold from every point of view. However, since there is usually no pressing need to take unnecessary risks, a reasonable threshold of 8 kg feed (0.14 kg P_2O_5 waste) per ha per day could be set as a safe, sustainable level at which ecological impact measurements would be evaluated before the threshold could be raised. The 8 kg feed per ha per day limit would be an average for the collective sum of feeds used over a multiple hectare area and not an actual amount for each specific hectare.

In addition to feeding limits per hectare based on the total area of the water environment, limits for designated specific cage-culture areas within the total environment may be imposed. Standing fish biomass of 20 and 60 mt in 1 and 10 ha designated areas, respectively, of large oligotrophic or mesotrophic reservoirs are the maximum standing crops recommended. These correspond to feed limits of 500 kg per day for a 1 ha designated area and 1,500 kg per day for a 10 ha designated area within the total environment. These numbers would need to be adjusted downward for more eutrophic environments, i.e., environments with visibilities less than about 80 cm. Cage fish culture should not be practiced in environments where visibilities would drop below 30 cm during the culture period or where DO level drops below 50 percent saturation in the top 70 percent of the water column.

BIBLIOGRAPHY

Beveridge, Malcolm C.M. (1996). *Cage aquaculture*. Cambridge, MA: Fishing News Books. Blackwell Scientific Publications, Inc.

Boyd, C.E. (1990). *Water quality in ponds for aquaculture*. Auburn, AL: Alabama Agricultural Experiment Station.

Boyd, C.E. (1995). *Bottom soils, sediment and pond aquacultures*. New York: Chapman & Hall.

Plumb, J.A. (1999). *Health maintenance and principal microbial diseases of cultured fishes*. Ames, IA: Iowa State University Press.

Schmittou, H.R., M.C. Cremer, and Jian Zhang. (1997). *Principles and practices of high density fish culture in low volume cages*. St. Louis, MO: American Soybean Association.

Chapter 10

Farming Tilapia in Saline Waters

Wade O. Watanabe
Kevin Fitzsimmons
Yang Yi

INTRODUCTION

Although tilapia culture has been limited primarily to freshwater and low-salinity brackishwater, a high degree of salt tolerance exhibited by certain species has suggested that they might be cultured in high-salinity brackishwater and marine systems, enabling their exploitation in tropical and arid coastal areas (Kuo and Neal 1982; Payne 1983; Hopkins et al. 1989; Watanabe, Burnett, et al. 1989; Watanabe 1991; Suresh and Kwei-Lin 1992; Watanabe et al. 1997). In many areas, limited fresh water supply is an important constraint to further expansion of the industry, which will therefore have to turn to mariculture. To date, the most comprehensive research on saltwater culture of tilapia has been conducted with the Florida red tilapia. The objectives of this chapter are to review the biotechnical and socioeconomic data for saltwater culture of the Florida red and other saline-tolerant tilapia, including the areas of hatchery design and management, broodstock husbandry and seedstock (eggs, yolksac fry, and free-swimming fry) production, nursery production of fingerlings, juvenile grow-out in land-based and sea cage systems, disease control, economics, and marketing. Although tilapia are being considered for culture in lagoonal systems where salinities under 15 ppt prevail (Legendre et al. 1989), the present

This project was supported by a grant from the National Undersea Research Program, National Oceanic and Atmospheric Administration, U.S. Department of Commerce, by the Perry Foundation, Inc., and by the University of North Carolina Wilmington.

doi:10.1300/5513_10

review is restricted to high-salinity culture systems of ≥15 ppt, conditions tolerated by relatively few species of commercial importance.

HISTORY OF SALTWATER TILAPIA CULTURE

Pioneering studies made in Hawaii in the late 1950s were associated with efforts to develop intensive tank culture methods for Mozambique tilapia *(Oreochromis mossambicus)* as a baitfish for the skipjack tuna industry (Hida et al. 1962; Uchida and King 1962). Results of small-scale experiments indicated that reproduction and growth were enhanced at elevated salinities (10 to 15 ppt), suggesting that a commercial facility should be operated on a brackishwater system (Uchida and King 1962). In Israel, small-scale experiments in aquaria, concrete tanks, and ponds were concurrently being initiated to study the adaptability of some commercial tilapia (e.g., blue tilapia *O. aureus* and redbelly tilapia *Tilapia zillii*) to various concentrations of seawater (see review by Chervinski 1982). These preliminary studies indicated that these tilapia could be acclimated and grown in brackishwater or seawater, but suggested that additional studies were needed to accurately assess their culture potential.

Experiments on a larger scale employing manmade ponds were also conducted in south Israel near the Dead Sea to determine the feasibility of utilizing salt marshes and saline waters of desert areas for fish culture (Fishelson and Popper 1968; Loya and Fishelson 1969). Results of these studies showed that hybrids of blue tilapia and Nile tilapia *O. niloticus* could be cultured in low-salinity brackishwater (3.6 to 14.5 ppt). The possibility for using saline-tolerant tilapia for culture in heated effluents of coastal power plants was later discussed (Kirk 1972).

Following the early studies, little additional information on saltwater tilapia culture was available for a number of years. However, saltwater tilapia culture regained considerable interest since the mid-1980s, perhaps due to the discovery of superior strains and species for culture and a greater awareness of their potential for widespread culture in arid and coastal areas (Kuo and Neal 1982; Hopkins 1983; Liao and Chen 1983; Payne 1983; Watanabe et al. 1984, 1985c; Stickney 1986; Hopkins et al. 1989; Watanabe, Wicklund et al. 1989; Watanabe 1991; Suresh and Kwei-Lin 1992; Watanabe et al. 1997).

During the 1990s, commercial saltwater tilapia culture in conjunction with marine shrimp production was initiated in the Caribbean (Head et al. 1996), Central America (Fitzsimmons 2000), and Thailand. In the Philippines, there is current interest in developing fast-growing, saline-tolerant strains for culture in brackishwater ponds and marine cages using semi-intensive methods (Romana-Eguia and Eguia 1999). In the Middle East, Florida red tilapia are being tested in saltwater tank and pond culture in Egypt (Maryut Fish Farming Company), whereas *O. spilurus* is being grown in cages and seapens in Saudi Arabia, both in monoculture and in polyculture with rabbit fish (*Siganus rivulatus;* S.A. Al-Thobaity, Fish Farming Center, personal communication). Tilapia, especially Mozambique tilapia, also remain the subject of basic research on mechanisms of euryhalinity in teleosts.

METHODS FOR MEASURING SALINITY TOLERANCE IN TILAPIA

There are numerous reports in the literature that describe the salinity tolerances of various species of tilapia, including descriptive accounts on ranges of salinities over which a species grows and reproduces naturally; small-scale experiments in brackishwater aquaria, concrete tanks, and earthen ponds; or production trials in seawater ponds; and on survival studies performed under laboratory conditions (see reviews by Chervinski 1982; Philippart and Ruwet 1982). Interpretation of these results, which represent a combination of descriptive and experimental evidence, is difficult. Descriptive data on natural distributions are often ambiguous since the distribution of a species depends upon many interacting factors (e.g., temperature, depth, oxygen, current velocity) including salinity. Interspecific comparisons are often confounded by the fact that the salinity limits for survival, reproduction, and growth may be quite different in a given species and that these limits have been defined in only a few species.

Salinity Tolerance Tests

The development of strains or hybrids suitable for culture under saline conditions requires a reliable index for routine monitoring of improved tolerance in experimental stocks. Provided that the assay conditions employed by an investigator are defined, the use of salinity tolerance tests in which survival is monitored at predetermined salinities under laboratory conditions provides a basis for accurate interspecific comparisons of tolerance limits (Watanabe et al. 1985c; Villegas 1990; Watanabe, Clark, et al. 1990b).

Several indices have been employed as practical measures of salinity tolerance in tilapia:

1. *Median lethal salinity-96 h (MLS-96),* defined as the salinity at which survival falls to 50 percent, 96 h following direct transfer from freshwater to varying salinities;

2. *Mean survival time (MST),* defined as the mean survival time over a 96-h period, following direct transfer from freshwater to seawater (32 ppt);
3. *Median survival time (ST-50),* defined as the time at which survival falls to 50 percent following direct transfer from freshwater to seawater.

These experiments are conducted in small aquaria under controlled environment conditions.

MSL-96

Table 10.1 compares MLS-96 values for different species of tilapia determined in several independent studies. *O. niloticus, O. aureus, O. mossambicus,* and *O. mossambicus* × *O. niloticus* (M × N) or N × M hybrids showed no significant age-specific differences in salinity tolerance between

TABLE 10.1. Median 96-h lethal salinity (MLS-96) of low-salinity (0-5 ppt) spawned and reared *O. niloticus* (N), *O. aureus* (A), and *O. mossambicus* (female) × *O. niloticus* (male) (M × N), *O. niloticus* (female) × *O. mossambicus* (male) (N × M) hybrids, and in Florida red tilapia (FRT) of different ages (days posthatching).

	MLS-96 (ppt)							
Age (dph)	**N (1)**	**N (2)**	**A (1)**	**M (2)**	**M × N (1)**	**M × N (2)**	**N × M (2)**	**FRT (3)**
7-10	19.7	nd	18.5	nd	21.5	nd	nd	25.2
15-20	17.0	19.5[a]	18.3	25.4[a]	24.0	23.0[a]	23.2[a]	nd
25-30	17.3	19.5[a]	19.0	25.4[a]	17.2	23.0[a]	23.2[a]	25.8
40-45	18.5	19.5[a]	20.3	25.4[a]	25.2	23.0[a]	23.2[a]	30.2
55-60	19.3	19.5[a]	19.7	25.4[a]	26.7	23.0[a]	23.2[a]	30.6
70-90	18.1	19.5[a]	18.5	25.4[a]	nd	23.2[a]	23.2[a]	>32
120	20.2	nd	20.0	nd	nd	nd	nd	nd
Overall	18.9	19.5	19.2	25.4	22.9	23.2	23.0	b

Note: Numbers in parentheses indicate sources: (1) Watanabe et al. 1985c; (2) Villegas 1990; and (3) Watanabe, Ellingson et al. 1990.

[a]Value determined from generalized survivorship curves for combined ages of 15-90 days posthatching.

[b]Age-specific differences in MLS-96 were evident.

nd = No data available.

7 and 120 days posthatching (dph). However, species-specific differences were evident. *Oreochromis niloticus* showed the lowest tolerance level, with overall MLS-96 values of 18.9 to 19.5 ppt (Watanabe et al. 1985c; Villegas 1990). These values were also very similar to that of *O. aureus* (19.2 ppt) (Watanabe et al. 1985a). In contrast, MLS-96 for *O. mossambicus* (25.4 ppt) was substantially higher than that for *O. niloticus* or *O. aureus* (Villegas 1990).

MLS-96 values for F1 hybrids of M × N and N × M (22.9 to 23.2 ppt) were intermediate to that for either parental species (Table 10.1). This indicates that salinity tolerance in *O. niloticus* was improved through hybridization with *O. mossambicus*, a more salt-tolerant species (Table 10.1). No studies to date have evaluated the salinity tolerance of hybrids produced by backcrossing F1 hybrids of M × N with *O. mossambicus* or *O. niloticus*.

In contrast to *O. niloticus, O. aureus, O. mossambicus,* M × N and N × M hybrids, Florida red tilapia displayed distinct age-specific differences in salinity tolerance, with MLS-96 increasing from 25.2 ppt at 7 to 10 dph to > 32.0 ppt by 70 to 90 dph (Watanabe, Ellingson et al. 1990). This suggests that the Florida red tilapia has higher salinity tolerance than *O. mossambicus.* A higher salinity tolerance of the Florida red tilapia could be attributed to the higher salinity (5 ppt) in which parental fish were maintained (Watanabe et al. 1985a). However, the clear trend toward increased MLS-96 with age probably reflects an exceptional tolerance of this hybrid strain attributable to genetic heritage (*O. hornorum* × *Oreochromics. mossambicus*). *O. hornorum,* like *O. mossambicus,* is a highly salt-tolerant species (Philippart and Ruwet 1982).

MST

Table 10.2 shows mean survival time (MST) values for the same species of tilapia shown in Table 10.1. In contrast to MLS-96, which did not clearly define age-specific differences in salinity tolerance (except for the Florida red tilapia), the MST index exhibited greater sensitivity in detecting such differences. All of the species showed distinct age-specific differences in salinity tolerance, with MST generally increasing with age.

Oreochromis niloticus and *O. aureus* showed similar age-specific differences in salinity tolerance, MST remaining at low levels (< 100 min) over the initial 45 to 60 dph, then increasing to moderate levels of 131 to 154 after 90 dph. In contrast, in *O. mossambicus* and its hybrids with *O. niloticus,* MST was relatively elevated (138 to 165 min) by 15 to 20 dph and reached much higher levels (225 to 292 min) by 70 to 90 dph. The

TABLE 10.2. Salinity tolerance (mean survival time = MST) of low-salinity (0 to 5 ppt) spawned and reared *O. niloticus* (N), *O. aureus* (A), and *O. mossambicus* (female) × *O. niloticus* (male) (M × N), *O. niloticus* (female) × *O. mossambicus* (male) (N × M) hybrids, and in Florida red tilapia (FRT) of different ages (days posthatching).

	MST (min)							
Age (dph)	N (1)	N (2)	A (1)	M (2)	M × N (1)	M × N (2)	N × M (2)	FRT (3)
7-10	50.2	nd	43.9	nd	52	nd	nd	190
15-20	29.8	52.3	51.8	165	140	156	138	nd
25-30	37.5	86.1	42.3	141 lethal	130	183	188	303
40-45	42.2	88.1	54.2	181 lethal	155	176	204	2,108
55-60	84.5	88.8	63.9	222	215	219	175	1,281
70-90	92.8	131	nd	292	nd	236	225	3,915
120	103	nd	136	nd	nd	nd	nd	nd
150	181	nd	nd	nd	nd	nd	nd	nd
180	154	nd	151	nd	nd	nd	nd	nd
395	148	nd	nd	nd	nd	nd	nd	nd

Note: Numbers in parentheses indicate sources: (1) Watanabe et al. 1985c; (2) Villegas 1990; and (3) Watanabe, Clark et al. 1990b.
nd = No data available.

Florida red tilapia exhibited the highest MST values, starting at 190 min at 7 to 10 dph and reaching very high levels of 3,916 min by 70 to 90 dph.

ST-50

The median survival time index (ST-50) provided a value closely similar to MST (Watanabe et al. 1985c; Villegas 1990) and will not be discussed in detail. A limitation of the ST-50 index is that it can underestimate tolerance when survivorship patterns involve a few relatively tolerant individuals tending to elevate MST.

The MLS-96 and ST-50 indices have limited comparative utility under high tolerance conditions when survival does not fall to 50 percent by 96 h (MLS-96 > 32 ppt; ST-50 > 96 h). The MST index was limited when there was full survival by 96 hours so that MST attained a ceiling value of 5,760 min. However, when MLS-96 was >32 ppt, the MST index still provided a relative tolerance value as long as survival remained less than 100 percent. The MST was generally the most sensitive index, providing relative tolerance values over the widest range of tolerance conditions. The ST-50 index, however, was more expedient and may be more practical for routine monitoring of improved tolerance in experimental stocks (Villegas 1990).

As discussed below, a number of abiotic and biotic factors have marked effects on salinity tolerance, including age/size (Watanabe et al. 1985c; Watanabe, Ellingson et al. 1990; Villegas 1990), early salinity exposure (i.e., through spawning under elevated salinities) (Watanabe et al. 1985a), ration level, nutritional status (Brett et al. 1982; Jurss et al. 1984), and temperature (Allanson et al. 1971; Tilney and Hocutt 1987), which must therefore be standardized when comparing salinity tolerance of tilapia. No studies have assessed the effects of other physical factors, including divalent ion concentrations, light intensity, photoperiod, and water turbulence on salinity tolerance. Potentially interactive effects of such factors should also be considered.

MORPHOLOGICAL AND PHYSIOLOGICAL MEASURES OF SALINITY TOLERANCE/ HYPOOSMOREGULATORY ABILITY

In tilapia, tolerance to salinity has also been assessed from gill structure, plasma sodium levels, and gill Na/K ATPase activity (Avella et al. 1993). Avella et al. noted in *O. niloticus* and *O. aureus* that an increase in external salinity induced the proliferation of chloride cells on gill filaments typical of seawater teleosts and an increase in gill Na/K ATPase. In salt tolerant strains, e.g., *O. mossambicus,* plasma Na did not change, indicating successful ion regulation in hypertonic media. These workers suggested that plasma Na levels (i.e., hypoosmoregulatory ability) and gill chloride cells proliferation upon transfer to isotonic medium should be considered as reliable indicators of salinity tolerance. Less-marked response of gill chloride cells to seawater in *O. niloticus* (Auperin et al. 1995) suggests a lower capacity of this species to modify osmoregulatory mechanisms in response to hypertonic environment.

In tilapia, differences in liver metabolic activity in hypertonic environment may also indicate relative osmoregulatory ability (Nakano et al. 1997). Whereas *O. niloticus* displayed similar growth and plasma ion concentrations in freshwater and in 50 percent seawater, seawater-adapted fish exhibited a pronounced alteration in carbohydrate metabolism. In contrast, *O. mossambicus* showed similar growth, plasma ion concentration, and carbohydrate metabolism in freshwater or seawater. In *O. niloticus,* the activities of key enzymes in carbohydrate metabolism in the liver were all higher in 50 percent seawater than in freshwater, whereas in *O. mossambicus,* no difference was seen among different salinities. Plasma glucose level was also higher in hyperosmotic environment than in freshwater only in

O. niloticus, which may therefore require an alteration in carbohydrate metabolism to produce energy required for osmoregulation. These results suggested a different efficiency in energy utilization for hypoosmoregulation between these species and a higher hypoosmoregulatory ability of *O. mossambicus* (Nakano et al. 1997).

REPRODUCTIVE PERFORMANCE AT VARIOUS SALINITIES

In tilapia, reproduction is generally inhibited by increasing salinity, with limits for reproduction varying among species. Reproductive performance at different salinities is, therefore, an alternative measure of relative salt tolerance. As measured by MST, ST-50, and MLS-96, *O. niloticus* has considerably lower salt tolerance than Florida red tilapia (Tables 10.1 and 10.2). Controlled laboratory studies showed that, while *O. niloticus* spawned at salinities ranging from freshwater to full seawater (32 ppt), fry production decreased above 5 ppt, with no fry produced in full seawater (32 ppt) (Watanabe and Kuo 1985). In contrast, in Florida red tilapia, fry production declined at salinities higher than 18 ppt, with viable fry produced in 36 ppt (Watanabe, Burnett et al. 1989).

To summarize, species-specific and age-related differences in salinity tolerance among tilapia have been clearly distinguished through tolerance indexing, which can be used for routine monitoring of improved tolerance in experimental stocks. Results of tolerance tests can be influenced by age/size, early salinity exposure, feeding and nutritional status, temperature, and other factors. Salinity tolerance and hypoosmoregulatory ability can also be correlated to morphological and physiological measures such as gill structure, plasma Na levels, gill Na/K ATPase, and carbohydrate metabolism. Although more difficult to quantify, relative reproductive performance (i.e., production of viable fry) at different salinities is a reliable indicator of salinity tolerance.

EXPERIMENTAL SALTWATER CULTURE OF FLORIDA RED TILAPIA

Red Hybrid Tilapia

Genetic Heritages of Red Hybrid Tilapia

Whereas the genetic heritages of the existing varieties of red tilapia are not well documented, their derivation is generally attributed to crossbreeding of mutant reddish-orange *O. mossambicus* (a normally black species) with other species, including *O. aureus, O. niloticus,* and *O. hornorum* (Fitzgerald 1979; Behrends et al. 1982; Galman and Avtalion 1983; Kuo 1984). The suitability of red hybrid tilapia for brackishwater and seawater culture is suggested by the salinity tolerance exhibited by these parental species, which are known to be moderately (*O. niloticus* and *O. aureus*) to highly (*O. mossambicus* and *O. hornorum*) euryhaline (Philippart and Ruwet 1982; Payne 1983; Wohlfarth and Hulata 1983; Stickney 1986). Genetic strain variations in growth and other economically important traits, like reproduction in high salinities, fecundity, feed conversion, red color, and the like have been demonstrated in earlier studies on red tilapia.

The feasibility of growing red hybrid tilapia in brackishwater and seawater was first studied by researchers in Taiwan (Liao and Chang 1983) who found good growth of Taiwanese red tilapia (*O. mossambicus* × *O. niloticus*) at salinities of 17 and 37 ppt, although fish appeared susceptible to handling stress. Good growth of Taiwanese red tilapia under intensive cage culture in brackishwater (11 to 17 ppt) shrimp ponds in Hawaii was also reported (Meriwether et al. 1984). Seawater-rearing studies of Taiwanese red tilapia in Kuwait showed that survival at 38 to 41 ppt was impaired at water temperatures below 24°C (Hopkins et al. 1989).

A red hybrid tilapia that originated in Florida (Florida red strain), descendants of an original cross between *O. urolepis hornorum* (female) and *O. mossambicus* (male) (Behrends et al. 1982), was found to have a high euryhaline capacity, as evidenced by faster growth and lower feed conversion ratio of juvenile, monosex males in brackish- and seawater than in freshwater (Watanabe, Ellingson et al. 1988). Detailed studies on saltwater culture methodology were then initiated for this strain (Watanabe et al. 1997).

Origin of the "Florida Red" Tilapia Strain

Red tilapia hybrids originating in the United States were first developed in the late 1970s by a commercial fish breeder (M. Sipe) in Florida (R. DeWandel and C. Harris, personal communication). Inbreeding of a population of *O. mossambicus* produced a mutant with reddish-yellow pigmentation, which was selectively bred to enhance the red and yellow coloration, resulting in a marked decline in growth and body conformation. To restore these qualities, mutant male *O. mossambicus* was crossbred with female *O. hornorum* (a black colored species) to produce a first generation (F1) red hybrid. F1 red hybrids were sold to producers for grow-out. In addition, black *O. hornorum* females and red *O. mossambicus* males were sold as broodstock to farmers, who could produce hybrids for grow-out, but could not generate additional pureline broodstock. In 1981, other commercial culturists began a selective breeding program with the F1 hybrids and, by the end of 1983, developed a relatively true breeding strain of red tilapia. Aquaculturists knew fish descended from F1 progeny as the "Florida red" strain. The Florida red tilapia strain has been introduced to the southeastern United States, the Caribbean, Latin America, Southeast Asia, East Africa, and the Middle East.

Intensive Tank Production of Seedstock

Numerous methods have been used for tilapia fry production including earthen ponds (Broussard et al. 1983; Rothbard et al. 1983; Little et al. 1987; Verdegem and McGinty 1989), tanks and pools (Balarin and Haller 1983; Berrios-Hernandez and Snow 1983; Snow et al. 1983; Guerrero and Guerrero 1985; Al-Ahmad et al. 1988b; Bautista et al. 1988; Ridha and Cruz 1989), hapas (net enclosures) (Hughes and Behrends 1983; Verdegem and McGinty 1987; Bautista et al. 1988; Costa-Pierce and Hadikusumah 1995), cages (Guerrero 1977; Guerrero and Garcia 1983), and recirculating tank systems (Uchida and King 1962; Ernst 1989; Watanabe et al. 1992). When freshwater and/or land resources are limited, intensive brackishwater production of tilapia seedstock (i.e., eggs, sacfry, and post-yolksac stage fry) in tanks may be advantageous since seed production per unit surface area is maximized and reliance on freshwater is minimized (Ernst et al. 1991).

Watanabe et al. (1997) described in detail the design and operation of a hatchery for intensive tank production of Florida red tilapia. The hatchery uses the "intensive tank method" of seedstock production, including high-density stocking of brood fish, clutch removal and artificial incubation of eggs, and high-density culture of fry. These workers used 34.2 m^3 rectangular broodfish tanks. Each broodfish spawning tank is stocked with 240 yearling breeders (6 fish per square meter) at a skewed sex ratio of 3:1 (females:males) (Watanabe et al. 1992). Fish are fed twice daily commercially prepared extruded diets containing 30 percent (Zeigler Bros. Inc., Pennsylvania) to 32 percent protein (Purina Mills Inc., Missouri) at a rate of approximately 1 to 3 percent body weight per day (Ernst et al. 1991).

Egg Hatching and Fry Culture Systems

Eggs are placed in 6.5 L upwelling incubators (Midland Plastics, Wisconsin) supplied by 12 ppt brackishwater that is continuously recirculated. Each incubator is stocked with up to 12,000 eggs (1,846 L^{-1}) (Watanabe et al.

1992). After hatching, yolksac fry swim up and out of the incubators into 126-L polyethylene troughs.

Fry are grown in 560 L culture tanks that have a 30°C cone-shaped bottom and a center drain. Each fry tank is supplied with aeration through three silica-glass air diffusers. Each tank is stocked with post-yolksac stage fry so that a biomass of 2.15 kg (3.84 $g \cdot L^{-1}$) is not exceeded by the end of a 40-day culture period (Watanabe, Smith et al. 1993).

Broodstock Management

The Effect of Salinity on Reproduction

When freshwater is limited, the production of seedstock in brackishwater or seawater reduces freshwater requirements for maintaining broodstock and for early rearing of fry (Uchida and King 1962; Watanabe and Kuo 1985). Although a number of species can tolerate full-strength seawater, relatively few are capable of reproduction in full-strength seawater. Among these are the Mozambique tilapia, which has been reported to reproduce in ponds under salinities as high as 50 ppt (Popper and Lichatowich 1975), and *O. hornorum,* which is capable of reproducing in full-strength seawater (Talbot and Newell 1957 in Wohlfarth and Hulata 1983). At the Seaphire Seawater Experimental Farm in Masawa, Eritrea, Florida red tilapia have been reared for three generations in seawater at 40 ppt. However, fry survival has been less than 10 percent. Controlled laboratory studies (Watanabe, Burnett et al. 1989) have shown that Florida red tilapia are capable of reproduction in full-strength seawater (36 ppt), but maximum production of seedstock requires water of lower salinities, with fertilization and hatching success and survival of prejuveniles declining markedly at salinities higher than 18 ppt. This suggests that Florida red tilapia broodstock may be maintained under salinities as high as 18 ppt without impairing fry production, suggesting further that a hatchery facility using brackishwater would be practical in areas where freshwater resources are limiting.

Few studies have compared reproductive performance of tilapia species at different salinities under controlled conditions. In Florida red tilapia, a marked decline in fertilization and hatching success at salinities of 27 and 36 ppt suggested that high salinities compromise gamete quality, either indirectly through osmotic stress on parental fish during gametogenesis or directly as gametes enter the external medium. In Taiwanese red tilapia, breeding behavior is also inhibited in brackishwater and seawater (Liao and Chang 1983).

Fry production, a practical measure of reproductive success, is a function of egg production, fertilization and hatching success, and prejuvenile survival. Fry production in full-strength seawater is impaired in all species studied to date. In Florida red tilapia, although egg production per unit female weight did not differ significantly under salinities of 9 to 36 ppt, relatively poor fertilization, hatching, and prejuvenile survival at 27 and 36 ppt resulted in a significantly lowered fry production per unit weight at these salinities, which was 30.5 percent of production in freshwater (Watanabe, Burnett et al. 1989). Clearly, egg production per unit weight alone is not a reliable measure of reproductive success. Fry production, a function of egg production, fertilization and hatching success, and prejuvenile survival is the best practical measure of reproductive success.

Watanabe, Burnett et al. (1989) noted that fertilization, hatching success, and survival of yolksac fry were higher at 9 ppt than at 1 ppt, suggesting that reproductive performance of the Florida red tilapia is optimized at salinities near isoosmotic (i.e., 12 ppt) (Febry and Lutz 1987). Spawning under salinities near isoosmotic is therefore beneficial, even when freshwater is not limiting. Uchida and King (1962) reported that *O. mossambicus* fry production was three times higher at salinities of 8.9 to 15.2 ppt than in freshwater and suggested that a commercial hatchery use brackishwater to reduce the high cost of freshwater. In Florida red tilapia, maintenance of broodfish in 12 ppt brackishwater optimized seed production, conserved low-salinity groundwater, and improved salinity tolerance of fry (Watanabe et al. 1984; 1985b; Watanabe, French et al. 1989).

A capacity for reproduction of the Florida red tilapia in full-strength seawater can be attributed to the high-salinity tolerance of the parental species, which originate from estuaries and lagoons along the east coast of Africa (Philippart and Ruwet 1982). The ability of *O. mossambicus* to reproduce in seawater is well known (Brock 1954; Chervinski 1961; Popper and Lichatowich 1975; Fishelson 1980; Stickney 1986). *Oreochromis hornorum* is also known to live and reproduce at salinities higher than 30 ppt (Philippart and Ruwet 1982). Ridha et al. (1985) reported that red tilapia originating from Taiwan did not reproduce in full seawater (37 to 40 ppt). This possibly reflects the different genetic heritages of the Taiwanese strains, which, in contrast to the Florida red strain, contain an *O. niloticus* component (Galman and Avtalion 1983), a species that does not reproduce in full-strength seawater (32 ppt) (Watanabe and Kuo 1985). Although fry production of Florida red tilapia in seawater was relatively low, high survival to yolksac absorption of broods from certain females suggests genotypic differences among individuals and the possibility of developing strains with improved capacity for reproduction in seawater through genetic selection (Watanabe, Burnett et al. 1989).

Collection of Seedstock: "Natural-Mouthbrooding" versus "Clutch-Removal" Methods

Seed (eggs, yolksac fry, and post-yolksac stage fry) production of Florida red tilapia broodstock was compared at 12 ppt using the natural-mouthbrooding and clutch-removal methods (Watanabe et al. 1992). Under the natural-mouthbrooding method, free-swimming (i.e., post-yolksac stage) fry were collected from each broodfish tank at 8- to 16-day intervals using a seine and dip net after crowding broodfish to one end of a tank. Under the clutch-removal method, seed was collected from broodfish tanks at 15- to 16-day intervals by crowding broodfish to one end of a tank and transferring fish into a floating container, where clutches (i.e., eggs and yolksac fry) were removed from mouthbrooding females for artificial incubation. Free-swimming fry were collected from the tank using a seine and dip net.

Seed production over a 3-month period under the natural-mouthbrooding method (3.3 seed per square meter per day) was markedly lower than under the clutch-removal method (91.7 seed per square meter per day), which is among the highest reported for tilapia hatcheries. Poor seed production under the natural-mouthbrooding method was attributable to cannibalism of eggs and fry by adults. Given an average hatching success of 64.8 percent obtained during artificial incubation of eggs (Watanabe et al. 1992), production of fry through the free-swimming stage under the clutch-removal method ($67.9 \cdot m^{2} \cdot day^{-1}$) was still substantially higher than that obtained under natural mouthbrooding ($3.3 \cdot m^{2} \cdot day^{-1}$).

Although of similar ages, post-yolksac stage fry obtained from artificial incubation of eggs and yolksac fry were larger and more robust than naturally incubated fry, exhibiting higher survival (73.9 versus 49.7 percent) over a 29-day culture period (Watanabe et al. 1992). This is partly attributable to the availability of prepared feeds to artificially incubated fry at first-feeding, whereas naturally incubated fry are dependent upon foods available in broodfish tanks (Rana 1986). In *Oreochromis* spp., a delay in initial feeding depressed growth rate of swim-up fry, reducing viability (Macintosh and De Silva 1984; Rana 1988). Early availability of prepared feeds may have also improved survival among artificially incubated fry by reducing cannibalism. Greater size-age uniformity among artificially incubated fry may have also minimized aggressive interactions and cannibalism (Ellis et al. 1993). These results, together with those showing higher rates of seed production under the clutch removal method (Watanabe et al. 1992), indicate that clutch-removal and artificial incubation of eggs and yolksac fry, while more labor intensive, yield markedly higher numbers of viable fry for growout.

Although poor fry production under natural mouthbrooding in tanks as practiced in this study would suggest the method to be inefficient, other workers have reported relatively good success with this method in tank culture. For example, Uchida and King (1962) obtained a fry production rate of $104 \cdot m^2 \cdot day^{-1}$ for *O. mossambicus* broodstock held in 11 to 13 ppt brackishwater in 4.48-m^2 plywood tanks, dip-netting schools of free-swimming fry daily. Berrios-Hernandez and Snow (1983) obtained similar rates of seed production under the clutch-removal (22.6 seed per square meter per day) or natural-mouthbrooding (21.7 seed per square meter per day) methods with *O. aureus* broodstock held in freshwater in 7.3-m^2 plastic pools, and they concluded that artificial incubation of eggs was not economically justifiable. High seed losses to cannibalism under the natural-mouthbrooding method observed by Watanabe et al. (1992) may have been related to a species-specific behavioral difference or to other conditions of broodstock management such as stocking density, feeding rate, frequency of collection, or broodtank design. For example, in crowded tanks, nonbreeding fish have been observed to eat young fry or feed on eggs as they are spawned (Balarin and Haller 1982). On the other hand, Pierce (1980) reported a dramatic reduction in cannibalism of *O. mossambicus* × *O. aureus* hybrid fry by parental fish when feeding rate was increased. The vertical walls and lack of cover in broodfish tanks, along with the 8- to 16-day collection intervals, may have made fry vulnerable to predation by adults (Berrios-Hernandez and Snow 1983).

Artificial Incubation of Eggs and Yolksac Fry

Following clutch removal, eggs and nonswimming yolksac fry, comprising 73.6 percent of total seed at each collection (Watanabe et al. 1992), are incubated in 6.5 L upwelling incubators supplied with recirculated 12 ppt brackishwater. Water flow rates averaging 2.73, 3.13, 3.86, and 5.27 $L \cdot min^{-1}$ are required for incubators stocked with 3,000, 6,000, 9,000, and 12,000 eggs (462, 923, 1385, and 1,846 eggs per liter), respectively. Within this range, survival of eggs and yolksac fry to the post-yolksac (i.e., free-swimming) stage averages 64.8 percent and does not differ significantly among densities (Watanabe et al. 1992). Thus, only four to five incubators (9,000 to 12,000 eggs per incubator) are required to incubate an average of 45,472 eggs produced by each broodfish tank every 16 days. Space and labor requirements for artificial incubation of eggs are therefore relatively small (Rana 1986; Head and Watanabe 1995).

Fry Culture: Survival and Growth of Prejuveniles

Post-yolksac stage fry, obtained from artificial incubation of eggs or collected from broodfish tanks, are stocked in 530 L culture tanks containing 12 ppt brackishwater and are fed diets containing the androgenic hormone 17 ∝-ethynyltestosterone (17-ET) for 28 days to transform genotypic females to phenotypic males (Guerrero 1975). Fry are fed a 50 percent protein flake diet for 7 days from first feeding, then weaned onto a 48 percent protein trout starter (Zeigler Bros., Inc., Pennsylvania). Feed is administered by hand four to five times daily. Sex reversal prevents unwanted reproduction at an early age, which results in overcrowding and stunting during growout.

Treatment of sexually undifferentiated fry with 30 to 60 $mg \cdot kg^{-1}$ diet ethynyltestosterone (ET) or methyltestosterone (MT) for a minimum of 21 to 28 days is considered necessary for producing all-male populations in most tilapia species (Phelps and Popma 2000). In Florida red tilapia, the minimum effective duration of treatment with ET for sex reversal is only 7 to 14 days (Watanabe, Mueller et al. 1993). This reduces hormone requirements for sex reversal to only 1.5 percent of the 28-day total, increases hatchery productivity by accelerating output, and allows transport to growout facilities at much smaller sizes to lower production/shipping costs. This improves cost and availability of sex-reversed fingerlings to farmers (Head and Watanabe 1995).

Considerable mortality and variable productivity of fry from the post-yolksac through juvenile stages may occur under high stocking densities used in tank culture (Al-Ahmad et al. 1988a; Hanley 1991; Watanabe et al. 1992; Watanabe, Smith et al. 1993; Ellis and Watanabe 1993a). Cannibalism is suspected as being a major cause of fry mortality in tilapia hatcheries (Uchida and King 1962; Macintosh and De Silva 1984; Pantastico et al. 1988; Watanabe, Smith et al. 1993; Ellis and Watanabe 1993a). High infrastructural and labor requirements of intensive tank culture necessitate that survival and growth are optimized in order to be cost effective (Head and Watanabe 1995).

Stocking Density

Stocking densities in fry culture tanks are often not quantified in tilapia hatcheries, and many culturists "overstock" to compensate for cannibalism (Macintosh and De Silva 1984), a practice that is wasteful of broodstock, facilities, feed, labor, and freshwater. Post-yolksac stage Florida red tilapia fry (avg. wt. = 0.010 g) stocked in 530-L culture tanks supplied with recirculated brackishwater (11 to 12 ppt) at densities of 5.7 to 13.2 L^{-1} (5,700-13,200 m^{-3}) and fed an ET-treated diet showed no significant differences in final weights (avg. wt. = 0.478 g) and survival (avg. wt. = 58.0 percent) after 30 days, suggesting that higher densities are feasible to increase production (Watanabe, Smith et al. 1993). Final biomass increased with stocking density from 1.25 to 3.84 $g \cdot L^{-1}$ ($kg \cdot m^{-3}$). Average survival (58 percent) of Florida red tilapia under these stocking densities (Watanabe, Smith et al. 1993) was comparable to that reported for *O. spilurus* (46 to 68 percent) reared at densities of 2.0 to 8.0 L^{-1} (2,000 to 8,000 m^{-3}) in 0.5 m^3 low-salinity (3 to 4 $g \cdot L^{-1}$) tanks (Al-Ahmad et al. 1988b). Higher survival rates have been reported for *O. aureus* (93.5 to 96.8 percent) (Snow et al. 1983), Taiwanese red tilapia (95 percent) (Liao and Chen 1983), and *O. niloticus* (78.1 percent) (Guerrero and Guerrero 1988) in freshwater tanks (7.3 to 32.5 m^3), but much lower stocking densities of 0.062 to 1.54 L^{-1} (62 to 1538 m^{-3}) were used.

Acclimation to Seawater: Optimum Age/Size

Age (or size) at time of seawater transfer influences survival of tilapia as has been demonstrated for several species (*O. mossambicus, O. niloticus, O. aureus*, and hybrid *O. mossambicus* × *O. niloticus*), with newly hatched fry being less tolerant than older individuals (Watanabe et al. 1985a,c; Perschbacher and McGeachin 1988; Villegas 1990). Optimal transfer age or size is therefore critical to the culturist to minimize freshwater requirements and to maximize survival and growth.

Studies have shown that none of the commercially important tilapia species tolerate direct transfer from freshwater (0 to 2 ppt) to full-strength seawater (≥32 ppt) (Watanabe et al. 1985a,c; Al-Amoudi 1987a; Villegas 1990; Perschbacher and McGeachin 1988). Fry and juvenile stages of Florida red tilapia survive direct transfer from 1.5 to 2.0 ppt to 19 ppt without apparent stress (Perschbacher and McGeachin 1988). This is similar to what was reported for *O. mossambicus* and hybrids with *O. niloticus,* where maximum salinities tolerated following direct transfer from freshwater were 15 to 20 ppt (Villegas 1990), 20 ppt (Dange 1985), 25.2 ppt (Al-Amoudi 1987a), and 18 to 23 ppt (Watanabe et al. 1985a,c).

When salinity tolerance was compared in Florida red tilapia fry, spawned at 5 ppt, at 10, 25, 40, 55, and 70 days posthatching, a trend toward increased tolerance with age (size) was observed (Watanabe, Ellingson et al. 1990). Mean survival time following abrupt transfer to 32 ppt increased from 190 min at 10 days posthatching (11.5 mm TL), to 3,915 min at 70 days posthatching (66.6 mm TL) (Table 10.2), and 96 hour median lethal salinity (MLS-96 h) increased from 24.8 ppt at 10 days to >32 ppt at 70 days posthatching (Table 10.1). Salinity tolerance remained low until 40 days posthatching (38.7 mm TL) when a marked increased was noted (Watanabe, Ellingson, et al. 1990). This is similar to what was reported in *O. niloticus*, *O. aureus* (Watanabe et al. 1985a,c), *O. mossambicus,* and *O. mossambicus* × *O. niloticus* hybrids (Villegas 1990), where salinity tolerance began to increase at 45 to 60 days posthatching (Tables 10.1 and 10.2).

Salinity tolerance varies ontogenetically in tilapia. When survival was compared for Florida red tilapia gradually (5 ppt·day^{-1}) acclimated to seawater (37 ppt) over 7 days beginning at different ages (11, 25, or 39 days posthatching), survival to 48 days posthatching improved from 20.0 to 55.9 percent as acclimation to seawater was delayed from 11 to 39 days posthatching (Watanabe, Ellingson et al. 1990). This demonstrated that premature acclimation to seawater impairs survival in this fish and that selection of proper transfer time, based on the knowledge of ontogenetic variation in salinity tolerance, improves survival.

For fry acclimated to seawater beginning 39 days posthatching, few mortalities were induced by acclimation, indicating that a level of tolerance was reached under which survival in seawater was not appreciably impaired (Watanabe, Ellingson et al. 1990). Thus, 39 days posthatching (38.7 mm TL) appears to be near a minimum age (size) to begin successful seawater acclimation, even though higher tolerance levels are reached in later development. This is important as costs associated with freshwater rearing beyond the fry stages may be prohibitive when freshwater is in scarce supply. Based on improving salinity tolerances with age, Villegas (1990) suggested that the optimum age (size) for transfer of *O. mossambicus* and F_1 hybrids with *O. niloticus* to full seawater (32 ppt) was 45 to 60 days posthatching (17.0 to 23.2 mm SL). In Florida red tilapia, tolerance improves markedly around 40 days posthatching, which is the recommended age for acclimating fry to full-strength seawater.

Gradual (5 ppt·day^{-1}) stepwise acclimation of Florida red tilapia from freshwater to seawater results in high survival, but single-step acclimation, where preacclimation at one intermediate salinity precedes direct transfer to seawater, may be more cost effective. The required salinty for preacclimation from freshwater may be 19 ppt, but additional studies are needed.

In *O. mossambicus,* a second hemoglobin appears at 47 days posthatching that has a higher affinity for O_2 than larval hemoglobin at higher osmotic pressure and temperature and that may enable the adults to tolerate both warmer and more saline environments (Perez and Maclean 1976). In Florida red tilapia, the marked improvement in salinity tolerance at 40 days posthatching may be related to this ontogenetic change, as was suggested for *O. niloticus* and *O. aureus* (Watanabe et al. 1985a,c). Improved salinity tolerance with age may also be related to other maturational events such as the functional development of the hypoosmoregulatory (Clarke 1982) or endocrinological (Helms et al. 1987) systems.

In *O. niloticus,* maximum salinity tolerance was acquired at a body length of about 52 mm (150 days posthatching), and salinity tolerance did not increase with a further increase in size (Watanabe et al. 1985a,c). In Florida red tilapia, maximum salinity tolerance is apparently attained at relatively large sizes, since adults (133.4 mm SL) were found to be more tolerant than juveniles (26.5 mm SL) and fry (6.3 mm SL), although the ages of these groups were not specified (Pershbacher and McGeachin 1988). However, since growth rates may vary widely depending upon culture conditions, interspecific or intraspecific comparisons of optimum/maximum transfer age and size must be made with caution. Due to age/size-related differences in salinity tolerances in tilapia, comparisons of broods of similar ages and sizes provide the most meaningful intra- or interspecific comparisons of salinity tolerance in tilapia.

Acclimation to Seawater: Optimum Rate

A gradual, stepwise increase in salinity until full-strength seawater is reached has been successfully used to acclimate a number of tilapia to seawater. Required acclimation periods range from 4 days (2 days at 18 ppt, 2 days at 27 ppt) *(O. aureus, O. mossambicus,* and *O. spilurus)* to 8 days (4 days at 18 ppt, 4 days at 27 ppt) (*O. niloticus* and *O. niloticus* × *O. aureus* hybrids), depending on the degree of euryhalinity (Al-Amoudi 1987a). For Florida red tilapia, acclimation from freshwater to 36 ppt seawater has been accomplished by increasing salinity at a rate of 5 ppt per day (Watanabe, Ellingson et al. 1988, 1990).

Following direct transfer of *O. mossambicus* from freshwater to 30 ppt seawater, a rapid elevation of plasma osmotic concentration to excessive levels occurs within 1 h, leading to death within 6 h (Hwang et al. 1989). On the other hand, transfer to a lower, sublethal salinity allows osmoregulatory mechanisms to adapt to and gradually reduce the rising plasma osmotic concentration to a new equilibrium level within 44 to 96 h (Assem and Hanke 1979; Al-Amoudi 1987a). This seawater adaptation process (see Prunet and Bornancin 1989 for review) involves the functional activation of the salt-secreting mitochondria-rich (MR) cells (i.e., chloride cells) of the gills (Dharmamba et al. 1975; Fishelson 1980; Foskett and Scheffey 1982; Hwang 1987, 1989; Cioni et al. 1991; Kultz et al. 1992; Uchida et al. 2000), including stimulation of mRNA of Na^+, K^+-ATPase in gills (Dange 1985; Hwang et al. 1989; Avella et al. 1993; Morgan et al. 1997;

Fontainhas-Fernandes et al. 2001), which enables subsequent transfer to higher salinities. Activation of chloride cells occurs within 12 to 24 h after acclimation to 20 ppt seawater (Hwang 1987; Hwang et al. 1989; Fontainhas-Fernandez et al. 2001). Under transmission electron microscopy, leaky junctions and intercellular digitations develop in branchial chloride cells of *O. mossambicus* adapted to seawater, but not in those adapted to freshwater (Hwang 1989), and this seems to be associated with the increase of ion permeability in the gill of teleosts adapted to seawater. Enhanced transcription and translation of the Na^+, K^+-ATPase gene was attributed to larger and more numerous MR cells in gill epithelium resulting in elevated Na^+, K^+-ATPase activity crucial to ion extrusion upon seawater challenge (Lee et al. 2000).

In addition to ion excretion, drinking rates are increased in seawater-adapted larvae. During acute salinity challenges, *O. mossambicus* larvae increased or decreased drinking rate within several hours when faced with hyper- or hypotonic challenges, respectively, to maintain constancy of body fluid (Lin et al. 2001). In *O. mossambicus,* drinking rates of seawater-adapted larvae were 4 to 9-fold higher than those of freshwater-adapted larvae from day 2 to day 5 posthatch (Lin et al. 2001). Increased drinking rate during larval development in seawater compensates for osmotic water loss due to increased surface:volume ratio and/or the increase of water permeability of larvae during development.

Acclimation to Seawater: Influence of Nutritional Status

Nutritional status of fish will likely influence seawater adaptation and salinity tolerance. In *O. mossambicus,* starvation reduces Na^+, K^+-ATPase, and survival during seawater acclimation (Jurss et al. 1984). Food deprivation also decreased the size and number of chloride cells (Kultz and Jurss 1991) and increased plasma Cl concentration relative to plasma Na concentration, suggesting that food deprivation directly or indirectly affects trans-epithelia branchial Cl transport mechanisms (Vijayan et al. 1996).

Endocrinology of Seawater Adaptation

The temporary elevation in plasma osmolality and Na and Cl ion concentrations in *O. mossambicus* during seawater adaptation (Assem and Hanke 1979; Hwang et al. 1989) is accompanied by a transient rise in plasma cortisol and growth hormone (GH) levels (Assem and Hanke 1979; Yada et al. 1994), whereas plasma prolactin levels ($tPRL_{177}$ and $tPRL_{188}$), which have a sodium retention effect in freshwater, decrease (Ayson et al. 1993;

Yada et al. 1994). These biochemical and hormonal responses result in a net efflux of Na^+ and Cl^- to maintain ionic balance in a hyperosmotic environment (Morgan et al. 1997).

There is evidence that GH plays a role during seawater acclimation in tilapia. In *O. mossambicus,* pituitary GH cell activity was greater in seawater-acclimated juveniles when compared to freshwater-acclimated juveniles (Borski et al. 1994), and plasma GH increased after 4 and 14 days in seawater (Morgan et al. 1997; Vijayan et al. 1996). The elevated activity of GH in seawater tilapia pituitaries and the ability of exogenous GH to stimulate Na^+, K^+-ATPase activity suggest that GH may be an important factor in seawater osmoregulation as well as in promoting greater growth of seawater tilapia (Borski et al. 1994). Furthermore, treatment of tilapia with GH increased chloride cell concentration in the opercular membrane, stimulated gill Na^+, K^+-ATPase, and increased salinity tolerance (Borski et al 1994; Sakamoto et al. 1997; Shepherd, Ron, et al. 1997; Shepherd, Sakamoto, et al. 1997) after transfer to seawater.

Acclimation to Seawater: Preacclimation Methods

Preacclimation has a crucial physiological significance for some euryhaline teleosts during seawater adaptation. The physiological processes of seawater adaptation in tilapia indicate that acclimation time (and hence, labor and facilities) can be minimized through single-step acclimation, where preacclimation at one intermediate salinity preceds direct transfer to seawater. This was shown possible in *O. mossambicus,* which can be successfully transferred to 36 ppt seawater following preacclimation to 12 ppt for 48 h (Foskett et al. 1981), or to 20 ppt for 24 h (Hwang 1987). Hwang et al. (1989) reported that tilapia *O. mossambicus,* when transferred to 30 percent seawater preacclimation to lower salinity (20 ppt) for 24 h, exhibited more rapid increase in gill Na^+, K^+-ATPase activity and less dehydration than fish transferred directly to 20 or 30 ppt seawater. *Oreochromis spilurus* was successfully acclimated from 2 to 40 ppt following preacclimation to 20 ppt for 24 h, and no benefit from slow, gradual acclimation was apparent (Osborne 1979). In Florida red tilapia, successful single-step transfer to full-strength seawater may require preacclimation at 19 ppt (Perschbacher and McGeachin 1988) but must be verified through further study.

Pretransfer feeding of a high salt diet (10 percent NaCl) in freshwater has been shown to stimulate the seawater adaptation process in *O. mossambicus, O. spilurus, O. aureus* × *O. niloticus* hybrids (Al-Amoudi 1987b; Fontainhas-Fernandes et al. 2001), and *O. niloticus* (Fontainhas-Fernandes et al. 2001). In *O. niloticus,* fish fed a high salt diet (8 percent NaCl) showed higher plasma levels of cortisol, MR cells in gills, and gill Na^+, K^+-ATPase activity after transfer to 15 and 20 ppt seawater than did control fish (Fontainhas-Fernandes et al. 2001). Fish fed dietary salt for 3 weeks showed some morphological features in freshwater similar to those described in seawater-adapted tilapia. Fish fed a control diet showed a more rapid increase in plasma osmolality and Cl concentration than fish fed dietary salt. However, this method failed to enable direct transfer to full-strength seawater, even after 4 weeks of feeding. Gradual salinity acclimation was therefore considered more effective, allowing successful transfer in less time.

Acclimation to Seawater: The Influence of Salinity Exposure During Early Development

Salinity exposure during early development also influences survival and growth in brackishwater and seawater. Early salinity exposure through spawning and hatching under elevated salinities enhances salinity tolerance of young tilapia fry and may facilitate acclimation to seawater (Watanabe et al. 1984; 1985b). In *O. niloticus,* MLS-96 h increased from 19.2 ppt for broods spawned in freshwater to >32 ppt for brood spawned at 15 ppt (Watanabe et al. 1984; 1985b). Since Florida red tilapia broodstock may be maintained under salinities as high as 18 ppt for effective reproduction (Watanabe, Burnett et al. 1989; Ernst et al. 1991), the influence of early salinity exposure on seawater transfer in Florida red tilapia should be determined. For fish spawned at 12 ppt, required salinity for preacclimation may be >19 ppt, and for fish spawned at 18 ppt, direct transfer to seawater may be possible.

After fertilization in tilapia eggs, the external medium passes through the egg chorion to form the perivitelline fluid (Peters 1983). This modifies the environment in which the embryo develops and may induce adjustments that persist through later development, a "nongenetic adaptation" (Kinne 1962) to environmental salinity. As oogenesis in fish is known to be sensitive to systemic stresses such as temperature and pH (Gerking 1979), the possibility that osmotic stress of parental fish may affect the salinity resistance of their progeny cannot be excluded.

A nongenetic adaptation to elevated salinity may be the differentiation of mitochondrion-rich (MR) cells. In adult fish, the gill epithelium represents a major portion of the body surface, and MR cells (chloride cells) in the gills are extrarenal sites of salt extrusion in seawater (Foskett et al. 1981; Foskett and Scheffey 1982; Kultz et al. 1992). In embryos and newly-hatched larvae, the yolksac membrane occupies a large part of body surface, since the developing body is small and gills, gut, and kidney are not fully developed. The yolksac membrane contains a rich population of MR cells, that play a major role in ion transport during the early developmental stages before the adult osmoregulatory organs become functional (Ayson et al. 1994a,b; Shiraishi et al. 1997). MR cells regulate salt balance and allow embryos and newly hatched larvae to survive transfer to different salinities at the early life stages.

Although endogenous hormone production may not have begun in embryos, they possess a large amount of yolk, which contain various hormones (Hwang et al. 1992). Tilapia embryos and larvae contain significant amounts of cortisol and thyroid hormones of maternal origin that have osmoregulatory functions. Maternal cortisol stimulates the activity of chloride cells in the yolksac membrane (Tagawa et al. 1997) and may be involved in osmoregulation during the early life stages of fish (Ayson et al. 1995). In hypoosmoregulating tilapia larvae acclimating to seawater, cortisol may enhance drinking and stimulate ion regulation and development of MR cells (Lin et al. 2000).

Juvenile Culture

In Florida red tilapia, fry are weaned from a 48 percent protein diet to a 30 percent diet beginning approximately 39 days posthatching when seawater acclimation is initiated. Following acclimation to seawater, fry (approximately 0.8 to 0.9 g in weight) are transferred to nursery tanks (Ernst et al. 1989) or floating sea cages (Watanabe, Clark et al. 1990b) and grown to large fingerling sizes (approximately 5 to 10 g) prior to stocking for grow-out in land-based or sea cage systems.

Production of Fingerlings in Flow-Through Tanks

Growth of Florida red tilapia fry was compared in seawater (37 ppt) tanks enriched with chicken manure (105 $kg \cdot ha^{-1} \cdot day^{-1}$) or a commercially prepared (30 percent protein) feed (Ernst et al. 1989). Sex-reversed males (avg. wt. = 1.3 g) were stocked in 23-m^3 tanks at a density of 25 fish per cubic meter and growth compared for 33 days. Fish were fed the prepared diet three to four times daily to satiation. Growth and survival were similar in manured and fed tanks through 20 days, and fingerling-size fish were obtained from manured pools by day 33. However, mean weights (11.3 g) and survival (84.5 percent) were lower than in fed tanks (18.0 g and 95.9 percent) (Ernst et al. 1989). Growth leveled off in manured tanks after 33 days whereas fish receiving prepared feeds continued rapid growth. The results demonstrate the feasibility of producing fingerling Florida red tilapia in seawater fertilized with chicken manure, which may therefore be used in lieu of prepared feeds during the first few weeks of growth.

Although tanks fertilized with chicken manure were managed identically, they developed substantially different food resources, and fish productivity varied greatly between the tanks. In tanks where primary production was dominated by macroalgae, the diet consisted largely of macroalgae whereas in tanks where phytoplankton was dominant, phytoplankton and particulate organic matter comprised the bulk of the diet (Grover et al. 1989). Food resources in phytoplankton-dominated pools were nutritionally superior to those in macroalgae-dominated pools, with zooplankton and larger phytoplankton selectively depleted by the fish. Hence, while nursery rearing with chicken manure is feasible, considerable variability in fish performance among culture units can be expected.

Production of Fingerlings in Floating Marine Cages

There is interest in raising salt-tolerant tilapia in marine cages in arid and tropical coastal regions where freshwater is in scarce supply or where there is competition for freshwater resources between aquaculture and agriculture (McGeachin et al. 1987; Al-Ahmad et al. 1988a,b; Clark, J.H. et al. 1990; Watanabe, Ernst et al. 1990; Watanabe, Clark et al. 1990 a,b). It is known that tilapia fry can be reared in cages in freshwater ponds and lakes (Coche 1982; Guerrero 1985; Campbell 1985), but the suitability of cage culture for tilapia fingerling production in marine environments had not been examined.

To test this possibility, growth of Florida red tilapia from fry through advanced fingerling stages was studied in floating cages near Great Exuma, Bahamas (Watanabe, Clark et al. 1990b). Cages (1 m^3) were stocked with sex-reversed male fry (avg. wt. = 1.79 g) at densities of 500 or 1000 m^{-3}, and fry were hand-fed a floating diet (Purina Tilapia Chow) containing 32 percent protein three to four times daily to satiation. After 30 days, survival averaged 88.8 percent and mean weights averaged 13.8 g, with 71.3 percent of the fingerlings ranging from 8.7 to 19.1 g (Watanabe, Clark et al. 1990b). There were no differences between densities in survival, growth rate, or feed consumption and conversion indicating that even higher densities may be possible. The results demonstrated that advanced fingerling Florida red tilapia of relatively uniform sizes can be produced in nursery cages in seawater stocked at high densities. Nursery rearing in marine cages can eliminate the costs of land-based nursery facilities.

Yields of fingerling Florida red tilapia obtained in floating marine cages (5.98 to 12.5 $kg \cdot m^{-3}$) are higher than those reported for Taiwanese red tilapia (2.31 $kg \cdot m^{-3}$) reared intensively in freshwater ponds at a density of 123 fish per cubic meter. However, these yields are lower than those obtained with *O. niloticus* (27.8 $kg \cdot m^{-3}$) reared in cages held in freshwater at densities of 1,300 to 3,000 fish per cubic meter (Campbell 1985) or for *O. spilurus* (14.1 to 30.7 $kg \cdot m^{-3}$) reared in marine cages at densities of 200 to 600 fish per cubic meter (Cruz and Ridha 1989). Cruz and Ridha (1989) also concluded that culture of *O. spilurus* fry in marine cages was an effective alternative to land-based nursery facilities. Maximum yields of fingerling Florida red tilapia in marine cages have not been determined, but yields as high as 52.2 $kg \cdot m^{-3}$ have been attained during later grow-out stages (Watanabe, Clark et al. 1990a).

Effects of Environmental and Other Factors on Growth of Juveniles in Saline Water

Salinity tolerance, as estimated by MST, ST_{50}, or MLS-96 indices, does not necessarily indicate relative growth at different salinities, which must be determined through controlled studies. Some of the potentially important factors that can influence the effects of salinity on growth in tilapia are discussed below.

Salinity

The effects of salinity on growth of tilapia, studied in relatively few species, are not well understood. The variety of conditions (e.g., temperature, photoperiod, sex ratios, and stocking densities) used in previous studies precludes meaningful comparisons of results. For many species, optimal ranges of salinities for growth have been inferred from data on natural distributions or fragmentary experimental evidence.

The effects of salinity on the growth of juvenile, sex-reversed male Florida red tilapia were studied under controlled photoperiod (12 L:12 D) and temperature (28°C). Fish (avg. wt. = 0.72 g) were grown in 200-L aquaria at salinities of 1, 10, 19, 28, and 36 ppt at a density of 15 fish per tank (0.075 fish per liter). Growth at 10 ppt and greater was significantly higher than that at 1 ppt, and there was a trend toward an increase in growth with increasing salinity (Watanabe, Ellingson et al. 1988). This was attributed to increased food consumption (appetite) and better feed conversion ratios with increasing salinity. Higher growth in brackishwater and seawater than in freshwater has also been observed in *O. mossambicus*

(Canagaratnam 1966; Jurss et al. 1984; Howerton et al. 1992; Kuwaye et al. 1993; Ron et al. 1995; Vonck et al. 1998) *O. mossambicus* × *O. hornorum* hybrid (Garcia and Sedjro 1987), and *O. spilurus* (Osborne 1979) and indicates an advantage of farming these tilapia in brackishwater or seawater.

Hybrids between *O. mossambicus* and other less-tolerant strains also show relatively fast growth rates in brackishwater and full-strength seawater. In *O. mossambicus* × *O. niloticus* hybrid, no significant differences in growth were observed among fingerlings (2.32 g mean wt.) grown under treatment salinities of 0.5, 17, and 32 ppt to around 25 to 35 g in 56 day at 27°C (Garcia-Ulloa et al. 2001). Feed consumption was best at 0.5 ppt, but FCR was better at 17 and 32 ppt (1.29 and 1.35, respectively). These authors concluded that this hybrid may perform equally well in freshwater and seawater.

In contrast to all-male Taiwanese red tilapia that exhibited faster growth in freshwater than in saltwater (Liao and Chang 1983), growth of sex-reversed male Florida red tilapia was faster in saltwater, suggesting a relatively high adaptability to seawater of the Florida red hybrid strain. This probably reflects the different genetic heritage of the Taiwanese strain, which, in contrast to the Florida red strain, contain an *O. niloticus* component (Galman and Avtalion 1983), a species that is less salt tolerant than *O. mossambicus* (Al-Amoudi 1987a; Villegas 1990).

In the Philippines, a number of red tilapia strains have recently been developed and tested for saltwater culture performance, including three genetically diverse strains derived from *O. mossambicus-hornorum* hybrid × *O. niloticus* (Philippine strains) and two imported Asian strains derived from *O. niloticus* × *O. mossambicus* (Taiwanese and Thai strains) (Romana-Eguia and Eguia 1999). When growth of these five Asian strains were compared at 0, 17, and 34 ppt, growth was best in brackishwater, consistent with the *O. niloticus* derivation of these strains. Saltwater-reared red tilapia grew at rates comparable to or even slightly higher than in freshwater-reared ones. These strains grew significantly different from each other under both freshwater and seawater conditions. In freshwater, the fastest growing was the Taiwanese strain whereas the Philippine strain grew the fastest in seawater. This indicated that both strain (genotype) and rearing salinity (environment) influenced growth in red tilapia. These authors concluded that the Taiwanese strain would be best for freshwater, any of the five strains for brackishwater, and the Philippine strain for seawater. They also concluded that red tilapia may be a practical alternative to milkfish or prawn for culture in areas where freshwater resources are limited. On Negros Island in central Philippines, FYD International has developed the "Jewel" tilapia, a hybrid of dark-skinned *O. mossambicus* and *O. urolepis hornorum*. These fish are grown in 18 to 25 ppt brackishwater ponds, often in polyculture with penaeid shrimp.

Interpretation of relative growth rates of *O. mossambicus* at different salinities with respect to metabolism is somewhat contradictory. Febry and Lutz (1987) reported that osmoregulation in tilapia hybrid (*O. mossambicus* × *O. hornorum*) in freshwater is more energetically expensive than in seawater, and least expensive in brackishwater. Rate of oxygen consumption of seawater tilapia was approximately half of that measured in freshwater fish (Febry and Lutz 1987). Iwama et al. (1997) compared oxygen consumption rates for *O. mossambicus* acclimated for 1 month in freshwater, seawater and 1.6 × seawater and found lowest oxygen consumption rates for fish acclimated to seawater. Ron et al. (1995) likewise found that the rate of oxygen consumption of seawater tilapia was approximately half of that measured in freshwater fish, suggesting that increased growth of the seawater animals over their freshwater counterparts is related to a reduction in routine metabolism in seawater (Ron et al. 1995). Vonck et al. (1998) noted the highest growth rates for *O. mossambicus* juveniles when they were raised in brackishwater ranging from 25 to 75 percent seawater. They further observed a low density of chloride cells in branchial epithelium and the lowest branchial Na^+, K^+-ATPase activity in 50 percent seawater, suggesting that food energy was spared for body growth rather than for osmoregulation.

Hormonal Enhancement of Growth in Seawater

Howerton et al. (1992) reported that both MT and thyroid hormone (T3 3,3′, 5 triiodo-L-thyronine), when administered orally and in appropriate doses, can significantly increase growth in seawater-adapted tilapia. Ron et al. (1995) reported that, while untreated *O. mossambicus* grow two to three times faster in seawater than in freshwater, concomitant exposure to MT and seawater leads to growth that is five to seven times higher than that seen in untreated freshwater fish. This was also attributed to enhanced appetite and aggressiveness in feeding in seawater and MT-treated tilapia. Tilapia in seawater may both consume more food and utilize that food more efficiently than fish in freshwater. Enhanced growth of seawater-reared tilapia is related to elevation in serum/pituitary GH levels and to lower routine metabolic rates (Borski et al. 1994; Ron et al. 1995). Once in the circulation, GH binds to specific binding proteins and stimulates, primarily in the liver, synthesis and secretion of insulinlike growth factor (IGF-I and IGF-II) to elicit the growth promoting action (Shepherd, Ron et al. 1997 Shepherd, Sakamoto et al. 1997; Mancera and McCormick 1998).

Divalent Ion Concentrations

Results of laboratory studies on the effects of salinity on growth of Florida red tilapia have been somewhat inconsistent. In contrast to earlier findings (Watanabe, Ellingson et al. 1988), where growth increased with salinity to a maximum in full-strength seawater, growth of juvenile, sex-reversed male Florida red tilapia was observed to be higher at 18 ppt than at 36 ppt in a later study (Watanabe, Ernst et al. 1993). It is known that the effects of salinity on growth in fish are related not only to total concentration of dissolved solids (i.e., salinity per se) but are also influenced by the concentrations of divalent ions (Ca^{2+} and Mg^{2+}) due to their effects on membrane permeability and osmoregulation (Wurts and Stickney 1989). In an earlier study by Watanabe, Ellingson et al. (1988), brackishwater was prepared from freshwater produced by reverse osmosis of seawater (hardness = 10.2 mg $CaCO_3$ L^{-1}), whereas in a latter study by Watanabe, Ernst et al. (1993), freshwater was obtained from groundwater sources (hardness = 157 to 162 mg $CaCO_3$ L^{-1}). Dissimilar results in these studies may therefore be attributed to marked differences in ionic composition of freshwater sources used. It is recommended that growth and tolerance studies use water of ionic composition similar to that to be used during culture. The available data is that the effects of salinity on metabolism and growth in tilapia are complex and can be modified by the potentially interactive effects of a number of factors, including concentrations of both total dissolved solids and divalent ions.

Temperature

Temperature tolerance in fish may be modified by salinity (Kinne 1960, 1963; Beamish 1970; Peters and Boyd 1972; Whitfield and Blaber 1976; Stauffer 1986) due to their interactive effects on osmoregulation (Allanson et al. 1971; Tilney and Hocutt 1987). Information on the combined effects of temperature and salinity on growth in tilapia is important to delineate environmental conditions suitable for culture and to optimize production in recirculating culture systems by environmental control.

Under salinities ranging from freshwater (0 ppt) to full-strength seawater (36 ppt), growth rates of juvenile, sex-reversed male Florida red tilapia (avg. wt. = 0.56 to 1.20 g) generally increased with increasing temperature within the range 22°C to 32°C, but were markedly lower at 22°C than at 27°C and 32°C demonstrating the thermophilic nature of this hybrid strain (Watanabe, Ernst et al. 1993). However, salinity modified the effects of temperature on growth: At 0 ppt, feed consumption and growth reached a maximum at 27°C whereas at 18 and 36 ppt, consumption and growth were highest at 32°C (Watanabe, Chan et al. 1993). The effect of temperature on growth depends upon the interaction between food consumption and metabolism (Brett 1979). In *O. mossambicus* × *O. hornorum* hybrid, energy requirements for osmoregulation have been shown to be higher in freshwater than in brackishwater (12 to 15 ppt) or in seawater (30 to 35 ppt) (Febry and Lutz 1987). This suggests that, in Florida red tilapia, a leveling of feed consumption and higher metabolic energy demands in freshwater result in an attenuation of growth at temperatures above 27°C.

Because rising temperatures increase metabolic energy demands, and high feeding rates are required to meet these demands, the optimum temperature for growth decreases if feed is restricted (Brett et al. 1982). Temperature optima for growth of Florida red tilapia may therefore be lower under a restricted feeding regime.

In terms of temperature unit (TU) requirements per gram of fish growth, maximum growth efficiency appeared to vary with salinity: at 0 ppt, TU requirements were lowest at 27°C whereas at 18 and 36 ppt, TU requirements were lowest over the range of 27°C to 32°C (Watanabe, Ernst et al. 1993). This suggests that, in freshwater, heating water to temperatures above 27°C would not be justifiable, whereas in brackish- and seawater, heating water to 32°C can increase growth rates without lowering growth efficiency.

Under all temperatures, TU requirements were lower at 18 ppt than at 0 or 36 ppt, although these differences were pronounced at 22°C (Watanabe, Ernst et al. 1993). This suggests an important advantage of brackishwater rearing to improve growth efficiency of the Florida red tilapia strain under suboptimum temperatures.

In Nile tilapia, salinity also modified the effects of temperature on growth: At 0 and 8 ppt, growth rates were highest at 32°C whereas at 12 ppt, growth rates were highest at 28°C (Likongwe et al. 1996). While the study showed that 32°C and 8 $g \cdot L^{-1}$ was most conducive to maximum growth and feed conversion efficiency of *O. niloticus,* the results suggested that a temperature range of 28°C to 32°C and 0 to 12 ppt produced rapid growth of juvenile Nile tilapia. On the other hand, a salinity close to 16 ppt combined with elevated water temperature of 32°C was injurious to the health of juvenile Nile tilapia.

The available data indicate that, in tilapia, salinity modifies the effects of temperature on growth (Stauffer 1986; Watanabe, Ernst et al. 1993; Likongwe et al. 1996). Hence, for practical purposes, the effects of salinity on growth should be evaluated under temperature ranges to be encountered during culture.

Overwintering

In areas where water temperatures approach lower lethal limits, production must be curtailed and stocks overwintered in covered ponds or heated tanks (Balarin and Haller 1982). In the subtropical Bahamas, evidence of disease and mortalities have been observed in Florida red tilapia maintained in seawater pools in association with seasonally declining temperatures to below 25°C (Ernst et al. 1989; Watanabe, French et al. 1989; Ellis and Watanabe 1993a). Other seawater-rearing studies of *Oreochromis* spp. have also reported poor growth and survival at temperatures below 25°C (Ting et al. 1984; Hopkins et al. 1989; McGeachin et al. 1987).

With the exception of one study (Jennings 1991), in which cold tolerance of *S. melanotheron* did not differ with salinity in the range of 5 to 35 ppt, studies with a number of tilapia species, including *O. aureus, S. melanotheron*, and *O. mossambicus,* showed that fish have better cold tolerance when maintained in low-salinity brackishwater (5 to 12 ppt) than in freshwater or full-strength seawater (Allanson et al. 1971; Stauffer et al. 1984; Zale and Gregory 1989) presumably because osmoregulatory stress is minimized at near isoosmotic salinities. Hence, osmoregulatory failure is prevented at temperatures that would be lethal in hypo- or hyperosmotic media (Zale and Gregory 1989).

Cage-culture overwintering experiments in Taiwan have shown that mortalities are considerably lower for tilapia maintained in brackishwater (8 to 16 ppt) than in seawater (30 ppt) at temperatures ranging from 14.5°C to 19.4°C and that growth rates were higher at 16 ppt than at either 8 or 30 ppt under these conditions (Ting et al. 1984). Based on lower TU requirements for Florida red tilapia at 18 ppt than at 0 or 36 ppt, brackishwater rearing can mitigate the effects of suboptimum temperature on osmoregulatory stress.

Florida red tilapia have survived seawater-rearing temperatures as low as 16°C without detrimental effects (unpublished data), suggesting that low-temperature tolerance may be related to the rate of decline (i.e., thermal history), as well as to a critical lower limit. Studies are required to determine lower lethal temperature tolerance in Florida red tilapia to determine the effect of salinity on low-temperature tolerance and to assess the practical feasibility of using brackishwater to overwinter fingerlings. The effects of fish size, thermal history, and other culture conditions (e.g. feeds, feeding regime, and water quality) on low-temperature tolerance should also be considered.

Early Salinity Exposure

When growth of juvenile, sex-reversed male Florida red tilapia (avg. wt. = 1.57 g), spawned at salinities of 4 and 18 ppt, was compared at rearing salinities of 18 and 36 ppt in 200-L aquaria under controlled photoperiod (12 light:12 dark) and temperature (28°C), growth at both rearing salinities was higher and feed conversion ratios lower for fish spawned at 18 ppt than those spawned at 4 ppt. This showed that fish spawned and reared through early ontogenetic development in brackishwater are better adapted for growth in brackish- and seawater than those spawned in freshwater (Watanabe, French et al. 1989).

In studies evaluating the effects of salinity on growth of fish, the conditions of salinity under which experimental animals are spawned and reared during early development are seldom specified. Brett (1979) concluded from the variability in reported salinity growth responses that uniformity of physiological state is often lacking. In Florida red tilapia, exposure to salinity during early development clearly influences survival and growth in brackishwater and seawater.

Photoperiod

Available evidence suggests that, in nonsalmonid fish species, long (or increasing) daylength stimulates growth whereas decreasing photoperiod inhibits growth (Gross et al. 1965, Brett 1979; Boehlert 1981; Woiwode and Adelman 1991). Information on the effects of photoperiod on growth in tilapia is limited. Watanabe et al. (1997) studied the effects of daylength and the rate and direction of change of daylength on growth of Florida red tilapia in seawater (36 ppt). Growth of sex-reversed male fingerlings (avg. wt. = 1.93 g) was compared for 56 days under six different photoperiod regimes:

1. 12 h L:12 h D (12 L:12 D);
2. 16 L:8 D;
3. 24 L:0 D;
4. daylength increasing from 12 to 16 h at a rate of 30 min·week^{-1} (12 to 16);
5. daylength increasing from 12 to 18 h at an accelerated rate of 1 h·week^{-1} (12 to 18);
6. daylength decreasing from 12 to 8 h at a rate of 30 min·week^{-1} (12 to 6).

Fish were fed twice daily a 32 percent protein diet to satiation and temperature was 28°C. In contrast to both salinity and temperature, which had pronounced effects on feeding and growth of juvenile sex-reversed male Florida red tilapia (Watanabe, Ellingson et al. 1988, Watanabe, Ernst et al. 1993), photoperiod produced little (albeit significant) effects on growth. Growth over 56 days was not influenced by rate or direction of change in daylength and was similar under increasing or decreasing daylength regimes (0.49 to 0.53 $g \cdot day^{-1}$) to that under a constant 12 L:12 D regime (0.54 $g \cdot day^{-1}$). Long daylength regimes of 16 L:8 D (0.47 $g \cdot day^{-1}$) and 24 L:0 D (0.43 $g \cdot day^{-1}$), however, produced slower growth than a constant 12 L:12 D regime (Watanabe et al. unpublished data). The results suggest that maximum growth can be obtained on a constant 12 L:12 D regime and that extended or continuous daylight regimes may be inhibitory. Possible interactive effects of photoperiod, salinity, and temperature on growth should not be ignored.

Ration

Effects of salinity and temperature on growth in fish depend upon the interaction between metabolism and food consumption (Brett 1979). Because salinity and temperature influence metabolic energy demands, and feeding rates must be adequate to meet these demands (Brett et al. 1982), the optimum combination of temperature and salinity for growth may be influenced by the level of feeding. Temperature and salinity optima for growth may therefore differ under restricted and satiation feeding regimes. The available data suggests that the interactive effects of many factors can modify the effects of salinity on metabolism and growth. Brett (1979) concluded from the variability in reported salinity on growth responses in fish that uniformity of physiological state is often lacking.

GROW-OUT FROM FINGERLING TO MARKETABLE SIZE

Flow-Through Seawater Tanks

Grow-out at high densities in flow-through tanks may be appropriate on tropical islands and other coastal areas where land for aquaculture may be costly, but where seawater resources are abundant and water temperatures favorable for tilapia culture year round. Ernst et al. (1989) first studied the feasibility of rearing Florida red tilapia in seawater from fingerlings through marketable sizes in plastic-lined, circular tanks (volume = 23 m^3). Sex-reversed male fry (avg. wt. = 1.3 g) were stocked in tanks at a density of 25 fish per cubic meter and fed three to four times daily to satiation a commercial diet containing 28.5 percent protein. Tanks were supplied with aeration and flow-through seawater (37 $g \cdot L^{-1}$) at an exchange rate of 250 to 500 percent·day^{-1}. Under an average temperature range of 27°C to 29°C, fish were grown to an average weight of 467 g in 170 days (weight gain = 2.74 g·day^{-1}) with a survival of 89.7 percent and a yield of 10.5 $kg \cdot m^{-3}$. Average feed conversion ratio was 1.6 (Ernst et al. 1989; Table 10.3). During the study, dissolved oxygen averaged 5.6 $mg \cdot L^{-1}$ under loading rates that increased from 0.01 $kg \cdot L^{-1} \cdot min^{-1}$ at stocking to 6.7 $kg \cdot L^{-1} \cdot min^{-1}$ when fish were harvested.

Effect of Stocking Density

Sex-reversed male Florida red tilapia fingerlings (avg. wt. = 5.36 g) were stocked into flow-through seawater tanks (10 m^3) at densities of 15, 25, and 35 fish per cubic meter and growth and survival compared for 150 days (Watanabe, Chan et al. 1993). Fish were fed commercial diets containing 32 and 20 percent protein and each tank was supplied with aeration and flow-through seawater at an exchange rate of 800 percent·day^{-1}. After 150 days, no significant differences were observed among treatments in fish weight (avg. = 462 g), coefficient of variation of weight (avg. = 22.8 percent), weight gain (avg. = 3.04 g·day^{-1}), survival (avg. = 94.4 percent), feed consumption (avg. = 5.17 percent·day^{-1}) and conversion ratio (avg. = 1.80) (Watanabe, Chan et al. 1993; Table 10.3). Individual weights ranged from 98.6 to 687.6 g, with 75.2 percent and 66.9 percent of these exceeding 393 and 430 g, respectively. Final biomass increased with stocking density from 6.69 to 15.4 $kg \cdot m^{-3}$. Since growth, survival, and feed conversion were not impaired under stocking densities of up to 35 fish per cubic meter, and no effect of stocking density on size variation was evident, even higher stocking densities appear feasible. Siddiqui et al. (1989) also reported no difference in growth or feed conversion of *O. niloticus* (avg. wt. = 40.3 g) grown in brackishwater (3.5 to 3.9 ppt) tanks over 164 days on a 34 percent protein diet at densities of 16, 32, and 42.6 m^{-3}.

Since rate of water exchange was constant (800 percent·day^{-1}) throughout the grow-out period, loading rate was low (0.015 to 0.033 $kg \cdot L^{-1} \cdot min^{-1}$) at stocking, increasing with fish growth to a maximum (1.20 to 2.78 $kg \cdot L^{-1} \cdot min^{-1}$) when the study was terminated, and no adverse effects on water quality were observed. Clearly, flow rates could be adjusted with growth to reduce pumping costs. Alternatively, loading rates can be optimized by the incorporation of nursery and grow-out phases, where smaller fry are reared at relatively high densities to large fingerling stages, which are then restocked at lower densities for grow-out. This also reduces space requirements and total grow-out time, thereby improving production. Although maximum loading rates for Florida red tilapia grown in flow-through tanks

TABLE 10.3. Summarized data on culture performance of Florida red tilapia and other species in saltwater (salinity $\geq 15\ g \cdot L^{-1}$) systems.

Species (sex)	Location	Culture system	Salinity/ temp. (ppt/°C)[a]	Initial wt. (g)	Stocking density (fish/m³)	Feed (type-percent protein)	Ration (percent bw/day)	Duration (day)	Final wt. (g)	Survival (percent)	Yield (kg/m³)	Wt. gain[b] (g/day)	FCR (wet/ dry)	Reference
Florida red tilapia (sex-reversed males)	Bahamas	23 m³ circular, plastic-lined pools; loading = 0.01-6.7 $kg \cdot L^{-1} \cdot min^{-1}$; aeration	37/27-29	1.3	25	CD[c]-28.9	(>30-1.5)[d]	170	467	89.7	10.5	2.74	1.6	Ernst et al. 1989
Florida red tilapia (sex-reversed males)	Bahamas	10 m³ circular, plastic-lined pools; loading = 0.01-2.7 $kg \cdot L^{-1} \cdot min^{-1}$; aeration	37/28	5.63	15	CD-32, 20	5.15	150	470	94.9	6.69	3.10	1.81	Watanabe, Mueller et al. 1993
				5.10	25		5.24		452	93.9	10.6	2.98	1.83	
				5.36	35		5.11		463	94.4	15.4	3.05	1.75	
Florida red tilapia (sex-reversed males)	Bahamas	10 m³ circular, plastic-lined pools; loading = 0.05-2.3 $kg \cdot L^{-1} \cdot min^{-1}$; aeration	37/28-32	10.2	25	CD-20	8.03	120	440	97.3	10.7	3.60	2.20	Clark, Watanabe et al. 1990
				11.2		CD-25	7.73		465	97.5	11.3	3.75	2.14	
				10.4		CD-30	7.60		453	97.0	11.0	3.70	2.01	
Florida red tilapia (sex-reversed males)	Bahamas	10 m³ circular, plastic-lined pools; loading = 0.12-2.09 $kg \cdot L^{-1} \cdot min^{-1}$ aeration	36-38/ 20.5-30.0	22	30	CD-20	satiation 2/day	210	592	84.2	11.6	2.71	ND	Ellis and Watanabe 1993[b]

TABLE 10.3 *(continued)*

Species (sex)	Location	Culture system	Salinity/ temp. (ppt/°C)[a]	Initial wt. (g)	Stocking density (fish/m^3)	Feed (type-percent protein)	Ration (percent bw/day)	Duration (day)	Final wt. (g)	Survival (percent)	Yield (kg/m^3)	Wt. gain[b] (g/day)	FCR (wet/dry)	Reference
Florida red tilapia (sex-reversed males)	Bahamas	1 m^3 floating cages	34-41/ 26-33	9.1	300	CD-32	13-3	84	149.7	98.2	35.2	1.82	2.04	Clark, Watanabe, et al. 1990
				9.2			11-3		121.8	99.3	30.0	1.47	1.68	
Florida red tilapia (sex-reversed males)	Bahamas	1 m^3 floating cages	34-39/ 29-31	8.51	100	CD-28	5.61	84	169.9	98.5	16.7	1.92	1.86	Watanabe, Ernst et al. 1990
				8.77	200		5.29		184.3	98.4	36.3	2.09	1.73	
				8.81	300		5.47		176.1	98.8	52.2	1.99	1.85	
				8.79	100	CD-32	5.39	84	165.9	97.0	16.1	1.87	1.91	
				8.82	200		5.53		170.4	97.8	33.3	1.92	1.88	
				9.00	300		5.78		162.9	96.8	47.3	1.83	2.02	
Florida red tilapia (sex-reversed males)	U.S. Virgin Islands	2 m^3 floating cages	35-38/ 27-32	152	20	CD-36	3	112	439	94.2	8.3	2.56	ND[e]	Hargreaves et al. 1992
Florida red tilapia (sex-reversed males)	Puerto Rico	0.2 ha earthen ponds	20-27/ 26-31	0.85	30,000/ ha	CD-25	satiation 2/day	160	452	70 (estimated)	9470 kg·ha^{-1}	2.82	1.8	Head et al. 1996
Taiwanese red tilapia (mixed-sex)	Taiwan	1.3-1.4 m^3 recirculating rectangular tank; partial exchange; aeration	34/25-28	17.5	22	CD-28.5	5-8	40	58.6	93.3	1.20	1.03	ND	Liao and Chang 1983

			17/25-28	18.4	22	CD-28.5	5-8	40	56.1	100	1.23	0.94	ND	
Taiwanese red tilapia (all-male)	Taiwan	1.3-1.4 m^3 recirculating rectangular tank; partial exchange; aeration	34/27	104	22	CD-28.5	5-8	40	153	83.3	2.92	1.35	ND	
			17/27	104	22	CD-28.5	5-8	40	148	86.7	2.82	1.10	ND	
Taiwanese red tilapia (mixed-sex)	Taiwan	20-m^2 octagonal pond; aeration	34/24-26	11.4	15 m^2	CD-24	3-5	84	104	94.7	1.47 $kg{\cdot}m^{-2}$	1.10	ND	
			20/24-26	11.4	15 m^2	CD-24	3-5	84	102	98.3	2.92 $kg{\cdot}m^{-2}$	1.08	ND	
Taiwanese red tilapia (mixed-sex)	Singapore	8 m^3 recirculating, circular fiberglass tanks with biodrum; loading = 0.1-3.4 $kg{\cdot}L^{-1}{\cdot}min^{-1}$	26-30/ 26-29	0.78	500	CD-28-31	8-3	58	38.6	44.2	8.53	0.65	1.55	Cheong et al. 1987
		Loading = 2.6-2.0 $kg{\cdot}L^{-1}{\cdot}min^{-1}$		38.6	250	CD-28-31	8-3	121	239	83.9	50.2	1.66	2.32	
		Loading = 8.8-19.7 $kg{\cdot}L^{-1}{\cdot}min^{-1}$		264	125	CD-28-31	8-3	60	438	90.2	49.3	2.89	3.41	

TABLE 10.3 *(continued)*

Species (sex)	Location	Culture system	Salinity/ temp. (ppt/°C)[a]	Initial wt. (g)	Stocking density (fish/m^3)	Feed (type-percent protein)	Ration (percent bw/day)	Duration (day)	Final wt. (g)	Survival (percent)	Yield (kg/m^3)	Wt. gain[b] (g/day)	FCR (wet/ dry)	Reference
Taiwanese red tilapia (mixed-sex)	Kuwait	2 m^3 circular, concrete tanks; loading = 0.1-1.0 $kg \cdot L^{-1} \cdot min^{-1}$ aeration	39-41/ 20-28	1	54	CD-40-45	20-2	226	132	38	2.71	0.58	ND	Hopkins et al.1989
Taiwanese red tilapia (mixed-sex)	Singapore	8 m^3 circular, fiberglass tanks; loading = 0.02-0.71 kg·L·min	28-30/ 26-29	0.72	500	CD-28-31	8-3	58	21.3	74.5	7.94	0.35	1.70	Cheong et al. 1987
		Loading = 0.4-2.7 Kg·L·min		21.3	250	CD-28-31	8-3	121	179	67.6	30.2	1.3	2.41	
Tiwanese red tilapia (90 percent males)	Hawaii	0.8 m^3 floating cages in 0.25-0.45 ha ponds	11-17/ 22-27	9.46	125	CD-24	3-5	56	34.0	97.6	4.14	0.44	1.76	Meriwether et al. 1984
O. spilurus (mixed-sex)	Kuwait	0.5 m^3 tanks; loading = 1.0-2.0 kg·L·min; aeration	38-41/ 27-30	71.5	100	CD-40	2	56	127	100	12.7	0.97	1.51	Al-Ahmad et al. 1988[b]
O. spilurus (mixed-sex)	Kuwait	0.5 m^3 tanks; loading = 1.0-2.0 kg·L·min; aeration	38-41/ 27-30	272	40	CD-40	1.5	56	427	98.3	16.8	2.71	1.53	

O. spilurus (mixed-sex)	Kuwait	7.7 m^3 plastic raceways; loading = 1.0-2.0 kg·L·min; aeration	38-41/ 22-29	142	130	CD-40	2-3	124	290	98.1	35.2	1.19	2.30	
O. spilurus (mixed-sex)	Kuwait	25.5 m^3 circular fiberglass tanks; loading = 2.0-3.3 kg·L·min; aeration	38-41/ 25-30	23.1	149	CD-40	3-1.5	180	259	94.2	36.4	1.28	1.37	
O. spilurus (mixed-sex)	Kuwait	5-6 m^3 fiberglass raceways; loading = 0.1-1.0 kg·L·min; aeration	39-41/ 19-30	1	216-260	CD-40-45	20-2	246	66	59	8.41-10.1	0.26	ND	Hopkins et al. 1989
O. spilurus (all males)	Kuwait	5-6 m^3 fiberglass raceways; loading = 0.1-1.0 kg·L·min; aeration	39-41/ 23-28	106	50	CD-40-45	3-1	175	371	83	15.4	1.51	ND	
O. spilurus (mixed-sex)	Kuwait	2 m^3 circular concrete tanks; loading = 0.1-1.0 kg·L·min; aeration	39-41/ 20-28	2	50	CD-40-45	20-2	226	70	82	2.87	0.3	ND	

TABLE 10.3 *(continued)*

Species (sex)	Location	Culture system	Salinity/ temp. (ppt/°C)[a]	Initial wt. (g)	Stocking density (fish/m^3)	Feed (type-percent protein)	Ration (percent bw/day)	Duration (day)	Final wt. (g)	Survival (percent)	Yield (kg/m^3)	Wt. gain[b] (g/day)	FCR (wet/ dry)	Reference
O. spilurus (sex-reversed males)	Kuwait	2 m^3 circular concrete tanks; loading = 0.1-1.0 kg·L·min; aeration	39-41/ 20-28	3	51	CD-40-45	20-2	226	80	71	2.90	0.34	ND	
O. spilurus (mixed-sex)	Kuwait	0.5 m^3 fiberglass tanks; loading = 4-5 kg·L·min	36/23-33	1.11	500	CD-46-55	20-2	301	175	78.4	68.6	0.58	1.98	Cruz and Ridha 1990
O. spilurus (mixed-sex)	Kuwait	100 m^3 circular concrete tanks; loading = 2.5-3.6 kg·L·min; aeration	33-35/ 24-27	230-395	15.4-26.7	CD-46	2-2.5	30-70	341-561	94.5-97.7	7.1-12.0	2.31-3.49	1.47-2.13	Cruz et al. 1990
O. spilurus (mixed-sex)	Kuwait	6 m^3 floating cages	38-41/ 23-33	118	167	CD-40	2-3	101	323	81.8	44	2.03	2.04	Al-Ahmad et al. 1988[b]
O. spilurus (mixed-sex)	Kuwait	1 m^3 floating cages	41/20-33	49.3	100	CD-46-49	Aff 8 h per day	75	144.3	97.5	14.1	1.27	1.96	Cruz and Ridha 1989
O. spilurus (mixed-sex)	Kuwait	1 m^3 floating cages	41/20-33	38.5	200	CD-46-49	Aff 8 h per day	75	124.7	95.9	23.9	1.15	2.32	
				36.6	300	CD-46-49	Aff 8 h per day	75	111.8	91.8	30.7	1.0	2.07	

O. spilurus (mixed-sex)	Kuwait	1 m^3 net cages	41/20-33	15.9	150	CD-46-55	20-2	160	324.9	83.3	40.6	1.93	1.98	
O. aureus x *O. spilurus* (mixed-sex)	Kuwait	2 m^3 circular concrete tanks; loading = 0.1-1.0 $kg \cdot L^{-1} \cdot min^{-1}$; aeration	39-41/ 20-28	1	73.5-8.0	CD-40-45	20-2	226	68	95	5.17	0.30	ND	Hopkins et al. 1989
O. aureus (mixed-sex)	Kuwait	5-6 m^3 fiberglass raceways; loading = 0.1-1.0 $kg \cdot L^{-1} \cdot min^{-1}$; aeration	39-41/ 20-28	1	113-136	CD-40-45	20-2	253	58	55	3.6-4.3	0.23	ND	
O. aureus (all males)	Kuwait	5 m^3 fiberglass raceways; loading = 0.1-1.0 $kg \cdot L^{-1} \cdot min^{-1}$; aeration	39-41 /20-28	67	50	CD-40	3-1	175	239	56	6.69	0.98	ND	
O. aureus (sex-reversed males)	Kuwait	2 m^3 circular concrete tanks; loading = 0.1-1.0 $kg \cdot L^{-1} \cdot min^{-1}$; aeration	39-41/ 20-28	1	50	CD-40-45	20-2	226	34	24	0.41	0.15	ND	
O. aureus (mixed-sex)	Israel	0.15 ha pond	41-45/ND	72	400 ha^{-1}	CD-25	2 $kg \cdot d^{-1}$	118	234	77	72.1 $kg \cdot ha^{-1}$	1.4	ND	Chervinski and Yashouv 1971
O. aureus (mixed-sex)	Israel	0.1 ha pond	39-44/ 23-29	70	1000 ha^{-1}	CD-25	ND	158	382	56	214 $kg \cdot ha^{-1}$	1.97	ND	Chervinski and Zorn 1974

TABLE 10.3 *(continued)*

Species (sex)	Location	Culture system	Salinity/ temp. (ppt/°C)[a]	Initial wt. (g)	Stocking density (fish/m^3)	Feed (type-percent protein)	Ration (percent bw/day)	Duration (day)	Final wt. (g)	Survival (percent)	Yield (kg/m^3)	Wt. gain[b] (g/day)	FCR (wet/dry)	Reference
O. aureus (mixed-sex)	Bahamas	1.0 m^3 floating cages in 0.02 ha concrete ponds	34-37/ 21-29	21.2	100-400	CD-36	6	90	55.4	ND	ND	0.34	ND	McGeachin et al. 1987
O. niloticus (mixed-sex)	Philippines	100 m^2 manured earthen pond	18-48/ 15-32	3.39	1.5/m^2	CD-30	5-3	120	104.6	94.5	0.148 kg·m^{-2}	0.84	1.68	Fineman-Kalio and Camacho 1987
O. niloticus (mixed-sex)	Philippines	100 m^2 manured earthen pond	17-50/ 16-32	2.8	1.5/m^2	CD-20	5-3	120	79.4	91.3	0.109 kg·m^{-2}	0.64	ND	Fineman-Kalio 1988
O. mossambicus (sex-reversed males)	U.S. Virgin Islands	Floating cages (2 m^3)	35-38/ 27-32	148	20	CD-36	3	112	381	83.3	6.5	2.08	ND	Hargreaves et al. 1992
T. zillii (mixed-sex)	Israel	0.1 ha pond	39-43/ 23-29	1.3	120 ha^{-1}	CD-25	ND	113	95.0	83.3	95.0 kg·ha^{-1}	0.83	ND	Chervinski and Zorn 1974

[a]Units are those in parentheses unless otherwise indicated; [b]DWG = Daily weight gain = [(ln final wt.(g)–ln initial wt.(g))/time(d)]100; [c]CD = Commercially prepared diet; [d]Beginning ration–ending ration; [e]ND = no data available; [f]AF = automatic feeder.

have not been determined, loading rates of 1.0 to 3.33 $kg \cdot L^{-1} \cdot min^{-1}$ have been reported to support tilapia yields ranging from 7.1 to 50 $kg \cdot m^{-3}$ in flow-through tanks (Balarin and Haller 1982, 1983; Al-Ahmad et al. 1988a; Siddiqui et al. 1989; Cruz et al. 1990). Yields of up to 52.2 $kg \cdot m^{-3}$ have been reached for Florida red tilapia grown in marine cages (Watanabe, Clark et al. 1990a).

Floating Sea Cages

Effects of Stocking Density

In Jamaica, where investment capital is scarce and startup costs are high, the industry will need to increase efficiency of production and markets in order for tilapia aquaculture to compete with land-based animal production. While there are sufficient land resources to expand traditional pond-based production, a diminishing supply of water for agriculture (Ryther et al. 1991; Chakallal and Noriega-Curtis 1992) is problematic. Diversification into brackish coastal sites or into marine cage or net pen culture in protected bays on the south coast, may be possible. Some areas of the Caribbean are characterized by sheltered embayments with high flushing rates that make them suitable for fish culture in marine cages.

The feasibility of raising Florida red tilapia fingerlings in marine cages was first studied by Watanabe, Clark et al. (1990a) in the Bahamas. Juvenile, sex-reversed male fingerlings (avg. wt. = 8.78 g) were stocked into 1-m^3 floating cages at densities of 100, 200, and 300 fish per cubic meter and reared on commercially prepared diets containing 28 or 32 percent protein. After 84 days, fish reached an average weight of 171.6 g, with no significant effects of stocking density observed on growth (avg. = 1.94 $g \cdot day^{-1}$), survival (97.9 percent), and feed conversion ratio (1.88; Table 10.3), suggesting that densities higher than 300 fish per cubic meter are possible. This is similar to what was seen for *O. aureus* (avg. wt. = 21.2 g) fed a 36 percent protein diet in marine cages, where stocking densities of up to 300 fish per cubic meter did not affect growth (McGeachin et al. 1987; Table 10.3). In freshwater, Wannigamma et al. (1985) found no differences in growth or feed conversion of *O. niloticus* (avg. wt. = 22 to 30 g) reared in cages over 120 days on 19 to 29 percent protein diets at densities of 400, 600, and 800 fish per cubic meter, causing these authors to suggest that stocking densities above 800 fish per cubic meter were possible.

Although Carro-Anzalotta and McGinty (1986) found that growth of *O. niloticus* in cages held in freshwater over 169 days on a 32 percent protein diet was negatively correlated with stocking density (250, 500, 750, and 1000 fish per cubic meter), growth rates did not diverge until late in the study, and final biomass densities ranged from 57 to 169 $kg \cdot m^{-3}$. This suggested that the inhibitory effects of high stocking densities emerged as cage biomass reached high densities. While comparisons of results are complicated by variations in experimental conditions (e.g., culture system, species, size of fish, duration, feed quality, and salinity) and differences in biomass densities attained, the results suggest clearly that stocking density can be varied widely for intensive culture of tilapia in cages, with negligible effects on growth or feed conversion.

Carrying capacities of 146 (Hargreaves et al. 1988) and 169 $kg \cdot m^{-3}$ (Carro-Anzalotta and McGinty 1986) have been reported for intensive cage culture of *O. aureus* and *O. niloticus,* respectively, in freshwater. For Florida red tilapia reared in marine cages, Hargreaves et al. (1992) obtained a yield of 8.3 $kg \cdot m^{-3}$ for fish stocked at a density of 20 fish per cubic meter (Table 10.3) whereas Watanabe, Clark et al. (1990a) obtained 52.2 $kg \cdot m^{-3}$ for fish stocked at a density of 300 fish per cubic meter (Table 10.3). This is comparable to maximum yields reported for *O. spilurus* reared in marine cages that range from 40.6 (Cruz and Ridha 1989) to 44 $kg \cdot m^{-3}$ (Al-Ahmad et al. 1988a). Since cage-rearing studies with Florida red tilapia were terminated when in-cage dissolved oxygen (DO) declined to adverse levels (<3 $mg \cdot L^{-1}$), due to fluctuating ambient levels as well as increasing biomass (Watanabe, Clark et al. 1990a), considerably higher biomass densities than achieved in these studies (52.2 $kg \cdot m^{-3}$) are attainable, given higher ambient DO. Proper site selection is therefore critical toward maximizing ambient levels of dissolved oxygen and production in cage culture.

Coche (1982) recommended 3.0 $mg \cdot L^{-1}$ as the minimum DO level below which adverse affects appear during cage culture of tilapia in freshwater. In Florida red tilapia, evidence of stress (i.e., reduced feeding) was observed while in-cage DO levels were still above 3 ppt (Clark, Ernst et al. 1990; Watanabe, Clark et al. 1990a). While circumstantial, the evidence suggests that tolerance to low DO is poorer in seawater than in freshwater and may be related to a higher oxygen requirement (i.e., metabolic rate) of Florida red tilapia reared in seawater (Febry and Lutz 1987). Furthermore, as saturation oxygen concentrations are lower in seawater than in freshwater (Stickney 1979), fish reared in seawater must obtain oxygen from a smaller supply. Under cage culture conditions where aeration is not used, maximum production of tilapia in seawater may therefore be lower than in freshwater. Selective cropping at an intermediate stage (Campbell 1985) and transfer to cages of larger mesh size may be advantageous to reduce biomass densities, improve circulation, and maintain in-cage DO levels.

A major drawback for producing marketable *O. niloticus* in freshwater cage culture was differential growth rates, which required frequent sorting

(Campbell 1985). Watanabe, Clark et al. (1990) found that higher stocking densities within the range of 100 to 300 fish per cubic meter lowered growth variation of Florida red tilapia reared in marine cages, possibly due to an inhibitory effect on aggression (Balarin and Haller 1982). This suggests that higher densities are advantageous not only for increasing production, but also for minimizing growth variation.

Earthen Ponds

Monoculture

A pilot-scale grow-out trial was conducted in northern Puerto Rico to evaluate saltwater pond production of Florida red tilapia. Sixty thousand sex-reversed male Florida red tilapia fingerlings (avg. wt. = 0.85 g) were stocked at a rate of 3 fish per cubic meter (30,000 ha^{-1}) into six 0.2 ha earthen ponds supplied with brackish well water (avg. = 22.7 $g \cdot L^{-1}$). Twice daily, fish were fed a sinking 25 percent protein tilapia grower (Zeigler Brothers, Gardners, Pennsylvania) to satiation. Because of limited groundwater supplies, only minimal water exchange (<10 $percent \cdot day^{-1}$) was maintained, and nighttime paddle wheel aeration was required during the late stages of growout.

After 160 days, fish reached an average weight of 452 g, with a feed conversion ratio of 1.8, results similar to those obtained in flow-through seawater tanks (Ernst et al. 1989; Watanabe, Chan et al. 1993; Table 10.3). Fifty percent of the fish had reached a marketable weight of 451 g or higher with an average of 545 g. Dress-out (gilled, gutted, and scaled) percentage increased with fish size from 75.5 percent in fish of 300 g to 83.5 percent in fish of 600 g.

Average maximum and minimum temperatures during the study were 31.2°C and 25.8°C, respectively. Morning (before 08:00) dissolved oxygen, NH_4-N, NO_2-N, alkalinity, and pH averaged 3.3 ± 0.2 $mg \cdot L^{-1}$, 0.60 ± 0.04 $mg \cdot L^{-1}$, 0.03 ± 0.01 $mg \cdot L^{-1}$, 185 ± 14 $mg \cdot L^{-1}$, and 7.38 (range = 6.15 to 9.86). Fish appeared healthy during the pilot trial, and few mortalities were observed; however, a final survival rate could not be determined as fish were lost during a major flood near the end of the study. Based on an estimated survival rate of 70 percent, a conservative estimate for pond culture of sex-reversed tilapia (Sin and Chiu 1983; Rakocy and McGinty 1989), each pond produced an average of 1,894 kg of whole fish (9,470 $kg \cdot ha^{-1}$).

Polyculture with Shrimp

Polyculture of tilapia and shrimp is rapidly being adopted as a primary management practice to improve economic return and to reduce shrimp losses due to infectious diseases. Farms in Thailand, the Philippines, Ecuador, Mexico, Eritrea, and the United States have adopted this particular polyculture practice. Some practitioners report higher yields of shrimp when tilapia are grown in a polyculture. There are also differences in microbial populations of bacteria and algae in production ponds utilizing polyculture. Most farms report improved economic returns per pond.

Examination of practices in the field found that three versions of tilapia-shrimp polyculture have been developed: simultaneous, sequential, and crop rotation systems (Fitzsimmons 2001). These practices are not exclusionary. Many farms will utilize these in conjunction with one another. Tilapia-shrimp farmers were interviewed regarding several aspects of why, how, when, and where they have adopted tilapia-shrimp polyculture. In the Philippines, 36 farmers from 13 separate provinces were visited and interviewed (Bolivar et al. in press). In Thailand, 61 farmers from 12 provinces were interviewed (Yang et al. 2002). In Peru, Ecuador, Eritrea, Mexico, and the United States, one or two farmers were visited and interviewed in each country.

Most of the farmers interviewed reported that the integration of tilapia and shrimp culture lessened disease problems, especially vibriosis. On-farm trials showed that luminous bacterial counts in the water and in the shrimp were less than 10 colony forming units (cfu) per milliliter and less than 10^3 cfu per hepatopancreas, respectively. These were conducted in ponds with the stocking of tilapia and shrimp directly in the ponds, as well as tilapia in hapas (simultaneous), and those whose water was taken from a tilapia pond or reservoir (sequential), and from ponds that had previously been used for tilapia culture (rotation). Several of the respondents also reported that the tilapia contributed to "conditioning" the water. Specifically, the density of green algae increased providing "greenwater." As the experience of farmers is added to lab and field studies, it appears that polyculture of tilapia and shrimp will become the norm for semi-intensive shrimp producers.

In a simultaneous polyculture setting, the tilapia and shrimp may be stocked and left unrestrained in the same pond, or the tilapia may be caged within the pond. In the case where the animals are unrestrained, the shrimp and tilapia can utilize different niches. In extensive culture, tilapia can filter feed on phytoplankton and zooplankton in the upper water column, while shrimp spend most of the time on the pond bottom grazing on bacterial films on the bottom substrate and on the detritus settling from above. In intensive culture receiving pelleted feeds, tilapia may monopolize the feed, especially floating feeds. However, some feed particles always get to the bottom where they are more available for the shrimp. More importantly, the fecal matter from the tilapia contributes to the detrital rain that supports the shrimp. Akiyama and Anggawati (1999) reported that yields of shrimp increased when red tilapia (*Oreochromis* spp.) were stocked into existing shrimp ponds. The authors reported that the stocking rate was 20 to 25 gram of fish per square meter and fish size at stocking was 50 to 100 g per fish. In addition, they reported that red tilapia assisted shrimp performance by improving and stabilizing the water quality by foraging and cleaning the pond bottom and by having a probiotic type effect in the pond environment. Tilapia, as a filter feeder, can reduce excessive phytoplankton biomass in later stages of pond culture and recycle nutrients effectively (Stickney et al. 1979).

In a simultaneous system with floating cages, the tilapia are confined, and uneaten feed and wastes fall out of the cage and are available to the shrimp below. Tilapia can also be stocked into hapa nets staked to the pond bottom. The fish can be fed or the hapas may be placed in a location within the shrimp pond where the fish will be allowed to consume materials that accumulate in that location due to the circulation pattern set up in the pond. Some producers prefer this system as it contributes to overall water quality.

In a sequential system, tilapia are typically grown in water that is then sent on to the shrimp ponds. This production scheme is one of the most popular. Tilapia have been stocked into the supply reservoir at some farms. At other locations, incoming water passes through one or more tilapia ponds before being introduced to shrimp production ponds. Most survey respondents reported, and field studies support (Saelee et al. 2002), that the tilapia culture changes water quality parameters. The most common result is an increase in density of chlorophytic phytoplankton (green algae). The density of other algae does not seem to change much. There is also an increase in suspended organic matter.

In crop rotation, tilapia are stocked between shrimp crops. Producers reported that incidences of shrimp disease are reduced, probably due to lack of hosts and changes in pond bottom ecosystem as the fish feed and build nests in the sediments. There may also be longer term benefits to the pond sediments as the tilapia typically leave the bottom pockmarked with nest sites. This may serve to reduce the anoxic and reducing conditions in bottom sediments, thereby improving pond bottom conditions for the next crop of shrimp.

In higher salinity situations, 20 ppt or above, red strains of tilapia are preferred. In intermediate levels of salinity, 10 to 20 ppt, various strains or hybrids of *O. mossambicus, O. aureus*, and *O. urolepis hornorum* are used. In the lowest salinities, *O. niloticus* seems to be the preferred tilapia.

The most highly developed polyculture system examined is the TIPS (Tilapia Introduction to Prawn System) developed by FYD International on Negros Island in the Philippines (Paclibare et al. 2001). The system was employed on 95 ha of ponds on Negros in 2002 and additional farms on Negros and other islands employed the system in 2003. TIPS utilizes tilapia in the supply canal to condition water, tilapia stocked into hapa nets staked into the center of shrimp ponds to maintain water quality, and a crop rotation with tilapia grown in monoculture after any shrimp crop that is affected with diseases.

Tilapia in the hapas consume much of the waste generated by the shrimp. Hapas are placed strategically over the spots that normally accumulate wastes based on the flow of four paddle wheel aerators operating in the corners of the rectangular ponds. Estuarine and well water sources are used with salinity range from 5 to 32 ppt during the year.

Species used for the TIPS polyculture system are *Penaeus monodon* and the locally developed "Jewel" hybrid, *O. aureus* × *O. hornorum*. Shrimp postlarvae are stocked at the rate of 15 to 25 m^{-2} and tilapia fingerlings at 3 m^{-2}. The ponds are fertilized with organic and inorganic fertilizers at the rate of 2,000 and 200 $kg \cdot ha^{-1}$, respectively. Chemicals such as agricultural lime, hydrated lime, and teaseed cake are used at the rates of 100, 25 $kg \cdot ha^{-1}$, and 20 to 30 ppm, respectively.

Water exchange is practiced as needed to maintain phytoplankton density of 500,000 to 700,000 cells per milliliter. Plankton and bacterial profiles are monitored every 2 weeks. Aeration is provided using paddle wheels. The shrimp are fed with a combination of fresh food and commercial pelleted feeds. The fresh food is normally crushed golden apple snails *(Pomacea canaliculatus)* fed daily to shrimp of 10 to 25 g in body weight. The commercial feed has a protein content of 37 to 38 percent, with feeding rates declining from 20 to 2 percent of the body weight. Feeding trays are used and checked regularly depending on the age of the shrimp. Trays are checked from 1 to 4 h when younger and every 45 min for older shrimp. The amount of feed is adjusted depending on the observation on the feeding trays.

Decreased incidence of parasitic infestation was noted in polyculture compared to monoculture of shrimp. Parasite problems were caused by protozoan parasites that were eradicated by either iodine bath or salinity change. Respondents had not used any antibiotics since they started their shrimp-tilapia polyculture. Shrimp yield was from 4,000 to 5,500 $kg \cdot ha^{-1}$ at about 30 shrimp per kilogram. Yield of tilapia was from 2,000 to 3,000 $kg \cdot ha^{-1}$ at 500 g average weight. Observed survival rate for shrimp in polyculture systems was from 70 to 90 percent and 90 percent for tilapia.

In Thailand, intensive culture of black tiger shrimp *(P. monodon)* has developed rapidly along the coastal area since 1987, and Thailand has been the world's leading marine shrimp producer since 1991 (Fast and Menasveta 2000). Some of the major problems for Thai shrimp farmers in coastal areas are disease outbreaks such as white spot that often cause the failure of marine shrimp production. Many shrimp ponds in coastal areas have been abandoned in Thailand and some other parts of the world due to diseases, poor management such as overstocking, and environmental degradation. The polyculture of shrimp-tilapia at relatively low stocking densities has provided an opportunity to develop a sustainable aquaculture system to best utilize abandoned shrimp ponds in coastal areas and low-salinity shrimp ponds in inland areas.

Nile tilapia *(O. niloticus)* is the most commonly cultured species among tilapia in Thailand. Although Nile tilapia is the least saline-tolerant species among the commercially important tilapia species, it may be the tilapia species of choice in brackishwater shrimp ponds, where the release of tilapia to estuaries is undesirable, because Nile tilapia is unlikely to reproduce at salinities higher than 25 ppt (Teichert-Coddington et al. 1997).

In the East African country of Eritrea, red tilapia are reared in the supply canal of a shrimp farm near Massawa. They are also stocked into cages in the shrimp ponds and into cages and ponds fed by the shrimp farm effluent. The water used is from the Red Sea and is hypersaline, 40 ppt. To date, slow growth and low reproduction rates have been observed in high salinity water used in the hatchery. Additional trials are continuing.

In Puerto Peñasco, Sonora, Mexico red tilapia have been reared in the effluent channel of a commercial shrimp farm for 12 years. The fish have repeatedly spawned at a salinity of 35 ppt. The tilapia were fed with commercial shrimp feed in addition to the wastes and algae they consumed from the effluent. The fish were provided to workers at the shrimp farm as extra compensation.

In Hyder, Arizona, Nile tilapia were reared in 0.04 ha ponds to condition water before entering 0.1 ha shrimp ponds. All ponds were lined with plastic and water supply was from a groundwater well at a temperature of 38°C and salinity of 4 ppt. The tilapia were fed shrimp diets, and the farm harvested an average of 2,400 kg from each 0.04 ha pond in 8 months. They reported that the effluent from the tilapia ponds improved survival and growth in the shrimp ponds.

In Peru, several shrimp producers have experimented with polyculture. The biggest trial used a crop rotation plan alternating red tilapia and shrimp. The water available was between 25 and 30 ppt. The tilapia needed repeated freshwater dips to control parasites and the farm eventually reverted to full-time shrimp production.

In Ecuador, several farms have adopted polyculture systems. Almost all utilize sequential systems that stock tilapia before shrimp to condition the water. Most farms use red tilapia with salinities that range from 12 to 30 ppt. The farms in the Guayaquil region have built a major international trade on tilapia produced in conjunction with the giant shrimp industry in that area.

Replicated Field Studies

Two experiments were conducted at the Asian Institute of Technology (AIT), Thailand for 65 and 75 days from February to April 2002 and from November to February 2003, respectively, to investigate growth performance of shrimp and Nile tilapia, and water quality at different stocking densities of Nile tilapia in tilapia-shrimp polyculture. Nine 200-m^2 earthen ponds were used for each experiment. There were three treatments in triplicate for both experiments: shrimp alone at 30 m^{-2} (monoculture, control); shrimp at 30 m^{-2} and Nile tilapia at 0.25 m^{-2} (low tilapia density polyculture); shrimp at 30 m^{-2} and Nile tilapia at 0.50 m^{-2} (high tilapia density polyculture). Feed rations varied in each pond and were estimated by feeding trays ($50 \times 50 \times 10$ cm^3) daily in experiment one. In experiment two, the same feed rations were used for all ponds and were determined by a feeding table (Lin 1995) and the estimated average shrimp survival in experiment one.

Juvenile shrimps (PL_{15}) (0.4 to 1.2 g) were stocked in all experimental ponds on February 20, 2002, for experiment one and November 27, 2002, for experiment two, while sex-reversed male Nile tilapia fingerings (5.5 to 8.0 g) were stocked 7 days after stocking shrimp. Prior to the start of the experiments, all ponds were drained completely, dried for 2 weeks, and filled with freshwater from a nearby canal to a depth of 80 cm. Then, hypersaline water (150 to 250 ppt) was added to all ponds to adjust the salinity level to 5 ppt. Water level in all ponds was maintained at 1.0 m by adding freshwater biweekly to replace water loss due to seepage and evaporation. There was no water exchange throughout the experimental periods. The ponds were fertilized once before stocking shrimp using urea at a rate of 28 kg $N \cdot ha^{-1} \cdot week^{-1}$ and triple superphosphate (TSP) at a rate of 7 kg $P \cdot ha^{-1} \cdot week^{-1}$. No other chemicals were used in the experiments.

One air blower (5 Hp) was used to supply air for all experimental ponds. Air manifolds were constructed of PVC pipes with holes drilled at regular intervals. Aerators were operated daily for 24 h except during feeding. Shrimp were fed commercial pelleted shrimp feed containing approximately 36 percent crude protein, 9 percent fat, 12 percent moisture, and 4 percent ash (Charoen Pokaphand Feed Mill Company, Thailand). Daily feed rations were divided into four equal portions and given at 600, 1,200, 1,800, and 2,200 h.

Dissolved oxygen (DO) and temperature were measured daily at 0600 h at 25 cm above bottom, middle, and 25 cm below water surface, while salinity and pH were monitored weekly at the three depths. Water samples were taken biweekly at 0900 to 0930 h for analyses of total Kjeldahl nitrogen (TKN), total ammonia nitrogen (TAN), nitrate nitrogen (nitrate-N), nitrite nitrogen (nitrite-N), total phosphorus (TP), soluble reactive phosphate (SRP), total alkalinity, and chlorophyll *a* following standard methods (Parsons et al. 1984; APHA et al. 1985).

Partial budget analyses were conducted to determine economic returns of shrimp monoculture and tilapia-shrimp polyculture systems tested (Shang 1990). The analyses were based on farm-gate prices in Thailand for harvested shrimp and tilapia, and current local market prices for all other items expressed in U.S. dollars (1US $ = 42 baht). Farm-gate prices of shrimp and tilapia varied with size: shrimp at \$4.29 kg^{-1} for size 15 to 17 g and \$3.81 kg^{-1} for size 12 to 14 g; tilapia at \$0.36 kg^{-1} for size 200 to 300 g, \$0.48 kg^{-1} for size 300 to 400 g. Market prices for shrimp PL60 (\$0.004/piece), tilapia (\$0.04 per piece), feed (\$0.78 kg^{-1}), urea (\$0.18 kg^{-1}), TSP (\$0.30 kg^{-1}), hypersaline water (\$8.66/m), and electricity (\$0.12 $W \cdot h^{-1}$) were applied to the analysis. The calculation for cost of working capital was based on an annual interest rate of 8 percent.

Shrimp growth was not significantly different among treatments in both experiments one and two (all significance determined at $p \leq 0.05$). Gross and net shrimp yields were not significantly different in experiment one, however, the shrimp yields in experiment two were significantly higher in the low tilapia density polyculture than those in the monoculture and the high tilapia density polyculture, between which there were no significant differences. In experiment one, shrimp survival in the high tilapia density polyculture was significantly higher than that in the low tilapia density polyculture whereas both were not significantly different from that in the monoculture. However, there was no significant difference in shrimp survival in experiment two. Feed input in experiment one increased significantly with increasing density of Nile tilapia whereas in experiment two it was fixed to be the same for all treatments. In experiment one, there was no significant difference in the apparent FCR for shrimps among all treatments whereas the apparent FCR for shrimps was significantly better in the low tilapia density polyculture than those in the monoculture and the high tilapia density polyculture.

Growth of Nile tilapia in both experiments was not significantly different between the low and high tilapia density polycultures, whereas yields were significantly higher in the high tilapia density polyculture than those in the low tilapia density polyculture. Survival of Nile tilapia was not significantly different between the low and high tilapia density polycultures in experiment one; however, it was significantly higher in the low tilapia density polyculture than that in the high tilapia density polyculture in experiment two.

There were no significant differences in both overall mean and final values of most water quality parameters among all polycultures during both experiments. DO concentrations at dawn fluctuated during the experimental periods and tended to be lower toward the end of the experimental period in both experiments. There were no significant differences in overall mean and final total alkalinity concentrations among polyculture in experiment one; however, total alkalinity concentrations were highest in the shrimp monoculture, intermediate in the low tilapia density polyculture, and lowest in the high tilapia density polyculture in experiment two. Overall mean SRP concentrations were significantly higher in the low tilapia density polyculture than those in the shrimp monoculture and the high tilapia density polyculture in experiment one; however, no significant differences were found in overall mean SRP concentrations in experiment two and final SRP concentrations in both experiments.

In both experiments, concentrations of TAN were higher at the beginning, then dramatically decreased in the first month, and remained stable for the rest of the experimental period. In experiment one, the overall mean concentrations of TAN were significantly higher in the shrimp monoculture and the low tilapia density polyculture than those in the high tilapia density polyculture whereas there were no significant differences in final concentrations of TAN among treatments. However, both overall mean and final concentrations of TAN were not significantly different among the treatments in experiment two. Concentrations of chlorophyll *a* fluctuated throughout the experimental period, and neither overall means nor final values were significantly different among treatments in both experiments. Secchi disk depths decreased gradually toward the end of the experimental period. The overall means of Secchi disk depths were not significantly different among treatments in both experiments; however, the final values of Secchi disk depth in the shrimp monoculture were significantly greater than those in the tilapia polyculture, which were not significantly different from each other. Salinity levels decreased rapidly from 5 to 0 ppt within the first 2 weeks in experiment one, but remained quite stable (2 to 5 ppt) for the entire culture period in experiment two.

Partial budget analyses for experiments one and two demonstrated that under varied feed input (experiment one), the highest net return was achieved in the shrimp monoculture, intermediate in the high tilapia density polyculture, and lowest in the low tilapia density polyculture. Under the fixed feed input (experiment two), the low tilapia density polyculture gave the highest net return, followed by the high tilapia density polyculture and the shrimp monoculture. Under varied feed input, the added cost produced negative added return in the low tilapia density polyculture, and the ratio of added return to added cost in the high tilapia density polyculture was 0.73. However, under fixed feed input, the ratio of added return to added cost in the low tilapia density polyculture reached 22.69, which is higher than that (5.04) in the high tilapia density polyculture.

The addition of Nile tilapia at both densities (0.25 and 0.5 tilapia per square meter) into intensive shrimp culture ponds did not significantly affect the growth and survival of shrimp under both varied and fixed feed input in the present study; however, the addition of Nile tilapia at 0.25 tilapia per square meter resulted in a significantly higher shrimp yield (21 percent) than the shrimp monoculture under the fixed feed input while there were no significant differences in shrimp yields between monoculture and polyculture under the varied feed input in the present study. Akiyama and Anggawati (1999) reported that the production and survival of shrimps was improved in an intensive polyculture system with red tilapia whereas the presence of Nile tilapia resulted in better growth and survival of shrimps at 0.4 tilapia per square meter but poorer shrimp performance at 0.6 tilapia per square meter in semi-intensive culture (Gonzales-Corre 1988). Similarly, Tian, Dong, Yan et al. (2001) reported that survival and net yield of shrimps in a polyculture system was higher by 3 to 16 percent and 5 to 17 percent than that in the monoculture, respectively, probably due to the better water quality in the polyculture system.

In the experiments conducted by Akiyama and Anggawati (1999), the stocked species was red tilapia of larger size (60 to 100 g) at densities of 0.2 and 0.3 tilapia per square meter, which resulted in the higher fish standing crop compared to Nile tilapia used in the present study. Akiyama and Anggawati (1999) attributed this positive effect to improving and stabilizing water quality, foraging and cleaning pond bottom, and having a probiotic type affect in the pond environment by red tilapia. In the semi-intensive culture, the positive effect of Nile tilapia at the low density on shrimp performance could be due to the addition of undigested food particles excreted by Nile tilapia that served directly as food for shrimps and simultaneously as fertilizer on the pond bottom, while the negative effect caused by Nile tilapia at the high density was probably due to competition for food and space (Gonzales-Corre 1988).

Tian, Dong, Yan, et al. (2001) reported that the best stocking rates were 7.2 shrimp per square meter, 0.08 tilapia per square meter, and 14 constricted tagelus *(Sinonovacula constricta)* per square meter in the polyculture of Chinese penaeid shrimp *(P. chinensis)*, Taiwanese red tilapia *(O. mossambicus* × *O. niloticus)*, and tagelus. Wang et al. (1998) reported that the optimum stocking density of Chinese shrimp and Taiwanese red tilapia was 6 shrimp per square meter and 0.32 tilapia per square meter (126.3 g in size), and shrimp growth and survival rate at all three stocking densities did not differ significantly among treatments. In comparison, Nile tilapia were stocked at average sizes of 5 to 8 g in the present experiment, thus Nile tilapia might be too small to effect any improvement in the pond environment. Compared to the above results, Thai farmers stock shrimps at much higher densities. Overstocking makes management more difficult and is not sustainable in the long run.

The growth of Nile tilapia at both low and high densities was fast and not significantly different in the present study. This density-independent growth indicated that the food availability in the polyculture ponds was sufficient to support Nile tilapia growth, and carrying capacity was not reached. It appears that the benefits reported by Akiyama and Anggawati (1999) can be achieved by stocking a higher biomass in the form of larger stocked fish or smaller fish in greater density. The polyculture of shrimp and tilapia augment the total production through nonreduction of the production of shrimp and additional tilapia production.

In experiment one of the present study, the daily feed ration was determined by the observation of feed consumption in the feeding trays, which is a normal practice in shrimp farms. The significantly higher feed input in the tilapia-shrimp polyculture than in the shrimp monoculture indicated that Nile tilapia consumed a considerable amount of costly shrimp feed. Thus, FCR of shrimp was higher in shrimp-tilapia polyculture than in shrimp monoculture; however, no significant difference in FCR was found among all treatments, due to the large variation of shrimp production and feed consumption within treatment especially in the low tilapia density polyculture. The difference of shrimp FCR between polyculture and monoculture in experiment one of the present study was higher than that reported by Akiyama and Anggawati (1999), probably due to the larger ponds and paddle wheel aerators used in their experiments, which might make more tilapia distribute in the central part of ponds and thus reduce consumption of shrimp feed. Gonzales-Corre (1988) reported that tilapia were found to compete with shrimp for food. To prevent Nile tilapia from eating costly shrimp feed in polyculture, more feed may be given during the nighttime, because shrimp can eat well during nighttime whereas Nile tilapia may not actively feed during nighttime. Alternatively, Nile tilapia could be confined in floating nets or cages to prevent them from accessing shrimp feed, and depend solely on natural foods (Fitzsimmons 2001).

In experiment two of the present study, feed input was fixed to be the same for the shrimp monoculture and tilapia-shrimp polyculture ponds. However, shrimp production was even higher in the tilapia-shrimp polyculture than in the shrimp monoculture, whereas the growth of Nile tilapia was not reduced, which was similar to that in experiment one with the higher feed input in polyculture ponds than in the monoculture ponds. This result indicated that either all ponds were overfed, or Nile tilapia preferred eating more natural foods, or shrimp were able to utilize fecal materials of Nile tilapia, which fed on the given pelleted feed. Thus, further research on optimizing feed input and detailed analysis of feeding behaviors of both Nile tilapia and shrimp is needed.

The issue of greatest concern for inland shrimp farms is the soil salinization caused by pond seepage; salinity increases in irrigation waters due to shrimp pond discharges, and sludge discharges from ponds into irrigation canals (Fast and Menasveta 2000). This concern resulted in the ban of all shrimp farming in freshwater inland areas by the Thai government in 1998 (Fast and Menasveta 2000). In the present study, the salinity level decreased to 0 ppt within the first 3 weeks in the first round of experiments, and shrimps showed good growth performance throughout the experimental period. Thus, it is possible to culture the marine shrimp species in freshwater in inland areas by stocking acclimated shrimp juveniles.

In intensive shrimp monoculture, wastes derived from feeding often stimulate phytoplankton growth and lead to dense blooms in ponds, and the collapses of phytoplankton can cause shrimp stress and mortality through disease, oxygen depletion, and increased metabolic toxicity (Briggs and Fung-Smith 1998; Fast and Menasveta 2000). The conventional solution to this situation has been increased water exchange causing environmental pollution. In the present study, the experiments were conducted in the closed system without any water exchange. However, the results of the present study showed that the concentrations of chlorophyll *a* in the tilapia-shrimp polyculture ponds were not lower than those in the shrimp monoculture ponds. Probably, the roles of Nile tilapia are not to reduce phytoplankton biomass, but to stabilize water quality in the tilapia-shrimp polyculture. Tian, Dong, Liu et al. (2001) investigated the water quality in a closed polyculture system containing Chinese penaeid shrimp with Taiwanese red tilapia and constricted tagelus. They found that bacteria and organic matter were significantly reduced in the polyculture system when compared to monoculture. In addition, nitrogen and phosphorus levels were measured in the sediments of the polyculture enclosure and found to be 39.76 and 51.26 percent lower than those of monoculture sediments, respectively. These results indicate that tilapia are useful in improving water quality in shrimp ponds.

The present study has demonstrated that the tilapia-shrimp polyculture is technically feasible and can be environmentally friendly and economically attractive with appropriate feeding strategy. The use of cost-effective diets and optimization of feeding inputs is therefore vital in sustainable shrimp farming and can make the shrimp-tilapia polyculture more attractive to shrimp farmers. The present study indicated that the addition of Nile tilapia into shrimp ponds can improve feed utilization efficiency, resulting in better economic returns and less environmental pollution.

Effect of Feeding Regimen

When Florida red tilapia were reared from small fingerlings to marketable sizes in flow-through seawater tanks on a 30 percent protein diet fed daily to satiation, defined as the percentage of body weight consumed in three 30-min feeding periods, feeding rate decreased as a function of size, as described by $F = 56.0\ W^{-0.566}$, where F is the feeding rate (percent $bw \cdot d^{-1}$) and W is the fish weight (g) (Ernst et al. 1989). Over the 170-day study period, feeding rate declined from >30 percent at stocking to 1.5 percent when the study was terminated. For fish >30 g, these feeding rates exceeded those calculated from published equations (Balarin and Haller 1982) and tables (Coche 1982; Granoth and Porath 1983; Liao and Chen 1983) for freshwater-reared tilapia. Two lines of evidence suggest that high feeding rates may have been due in part to salinity. Feeding and growth rates of juvenile male Florida red tilapia increased with salinity (to 36 ppt) (Watanabe, Ellingson et al. 1988; Watanabe, French et al., 1988). Additionally, oxygen consumption rates (i.e., metabolic rates) of *Oreochromis* spp. are reported to be greater in seawater (>30 ppt) than in freshwater (<1 ppt) (Farmer and Beamish 1969; Job 1969; Febry and Lutz 1987).

To further examine the effect of feeding regimen, growth and feed conversion were monitored for sex-reversed male Florida red tilapia (avg. wt. = 10 g) held in floating marine cages at a density of 300 fish per cubic meter and fed a floating pelleted diet (32 percent protein) under six different regimens: Four programmed rates representing 50, 70, 90, and 110 percent of an estimated satiation rate, apparent satiation (based on 30-min feeding period) and by demand feeders (Clark, Watanabe, Ernst et al. 1990).

Growth rates, expressed as mean daily weight gain, were significantly lower in the 50 percent programmed feeding treatment (0.98 $g \cdot day^{-1}$) than in all other treatments (1.47 to 1.82 $g \cdot day^{-1}$), while feed conversion ratios were significantly lower under the 50 percent (1.57) and 70 percent (1.68) programmed feeding treatments than under all other treatments (range = 1.98 to 2.26) (Clark, Watanabe, Ernst et al. 1990). Thus, maximum growth was achieved at feeding rates of 70 percent or above of an estimated satiation rate whereas feed conversion was improved at lower feeding rates. This is similar to what was observed in *O. niloticus* grown in cages held in freshwater in the Ivory Coast (Campbell 1985). Lowered conversion efficiency at higher than optimum feeding rates may result from decreased digestion of feed (Balarin and Haller 1982).

The use of demand feeders produced growth and feed conversion similar to manual feeding at near-satiation rates and may be the most cost-effective method for feeding tilapia in marine cages since labor is reduced (Clark, Watanabe, Ernst et al. 1990). Use of demand feeders for *O. aureus* raised in cages in freshwater reduced labor requirements by 88 to 94 percent (Hargreaves et al. 1988) and produced fish that were 72 percent heavier than manually fed fish after 11 weeks (Meriwether 1986). Feed conversion under demand feeding in marine cages can be further improved through feeder designs to minimize wind- and wave-induced release.

Dietary Protein Requirements During Grow-Out

Economic analyses of intensive saltwater pond (Head et al. 1996) and tank culture operations (this chapter) for Florida red tilapia have revealed that feed costs represent from 32.4 to 34.2 percent of annual operating expenditures (Tables 10.4 and 10.9). As protein represents the most expensive component in a prepared feed, it is of considerable practical importance to determine the lowest level supporting optimal growth and survival.

Although many studies have assessed the dietary protein requirement in tilapia, most have been conducted with juveniles or fry and for relatively short periods (see Jauncey and Ross 1982 for review). Relatively little information is available on dietary protein requirements from the fingerling to marketable sizes, a period when most of the feed costs are incurred during intensive culture. Furthermore, there is few or no information on dietary protein requirements of tilapia reared in full-strength seawater.

Growth, feed conversion, and protein utilization were compared for Florida red tilapia reared in flow-through seawater pools from fingerling to marketable size on prepared diets containing different protein levels (Clark, Watanabe, Olla, and Wicklund 1990). Juvenile sex-reversed males (10.6 g avg. wt.) were stocked into 10 m^3 pools at a density of 25 fish per cubic meter. Fish were fed three isocaloric diets containing 20, 25, or 30 percent protein of equal quality, with protein from fishmeal representing approximately 26 percent of the total protein in all diets. The gross caloric content of the diets was approximately 4.08 $kcal \cdot g^{-1}$. After 120 days, fish reached an average weight of 440 g, and there were no treatment differences observed in growth rate (3.68 $g \cdot day^{-1}$), survival (97.3 percent), feed conversion ratio (2.12), feed consumption (7.79 $percent \cdot day^{-1}$), or carcass composition (Clark, Watanabe, Olla, and Wicklund 1990; Table 10.3). Hence, a dietary protein level of 20 percent supported growth at rates comparable to

TABLE 10.4. Parameters used for the economic analysis of a saltwater pond aquaculture facility for Florida red tilapia in Puerto Rico.

Parameter	Value
Farm size	
Total farm area (ha)	30
Pond size (ha)	0.8
Total pond area (ha)	20
Stocking data	
Average initial wt. of reversed males (g)	0.85
Stocking density (fish per m^2)	3.5
(fish per ha)	35,000
Harvest data	
Duration of grow-out (days)	220
Pond preparation time (days)	20
Avg. number of crops per year per pond	1.5
Average final weight (g)	545
Survival (percentage)	70
Feed conversion ratio (dry wt. per wet wt.)	1.8
Yield per crop (kg per ha)	13,353
Average annual yield (kg per ha per year)	20,030
Total farm yield (even years) (kg)	427,296
Total farm yield (odd years) (kg)	373,884
Unmarketable fish (percentage)	4
Dress-out (percentage)[a]	82.7

Source: Data from Head et al. 1996.
[a] Gilled, gutted, and scaled.

those obtained at higher protein levels. These workers observed no differences in body protein content among Florida red tilapia fed at the three dietary protein levels. Other studies with tilapia similarly have reported that body protein content was not greatly affected by changing dietary protein level (Mazid et al. 1979; Winfree and Stickney 1981; Jauncey 1982).

Studies with other species have also reported a relatively low protein requirement for tilapia reared in seawater. In sea cage-rearing studies with Florida red tilapia, a 28 percent protein diet produced higher survival and growth than a 32 percent protein diet, suggesting that excess dietary protein inhibited growth (Watanabe, Clark et al. 1990a). These findings are similar to what Shiau and Huang (1989) found in hybrid tilapia (*O. niloticus* × *O. aureus*) reared in 32 to 34 ppt seawater, in which optimal dietary protein level within the range of 0 to 56 percent protein (in 8 percent increments) was 24 ppt. Shiau and Huang (1990) reported that, when energy level of the diets was 310 kcal·100 g^{-1}, the dietary protein level for hybrid tilapia grown in seawater can be lowered from 24 to 21 percent. These data, together with those of Clark, Watanaber, Olla, and Wicklund (1990), may suggest that tilapia grown in seawater may have lower protein requirement than those grown in freshwater. In contrast, Fineman-Kalio and Camacho (1987) reported that a diet containing 30 percent protein produced better growth than 20 percent or 25 percent protein when fed to *O. niloticus* fingerlings grown in brackishwater (29 ppt) ponds. These dissimilar results may be due to differences in protein sources, dietary energy levels, feeding management, salinity differences, differences in salinity tolerance among various species, or other conditions of culture. More studies are needed to determine species-specific differences in protein requirements of tilapia relative to salinity, but the results to date indicate that tilapia requires about 24 percent protein to produce maximum growth when grown in seawater.

Based on the ingredient costs of the different diets, Clark, Watanabe, Olla, and Wicklund (1990) found that relative feed costs per unit fish weight produced decreased with decreasing dietary protein from US$2.39 in the 30 percent protein diet to US$2.20 in the 20 percent diet. This demonstrated that feed costs might be reduced during growout of fingerling Florida red tilapia to marketable sizes by using diets of relatively low protein content. As growth was not impaired at the 20 percent protein level, it is possible that reasonable growth rates may be sustained at even lower protein levels. To further reduce costs, the maximum rate of inclusion of inexpensive plant protein sources as a substitute for fishmeal in a 20 percent protein diet should be determined.

COMPARISON OF SALINE-TOLERANT SPECIES UNDER SALTWATER CULTURE

Considerable data has been obtained in recent years on the saltwater culture performance of a number of tilapia species (Table 10.3). However, complete data on growth in saltwater from small fingerlings to marketable sizes is often lacking, and differences in both abiotic (salinity, temperature, and tank type) and biotic (feeding, feeds, sex ratios, and stocking densities) conditions used in these studies make inter- and intraspecific comparisons difficult.

Despite these inconsistencies, the relative saltwater culture performance of the different species may be discerned. Growth rates of Florida red tilapia in flow-through seawater pools from small fingerling (1 to 10 g) to marketable size (>450 g) range from 2.71 to 3.75 g·day^{-1} (Ernst et al. 1989; Clark,

Watanabe, Olla, and Wicklund 1990; Watanabe, Chan et al. 1993; Ellis and Watanabe 1993b) (Table 10.3). Irrespective of the culture system, these values are the highest reported for tilapia cultured in seawater when studies beginning with fingerling-stage fish are considered. Growth rates of Florida red tilapia in seawater are comparable to those reported for Taiwanese red tilapia *(O. mossambicus* × *O. niloticus)* under intensive freshwater culture where fish are grown from 1 to 500 g in 150 to 180 days (2.77 to 3.33 g·day^{-1}) (Liao and Chen 1983). An exceptional growth capacity of the Florida red tilapia in seawater is evident from these results.

Next to the Florida red tilapia, the species studied most extensively for saltwater culture is the *O. spilurus* (Al-Ahmad et al. 1988a,b; Hopkins et al. 1989; Cruz and Ridha 1989; Cruz and Ridha 1990). Available data indicates that *O. spilurus* grows slower in seawater than the Florida red tilapia under moderate temperatures. For example, growth of sex-reversed males from 3 to 80 g in concrete tanks in 39 to 41 ppt seawater required 226 days (0.26 g·day^{-1}) at temperatures ranging from 20°C to 28°C (Hopkins et al. 1989), whereas growth of mixed-sex fingerlings from 1.1 to 175 g in fiberglass tanks in 36 ppt seawater required 301 days (0.58 g·day^{-1}) at temperatures ranging from 22.5°C to 33.0°C (Cruz and Ridha 1990) (Table 10.3). In contrast, Florida red tilapia were grown in 37 g·L^{-1} seawater from 1 to 467 g in 170 days (2.7 g·d^{-1}) at temperatures ranging from 22.0 to 29.5°C (Ernst et al. 1989) (Table 10.3).

Although relatively slow growing, *O. spilurus* (as well as its hybrids with *O. aureus*) appear to be very cold tolerant, surviving much better than Taiwanese red tilapia or *O. aureus* under winter conditions in Kuwait when seawater temperatures declined to 19°C to 25°C (Hopkins et al. 1986, 1989). Hence, *O. spilurus* and its hybrids with *O. aureus* may be best suited for seawater culture under the low-temperature conditions that prevail in arid or sub-tropical regions during winter (Hopkins et al. 1989). Yields of up to 68.6 kg·m^{-3} attained for *O. spilurus* grown in 0.5 m^3 tanks in seawater (Cruz and Ridha 1990) (Table 10.3) are the highest reported to date for tilapia grown in seawater.

Taiwanese red tilapia have also received considerable attention for saltwater culture (Liao and Chang 1983; Meriwether et al. 1984; Hopkins et al. 1989; Cheong et al. 1987), but growth from small fingerlings through marketable sizes under these conditions has been demonstrated in only one study (Cheong et al. 1987) where fish were reared from 0.78 to 438 g in 239 days (1.82 g·day^{-1}) in recirculating seawater (26 to 30 ppt) tanks, with yields of up to 49.3-50.2 kg·m^{-3} attained (Table 10.5). These growth rates are moderate when compared to growth of Taiwanese red tilapia in freshwater (2.77 to 3.33 g·day^{-1}) (Liao and Chen 1983) and when compared to the Florida red strain in seawater (2.74 to 3.75 g·day^{-1}) (Ernst et al. 1989; Clark,

Watanabe, Olla, and Wicklund 1990; Ellis and Watanabe 1993b; Watanabe, Chan et al. 1993) (Table 10.3). Although Taiwanese red tilapia grew much faster than *O. spilurus* or *O. aureus* in seawater, survival was poorer during the winter when temperatures declined below 25°C (Hopkins et al. 1989).

Oreochromis aureus was the first species studied for seawater culture and has been grown in seawater (39 to 45 ppt) ponds to relatively large sizes of 239 to 382 g, albeit at low stocking densities (400 to 1,000 ha^{-1}) (Chervinski and Yashouv 1971; Chervinski and Zorn 1974; Hopkins et al. 1989) (Table 10.3). Under intensive culture in tanks or floating cages, however, growth rates (0.15 to 0.34 $g \cdot day^{-1}$) and survival rates have been disappointing (McGeachin et al. 1987; Hopkins et al. 1989) for fish reared through smaller final sizes (55.4 to 58 g), and a maximum yield of only 6.69 $kg \cdot m^{-3}$ has been attained with this species (Hopkins et al. 1989; Table 10.3). In Kuwait, *O. aureus* grew more slowly than *O. spilurus* or Taiwanese red tilapia during seawater grow-out trials and also exhibited skin lesions under cold conditions, indicating a relatively poor tolerance to high salinity and low temperature conditions simultaneously (Hopkins et al. 1989).

Saltwater pond culture of *O. niloticus* (Fineman Kalio and Camacho 1987; Fineman Kalio 1988) and *T. zillii* (Chervinski and Zorn 1974) has been attempted in the Philippines and in Israel, respectively, although growout of these species to large marketable sizes was not demonstrated (Table 10.3).

In the U.S. Virgin Islands, *O. mossambicus* survived well in floating marine cages at 35 to 38 ppt, but grew considerably slower (2.08 $g \cdot day^{-1}$) than the Florida red tilapia (2.56 $g \cdot day^{-1}$) under these conditions (Hargreaves et al. 1992).

To summarize, available information suggests the Florida red tilapia to be best suited for saltwater culture under tropical conditions, exhibiting the highest growth rates of the various species studied for culture in high-salinity systems. *O. spilurus,* while slower growing than the Florida red tilapia in seawater, possesses better low temperature tolerance and is better suited for culture under arid or subtropical conditions where temperatures below 25°C prevail. Taiwanese red tilapia, *O. aureus, O. niloticus, T. zillii,* and *O. mossambicus* can be acclimated to high-salinity brackish- or seawater, but grow-out performance for these species has been relatively poor.

DISEASES ENCOUNTERED DURING CULTURE IN SALTWATER

A potential constraint to commercial culture of tilapia in full-strength seawater is susceptibility to infection by the marine monogenean,

Neobenedenia melleni Yamaguti 1968 (Monogenea: Capsalidae) (Kaneko et al. 1988; Gallet de Saint Aurin 1990; Mueller et al. 1992; Robinson et al. 1992; Ellis and Watanabe 1993b; Watanabe, Chan et al. 1993), an ectoparasitic flatworm that has been reported to infect fish in over 20 different families (Jahn and Kuhn 1932; Nigrelli and Breder 1934; Mueller et al. 1994). The reproductive biology and life history of *N. melleni* was first described by Jahn and Kuhn (1932). Adults, ranging from 2 to 7 mm in length, attach to the sides, head, and eyes of fish by means of a spined posterior sucker or opisthaptor. Using anterior, disk-like feeding organs, collectively known as the prohaptor, they feed on mucus and epithelium, causing external lesions, blindness, and exposing the dermis to secondary infections of bacteria, viruses, and fungi, leading to death (Paperna 1987; Thoney and Hargis 1991).

The life cycle of *N. melleni* is short and direct, involving only a fish host. The hermaphroditic adults produce hundreds of eggs, which are tetrahedral in form (approximately 120 μm on a side) and usually bearing a single long filament and two shorter filaments. Eggs are produced singularly at oviposition and form long strings or clusters, which fall to the substrate and hatch, or are deposited and developed directly on the surface of the host fish. At a temperature of 25°C, a ciliated onchomiracidium larva (approximately 225 μm in length) emerges approximately 24 to 48 h after the eggs are laid (Mueller et al. 1992) and swims in search of a host. If no host is found within 6 h, the larva sinks to the bottom and dies. A larva that finds a host attaches itself by the posterior sucker, loses its cilia, and begins to feed and develop into an adult. Although the generation time is not well known and likely varies with temperature, it is estimated to be around 7 to 10 days at 25°C. A short generation time, high fecundity, and the adhesive nature of the eggs create a high potential for proliferation under intensive fish culture conditions.

Kaneko et al. (1988) first reported parasitosis by *N. melleni* in *O. mossambicus* reared in seawater cages in Hawaii. Ernst et al. (1989) later noted that Florida red tilapia raised in flow-through seawater tanks in the Bahamas were susceptible to a marine monogenean, later identified by Watanabe, Chan et al. (1993) as *N. melleni.* Gallet de St. Aurin et al. (1990) also reported that Florida red tilapia grown in seawater cages in the French West Indies were susceptible to infection by this parasite. *Neobenedenia melleni* was also found on red tilapia grown in sea cages in Jamaica (Hall 1992; Robinson et al. 1992). In Florida red tilapia, early symptoms of parasitosis include reduced feeding, frayed fins, resting on tank bottoms, hyperproduction of skin mucus, exophthalmia, and corneal opacity (Plate 15) (Ernst 1989; Watanabe 1991; Mueller et al. 1992; Ellis and Watanabe 1993b).

Hargreaves et al. (1992) reported that both Florida red tilapia and *O. mossambicus* cultured in floating marine cages in the U.S. Virgin Islands showed disoriented swimming behavior, epidermal hyperplasia, exophthalmia, and pale livers, but without evidence of internal and external parasites. Cruz and Ridha (1989) also observed exophthalmia in *O. spilurus* reared in seawater cages, although an etiological agent was not identified. The possibility that *N. melleni* could have been a primary pathogen in both these studies should not be dismissed.

Chemical Control of N. melleni Parasitosis

Formalin has been frequently used in the treatment of monogeneans, including *N. melleni* (Thoney and Hargis 1991). Robinson et al. (1992) reported that treatment of Jamaican red tilapia (*Oreochromis* spp.) parasitized by *N. melleni* with 60 ppm formaldehyde (formalin) for 30 s resulted in the dislodgement of both juveniles and adult parasites and the recovery of treated fish following transfer to a parasite-free environment. On the other hand, Watanabe and Smith (unpublished data) found that treatment of infected Florida red tilapia in seawater with Paracide-F, a formalin-based product (Argent Chemical Co., Washington) at concentrations ranging from 50 to 150 ppm for 1 h had minimal effects on *N. melleni,* although fish were visibly stressed at 150 ppm. It must be considered that, due to the effects of hyposalinity on *N. melleni,* the salinity of the treatment medium will heavily influence the efficacy of a chemical agent.

Potassium permanganate ($KMnO_4$) has been used to treat monogenean infestations in cultured fish (Thoney and Hargis 1991). Treatment of *N. melleni*-infected Florida red tilapia with $KMnO_4$ at concentrations of 0.25 to 1.0 ppm (in seawater) for 10 min killed juveniles and adult monogeneans, and adverse effects on fish were only observed at $KMnO_4$ concentrations of ≥2 ppm (Watanabe and Smith unpublished data). Based on these laboratory-scale results, $KMnO_4$ was tested at concentrations of 0.25 and 0.5 ppm (applied for 1 hour, weekly) as a prophylaxis against parasitosis for Florida red tilapia reared at a density of 25 fish per cubic meter for 175 days in 10 m^3 flow-through seawater tanks. Fish remained robust and survival was high (95 percent) in both $KMnO_4$-treated and untreated (control) groups through 120 days, but survival declined thereafter as parasitosis by *N. melleni* intensified in both groups. By day 175, survival of control fish (60.7 percent) was not significantly different from those treated with $KMnO_4$ (78.0 percent) (Watanabe et al. 1991). Hence, $KMnO_4$, at concentrations lethal to adult *N. melleni* in aquaria, did not prevent parasitosis under practical culture conditions. Oxidation of attached algae and detritus on the bottom and sides of the tanks probably lowered the effectiveness of $KMnO_4$. Furthermore, egg stages are much more resistant than adult *N. melleni* to $KMnO_4$, with 3.6 percent hatching success observed after a 24-h exposure to 100 ppm $KMnO_4$ in freshwater (Watanabe et al. unpublished data). The excessive concentrations required for destruction of both egg and adult stages would make this chemical impractical for commercial application.

Although the organophosphate insecticide trichlorfon is considered to be the most cost-effective chemical method to treat monogenean infections in large fish-culture systems, it can produce harmful effects on mammals and is a potential teratogen, requiring extreme caution in handling (see Thoney and Hargis 1991 for review). Administration of trichlorfon in feed (50 $mg \cdot kg^{-1}$ live weight, four times per month) is recommended for treatment of *N. melleni* parasitosis of Florida red tilapia reared in floating sea cages (Gallet de St. Aurin et al. 1990). However, when tested on Florida red tilapia reared in flow-through seawater tanks, the effectiveness of this treatment appeared to diminish over a 150-day grow-out period (Watanabe et al. unpublished data). Thoney (1990) suggested that greater monogenean survival during repeated treatment with trichlorfon may be due to more resistant individuals that produce resistant offspring.

Other chemicals used to treat monogenean infections in fish include copper sulfate, mebendazole, praziquantel, and benzocaine (Thoney 1990; Svendsen and Haug 1991; see Thoney and Hargis 1991 for review). However, their specific effects on treatment of *N. melleni* infection in tilapia have not been examined.

Although chemical treatments may be effective in the laboratory, they are often impractical for treating food fish on a commercial scale due to high costs, potentially harmful effects to consumers and the environment, and the development of resistant strains of parasites and related governmental restrictions. These factors presently limit the widespread use of chemical treatments on food fish (Thoney and Hargis 1991).

Effects of Hyposalinity on Eggs, Juveniles, and Adults of N. melleni

Short-term (2 to 3 min) immersion of infected fish in freshwater was reported to be effective in removing adult *Benedenia seriolae* (Monogenea) and *N. melleni* from seawater-cultured yellowtail *Seriolae quinqueradiata* (Hoshina 1968) and tilapia *O. mossambicus* (Kaneko et al. 1988), respectively, and is a safe, environmentally sound alternative to chemical treatments. On a commercial scale, treatment with freshwater can be impractical when freshwater is in limited supply and can be stressful to seawater-acclimated fish. Brackishwater is often available and can be used to conserve freshwater supplies by mixing with seawater, and is less stressful to fish.

Egg Stages

The effects of hyposalinity on survival of eggs stages of *N. melleni* were studied (Mueller et al. 1992; Ellis and Watanabe 1993b) as a basis for treatment of ectoparasitosis in seawater-cultured tilapia. Eggs, collected from adult monogeneans at 37 to 38 ppt, were exposed in vitro to salinities of 0, 3, 6, 12, 15, 18, 24, 30, and 37 to 38 ppt (control) for periods ranging from 2 to 14 days, then returned to 38 ppt and posttreatment development and hatching observed. Hatching success generally declined with decreasing salinity and increasing duration of exposure. Under all durations of exposure, hatching success remained relatively high (≥69.6 percent) at salinities of 24 ppt and above, but declined markedly (≤32.5 percent) at lower salinities. Hatching was prevented by exposure to salinities of 18 ppt for 7 days and 15 ppt and below for 5 days. Although considerable embryonic development (22.5 percent eyed-eggs) was still observed after exposure to 18 ppt for 7 days, exposure to 15 ppt for only 5 days nearly eliminated embryonic development (0.5 percent eyed-eggs). Eggs exposed to freshwater (0 ppt) for 48 h still showed a hatching of 13.3 percent, while hatching did not occur in eggs exposed to freshwater for 72 h and longer.

Juvenile and Adult Stages

To determine the effects of hyposalinity on juveniles and adults of *N. melleni,* infected Florida red tilapia were exposed to salinities of 12, 15, 18, and 36 to 38 ppt (control) for periods ranging from 2 to 7 days (Ellis and Watanabe 1993b). Exposure to salinities of 15 ppt or below for a minimum of 2 days eliminated juvenile and adult monogeneans *in situ*. Although exposure of eggs to 18 ppt in vitro for 7 days eliminated hatching, this treatment was not 100 percent effective on juveniles and adults *in situ*. This result was unexpected, since it was presumed that the egg, with its sclerotized protein shell (Kearn 1986), is the more resistant life stage (Mueller et al. 1992; Watanabe, Chan et al. 1993). However, *in vitro* studies on eggs were conducted in June under temperatures of 22.6°C to 29.2°C, whereas *in situ* studies on juveniles and adults were conducted in November and December, when ambient temperature fell as low as 20.5°C. While circumstantial, the evidence suggests that *N. melleni* may be more resistant to hyposaline conditions under lower temperatures prevailing in autumn and winter. Seasonality of *N. melleni* infections has been previously reported for seawater-cultured tilapia in Martinique (Loyau 1985, cited by Gallet de Saint Aurin et al. 1990), where infestation peaked from March to May and from August-October. A higher incidence and severity of infection levels during the months of October and November during grow-out of Florida red tilapia in seawater tanks (Ernst et al. 1989; Ellis and Watanabe 1993b) also support a higher virulence of this parasite under seasonally declining water temperatures.

Increased severity of infection during the fall, on the other hand, may be related to a lowered resistance of tilapia to infection under declining temperatures (Roberts and Sommerville 1982). Ernst et al. (1989) observed evidence of trematode parasitosis in Florida red tilapia reared in seawater pools when water temperature rapidly dropped from 28-29°C to 24-25°C. The Florida red strain is known to be thermophilic, with growth rates in seawater being markedly lower at 22°C than at 27°C to 32°C (Watanabe, Ernst et al. 1993). The interactive effects of temperature and salinity on the survival of eggs and adults of *N. melleni* as well as on the resistance of Florida red tilapia to infection require further study.

Treatment of N. melleni Ectoparasitosis in Seawater-Cultured Tilapia

In a preliminary study (Watanabe et al. 1991), survival of nonparasitized Florida red tilapia grown in seawater tanks receiving prophylactic treatment with brackishwater (i.e., 18 ppt for 72 h, twice monthly) remained high (93.6 percent) after 175 days whereas survival of untreated fish declined to 60.6 percent as a result of infection by *N. melleni.* In a subsequent study (Ellis and Watanabe 1993b), hyposalinity was tested as a therapeutant for *N. melleni* parasitosis of Florida red tilapia during seawater grow-out under commercial-scale conditions. Fingerlings (avg. wt. = 22 g) were stocked into 10 m^3 flow-through seawater pools at a density of 30 fish/m^3 and infection levels monitored over 210 days. Fish were treated with brackishwater (18 ppt for 7 days) to prevent hatching of *N. melleni* eggs when average infection level reached ≥5 monogeneans per fish. An average of only 1.3 (range = 0 to 2) treatments was sufficient to maintain good growth, normal appearance, and high survival (avg. = 84.2 percent) over an extended grow-out period of 210 days, with fish reaching a final average weight of 592 g. Frequency of occurrence of parasitosis during this study suggests that, if fish are harvested earlier at an average weight of 454 g (day 140), one treatment with brackishwater (18 ppt for 7 days) would suffice.

Fish treated with brackishwater usually showed a very low level of infection a month earlier. This suggests that, by monthly monitoring of fish stocks, a farmer can detect infection at the early stages, with ample time to initiate successful treatment. Treatment begun at the incipient stages would also minimize proliferation and the possibility of autoinfection following treatment. Since exposure to 15 ppt for 5 days was found to eliminate hatching of eggs in vitro, as well as juvenile and adult stages in situ, these conditions (15 ppt for 5 days) should provide an even more effective therapeutant, particularly at lower temperatures, but require verification on a commercial scale.

As prevention is the key to effective disease management, it is recommended that land-based commercial growers of tilapia filter influent seawater to minimize introduction of parasites including *N. melleni* (Watanabe 1991; Robinson et al. 1992). Considering the size of *N. melleni* eggs (120 μm) (Jahn and Kuhn 1932; Mueller et al. 1992), this may be achieved by using a buried intake or seawater well (Watanabe 1991). Parasite load of fish stocks should be monitored each month by bathing a sample of fish for a minimum of 10 min in freshwater to kill *N. melleni,* which turn opaque in death. If necessary, fish may be treated therapeutically by reducing salinities for a sufficient duration to kill *N. melleni*. Periodic brackishwater treatment of nonparasitized fish can also be highly effective as a prophylactic measure (Watanabe et al. 1991).

Hyposaline control of *N. melleni* parasitosis is impractical during culture of tilapia in marine cages. Robinson et al. (1992) reported that relocation of cages containing formaldehyde-treated, previously infected Jamaican red tilapia to a site with a slight (<0.5 knots) current, and at least 3 m above the substratum, allowed parasite-free culture for a period exceeding 14 weeks, presumably by preventing onchomiracidia from reaching the cages and becoming established. This simple, inexpensive method for preventing *N. melleni* parasitosis in marine cages merits further evaluation.

Biological Control Methods: Cleaning Symbiosis

Recent studies suggest that tropical cleaner fish may be a viable biological method for controlling monogenean parasitosis in seawater-cultured tilapia. The abilities of three well-known tropical species, the cleaning goby *(Gobiosoma genie),* the neon goby *(G. oceanops),* and the juvenile bluehead wrasse *(Thalassoma bifasciatum),* to remove *N. melleni* from seawater-cultured Florida red tilapia were compared (Cowell et al. 1993). The cleaning goby and the bluehead wrasse were wild-caught whereas the neon goby was artificially cultured. Infection levels (number of monogeneans per fish) were monitored for individual tilapia maintained with and without cleaner fish in three 8-day trials.

Although all the three species ingested monogeneans, the cleaning and neon gobies displayed cleaning abilities that were superior to the bluehead wrasse. Both goby species significantly reduced monogenean infection levels in two trials, with nearly complete removal of monogeneans observed when initial infection levels were moderate. This suggests that the gobies are capable of rapidly eliminating monogeneans from light to moderately infected fish. On the other hand, the bluehead wrasse, which is a facultative cleaner (Feddern 1965; Itzkowitz 1979), consumed more planktonic foods than did the gobies and appeared incapable of rapidly reducing or eliminating monogenean populations on infected tilapia. The cleaning performance of the neon gobies is especially noteworthy, given that they were aquarium raised and unaccustomed to symbiotic cleaning.

The widespread use of cleaner fish for control of ectoparasites on cultured fish will be dependent on their availability. Many members of the genus *Gobiosoma* are easy to spawn and rear in captivity (Colin 1975; Thresher 1980). Techniques for raising neon gobies are well established (Moe 1975; Walker 1977), and they are currently being produced commercially (G. Waugh, Aqualife Research Corporation, personal communication).

Given their effectiveness in removing *N. melleni* and their suitability for artificial propagation, neon and cleaning gobies are recommended for future studies on the control of monogenean parasitosis in seawater-cultured tilapia. These studies should determine the ratios of cleaners to tilapia required to prevent or eliminate infection. Long-term experiments conducted in larger culture units, with larger numbers of tilapia and cleaners, are needed to evaluate the practicality of this method. The cost of supplying and maintaining the cleaners, their possible predation by tilapia, and the effect of their presence on harvesting procedures must also be evaluated.

Other Diseases

Under static brackishwater (18 $g \cdot L^{-1}$) conditions used to treat *N. melleni*-infected Florida red tilapia in seawater pools, Ellis and Watanabe (1993b) observed secondary infection with the dinoflagellate *Amyloodinium ocellatum,* leading to poor survival. These symptoms, however, were eliminated by restoring flow-through seawater conditions (Watanabe and Ellis unpublished data). *Amyloodinium ocellatum* should therefore not represent a problem under normal culture conditions.

Information on other diseases affecting tilapia reared in full-strength seawater is sparse. Tilapia cultured in full-strength seawater are more prone to stress caused by handling. Liao and Chang (1983) noted that Taiwanese red tilapia were easily bruised and attacked by fish lice when cultured in saline water (17 to 34 $g \cdot L^{-1}$). Al-Ahmad et al. (1986; 1988a) found that *O. spilurus* raised in seawater were less resistant to bacterial infections, with bigger fish being more susceptible, indicating that fish size may affect handling stress in seawater. These authors recommended stocking fish of not more than 60 g for optimal production.

Available evidence suggests that in tilapia, cold stress is exacerbated under high salinities and appears to be based on the interactive effects of temperature and salinity on osmoregulation (Allanson et al. 1971; Tilney and Hocutt 1987). Hence, under low-temperature conditions, tilapia reared in seawater are particularly susceptible to handling stress. Cruz et al. (1990) noted that transfer of *O. spilurus* from net cages under low seawater temperatures (18°C) induced injury and associated bacterial infections and recommended transferring fish when seawater temperature is above 23°C. *Omyloodinium aureus* is known to be relatively cold tolerant (Chervinski 1982) and can be overwintered in freshwater at 15°C to 16°C for 5 to 6 months (Behrends and Smitherman 1984). However, McGeachin et al. (1987) reported a high incidence of disease and mortalities in *O. aureus* reared in seawater (34 to 37 $g \cdot L^{-1}$) cages when temperatures declined to 21 to 22°C during winter. They noted that handling of seawater-reared *O. aureus* at these temperatures induced external lesions similar to those reported by other workers (Chervinski and Yashouv 1971; Chervinski and Zorn 1974) for this species raised in seawater ponds. Cultures from external lesions and internal organs revealed the presence of four strains of Gram variable, spore-forming bacteria identified as *Bacillus* sp., probably opportunistic pathogens that became established following stress of high salinity and low temperature conditions (McGeachin et al. 1987).

ECONOMICS AND MARKETING

Economic Analysis of a Recirculating, Brackishwater Tilapia Hatchery

Although there have been economic studies on open pond (Broussard and Reyes 1985; Yater and Smith 1985; Little et al. 1987), hapa (Yater and Smith 1985), and tank (Hida et al. 1962) tilapia hatchery systems, little or no information is available for intensive, recirculating systems. Using actual design criteria and production data from a commercial-scale research hatchery in the Bahamas, Head and Watanabe (1995) performed an economic analysis of a recirculating, brackishwater hatchery for Florida red tilapia. This analysis showed that, at annual production levels of 1.0 to 2.5 million sex-reversed fry, breakeven prices ranged from $0.07 to $0.16 per fry. At annual sales of 2 million sex-reversed fry, internal rates of return greater than 15 percent were projected at a price of $0.11 per fry, indicating economic feasibility at competitive pricing, if fry can be sold year round. The vertical integration of hatchery and grow-out operations offers best prospects for economic success, allowing timing of production to meet inhouse needs as well as external sales demands for sex-reversed fry or broodstock.

A minimum internal rate of return (IRR) (Shang 1990) of 15 percent is widely used as a criterion for business investment, with a higher IRR required as investment risk and market uncertainty increases (Newnan 1991). In a survey of eight commercial freshwater tilapia hatcheries, 30-day old sex-reversed fry sold for prices that ranged from $0.10 to $0.18 per fry. In this study, IRRs greater than 15 percent occur at annual sales levels of 2 to 2.5 million fry at competitive prices ranging from $0.09 to $0.12 per fry, indicating that a recirculating, brackishwater hatchery for Florida red tilapia is economically feasible if fry can be sold year round. However, economic return would be adversely affected if market demand is seasonal and timing of sales as well as potential sales volume must be considered.

Marketability of Saltwater-Cultured Florida Red Tilapia in Puerto Rico

A cultural bias against freshwater fish and against fish with the silver-black appearance of most common varieties of tilapia (primarily *O. mossambicus*) has likely limited market demand and commercial production in some areas of the Caribbean (Sandifer 1991; Chakalall and Noriega-Curtis 1992). However, culture of red tilapia hybrids in saltwater could produce a more broadly accepted product with good commercial potential (Watanabe, Wicklund et al. 1989; Sandifer 1991; Watanabe 1991; Chakalall and Noriega-Curtis 1992; Head et al. 1996).

In conjunction with pilot-scale saltwater pond grow-out trials in Puerto Rico (Head et al. 1996), fish were harvested and sold at a farm-site retail store and at nine restaurants in Puerto Rico. Consumer surveys were conducted to assess marketability of saltwater-cultured Florida red tilapia at these distribution channels (Head et al. 1994).

At the farm-site store in Dorado, customers preferred dressed-out fish in the 454 to 567 g size range and paid a retail price of $7.70 kg^{-1}. During the 4-month study period, the equivalent of 5,380 kg of dressed-out red tilapia were sold at the farm-site store. Located just outside the major population center of San Juan, the farm-site store attracted customers with little advertising. Although sales could be increased through advertising and product promotion, it is likely that retail prices would decrease as competition from other producers increased (Nelson et al. 1983). However, the results suggest that farm-site retail sales could represent a significant marketing channel if the farm is located near a population center.

Due to logistical constraints and availability of product, restaurants were the only wholesale distribution channels considered in this study. The nine restaurants purchased 1,071 kg of Florida red tilapia (gilled, gutted, and scaled) during the study period and managers preferred fish in the 567 to 680 g range, which was optimum for common preparation methods that used the whole fish with head and skin. Most managers believed that red saltwater-cultured tilapia provided a prepared product of similar or better taste and presentation than silk snapper. Thus restaurant managers paid a wholesale price (\$4.96 to 5.18 kg^{-1}) for Florida red tilapia that was within the wholesale price range for imported silk snapper (\$3.96 to \$6.05 kg^{-1}) and the price per serving for whole fish paid by restaurant customers was similar for both species (\$7.00 to \$25.00). Although restaurants participating in the marketing study each purchased from 7 to 14 kg of red tilapia per week, less than the 20 to 50 kg of silk snapper they purchased weekly, all of the managers stated that they would purchase more red tilapia as promotion and customer awareness increased. The restaurant sector could become economically important to local red tilapia aquaculture enterprises.

The overall appraisal of saltwater-cultured Florida red tilapia by retail and restaurant customers and restaurant managers was high, with all product attributes, including taste, texture, freshness, and presentation rated above average to very good, and equal to or better than silk snapper *(Lutjanus vivanus),* a marine food fish popular in restaurants and supermarkets and the most important commercial fishery product in Puerto Rico (Caraballo and Sadovy 1990). Seventy-five to eighty-one percent of the respondents were new consumers of red tilapia, and appearance (presumably related to body conformation and color), dietary and health reasons, and curiosity were primary factors prompting them to try this new product. Red coloration was clearly an important factor in the initial acceptance of this product.

A problem area identified by restaurant managers was the number of bones in smaller (<454 g) fish. In addition, retail customers and restaurant managers indicated that the name "Florida red tilapia" presented marketing obstacles because of its association with wild-caught, freshwater tilapia *(O. mossambicus)* introduced to Puerto Rico in 1958 and found in lowland areas throughout the island (Erdman 1984). The dark color and reputed earthy-musty flavor of wild-caught *O. mossambicus* has made it a low-priced commercial species (Romaguera et al. 1987). Survey participants emphasized the need for a market-oriented name that reduced the association with wild-caught tilapia. They stressed the importance of product promotion, appealing to cultural preferences for red over dark-colored fish (PR-SBDC 1989) and for a local product farm-raised in saltwater.

The results of this preliminary study indicate a favorable potential for the marketability of saltwater-cultured Florida red tilapia in Puerto Rico if producers can meet the demand for quality, availability, and price. A more comprehensive marketing study over a wide range of distribution channels is warranted.

Economic Analysis of Intensive, Saltwater Pond Culture in Puerto Rico

An economic analysis was performed of a proposed commercial-scale (20 ha) saltwater pond culture operation for Florida red tilapia in Dorado, northern Puerto Rico (Head et al. 1996). The analysis was based on actual cost and production data from a commercial-scale hatchery, from pilot-scale grow-out trials conducted at the Dorado facilities (Table 10.4) and on wholesale market prices received for dressed-out product in Puerto Rico. Major capital expenses for the proposed 20 ha red tilapia aquaculture enterprise include water supply and drainage, pond construction, buildings, and aeration that represent 27.2, 23.0, 12.6, and 10.6 percent, respectively, of the total investment cost (Table 10.5).

Enterprise and Cash Flow Budgets

Imported feed, processing and distribution, and sex-reversed fry are the highest variable costs, accounting for 32.4, 16.3, and 14.7 percent, respectively, of the total annual costs (Table 10.6). Salaries and benefits, and depreciation represent the highest fixed costs, accounting for 7.6 and 5.0 percent, respectively, of the annual costs. Under the basic assumptions used in this study (fry price of \$0.11 per fry, production feed price of \$0.55 per kg^{-1}, and stocking density of 3.5 fish per square meter), a wholesale price of \$4.55 kg^{-1} resulted in an IRR of 18.2 percent, a discounted payback period (DPP) of 6.8 years, and a breakeven price of \$3.86 kg^{-1}, suggesting that the proposed 20 ha operation would be marginally feasible.

IRR and DPP were very sensitive to changes in wholesale price; a slight increase to \$4.70 kg^{-1} boosted IRR to 23.5 percent and reduced DPP to 5.6 years, but a decrease in wholesale price to \$4.40 kg^{-1} reduced the IRR to 12.4 percent and increased DPP to 9.1 years, making the proposed

TABLE 10.5. Initial capital investment requirements for a 20 ha saltwater pond aquaculture facility for Florida red tilapia.

Item	Cost ($)	Percentage of total
Water supply and drainage	179,000	27.2
Pond construction	151,160	23.0
Buildings	83,000	12.6
Aeration	70,000	10.6
Operations equipment	50,500	7.7
Backup systems	40,100	6.1
Electric service	40,000	6.1
Lab and conditioning equipment	22,000	3.3
Feed distribution and storage	9,600	1.5
Office equipment	6,500	1.0
Harvest equipment	5,965	0.9
Total cost	657,825	100

Source: Data from Head et al. 1996. Used with permission.

operation uneconomical. Consumer demand and wholesale prices are likely to be greatly influenced by product promotion and the ability of producers to maintain quality and availability.

Early producers can command premium prices for a short time, but competition from other tilapia producers entering the market may lower the wholesale price (Nelson et al. 1983). It is, therefore, important to minimize production costs by targeting areas that have the greatest impact on IRR (Rhodes and Hollin 1990; Rhodes 1991). After wholesale price, the IRR was most sensitive to changes in stocking density, followed by feed and fry prices. A marginal increase in stocking density from 3.5 to 4.0 fish per square meter raises the IRR from 18.2 to 27.0 percent, considerably enhancing economic outlook. The stocking densities used in the sensitivity analysis (2.5 to 4.5 fish per square meter) and their respective projected annual production values (14.3 to 25.8 $t \cdot ha^{-1} \cdot year^{-1}$) are within the ranges (2 to 11 fish per square meter; 5 to 28 $t \cdot ha^{-1} \cdot year^{-1}$) reported for intensive pond culture of tilapia (Sarig and Marek 1974; Hepher and Pruginin 1982; Balarin and Haller 1983; Liao and Chen 1983; Sin and Chiu 1983; Rakocy and McGinty 1989). Aeration and/or water exchange are necessary at these higher stocking densities (Sin and Chiu 1983; Rakocy and McGinty 1989) and are accounted for in the present analysis.

Since imported feed accounts for 39.1 percent of the variable costs and shipping represents 32 percent of the feed cost, profitability can be

TABLE 10.6. Annual even- and odd-year enterprise budgets for a 20 ha saltwater pond aquaculture facility for Florida red tilapia in Puerto Rico. Fish are stocked at a density of 3.5 fish per m^2.

	Even years	Odd years	Percentage of total variable cost	Percentage of total cost
Production				
Dress-out yield (kg)	339,239	296,834		
Variable costs				
Sex-reversed fry	$168,438	$192,500	17.7	14.7
Feed	424,539	371,878	39.1	32.4
Processing and distribution	213,648	186,942	19.6	16.3
Electricity	100,500	100,500	9.9	8.2
Labor	53,820	53,820	5.3	4.4
Repair and maintenance (equipment, levees)	13,000	13,000	1.3	1.1
Fuel, oil, lubrication (vehicles)	4,200	4,200	0.4	0.3
Advertising	6,000	6,000	0.6	0.5
Chemicals	2,400	2,400	0.2	0.2
Misc. (telephone, materials, and supplies)	10,000	10,000	1.0	0.8
Interest on operating capital (@9.5 percent)	51,789	49,162	4.9	4.1
Total variable costs	1,048,334	990,402	100	83.0
Fixed costs				
Salaries and benefits	93,750	93,750	44.7	7.6
Overhead (insurance, taxes, fees, and permits)	14,000	14,000	6.7	1.1
Land lease	11,250	11,250	5.4	0.9
Depreciation	61,434	61,434	29.3	5.0
Interest on investment capital[a]	29,100	29,100	13.9	2.4
Total fixed costs	209,534	209,534	100	17.0
Total annual costs	$1,257,868	$1,199,936		100

Source: Data from Head et al. 1996.
[a]Equipment at 9.5 percent; buildings, ponds, water supply and drainage, and electric service at 8.5 percent.

significantly improved by using a locally produced feed. Reducing feed costs by 18.2 percent (from \$0.55 to \$0.45 per kg^{-1}) boosts the IRR from 18.2 to 27.3 percent and reduces the DPP from 6.8 to 5.1 years.

Vertical integration of a hatchery operation with the grow-out phase can substantially improve economic feasibility. At a stocking density of 3.5 fish per square meter, the proposed 20 ha facility has a sex-reversed fry requirement of approximately 1.7 million each year, which can be produced by a commercial hatchery facility at a breakeven price of \$0.06 per fry (Head and Watanabe 1995). At a stocking density of 3.5 fish per square meter this increases the IRR from 18.2 to 29.4 percent and reduces the DPP from 6.8 to 4.8 years.

Economic Analysis of Intensive, Flow-Through Tank Culture in Puerto Rico

Intensive culture in tanks may be more practical than ponds on Caribbean islands where coastal land can be difficult to obtain due to competing needs from tourism and other economic sectors (Clark 1991). Intensive culture systems also offer a high degree of production control compared to other systems (Parker 1987), which may enable a grower to competitively meet market demands for fresh, high-quality product.

An economic analysis was performed of a proposed commercial-scale (6,240 m^3) saltwater tank culture operation for Florida red tilapia, based in Dorado, Puerto Rico (Head et al. unpublished data). The analysis is based on cost and production data from pilot-scale tank grow-out trials with Florida red tilapia conducted in the Bahamas (Ernst et al. 1989; Clark, A.E. et al. 1990; Ellis and Watanabe 1993b; Watanabe, Chan et al. 1993) (Table 10.7).

Major capital expenses for the proposed 6,240 m^3 intensive tank operation include tanks and water supply/drainage that represent 41.6 and 22.5 percent, respectively, of the total investment cost (Table 10.8). Startup costs for the intensive tank system (\$834,115) are 27 percent greater than for the 20 ha pond system (\$657,825).

Stocking, Production, and Harvesting

The intensive tank production system consists of three phases: hatchery, nursery, and grow-out. The hatchery phase assumes a 14-day sex-reversal period (Watanabe, Mueller, et al. 1993; Head and Watanabe 1995). Sex-reversed fry (0.04 g) are stocked in nursery tanks at a rate of 1,000 fish/m^3 (Balarin and Haller 1983) and reared for 60 days to a weight of 22 g

TABLE 10.7. Parameters used for the economic analysis of a proposed 6,240 m^3 intensive saltwater tank aquaculture facility for Florida red tilapia in Puerto Rico.

Parameter	Value
Farm size	
Total farm area (ha)	5
Nursery tank size (m^3)	30
Number of tanks	9
Total nursery tank area (ha)	0.03
Grow-out tank size (m^3)	120
Number of grow-out tanks	52
Total grow-out tank area (ha)	0.54
Sedimentation basin area (ha)	0.05
Stabilization pond area (ha)	1.0
Nursery phase	
Average initial wt. of reversed males (g)	0.04
Stocking density (fish/m^3)	1,000
Duration of nursery phase (d)	60
Average final weight (g)	22.0
Survival (percent)	60.0
Feed conversion ratio (dry wt./wet wt.)	1.4-1.6
Fish biomass at harvest (kg/m^3)	13.2
Grow-out phase	
Average initial weight at stocking (g)	22.0
Stocking density (fish/m^3)	75
Duration of grow-out phase (d)	182
Number of crops per year per tank	2
Average final weight (g)	572
Survival (percent)	91
Feed conversion ratio (dry wt./wet wt.)	1.8
Fish biomass at harvest (kg/m^3)	39.0
Total annual yield (kg)	487,207
Unmarketable fish (percent)	4
Dress-out (percent)	83.1

Source: Data from Watanabe et al. 1997. Used with permission.

(Watanabe, Smith et al. 1993) with 60 percent survival (Ernst et al. 1989; Head et al. 1996). Sex-reversed fry are fed #3 salmon starter for the first 22 days at a food conversion ratio (FCR) of 1.4 (Head et al. 1996) and then tilapia grower at an FCR of 1.6 (Watanabe, Clark et al. 1990b) until transfer

TABLE 10.8. Initial capital investment requirements for a proposed 6,240 m^3 intensive saltwater tank aquaculture facility for Florida red tilapia in Puerto Rico.

Item	Cost ($)	Percentage of total
Tanks	347,100	41.6
Water supply and drainage	188,000	22.5
Backup systems	61,900	7.4
Buildings	48,000	5.8
Operations equipment	47,000	5.6
Feed distribution and storage	33,850	4.1
Water treatment	32,500	3.9
Electric service	30,000	3.6
Aeration	24,500	2.9
Lab equipment	8,800	1.1
Office equipment	6,500	0.8
Harvest equipment	5,965	0.7
Total cost	834,115	100

Source: Data from Watanabe et al. 1997. Used with permission.

to growout tanks. Fingerlings are stocked in growout tanks at a rate of 75 fish per cubic meter and reared to a final average weight of 572 g in 182 days (Ellis and Watanabe 1993b) with a survival of 91 percent (Ernst et al. 1989; Clark, Watanabe, Olla, and Wicklund et al. 1990; Ellis and Watanabe 1993b; Watanabe, Chan, et al. 1993) and a final standing crop of 39 $kg \cdot m^{-3}$. In grow-out tanks, tilapia grower is fed at a conversion ratio of 1.8 (Ernst et al. 1989; Watanabe, Chan et al. 1993).

To permit continuous year-round production, one nursery tank is stocked on day 122 of the first year and at weekly intervals thereafter. Beginning on week 26 of the first year, two grow-out tanks are stocked at weekly intervals until all 52 tanks are stocked by the end of the first year. Two growout tanks are harvested weekly, with each tank producing two crops annually. Total annual farm harvest is 487,207 kg, similar to that of the 20 ha pond culture system (373,884 to 427,296 kg).

Variable and Fixed Costs

Variable and fixed costs for the tank-culture operation (Table 10.9) are similar to those of the pond-culture operation (Table 10.6).

TABLE 10.9. Annual enterprise budget for a proposed 6,240 m^3 intensive saltwater tank aquaculture facility for Florida red tilapia in Puerto Rico. Fish are stocked at a density of 75 fish per m^3.

	Amount	Percentage	Percent of total
Production			
Dress-out yield (kg)	388,674		
Variable costs			
Sex-reversed fry	$171,600	14.8	12.5
Feed	470,783	40.6	34.2
Processing and distribution	243,604	21.0	17.7
Electricity	123,566	10.7	9.0
Labor	53,820	4.6	3.9
Waste removal	10,000	0.9	0.7
Repair and maintanence	6,690	0.6	0.5
Fuel, oil, lubrication (vehicles)	4,200	0.4	0.3
Advertising	6,000	0.5	0.4
Chemicals	2,000	0.2	0.2
Misc. (telephone, materials, and supplies)	10,000	0.9	0.7
Interest on operating capital (@9.5 percent)	56,811	4.9	4.1
Total variable costs	1,159,074	100	84.3
Fixed costs			
Salaries and benefits	93,750	43.0	6.8
Overhead (insurance, taxes, fees, and permits)	14,375	6.6	1.0
Land lease	1,875	0.9	0.1
Depreciation	71,829	32.9	5.2
Interest on investment capital[a]	36,450	16.7	2.6
Total fixed costs	218,279	100	15.7
Total annual costs	1,377,353		100

[a]Equipment at 9.5 percent; buildings, tanks, water supply and drainage, and electric service at 8.5 percent.
Source: Data from Watanabe et al. 1997. Used with permission.

Enterprise and Cash Flow Budgets

Feed, processing/distribution, and sex-reversed fry are the highest variable cost representing 34.2, 17.7, and 12.5 percent, respectively, of the total annual costs (Table 10.9). Salaries and benefits, and depreciation are the largest fixed costs representing 6.8 and 5.2 percent of the annual costs, respectively. Under these conditions, a wholesale price of $4.55 kg^{-1} results in a positive cash flow by year five, while the breakeven price, IRR, and DPP are $3.54 kg^{-1}, 27.8 percent, and 5.0 years, respectively.

Projected IRR and DPP are most sensitive to changes in wholesale price, followed by feed and fry price. A 6.6 percent increase in wholesale price (from $4.55 to $4.84 kg^{-1}) increases IRR by 9.3 percent and reduces DPP by 1.0 year. A decrease in feed price to $0.45 kg^{-1} by use of locally prepared feeds, or in fry prices to $0.06 per fry (by vertical integration of a hatchery operation), raises the IRR to 36.3 and 35.6 percent, and lowers the DPP to 4.1 and 4.2 years, respectively.

Stocking density shows a variable effect on IRR and DPP; decreasing the stocking density from 75 to 50 fish per cubic meter reduces the IRR by 20.3 percent and increases the DPP to greater than 10 years, whereas raising the stocking density to 100 fish per cubic meter increases the IRR by only 9.1 percent and decreases the DPP by 1.0 years. At very high stocking density, increases in capital and operating costs offset increased returns.

Results of these analyses suggest that the proposed saltwater tank culture operation for Florida red tilapia is economically feasible at a stocking density of 75 fish per cubic meter and at a wholesale price of $4.55 kg^{-1}. Production costs can be lowered considerably by using locally prepared feeds and vertical integration of a hatchery operation. The advantages of an intensive tank culture system, which include a high degree of production control and small land area requirement, must be weighed against higher startup costs and management and marketing requirements.

POTENTIAL FOR UNWANTED INTRODUCTION OF TILAPIA INTO MARINE WATERS

Concerns regarding the possible introduction of tilapia into marine waters and of the potentially adverse ecological effects (Knaggs 1977; Lobel 1980; Dial and Wainwright 1983) can constrain the commercial development of saltwater tilapia culture in some areas. In this section, the relevant scientific literature is reviewed as a basis for identifying required research and for developing sound public policy.

It is well known that certain tilapia are highly salt tolerant (see Stickney 1986 for review). The Mozambique tilapia, *O. mossambicus,* can grow in ponds at salinities from 32 to 40 ppt, reproduce at salinities as high as 49 ppt (Popper and Lichatowich 1975), and adapt to salinities as high as 120 ppt (Whitfield and Blaber 1979). The ability of tilapia to become established and to thrive in a variety of nonnative, fresh- and saline-water habitats, including mixohaline estuaries, is well documented. In the Caribbean, the

Mozambique tilapia was introduced to Puerto Rico in 1958 by the Puerto Rico Department of Agriculture to control algae in sugar cane irrigation canals (Erdman 1984) and has since become widely distributed in lowland areas throughout the island. Their presence in an estuarine location in Puerto Rico (salinities of ≤20 ppt) was first reported by Austin (1971). Burger et al. (1992) found that *O. mossambicus* accounted for 55 to 79 percent of fish sampled at three brackish, coastal marshes in Puerto Rico, and were more common in open lagoons (6 to 14 ppt) than creeks (0 to 4 ppt) or bays (36 ppt).

In Florida, *O. mossambicus* and *Sarotherodon melanotheron* were reported to occur in shallow estuarine waters, coastal lagoons, and canal systems (salinity = 8 to 23 ppt) (Dial and Wainwright 1983). *Sarotherodon melanotheron* has been established in estuaries along the eastern side of Tampa Bay since 1958 and is sufficiently abundant to support a small commercial fishery (Dial and Wainwright 1983). In California, *O. mossambicus* and *T. zillii,* introduced in 1973 into tributaries of the San Gabriel and Santa Ana Rivers, were collected in a flood control canal and a coastal lagoon where salinities ranged as high as 34.5 $g \cdot L^{-1}$ (Knaggs 1977). In Hawaii, *O. mossambicus, T. zillii,* and *T. melanopleura* have also been reported to inhabit coastal estuaries (Maciolek 1984).

Following the introduction of *O. mossambicus* to Fanning Atoll (Pacific Ocean) in 1958, they have become established in shallow, enclosed areas at the periphery of the central lagoon (Lobel 1980). Although Lobel provided no data on environmental salinity, data from an earlier study (Guinther 1971) suggests that these areas were estuarine rather than marine.

The available evidence suggests, therefore, that certain tilapia can become established under nonnative estuarine conditions similar to those of their native range. However, apart from occasional forays of individuals from estuarine populations into open marine waters (Lobel 1980), there have been no reports to date on established populations of tilapia in strict marine ecosystems. Given the high salinity tolerances of these tilapia species, the direct accessibility of open coastal marine waters to these fish via rivers and canals and estuaries, and the long duration of time since their introduction to many areas, the absence of reports documenting their presence in open marine waters is a strong indication that other ecological factors have prevented the colonization of such ecosystems.

Studies suggest that the ability of *O. mossambicus* to colonize a marine environment involves the interplay of many factors, only one of which is salinity tolerance (Whitfield and Blaber 1979). In native Africa, *O. mossambicus* is found in coastal lagoons and estuarine areas that are closed off from the sea for most of the year, but is absent from open estuaries that are strongly affected by tides and currents and subject to rapid salinity fluctuations (Whitfield and Blaber 1979). A number of factors were hypothesized to be responsible for their absence in open estuaries. While *O. mossambicus* can adapt to a gradual change in salinity, they may not tolerate rapid salinity fluctuations of open estuaries. *Oreochromis mossambicus* avoids strong water currents that cause substrate movement and hampers nesting and reproduction. The absence of suitable marginal vegetation in open estuaries and the exposure of littoral areas at low tide also prevent nest establishment. The presence of marine piscivorous species may further reduce their ability to persist in a marine environment (Whitfield and Blaber 1979).

In the Bahamas, large populations of Florida red tilapia broodstock were maintained on Lee Stocking Island from 1986 through 1995. Although larvae have accidentally escaped via the effluent system, no sightings of juvenile or adult tilapia have been reported in waters around the island despite many research dives made each year (R. Wicklund and W. Head, personal communication). Underwater surveys of potential shallow-water habitats for tilapia in the vicinity of Lee Stocking Island, including mangroves swamps, have also failed to detect the presence of tilapia (W. Head and S. Ellis, personal communication).

On Lee Stocking Island, Florida red tilapia that escape through the effluent system are often observed being preyed upon by piscivorous species, including barracuda (*Sphyraena* sp.) and jacks (*Caranx* sp.), which therefore represent a biological safeguard against unwanted introduction. Furthermore, strong tidal movement of water probably interferes with nest building and reproduction. Hence, both biological and environmental conditions minimize the likelihood that a founder population could persist in open marine waters around Lee Stocking Island.

The available evidence indicates that the potential for colonization of open marine waters by tilapia is low. Given their ability to inhabit nonnative, estuarine areas, the possibility of detrimental impact on commercially-important native species that use the estuaries as nursery grounds appears to be of primary importance, but data are lacking. Hence, the importance of tilapia as competitors for food and shelter, or as predators of young is unknown, and definitive ecological studies are needed (Maciolek 1984).

While the use of nonnative species without careful evaluation of potential ecological consequences should not be advocated, it is recommended that government policy-makers weigh the risks against the potential economic benefits of saltwater tilapia aquaculture to their specific regions. As environmental conditions vary with location, such evaluations should be made on a location-specific basis.

GENETIC ISSUES IN THE USE OF RED HYBRID TILAPIA

Perpetuation of Florida red tilapia broodstock at the Caribbean Marine Research Center was aimed primarily at maintaining a "standard reference" strain by preventing inbreeding. When replacing broodstock, nonsex-reversed fingerlings were initially obtained by collection of clutches from as many different parental females as possible from a population of 540 to 1,080 individuals. Broodstock are selected as those showing optimum coloration, body conformation, and size. The Florida red tilapia displays a variety of color pattern including reddish-orange with or without black mottling, pink, and white (albino). Inheritance of red coloration in this strain is complex and is believed to be a dominant trait controlled by two or three loci (Behrends et al. 1982). At CMRC, individuals with excessive mottling are culled. White (albino) and pink phenotypes (no black pigmentation) are considered less viable than other color types (Behrends et al. 1982; Gamal et al. 1988; Wohlfarth et al. 1990) and are also culled.

When broodstock numbers are relatively small, the advantages of replacing broodstock with more productive, yearling broodstock each season must be weighed against the risk of inbreeding that can result from frequent replacement of broodstock with next-generation progeny. To minimize this potential problem, it is recommended that a separate "genetic" broodstock be maintained for at least two years as a source of replacement brooders. Furthermore, effective breeding number among "genetic broodstock" may be maximized by maintaining a 1:1 rather than a skewed sex ratio (Tave 1986) that was used to maximize seed production.

In addition to the Florida red, other strains of red tilapia with distinct genetic heritages have been introduced into the Caribbean and the United States, including Taiwanese red tilapia *(O. mossambicus × O. niloticus)* (Liao and Chang 1983; Galman and Avtalion 1983). Researchers and commercial growers in the United States and Jamaica have cross-bred the Florida red strain with other species including *O. aureus* to improve cold tolerance (Behrends and Smitherman 1984) and/or *O. niloticus* to improve growth rates (R. DeWandel, American Tilapia Association, personal communication). Florida red tilapia, as well as their three- and four-way hybrid crosses with *O. aureus* and *O. niloticus,* were introduced into various parts of the Caribbean including the Bahamas, U.S. Virgin Islands, Martinique, Curacao, Bonaire, and Jamaica for commercial culture trials in both freshwater and saltwater. Clearly, the maintenance of the genetic integrity of a commercial strain will depend upon many factors including market demand, adequate facilities, skills of hatchery biologists, and accidental contamination by other strains.

Exceptional culture performance of the Florida red strain in saltwater is attributed to the high salt tolerance of both its parental species; indiscriminate hybridization with less salt-tolerant red strains could cause a deterioration in performance. It is critical, therefore, that standard reference lines of Florida red tilapia be maintained by fish culturists as a source of replacement broodstock.

Maintaining a minimum effective population size and random mating are critical toward maintaining genetic variance in perpetuating a commercial broodstock (Tave 1986). Periodic outcrossing of broodstock with unrelated standard reference lines can also increase genetic variance and prevent inbreeding (Tave 1986). Prior to outcrossing, saltwater performance of newly acquired broodstock should be ascertained by saltwater yield trials. Salinity tolerance testing may also be used to rapidly screen potential broodstock (Watanabe et al. 1985a,c). It is recommended that a commercial culturist continue a regular program of strain comparisons and outcrossing as a basis for genetic improvement.

Periodic outcrossing is only practical if high-quality standard reference lines are obtainable from other commercial and/or research hatcheries. Therefore, it is further recommended that such hatcheries re-derive the F1 hybrid cross (*O. mossambicus* × *O. hornorum*) as a basis for the development of new standard reference lines of the Florida red tilapia strain.

COLD TOLERANCE

A basic constraint to saltwater tilapia culture in subtropical areas is poor cold tolerance that precludes production during the winter months and that is exacerbated under high salinities. Studies have demonstrated the feasibility of improving the cold tolerance of a red tilapia strain in freshwater through introgressive hybridization with *O. aureus,* a relatively cold-tolerant species (Behrends and Smitherman 1984). Seawater culture studies in Kuwait have shown that while *O. aureus* was unable to tolerate both low temperature and high salinity simultaneously, *O. spilurus* and *O. aureus* × *O. spilurus* hybrid survived cold water temperatures during winter better than did *O. aureus* or Taiwanese red tilapia. This suggests the feasibility of developing a Florida red tilapia strain with improved cold tolerance in seawater through introgressive hybridization with *O. spilurus.*

Improved Environmental Tolerance

The Florida red tilapia is a hybrid strain originally developed by commercial culturists for its color characteristics as well as productivity in freshwater. Exceptional culture performance of this hybrid strain under high salinities demonstrates that varieties capable of fast growth in seawater can be produced through cross-breeding.

The Florida red tilapia is a hybrid strain originally developed by commercial culturists for its color characteristics as well as productivity in freshwater. Exceptional culture performance of this hybrid strain under high salinities demonstrates that varieties capable of fast growth in seawater can be produced through cross-breeding. It has been shown that salinity tolerance can be improved in a moderately tolerant species *(O. niloticus)* through hybridization with a highly tolerant species *(O. mossambicus)*. Other salt-tolerant species that may be considered for cross-breeding include *O. hornorum* and *O. spilurus*. Salinity tolerance and growth of progeny produced by backcrossing F1 hybrids with either parental species have not been assessed.

Tilapia are known for wide environmental tolerances, but rapid growth occurs only under favorable conditions. Future research will aim to improve environmental tolerance to expand culture into areas of lower temperature and higher salinities. With new genetic mixes, optimal environmental conditions are likely to change and optimal temperatures, salinity, pH, DO, and acceptable levels of other water quality parameters will need to be determined. Red tilapia hybrids with improved cold tolerance have been developed by introgressive breeding with *O. aureus* (a cold-tolerant species), followed by two generations of backcrossing red hybrid males to female *O. aureus* (Behrends et al. 1982; Behrends and Smitherman 1984). The potential for improving growth rate and environmental tolerance of tilapia by selective breeding or other genetic manipulations has not yet been fully exploited.

Future development of stocks with improved environmental tolerances can be accelerated through greater use of genetic markers for broodstock management, identification of loci controlling quantitative traits, and development of superior strains through marker-assisted selection (Kocher 1997; Agresti et al. 2000). The application of transgenics to such production characteristics will require the identification of genes responsible for these traits. With new genetic combinations, optimal environmental conditions may change and optimal temperatures, salinity, pH, dissolved oxygen, and acceptable levels of other water quality parameters will need to be determined.

REFERENCES

Agresti, J.J., S. Seki, A. Cnaani, S. Poompuagn, E.M. Hallerman, N. Umiel, G. Hulata, G.A.E. Gall, and B. May (2000). Breeding new strains of tilapia: Development of an artificial center of origin and linkage map based on AFLP and microsatellite loci. *Aquaculture* 185:43-56.

Akiyama, D.M. and A.M. Anggawati (1999). Polyculture of shrimp and tilapia in East Java. American Soybean Association (ASA). *Technical Bulletin* AQ 47-1999.

Al-Ahmad, T.A., K.D. Hopkins, M. Ridha, A.A. Al-Ahmed, and M.C. Hopkins (1986). *Tilapia culture in Kuwait.* Kuwait Institute for Scientific Research/International Center for Living Aquatic Resources Management, Report No. KISR2122, Kuwait.

Al-Ahmad, T.A., M. Ridha, and A.A. Al-Ahmed (1988a). Production and feed ration of the tilapia *Oreochromis spilurus* in seawater. *Aquaculture* 73:111-118.

———. (1988b). Reproductive performance of the tilapia *Oreochromis spilurus* in seawater and brackish groundwater. *Aquaculture* 73:323-332.

Al-Amoudi, M.M. (1987a). Acclimation of commercially cultured *Oreochromis* species to seawater—An experimental study. *Aquaculture* 65:333-342.

———. (1987b). The effect of high salt diet on the direct transfer of *Oreochromis mossambicus, O. spilurus* and *O. aureus/O. niloticus* hybrids to sea water. *Aquaculture* 64:333-338.

Allanson, B.R., A. Bok, and N.I. van Wyk (1971). The influence of exposure to low temperature on *Tilapia mossambica Peters* (Cichlidae). II. Changes in serum osmolarity, sodium and chloride ion concentrations. *Journal of Fish Biology* 3:181-185.

APHA, AWWA, WPCF (1985). *Standard methods for the examination of water and wastewater* (16th edn.). Washington, DC: American Public Health Association, American Water Works Association and Water Pollution Control Federation.

Assem, H. and W. Hanke (1979). Volume regulation of muscle cells in the euryhaline teleost, *Tilapia mossambica. Comparative Biochemistry and Physiology* 64(A):17-23.

Auperin, B., I. Leguen, F. Rentier-Delrue, J. Small, and P. Prunet (1995). Absence of a tGH effect on adaptability to brackish water in tilapia *(Oreochromis niloticus). General and Comparative Endocrinology* 97:145-159.

Austin, H.M. (1971). A survey of the ichthyofauna of the mangroves of western Puerto Rico during December 1967-August 1968. *Caribbean Journal of Science* 11:27-39.

Avella, M., J. Berhaut, and M. Bornancin (1993). Salinity tolerance of two tropical fishes, *Oreochromis aureus* and *O. niloticus,* 1. Biochemical and morphological changes in the gill epithelium. *Journal of Fish Biology* 42:243-254.

Ayson, F.G., T. Kaneko, S. Hasegawa, and T. Hirano (1994a). Development of mitochondrion-rich cells in the yolk-sac membrane of embryos and larvae of tilapia *(Oreochromis mossambicus),* in fresh water and seawater. *Journal Experimental Zoology* 270:129-135.

Ayson, F.G., T. Kaneko, S. Hasegawa, and T. Hirano (1994b). Differential expression of two prolactin and growth hormone genes during early development of tilapia *(Oreochromis mossambicus)* in fresh water and seawater: Implications for possible involvement in osmoregulation during early life stages. *General Comparative Endocrinology* 95:143-152.

Ayson, F.G., T. Kaneko, S. Hasegawa, and T. Hirano (1995). Cortisol stimulates the size and number of mitochondrion-rich cells in the yolk-sac membranes of embryos and larvae of tilapia *(Oreochromis mossambicus)* in vitro and in vivo. *Journal Experimental Zoology* 272:419-425.

Ayson, F.G., T. Kaneko, M. Tagawa, S. Hasegawa, E.G. Grau, R.S. Nishioka, D.S. King, H.A. Berna, and T. Hirano (1993). Effects of acclimation to hypertonic environment on plasma and pituitary levels of two prolactins and growth hormone in two species of tilapia, *Oreochromis mossambicus* and *Oreochromis niloticus. General and Comparative Endocrinology* 89:138-148.

Balarin, J.D. and R.D. Haller (1982). The intensive culture of tilapia in tanks, raceways and cages. In J.F. Muir and R.J. Roberts (Eds.), *Recent advances in aquaculture* (pp. 267-335). Boulder, CO: Westview Press.

———. (1983). Commercial tank culture of tilapia. In L. Fishelson and Z. Yaron (compilers), *International Symposium on Tilapia in Aquaculture, Nazareth, Israel, 8-13 May 1983* (pp. 473-483). Tel Aviv, Israel: Tel Aviv University.

Bautista, A. M., M.H. Carlos, and A.I. San Antonio (1988). Hatchery production of *Oreochromis niloticus* L. at different sex ratios and stocking densities. *Aquaculture* 73:85-95.

Beamish, F.W.H. (1970). Influence of temperature and salinity acclimation on temperature preferenda of the euryhaline fish *Tilapia nilotica. Journal of the Fisheries Research Board Canada* 27:1209-1214.

Behrends, L.L., R.G. Nelson, R.O. Smitherman, and N.M. Stone (1982). Breeding and culture of the red-gold color phase of tilapia. *Journal of the World Mariculture Society* 13:210-220.

Behrends, L.L. and R.O. Smitherman (1984). Development of a cold-tolerant population of red tilapia through introgressive hybridization. *Journal of the World Mariculture Society* 15:172-178.

Berrios-Hernandez, J.M. and J.R. Snow (1983). Comparison of methods for reducing fry losses due to cannibalism in tilapia production. *Progressive Fish-Culturist* 45:116-118.

Boehlert, G.W. (1981). The effects of photoperiod and temperature on laboratory growth of juvenile *Sebastes diploproa* and a comparison with growth in the field. *Fishery Bulletin, U.S.* 79:789-794.

Borski, R.J., J. Yoshikawa, S. Madsen, R.S. Nishoka, C. Zabetian, H.A. Bern, and E.G. Grau (1994). Effects of environmental salinity on pituitary growth hormone content and cell activity in the euryhaline tilapia, *Oreochromis mossambicus. General and Comparative Endocrinology* 95:483-494.

Brett, J.R. (1979). Environmental factors and growth. In W.S. Hoar, D.J. Randall, and J.R. Brett (Eds.), *Fish physiology* (Vol. 8, pp. 599-675). London, England: Academic Press.

Brett, J.R., W.C. Clarke, and J.E. Shelbourn (1982). *Experiments on thermal requirements for growth and food conversion efficiency of juvenile chinook salmon Oncorhynchus tshawytscha.* Canadian Technical Report of Fisheries and Aquatic Sciences No. 1127.

Briggs, M.R.P. and S.J. Funge-Smith (1994). A nutrient budget of some intensive marine shrimp ponds in Thailand. *Aquaculture and Fisheries Management* 25:789-811.

Brock, V.E. (1954). A note on the spawning of *Tilapia mossambica* in seawater. *Copeia* 1954(1):72.

Broussard, M.C. and C.G. Reyes (1985). Cost Analysis of a large-scale hatchery for the production of *Oreochromis niloticus* fingerlings in Central Luzon, Philippines. In I.R. Smith, E.B. Torres, and E.O. Tan (Eds.). *Philippine Tilapia Economics ICLARM Conference Proceedings 12* (pp. 33-43). Manila, Philippines: Philippine Council for Agriculture and Resources Research and Development, Los Banos, Launa and International Center for Living Aquatic Resources Management.

Broussard, M.C., R. Reyes, and F. Raguidin (1983). Evaluation of hatchery management schemes for large scale production of *Oreochromis niloticus* fingerlings in central Luzon, Philippines. In L. Fishelson and Z. Yaron (compilers), *Proceedings of the International Symposium on Tilapia in Aquaculture, Nazareth, Israel, 8-13 May 1983* (pp. 414-424). Tel Aviv, Israel: Tel Aviv University.

Burger, J., K. Cooper, D.J. Gochfeld, J.E. Saliva, C. Safina, D. Lipsky, and M. Gochfeld (1992). Dominance of *Tilapia mossambica,* an introduced fish species, in three Puerto Rican estuaries. *Estuaries* 15(2):239-245.

Campbell, D. (1985). Large scale cage farming of *Sarotherodon niloticus. Aquaculture* 48:57-69.

Canagaratnam, P. (1966). Growth of *Tilapia mossambica* (Peters) at different salinities. *Bulletin of the Fisheries Research Station, Ceylon* 19:47-50.

Caraballo, D.M. and Y. Sadovy (1990). Overview of Puerto Rico's small-scale fisheries statistics, 1988-1989. Corporation for the Development and Administration of the Marine, Lacustrine and Fluvial Resources of Puerto Rico. *Fisheries Research Laboratory Technical Report* 1(4):1-17.

Carro-Anzalotta, A.E. and A.S. McGinty (1986). Effects of stocking density on growth of *Tilapia nilotica* cultured in cages in ponds. *Journal of the World Aquaculture Society* 17:52-57.

Chakalall, B. and P. Noriega-Curtis (1992). Tilapia farming in Jamaica. *Gulf and Caribbean Fisheries Institute* 41:545-569.

Cheong, L., F.K. Chan, F.J. Wong, and R. Chou (1987). Observations on the culture of red tilapia (*Oreochromis niloticus* hybrid) in seawater under intensive tank condition using a biodrum. *Singapore Journal of Primary Industries* 15(1):42-56.

Chervinski, J. (1961). On the spawning of *Tilapia nilotica* in brackish water during experiments in concrete tanks. *Bamidgeh* 13(1):30.

———. (1982). Environmental physiology of tilapias. In R.S.V. Pullin and R.H. Low-McConnell (Eds.), *The biology and culture of tilapias* (pp. 119-128). Manila, Philippines: ICLARM Conference Proceedings 7, International Center for Living Aquatic Resources Management.

Chervinski, J. and A. Yashouv (1971). Preliminary experiments on the growth of *Tilapia aurea,* Steindachner (Pisces, cichlidae) in seawater ponds. *Bamidgeh* 23(4):125-129.

Chervinski, J. and M. Zorn (1974). Note on the growth of *Tilapia aurea* (Steindachner) and *Tilapia zillii* (Gervais) in sea-water ponds. *Aquaculture* 4:249-255.

Cioni, C., D. de Merich, E. Cataldi, and S. Cataudella (1991). Fine structure of chloride cells in freshwater- and seawater-adapted *Oreochromis niloticus* (Linnaeus) and *Oreochromis mossambicus* (Peters). *Journal of Fish Biology* 39:197-209.

Clark, A.E., W.O. Watanabe, B.L. Olla, and R.I. Wicklund (1990). Growth, feed conversion, and protein utilization of Florida red tilapia fed isocaloric diets with different protein levels in seawater pools. *Aquaculture* 87:75-85.

Clark, J.H., W.O. Watanabe, D.H. Ernst, R.I. Wicklund, and B.L. Olla (1990). Effects of feed rate on growth and feed conversion of Florida red tilapia reared in floating marine cages. *Journal of the World Aquaculture Society* 21(1):16-24.

Clark, J.R. (1991). Environmental planning for aquaculture in the coastal zone. In J.A. Hargreaves and D.E. Alston (Eds.), *Status and potential of aquaculture in the Caribbean* (pp. 109-134). Baton Rouge, LA: The World Aquaculture Society.

Clarke, W.C. (1982). Evaluation of the seawater challenge test as an index of marine survival. *Aquaculture* 18:177-183.

Coche, A.G. (1982). Cage culture of tilapias. In R.S.V. Pullin and R.H. Lowe-McConnell (Eds.), *The biology and culture of tilapias* (pp. 205-246). Manila, Philippines: International Center for Living Aquatic Resources Management (ICLARM).

Colin, P. (1975). *The neon gobies.* Neptune City, NJ: T.F.H. Publications, Inc.

Costa-Pierce, B.A. and H. Hadikusumah (1995). Production management of double-net tilapia *Oreochromis* spp. Hatcheries in a eutrophic tropical reservoir. *Journal of the World Aquaculture Society* 26:453-459.

Cowell, L.E, W.O. Watanabe, W.D. Head, J.J. Grover, and J.M. Shenker (1993). Use of tropical cleaner fish to control the ectoparasite *Neobenedenia melleni* (Monogenea: Capsalidae) on seawater-cultured Florida red tilapia. *Aquaculture* 113:189-200.

Cruz, E.M. and M. Ridha (1989). Preliminary study on the production of the tilapia, *Oreochromis spilurus* (Gunther), cultured in seawater cages. *Aquaculture and Fisheries Management* 20:381-388.

———. (1990). Production of marketable-size tilapia, *Oreochromis spilurus* (Gunther), in seawater cages using different production schedules. *Aquaculture and Fisheries Management* 21:187-194.

Cruz, E.M., M. Ridha, and M.S. Abdullah (1990). Production of the African freshwater tilapia *Oreochromis spilurus* (Gunther) in seawater. *Aquaculture* 84:41-48.

Dange, A.D. (1985). Branchial Na^+-K^+-ATPase activity during osmotic adjustments in two freshwater euryhaline teleosts, tilapia *(Sarotherodon mossambicus)* and orange chromide *(Etroplus maculatus). Marine Biology* 87:101-107.

Dharmamba, M., M. Bornancin, and J. Maetz (1975). Environmental salinity and sodium and chloride exchanges across the gill of *Tilapia mossambica. Journal of Physiology* 70:627-636.

Dial, R.S. and S.C. Wainright (1983). New distributional records for non-native fishes in Florida. *Florida Scientist* 46:8-15.

Ellis, S.C. and W.O. Watanabe (1993a). Comparison of raceway and cylindro-conical tanks for brackishwater production of juvenile Florida red tilapia under high stocking densities. *Aquacultural Engineering* 13:59-69.

———. (1993b). The effects of hyposalinity on eggs, juveniles and adults of the marine monogenean, *Neobenedenia melleni.* Treatment of ecto-parasitosis in seawater-cultured tilapia. *Aquaculture* 117:15-27.

Ellis, S.C., W.O. Watanabe, and W.D. Head (1993). The effect of initial age variation on production of Florida red tilapia fry under intensive, brackishwater tank culture. *Aquaculture and Fisheries Management* 24:465-471.

Erdman, D.S. (1984). Exotic fishes in Puerto Rico. In W.R. Courtenay Jr. and J.R. Stauffer Jr. (Eds.), *Distribution, biology, and management of exotic fishes* (pp. 162-176). Baltimore, MD: John Hopkins University Press.

Ernst, D.H. (1989). Design and operation of a hatchery for seawater production of tilapia in the Caribbean. In G.T. Waugh and M.H. Goodwin (Eds.), *Proceedings of the 39th Annual Gulf and Caribbean Fisheries Institute, November 1986, Hamilton, Bermuda* (pp. 420-434). Charleston, SC: Gulf and Caribbean Fisheries Institute.

Ernst, D.H., L.J. Ellingson, B.L. Olla, R.I. Wicklund, W.O. Watanabe, and J.J. Grover (1989). Production of Florida red tilapia in seawater pools: Nursery rearing with chicken manure and growout with prepared feed. *Aquaculture* 80:247-260.

Ernst, D.H., W.O. Watanabe, L.J. Ellingson, R.I. Wicklund, and B.L. Olla (1991). Commercial-scale production of Florida red tilapia seed in low- and brackish-salinity tanks. *Journal of the World Aquaculture Society* 22(1):36-44.

Farmer, G.J. and F.W.H. Beamish (1969). Oxygen consumption of *Tilapia nilotica* in relation to swimming speed and salinity. *Journal of Fisheries Research Board of Canada* 26:2807-2821.

Fast, A.W. and P. Menasveta (2000). Some recent issues and innovations in marine shrimp pond culture. *Reviews in Fisheries Science* 8(3):151-233.

Febry, R. and P. Lutz. (1987). Energy partitioning in fish: the activity-related cost of osmoregulation in a euryhaline cichlid. *Journal of Experimental Biology* 128: 63-85.

Feddern, H.A. (1965). The spawning, growth, and general behavior of the bluehead wrasse, *Thalassoma bifasciatum* (Pisces: Labridae). *Bulletin of Marine Science* 15:896-941.

Fineman-Kalio, A.S. (1988). Preliminary observations on the effect of salinity on the reproduction and growth of freshwater Nile tilapia, *Oreochromis niloticus* (L.), cultured in brackishwater ponds. *Aquaculture and Fisheries Management* 19:313-320.

Fineman-Kalio, A.S. and A.S. Camacho (1987). The effects of supplemental feeds containing different protein:energy ratios on the growth and survival of *Oreochromis niloticus* (L.) in brackishwater ponds. *Aquaculture and Fisheries Management* 18:139-149.

Fishelson, L. (1980). Scanning and transmission electron microscopy of the squamose gill-filament epithelium from fresh- and seawater-adapted tilapia. *Environmental Biology of Fishes* 5:161-165.

Fishelson, L. and D. Popper (1968). Experiments on rearing fish in saltwater near the Dead Sea, Israel. *FAO Fisheries Report* 44:244-245.

Fitzgerald, W.J. (1979). The red-orange tilapia: A hybrid that could become a world favorite. *Fish Farming International* 6(1):26-27.

Fitzsimmons, K. (2000). Tilapia and penaeid shrimp polycultures. Pond Dynamics/ Aquaculture CRSP, *Aquanews,* Fall 2000.

——— (2001). Polyculture of tilapia and penaeid shrimp. *Global Aquaculture Advocate* 4(3):43-44.

Fontainhas-Fernandes, A., F. Russell-Pinto, E. Gomes, M.A. Reis-Henriques, and J. Coimbra (2001). The effect of dietary sodium chloride on some osmoregulatory parameters of the teleost, *Oreochromis niloticus,* after transfer from freshwater to seawater. *Fish Physiology and Biochemistry* 23:307-316.

Foskett, J.K., C.D. Logsdon, T. Turner, T.E. Machen, and H.A. Bern (1981). Differentiation of the chloride extrusion mechanism during seawater adaptation of the teleost fish, the cichlid *Sarotherodon mossambicus. Journal of Experimental Marine Biology* 43:209-224.

Foskett, K.J. and C. Scheffey (1982). The chloride cells: Definitive identification as the salt secreting cell in teleosts. *Science* 215:164-166.

Gallet de Saint Aurin, D., J.C. Raymond, and J. Vianas (1990). Marine finfish pathology: Specific problems and research in the French West Indies. In *Advances in tropical aquaculture. Tahiti, February 20-March 4, 1989* (pp. 143-160). Paris, France: Actes de Colloque 9. AQUACOP. IFREMER.

Galman, O.R. and R.R. Avtalion (1983). A preliminary investigation of the characteristics of red tilapias from the Philippines and Taiwan. In L. Fishelson and Z. Yaron (compilers), *Proceedings of the International Symposium on Tilapia in Aquaculture, Nazareth, Israel, 8-13 May 1983* (pp. 291-301). Tel Aviv, Israel: Tel Aviv University.

Gamal, A.A., R.O. Smitherman, and L.L. Behrends. (1988). Viability of red and normal-colored *Oreochromis aureus* and *O. niloticus hybrids.* In R.S.V. Pullin, T. Bhukaswan, K. Tonguthai, and J.L. Maclean (Eds.), *Second International Symposium on Tilapia in Aquaculture* (pp. 153-157). ICLARM Conference Proceeding 15. Department of Fisheries, Bangkok, Thailand, and ICLARM, Manila, Philippines.

Garcia, T. and K. Sedjro (1987). Estudio comparativo del crecimiento de la perca dorada (*Oreochromis mossambicus* × *reochromis hornorum*) en agua dulce y salada. *Revista de Investigaciones Marinas* 8(2):61-65.

Garcia-Ulloa, R. L. Villa, and T. M. Martinez. (2001). Growth and feed utilization of the tilapia hybrid *Oreochromis mossambicus* × *O. niloticus* cultured at different salinities under controlled laboratory conditions. *Journal of the World Aquaculture Society* 32:117-121.

Gerking, S.D. (1979). Fish reproduction and stress. In M.A. Ali (Ed.), *Environmental physiology of fishes* (pp. 569-588). New York: Plenum Press.

Gonzales-Corre, K. (1988). Polyculture of the tiger shrimp *(Penaeus monodon)* with the Nile tilapia *(Oreochromis niloticus)* in brackish water fish ponds. In R.S.V. Pullin, T. Bhukaswan, and K. Tonguthai (Eds.), *Proceedings of the*

Second International Symposium on Tilapia in Aquaculture, 16-20 March 1987 (pp. 569-588). Manila, Philippines.

Granoth, G. and D. Porath (1983). An attempt to optimize feed utilization by tilapia in a flow-through aquaculture. *Symposium on Tilapia in Aquaculture, Nazareth, Israel, 8-13 May 1983* (pp. 550-558). Tel Aviv, Israel: Tel Aviv University.

Gross, W.L., E.W. Roelofs, and P.O. Fromm (1965). Influence of photoperiod on growth of green sunfish, *Lepomis cyanellus* (Rafinesque). *Transactions of the American Fisheries Society* 92:401-408.

Grover, J.J., B.L. Olla, M. O'Brien, and R.I. Wicklund (1989). Food habits of Florida red tilapia fry in manured seawater pools in the Bahamas. *Progressive Fish-Culturist* 51:152-156.

Guerrero, R.D. (1975). Use of androgens for the production of all-male *Tilapia aurea* (Steindachner). *Transactions of the American Fisheries Society* 104:342-348.

———. (1977). Production of tilapia fry in floating net enclosures. *FAO Aquaculture Bulletin* 8:4.

Guerrero, R.D. III (1985). Tilapia farming in the Philippines: Practices, problems and prospects. In I.R. Smith, E.B. Torres, and E.O. Tan (Eds.), *Philippines Tilapia Economics: Proceedings of a PCARRD-ICLARM Workshop* (pp. 3-14). Manila, Philippines: International Center for Living Aquatic Resources Management, ICLARM Conference Proceedings 12.

Guerrero, R.D. and A.M. Garcia (1983). Studies on the fry production of *Sarotherodon niloticus* in a lake-based hatchery. In L. Fishelson and Z. Yaron (compilers), *International Symposium on Tilapia in Aquaculture, Nazareth, Israel, 8-13 May 1983*, pp. 388-393). Tel Aviv, Israel: Tel Aviv University.

Guerrero, R.D. and L.A. Guerrero (1985). Further observations on the fry production of *Oreochromis niloticus* in concrete tanks. *Aquaculture* 47:257-261.

——— (1988). Feasibility of commercial production of sex-reversed Nile tilapia fingerlings in the Philippines. In R.S.V. Pullin, T. Bhukaswan, K. Tonguthai, and J.L. Maclean (Eds.), *The Second International Symposium on Tilapia in Aquaculture* (pp. 183-186). ICLARM Conference Proceedings 15, Department of Fisheries, Bangkok, Thailand and International Center for Living Aquatic Resources Management, Manila, Philippines.

Guinther, E.B. (1971). Ecologic observations on an estuarine environment at Fanning Atoll. *Pacific Science* 25:249-259.

Hall, R.N. (1992). Preliminary investigations of marine cage culture of red hybrid tilapia in Jamaica. *Proceedings of the Gulf and Caribbean Fisheries Institute* 42:440.

Hanley, F. (1991). Freshwater tilapia culture in Jamaica. *World Aquaculture* 22(1): 42-48.

Hargreaves, J.A., J.E. Rakocy, D.S. Bailey, and D.J. Miller (1992). An evaluation of three cage designs and two tilapias for mariculture. *Proceedings of the Gulf and Caribbean Fisheries Institute* 42:103-113.

Hargreaves, J.A., J.E. Rakocy, and A. Nair (1988). An evaluation of fixed and demand feeding regimes for cage culture of *Oreochromis aureus*. In R.S.V. Pullin, T. Bhukaswan, K. Tonguthai, and J.L. Maclean (Eds.), *The Second International*

Symposium on Tilapia in Aquaculture (pp. 335-340). ICLARM Conference Proceedings 15, Department of Fisheries, Bangkok, Thailand and International Center for Living Aquatic Resources Management, Manila, Philippines.

Head, W.D. and W.O. Watanabe (1995). Economic analysis of a commercial-scale recirculating, brackishwater hatchery for Florida red tilapia. *Journal of Applied Aquaculture* 5:1-23.

Head, W.D., A. Zerbi, and W.O. Watanabe (1994). Preliminary observations on the marketability of saltwater-cultured Florida red tilapia in Puerto Rico. *Journal of the World Aquaculture Society* 25:432-441.

———. (1996). Economic evaluation of commercial-scale, saltwater pond production of Florida red tilapia in Puerto Rico. *Journal of the World Aquaculture Society* 27:275-289.

Helms, L.M.H., E.G. Grau, S.K. Shimoda, R.S. Nishioka, and H.A. Bern (1987). Studies on the regulation of growth hormone release from the proximal pars distalis of male tilapia, *Oreochromis mossambicus,* in vitro. *General and Comparative Endocrinology* 65:48-55.

Hepher, B. and Y. Pruginin (1982). Tilapia culture in ponds under controlled conditions. In R.S.V. Pullin and R.H. Lowe-McConnell (Eds.), *The biology and culture of tilapias* (pp. 185-203). ICLARM Conference Proceedings 7, International Center for Living Aquatic Resources Management, Manila, Philippines.

Hida, T.S., J.R. Harada, and J.E. King (1962). Rearing tilapia for tuna bait. *Fishery Bulletin of the Fish and Wildlife Service* 62:1-20.

Hopkins, K.D. (1983). Tilapia culture in arid lands. *ICLARM Newsletter* 6(1):8-9.

Hopkins, K., M. Hopkins, D. Leclercq, and A.A. Al-Ameeri (1986). Tilapia culture in Kuwait: A preliminary economic analysis of production systems. *Kuwait Bulletin of Marine Science* 7:45-64.

Hopkins, K., M. Ridha, D. Leclercq, A.A. Al-Meeri, and T. Al-Ahmad (1989). Screening tilapias for culture in sea water in Kuwait. *Aquaculture and Fisheries Management* 20:389-397.

Hoshina, T. (1968). On the monogenetic trematode, *Benedenia seriolae,* parasitic on yellow-tail, *Seriola quinqueradiata.* In *Proceedings of the Third Symposium de La Commission de L'Office International des Epizooties pour L'Etude des Maladies des Poissons, September 23-27, 1968* (pp. 1-11). Stockholm, Sweden.

Howerton, R.D., D.K. Okimoto, and E.G. Grau (1992). The effect of orally administered 17a-methyltestosterone and triiodothyronine on growth and proximate body composition of seawater-adapted tilapia *(Oreochromis mossambicus). Aquaculture and Fisheries Management* 23:123-128.

Hughes, D.G. and L.L. Behrends (1983). Mass production of Tilapia nilotica seed in suspended net enclosures. In L. Fishelson and Z. Yaron (compilers), *Proceedings of the International Symposium on Tilapia in Aquaculture, Nazareth, Israel, 8-13 May, 1983* (pp. 394-401). Tel Aviv, Israel: Tel Aviv University.

Hwang, P.P. (1987). Tolerance and ultrastructural responses of branchial chloride cells to salinity changes in the euryhaline teleost *Oreochromis mossambicus. Marine Biology* 94:643-649.

———. (1989). Distribution of chloride cells in teleost larvae. *Journal of Morphology* 200:1-8.

Hwang, P.P., C.M. Sun, and S.M. Wu (1989). Changes of plasma osmolality, chloride concentration and gill Na-K-ATPase activity in tilapia *Oreochromis mossambicus* during seawater acclimation. *Marine Biology* 100(3):295-299.

Hwang, P.P., S.M. Wu, J.H. Lin, and L.H. Wy (1992). Cortisol content of eggs and larvae of teleosts. *General and Comparative Endocrinology* 86:189-196.

Itzkowitz, M. (1979). The feeding strategies of a facultative cleanerfish, *Thalassoma bifasciatum* (Pisces: Labridae). *Journal of Zoology (London)* 187:403-413.

Iwama, G.K., A. Takemura, and K. Takano (1997). Oxygen consumption rates of tilapia in fresh water, sea water, and hypersaline sea water. *Journal of Fish Biology* 51:886-894.

Jahn, T.L. and L.R. Kuhn (1932). The life history of *Epibdella melleni* (MacCallum, 1927), a monogenetic trematode parasitic on marine fishes. *Biological Bulletin* 62:89-111.

Jauncey, K. (1982). The effects of varying dietary protein level on the growth, food conversion, protein utilization and body composition of juvenile, tilapias *(Sarotherodon mossambicus). Aquaculture* 27:43-54.

Jauncey, K. and B. Ross (1982). *A guide to tilapia feeds and feeding.* Institute of Aquaculture, University of Stirling, United Kingdom.

Jennings, D.P. (1991). Behavioral aspects of cold tolerance in blackchin tilapia, *Sarotherodon melanotheron,* at different salinities. *Environmental Biology of Fishes* 31:185-195.

Job, S.V. (1969). The respiratory metabolism of *Tilapia mossambica* (Teleostei). II. The effect of size, temperature, salinity and partial pressure of oxygen. *Marine Biology* 3:222-226.

Jurss, K., T. Bittorf, T. Vokler, and R. Wacke (1984). Biochemical investigations into the influence of environmental salinity on starvation of the tilapia, *Oreochromis mossambicus. Aquaculture* 40:171-182.

Kaneko, J.J., R. Yamada, J.A. Brock, and R.M. Nakamura (1988). Infection of tilapia, *Oreochromis mossambicus* (Trewavas), by a marine monogenean, *Neobenedenia melleni* (MacCallum, 1927) Yamaguti, 1963 in Kaneohe Bay, Hawaii, USA, and its treatment. *Journal of Fish Diseases* 11:295-300.

Kearn, G.C. (1986). The eggs of monogeneans. *Advances in Parasitology* 25:175-273.

Kinne, O. (1960). Growth, food intake, and food conversion in a euryplastic fish exposed to different temperatures and salinities. *Physiological Zoology* 33:288-317.

———. (1962). Irreversible nongenetic adaptation. *Comparative Biochemistry and Physiology* 5:265-282.

———. (1963). The effects of temperature and salinity on marine and brackish water animals. *Oceanography and Marine Biology, An Annual Review* 1:301-340.

Kirk, R.G. (1972). A review of recent developments in tilapia culture, with special reference to fish farming in heated effluents of power stations. *Aquaculture* 1:45-60.

Knaggs, E.H. (1977). Status of the genus tilapia in California's estuarine and marine waters. *Cal-Neva Wildlife Transactions* 1977:60-67.

Kocher, T.D. (1997). Introduction to the genetics of tilapias. In K. Fitzsimmons (Ed.), *Proceedings from the Fourth International Symposium on Tilapia in*

Aquaculture (pp. 61-63). Ithaca, NY: Northeast Regional Agricultural Engineering Service.

Kultz, D., R. Bastrop, K. Jurss, and D. Siebers (1992). Mitochondria-rich (MR) cells and the activities of the Na^{+}/K^{+}-ATPase and carbonic anhydrase in the gill and opercular epithelium of *Oreochromis mossambicus* adapted to various salinities. *Comparative Biochemistry and Physiology* 102B:293-301.

Kultz, D. and K. Jurss (1991). Acclimation of chloride cells and Na^{+}/K^{+}-ATPase to energy deficiency in tilapia *(Oreochromis mossambicus). Zoologische jahrbuecher Abteiling fuer Allgemeine Zoologie und Physiologie der Tiere* 95:39-50.

Kuo, C.-M. (1984). The development of tilapia culture in Taiwan. *ICLARM Newsletter* 7(1):12-14.

Kuo, C.-M. and R.A. Neal (1982). ICLARM'S tilapia research. *ICLARM Newsletter* 5:11-13.

Kuwaye, T.T., D.K. Okimoto, S.K. Shimoda, R.D. Howerton, H.-R. Lin, P.K.T. Pang, and E. G. Grau (1993). Effect of 17α-methyltestosterone on the growth of the euryhaline tilapia, *Oreochromis mossambicus,* in fresh water and in sea water. *Aquaculture* 113:137-152.

Lee, T.H., P.P. Hwang, Y.E. Shieh, and C.H. Lin (2000). The relationship between "deep-hole" mitochondria-rich cells and salinity adaptation in the euryhaline teleost, *Oreochromis mossambicus. Fish Physiology and Biochemistry* 23:133-140.

Legendre, M., S. Hem, and A. Cisse (1989). Suitability of brackish water tilapia species from the Ivory Coast for lagoon aquaculture. II—Growth and rearing methods. *Aquatic Living Resources* 2:81-89.

Liao, I.-C. and S.-L. Chang (1983). Studies on the feasibility of red tilapia culture in saline water. In L. Fishelson and Z. Yaron (compilers), *International Symposium on Tilapia in Aquaculture, Nazareth, Israel, 8-13 May 1983* (pp. 524-533). Tel Aviv, Israel: Tel Aviv University.

Liao, I.-C. and T.-P. Chen (1983). Status and prospects of tilapia culture in Taiwan. In L. Fishelson and Z. Yaron (Eds.), *International Symposium on Tilapia in Aquaculture* (pp. 588-596). Tel Aviv, Israel: Tel Aviv University.

Likongwe, J.S., T.D. Stecko, J.R. Stauffer Jr., and R.F. Carline (1996). Combined effects of water temperature and salinity on growth and feed utilization of juvenile Nile tilapia *Oreochromis niloticus* (Linneaus). *Aquaculture* 146:37-46.

Lin, C.K. (1995). Progression of intensive marine shrimp farm in Thailand. In C.L. Browdy and J.S. Hopkins (Eds.), *Swimming through troubled water, Proceedings of the Special Session on Shrimp Farming, Aquaculture '95* (pp. 13-22). Baton Rouge, LA: World Aquaculture Society.

Lin, L.-Y., C.-F. Weng, and P.-P. Hwang (2000). Effects of cortisol and salinity challenge on water balance in developing larvae of tilapia *(Oreochromis mossambicus). Physiological and Biochemical Zoology* 73(3):283-289.

———. (2001). Regulation of drinking rate in euryhaline tilapia larvae *(Oreochromis mossambicus)* during salinity challenges. *Physiological and Biochemical Zoology* 74:171-177.

Little, D., M. Skladany, and R. Rode (1987). Small-scale hatcheries in north-east Thailand. *Aquaculture and Fisheries Management* 18:15-31.

Lobel, P.S. (1980). Invasion by the Mozambique tilapia (*Sarotherodon mossambicus;* Pisces; Cichlidae) of a Pacific atoll marine ecosystem. *Micronesica* 16(2):349-355.

Loya, Y. and L. Fishelson (1969). Ecology of fish breeding in brackish water ponds near the Dead Sea (Israel). *Journal of Fish Biology* 1:261-278.

Macintosh, D.J. and S.S. De Silva (1984). The influence of stocking density and food ration on fry survival and growth in *Oreochromis mossambicus* and *O. niloticus* female × *O. aureus* male hybrids reared in a closed circulated system. *Aquaculture* 41:345-358.

Maciolek, J.A. (1984). Exotic fishes in Hawaii and other Islands of Oceania. In W.R. Courtenay Jr. and J.R. Stauffer Jr. (Eds.), *Distribution, biology, and management of exotic fishes* (pp. 131-161). Baltimore, MD: Johns Hopkins University Press.

Mancera, J.M. and S.D. McCormick (1998). Osmoregulatory actions of the GH/IGF axis in non-salmonid teleosts. *Comparative Biochemistry and Physiology* 121B: 43-48.

Mazid, M.A., Y. Tanaka, T. Katayama, M. Asadur Rahman, K.L. Simpson, and C.O. Chichester (1979). Growth response of *Tilapia zillii* fingerlings fed isocaloric diets with variable protein levels. *Aquaculture* 18:115-122.

McGeachin, R.B., R.I. Wicklund, B.L. Olla, and J.R. Winton (1987). Growth of *Tilapia aurea* in seawater cages. *Journal of the World Aquaculture Society* 18(1):31-34.

Meriwether, F.H. (1986). An inexpensive demand feeder for cage-reared tilapia. *Progressive Fish-Culturist* 48:226-228.

Meriwether, F.H. II, E.D. Scura, and W.Y. Okamura (1984). Cage culture of red tilapia in prawn and shrimp ponds. *Journal of the World Mariculture Society* 15:254-265.

Moe, M.A. Jr. (1975). Propagating the Atlantic neon goby. *Marine Aquarist* 6(2):4-10.

Morgan, J.D., T. Sakamoto, E.G. Grau, and G.K Iwama (1997). Physiological and respiratory responses of the Mozambique tilapia *(Oreochromis mossambicus)* to salinity acclimation. *Comparative Biochemistry and Physiology* 117A:391-398.

Mueller, K.W., W.O. Watanabe, and W.D. Head (1992). Effect of salinity on hatching in *Neobenedenia melleni,* a monogenean ectoparasite of seawater-cultured tilapia. *Journal of the World Aquaculture Society* 23(3):199-204.

———. (1994). Occurrence and control of *Neobenedenia melleni* (Monogenea: Capsalidae) in cultured marine fish, including three new host records. *Progressive Fish-Culturist* 56:140-142.

Nakano, K., M. Tagawa, A. Takemura, and T. Hirano (1997). Effects of ambient salinities on carbohydrate metabolism in two species of tilapia *Oreochromis mossambicus* and *O. niloticus. Fisheries Science* 63:338-343.

Nelson, R.G., L.L. Behrends, E.L. Waddell, and D.W. Burch (1983). Estimating relative sales potential of tilapia in supermarkets. *Proceedings Annual Conference Southeast Association Fish and Wildlife Agencies* 37:314-326.

Newnan, D.G. (1991). *Engineering economic analysis.* San Jose, CA: Engineering Press, Inc.

Nigrelli, R.F. and C.M. Breder Jr. (1934). The susceptibility and immunity of certain marine fishes to *Epibdella melleni,* a monogenetic trematode. *Journal of Parasitology* 20(5):259-269.

Osborne, T.S. (1979). Some aspects of salinity tolerances and subsequent growth of three tilapia species: *Sarotherodon aureus, S. spilurus* and *Tilapia zillii.* Field Report White Fish Authority, Fishery Development Project Saudi Arabia 48, Ministry of Agriculture and Water, Kingdom of Saudi Arabia.

Paclibare, J.O., R.C. Usero, J.R. Somga, and R. Visitacion (2000). Integration of finfish in shrimps *(Penaeus monodon)* culture: An effective disease prevention strategy. In Y. Inui and E. R. Cruz-Lacierda (Eds.), *Proceedings of Disease Control in Fish and Shrimp Aquaculture in Southeast Asia—Diagnosis and Husbandry Techniques 4-6 Dec 2001* (pp. 152-180). SEAFDEC and OIE. Tigbauan, Iloilo, Philippines.

Pantastico, J.B., M.M.A. Dangilan, and R.V. Eguia (1988). Cannibalism among different sizes of tilapia *(Oreochromis niloticus)* fry/fingerlings and the effect of natural food. In R.S.V. Pullin, T. Bhukaswan, K. Tonguthai, and J.L. Maclean, (Eds.), *The Second International Symposium on Tilapia in Aquaculture* (pp. 465-468). ICLARM Conference Proceedings 15, Department of Fisheries, Bangkok, Thailand and International Center for Living Aquatic Resources Management, Manila, Philippines.

Paperna, I. (1987). Solving parasite-related problems in cultured marine fish. *International Journal of Parasitology* 17:327-336.

Parker, N.C. (1987). Intensive fish culture in raceways, silos and tanks. *Proceedings of the 1987 Fish Farming Conference and Annual Convention, Fish Farmers of Texas.* Texas A&M University, College Station, Texas.

Parsons, T.R., Y. Maita, and C.M. Lalli (1984). *A manual of chemical and biological methods for seawater analysis.* New York: Pergamon Press.

Payne, A.I. (1983). Estuarine and salt tolerant tilapias. In L. Fishelson and Z. Yaron (compilers), *Proceeding of the International Symposium on Tilapia in Aquaculture, Nazareth, Israel, 8-13 May 1983* (pp. 534-543). Tel Aviv, Israel: Tel Aviv University.

Perez, J.E. and N. Maclean (1976). The haemoglobins of the fish *Sarotherodon mossambicus* (Peters): Functional significance and ontogenetic changes. *Journal of Fish Biology* 9:447-455.

Perschbacher, P.W. and R.B. McGeachin (1988). Salinity tolerances of red hybrid tilapia fry, juveniles and adults. In R.S.V. Pullin, T. Bhukaswan, K. Tonguthai, and J.L. Maclean (Eds.), *The Second International Symposium on Tilapia in Aquaculture* (pp. 415-420). ICLARM Conference Proceedings 15. Department of Fisheries, Bangkok, Thailand and International Center for Living Aquatic Resources Management, Manila, Philippines.

Peters, D.S. and M.T. Boyd (1972). The effect of temperature, salinity, and availability of food on the feeding and growth of the hogchoker, *Trinectes maculatus* (Bloch & Schneider). *Journal of Experimental Marine Biology* 7:201-207.

Peters, H.M. (1983). Fecundity, egg weight, oocyte development in tilapias. ICLARM Translations 2, International Center for Living Aquatic Resources Management, Manila, Philippines.

Phelps, R. and T.J. Popma (2000). Sex reversal of tilapia. In B.A. Costa-Pierce and J.E. Rakocy (Eds.), *Tilapia aquaculture in the Americas* (Vol. 2, pp. 34-59). Baton Rouge, LA: The World Aquaculture Society.

PR-SBDC (1989). *Estudio de mercadeo sobre la tilapia roja en el mercado de Puerto Rico.* Mayagüez: Puerto Rico Small Business Development Center.

Philippart, J.-C. and J.-C. Ruwet (1982). Ecology and distribution of tilapias. In R.S.V. Pullin and R.H. Lowe-McConnell (Eds.), *The biology and culture of tilapias* (pp. 15-59). ICLARM Conference Proceedings 7, International Center for Living Aquatic Resources Management, Manila, Philippines.

Pierce, B.A. (1980). Production of hybrid tilapia in indoor aquaria. *Progressive Fish-Culturist* 42:233-234.

Popper D. and T. Lichatowich (1975). Preliminary success in predator contact of *Tilapia mossambica. Aquaculture* 5:213-214.

Prunet, P. and M. Bornancin (1989). Physiology of salinity tolerance in tilapia: An update of basic and applied aspects. *Aquatic Living Resources* 2:91-97.

Rakocy, J.E. and A.S. McGinty (1989). *Pond culture of tilapia.* Florida Cooperative Extension Service. Southern Regional Aquaculture Center Pub. No. 280. University of Florida, Gainesville, Florida.

Rana, K.J. (1986). An evaluation of two types of containers for the artificial incubation of *Oreochromis* eggs. *Aquaculture and Fisheries Management* 17:139-145.

———. (1988). Reproductive biology and hatchery rearing of tilapia eggs and fry. In J.J. Muir and R.J. Roberts (Eds.), *Recent advances in aquaculture* (Vol. 3, pp. 343-406). London, England: Croom Helm.

Rhodes, R.J. (1991). Economics of aquaculture production: Financial feasibility. In J.A. Hargreaves and D.E. Alston (Eds.), *Status and potential of aquaculture in the Caribbean* (pp. 192-208). Baton Rouge, LA: The World Aquaculture Society.

Rhodes, R.J. and D. Hollin (1990). Financial analysis of commercial red drum aquaculture enterprise. In G.W. Chamberlain, R.J. Miget, and M.G. Haby (compilers), *Red drum aquaculture* (pp. 189-208). College Station: Texas A&M Sea Grant Publication 90-603.

Ridha, M., T.-A. Al-Ahmad, and A.-A. Al-Ahmad (1985). *Tilapia culture in Kuwait: Spawning experiments, 1984.* KISR technical report 1875, Kuwait Institute for Scientific Research, Safat, Kuwait.

Ridha, M. and E.M. Cruz (1989). Effect of age on the fecundity of the tilapia *Oreochromis spilurus. Asian Fisheries Science* 2:239-247.

Roberts, R.J. and C. Sommerville (1982). Diseases of tilapias. In R.S.V. Pullin and R.H. Lowe-McConnell (Eds.), *The biology and culture of tilapias* (pp. 247-263). ICLARM Conference Proceedings, 7. International Center for Living Aquatic Resources Management, Manila, Philippines.

Robinson, R.D., L.F. Khalil, R.N. Hall, and R.D. Steele (1992). Infection of red hybrid tilapia with a monogenean in coastal waters off Southern Jamaica. *Proceedings of the Gulf and Caribbean Fisheries Institute* 42:441-447.

Romaguera, J.M., P.D. Molina, and J.L. Vega (1987). *Puerto Rico's fishing centers: An assessment for development.* Washington, DC: U.S. Department of Commerce Economic Development Administration Technical Assistance Program Report 01-06-0244.

Romana-Equia, M.R. and R.V. Eguia (1999). Growth of five Asian strains in saline environments. *Aquaculture* 173:161-170.

Ron, B., S.K. Shimoda, G.K. Iwama, and E.G. Grau (1995). Relationships among ration, salinity, 17α-methyltestosterone and growth in the euryhaline tilapia, *Oreochromis mossambicus. Aquaculture* 135:185-193.

Rothbard, S., E. Solnik, S. Shabbath, R. Amado, and I. Grabie (1983). The technology of mass production of hormonally sex-inversed all-male tilapias. In L. Fishelson and Z. Yaron (compilers), *International Symposium on Tilapia in Aquaculture, Nazareth, Israel, 8-13 May 1983* (pp. 425-432). Tel Aviv, Israel: Tel Aviv University.

Ryther, J.H., R.L. Creswell, and D.E. Alston (1991). Historical overview: Aquaculture in the Caribbean. In J.A. Hargreaves and D.E. Alston (Eds.), *Status and potential of aquaculture in the Caribbean* (pp. 4-8). Baton Rouge, LA: World Aquaculture Society.

Saelee W., Yang Yi, and K. Fitzsimmons (2002). Stocking densities of Nile tilapia in tilapia-shrimp polyculture at low salinity. In *Proceedings of the 4th National Symposium on Marine Shrimp* (pp. 93-107). BIOTECH, Thailand.

Sakamoto, T., Shepherd, B.S, and Madsen, S.S. (1997). Osmoregulatory action of growth hormone and prolactin in an advanced teleost. *General and Comparative Endocrinology* 106:95-101.

Sandifer, P.A. (1991). Species with aquaculture potential for the Caribbean. In J.A. Hargreaves and D.E. Alston (Eds.), *Status and potential of aquaculture in the Caribbean* (pp. 30-60). Baton Rouge, LA: World Aquaculture Society.

Sarig, S. and M. Marek (1974). Results of intensive and semi-intensive fish breeding techniques in Israel in 1971-1973. *Bamidgeh* 26:28-48.

Shang, Y.C. (1990). Aquaculture economic analysis: an introduction. *Advances in World Aquaculture* 2. The World Aquaculture Society, Baton Rouge, Louisiana.

Shepherd, B.S., B. Ron, A. Burch, R. Sparks, N.H. Richman, III, S.K. Shimoda, M.H. Stetson, C. Lim, and E.G. Grau (1997). Effects of salinity, dietary level of protein and 17α-methyltestosterone on growth hormone (GH) and prolactin (tPRL177 and tPRL188) levels in the tilapia, *Oreochromis mossambicus. Fish Physiology and Biochemistry* 17:279-288.

Shepherd, B.S., T. Sakamoto, R.S. Nishioka, N.H., Richman III, I. Mori, SS. Madsen, TT. Chen, T. Hirano, H.A. Bern, and E.G. Grau (1997). Somatotropic actions of the homologous growth hormone and prolactins in the euryhaline teleost, the tilapia, *Oreochromis mossambicus. Proceedings National Academy of Sciences* 94:2068-2072.

Shiau, S.-Y. and S.L. Huang (1989). Optimal dietary protein level for hybrid tilapia *Oreochromis niloticus* × *O. aureus*) reared in seawater. *Aquaculture* 81:119-127.

———. (1990). Influence of varying energy levels with two protein concentrations in diets for hybrid tilapia (*Oreochromis niloticus* × *O. aureus*) reared in seawater. *Aquaculture* 91:143-152.

Shiraishi, K., T. Kaneko, S. Hasegawa, and T. Hirano (1997). Development of multicellular complexes of chloride cells in the yolk-sac membrane of tilapia (*Oreochromis mossambicus*) embryos and larvae in seawater. *Cell Tissue Research* 288:583-590.

Siddiqui, A.Q., M.S. Howlader, and A.B. Adam (1989). Culture of Nile tilapia, *Oreochromis niloticus* (L.), at three stocking densities in outdoor concrete tanks using drainage water. *Aquaculture and Fisheries Management* 20:49-57.

Sin A.W. and M.T. Chiu (1983). The intensive monoculture of the tilapia hybrid, *Sarotherodon nilotica* (male) × *S. mossambica* (female) in Hong Kong. In L. Fishelson and Z. Yaron (Eds.), *International Symposium on Tilapia in Aquaculture* (pp. 506-516). Tel Aviv, Israel: Tel Aviv University.

Snow, J.R., J.M. Berrios-Hernandez, and H.Y. Ye (1983). A modular system for producing tilapia seed using simple facilities. In L. Fishelson and Z. Yaron (compilers), *International Symposium on Tilapia in Aquaculture* (pp. 402-413). Tel Aviv, Israel: Tel Aviv University.

Stauffer, J.R., Jr. (1986). Effects of salinity on preferred and lethal temperatures of the Mozambique tilapia, *Oreochromis mossambicus* (Peters). *Water Resources Bulletin* 22(2):205-208.

Stauffer, J.R., Jr., D.K. Vann, and C.H. Hocutt (1984). Effects of salinity on preferred and lethal temperatures of the blackchin tilapia, *Saratherodon melanotheron. Water Resources Bulletin. American Water Resources Association* 20(5): 771-775.

Stickney, R.R. (1979). *Principles of warmwater aquaculture.* New York: John Wiley and Sons.

——— (1986). Tilapia tolerance of saline waters: A review. *Progressive Fish-Culturist* 48:161-167.

Stickney, R.R., J.H. Hesby, R.B. Mcgeachin, and W.A. Isbell (1979). Growth of *Tilapia niloticus* in ponds with differing histories of organic fertilization. *Aquaculture* 17:189-194.

Suresh, A.V. and C. Kwei-Lin (1992). Tilapia culture in saline waters: A review. *Aquaculture* 106:201-226.

Svendsen, Y.S. and T. Haug (1991). Effects of formalin, benzocaine, and hypo- and hypersaline exposures against adults and eggs of *Entobdella hippoglossi* (Muller), an ectoparasite on Atlantic halibut (*Hippoglossus hippoglossus* L.). Laboratory studies. *Aquaculture* 94:279-289.

Tagawa, M., H. Hagiwara, A. Takemura, S. Hirose, and T. Hirano (1997). Partial cloning of the hormone-binding domain of the cortisol receptor in tilapia, *Oreochromis mossambicus* and changes in the mRNA level during embryonic development. *General and Comparative Endocrinology* 108:132-140.

Tave, D. (1986). *Genetics for fish hatchery managers.* Westport, CT: AVI Publishing Company, Inc.

Teichert-Coddington, D.R., T.J. Popma, and L.L. Lovshin (1997). Attributes of tropical pond-cultured fish. In H.S. Egna and C.E. Boyd (Eds.), *Dynamics of pond aquaculture* (pp. 183-198). Boca Raton, FL: CRC Press.

Thoney, D.A. (1990). The effects of trichlorfon, praziquantel and copper sulphate on various stages of the monogenean *Benedeniella posterocolpa,* a skin parasite of the cownose ray, *Rhinoptera bonasus* (Mitchill). *Journal of Fish Diseases* 13:385-389.

Thoney, D.A. and W.J. Hargis (1991). Monogenea (Platyhelminthes) as hazards for fish in confinement. *Annual Review of Fish Diseases* 133-153.

Thresher, R.E. (1980). *Reeffish.* St. Petersburg, FL: Palmetto Publishing Company.

Tian, X., D. Li, S. Dong, G. Liu, Z. Qi, and J. Lu (2001). Water quality of closed polyculture of penaeid shrimp with tilapia and constricted tagelus. *Chinese Journal of Applied Ecology* 12(2):287-292.

Tian, X., D. Li, S. Dong, X. Yan, Z. Qi, G. Liu, and J. Lu (2001). An experimental study on closed-polyculture of penaeid shrimp with tilapia and constricted tagelus. *Aquaculture* 202(1-2):57-71.

Tilney, R.L. and C.H. Hocutt (1987). Changes in gill epithelia of *Oreochromis mossambicus* subjected to cold shock. *Environmental Biology of Fishes* 9(1):35-44.

Ting, Y.-Y., M.-H. Chang, S.-H. Chen, Y.-S. Wang, and W.-H. Cherng (1984). Three kinds of *Tilapia* spp. reared in three different salinities water for cage- cultural wintering experiment. *Bulletin of Taiwan Fisheries Research Institute* 37:101-115.

Uchida, K., T. Kaneko, H. Miyazaki, S. Hasegawa, and T. Hirano (2000). Excellent salinity tolerance of Mozambique tilapia *(Oreochromis mossambicus):* Elevated chloride cell activity in the branchial and opercular epithelia of the fish adapted to concentrated seawater. *Zoological Science* 17:149-160.

Uchida, R.N. and J.E. King (1962). Tank culture of tilapia. *U.S. Fish and Wildlife Service Fishery Bulletin* 199(62):21-47.

Verdegem, M.C. and A.S. McGinty (1987). Effects of frequency of removal of eggs and fry from *Tilapia nilotica* spawned in hapas. *Progressive Fish-Culturist* 49:129-131.

———. (1989). Evaluation of edge seining for harvesting *Oreochromis niloticus* fry from spawning ponds. *Aquaculture* 80:195-200.

Vijayan, M.M., J.D. Morgan, T. Sakamoto, E.G. Grau, and G.K. Iwama (1996). Food-deprivation affects seawater acclimation in tilapia: Hormonal and metabolic changes. *Journal Experimental Biology* 199:2467-2474.

Villegas, C.T. (1990). Evaluation of the salinity tolerance of *Oreochromis mossambicus, O. niloticus* and their F1 hybrids. *Aquaculture* 85:281-292.

Vonck, A.P.M.A., S.E. Wendelaar Bonga, and G. Glik (1998). Sodium and calcium balance in Mozambique tilapia, *Oreochromis mossambicus,* raised at different salinities. *Comparative Biochemistry and Physiology* 199A:441-449.

Walker, S. (1977). Walker successfully spawns five species. *Marine Hobbyist News* 5(9):1-4.

Wang, J., D. Li, S. Dong, K. Wang, and X. Tian (1998). Experimental studies on polyculture in closed shrimp ponds: I. Intensive polyculture of Chinese shrimp *(Penaeus chinensis)* with tilapia hybrids. *Aquaculture* 163(1-2):11-27.

Wannigamma, N.D., D.E.M. Weerakoon, and G. Muthukumarana (1985). Cage culture of *S. niloticus* in Sri Lanka: Effect of stocking density and dietary crude protein levels on growth. In C.Y. Cho, C.B Cowey, and T. Watanabe (compilers), *Finfish nutrition in Asia: Methodological approaches to research and development* (pp. 113-117). Ottawa, Ontario, Canada: IDRC.

Watanabe, W.O. (1991). Saltwater culture of tilapia in the Caribbean. *World Aquaculture* 22:49-54.

Watanabe, W.O., K.M. Burnett, B.L. Olla, and R.I. Wicklund (1989). The effects of salinity on reproductive performance in Florida red tilapia. *Journal of the World Aquaculture Society* 20(4):223-229.

Watanabe, W.O., J.H. Clark, J.B. Dunham, R.I. Wicklund, and B.L. Olla (1990a). Culture of Florida red tilapia in marine cages: The effects of stocking density and dietary protein on growth. *Aquaculture* 90:123-124.

———. (1990b). Production of fingerling Florida red tilapia (*T. hornorum* × *T. mossambica*) in floating marine cages. *Progressive Fish-Culturist* 52(3):158-161.

Watanabe, W.O., J.R. Chan, S.J. Smith, R.I. Wicklund, and B.L. Olla (1993). Production of Florida red tilapia in flowthrough seawater pools at three stocking densities. In R.S.V. Pullin, J. Lazard, M. Legendre, J.B. Amon Kothias, and D. Pauly (Eds.), *The Third International Symposium on Tilapia in Aquaculture* (pp. 168-174). Makati City, Philippines: ICLARM Conference Proceedings 41, ICLARM.

Watanabe, W.O., L.J. Ellingson, B.L. Olla, D.H. Ernst, and R.I. Wicklund (1990). Salinity tolerance and seawater survival vary ontogenetically in Florida red tilapia. *Aquaculture* 87:311-321.

Watanabe, W.O., L.J. Ellingson, R.I. Wicklund, and B.L. Olla (1988). The effects of salinity on growth, food consumption and conversion in juvenile, monosex male Florida red tilapia. In R.S.V. Pullin, T. Bhukaswan, K. Tonguthai, and J.L. Maclean (Eds.), *The Second International Symposium on Tilapia in Aquaculture* (pp. 515-523). ICLARM Conference Proceedings 15. Department of Fisheries, Bangkok, Thailand and International Center for Living Aquatic Resources Management, Manila, Philippines.

Watanabe, W.O., D.H. Ernst, M.P. Chasar, R.I. Wicklund and B.L. Olla (1993). The effects of temperature and salinity on growth and feed utilization of juvenile, sex-reversed male Florida red tilapia cultured in a recirculating system. *Aquaculture* 112:309-320.

Watanabe, W.O., D.H. Ernst, B.L. Olla, and R.I. Wicklund (1990). Aquaculture of red tilapia (*Oreochromis* sp.) in marine environments: State of the art. Advances in Tropical Aquaculture, 20 February-4 March, 1989, Tahiti, French Polynesia, *Actes de Colloque* 9:487-499.

Watanabe, W.O., K.E. French, L.J. Ellingson, R.I. Wicklund, and B.L. Olla (1988). Further investigations on the effects of salinity on growth of Florida red tilapia: Evidence for the influence of behavior. In R.S.V. Pullin, T. Bhukaswan, K. Tonguthai, and J.L. Maclean (Eds.), *The Second International Symposium on Tilapia in Aquaculture* (pp. 525-530). ICLARM Conference Proceedings 15, Department of Fisheries, Bangkok, Thailand and International Center for Living Aquatic Resources Management, Manila, Philippines.

Watanabe, W.O., K.E. French, D.H. Ernst, B.L Olla, and R.I. Wicklund (1989). Salinity during early development influences survival and growth of Florida red tilapia in brackish- and seawater. *Journal of the World Aquaculture Society* 20(3):134-142.

Watanabe, W.O. and C.-M. Kuo (1985). Observations on the reproductive performance of Nile tilapia *(Oreochromis niloticus)* in laboratory aquaria at various salinities. *Aquaculture* 49:315-323.

Watanabe, W.O., C.-M. Kuo, and H.-C. Huang (1984). Experimental rearing of Nile tilapia fry *(Oreochromis niloticus)* for saltwater culture. ICLARM Technical Reports 14, Council for Agricultural Planning and Development, Taipei, Taiwan and International Center for Living Aquatic Resources Management, Manila, Philippines.

———. (1985a). The ontogeny of salinity tolerance in the tilapias *Oreochromis niloticus, O. aureus,* and *O. mossambicus* × *O. niloticus* hybrid, spawned and hatched in freshwater. *Aquaculture* 47:353-367.

———. (1985b). Salinity tolerance of Nile tilapia fry *(Oreochromis niloticus)* spawned and hatched at various salinities. *Aquaculture* 48:159-176.

———. (1985c). Salinity tolerance of tilapias *Oreochromis aureus* (Steindachner), *O. niloticus* (L.), and *O. mossambicus* (Peters) × *O. niloticus* hybrid. ICLARM Technical Reports 16, Council for Agricultural Planning and Development, Taipei, Taiwan and International Center for Living Aquatic Resources Management, Manila, Philippines.

Watanabe, W.O., K.W. Mueller, W.D. Head, and S.C. Ellis (1993). Sex reversal of Florida red tilapia in brackish water tanks under different treatment durations of 17α-ethynyltestosterone administered in feed. *Journal of Applied Aquaculture* 2(1):29-41.

Watanabe, W.O., S.J. Smith, W.D. Head, and K.W. Mueller (1993). Production of Florida red tilapia fry in brackishwater tanks under different stocking densities and feeding regimes. In R.S.V. Pullin, J. Lazard, M. Legendre, J.B. Amon Kothias, and D. Pauly (Eds.), *The Third International Symposium on Tilapia in Aquaculture* (pp. 168-174). Makati City, Philippines: ICLARM Conference Proceedings 41, ICLARM.

Watanabe, W.O., S.J. Smith, R.I. Wicklund, and B.L. Olla (1991). Evaluation of methods for prevention and control of monogenetic trematode *(Neobenedenia melleni)* parasitosis of Florida red tilapia reared in seawater pools. *Journal of the World Aquaculture Society* 22:63A.

———. (1992). Hatchery production of Florida red tilapia seed in brackishwater tanks under natural-mouthbrooding and clutch-removal methods. *Aquaculture* 102:77-88.

Watanabe, W.O., R.I. Wicklund, B.L. Olla, D.H. Ernst, and L.J. Ellingson (1989). Potential for saltwater tilapia culture in the Caribbean. *Gulf and Caribbean Fisheries Institute* 39:435-445.

Watanabe, W.O., R.I. Wicklund, B.L. Olla, and W.D. Head (1997). Saltwater culture of the Florida red and other saline tolerant tilapias: A review. In B.A.

Costa-Pierce and J.E. Rakocy (Eds.), *Tilapia aquaculture in the Americas* (Vol. 1, pp. 54-141). Baton Rouge, LA: World Aquaculture Society.

Whitfield, A.K. and S.J.M. Blaber (1976). The effects of temperature and salinity on *Tilapia rendalli* Boulenger 1896. *Journal of Fish Biology* 9:99-104.

———. (1979). The distribution of the freshwater cichlid *Sarotherodon mossambicus* in estuarine systems. *Environmental Biology of Fishes* 4:77-81.

Winfree, R.A. and R.R. Stickney (1981). Effects of dietary protein and energy on growth, feed conversion efficiency and body composition of *Tilapia aurea. Journal of Nutrition* 3:1001-1012.

Wohlfarth, G.W. and G. Hulata (1983). *Applied genetics of tilapias. ICLARM Studies and Reviews 6* (2nd edn.). International Center for Living Aquatic Resources Management, Manila, Philippines.

Wohlfarth, G.W., S. Rothbard, G. Hulata, and D. Szweigman (1990). Inheritance of red body coloration in Taiwanese tilapias and in *Oreochromis mossambicus. Aquaculture* 84:219-234.

Woiwode, J.G. and I.R. Adelman (1991). Effects of temperature, photoperiod, and ration size on growth of hybrid striped bass × white bass.*Transaction of the American Fisheries Society* 120:217-229.

Wurts, W.A. and R.R. Stickney (1989). Responses of red drum *(Sciaenops ocellatus)* to calcium and magnesium concentrations in fresh and salt water. *Aquaculture* 76:21-35.

Yada, T., T. Hirano, and E.G. Grau (1994). Changes in plasma levels of the two prolactins and growth hormone during adaptation to different salinities in the euryhaline tilapia, *Oreochromis mossambicus. General Comparative Endocrinology* 93:214-223.

Yang, Yi, P. Nadtirom, V. Tansakul, and K. Fitzsimmons (2002). Current status of tilapia-shrimp polyculture in Thailand. In *Proceedings of the 4th National Symposium on Marine Shrimp* (pp. 77-92). BIOTECH, Thailand.

Yater, L.R. and I.R. Smith (1985). Economics of private tilapia hatcheries in Laguna and Rizal Provinces, Philippines. In I.R. Smith, E.B. Torres, and E.O. Tan (Eds.), *Philippine tilapia economics* (pp. 15-32). ICLARM Conference Proceedings 12, Philippine Council for Agriculture and Resources Research and Development, Los Banos, Launa and International Center for Living Aquatic Resources Management, Manila, Philippines.

Zale, A.V. and R.W. Gregory (1989). Effect of salinity on cold tolerance of juvenile blue tilapias. *Transactions of the American Fisheries Society* 118:718-720.

Chapter 11

Management of Bottom Soil Condition and Pond Water and Effluent Quality

Claude E. Boyd

INTRODUCTION

Good bottom soil condition and high-quality water are essential ingredients for successful pond aquaculture of tilapia and other species. Some problems with pond soil and water quality are related to site characteristics (Hajek and Boyd 1994). Soils may have undesirable properties such as acidity, high organic matter content, or excessive porosity. Water supplies may not be large enough or the source water naturally may be of poor quality or polluted with domestic, industrial, or agricultural wastes. Even if a good site is available, large inputs of nutrients and organic matter in feeds to enhance aquacultural production can lead to excessive phytoplankton, low dissolved oxygen concentration, high ammonia concentration, poor bottom soil condition, and other problems (Boyd and Tucker 1998).

Many soil and water quality problems can be avoided by attention to site selection, pond design, and pond construction and by the use of moderate stocking and feeding rates. Nevertheless, sites are seldom perfect, and often, site limitations are not adequately mitigated during design and construction. Pond managers also may strive for unrealistically high production. Thus, soil and water quality problems are not uncommon in pond culture of tilapia. When soil and water quality in ponds are impaired, fish suffer stress. This makes them more susceptible to disease, and they do not consume feed efficiently or grow as well as they should.

Effluents from ponds with poor-quality water may have low dissolved oxygen concentration and high concentrations of nutrients, organic matter, and suspended solids. Release of such effluents into natural waters can cause pollution that harms aquatic communities and lessen the quality of water for other beneficial uses.

doi:10.1300/5513_11

The purpose of this chapter is to discuss management of soil and water in ponds and to present suggestions for reducing the volume and improving the quality of pond effluents.

BOTTOM SOILS

Liming

The reason for liming aquaculture ponds is to neutralize soil acidity and increase total alkalinity and total hardness concentrations in water (Hickling 1962). This can enhance conditions for productivity of fish food organisms and increase aquatic animal production. Freshwater ponds with less than 40 or 50 $mg \cdot L^{-1}$ total alkalinity, brackishwater ponds with total alkalinity below 60 $mg \cdot L^{-1}$, and any pond with soil pH below 7 usually will benefit from liming (Boyd and Tucker 1998).

Samples of bottom soil may be analyzed for lime requirement (Boyd 1974; Pillai and Boyd 1985), but if this is not possible, data on soil pH and rough approximations of soil texture (Boyd 1990) may be used to estimate lime requirement (Table 11.1).

Agricultural limestone should be spread uniformly over the bottoms of empty ponds, or alternatively, it may be spread uniformly over water surfaces. Agricultural limestone should be applied at the beginning of the crop. It should also be applied at least 1 week before fertilization is initiated because phosphorus in water may be precipitated by the initial reaction of the liming material.

Agricultural limestone will not react with dry soil, so during application, it should be applied while the soil is still visibly moist but dry enough to walk on without soiling your shoes. Tilling after liming can improve the

TABLE 11.1. Lime requirements of pond bottom soils based on pH and texture of mud.

	Lime requirement ($kg \cdot ha^{-1}$ as $CaCO_3$)		
Mud pH	**Heavy loam or clay**	**Sandy loam**	**Sand**
<4.0	14,320	7,160	4,475
4.0-4.5	10,740	5,370	4,475
4.6-5.0	8,950	4,475	3,580
5.1-5.5	5,370	3,580	1,790
5.6-6.0	3,580	1,790	895
6.1-6.5	1,790	1,790	0
>6.5	0	0	0

reaction of agricultural limestone with soil, but no studies have yet been made to verify the benefits of tilling.

Drying

The purpose of drying pond bottoms between crops is to reduce the moisture content of soil so that air can enter the pore spaces between soil particles. Better aeration will improve the supply of oxygen and enhance aerobic decomposition of organic matter (Wurtz 1960). By drying for 2 to 3 weeks, most of the labile organic matter remaining in the bottom soil from the previous crop will decompose and reduced inorganic compounds will be oxidized (Boyd and Pippopinyo 1994). The main benefit of this practice is to reduce the oxygen demand of bottom soil as much as possible before beginning the next crop (Seo and Boyd 2001a,b).

The time required to dry a pond bottom depends upon the soil texture, air temperature, wind conditions, rainfall, and infiltration of water from adjacent ponds or shallow water tables. Light-textured soils dry faster than heavy-textured soils. Warm, dry weather and windy conditions hasten drying whereas rainy weather or infiltration into ponds retards drying. Intensive tilapia ponds are especially difficult to dry because the fish make many redds (depressions) in pond bottoms. Fine soil collects in the redds and the bottoms of the redds do not dry well. The decomposition rate in soil increases up to the optimum moisture content and then declines if soils are dried more. The optimum moisture content is about 30 to 40 percent for heavy clays, 20 to 30 percent for loams, and 10 to 20 percent for sandy soils (Boyd and Pipoppinyo 1994; Boyd and Teichert-Coddington 1994). It is usually not useful to dry pond bottoms for periods of many weeks.

In soils with a high clay content or in deep layers of silty or clayey sediment, the soil will crack into columnar blocks upon drying. Surfaces of blocks of soil may appear oxidized and quite dry, but if blocks are broken, the soil inside will still be black and wet. Additional drying of the soil usually will not be of much benefit because dry surfaces serve as a barrier to further evaporation. Tilling with a disc harrow can break up blocks of soil or penetrate dense soils to enhance drying and aeration.

Tilling

Tilling bottom soils can enhance drying to increase aeration and accelerate decomposition of organic matter and oxidation of reduced compounds. Soil amendments such as agricultural limestone or burnt lime can be mixed into soil by tilling. Accumulation of organic matter or other substances in the surface layer of soil also can be mixed with deeper soils to reduce concentrations of the substances in the surface layer.

Pond bottoms should not be tilled when they are too wet to support tillage machinery. Ruts caused by machinery will fill with soft sediment and be likely sites for anaerobic conditions. Ruts also interfere with draining and increase the difficulty of drying pond bottoms. Where tractors are used for tilling, dual tires or extra-wide tires are recommended to prevent the occurrence of ruts.

Depth of tillage usually should be 5 to 10 cm, so a disk harrow can be used. Roto-tillers require much more energy than disk harrows and are destructive of soil texture. Mould board plows, often called turning plows, can be used to turn soil over. They can be useful if surface soil has unacceptably high concentrations of one or more substances and deeper soils are of better quality. A mould board plow should not be used for routine tilling as it requires more energy than a disk harrow.

Tilling can be counterproductive in ponds where heavy mechanical aeration is used. Tilling loosens the soil particles and aerator-induced water currents cause severe erosion of the pond bottom. If bottoms of heavily aerated ponds are tilled, they should be compacted with a heavy roller before refilling.

Sediment Removal

Sediment accumulates in ponds due to several reasons (Boyd 1995). There may be a large external sediment load from turbid water supplies. Erosion of embankments can result in large amounts of sediment in deep water areas even where there is not a large external sediment input. If ponds are left empty between crops, rain falling on bottoms can cause the insides of embankments and shallow water edges to erode and the eroded material will settle in the deep parts of the pond bottom. Mechanical aeration can cause erosion in front of aerators where water currents are strong, and deposition of eroded particles will occur in areas of the pond with weaker water currents.

Accumulation of soft sediment in ponds is undesirable as it fills deeper areas and can cause ponds to lose volume. Soft sediment can trap feed pellets and fertilizer granules. Anaerobic zones often occur in soft sediment (Avnimelech and Zohar 1986); soft sediment is not good habitat for benthic organisms. Dissolved oxygen concentrations are often lower in older ponds with deep sediment than in newer ponds with less sediment (Steeby et al. 2001). Fish harvest is also hampered by soft sediment because it impedes seining operations. Soft sediment should be removed periodically before it reaches a troublesome thickness.

Sediment can be excavated with a variety of equipment ranging from shovels to bulldozers. Sediment in pond bottoms does not contain as much organic matter as farmers often think. There normally is no valid reason for disposing of sediment outside of ponds. Sediment usually can be put back on the areas from which it eroded. Of course, the loose material should be compacted or protected from erosion by covering it with vegetation, stone, or other barriers. When sediment must be disposed of outside of ponds, disposal should be done in a responsible way to prevent unsightly, ecologically degrading spoil piles and erosion.

Fertilization

There are places where ponds are constructed on soils with high concentrations of fibrous organic matter. Decomposition in organic soils is slow because the pH usually is low and the amount of carbon relative to nitrogen (carbon:nitrogen ratio) is high. Nevertheless, because of high organic matter content, such soil often becomes anaerobic during fish or shrimp culture. Application of agricultural limestone to increase pH and inorganic nitrogen fertilizers to supply nitrogen will increase soil organic matter degradation during fallow periods between crops. Nitrate can be used to oxidize wet soils that cannot be dried.

Urea can be spread over pond bottoms at 200 to 400 $kg \cdot ha^{-1}$ at the beginning of the fallow period to accelerate decomposition of organic soil. Agricultural limestone should not be applied until a few days after urea is applied to prevent a high pH. Urea hydrolyzes to ammonia, and if pH is above 8, much of the ammonia will diffuse into the air. Bottoms may be tilled to incorporated lime and urea into soil to avoid ammonia volatilization. Tilling also provides better aeration of the soil mass to encourage bacterial activity.

In some ponds, there will be areas that will not dry sufficiently to enhance decomposition of organic matter and oxidation of reduced inorganic compounds. Sodium, potassium, or sodium nitrate can be applied to wet soil to encourage organic matter decomposition by denitrifying bacteria and to oxidize ferrous iron, manganous manganese, and hydrogen sulfide. The usual application rate is 20 to 40 $g \cdot m^{-2}$ over wet areas. Nitrate fertilizers are more expensive than urea and are not recommended where soils can be adequately dried.

Productivity of benthic organisms may be low in ponds with concentrations of organic carbon below 0.5 to 1.0 percent. Organic fertilizer can be applied to such soils to enhance organic matter concentration. Chicken manure and other animal manures have been applied at 1,000 to 2,000 $kg \cdot ha^{-1}$ to pond bottoms during the fallow period. However, application of a higher quality organic matter such as plant meals, e.g., rice bran, soybean meal, crushed corn, and the like, or low-protein content animal feed, at 500 to 1,000 $kg \cdot ha^{-1}$ is more efficient. When organic fertilization of pond bottoms is practiced, ponds should be filled with 10 to 20 cm of water and a dense plankton bloom must be allowed to develop. Water level should be increased and 1 or 2 weeks allowed for the development of benthic community before stocking ponds.

Bottom Raking

During the culture period when the pond bottom is covered with water, the most common bottom soil problem is loss of the oxidized layer. Stirring of the sediment surface can improve contact with oxygenated water and help maintain the oxidized layer (Beveridge et al. 1994).

Several methods have been used to introduce oxygenated water into the surface sediment, but the two most practical techniques appear to be manual raking in small ponds and dragging a chain across the bottom of larger ponds. A 1 cm chain (refers to the diameter of the metal composing the links of the chain) is heavy enough for this purpose. For ponds of 30 to 50 m in width, two workers, one on each side of the pond, can drag a chain over the pond bottom. This practice should be applied at 1 or 2-day intervals to be effective. The benefits of bottom raking in tilapia ponds may be less than in other types of pond fish culture. Tilapia tend to stir the bottom when hunting for benthic organisms, and this can effect considerable mixing and aeration.

Organic matter originating from dead algae, manure particles, or uneaten feed often accumulates in the windward corners of the ponds and settles to the bottom to spoil the sediment surface. Where feasible, this material should be removed with nets, dippers, or other hand tools and the bottom in the corner raked thoroughly.

Disinfection

Bottom soils can harbor aquatic animal pathogens or their vectors between crops and cause diseases in the succeeding crop. It is common practice to attempt to disinfect pond bottoms following disease outbreaks. Drying can eliminate most disease organisms, but the combination of drying and application of chemical disinfectant is more effective. The two most common treatments are chlorination with calcium hypochlorite to kill organisms by chlorine contact and lime (calcium oxide or calcium hydroxide) to cause a high soil pH and kill disease organisms and their vectors.

Calcium hypochlorite is expensive, so lime treatment is more feasible. Application of 1,000 $kg \cdot ha^{-1}$ of lime is the minimum amount necessary to raise pH high enough for disinfection, and 1,500 to 2,000 $kg \cdot ha^{-1}$ is a more reliable dose. Lime should not be applied after pond bottoms are very dry, for it will not dissolve and increase pH. Uniform coverage of bottom soil is necessary, and the distribution of lime over the bottom and its penetration into the soil mass can be facilitated by adding a few centimeters of water over the bottom.

Liming for disinfection will also improve pH in acidic soils, but it kills beneficial bacteria as well as pathogenic ones. Where pond soils are acidic, lime application will not enhance bacterial activity unless time is allowed for the pH to decline to 8 or 8.5 so that reestablishment of beneficial communities of soil microorganisms will occur. This usually takes only 3 or 4 days, but ponds should be left fallow for another 2 or 3 weeks to promote organic matter degradation.

Probiotics

A number of products are promoted to enhance beneficial chemical and biological processes and to improve soil quality. These products include cultures of living bacteria, enzyme preparations, composted or fermented residues, plant extracts, and other concoctions. There is no evidence from research that any of these products will improve soil quality. Nevertheless, they are not harmful to the culture species, surrounding environment, workers, or quality of aquaculture products.

WATER QUALITY

Feed and Nutrients

Feed is an expensive but necessary ingredient in commercial aquaculture. It is also the source of nutrients that causes excessive phytoplankton blooms and water quality deterioration in ponds. By using high-quality feed and proper feeding practices to assure that most of the feed is consumed by the fish, a feed conversion ratio (FCR; weight of feed applied:net production) of 1.5 to 1.8 can be achieved with many species. One should remember that feed is about 90 percent dry matter and fish are about 25 percent dry matter. Thus, at an FCR of 2.0, it requires 2,000 kg feed to produce 1,000 kg fish. In this example, in terms of dry matter (constituents other than water), it requires 1,800 kg dry feed to produce 250 kg dry fish, and the waste load is 1,550 kg. The feed waste enters the water as uneaten feed, feces, carbon dioxide, and ammonia.

A specific example of tilapia production will be used to calculate nitrogen and phosphorus removal in tilapia. Diana et al. (1994) reported the production of 7,267 kg·ha^{-1} of Nile tilapia *(Oreochromis niloticus)* in ponds receiving 10,319 kg·ha^{-1} of feed. This species contains 26.5 percent dry matter of which 8.5 percent is nitrogen and 3.01 percent is phosphorus (Boyd and Green 1998). The feed had 30 percent crude protein (4.89 percent nitrogen), and it will be assumed that the feed contained 1.2 percent phosphorus and 90 percent dry matter. The amounts of nitrogen and phosphorus added to ponds in feed and removed in fish are presented in Table 11.2. In this example, 20.7 percent of dry matter, 32.5 percent of nitrogen, and 46.8 percent of phosphorus added in feed was recovered at harvest in fish.

The differences in the amounts of dry matter and nutrients applied in feed and recovered in fish is not equal to the amounts of dry matter and nutrients from feed contained in effluent. Ponds have a remarkable ability to assimilate wastes through microbial degradation of organic carbon to carbon dioxide and water, conversion of ammonia to nitrate, and transformation of nitrate to gaseous nitrogen by bacteria, ammonia volatilization, and accumulation of phosphorus and organic nitrogen in bottom soil (Boyd 1995; Gross et al. 2000).

As production and feeding rate increase, ponds become polluted with nutrients and organic matter. Heavy phytoplankton blooms may develop, and ammonia concentrations increase. When feeding rates exceed 30 to 40 kg·ha^{-1}·day^{-1}, mechanical aeration is usually required in the culture of many species to prevent low dissolved oxygen concentrations and fish stress at night (Boyd 1990; Boyd and Tucker 1998). Tilapia are more tolerant to low dissolved oxygen than most other species, and aeration may not be required until feeding rates exceed 100 kg·ha^{-1}·day^{-1}.

TABLE 11.2. Mass balance for dry matter, nitrogen, and phosphorus in a tilapia pond.

Variable	Dry matter	Nitrogen	Phosphorus
Feed (10,319 kg·ha^{-1})	9,287	505	124
Fish (7,267 kg·ha^{-1} live weight)	1,926	164	58
Waste load to pond (kg·ha^{-1})	7,361	341	66
Removal in fish (percentage of input)	20.7	32.5	46.8

Fertilization

Although pond fertilization is a proven technique for increasing fish production, the tendency in commercial aquaculture is to use more feeds. Feeding gives greater production than can be achieved with fertilization, but there may be benefits from fertilization when feed inputs are low. Chemical fertilizers are more dependable than manures, and their use is increasing relative to manures in most nations (Boyd and Tucker 1998). Research showed that liquid fertilizers (Metzger and Boyd 1980) or finely pulverized, instantly soluble fertilizers (Rushton and Boyd 2000) are effective at lower application rates than traditional prilled or granular fertilizers. A substitute for liquid fertilizer can be made by mixing traditional fertilizers with water and splashing the resulting slurry over pond surfaces. Controlled-release fertilizers also may have potential in aquaculture (Kastner and Boyd 1996). Fertilizer prills are coated with a polymer shell, and they gradually release nutrients over several months. Controlled-release fertilizers need not be applied as frequently as other kinds of fertilizers, but they are very expensive.

The main nutrient limiting phytoplankton production in most freshwater ponds is phosphorus. Nitrogen is also a limiting nutrient in many ponds, but fertilizers should contain at least as much P_2O_5 as N. Optimum ratios of $N:P_2O_5$ in fertilizers are about 1:3 for freshwater ponds. Fertilizers usually should be applied at rates of 5 to 10 kg $P_2O_5 \cdot ha^{-1}$ per application at 2- to 4-week intervals. High application rates of nitrogen fertilizers are sometimes recommended (Lin et al. 1997), but they are usually not necessary. Urea and ammonium fertilizers are popular and inexpensive, but if used at high rates, they can cause ammonia toxicity. Microbial oxidation of ammonia causes an oxygen demand and is a source of acidity (Boyd and Tucker 1998). Nitrate fertilizers are more expensive, but they have advantages over other nitrogen fertilizers; nitrate is not toxic and is fully oxidized (Tepe and Boyd 2001).

Liming

The most common liming materials are agricultural limestone ($CaCO_3$), burnt lime (CaO), and hydrated lime [$Ca(OH)_2$]. These materials often contain some magnesium in addition to calcium. Liming neutralizes acidity and increases total alkalinity and total hardness in ponds. Liming compounds also react with carbon dioxide and convert it to bicarbonate or carbonate. Usual application rates are 1,000 to 2,000 $kg \cdot ha^{-1}$. Liming materials are not very soluble. For example, the equilibrium concentration of both total alkalinity and total hardness in pure water is about 60 $mg{\cdot}L^{-1}$. Carbon dioxide from decomposition of organic matter in ponds increases the solubility of liming materials, but it is difficult to increase the alkalinity of pond water above 80 to 100 $mg{\cdot}L^{-1}$ by liming.

Fish farmers in Asia tend to lime ponds as a general practice, but if the total alkalinity is above 50 to 100 $mg{\cdot}L^{-1}$, liming materials will not dissolve well. The common practice of adding 5 to 10 $kg{\cdot}ha^{-1}$ of liming materials to ponds on a daily basis also is not advisable for it has little effect on pond water quality (Giri and Boyd 2000). Excessive lime can be harmful by removing carbon dioxide, raising pH too much, and precipitating dissolved phosphate (Boyd and Tucker 1998). Phytoplankton will be starved of carbon dioxide and phosphorus, and a high pH favors ammonia toxicity to aquatic animals. Liming can be an effective practice in ponds with low alkalinity water, but excessive liming should be avoided.

Mechanical Aeration

Mechanical aeration is a proven technique for improving dissolved oxygen availability in ponds. There are many types of aerators, but propeller-aspirator-pump aerators, such as the Aire-O_2 (Aeration Industries, Chaska, Minnesota[1]), and paddle-wheel aerators are used most widely. Each horsepower of aeration by these devices normally will allow about 500 $kg{\cdot}ha^{-1}$ more fish production than can be obtained in unaerated ponds (Boyd 1997). In fish ponds, it is usually not necessary to operate aerators during the day, because dissolved oxygen concentrations are usually high. Dissolved oxygen concentrations decline during the night, and the period from midnight until 0700 to 0800 h is usually the most critical time. Some farmers install timers that start aerators during the critical period and turn them off in the early morning. Automated dissolved oxygen equipment can be used to turn aerators on and off in response to dissolved oxygen concentrations, but this equipment is not yet reliable enough to be recommended for use.

In heavily aerated ponds where aerators are positioned around the peripheries to create circular water flow, strong water currents cause severe erosion of pond bottoms. Mineral soil and organic matter particles eroded from peripheral areas settle in the central part of the pond where water currents are weaker. Farmers think that accumulation of waste in one place isolates the effect of the waste on soil and water quality. However, they are badly mistaken. Organic matter in sediment mounds in the center of ponds decomposes and causes anaerobic conditions at the soil surface with release of toxic metabolites such as hydrogen sulfide into the water. A method of aeration that does not erode mineral soil and produces water movement over the entire pond bottom, instead of just around the periphery, is needed. This method of aeration would produce water currents strong enough to suspend organic particles, but it would not suspend the heavier, mineral soil particles. If suspended in well-oxygenated water, organic particles would decompose aerobically without the associated production of toxic metabolites. Unfortunately, such aeration equipment is not available in the market.

Mounds of sediment in heavily aerated ponds usually do not contain large amounts of organic matter as is commonly thought. They consist primarily of mineral soil (95 to 98 percent) and contain only a little organic matter (2 to 5 percent) (Boyd et al. 1994). Removal of sediment using water jets to wash it into water supply canals is a bad practice as it contaminates the canals. Sediment mounds should be prevented by better aeration techniques. If mounds are allowed to form, they should be dried between crops and must be spread back over eroded areas in ponds. Bottoms should be compacted to minimize erosion by aerator-induced currents during the next crop.

Turbidity

The water supply for some farms may be highly turbid with suspended soil particles. High external sediment loads can rapidly fill ponds and reduce water volume. This problem can be solved with sedimentation ponds for removing suspended solids before they enter ponds. Sedimentation ponds must be dredged occasionally, or they will fill in and shorten the water retention time.

Pond bottoms erode naturally with net movement of soil from shallow to deeper water areas. After a number of years, pond bottoms usually must be reshaped, and some sediment removal may be necessary.

Pond waters turbid with suspended soil particles have been cleared by manure applications of 500 to 1,000 $kg \cdot ha^{-1}$, gypsum applications of 250 to 500 ppm, or alum applications of 25 to 50 ppm (Boyd 1979). Unless the source of turbidity is eliminated, no lasting benefit can be expected.

Water Exchange

Routine water exchange in aquaculture ponds is an example of inefficiency. There are reasons to exchange water in specific instances, i.e., to flush out excessive nutrients and plankton and to reduce ammonia concentrations. However, daily water exchange usually does not improve water quality in ponds, and pumping costs are a liability. Ponds are highly efficient in assimilating carbon, nitrogen, and phosphorus inputs not converted to fish or shrimp flesh, but if water exchange is great, these substances are discharged from ponds before they can be assimilated (Boyd and Tucker 1995). The pollution potential of aquaculture ponds increases as a function of increasing water exchange. From both economic and environmental perspectives, water exchange should be used only when necessary.

In channel catfish farming in the United States, water exchange is not used and fish are harvested with long seines without draining ponds. A recent study (National Animal Health Monitoring System 1997) revealed that ponds are drained every 6.1 years on average.

Chlorination

Hypochlorous acid and hypochlorite (free chlorine residuals) are responsible for the disinfecting power of chlorine products in pond water (White 1992). Chlorine compounds are strong oxidizing agents, and enough free chlorine residual must be applied to overcome the chlorine demand of organic matter and other substances that react with free chlorine residuals to convert them to nontoxic chloride or less-toxic combined chlorine residuals (White 1992). Free chlorine residuals have similar toxicities to pathogenic and nonpathogenic microorganisms and fish. The fact that chlorine products are often applied to ponds without killing fish proves that chlorine doses were not high enough to kill target microorganisms. If enough chlorine is applied to kill target microorganisms, fish will be killed. Chlorination of waters containing fish is both dangerous to the fish and provides no benefits (Potts and Boyd 1998).

It is possible to sterilize water in newly filled, but unstocked, ponds by applying chlorine products. When this is done, enough chlorine should be applied to overcome the chlorine demand and provide 2 to 3 $mg \cdot L^{-1}$ or more of free chlorine residual. Because of the reduction of chlorine by organic matter, 20 to 30 $mg \cdot L^{-1}$ of commercial calcium hypochlorite may be needed to provide 2 to 3 $mg \cdot L^{-1}$ of residual chlorine. These residuals will detoxify naturally in a few days so that ponds can be stocked safely.

Nutrient Removal

Addition of nutrients, especially phosphorus, to aquaculture ponds in feed are responsible for heavy phytoplankton bloom and are associated with water quality problems. Moreover, blue-green algae often dominate the phytoplankton in aquaculture ponds, and odorous compounds produced by blue-green algae can cause off-flavor in fish flesh (Tucker 2000). There has been much interest in techniques for removing nutrients from water to reduce plankton bloom and blue-green algal abundance in fish ponds. It is possible to precipitate phosphorus from pond water by applying sources of iron, aluminum, or calcium ions. These ions precipitate phosphate as insoluble iron, aluminum, or calcium phosphates. Alum (aluminum sulfate) and ferric chloride are commercially available sources of aluminum and iron, respectively. Alum is generally cheaper and more widely available than ferric chloride. Gypsum (calcium sulfate) is a good source of calcium because it is more soluble than liming materials (agricultural limestone, burnt lime, and hydrated lime). Treatment rates of 20 to 30 $mg{\cdot}L^{-1}$ of alum and 100 to 200 $mg{\cdot}L^{-1}$ of gypsum have lowered phosphorus concentrations in pond waters (Rowan 2001). Alum is acidic and more suitable for use in waters of 50 $mg{\cdot}L^{-1}$ total alkalinity and above. Gypsum is better for use in low alkalinity waters. Further research is needed to determine how best to use these materials for precipitating phosphorus and limiting phytoplankton growth.

Algicides

Algicides have been used in attempts to reduce the abundance of phytoplankton in intensive culture ponds. Synthetic algicides usually have a long residual life and can cause chronically low dissolved oxygen concentrations. Copper sulfate has a shorter residual life, and some workers recommend its use for reducing phytoplankton abundance, in general, and in reducing the abundance of blue-green algae, in particular. The toxic form of copper is the cupric ion, and the concentration of cupric ion depends upon the pH. The greater the pH, the larger the copper dose required to kill algae. The usual recommendation is to apply a dose of copper sulfate equal to 1/100 of the total alkalinity. The advantages of chelated copper algicides over copper sulfate have never been conclusively demonstrated, and chelated copper compounds are more costly.

Various dyes have been added to pond water for the purpose of limiting light and reducing phytoplankton growth. There is no clear evidence that dyes are effective for this purpose.

With the exception of copper sulfate, there is little demonstrated success in algicide treatments to limit phytoplankton growth in ponds (Boyd and Tucker 1998). The best approach to phytoplankton control is to regulate nutrient inputs by moderate stocking and feeding rates.

Probiotics

It is popular to apply commercial bacterial inocula or enzyme preparations to ponds. These preparations are often called probiotics, and they are advertized to improve water quality by enhancing nutrient removal, stimulating organic matter oxidation, reducing ammonia concentrations, and the like. Because bacteria and enzymes in these products already occur naturally in ponds, application of these products may not be necessary (Boyd et al. 1984; Queiroz and Boyd 1998; Queiroz et al. 1998). Much additional research is needed to determine whether probiotics can provide benefits and to define the conditions under which they should be used.

Zeolite

Zeolite, an aluminosilicate mineral with ion exchange properties, can adsorb ammonium (Mumpton 1984). Zeolite is both mined and produced synthetically for many industrial purposes, and some fish farmers believe that zeolite can remove ammonia from pond water. While this is technically true, a very large amount of zeolite would be required to significantly lower ammonia concentrations.

EFFLUENTS

Many countries are beginning to impose regulations on effluents from aquaculture farms. Presently, coastal aquaculture is more likely to be regulated than inland, freshwater aquaculture. However, because environmental groups are pressuring governments to regulate aquaculture, most types of aquaculture will eventually be subjected to effluent regulations. The U.S. Environmental Protection Agency (USEPA) announced effluent regulations for U.S. aquaculture (Federal Register 2004). For warmwater aquaculture, facilities that produce over 45,454 $kg \cdot year^{-1}$ and discharge 30 days or more per year other than excess runoff require a National Pollution Dischrge Elimination System (NPDES) permit. However, USEPA did not specify effluent limitation guidelines in the aquaculture effluent rule (Federal Register 2004). Individual NPDES-delegated states will be responsible for the requirements in NPDES permits for concentrated aquatic animal production facilities. The most likely form of effluent regulations in pond aquaculture is the application of best management practices (BMP) to reduce effluent volume, improve effluent quality, and to minimize pollution loads to natural waters (Federal Register 2004). Of course, the regulations may also have numerical standards in some states.

Boyd et al. (2000) and Boyd and Queiroz (2001) made an environmental impact assessment of channel catfish farming in Alabama (United States) and recommended BMP for preventing or mitigating the negative impacts. These BMP can also be easily applied to pond culture of tilapia, and are listed below.

BMP to Reduce Effluent Volume

- Use embankment ponds where possible, and new watershed ponds should be designed to have watershed area to pond area ratio of 10:1 or less.
- Use terraces to divert excess runoff around watershed ponds or an additional pond may sometimes be built to increase storage on the watershed.
- Maintain good vegetative cover on all parts of watersheds, and where feasible, replace short grass with evergreen trees.
- Harvest fish by seining and without partially or completely draining ponds unless it is necessary to renovate fish stocks or repair pond earthwork.
- Maintain at least 20 cm of storage capacity in ponds to capture rainfall.
- Do not flush well or stream water through ponds. This practice usually does not improve water quality in ponds.

BMP to Minimize Suspended Solids Through Erosion Control

- Control erosion on watersheds by providing vegetative cover, eliminating gully erosion, and using terraces to route water from areas of high erosion potential.
- Restrict livestock from watersheds and embankments of ponds.
- Eliminate steep slopes on farm roads and cover these roads with gravel.
- Provide grass cover on the sides of pond dams or embankments and grass or gravel on tops of dams or embankments.
- Do not leave ponds partially or completely empty in winter and spring, and immediately close drains in empty ponds.
- Mechanical aerators should be installed so that water currents caused by these devices do not cause erosion of pond earthwork.
- Sediment should not be disposed of outside of ponds.
- Install structures to prevent drainpipe discharge from impacting and eroding earthwork.
- Construct ditches with adequate hydraulic cross-section and provide grass cover on the sides of ditches.
- Settling basins are alternative methods for improving the quality of final draining effluent from ponds where space is available.
- Trees or shrubs could be used in critical areas to shelter ponds from excessive wind velocities and reduce wave erosion of embankments.

BMP to Improve Pond Water and Effluent Quality

- Select high-quality feeds that contain adequate, but not excessive, nitrogen and phosphorous.
- Store feed in well-ventilated, dry bins, or if bagged, in a well-ventilated, dry room. The feed should be used by the expiration date suggested by the manufacturer.
- Apply feed uniformly.
- Do not apply more feed than fish will eat.
- When uneaten feed accumulates in the corners of ponds, it should be manually removed.
- Apply fertilizers only when necessary to promote phytoplankton blooms.
- Use chemical fertilizers and avoid the use of animal manures.
- Avoid excessive fertilization by using moderate doses and relying on the Secchi disk visibility to determine whether fertilization is needed.
- Apply agricultural limestone to ponds with total alkalinity below 20 $mg \cdot L^{-1}$.
- Store fertilizers under a roof in a dry place to prevent rain from washing them into surface waters.
- Avoid deep water intake structures in ponds.
- Restrict livestock from watersheds of ponds.
- Avoid discharge when harvesting fish, but if ponds must be drained completely, hold the final 20 to 25 percent of pond volume for 2 or 3 days and then discharge it slowly.

BMP for Use of Therapeutic Agents and Other Chemicals

- Store therapeutants so that they cannot be accidentally spilled to enter the environment.
- Use good water quality management procedures to prevent unnecessary stress to fish.
- Obtain a definite diagnosis for diseases and a recommendation for disease treatment before applying therapeutic agents.
- Follow instructions on labels of therapeutic agents for dose application method, safety precautions, and the like.
- Copper sulfate applications in milligrams per liter should not exceed 1 percent of total alkalinity, also measured in milligrams per liter, or a maximum dose of 1.0 $mg \cdot L^{-1}$. Pond water should not be released for 72 h after application of copper sulfate.
- Sodium chloride applications should not exceed 100 $mg \cdot L^{-1}$.

- Lime (calcium oxide or calcium hydroxide) applications should not exceed 100 $mg \cdot L^{-1}$.
- Agricultural limestone and gypsum (calcium sulfate) applications should not exceed 5,000 $kg \cdot ha^{-1}$ and 2,000 $kg \cdot ha^{-1}$, respectively.
- Calcium hypochlorite or other chlorine compounds should not be applied to ponds during the culture period.

BMP for New Ponds or Farms

- New ponds should be constructed according to National Resource Conservation Service (NRCS) or similar standards. Riparian vegetation of trees or shrubs should be preserved or established to provide a vegetative buffer zone along streams.
- New ponds should not be located on watersheds that are already impacted by subdivisions, industrial activities, or row-crops.
- Design of new ponds should conform to NRCS or similar standards and be compatible with implementation of BMP outlined above.

ENDNOTE

1. Use of trade or manufacture's name does not imply endorsement.

REFERENCES

Avnimelech, Y., and G. Zohar (1986). The effect of local anaerobic conditions on growth retardation in aquaculture systems. *Aquaculture* 58:167-174.

Beveridge, M.C.M., A. Wahab, and S. Dewan (1994). Effects of daily harrowing on pond soil and water nutrient levels and on Roho fingerling production. *The Progressive Fish Culturist* 57:282-287.

Boyd, C.E. (1974). *Lime requirements of Alabama fish ponds.* Bulletin 459, Alabama Agricultural Experiment Station, Auburn University, Alabama.

———. (1979). Aluminum sulfate (alum) for precipitating clay turbidity from fish ponds. *Transactions of the American Fisheries Society* 108:307-313.

———. (1990). *Water quality in ponds for aquaculture.* Auburn: Alabama Agricultural Experiment Station, Auburn University.

———. (1995). *Bottom soils, sediment, and pond aquaculture.* Chapman and Hall, New York.

———. (1997). Advances in pond aeration technology and practices. *INFOFISH* 2/97:24-28.

Boyd, C.E., and B. Green (1998). Dry matter, ash, and elemental composition of pond-cultured tilapia (*Oreochromis aureus* and *O. niloticus*). *Journal of the World Aquaculture Society* 29:125-128.

Boyd, C.E., W.D. Hollerman, J.A. Plumb, and M. Saeed (1984). Effect of treatment with a commercial bacterial suspension on water quality in channel catfish ponds. *Progressive Fish-Culturist* 46:36-40.

Boyd, C.E., P. Munsiri, and B.F. Hajek (1994). Composition of sediment from intensive shrimp ponds in Thailand. *World Aquaculture* 25:53-55.

Boyd, C.E., and S. Pippopinyo (1994). Factors affecting respiration in dry pond bottom soils. *Aquaculture* 120:283-293.

Boyd, C.E., and J. Queiroz (2001). Feasibility of retention structures, settling basins, and best management practices in effluent regulation for Alabama channel catfish farming. *Reviews in Fisheries Science* 9:43-67.

Boyd, C.E., J. Queiroz, J. Lee, M. Rowan, G.N. Whitis, and A. Gross (2000). Environmental assessment of channel catfish, *Ictalurus punctatus,* farming in Alabama. *Journal of the World Aquaculture Society* 31(4):511-544.

Boyd, C.E., and D. Teichert-Coddington (1994). Pond bottom soil respiration during fallow and culture periods in heavily fertilized tropical fish ponds. *Journal of the World Aquaculture Society* 25:417-423.

Boyd, C.E., and C.S. Tucker (1995). Sustainability of channel catfish farming. *World Aquaculture* 26:45-53.

———. (1998). *Pond aquaculture water quality management.* Boston: Kluwer Academic Publishers.

Diana, J.S., C.K. Lin, and K. Jaiyan (1994). Supplemental feeding of tilapia in fertilized ponds. *Journal of the World Aquaculture Society* 25:497-506.

Federal Register (2004). Environmental Protection Agency, 40 CFR Part 451. Effluents limitations guidelines and new source performance standards for the concentrated aquatic animal production point source category: final rule. *Federal Register* 69(162):51892-51930 (August 23). Washington, DC: Office of the Federal Register, National Archives and Records Administration.

Giri, B.J., III, and C.E. Boyd (2000). Effects of frequent, small doses of calcium carbonate on water quality and phytoplankton in channel catfish ponds. *North American Journal of Aquaculture* 62:225-228.

Gross, A., C.E. Boyd, and C.W. Wood (2000). Nitrogen transformations and balance in channel catfish ponds. *Aquacultural Engineering* 24:1-14.

Hajek, B.F., and C.E. Boyd (1994). Rating soil and water information for aquaculture. *Aquacultural Engineering* 13:115-128.

Hickling, C.F. (1962). *Fish cultures.* London: Faber and Faber.

Kastner, R.J., and C.E. Boyd (1996). Production of sunfish (*Lepomis* spp.) in ponds treated with controlled-release fertilizers. *Journal of the World Aquaculture Society* 27:228-234.

Lin, C.K., D.R. Teichert-Coddington, B.W. Green, and K.L. Veverica (1997). Fertilization regimes. In H.S. Egna, and C.E. Boyd (Eds.), *Dynamics of pond aquaculture* (pp. 73-106). Boca Raton, FL: CRC Press.

Metzger, R.J., and C.E. Boyd (1980). Liquid ammonium polyphosphate as a fish pond fertilizer. *Transactions of the American Fisheries Society* 109:563-570.

Mumpton, F.A. (1984). Natural zeolites. In W.G. Pond, and F.A. Mumpton (Eds.), *Zeo-agriculture* (pp. 33-43). Boulder, CO: Westview Press.

National Animal Health Monitoring System (1997). Catfish part II: Reference of 1996 U.S. catfish management practices. Fort Collins, CO: United States Department of Agriculture, Animal and Plant Health Inspection Service, Veterinary Services.

Pillai, V.K., and C.E. Boyd (1985). A simple method for calculating liming rates for fish ponds. *Aquaculture* 46:157-162.

Potts, A.C., and C.E. Boyd (1998). Chlorination of channel catfish ponds. *Journal of the World Aquaculture Society* 29:432-440.

Queiroz, J.F., and C.E. Boyd (1998). Effects of a bacterial inoculum in channel catfish ponds. *Journal of the World Aquaculture Society* 29:67-73.

Queiroz, J.F., C.E. Boyd, and A. Gross (1998). Evaluation of a bio-organic catalyst in channel catfish, *Ictalurus punctatus,* ponds. *Journal of Applied Aquaculture* 8:49-61.

Rowan, M. (2001). Chemical phosphorus removal from aquaculture pond water and effluent. Doctoral dissertation, Auburn University, Auburn Alabama.

Rushton, Y.G., and C.E. Boyd (2000). A comparison of water-soluble fertilizer with liquid fertilizer for sportfish pond fertilization. *North American Journal of Aquaculture* 62:212-218.

Seo, J., and C.E. Boyd (2001a). Dry-tilling of pond bottoms and calcium sulfate treatment for water quality improvement. *Journal of the World Aquaculture Society* 32:257-268.

———. (2001b). Effects of bottom soil management practices on water quality improvement in channel catfish, *Ictalurus punctatus,* ponds. *Aquacultural Engineering* 25:83-97.

Steeby, J., S. Kingsbury, C. Tucker, and J. Hargreaves (2001). Sediment accumulation in channel catfish production ponds. *Global Aquaculture Advocate* 4(3):54-56.

Tepe, Y., and C.E. Boyd (2001). A sodium nitrate-based, water-soluble fertilizer for sportfish ponds. *North American Journal of Aquaculture* 63:328-332.

Tucker, C.S. (2000). Off-flavor problems in aquaculture. *Reviews in Fisheries Science* 8:45-88.

White, G.C. (1992). *The handbook of chlorination and alternative disinfectants.* New York: Van Nostrand Reinhold.

Wurtz, A.G. (1960). *Methods of treating the bottom of fish ponds and their effects on productivity. General Fisheries Council in the Mediterranean,* Studies and Reviews No. 11. Rome, Italy: FAO.

Chapter 12

Nutrient Requirements

Chhorn E. Lim
Carl D. Webster

INTRODUCTION

Tilapia are the most adaptable and successful aquaculture species worldwide. These are tropical species endemic to freshwater in Africa, Jordan, and Israel. They are, however, being cultured in virtually all types of production systems in both fresh and saltwater in tropical, subtropical, and temperate climates. Early tilapia production relied on natural food organisms as the source of nutrients for growth and well-being. In the past two decades, as a result of technological development and improvements, particularly those concerned with seed production, culture system designs, water management, disease prevention and control, feeds and feeding practices, tilapia culture has expanded rapidly worldwide. Global tilapia production through aquaculture reached 1.25 million metric tons (MMT) in 2000 as compared with only 800,000 metric tons (MT) in 1996 (FAO 2002) and 200,000 MT in 1986 (Luquet 1991). This trend is expected to continue due to increased demand of tilapia in both domestic and international markets. As the industry expands and technology development continues, traditional extensive culture is being replaced by semi-intensive and intensive production systems. In semi-intensive farming systems, supplemental feeds that consist of locally available, low-cost single feedstuffs such as rice bran, corn meal, copra meal, coffee pulp, brewery by-product and/or their combination are generally used as supplements to natural food (Lim 1989). These feeds are high in energy, low in protein and deficient in micronutrients such as vitamins and minerals. It is assumed that the deficient nutrients will be provided by natural food organisms. As stocking rate increases, the contribution of natural food decreases and more nutritionally complete feeds are needed. In intensive culture systems such as in ponds, raceways, cages, and tanks, feed is the most expensive item, often ranging from 30 to 60 percent

doi:10.1300/5513_12

of the total variable expenses, depending on the intensity of the culture operation. Thus, the availability of least-cost, nutritionally well-balanced feeds is one of the most important requisites for successful and sustainable tilapia production. Data on nutrient requirements, among other information, are needed for least-cost feed formulation.

This chapter provides an overview of the nutritional requirements of tilapia with respect to protein, energy, lipids, carbohydrates, vitamins, and minerals.

PROTEINS AND AMINO ACIDS

Proteins are the principal organic constituent of animal tissue and are the most expensive component in fish diets. Since body protein is constantly undergoing two major processes: protein synthesis (anabolism) and protein breakdown (catabolism), animals, including tilapia, need a continuous supply of protein throughout life for maintenance, growth, and other physiological functions. Inadequate intake of protein will result in retardation or cessation of growth, or loss of weight due to the withdrawal of protein from less vital tissues to maintain the function of more vital ones. If too much protein is supplied, however, only part will be used to synthesize new tissues and the remainder will be converted to energy (NRC 1983).

Proteins vary greatly in chemical structure, physical properties, size, shape, solubility, and biological functions but always contain carbon (51 to 55 percent), hydrogen (6.5 to 7.3 percent), oxygen (21.5 to 23.5 percent), nitrogen (15.5 to 18.0 percent), sulfur (0.5 to 2.0 percent), phosphorus, (0.0 to 1.5 percent), and a small amount of iron. The proportion of nitrogen varies depending on the nature of the protein. For the purposes of quantitative measurement of protein, however, an average value of 16 percent has been universally adapted. Thus, the approximate protein (crude protein) content of a product is obtained by determining the nitrogen content and then multiplied by 6.25 (100/16 = 6.25).

Proteins are made up of simple units, amino acids, linked together by peptide bonds. There are 18 amino acids that can be found in most protein sources, although proteins usually contain 22 to 26 amino acids. The amino acid content of protein (relative amount and kind) vary among protein sources. A number of amino acids cannot be synthesized by the animals or are not synthesized in sufficient amounts and must be supplied in diets; they are referred to as essential or indispensable amino acids. Those that can be synthesized in adequate quantity are termed nonessential or dispensable amino acids.

Tilapia, like other fish, do not have an absolute protein requirement per se but have a requirement for a well-balanced mixture of essential and nonessential amino acids. Numerous investigators have utilized semipurified, purified, and practical diets to determine the protein requirements of various tilapia species. These values have been obtained mostly by measuring the growth response of fish that were fed diets containing graded levels of good quality protein. The minimum amount of dietary protein for maximum growth of tilapia ranges from about 56 percent for fry to 30 percent for 40 g fish (Table 12.1). A study evaluating the least-cost dietary protein level using the available data on dietary protein requirements of four species of tilapia *(O. mossambicus, O. niloticus, O. aureus,* and *Tilapia zillii)* showed that the dietary protein level of 34 to 36 percent provided maximum growth of young tilapia (1 to 5 g), but the most cost-effective protein level was 25 to 28 percent (De Silva et al. 1989). It has been reported that the dietary protein requirement of tilapia is affected by a number of factors such as size or age, protein quality, nonprotein energy level, water temperature and salinity, presence of natural food, and feed allowance (NRC 1993), but limited studies have been conducted to determine the effect of these parameters on protein requirement of tilapia.

TABLE 12.1. Dietary protein requirements of some species of tilapia cultured in freshwater.

Species	Protein source	Size (g)	Protein requirement (percent)	Reference
O. aureus	Casein/egg albumin	Fry-2.5 2.5-7.5	56 34	Winfree and Stickney (1981)
	Soybean or fish meal	0.3-0.5	36	Davis and Stickney (1978)
O. niloticus	Fish meal	0.8 40	40 30	Siddiqui et al. (1988)
	Casein	0.012	45	El-Sayed and Teshima (1992)
	Casein	3.5 9.0	30 25	Wang et al. (1985)
O. mossambicus	Fish meal	Fry 0.5-1.0 6-30	50 40 30-35	Jauncey and Ross (1982)
T. zillii	Casein	1.7	35-40	Teshima et al. (1978)
	Casein	1.3-3.5	35	Mazid et al. (1979)
Red hybrid tilapia	Casein/ gelatin	0.2	35-40	Santiago and Laron (1991)

Younger or smaller fish have a higher protein requirement than older or larger fish. Siddiqui et al. (1988) showed that the protein requirement of 0.8 g Nile tilapia (*Oreochromis niloticus*) was 40 percent as compared with only 30 percent for 40 g fish. Higher dietary levels of protein are required if poor quality protein sources are used. Insufficient nonprotein energy in the diet will lead to higher dietary protein requirements because fish will utilize part of the protein as energy to meet their metabolic energy needs. Water quality, such as temperature and dissolved oxygen (DO), has considerable effect on the metabolic rate of fish and thus their dietary protein requirements. A higher dietary protein level is required at water temperature and DO optimum for growth than at lower temperature and DO. Water salinity also appears to influence dietary protein requirements of tilapia, being lower at high salinity (Table 12.2). The decrease of dietary protein requirements (20 to 24 percent) of red hybrid tilapia grown in sea water with 32-37 ppt salinity (Shiau and Huang 1989; Clark et al. 1990) may be due to increased feed consumption at higher salinity. Watanabe et al. (1988) observed that the average daily feed consumption of juvenile, male Florida red tilapia over a 43-day feeding period increased from 5.93 percent at 1 ppt to 8.71 percent at 36 ppt. Tilapia are efficient consumers of natural food. At low stocking densities in ponds, they obtain significant amount of protein from natural food, and under these conditions, lower protein diets can be used. Lower dietary protein levels are required for fish that are fed to satiation as compared with those receiving restricted feeding.

TABLE 12.2. Dietary protein requirements of some species of tilapia at different salinities.

Species	Size/age	Salinity (ppt)	Protein requirement (percent)	References
O. niloticus	2-3 weeks (24 mg)	0	30.4	De Silva and Perera (1985)
		5	30.4	
		10	28.0	
		15	28.0	
	2.9 g	18-50	30.0	Fineman-Kalio and Camacho (1987)
Male hybrid tilapia *O. niloticus* x *O. aureus*	2.9 g	32-34	24.0	Shiau and Huang (1989)
Male Florida red tilapia *O. urolepis honorum* x *O. mossambicus*	10.6	37	20	Clark et al. (1990)

Tilapia require the same ten essential amino acids (arginine, histidine, isoleucine, leucine, lysine, methionine, phenylalanine, threonine, tryptophan, and valine) as other fish and terrestrial animals. The quantitative requirements for these essential amino acids have been determined for young Mozambique (Jauncey et al. 1983) and Nile (Santiago and Lovell 1988) tilapia using different methods. Jauncey et al. (1983) employed the whole body and muscle amino acid profile methods (Cowey and Tacon 1983) and the daily deposition rate of essential amino acid (Ogino1980); the average value obtained from these methods are given in Table 12.3. The procedure used by Santiago and Lovell (1988) involved feeding the fish test diets containing graded levels of each essential amino acid (Mertz 1972). The basal diet contained casein-gelatin supplemented with a mixture of crystalline amino acids to provide an amino acid profile similar to that of the 28 percent whole egg protein except for the test amino acid. The pH of the crystalline amino acid mixture and mineral mix was adjusted to 7.0 by adding 6N NaOH. The optimum quantitative requirements of Nile tilapia for the ten essential amino acids determined by break points in the growth response curves are also presented in Table 12.3. The requirements reported for Nile tilapia are generally similar to those reported for other fish species. However, the values reported for Mozambique tilapia (Jauncey et al. 1983) expressed on a dietary protein basis are considerably lower than those reported for Nile

TABLE 12.3. Essential amino acid requirements of young *O. mossambicus and O. niloticus.*

	Percent of dietary protein	
Amino acid	***O. mossambicus*[a]**	***O. niloticus*[b]**
Arginine	2.82	4.20
Histidine	1.05	1.72
Isoleucine	2.01	3.11
Leucine	3.40	3.39
Lysine	3.78	5.12
Methionine	0.99	2.68[c]
Phenylalanine	2.50	3.75[d]
Threonine	2.93	3.75
Tryptophane	0.43	1.00
Valine	2.20	2.80

Notes: [a]Jauncey et al. 1983. [b]Santiago and Lovell 1988. [c]In the presence of cystine at 0.54 percent of dietary protein. [d]In the presence of tyrosine at 1.79 percent of dietary protein.

tilapia (Santiago and Lovell 1988), except for leucine. It is unlikely that the requirements differ so much between these two tilapia species.

The nonessential amino acids can be adequately synthesized by fish but their presence in the diet have nutritional significance because the need for fish to synthesize them is reduced. Two special examples of sparing action are the conversion of methionine to cystine and phenylalanine to tyrosine. These nonessential amino acids (cystine and tyrosine) can only be synthesized from the essential amino acid precursors (NRC 1983). Tilapia actually have a requirement for sulfur-containing amino acids which can be met by either methionine alone or the proper mixture of methionine and cystine. Dietary cystine can replace up to 50 percent of the total sulfur acid requirement for *O. mossambicus* (Jauncey and Ross 1982) and *O. niloticus* (Abdelghany 2000). A similar relationship exists between aromatic amino acids. The presence of tyrosine in the diet will reduce some of the requirement for phenylalanine (NRC 1993). Since most practical feeds contain adequate levels of phenylalanine and tyrosine, the sum of these two amino acids normally exceeds the dietary needs of fish.

The nutritional value of protein sources, commonly referred to as protein quality, is determined based on the essential amino acids content and their digestibility or bioavailability. A protein source having an essential amino acid profile that closely matches the essential amino acid requirements of fish are likely to have high nutritional value. Fish meal has customarily been used as a major protein source in aquaculture feeds, including those for tilapia, because of its high protein content, good amino acid profile, and palatability. Protein and amino acid digestibility of good quality fish meal is relatively high for most fish species. Fish meal is also a good source of essential fatty acids and minerals. Fish meal currently constitutes a substantial part of the feed formulae of aquaculture species. However, the increased demand for fish meal coupled with the rising cost and uncertain availability have initiated much interest to evaluate alternative protein ingredients to partially or totally replace fish meal in diets of various aquaculture species.

Animal by-product meals such as poultry by-product meal (PBM), hydrolyzed feather meal (HFM), meat and bone meal (M and BM), and blood meal (BM) have been evaluated as substitutes for fish meal in tilapia diets with varying results. Poultry by-product meal at a level of 10 percent in diets, as a replacement of half of the fish meal on an equal protein basis, had no effect on growth performance and body composition of hybrid tilapia (*O. niloticus* × *O. aureus*). Replacement of the same amount of fish meal by 7.5 percent feather meal reduced weight gain and body fat content (Viola and Zohar 1984). The poor performance of fish that are fed feather meal diet was attributed to the deficiency of essential amino acids and energy.

Tacon et al. (1983) reported that, even with supplementation of lysine, methionine, and histidine, only about 30 percent of fish meal protein in diets of Nile tilapia could be replaced by HFM. El-sayed (1998) obtained similar growth of Nile tilapia fed diets containing 47, 50, and 40 percent of PBM, shrimp meal, and M and BM, respectively, as replacements of 30 percent fish meal. Feed efficiency and protein efficiency ratio, however, were significantly poorer than those fed the fish meal diets. Poor growth performance was observed in tilapia fed diets containing 30 percent BM and 20 percent M and BM + 15 percent BM. Wu et al. (1999) showed that inclusion of 6 percent M and BM as a total replacement of menhaden fish meal had no effect on the growth performance of Nile tilapia. Good-quality sprayed-dried BM can be used up to 7.45 percent as a substitute for 50 percent fish meal in diets of Nile tilapia (Lee and Bai 1997). Total replacement of fish meal by 14.9 percent BM, with or without addition of L-arginine, L-isoleucine, and methionine, adversely affected growth performance.

Many conventional and unconventional plant protein sources have been evaluated as substitutes for fish meal in tilapia diets. Soybean meal (SBM), because of its availability, consistent quality, high protein content with good amino acid profile, and low cost, is the most studied plant feedstuff in aquaculture diets (Lim and Dominy 1989). Shiau et al. (1987) showed that fish meal in *O. niloticus* × *O. aureus* diets can be partially replaced by SBM when the dietary protein level is below the optimum level for growth (24 percent). At the optimum protein level (32 percent), replacement of 30 percent fish meal with SBM resulted in depressed growth and poor feed efficiency. These, however, were restored by addition of methionine to the level of the control diet. Supplementation of 0.8 percent D,L-methionine to a diet in which 75 percent of brown fish meal was replaced by SBM improved the growth of Nile tilapia to a level comparable to that obtained with the fish meal diet (Tacon et al. 1983). Viola and Arieli (1983) found that SBM could be used to replace up to half of the fish meal in tilapia diets having 25 percent crude protein without requiring any supplementation. Complete substitution resulted in reduction of weight gain and feed efficiency that were not overcome by supplementation of oil, lysine, methionine, and vitamins. For a 30 percent protein diet, isonitrogenous substitution of fish meal with 24 percent SBM reduced the growth of *O. aureus* × *O. niloticus* hybrid. When 2 to 3 percent dicalcium phosphate was added, the growth was comparable to that of the all-fish meal control (Viola et al. 1986). A growth reduction was observed in *O. mossambicus* that were fed a diet in which 50 percent or more fish meal was replaced by SBM (Jackson et al. 1982). This growth reduction was attributed to the deficiency of methionine and the presence of antinutritional factors such as trypsin inhibitors and hemagglutinin (lectin). Roasted full-fat soybean meal (FFSBM) has similar nutritional value as that of SBM with added oil (Lim and Dominy 1989). A study with Nile tilapia showed that, supplementation of 0.5 percent D,L-methionine to a diet containing 50 percent puffed FFSBM as a replacement of 75 percent brown fish meal provided the same growth and feed conversion as those of the fish meal diet (Tacon et al. 1983).

Cottonseed meal (CSM) is an important protein source for domestic animals, but its use in fish feed is limited due mainly to the presence of gossypol and low available lysine. Prepress solvent-extracted CSM containing 300 mg gossypol·kg^{-1} has been shown to be a good protein source for *O. mossambicus*. The growth of fish was improved when 50 percent of fish meal was replaced by CSM. The growth was essentially the same even at 100 percent substitution levels (Jackson et al. 1982). Viola and Zohar (1984) reported that low-gossypol CSM can be included at the same level as SBM in the diet of market-size hybrid tilapia (*O. niloticus* × *O. aureus*) reared in ponds.

For juvenile blue tilapia, Robinson et al. (1984a) found that inclusion of 26.5 percent solvent-extracted CSM (450 mg free gossypol·kg^{-1}) with lysine supplementation as a replacement for 25 percent peanut meal and 5 percent fish meal in a 35 percent crude protein diet depressed growth and feed efficiency. Lim et al. (2002) reported that substitution of one-third of SBM by 19 percent CSM and lysine supplementation had no effect on the growth performance of juvenile Nile tilapia; increasing CSM levels to 38 percent or higher adversely affected weight gain and hematological parameters. Since Nile and blue tilapia can tolerate at least up to 1,600 and 1,800 mg gossypol·kg^{-1} diet from gossypol-acetic acid, respectively, and diets used were supplemented with lysine, the negative effect of CSM was thought to be due to the deficiency of essential amino acids other than lysine, or the toxicity of compounds other than gossypol, such as cyclopropenoid fatty acids that occur on cottonseed lipid (Robinson et al. 1984a; Lim et al. 2002, 2003).

Ofojekwu and Ejike (1984) reported reduced weight gain and feed efficiency of Nile tilapia juveniles fed a diet containing 19.4 percent cottonseed cake, as compared with the groups fed a control diet containing fish meal. With the same species, El-Sayed (1990) obtained a 24 and 35 percent reduction in weight gain and feed efficiency, respectively, in fish fed a diet containing 65 percent CSM compared to fish fed a diet containing fish meal. Mbahinzireki et al. (2001) found that 50 percent of fish meal in the diet of juvenile *Oreochomis* sp. can be replaced by 29.4 percent CSM without affecting weight gain and feed efficiency.

Rapeseed/canola meal has a relatively good amino acid profile but is slightly lower in lysine than solvent-extracted SBM. It has been used primarily as a protein supplement for livestock and poultry, but its use in fish feed is limited due to the presence of high fiber and other indigestible carbohydrates, antinutritional factors (protease inhibitors), and phenolic compounds (such as glucosinolates, sinapine, tannins, and phytic acid). Jackson et al. (1982) obtained good growth of juvenile Mozambique tilapia fed a diet containing 41.8 percent rapeseed meal by removal of 50 percent of fish meal in the control diet. Increasing the level of rapeseed meal to 62.8 percent as a replacement of 75 percent of dietary fish meal significantly reduced weight gain. Jackson et al. indicated that the essential amino acids were not a limiting factor in this diet, but the limitation was due to the toxic effect of glucosinolates. Davies et al. (1990), however, reported that rapeseed meal at a level higher than 15 percent of the diet as a replacement of soybean meal, on an equal protein basis, resulted in poor growth and feed efficiency of *O. mossambicus*. The poor growth performance was attributed to either an inadequate nutritional profile or the presence of antinutritional factors.

Other plant proteins such copra, peanut, sunflower seed meal (Jackson et al. 1982), *Leucaena leucocephala* leaf meal (Jackson et al. 1982; Wee and Wang 1987; Santiago et al. 1988a), corn gluten feed, corn distiller's grain with solubles (Wu et al. 1994, 1996), azolla meal (Santiago et al. 1988b; El-sayed 1992), and cassava leaf meal (Ng and Wee 1989) have also been evaluated as substitutes for fish meal or other more costly protein ingredients in diets for various species of tilapia, but with varying degrees of success.

The reasons for low nutritional value of plant proteins as compared with fish meal have not been studied thoroughly, but several hypotheses have been suggested:

1. Presence of antinutritional factors or toxic substances;
2. Improper balance of essential nutrients such as amino acid, energy, and mineral;
3. Presence of high amount of fiber and indigestible carbohydrates;
4. Decrease palatability of the feed;
5. Reduction of pellet quality especially its water stability (Lim and Dominy 1989).

Digestibility data of selected feedstuffs for *O. aureus*, *O. niloticus,* and *O. mossambicus* are given in Tables 12.4, 12.5, and 12.6, respectively.

TABLE 12.4. Digestibility coefficients for protein, fat, carbohydrate, and gross energy of selected feedstuffs for *O. aureus*.

	Percent digestibility			
Feedstuff	**Protein**	**Fat**	**Carbohydrate**	**Gross energy**
Fish meal	84.8	97.8	—	87.4
Fish meal plus corn	84.9	—	—	—
Meat and bone meal	77.7	—	—	68.7
Soybean meal	94.4	—	53.5	72.5
Corn (uncooked)	83.8	89.9	45.4	55.5
Corn (uncooked mixed with fish meal)	—	—	65.4	—
Corn (cooked)	78.6	—	72.2	67.8
Wheat	89.6	84.9	60.8	65.3
Wheat bran	70.7	—	—	—
Alfalfa meal	65.7	—	27.7	22.9

Source: Data adapted from Popma (1982).

TABLE 12.5. Digestibility coefficients for protein and gross energy, and digestible energy values of selected feedstuffs for *O. niloticus*.

	Percent digestibility		Digestible energy
Feedstuff	**Protein**	**Gross energy**	**(kcal·kg^{-1})**
Fish meal	86.5	79.8	3,840
Poultry-offal meal	73.9	58.8	3,624
Soybean meal	90.7	56.6	2,678
Wheat middlings	75.6	57.6	2,746
Brewers grain	62.6	30.5	1,416
Ground corn	83.3	76.0	3,099
Gelatin	72.8	50.7	2,187
Animal oil	—	93.0	8,676

Source: Data adapted from Hanley (1987).

TABLE 12.6. Digestibility coefficients for protein and fat of selected feedstuffs for *O. mossambicus.*

	Apparent digestibility		True protein digestibility
Feedstuff	**Protein**	**Lipid**	
Fish meal	82.0	90.0	85.8
Silkworm pupae	85.7	94.5	89.3
Deoiled silkworm pupae	85.0	93.0	88.7
Fish silage (HCOOH)	88.2	89.3	92.0
Fish silage (H_2SO_4)	88.9	90.7	92.7
Mustard	78.9	90.4	82.9
Linseed	81.4	90.9	85.4
Sesame	81.9	87.2	85.8
Soybean	84.7	93.2	88.3

Source: Data adapted from Hossain et al. (1992).

ENERGY

Energy is not a nutrient but is a property of nutrients that are released during metabolic oxidation of proteins, carbohydrates, and lipids. Generally, protein is given the first priority in formulating fish feeds because it is the most expensive component of the prepared feeds. However, energy should be the first nutritional consideration in feed formulation since tilapias, like other fish, eat to satisfy their energy needs. If insufficient nonprotein energy is available, part of the protein will be broken down into energy. An excess of energy in the diet can limit feed consumption, thus reducing the intake of protein and other nutrients (NRC 1993). Kubaryk (1980) observed that, as the digestible energy (DE) content in the diet increased, feed consumption by *O. niloticus* decreased, but the amount of protein in the diet did not affect feed consumption.

Available data on energy requirements of tilapias have been reported in terms of gross energy (GE), DE or metabolizable energy (ME) in relation to the dietary level of protein. The dietary protein to energy ratio required for maximum growth decreases with increasing size of fish (Table 12.7). Young *O. aureus* (2.5 g) grew best when fed the 56 percent protein diet with a P/DE ratio of 123 mg·kcal^{-1}. Larger fish (7.5 g) grew maximally when fed a diet containing 106 mg protein/kcal of DE (Winfree and Stickney 1981). Fry (12 mg) of Nile tilapia *(O. niloticus)* grew best on a 45 percent protein diet with a P/GE ratio of 110 mg·kcal^{-1} (El-Sayed and Teshima 1992). Kubaryk (1980) obtained best performance of 1.7 g Nile tilapia with a diet

TABLE 12.7. Optimum dietary protein to energy ratio for various tilapia species.

Species	Fish size (g)	Protein/energy ratio	References
O. aureus	2.5	123 mg·kcal DE	Winfree and Stickney (1981)
	7.5	106 mg·kcal DE	
O. niloticus	0.012	110 mg·kcal GE	El-Sayed and Teshima (1992)
	1.7	120 mg·kcal DE	Kubaryk (1980)
Tilapia zilli	50	103 mg·kcal DE	El-Sayed (1987)
Red hybrid *(O. niloticus* × *O. aureus)*	0.6	111 mg·kcal DE	Santiago and Laron (1991)
O. niloticus[a]	2.9	75 mg·kcal ME	Fineman-kalao and Camacho (1987)
O. niloticus × *O. aureus* hybrid[b]	1.6	68-104 mg·kcal ME	Shiau and Huang (1990)

Notes: [a]Cultured in seawater at 18 to 50 ppt salinity. [b]Cultured in seawater at 32 to 34 ppt salinity.

containing 36 percent protein with a P/DE ratio of 120 mg·kcal^{-1}. For redbelly tilapia, *Tilapia zillii* (50 g), good growth was obtained with a 30 percent protein diet having a P/DE ratio of 103 mg·kcal^{-1} (El-Sayed 1987). Red hybrid tilapia (*O. niloticus* × *O. aureus*) fry (0.16 g) performed best on a 40 percent protein diet with a P/ME ratio of 111 mg·kcal^{-1} (Santiago and Laron 1991). Fineman-Kalio and Camacho (1987) reported maximum growth of *O. niloticus* (2.9 g) reared in brackishwater (18-50 ppt salinity) when fed a 30 percent protein diet with a P/ME (mammalian physiological fuel values) ratio of 75 mg·kcal^{-1}. With 1.60 g hybrid tilapia *(O. niloticus* × *O. aureus)* cultured in seawater (32-34 ppt salinity), good growth was obtained in fish fed 21 and 24 percent crude protein diets with P/ME ratios of approximately 68 and 104 mg·kcal^{-1}, respectively (Shiau and Huang 1990). Improper balance of P/E ratio will lead to poor growth; however, the growth reduction of diets with P/E ratios lower than the optimum value was not as pronounced as that of diets with excessive P/E ratios (Santiago and Laron 1991).

LIPIDS AND FATTY ACIDS

Dietary lipids are important sources of highly digestible energy and are the only source of essential fatty acids needed by fish for normal growth and development. They are also important carriers, and assist in the absorption, of fat soluble vitamins (A, D, E, and K). Lipids, especially phospholipids, are important for cellular structure and maintenance of membrane flexibility and permeability. They serve as precursors of steroid hormones and prostaglandins, improve the flavor of feeds, and affect feed texture and fatty acid composition of fish.

Dietary lipids have been shown to have sparing effect on the utilization of dietary protein. The level of protein in the diet of *O. niloticus* can be reduced from 33.2 to 25.7 percent by increasing dietary lipid from 5.7 to 9.4 percent and carbohydrate from 31.9 to 36.9 percent (Li et al. 1991). The sparing effect of dietary protein by increasing dietary lipid levels has also been reported in hybrid tilapia (*O. niloticus* × *O. aureus*) (Jauncey 2000). However, tilapia do not tolerate as high a dietary lipid as do salmonids. A dietary lipid level in excess of 12 percent depressed the growth of juvenile *O. aureus* × *O. niloticus* hybrids and increased the accumulation of carcass lipid (Jauncey and Ross 1982; Jauncey 2000). For juvenile hybrid tilapia *(O. niloticus* × *O. aureus)*, Chou and Shiau (1996) obtained good growth and feed efficiency in fish fed the 10 and 15 percent lipid diets. They suggested, however, that a level of 5 percent dietary lipid appeared to be sufficient to meet the minimal requirement of this tilapia hybrid but a level of about 12 percent was needed for maximum growth. The growth of *O. aureus* can be substantially improved when menhaden oil or fish oil was provided at 7.5 to 10 percent of the diet as compared with lower levels; however, best performance was obtained with menhaden oil at 10 percent of the diet (Stickney and Wurts 1986).

Research evaluating dietary lipid sources showed that *O. niloticus* that were fed diets supplemented with soybean or corn oil rich in linoleic acid (18:2 n–6) showed better growth performance than those that were fed diets containing fish oil rich in 20:5 n–3 polyunsaturated fatty acids (PUFA) and beef tallow rich in 18:1 n–9 (Takeuchi et al. 1983a). Dietary lipid sources also had an effect on the reproductive performance of Nile tilapia. The number of female that spawn, spawning frequency, number of fry per spawn, and total fry production were increased at varying degrees by supplementation of broodstock diets with soybean, corn, or coconut oils, but not with cod liver oil (Santiago and Reyes 1993). They reported that the overall performance over a 24-week period was best for fish fed the soybean oil diet. Fish fed the cod liver oil diet had the highest weight gain but the poorest reproductive performance. Extrapolation of these results suggests that linoleic series (n–6) fatty acids are a dietary essential for Nile tilapia. Linolenic series (n–3) fatty acids may also be dietary essential for Nile tilapia because soybean oil that provided good growth and reproductive performance is also high in linolenic acid (18:3 n–3). However, Kanazawa et al. (1980) demonstrated that the growth-promoting effects of n–6 series fatty acids (18:2 n–6 and 20:4 n–6) for *T. zillii* were superior to those of the n–3 series (18:3 n–3 and 20:5 n–3). The optimum dietary levels of n–6 fatty acids (18:2 n–6 or 20:4 n–6) have been has been estimated to be about 1 percent for *T. zillii* (Kanazawa et al. 1980) and 0.5 percent for *O. niloticus* (Takeuchi et al. 1983b). The essential fatty acid requirements of *O. aureus* has not been defined. Stickney and McGeachin (1983) indicated that neither linoleic nor linolenic acids appear to be required at above 1 percent of the diet of *O. aureus.* Stickney and Hardy (1989), however, reported that *O. aureus* have a requirement for relatively high level of n–6 fatty acids, although the requirement can be reduced when n–3 fatty acids are present. Chou and Shiau (1999) and Chou et al. (2001) also suggested that n–3 fatty acids, as well as n–6 fatty acids, are essential for maximum growth of hybrid tilapia *(O. niloticus* × *O. aureus).* However, the optimum dietary requirement of tilapias for n–3 fatty acids has not been determined. Deficiency signs observed in tilapia that were fed diets deficient in n–6 and n–3 fatty acids were anorexia; poor growth; increased body content of 18:1 n–9 and 20:3 n–9; and swollen, pale, and fatty livers. Fatty acid requirements of various tilapia species are given in Table 12.8.

Nile, blue, and redbelly tilapia appear to possess the ability to desaturate and chain elongate 18:2 n–6 and 18:3 n–3 to longer chain n–6 and n–3 PUFA as increased levels of these fatty acids were observed in fish that were fed diets supplemented with linoleic or linolenic acids or injected with ^{14}C-labeled fatty acids (Kanazawa et al. 1980; Stickney and McGeachin 1983; Olsen et al. 1990). In Nile tilapia, the rate of conversion is thought to be dependent on the fatty acid composition of the diets. In tilapia that were fed diets containing sufficient levels of preformed C20 and C22 PUFA, the rate of conversion was low (Olsen et al. 1990). They suggested that the enzymes

TABLE 12.8. Essential fatty acid requirements of juvenile tilapia.

Species	Essential fatty acid	Requirement (percent in diet)	References
T. zilli	18:2 n–6 or 20:4 n–6	1.0	Kanazawa et al. (1980)
O. niloticus	18:2 n–6 or 20:4 n–6	0.5	Takeuchi et al. (1983b)
	18:3 n–3	Unknown	
O. aureus	18:2 n–6 and 18:3 n–3	≥1.0	Stickney and McGeachin (1983)
O. niloticus × *O. aureus*	18:2 n–6 and 18:3 n–3	Unknown	Chou et al. (2001)

involved in this conversion may always be present in the tilapia but be inhibited by long chain PUFA. In contrast, production of the enzymes may be induced when the levels of C20 and C22 PUFA fall below those required for phospholipid synthesis. Based on this information, tilapia can probably utilize 18:2 n–6 and 18:3 n–3 equally well as 20:4 n–6, and 20:5 n–3 and 22:6 n–3, as essential fatty acids.

CARBOHYDRATES

Tilapia, like other finfish, do not have specific requirements for carbohydrates. Several studies have shown that fish grew satisfactorily and without any pathological signs when fed carbohydrate-free diets. However, carbohydrates are always included in fish feeds because they are the most abundant and least expensive source of energy. They function as pellet binder, serve as precursor for the formation of various metabolic intermediates essential for growth, and have a sparing effect on the utilization of dietary protein (NRC 1993). Shiau and Peng (1993) reported that the protein-sparing effect of carbohydrates (dextrin or starch) in hybrid tilapia *(O. niloticus* × *O. aureus)* only occurred when the dietary protein level was suboptimal. A study on the effect of carbohydrate:lipid ratios on intermediary metabolism in Nile tilapia by Shimeno et al. (1993) appears to also suggest the protein-sparing action of dietary carbohydrates. They reported that increasing dietary carbohydrate:lipid ratios lead to increased glycolysis and lipogenesis, but decreased gluconeogenesis and amino acid degradation in the liver. Provided that the dietary essential fatty acid requirement are met, digestible carbohydrates can also be used to replace dietary lipid as an energy source. El-sayed and Garling (1988) observed that dextrin can be used as a substitute for lipid in diets of redbelly tilapia at a rate of 2.25:1 dextrin:lipid (based on mammalian metabolizable energy).

Tilapia appear to digest carbohydrates in most feedstuffs relatively well (Table 12.4). Complex carbohydrates such as dextrin or starch were more readily utilized by tilapia than simple sugars such as glucose (Shiau and Peng 1993). Cassava, sorghum, corn, wheat, and rice are good carbohydrate sources for Nile tilapia (Wee and Ng 1986; El-sayed and Garling 1988), whereas barley is poorly utilized by Nile tilapia (El-sayed and Garling 1988).

Tilapia, like other fish species, do not digest highly fibrous feedstuffs, such as alfalfa and coffee pulp, well for energy (Table 12.4). In Nile tilapia, dietary fiber at levels >5 percent reduced diet utilization and digestibility, and levels >10 percent reduced protein utilization (Anderson et al. 1984). A level of 15 percent alpha-cellulose in the diet has also been suspected to adversely affect the growth and feed efficiency of Nile tilapia (Wee and Ng 1986). Tilapia, however, can utilize carbohydrates as an energy source more efficiently than salmonids, seabass, seabream, and yellowtail. Anderson et al. (1984) reported growth improvement of *O. niloticus* when dietary level of carbohydrates was increased from 0.0 to 40 percent. Wee and Ng (1986) also observed a trend of improved growth and feed efficiency with increased level of cassava in the diet up to 60 percent. *Tilapia zillii* appear to utilize carbohydrates as efficiently as *O. niloticus* (El-sayed and Garling 1988).

VITAMINS

Vitamins are organic compounds that are required in small amounts for normal growth, reproduction, and health. However, vitamin supplements are often not included in practical feeds for tilapias stocked at moderate densities in fertilized ponds. It is assumed that, under these conditions, tilapia consume a sufficient quantity of natural food organisms to meet their vitamin needs. In intensive systems where limited or no natural foods are available, supplemental vitamins must be added to the diets to sustain normal growth and health.

Water-Soluble Vitamins

Water-soluble vitamin requirements of tilapia that have been studied are: thiamin, riboflavin, pyridoxine, pantothenic acid, niacin, biotin, folic acid, vitamin B_{12}, inositol, choline, and vitamin C. These vitamin requirements for various tilapia species are presented in Table 12.9.

Thiamin (vitamin B_1) deficiency signs observed in red hybrid tilapia *(O. mossambicus* × *O. niloticus)* fingerlings cultured in seawater (32 ppt salinity) were reduced growth, poor feed efficiency, and low hematocrit. A dietary thiamin level of 2.5 $mg{\cdot}kg^{-1}$ was sufficient for maximum growth and prevention of deficiency signs (Lim and LeaMaster 1991). Reductions in weight gain, feed intake, feed efficiency, total and red blood cell counts, and increased serum pyruvate were observed in Nile tilapia that were fed thiamin-deficient diets. A dietary thiamin level of about 4 $mg{\cdot}kg^{-1}$ was adequate for optimum growth and prevention of deficiency signs (Lim et al. 2000).

Typical deficiency signs reported for tilapia that were fed a riboflavin (vitamin B_2)-free diet were anorexia, poor growth, high mortality, fin erosion, loss of normal body color, short body dwarfism, and lens cataracts. The dietary riboflavin requirements were 6 $mg{\cdot}kg^{-1}$ for juvenile *O. aureus*

TABLE 12.9. Vitamin requirements of juvenile tilapia.

Vitamin/fish	Requirement (mg·kg^{-1} diet)	Deficiency signs	Reference
Thiamin			
O. mossambicus × *O. niloticus*	2.5	Anorexia, light coloration, nervous disorder, poor growth and poor feed efficiency, low hematicrit and red blood cell count, and increased serum pyruvate	Lim et al. (1991)
O. niloticus	4.0		Lim et al. (2000)
Riboflavin			
O. aureus	6.0	Lethargy, anorexia, poor growth, high mortality, loss of color, fin erosion, short body dwarfism, and lens cataracts	Soliman and Wilson (1992)
O. mossambicus × *O. niloticus*	5.0		Lim et al. (1993)
Pyridoxine			
O. mossambicus × *O. niloticus*	3.0	Anorexia, nervousness, convulsion, caudal fin erosion, mouth lesions, poor growth and feed efficiency, high mortality, reduced hepatic alanine aminotransferase	Lim et al. (1995)
O. niloticus × *O. aureus*	1.7-9.5 and 15.0-16.5 for 28 and 36 percent protein diets, respectively		Shiau and Hsieh (1997)
O. mossambicus	5.0-11.7		Oyetayo et al. (1985)
Pantothenic acid			
O. aureus	10.0	Poor growth, hemorrhage, sluggishness, high mortality, anemia, and hyperplasia of gill lamellae epithelial cell	Soliman and Wilson (1992)
Niacin			
O. niloticus × *O. aureus*	26.0 (for glucose diet) 121.0 (for dextrin diet)	Hemorrhages, deformed snout, gill edema, and skin, fin, and mouth lesions	Shiau and Suen (1992)
Biotin			
O. niloticus × *O. aureus*	0.06	Poor growth, low tissue biotin, and reduced hepatic pyruvate carboxylase and acetyl CoA carboxylase activities	Shiau and Chin (1999)

TABLE 12.9 *(continued)*

Vitamin/fish	Requirement (mg·kg^{-1} diet)	Deficiency signs	Reference
Folic acid			
O. niloticus	0.5	Poor growth and reduced feed intake and efficiency	Lim and Klesius (2001)
Cyanocobalamin			
O. niloticus	Not required		Lovell and Limsuwan (1982)
			Sugita et al. (1990)
O. niloticus × *O. aureus*	Not required		Shiau and Lung (1993)
Inositol			
O. niloticus	Not required		Perres et al. (2004)
Choline			
O. niloticus × *O. aureus*	1,000	Poor growth and survival, and reduced blood triglyceride, cholesterol, and phospholipid concentration	Shiau and Lo (2000)
Ascorbic acid[a]			
O. niloticus	50.0	Lordosis, scoliosis, poor growth, feed efficiency, poor wound healing, hemorrhage, anemia, exophthalmia and gill, and operculum deformity	Stckney et al. (1984)
O. aureus	79.0		Shiau and Jan (1992)
O. spilurus	50.0		Al-Amoudi et al. (1992)
O. niloticus × *O. aureus*	19.0		Shiau and Hsu (1999)
Vitamin A			
O. niloticus	5,000 IU	Poor growth and poor feed efficiency, high mortality, restlessness, abnormal swimming, blindness, skin, fin, and eye hemorrhages and reduced mucus secretion	Saleh et al. (1995)
Cholecalciferol			
O. niloticus × *O. aureus*	375 IU	Poor growth and poor feed efficiency, and reduced hemoglobin, hepatosomatic index, and alkaline phosphatase activity	Shiau and Hwang (1993)
O. aureus	Not required		O'Connel and Gatlin (1994)

TABLE 12.9 *(continued)*

Vitamin/fish	Requirement (mg·kg⁻¹ diet)	Deficiency signs	Reference
Vitamin E[b]			
O. aureus	10-25	Anorexia, poor growth, and poor feed efficiency, skin hemorrhage, impaired erythropoiesis, muscle degeneration, seroid in liver and spleen, and abnormal skin coloration	Roem, Kohler, and Stickney (1990b)
O. niloticus	50-100		Satoh et al. (1987)
O. niloticus × *O. aureus*	42-44		Shiau and Shiau (2001)

Notes: [a]Since ascorbic acid or its derivatives were used, the requirement values reported are expressed in ascorbic acid equivalent. [b]Requirement values at 5-6 percent dietary lipid. The requirements increase with increasing dietary lipid.

grown in freshwater (Soliman and Wilson 1992) and 5 mg·kg^{-1} for red hybrid *(O. mossambicus* × *O. niloticus)* grown in 32 ppt seawater (Lim et al. 1993).

Tilapia are very sensitive to pyridoxine (vitamin B_6) deficiency. Red hybrid tilapia *(O. mossambicus* × *O. niloticus)* grown in seawater and fed a diet without pyridoxine supplementation developed abnormal neurological signs, anorexia, convulsion, caudal fin erosion, mouth lesions, poor growth, and high mortality within 2 to 3 weeks. A dietary level of 3 mg pyridoxine·kg^{-1} was adequate for maximum growth and prevention of various deficiency symptoms (Lim et al. 1995). The requirement of this vitamin in *O. niloticus* × *O. aureus* hybrid has been shown to be affected by dietary levels of protein. Dietary levels of pyridoxine necessary for optimum growth and maintaining high levels of liver alanine aminotransferase activity were 1.7 to 9.5 and 15 to 16.5 mg pyridoxine·kg^{-1} in diets containing 28 and 36 percent crude protein, respectively (Shiau and Hsieh 1997). The requirement of this vitamin reported for Mozambique tilapia ranged from 5.0 to 11.7 mg·kg^{-1} diet (Oyetayo et al. 1985).

Niacin is essential in diet for *O. niloticus* × *O. aureus* hybrid but the level required varies depending on the source of dietary carbohydrate. Optimum dietary levels for maximum growth have been reported to be 26 mg·kg^{-1} in fish that are fed a glucose diet and 121 mg·kg^{-1} in fish that are fed the dextrin diet. Fish deprived of dietary niacin developed hemorrhages; deformed snout; gill edema; and skin, fin, and mouth lesions (Shiau and Suen 1992).

Pantothenic acid is essential in diet for tilapia. Blue tilapia that are fed pantothenic acid-in deficient quality in diet had poor growth, hemorrhage, sluggishness, high mortality, anemia, and severe hyperplasia of the epithelial cells of gill lamellae. A dietary level of 10 mg of calcium *d*-pantothenate per kilogram diet was sufficient to prevent these deficiency symptoms (Soliman and Wilson 1992). Roem, Stickney, and Kohler (1990) reported that *O. aureus* could satisfy their requirement for pantothenic acid, choline, and possibly some other vitamins by feeding on bacteria in a recirculating system.

Biotin is dietary essential for *O. niloticus* × *O. aureus* hybrid. Fish that are fed biotin-deficient diets resulted in poor growth, low body biotin concentrations, and reduced hepatic pyruvate carboxylase and acetyl CoA carboxylase activities. A dietary level of 0.06 mg biotin·kg^{-1} was adequate to prevent these signs of deficiency (Shiau and Chin 1999).

Nile tilapia appear to have a requirement for dietary folic acid. Fish fed a diet without folic acid supplementation had reduced weight gain, reduced feed intake, and poor feed efficiency as compared to fish fed the diet with 0.5 mg folic acid·kg^{-1}. Total and red blood cell counts, however, were lower in fish that were fed 0.5 mg folic acid diet as compared to the groups fed the 1.0 mg or higher folic acid·kg^{-1}. A level of folic acid of about 0.5-1.0 mg·kg^{-1} diet was suggested (Lim and Klesius 2001).

Vitamin B_{12} (cyanocobalamin) is not a dietary essential for tilapia. It has been shown that Nile tilapia produced vitamin B_{12} in their gastrointestinal tract through bacterial synthesis to meet their metabolic requirements (Lovell and Limsuwan 1982; Sugita et al. 1990). Vitamin B_{12} is likewise not a dietary essential for juvenile hybrid tilapia *(O. niloticus* × *O. aureus)* since no deficiency symptoms were detected in fish fed a vitamin B_{12}-deficient diet for 16 weeks (Shiau and Lung 1993).

Inositol synthesis has been reported to occur, to some degree, in liver, kidney, and other tissues of several fish species. For some species, however, de novo synthesis is inadequate to support their metabolic needs and, thus, require an exogenous source of this vitamin (NRC 1993). A study by Peres et al. (2004) showed that supplementation of inositol to purified diets had no effect on growth, feed efficiency, and hematological parameters, but reduced-lipid accumulation in liver and muscle were observed in fish that were fed diets supplemented with 100 and 400 mg inositol·kg^{-1}, respectively. They suggested that Nile tilapia juveniles can probably synthesize inositol in sufficient quantity to meet their need for normal growth, but insufficient to prevent alteration of lipid metabolism. They concluded that, since inositol is widely distributed in common feed ingredients, practical diets should contain sufficient levels of this vitamin to meet various metabolic needs of Nile tilapia.

Roem et al. (1990a) were unable to determine the choline requirement of blue tilapia when held in a recirculating water system. It was suggested that dietary methionine may have been present in excess of the fish's metabolic requirement and may have provided sufficient methyl groups for choline synthesis and thus, satisfied the choline requirement in *O. aureus* (Roem et al. 1990a,b). Several studies have suggested that most fish species can utilize dietary methionine to spare at least some choline (NRC 1993). In a study, Shiau and Lo (2000) showed that choline is essential in diet for juvenile *O. niloticus* × *O. aureus* and a level of about 1,000 mg·kg^{-1} was sufficient to maintain high growth rate, survival, and maintain high blood triglyceride, cholesterol, and phospholipid concentrations.

Tilapia exhibit classical vitamin C deficiency signs when fed a vitamin-deficient diet in the absence of natural foods. The requirement for normal growth of *O. niloticus* has been reported to be 50 mg ascorbic acid (AA)·kg^{-1} diet (Stickney et al. 1984). In juvenile *O. niloticus* × *O. aureus,* a dietary level of 79 mg AA·kg^{-1} is needed for maximum growth (Shiau and Jan 1992). In these studies, AA was used as the source of vitamin C. Since AA is relatively unstable and large amount is lost during feed processing and storage, a large margin of allowance must be provided. Several AA derivatives, such as L-ascorbyl-2-sulfate (AS), L-ascorbyl-2-monophosphate-magnesium (AMP-Mg), L-ascorbyl-2-monophosphate-sodium (APM-Na), and L-ascorbyl-2-polyphosphate (APP), which have been shown to have antiscorbutic activity for tilapia, are relatively stable. The requirement for optimum growth of *O. spilurus* in seawater has been determined to be between, 100 and 200 mg of AS·kg^{-1} diet (Al-Amoudi et al. 1992). L-ascorbyl-2-sulfate, AMP-Mg, and APP had approximately equal ascorbic acid activity (Shiau and Hsu 1995; Abdelghany 1996) and a dietary level of 50 mg AA equivalent per kilogram was sufficient for maximum growth and AA liver storage in Nile tilapia (Abdelghany 1996). The requirements of hybrid tilapia *(O. niloticus* × *O. aureus)* are about 41 mg APM-Mg·kg^{-1} and 63 mg AMP-Na·kg^{-1}, both equivalent to approximately 19 mg AA activity·kg^{-1} diet (Shiau and Hsu 1999).

Fat-Soluble Vitamins

Available information on fat-soluble vitamins for tilapia concerns only vitamins A, D, and E (Table 12.9). No studies have been conducted on the requirements of vitamin K for tilapia.

The requirement of vitamin A (retinol) for juvenile Nile tilapia have been reported to be about 5,000 IU·kg^{-1} diets (Saleh et al. 1995). Severe deficiency symptoms including poor growth and poor feed efficiency, high mortality, restlessness, abnormal swimming, exophthalmia, blindness, hemorrhages of eyes, fins, and skin, and reduced mucus secretion were observed in fish that were fed the vitamin A unsupplemented diet. Hypervitaminosis A, characterized by reduced weight gain, high mortality, impaired skeletal formation, skin hemorrhages, necrosis of caudal fin, and enlargements of liver and spleen, were observed in the groups fed 40,000 IU vitamin A per kilogram diet.

Sunita Rao and Raghuramulu (1996) reported that intravenous injection or incubation of crude liver homogenate of Mozambique tilapia with ^{14}C-labeled cholesterol or intraperitoneal injection of ^{14}C-labeled acetate did not result in the formation of ^{14}C-labeled vitamin D_3. They suggested the absence of nonphotochemical pathway of vitamin D_3 synthesis in Mozambique tilapia, or even if it existed, cholesterol may not be the substrate. Moreover, fish may not be able to synthize cholesterol from 2-carbon units like acetate. The optimum dietary level of vitamin D_3 (cholecalciferol) for maximum growth of juvenile tilapia hybrid (*O. niloticus* × *O. aureus*) is about 375 $IU \cdot kg^{-1}$ (Shiau and Hwang 1993). Fish that fed diets without vitamin D_3 supplementation had poor growth and poor feed efficiency, and low hemoglobin, hepatosomatic index, and alkaline phosphatase activity. Blue tilapia, however, appear not to have a dietary requirement for cholecalciferol (vitamin D_3) for normal growth and tissue mineralization (O'Connel and Gatlin 1994).

Tilapia have been shown to have a dietary requirement for vitamin E (∝-tocopherol). Vitamin E-deficient *O. aureus* exhibited anorexia; reduced weight gain, and poor feed efficiency; skin hemorrhages; impaired erythropoiesis; muscle degeneration; ceroid in liver and spleen; and abnormal skin coloration. The dietary vitamin E requirement increased with increasing levels of dietary lipid. The requirement of *O. aureus* was estimated at 10 mg and 25 mg dl-α-tocopheryl acetate per kilogram of diet at 3 and 6 percent dietary lipid, respectively, or 3 to 4 mg α-tocopheryl acetate per percent of corn oil (Roem, Kohler, and Stickney 1990b). The dietary vitamin E requirement of *O. niloticus* was reported to be 50 to 100 $mg \cdot kg^{-1}$ for a diet containing 5 percent lipid and increased to 500 $mg \cdot kg^{-1}$ for a diet containing 10 to 15 percent lipid (Satoh et al. 1987). The dietary vitamin E requirement of tilapia hybrid *(O. niloticus × O. aureus)* was 42 to 44 $mg \cdot kg^{-1}$ and 60 to 66 $mg \cdot kg^{-1}$ in diets containing 5 and 12 percent lipid, respectively (Shiau and Shiau 2001).

MINERALS

There is little information on mineral requirements of tilapia. However, tilapia probably require the same minerals as do other fish for tissue formation and various metabolic functions such as osmoregulation, acid-base balance, and proper functioning of muscles and nerves. Like other fish, they probably can absorb several minerals from the surrounding water to meet part of their metabolic needs. Luquet (1991) reported that *O. mossambicus* can absorb calcium efficiently via the gills and phosphorus via the gut. Although phosphorus can be absorbed by fish, unlike calcium, phosphorus is usually a limiting mineral in most natural waters. Thus, diets must contain sufficient levels of phosphorus to meet the fish needs. Since phosphorus is an important component for the eutrophication of water, the amount of phosphorus in culture water and aquaculture effluent is of major concern. Thus, requirement for phosphorus in fish must be met without adding excess phosphorus to the water.

In low-calcium water, a supplementation of about 7.0 g (0.7 percent) Ca and 5 g (0.5 percent) $P{\cdot}kg^{-1}$ purified diet are required for normal growth and bone mineralization of blue tilapia (Robinson et al. 1984b, 1987). O'Connell and Gatlin (1994) obtained best growth and high concentrations of minerals in bone and scale of blue tilapia that were reared in water with <0.1 $g\ Ca{\cdot}L^{-1}$ and fed purified diets supplemented with 7.5 g (0.75 percent) $Ca{\cdot}kg^{-1}$. Watanabe et al. (1988) recommended an available phosphorus level of 0.8 to 1.0 percent in diets of Nile tilapia. The availability of phosphorus varies greatly depending on the source. More water-soluble source such as monocalcium, monoammonium, and monosodium phosphate are more available than dicalcium phosphate. Tricalcium phosphate and finely ground defluorinated rock phosphate are less available. About 70 percent of phosphorus in plant feedstuffs is in the form of phytin that is unavailable to fish. Moreover, phytin-phosphorus readily forms complexes with di- and trivalent cations such as calcium, magnesium, iron, zinc, and copper in the gastrointestinal tract, thus reducing the availability of these minerals. Treatments of plant feedstuffs with microbial enzyme phytase or addition of phytase to the diets have been shown to greatly improve the bioavailability of phytin-phosphorus.

Other minerals in which dietary requirement have been determined are magnesium, manganese, potassium, iron, copper, and zinc (see Table 12.10). Dietary magnesium at levels of 0.59 to 0.77 g and 0.50 to 0.65 $g{\cdot}kg^{-1}$ diet have been reported to be adequate for optimum performance of Nile tilapia (Dabrowska et al. 1989) and blue tilapia (Reigh et al. 1991), respectively. Excessive levels of Mg (3.2 $g{\cdot}kg^{-1}$) resulted in depression in growth, but the effect was more pronounced in fish that were fed low protein (24 percent) diets than those fed with high protein (44 percent) diets (Dabrowska et al. 1989). Low hematocrit and hemoglobin content and sluggishness were also observed in tilapia that were fed low-protein and high-magnesium diet. Fish that were fed magnesium-deficient diets had poor growth, low tissue magnesium concentrations and abnormal tissue mineralization.

TABLE 12.10. Mineral requirements of juvenile tilapia.

Mineral	Species	Requirement (/kg diet)	References
Calcium	*O. aureus*	7.0 g	Robinson et al. (1987)
		7.5 g	O'Connell and Gatlin (1994)
Phosphorus	*O. aureus*	5.0 g	Robinson et al. (1987)
	O. niloticus	0.8-1.0 g	Watanabe, Satoh et al. (1988)
Potassium	*O. niloticus* × *O. aureus*	2.0-3.0 g	Shiau and Hsieh (2001)
Magnesium	*O. niloticus*	0.59-0.77 g	Dabrowska et al. (1989)
	O. aureus	0.50-0.65 g	Reigh et al. (1991)
Manganese	*O. mossambicus*	1.7 mg	Ishac and Dollar (1967)
	O. aureus	12.0 mg	Watanabe et al. (1988)
Iron	*O. niloticus* × *O. aureus*	150-160 (Fe citrate)	Shiau and Su (2003) Kleemann et al. (2003)
	O. niloticus	85 (Fe sulfate)	
		60 (available Fe)	
Zinc	*O. niloticus*	30.0 mg	Elhamid Eid and Ghomin (1994)
Copper	*O. niloticus*	2.0-3.0 mg	Watanabe et al. (1988)

Ishac and Dollar (1967) reported that *O. mossambicus* required manganese in both culture water and diet to meet the daily requirement of approximately 1.7 mg·kg^{-1} of fish per day. Watanabe et al. (1988b) recommended a dietary level of 12 mg Mn·kg^{-1} for *O. niloticus.* Deficiency of manganese resulted in anorexia, poor growth, loss of equilibrium, and high mortality.

Iron has also been found to be essential in diet for tilapia. Deficiency signs that were reported were poor growth, hypochromic microcytic anemia, and low hepatic iron content. The dietary iron requirement is affected by the source of dietary iron. Shiau and Su (2003) observed that ferric citrate was about 50 percent as effective as iron sulfate in meeting the iron requirements of hybrid tilapia *(O. niloticus × O. aureus).* A dietary level of 150 to 160 mg Fe·kg^{-1} from iron citrate or 85 mg Fe·kg^{-1} from iron sulfate was recommended. For Nile tilapia, a minimum available iron of 60 mg·kg^{-1} diet was suggested to maintain normal erythropoiesis (Kleemann et al. 2003).

Other minerals of importance are potassium, zinc, and copper. Potassium at a level of 2 to 3 g (0.2 to 0.3)·kg^{-1} diet are required for optimum growth, gill Na^{+}-K^{+} ATPase activity, and whole-body potassium retention of hybrid tilapia *(O. niloticus × O. aureus)* (Shiau and Hsieh 2001). The requirements of *O. niloticus* for zinc have been reported to be 30 mg·kg^{-1} diet (Elhamid Eid and Ghonim 1994). A dietary copper level of 2 to 3 mg·kg^{-1} has been suggested by Watanabe et al. (1988b) for *O. niloticus,* although the requirement of this mineral has not been determined.

CONCLUSION

The development of least-cost and nutritionally balanced feeds relies heavily, among other things, on information about nutrient requirements and nutrient availability of feedstuffs. This information for tilapia is still limited, although considerable progress has been made in the past decade. Observation of the available data suggests that the qualitative nutrient requirements of different tilapia species are similar, but that the quantitative requirements appear to vary considerably even for the same species. These differences are probably related to differences in fish size or age, composition of basal diet, ingredient quality, dietary nutrient levels, physical quality of diet, feeding management, as well as culture conditions. Moreover, most of the information available is confined to early juvenile stage cultured in the laboratory under well-controlled environmental conditions. Thus, the data generated are well defined and oftentimes cannot be directly applied to large-scale commercial production.

REFERENCES

Abdelghany, A.E. (1996). Growth Response of Nile Tilapia, *Oreochromis niloticus* to Dietary L-Ascorbic Acid, L-Ascorbyl-2-Sulfate, and L-Ascorbyl-2-Polyphosphate. *Journal of the World Aquaculture Society* 27:449-455.

———. (1998). Feed efficiency, nutrient retention and body composition of Nile tilapia, *Oreochromis niloticus* L., fed diets containing L-ascorbic acid, L-ascorbyl-2-sulphate or L-ascorbyl-2-polyphosphate. *Aquaculture Research* 29: 503-510.

———. (2000). Replacement value of cystine for methionine in semi-purified diets supplemented with free amino acids for Nile tilapia (*Oreochromis niloticus* L.). In K. Fitzimmons (Ed.), pp. 31-39. *Proceedings, International Symposium on Tilapia in Aquaculture*, 9-11 November 1997, Orlando, FL.

Al-Amoudi, M.M., A.M.N. El-Nagar, and B.M. El-Nouman (1992). Evaluation of optimum dietary requirement of vitamin C for the growth of *Oreochromis spifurees* fingerlings in water from the red sea. *Aquaculture* 105:165-173.

Al-Ogaily, S.M., N.A. Al-Asgah, and A. Ali (1996). Effect of feeding different grain sources on the growth performance and body composition of tilapia, *Oreochromis niloticus* (L.). *Aquaculture Research* 27:523-529.

Anderson, J., A.J. Jackson, A.J. Matty, and B.S. Capper (1984). Effects of dietary carbohydrate and fiber on the tilapia, *Oreochromis niloticus* (Linn). *Aquaculture* 37:303-314.

Chou, B.S., and S.Y. Shiau (1996). Optimum dietary lipid level for growth of juvenile hybrid tilapia, *Oreochromis niloticus* × *Oreochromis aureus*. *Aquaculture* 143:185-195.

———. (1999). Both n–6 and n–3 fatty acids are required for maximal growth of juvenile hybrid tilapia. *North American Journal of Aquaculture* 61:13-20.

Chou, B.S., S.Y. Shiau, and S.S.O. Hung (2001). Effect of dietary cod liver oil on growth and fatty acids of juvenile hybrid tilapia. *North American Journal of Aquaculture* 63:277-284.

Clark, J.H., W.O. Watanabe, and D.H. Ernest (1990). Effect of feeding rate on growth and feed conversion of Florida red tilapia reared in floating marine cages. *Journal of the World Aquaculture Society* 21:16-24.

Cowey, C.B., and A.G.J. Tacon (1983). Fish nutrition- relevance to invertebrates. In G.D. Pruder, C.J. Langdon, and D.E. Conklin (Eds.), *Proceeding, Second International Conference on Aquaculture Nutrition: Biochemical and Physiological Approaches to Shellfish Nutrition* (pp. 13-30) Baton Rouge, LL: Louisiana State University.

Dabrowska, H., K. Meyer-Burgdorff, and K.D. Gunther (1989). Interaction between dietary protein and magnesium level in tilapia (*Oreochromis niloticus*). *Aquaculture* 76:277-291.

Davies, S.J., S. McConnell, and R.I. Bateson (1990). Potential of rapeseed meal as an alternative protein source in complete diets for tilapia (*Oreochromis mossambicus* Peters). *Aquaculture* 87:145-154.

Davis, A.T., and R.R. Stickney (1978). Growth response of tilapia aurea to dietary protein quality and quantity. *Transactions of the American Fisheries Society* 107:479-483.

De Silva, S., and M.K. Perera (1985). Effects of dietary protein level on growth, food conversion, and protein use in young *Tilapia nilotica* at four salinities. *Transactions of the American Fisheries Society* 114:584-589.

De Silva, S.S., R.M. Gunasekera, and D. Atapattu (1989). The dietary protein requirements of young tilapia and an evaluation of the least-cost dietary protein level. *Aquaculture* 80:271-284.

Elhamid Eid, A., and S.I. Ghonim (1994). Dietary zinc requirement of fingerling *Oreochromis niloticus*. *Aquaculture* 119:259-264.

El-Sayed, A.F.M. (1987). Protein and Energy Requirement of *Tilapia zillii*. Doctoral Dissertation, Michigan State University, East Lansing, MI.

———. (1990). Long-term evaluation of cottonseed meal as a protein source for Nila tilapia, *Oreochromis niloticus* (Linn.). *Aquaculture* 84:315-320.

———. (1992). Effects of substituting fish meal with Azolla pinnata in practical diets for fingerling and adult Nile tilapia *Oreochromis niloticus* (L.). *Aquaculture and Fisheries Management* 3:167-173.

———. (1998). Total replacement of fish meal with animal protein sources in Nile tilapia, *Oreochromis niloticus* (L.), feeds. *Aquaculture Research* 29:275-280.

El-Sayed, A.F.M, and D.L. Garling, Jr (1988). Carbohydrate-to-lipid ratios in diets for *Tilapia* fingerlings. *Aquaculture* 73:157-163.

El-Sayed, A.F.M., and S. Teshima (1992). Protein and energy requirements of Nile tilapia, *Oreochromis niloticus*, fry. *Aquaculture* 103:55-63.

FAO (2002). State of the World Fisheries and Aquaculture. Food and Agricultural Organization, Part 1. Rome, Italy.

Fineman-Kalio., A.S., and A.S. Camacho (1987). The effects of supplemental feeds containing different protein: energy ratios on the growth and survival of *Oreochromis niloticus* (L.) in brackish water ponds. *Aquaculture and Fisheries Management* 18:139-149.

Hanley, F. (1987). The digestibility of foodstuffs and the effects of feeding selectivity on digestibility determinations in tilapia, *Oreochromis niloticus* (L.). *Aquaculture* 66:163-179.

Hossain, M.A., N. Nahar, M. Kamal, and M.N. Islam (1992). Nutrient digestibility coefficients of some plant and animal proteins for tilapia (*Oreochromis mossambicus*). *Journal of Aquaculture in the Tropics* 7:257-266.

Ishac, M.M., and A.M. Dollar (1967). Studies on manganese uptake in *Tilapia mossambica* and *Salmo gairdnerii*. I-Growth and survival of *Tilapia mossambica* in response to manganese. *Hydrobiolobia* 31:572-584.

Jackson, A.J., B.S. Capper, and A.J. Matty (1982). Evaluation of some plant proteins in complete diets for the tilapia *Saratherodon mossambicus*. *Aquaculture* 27:97-109.

Jauncey, K. (2000). Nutritional requirement. In M.C.M. Beveridge and B.J. McAndrew (Eds.), *Tilapias: Biology and Exploitation* (pp. 327-375). London, UK: Kluwer Academic Publishers.

Jauncey, K., and B. Ross (1982). *A guide to tilapia feed and feeding*. Stirling, Scotland: Institute of Aquaculture, University of Stirling.

Jauncey, K., A.G.J. Tacon, and A.J. Jackson (1983). The quantitative essential amino acid requirements of *Oreochromis (=Sarotherodon) mossambicus*. In L. Fishelson and Z. Yaron (Eds.), *Proceedings, International Symposium on Tilapia in Aquaculture* (pp. 328-337), Tel Aviv, Israel: Tel Aviv University.

Kanazawa, A., S. Teshima, M. Sakamoto, and M.A. Awal (1980). Requirements of *Tilapia zillii* for essential fatty acids. *Bulletin of the Japanese Society of Scientific Fisheries* 46:1353-1356.

Kleeman, G.A., M.M. Barros, L.E. Pezzato, F.G. Sampaio, J.C. Ferrari, J.B. Valle, E.S. Freire, and J.A. Zuanon (2003). Iron requirement for Nile tilapia, *Oreochromis niloticus*. In Book of Abstracts, World Aquaculture, May 19-23, 2003 (p. 82). Salvador, Brazil. Baton Rouge, LA: World Aquaculture Society.

Kubaryk, J.M. (1980). Effect of Diet, Feeding Schedule and Sex on Food Consumption, Growth and Retention of Protein and Energy by Tilapia. Doctoral Dissertation, Auburn University, Auburn, AL.

Lee, K.J., and S.C. Bai (1997). Hemoglobin powder as a dietary animal protein source for juvenile Nile tilapia. *Progressive Fish-Culturist* 59:266-271.

Li, Z., W. Lei, J. Ye, and X. He (1991). The nutritional value of commercial feed ingredients for Nile tilapia (*Oreochromis* L.) in China. In S.S. De Silva (Ed.), *Fish*

Nutrition Research in Asia (pp. 101-106), Asian Fisheries Society Special Publication No. 5, Manila, Philippines.

Lim, C. (1989). Practical feeding- Tilapias. In T. Lovell (Ed.), *Nutrition and Feeding of Fish* (pp. 163-183) New York, NY: Van Nostrand Reinhold.

Lim, C., M.M. Barros, P.H. Klesius, and C.A. Shoemaker (2000). Thiamin requirement of Nile tilapia, *Oreochromis niloticus*. In Book of Abstracts, Aquaculture America 2000. 2-5 February 2000 (p. 201), New Orleans, LA. Baton Rouge, LA: World Aquaculture Society.

Lim, C., and W. Dominy (1989). Utilization of plant proteins by warmwater fish. In T.H. Applewhite (Ed.), *Proceedings of the World Congress on Vegetable Protein Utilization in Human Foods and Animal Feedstuffs* (pp. 245-251), Champaign, IL: American Oil Chemists' Society.

Lim, C., and P.H. Klesius (2001). Influence of dietary levels of folic acid on growth response and resistance of Nile tilapia *(Oreochromis niloticus)* to *Streptococcus iniae*. In Book of Abstract, 6th Asian Fisheries Forum, 25-30 November, 2001 (p. 150), Kaohsiung, Taiwan.

Lim, C., B. LeaMaster, and J.A. Brock (1991). Thiamin requirement of red hybrid tilapia grown in seawater. In Abstracts. World Aquaculture Society 22nd Annual Conference and Exposition, 16-20 June 1991 (p. 39) San Juan, Puerto Rico. Baton Rouge, LA: World Aquaculture Society.

———. (1993). Riboflavin requirement of fingerling red hybrid tilapia grown in seawater. *Journal of the World Aquaculture Society* 24:451-458.

———. (1995). Pyridoxine Requirement of fingerling red hybrid tilapia grown in seawater. *Journal of Applied Aquaculture* 5:49-60.

Lim, C., M. Yildirim, and P.H. Klesius (2002). Effect of substitution of cottonseed meal for soybean meal on growth, hematology and immune response of tilapia *(Oreochromis niloticus). Global Aquaculture Advocate* 5:28-32.

Lim, C., M. Yildirim, P.H. Klesius, and P.J. Wan (2003). Levels of dietary gossypol affect growth, bacterial resistance of Nile tilapia. *Global Aquaculture Advocate* 6:42-43.

Lovell, R.T, and T. Limsuwan (1982). Intestinal synthesis and dietary nonessentiality of vitamin B_{12} for *Tilapia nilotica*. *Transaction of the American Fisheries Society* 11:485-490.

Luquet, P. (1991). Tilapia, *Oreochromis* spp. In R.P. Wilson (Ed.), *Handbook of Nutrient Requirement of Finfish* (pp. 169-179). CRC Press, Boca Raton, FL.

Mazid, A.M., Y. Tanaka, T. Katayama, A.M. Rahman, K.L. Simpson, and C.O. Chichester (1979). Growth response of *Tilapia zillii* fingerlings fed isocaloric diets with variable protein levels. *Aquaculture* 18:115-122.

Mbahinzeriki, G.B., K. Dabrowshi, K.J. Lee, D. El-Saidy, and E.R. Wisner (2001). Growth, feed utilization and body composition of tilapia (*Oreochromis* sp.) feet cottonseed meal-based diets in a recirculating system. *Aquaculture Nutrition* 7:189-200.

Mertz, E.T. (1972). The protein and amino acid needs. In J.E. Halver (Ed.), *Fish Nutrition* (pp. 105-143). New York, NY: Academic Press.

Ng, W.K., and K.L. Wee (1989). The nutritive value of cassava leaf meal in pelleted feed for Nile tilapia. *Aquaculture* 83:45-58.

NRC (National Research Council) (1983). *Nutrient Requirements of Warmwater Fishes and Shellfishes.* Washington, D.C.: National Academy Press.

———. (1993). Nutrient requirements of fish. Washington, D.C.: National Academy Press.

O'Connell, J.P., and D.M. Gatlin (1994). Effects of dietary calcium and vitamin D_3 on weight gain and mineral composition of the blue tilapia *(Oreochromis aureus)* in-calcium water. *Aquaculture* 125:107-117.

Ofojekwu, P.C., and C. Ejike (1984). Growth response and feed utilization in the tropical cichlid *Oreochromis niloticus* (Linn.). Fed on cottonseed-based artificial diets. *Aquaculture* 42:27-36.

Ogino, C. (1980). Requirements of carp and rainbow trout for essential amino acids. *Bulletin of the Japanese Society of Scientific Fisheries* 46:171-174.

Olsen, R.E., R.J. Henderson, and B.J. McAndrew (1990). The Conversion of linoleic acid and linolenic acid to longer chain polyunsaturated fatty acids by tilapia *(Oreochromis* niloticus*) in vivo. Fish Physiology and Biochemistry* 8: 261-270.

Oyetayo, A.S., C.C. Thornburn, A.J. Matty, and A. Jackson (1985). Pyridoxine and survival of tilapia *(Sarotherodon mossambicus Peters).* In Proceedings of the 4th Annual Conference of the Fisheries Society of Nigeria (Fison), 26-29 November, 1985 (pp. 223-230). Port Harcourt, Nigeria.

Perres, H., C. Lim, and P.H. Klesius (2004). Growth, chemical composition and resistance to *Streptococcus iniae* challenge of Nile tilapia *(Oreochromis niloticus)* fed graded levels of dietary inositol. *Aquaculture* 235:423-432.

Popma, T.J. (1982). Digestibility of Selected Feedstuffs and Naturally Occurring Algae by Tilapia. Doctoral Dissertation, Auburn University, Auburn, AL.

Reigh, R.C., E.H. Robinson, and P.B. Brown (1991). Effects of dietary magnesium on growth and tissue magnesium content of blue tilapia, *Oreochromis aureus. Journal of the World. Aquaculture Society* 22:192-200.

Robinson, E.H., D. LaBomascus, P.B. Brown, and T.L. Linton (1987). Dietary calcium and phosphorus requirements of *Oreochromis aureus* reared in calcium-free water. *Aquaculture* 64:267-276.

Robinson, E.H., S.D. Rawles, P.W. Oldenburg, and R.R. Stickney (1984a). Effects of feeding glandless or glanded cottonseed products and gossypol to *Tilapia aurea. Aquaculture* 38:145-154.

Robinson, E.H., S.D. Rawles, H.E. Yette, and L.W. Greene (1984b). An estimate of the dietary calcium requirement of fingerling *Tilapia aureus* reared in calcium-free water. *Aquaculture* 41:389-393.

Roem, A.J., C.C. Kohler, and R.R. Stickney (1990a). Inability to detect a choline requirements for the blue tilapia, *Oreochromis aureus. Journal of the World Aquaculture Society* 21:238-240.

———. (1990b). Vitamin E requirements of the blue tilapia, *Oreochromis aureus* (Steindachner), in relation to dietary lipid level. *Aquaculture* 87:115-164.

Roem, A.J., R.R. Stickney, and C.C. Kohler (1990). Vitamin requirements of blue tilapias in a recirculating water system. *Progressive Fish-Culturist* 52:15-18.

Saleh, G., W. Eleraky, and J.M. Gropp (1995). A short note on the effects of vitamin A hypervitaminosis and hypovitaminosis on health and growth of Tilapia nilotica *(Oreochromis niloticus). Journal of Applied Ichthyology* 11:382-385.

Santiago, C.B., M.B. Aldaba, M.A. Laron, and O.S. Reyes (1988a). Reproductive performance and growth of Nile tilapia *(Oreochromis niloticus)* broodstock fed diets containing *Leucaena leucocephala* leaf meal. *Aquaculture* 70:53-61.

Santiago, C.B., M.B. Aldaba, O.S. Reyes, and M.A. Laron (1988b). Response of Nile tilapia *(Oreochromis niloticus)* fry to diets containing azolla meal. In R.S.V. Pullin, T. Bhukaswan, K. Tonguthai and J.L. Maclean (Eds.), *The Second International Symposium on Tilapia in Aquaculture*. ICLARM Conference Proceedings 15 (pp. 341-345), Department of Fisheries, Bangkok, Thailand, and International Center for Living Aquatic Resources Management, Manila, Philippines.

Santiago, C.B., and M.A. Laron (1991). Growth response and carcass composition of red tilapia fry fed diets with varying protein levels and protein to energy ratio. In S.S. De Silva (Ed.), *Fish Nutrition Research in Asia*. Asian Special Publication No. 5 (pp. 55-62), Manila, Philippines.

Santiago, C.B., and R.T. Lovell (1988). Amino acid requirements for growth of Nile tilapia. *Journal of Nutrition* 118:1540-1546.

Santiago C.B., and O.S. Reyes (1993). Effects of dietary lipid source on reproductive performance and tissue lipid levels of Nile tilapia *Oreochromis niloticus* (Linnaeus) broodstock. *Journal of Applied Ichthyology* 9:33-40.

Satoh, S., T. Takeuchi, and T. Watanabe (1987). Requirement of tilapia for alpha tocopherol. *Nippon Suisan Gakkaishi* 53:119-124.

Shiau, S.Y., and Y.H. Chin (1999). Estimation of the dietary biotin requirement of juvenile hybrid tilapia, *Oreochromis niloticus* × *O. aureus*. *Aquaculture* 170:71-78.

Shiau, S.Y., J.L. Chuang, and C.L. Sun (1987). Inclusion of soybean meal in tilapia *(Oreochromis niloticus* × *O. aureus)* diets at two protein levels. *Aquaculture* 65:251-261.

Shiau, S.Y., and H.L. Hsieh (1997). Vitamin B sub(6) requirements of tilapia *Oreochromis niloticus* × *O. aureus* fed two dietary protein concentrations. *Fisheries Science* 63:1002-1007.

Shiau, S.Y., and J.F. Hsieh (2001). Quantifying the dietary potassium requirement of juvenile hybrid tilapia *(Oreochromis niloticus* × *O. aureus). British Journal of Nutrition* 85:213-218.

Shiau, S.Y., and T.S. Hsu (1995). L-ascorbyl-2-sulfate has equal antiscorbutic activity as L-ascorbyl-2-monophosphate for tilapia, *Oreochromis niloticus* × *O. aureus*. *Aquaculture* 133:147-157.

———. (1999). Quantification of vitamin C requirement for juvenile hybrid tilapia, Oreoxhromis niloticus × *Oreochromis aureus*, with L-ascorbyl-2-monophosphate-Na and L-ascorbyl-2-monophosphate-Mg. *Aquaculture* 175:317-326.

Shiau, S.Y., and S.L. Huang (1989). Optimal dietary protein level for hybrid tilapia *(Oreochromis niloticus* × *O. aureus)* reared in seawater. *Aquaculture* 81:119-127.

———. (1990). Influence of varying energy levels with two protein concentrations in diets for hybrid tilapia *(Oreochromis niloticus × O. aureus)* reared in seawater. *Aquaculture* 91:143-152.

Shiau, S.Y., and J.Y. Hwang (1993). Vitamin D requirements of juvenile hybrid tilapia, *Oreochromis niloticus × O. aureus. Nippon Suisan Gakkaishi* 59:553-558.

Shiau, S.Y., and F.L. Jan (1992). Dietary ascorbic acid requirement of juvenile tilapia *Oreochromis niloticus × O. aureus. Nippon Suisan Gakkaishi* 58:671-675.

Shiau, S.Y., and P.S. Lo (2000). Dietary choline requirements of juvenile hybrid tilapia, *Oreochromis niloticus × O. aureus. Journal of Nutrition* 130:100-103.

Shiau, S.Y., and C.Q. Lung (1993). No dietary vitamin B_{12} required for juvenile tilapia *Oreochromis niloticus × O. aureus. Comparative Biochemistry and Physiology* 105:147-150.

Shiau, S.Y., and C.Y. Peng (1993). Protein-sparing effect by carbohydrates in diets for tilapia, *Oreochromis niloticus × O. aureus. Aquaculture* 117:327-334.

Shiau, S.Y., and L.F. Shiau (2001). Re-evaluation of the vitamin E requirements of juvenile tilapia *(Oreochromis niloticus × O. aureus). Animal Science* 72:529-534.

Shiau, S.Y., and L.W. Su (2003). Ferric sulfate is half effective as ferrous sulfate in meeting iron requirement for juvenile tilapia, *Oreochromis niloticus + O. aureus.* In Book of Abstracts, Wold Aquaculture, May 19-23, 2003 (p. 719). Salvador, Brazil. Baton Rouge, LA: World Aquaculture Society.

Shiau, S.V., and G.S. Suen (1992). Estimation of the niacin requirements for tilapia fed diets containing glucose or dextrin. *Journal of Nutrition* 122:2030-2036.

Shimeno, S., D.C. Ming, and M. Takeda (1993). Metabolic response to dietary carbohydrate to lipid ratios in *Oreochromis niloticus. Bulletin of the Japanese Society of Scientific Fisheries* 59:827-833.

Siddiqui, A.Q., M.S. Howlander, and A.A. Adam (1988). Effects of dietary protein levels on growth, feed conversion and protein utilization in fry and young Nile tilapia, *Oreochromis niloticus. Aquaculture* 70:63-73.

Soliman, A.K., and R.P. Wilson (1992). Water-soluble vitamin requirements of tilapia.2. Riboflavin requirement of blue tilapia, *Oreochromis aureus. Aquaculture* 104:309-314.

Stickney, R., and R.W. Hardy (1989). Lipid requirements of some warmwater species. *Aquaculture* 79:145-156.

Stickney, R.R., and R.B. McGeachin (1983). Responses of *Tilapia aurea* to semi-purified diets of differing fatty acid composition. In L. Fishelson and Z. Yaron (Eds.), *Proceedings, International Symposium on Tilapia in Aquaculture* (pp. 346-355), Tel Aviv, Israel: Tel Aviv University.

Stickney, R.R., R.B. McGeachin, D.H. Lewis, J. Marks, R.S. Sis, E.H. Robin, and W. Wurts (1984). Response of *Tilapia aurea* to dietary vitamin C. *Journal of the World Aquaculture Society* 15:179-185.

Stickney, R.R., and W.A. Wurts (1986). Growth response of blue tilapias to selected levels of dietary menhaden and catfish oils. *Progressive Fish-Culturist* 48:107-109.

Sugita, H., C. Miyajima, and Y. Deguchi (1990). The vitamin B_{12}-producing ability of intestinal bacteria isolated from tilapia and channel catfish. *Nippon Suisan Gakkaishi* 56:701.

Sunita Rao, D., and N. Raghuramulu (1996). Lack of vitamin D sub(3) synthesis in *Tilapia mossambica* from cholesterol and acetate. *Comparative Biochemistry and Physiology* 114A:21-25.

Tacon, A.G.J., K. Jauncey, A. Falaye, M. Pentah, I. MacGowen, and E. Stafford (1983). The use of meat and bone meal and hydrolyzed feather meal and soybean meal in practical fry and fingerling diets for *Oreochromis niloticus*. In J. Fishelton and Z. Yaron (Eds.), *Proceedings, First International Symposium on Tilapia in Aquaculture* (pp. 356-365), Israel: Tel Aviv University Press.

Takeuchi, T., S. Satoh, and W. Watanabe (1983a). Dietary lipid suitable for practical feed of *Tilapia nilotica*. *Bulletin of the Japanese Society of Scientific Fisheries* 49:1361-1365.

———. (1983b). Requirement of *Tilapia nilotica* for essential fatty acids. *Bulletin of the Japanese Society of Scientific Fisheries* 49:1127-1134.

Teshima, S., G. Gonzalez, and A. Kanazawa (1978). Nutritional requirements of tilapia: Utilization of dietary protein by *Tilapia zillii*. Memoirs of the Faculty of Fisheries, Kagoshima University 27:49-57.

Viola, S., and Y. Arieli (1983). Nutrition studies with tilapia *(Sarotherodon)*. 1- Replacement of fish meal by soybean meal in feeds for intensive tilapia culture. *Israeli Journal of Aquaculture (Bamidgeh)* 35:9-17.

Viola, S., and G. Zohar (1984). Nutritional study with market size tilapia hybrid *Oreochomis* in intensive culture. Protein levels and sources. *Israeli Journal of Aquaculture (Bamidgeh)* 36:3-15.

Viola, S., G. Zohar, and Y. Arieli (1986). Phosphorus requirements and its availability from different sources for intensive pond culture species in Israel. Part 1. Tilapia. *Bamidgeh* 38:3-12.

Wang, R., T. Takeuchi, and T. Watanabe (1985). Effect of dietary protein levels on growth of *Tilapia nilotica*. *Bulletin of the Japanese Society of Scientific Fisheries* 51:133-140.

Watanabe, W.O., L.J. Ellingson, R.I. Wiklum, and B.L. Olla (1988a). The effects of salinity on growth, food consumption and conversion in juvenile, monosex male Florida red tilaipa. In R.S.V. Pullin, T. Bhukaswan, K. Tonguthai and J.L. Maclean (Eds.), *The Second International Symposium on Tilapia in Aquaculture* (pp. 515-523), ICLARM Conference Proceedings 15, Department of Fisheries, Bangkok, Thailand, and International Center for Living Aquatic Resources Management, Manila, Philippines.

Watanabe, T., S. Satoh, and T. Takeuchi (1988b). Availability of minerals in fish meal to fish. *Asian Fisheries Science* 1:75-195.

Wee, K.L., and L.T. Ng (1986). Use of cassava as an energy source in a pelleted feed for the tilapia, *Oreochromis niloticus* L. *Aquaculture and Fisheries Management* 17:129-138.

Wee, K.L., and S.S. Wang (1987). Nutritive value of *Leucaena* leaf meal in pelleted feed for Nile tilapia. *Aquaculture* 62:97-108.

Winfree, R.A., and R.R. Stickney (1981). Effect of dietary protein and energy on growth, feed conversion efficiency and body composition of *Tilapia aurea*. *Journal of Nutrition* 111:1001-1012.

Wu, Y.V., R. Rosati, and P. Brown (1996). Effects of diets containing various levels of protein and ethanol co-products from corn on growth of tilapia fry. *Journal of Agricultural and Food Chemistry* 44:1491-1493.

Wu, Y.V., R. Rosati, D.J. Sessa, and P. Brown (1994). Utilization of protein-rich ethanol co-products from corn in tilapia feed. *Journal of American Oil Chemists Society* 71:1041-1043.

Wu, Y.V., K.Y. Tudor, P. Brown, and R.R. Rosati (1999). Substitution of plant protein and meat and bone meal for fish meal in diets for Nile tilapia. *North American Journal of Aquaculture* 61:58-63.

Chapter 13

Nonnutrient Components of Fish Diets

Steven G. Hughes
Chhorn E. Lim
Carl D. Webster

INTRODUCTION

Though the various nutrients are the primary concerns of nutritionists when formulating diets for intensively cultured tilapia, the inclusion or exclusion of other dietary components, which do not have nutritional value, can have profound impacts on the performance of the fish fed these diets. These components may be added to address physiological or health concerns of the aquaculturist (e.g., hormones, other modifiers of metabolism, and antibiotics), items to improve pellet quality (fiber and pellet binders), or to lengthen the shelf-life of the diet (antioxidants). Though the general concept is that these components are diet additives, many are naturally occurring parts of common feedstuffs and, therefore, do not appear in many diet formulation programs.

ANTIOXIDANTS

Antioxidants, either as natural components of feedstuffs or as discrete chemical compounds, are routinely added to all aquatic animal diets to prevent the oxidation of dietary lipids. Though little research has been conducted in this area using tilapia, free radicals, resulting from the oxidation of dietary lipids, have been associated with decreasing the availability of several amino acids, particularly lysine; with lipoid degeneration of the liver; and with hematological abnormalities in other fish species (Murai and Andrews 1974; Smith 1979; Tacon 1985). Data have also been published that would indicate younger fish to be more sensitive to oxidized lipids than older individuals (Hung and Slinger et al. 1980; Murai et al. 1988; Ketola et al. 1989).

doi:10.1300/5513_13

Though a wide variety of nutrients have antioxidant properties, the primary natural antioxidant sources in most feedstuffs are represented by vitamin E, selenium, ascorbic acid (vitamin C), and glutathione. It has been reported that vitamin E stores are definitely affected by the level of lipid in the diet (Satoh et al. 1987). Although the results of that study indicate a strong linkage between dietary levels of lipid and vitamin E, it should be remembered that it is only the alcohol form, not the acetate form, of vitamin E that acts as an antioxidant. Therefore, the vitamin E acetate that is normally added as a supplement to the diet does little or nothing to prevent lipid oxidation until it is transformed to the alcohol form in the fish's gut. For this reason, synthetic antioxidants are usually added to the feedstuffs or the diet to prevent dietary oils from going rancid during storage.

Ethoxyquin, BHT (butylated hydroxtoluene), and BHA (butylated hydroxyanisole) have been approved by the Food and Drug Administration (FDA) (21 C. F. R. 573.380, 582.3169, 582.3173 [1987]) for use in fish diets. BHA and BHT are limited to 0.02 percent of the fat content of the diet or feedstuff, and ethoxyquin can be used at levels up to 150 $mg \cdot kg^{-1}$ of diet. All three compounds are effective in protecting dietary lipids from oxidation, but no research has been reported on any adverse effects of feeding these compounds to tilapia.

FEEDING STIMULANTS AND METABOLIC MODIFIERS

The selection of food items by fish is mitigated by the chemosensitivity of the fish to certain chemical cues and the presence (or absence) of these chemical constituents or cues in the food item or environment. The primary modes of food detection by fish are through either olfaction or sight, but the taste of the item is the key factor in determining whether or not the item is swallowed or rejected once it has entered the mouth (Adron and Mackie 1978; Mackie and Adron 1978). There appears to be a well-defined and species-specific "tuning" of the taste receptors of fish for the particular cues present in their food items (Goh and Tamura 1980). Carr (1982) identified four major characteristics of feeding stimulants for fish that were derived from animal tissues:

1. they have a low molecular weight (<1,000);
2. they contain nitrogen;
3. they are nonvolatile and water soluble;
4. they are amphoteric (i.e., having both acid and base properties simultaneously).

Several substances, or groups of substances, for which these generalizations apply (i.e., amino acids, betaine, and inosine) have been shown to effectively alter feeding behavior in both carnivorous and omnivorous species (as reviewed by Atema 1980; Carr 1982; Mackie 1982; Adams and Johnsen 1986a; Rumsey 1986) when presented either individually or in combination.

Little data exists on feeding stimulants for herbivorous species such as tilapia, but in four studies using red belly tilapia, *Tilapia zilli* (Adams and Johnsen 1986a,b; Johnsen and Adams 1986; Adams et al. 1988), it was found that, along with certain acidic amino acids (e.g., alanine, aspartic acid, and glutamic acid) and lysine, organic acids were the most stimulatory. This is very much in line with generalizations on trophic level chemosensitivities that assume that the fish respond most strongly to the compounds that are found in the greatest amounts in their primary food items (Mackie 1982; Adams and Johnsen 1986a).

When the data on the effectiveness of the various ammonia-containing compounds for stimulating feeding are taken into consideration, a pattern emerges, relating the trophic level with which general classes of compounds are stimulatory. In general, carnivores show the greatest positive response to alkaline or neutral substances (e.g., glycine, proline, taurine, valine, and betaine), whereas herbivorous fish respond to acidic substances. This pattern is reasonable when one considers that the fish are responding best to the substances that are in the highest concentrations in the food items that they would be seeking in the natural environment.

In addition to enhancing feeding, the addition or presence of certain compounds can also act as feeding deterrents. Though this has been shown to occur in carnivorous fish with certain combinations of amino acids or rancid oils (reviewed in Hughes 1991), these circumstances have not been identified as occurring in tilapia and are not viewed as having practical significance. The presence of aflatoxins in the diet has been shown to reduce diet consumption (Chavez-Sanchez et al. 1994). It is not completely clear, however, as to whether this was due to the impact of aflatoxin on olfaction and/or gustation, or if the reduced diet consumption was a secondary effect of the toxin on the physiology of the fish.

Few studies, other than those involving hormones (see below), have been published wherein the addition of metabolic modifiers have been incorporated into the diets of tilapia. Carnitine, a promoter of fatty acid oxidation, has been added to tilapia diets with mixed results. Huang et al. (1998) found that the addition of carnitine at levels up to 3,000 $mg{\cdot}kg^{-1}$ to diets of juvenile (21 g) hybrid tilapia (*Oreochromis niloticus* × *O. aureus*) had no direct impact on the growth, feed conversion efficiency, or fillet protein and lipid content. Liver lipid levels were reduced in fish fed a high (6 percent) lipid diet, but the impact was not seen as having any economic significance to tilapia producers. Kumar and Jayaprakas (1996), however, found that 900 $mg \cdot kg^{-1}$ of carnitine added to the diet of male *(O. mossambicus)* fry (2.0 to 2.6 g) was very beneficial and led to significant improvements in weight gain, feed efficiency, digestibility, and carcass composition. Both studies attribute their results to modifications in lipid metabolism, but whether the profound differences in the findings are due to species, size, or gender differences is not clear.

FIBER AND PELLET BINDERS

When one discusses pelleted diets, it is realized that fiber serves both as a binder or structural component that gives stability to the pellet and as a carbohydrate source that may have some nutritional value for tilapia or may act as a diluent for other nutrients in the diets. Because of a predominantly herbivorous diet in nature, it is commonly believed that the various species of tilapia could utilize large amounts of dietary fiber as an energy source. However, the opposite appears to be true. The common conclusion of a number of studies (Teshima et al. 1987; Shiau, Kwok et al. 1988; Shiau, Yu et al. 1988; Shiau and Kwok 1989; Yong et al. 1989; Dioundick and Stom 1990) is that the optimum level of dietary fiber and/or other poorly digested complex carbohydrates (e.g., guar gums, carrageenans, chitin, etc.) is between 2 and 5 percent of the diet. This does make sense when it is taken into account that aquatic vegetation is higher in water content and lower in structural components (e.g., cellulose fiber) than its terrestrial counterparts. As with other vertebrates, high levels of dietary fiber have also been shown to decrease diet intake (Yong et al. 1989), and the digestibility and absorption of other dietary nutrients (Shiau, Kwok et al. 1988).

Under practical conditions, the primary source of fiber and poorly digested complex carbohydrates in tilapia diets comes from plant-based feedstuffs. The differences in quantity and quality of these substances can have profound effects, both on the overall digestibility of the diet and on the water stability of the manufactured pellet. Viola et al. (1985) found that pellet stability and tilapia growth was improved by replacing wheat with sorghum and then using steam to heat the mixed diet during pelleting. LaVorgna (1998), however, reported no difference in either digestibility by, or growth of, hybrid tilapia presented with the same plant-based diet that had either been cold-pelleted or stream-extruded. Additional research is needed in this area especially when a wide variety of milling methods used to manufacture tilapia diets is considered.

Pelleted fish diets must remain intact during processing, shipping, storage, and feeding. Since fish are fed in an aquatic environment, it is vital that the diet does not disintegrate before the fish has time to consume it. Generally, this is not a problem since tilapia, as well as most other finfish, are aggressive feeders and readily consume prepared diets. However, it is important that diets, especially sinking pellets, be water stable. Use of pellet binders can accomplish this goal. Commonly used binders include bentonite, lignin sulfonate, and hemicellulose. All three do not add any nutritional value to the diet. Bentonite is a naturally occurring clay and should be added at 2 to 3 percent of the diet. Lignin sulfonate is a product of the wood pulp industry and can be added at levels up to 4 to 5 percent of the diet. Hemicellulose is not often used as a binder and is a by-product of the pressed wood manufacturing process.

HORMONES

Numerous natural and synthetic hormones have been incorporated into fish diets in an attempt to regulate growth, feed conversion, and reproduction. Prolonged feeding of androgens and steroids for accelerated growth, however, has been shown to cause numerous physiological and physical abnormalities (Zohar 1989; Gannam and Lovell 1991a,b).

The use of dietary hormones to alter the gender of tilapia is discussed in detail in Chapter 6 "Hormone Manipulation of Sex" in this book. Thus, the following represents only a brief overview of that information. Hormone supplemented diets have been used to create either all-male or all-female population of tilapia. Studies on the use of natural and synthetic androgens, however, have shown that only those with chemical modifications at the 17-alpha position allow for the compound to make it through the digestive tract without being catabolized.

Probably the most commonly added hormone in tilapia diets is 17-alpha-methyltestosterone (MT) that is currently being used by selected producers who are participating in an Investigational New Animal Drug (INAD) program. MT is added to the diet of very young fry at a rate of 60 $mg \cdot kg^{-1}$ of diet to create 100 percent male populations for the purpose of reducing growth losses related to reproduction (Ridha and Lone 1990; Phelps and Popma 2000; Teichert-Coddington et al. 2000). Reports of increased growth in Nile tilapia are mixed with some studies indicating a significant increase in the growth rate of MT-treated fry (e.g., Sikoki and Ekwu 2000) and others stating no increase in growth rate (e.g., Green and Teichert-Coddington 1994). An extensive study with Mozambique tilapia indicated that any factor that negatively affects diet intake decreases the effectiveness of MT-treated diets and that high temperatures (38°C) may actually lead to MT causing feminization rather than masculinization (Varadaraj et al. 1994).

It was postulated that this was due to the aromatization of excess androgen to estrogen, or the combination of the process with temperature-related inhibition of the *in vivo* production of androgen.

Successful use of the yeast *Saccharomyces cerevisiae* to carry recombinant rainbow trout *(Oncorhynchus mykiss)* growth hormone has been reported and significant increases in tilapia growth were obtained (Tsai et al. 1993). Recent public opinion on the use of genetically modified organisms, however, has made the probability of using compounds either produced by or delivered by these sources less than economically feasible.

PIGMENTS

One of the most important group of natural pigments is the carotenoids that is used by fish and birds to impart color to flesh, skin, and eggs. It is generally regarded that since fish cannot synthesize these pigments, they must be provided in the diet (NRC 1993). Aside from coloration, the primary function of carotenoids seems to be as a precursor for vitamin A (Schiedt et al. 1985), but a number of other uses, ranging from its value as an antioxidant to metabolic functions in developing eggs, have been postulated (Tacon 1981).

Reviews of the literature on the absorption, metabolism, deposition, and excretion of carotenoids by fish indicate that there are wide differences in the utilization of the various carotenoids and that the primary limiting factors on absorption and deposition are the form of the carotenoid and the species of fish (Torrissen et al. 1989; NRC 1993). Most of the research has involved astaxanthin and canthaxanthin (derived predominantly from animal or synthetic sources) that are primary pigments for salmonid coloration and that impart a red or orange color to flesh and eggs (NRC 1993). Major plant-based carotenoids, such as lutein and zeaxanthin, have also received some interest, but their tendency to impart a yellow color to fish flesh has made them less desirable (Lee 1987).

Zeaxanthin is metabolized to astaxanthin by goldfish and red carp and is used to provide the red colors seen in these fish (Hata and Hata 1972, 1973, 1976). These fish also effectively utilize zeaxanthin giving evidence that a plant-based pigment is a natural path for attaining carotenoid pigments for these species. In the one reported study where different pigment sources were fed to red tilapia, the xanthophylls from *Spirulina* and marigold petals were readily converted to red pigments (presumably astaxanthin) that gave higher intensities than that derived from the astaxanthin present in shrimp waste meal. Once again, this indicates that herbivorous fish metabolize plant-based pigments effectively (Boonyaratpalin and Unprasert 1989).

PROBIOTICS AND NEUTRACEUTICALS

Probiotics are live microorganisms, such as bacteria and yeast, administered in the form of an injection, a water treatment, or a supplement to artificial diet or live food that are thought to stimulate and enhance growth and health by altering the microbial population of the fish's digestive tract. Probiotics can be either a single species or a mixture of species that could outcompete pathogenic microorganisms in the digestive tract, compete for nutrients and space in the mucosal lining, and stimulate the host immunity thereby improving disease resistance in fish. This could also allow the fish to utilize metabolic energy on growth, instead of having to direct energy to fight the effects of detrimental microorganisms.

Neutraceuticals are nonspecific immune stimulants and unregulated dietary additives that might enhance the health of the fish by making the fish less susceptible to infectious diseases. Beta-glucans are the most common of these neutriceuticals, which are fragments of the cell walls of yeast and mycelial fungi. Glucans have been reported to be sometimes effective while other times not (see review by Gannam and Schrock 2001). There are still many unanswered questions regarding the use of beta-glucans including effective dose, chemical form, administration route, feeding duration, and size of fish that can be treated effectively.

ENZYMES

Enzyme supplements are added to diets to enhance the digestion of one or more dietary components that the fish cannot digest at all or cannot digest efficiently. Enzymes are generally denatured at temperatures above 65°C so they must be added after the diet has been made (sprayed onto the diet). Phytase is an example of an enzyme supplement that is widely used in poultry and hog diets, and it is being considered more and more for use in aquaculture diets. Fish cannot effectively utilize the storage form of phosphorous (phytate) in plant feedstuffs (grains and oilseeds), but fish require phosphorus for proper growth and health. Phytase makes the phosphorus in phytate available to the fish so that it can be utilized. This reduces the amount of supplemental phosphorus that has to be added to the diet, which in turn, reduces the amount of phosphorus that enters the culture environment.

WATER

Water (moisture) is found in diets both as an additive and as a natural component of feedstuffs. Moisture may be added to facilitate pelleting or extrusion, but the resulting diet is then dried to bring the moisture level back to the ambient levels (about 10 percent). Diets of low moisture content are readily consumed by tilapia at various live stages.

Research on the role of varying moisture levels in diets for tilapia has not been reported to date. However, the excellent performance of all life-stages of tilapia on both standard pelleted and extruded diets that contain 8 to 12 percent moisture would seem to indicate that these fish do not require high levels of dietary moisture to efficiently utilize dietary nutrients. Though the addition of moisture to the diet has been shown to impact the consistency of the food bolus in the stomach of certain carnivorous fishes (Hughes and Barrows 1990), it is assumed that the gastrointestinal physiology of tilapia and other herbivores is such that they are able to impart sufficient moisture to their digesta to allow for efficient function of the digestive enzymes (Steffens 1989). Further, maintaining high percentages of moisture in the diets could lead to more rapid spoilage of the diet due to microbial and/or fungal growth.

TOXINS AND ANTINUTRITIONAL FACTORS

Microbial toxins (mycotoxins) and other toxic substances, if present in the diets in sufficient concentrations, can severely affect the growth and health of fish. Mycotoxins are produced from various genera of molds, such as *Asperigillus, Penicillium,* and *Fusarium.* While mycotoxins are generally produced in the plant ingredient crops prior to their harvest, they can be produced in the finished feeds if the diets are improperly stored (damp, humid, and warm conditions), or the diets have a high moisture content (>15 percent) and are stored in a hot, humid environment. Diet manufacturers should avoid using ingredients that are suspected to have any trace of mycotoxins since steam pelleting or extrusion processing does not destroy most of the mold toxins. Use of antimold agents in diets will reduce the potential threat of harmful molds that might grow on the diet after processing. Propionic acid at a level of 0.25 percent of the diet may be used to inhibit the growth of molds (NRC 1977).

In the United States, aflatoxins are of the greatest concern due to their prevalence (especially in southeastern United States) and their toxicity. Mortality, reduced growth, and liver damage are some of the adverse effects caused by the consumption of aflatoxins. Corn, cottonseed, and peanut meals are three ingredients that are most likely to be contaminated with aflatoxins. Generally, coldwater fish are more sensitive to aflatoxins than warmwater fish. The dose of aflatoxin B_1 that causes death in 50 percent (LD50) in rainbow trout *(O. mykiss)* is between 0.5 and 1.0 $mg \cdot kg^{-1}$ of diet. However, Jantrarotai and Lovell (1990) were unable to determine an LD50 for channel catfish *(Ictalarus punctatus)* fed various levels of alfatoxin B_1 because the fish regurgitated their diet when aflatoxin levels were too high. Intraperitoneal LD50 was found to be 11.5 $mg \cdot kg^{-1}$ of body weight. In Nile tilapia, the growth performance and hematocrit were not affected if dietary concentrations of aflatoxin B_1 are 0.25 $mg \cdot kg^{-1}$ or less. A dietary aflatoxin B_1 of 10 $mg \cdot kg^{-1}$ significantly reduced weight gain and hematocrit and caused liver histopathological abnormality, such as lipofuscin and irregular-sized hepatocellular nuclei (Tuan et al. 2002).

Fusarium toxins that are most detrimental to fish health are zearalenones, tricothecenes, vomitoxin, fumonsins, and moniliformin. Fumonsin B_1, produced from the mold *Fusarium moniliforme,* is very toxic and is found primarily in corn and corn screenings. Two toxins produced from various species of fungi from the genea *Aspergillus* and *Penicillium* are cyclopiazonic acid (CPA) and ochratoxins. Cyclopiazonic acid is more toxic to channel catfish than aflatoxin B_1 with an LD50 of 2.8 $mg \cdot kg^{-1}$ of body weight, compared to 11.5 $mg \cdot kg^{-1}$ of body weight for aflatoxin B_1. Ochratoxins are not generally considered a major cause of problems to fish health; however, they can cause kidney and liver necrosis, and if consumed by fish, can result in death. The LD50 for rainbow trout is 4.7 mg ochratoxin per 1 kg of body weight.

There are other substances that might not be as toxic to fish as mycotoxins; however, their presence in a diet can result in reduced growth, internal organ damage, and possibly death. Two compounds that can be toxic are histamine and gizzerosine. Both substances are found in fish meals that are produced when fish meal is improperly produced or stored. Histamine is produced from autolysis or actions of microbial enzymes on the amino acid histidine when fish are improperly stored prior to the production of fish meal. Gizzerosine is produced when fish meal is heated at too high a temperature resulting in a reaction between free histidine and some protein side chain. Both substances can result in reduced growth in fish.

Two antinutritional factors present in cottonseeds are gossypol and cyclopropenoic (fatty) acids. Free gossypol can decrease growth rate in fish at levels as low as 0.01 percent; however, the affected level will be dependent upon fish species and fish size. Nile and blue tilapia juvenile have been fed diets containing up to 0.16 and 0.18 percent gossypol per kg diet from gossypol-acetic acid, respectively without any observed effects (Lovell 1998; Lim et al. 2003). Lysine content in cottonseed meal that is available for use by fish is inversely related to the level of free gossypol. This is due to the fact that during processing into cottonseed meal, some free gossypol in whole cottonseed is bound to the protein, primarily on the epsilon carbon of lysine, rendering it nontoxic and reducing the bioavailability of lysine. Cyclopropenoic fatty acids can increase cholesterol levels, increase saturated fatty acid levels, and result in liver damage; however, there does not seem to be any specific problems associated with feeding fish diets containing cottonseed meal attributable to cyclopropenoic acids.

Oxidized fish oil, thiaminase, and protease inhibitors (anitproteases) are three other groups of antinutritional factors that generally are not of importance in most commercial fish diets, but should be avoided. Use of rancid marine fish oils can result in reduced growth, liver damage, and anemia in fish. Use of antioxidants in the diet can reduce the effects of oxidized lipids or prevent oxidation from occurring. Protease inhibitors are generally found in unheated or insufficiently-heated legume seed meals, such as soybean meal. These globulin proteins combine with, and render inactive, the proteolytic enzymes trypsin and chymotrypsin. Heat treatment of soybean meal, or other legume seed meal where antiproteases are present, reduces or eliminates the activity of protease inhibitors. Thiaminase, a heat-sensitive enzyme is present in raw fish tissues that can reduce the level of the vitamin B_1 (thiamine) in the diet. Raw fish flesh should not be added to a prepared diet.

REFERENCES

Adams, M.A., and P.B. Johnsen (1986a). Chemical control of feeding in herbivorous and carnivorous fish. In D. Duvall, D. Muller-Schwarze, and R. Silverstein (Eds.), *Chemical signals in vertebrates 4* (pp. 45-61). New York: Plenum Publishing Corporation.

———. (1986b). A solid matrix bioassay for determining chemical feeding stimulants. *Progressive Fish-Culturist* 48:147-149.

Adams, M.A., P.B. Johnsen, and Z. Hong-Qi (1988). Chemical enhancement of feeding for the herbivorous fish *Tilapia zilla. Aquaculture* 72:95-107.

Adron, J.W., and A.M. Mackie (1978). Studies on the chemical nature of feeding stimulants for rainbow trout, *Salmo gairdneri* Richardson. *Journal of Fish Biology* 12:303-310.

Atema, J. (1980). Chemical senses, chemical signals, and feeding behavior in fishes. In J.E. Bardach, J.J. Magnuson, R.C. May, and J.M. Reinhart (Eds.), *Fish behavior and its use in the capture and culture of fishes* (pp. 57-101). Manila, Philippines: International Center for Living Aquatic Resources Management.

Boonyaratpalin, M., and N. Unprasert (1989). Effects of pigments from different sources on color changes and growth of red tilapia, *Oreochromis niloticus. Aquaculture* 79:375-380.

Chavez-Sanchez, M.C., C.A.M. Palacios, and I.O. Moreno (1994). Pathological effects of feeding young *Oreochromis niloticus* diets supplemented with different levels of aflatoxin B1. *Aquaculture* 127:49-60.

Carr, W.E.S. (1982). Chemical stimulation of feeding behavior. In T.J. Hara (Ed.), *Chemoreception in fishes* (pp. 259-273). Amsterdam, the Netherlands: Elsevier Scientific Publishing Company.

Dioundick, O.B., and D.I. Stom (1980). Effects of dietary α-cellulose levels in the juvenile tilapia. *Oreochromis mossambicus. Aquaculture* 91:311-315.

Gannam, A.L., and R. T. Lovell (1991a). Effects of feeding 17α-methyltestosterone, 11-ketosterone, 17β-estradiol, and 3, 5,3′-griiodothyronine to channel catfish, *Ictalurus punctatus. Aquaculture* 92:377-388.

———. (1991b). Growth and bone development in channel catfish fed 17α-methyltestosterone in production ponds. *Journal of the World Aquaculture Society* 22:95-100.

Gannam, A.L., and R.M. Schrock (2001). Immunostimulants in fish diets. In C. Lim, and C.D. Webster (Eds.), *Nutrition and fish health* (pp. 235-266). Binghamton, NY: The Haworth Press.

Goh, Y. and T. Tamura (1980). Effect of amino acids on the feeding behavior of red sea bream. *Comparative Biochemistry and Physiology* 66C: 225-229.

Green, B.W., and D.R. Teichert-Coddington (1994). Growth of control and androgen-treated Nile tilapia, *Oreochromis niloticus* (L.), during treatment, nursery and grow-out phases in tropical fish ponds. *Aquaculture and Fisheries Management* 25:613-621.

Hata, M., and M. Hata (1972a). Carotenoid pigments in goldfish. IV. Carotenoid metabolism. *Bulletin of the Japanese Society of Scientific Fisheries* 38:331-338.

———. (1972b). Carotenoid pigments in goldfish. V. Conversion of zeaxanthin to astaxanthin. *Bulletin of the Japanese Society of Scientific Fisheries* 38:339-343.

———. (1973). Studies on astaxanthin formation in some fresh-water fishes. *Tohoku Journal of Agricultural Research* 24:192-196.

———. (1976). Carotenoid metabolism in fancy red carp, *Cyprinus carpio* 2. Metabolism of 14C-zeaxanthin. *Bulletin of the Japanese Society of Scientific Fisheries* 42:203-205.

Hughes, S.G. (1991). Response of first-feeding spring chinook salmon to some potential feeding attractants. *Progressive Fish-Culturist* 53:15-17.

Hughes, S.G., and F.T. Barrows (1990). Measurements of the abilities of cultured fishes to impart moisture to their digesta. *Comparative Biochemistry and Physiology* 96A:109-112.

Huang, C.-H., M.-C. Huang, and P.-C. Hou (1998). Effect of dietary lipids on fatty acid composition and lipid peroxidation in sacroplasmic reticulum of hybrid tilapia, *Oreochromis niloticus* × *O. aureus. Comparative Biochemistry and Physiology* B 120: 331-336.

Hung, S.S.O., and S.J. Slinger (1980). Measurement of oxidation in fish oil and its effect on rainbow trout *(Salmo gairdneri). Canadian Journal of Fisheries and Aquatic Sciences* 37:1248-1253.

Jantrarotai, W., and R.T. Lovell (1990). Subchronic toxicity of aflatoxin B, to channel catfish. *Journal of Aquatic Animal Health* 2:248-254.

Johnsen, P.B., and M.A. Adams (1986). Chemical feeding stimulants for the herbivorous fish, *Tilapia zilli. Comparative Biochemistry and Physiology* 68A:109-112.

Ketola, H.G., C.E. Smith, and G.A. Kindschi (1989). Influence of diet and peroxidative rancidity on fry of Atlantic and coho salmon. *Aquaculture* 79:417-423.

Kumar, S.S., and V. Jayaprakas (1996). Role of dietary L-carnitine on the monosex culture of male Mozambique tilapia *Oreochromis mossambicus* (Peters). *Proceedings of the Indian National Science Academy Part B, Biological Sciences* 62(4): 247-258.

LaVorgna, M.W. (1998). Utilization of phytate phosphorous by tilapia. Doctoral dissertation, University of Maryland Eastern Shore, Princess Anne, Maryland.

Lee, P.H. (1987). Carotenoids in cultured channel catfish. Doctoral dissertation, Auburn University, Auburn, Alabama.

Lim, C., M. Yildirim, P.H. Klesius, and P.J. Wan (2003). Levels of dietary gossypol affect growth, bacterial resistance of Nile tilapia. *Global Aquaculture Advocate* 6:42-43.

Lovell, T. (1998). *Nutrition and feeding of fish.* Boston: Kluwer Academic Publishers.

Mackie, A.M. (1982). Identification of the gustatory feeding stimulants. In T.J. Hara (Ed.), *Chemoreception in fishes* (pp. 275-291). Amsterdam, The Netherlands: Elsevier Scientific Publishing Company.

Mackie, A.M., and J.W. Adron (1978). Identification of inosine and inosine-5'-mono-phosphate as the gustatory feeding stimulants for the turbot, *Scophthalmus maximus. Comparative Biochemistry and Physiology* 60A:79-83.

Murai, T., T. Akiyanma, H. Ogata, and T. Suziki (1988). Interaction of dietary oxidized fish oil and glutathione on fingerling yellowtail *Seriola quinqueradiata. Nippon Suisan Gakkaishi* 54:145-149.

Murai, T., and J. W. Andrews (1974). Interactions of dietary a-tocopherol, oxidized menhaden oil and ethoxyquin on channel catfish *(Ictalurus punctatus). Journal of Nutrition* 104:1416-1431.

NRC (National Research Council) (1977). *Nutrient requirements of warmwater fishes.* Washington, DC: National Academy Press.

———. (1993). *Nutrient requirements of finfish.* Washington, DC: National Academy Press.

Phelps, R.P. and T.J. Popma (2000). Sex reversal of tilapia. In B.A. Costa-Pierce and J.E. Rakocy (Eds.), *Tilapia aquaculture in the Americas,* (Volume 2, pp. 34-59). Baton Rouge, LA: The World Aquaculture Society.

Ridha, M.T. and K.P. Lone (1990). Effect of oral administration of different levels of 17-a-methyltestosterone on the sex reversal, growth and food conversion efficiency of the tilapia *Oreochromis spilurus* (Guenther) in brackish water. *Aquaculture and Fisheries Management* 21:391-397.

Rumsey, G.L. (1986). Chemical control of feed intake in fishes. In *Proceedings of the 1986 Cornell Nutrition Conference* (pp. 40-45). Ithaca, NY: Cornell University.

Satoh, S., T. Takeuchi, and T. Watanabe (1987). Requirement of tilapia for alpha-tocopherol. *Bulletin of the Japanese Society of Scientific Fisheries* 53(1):119-124.

Scheidt, K., F.J. Leuenberger, M. Vecchi, and E. Glinz (1985). Absorption, retention and metabolic transformations of carotenoids in rainbow trout, salmon, and chicken. *Pure and Applied Chemistry* 57:685-692.

Shiau, S.-Y., and C.-C. Kwok (1989). Effects of cellulose, agar, carrageenan, guar gum and carboxymethylcellulose on tilapia growth. *World Aquaculture* 20(2):60.

Shiau, S.-Y., C.-C. Kwok, C.-J. Chen, H.-T. Hong, and H.-B. Hsigh (1988). Effects of different dietary fiber on the utilization of dextrin in tilapia *(Oreochromis niloticus × O. aureus)*. In *Proceedings of the Aquaculture International Congress and Exposition* (p. 66). Vancouver, British Columbia, Canada: Aquaculture International Congress.

Shiau, S.-Y., H.-L. Yu, S. Hwa, S.-Y. Chen, and S.-I. Hsu (1988). The influence of carboxymethylcellulose on growth, digestion, gastric emptying time and body composition of tilapia. *Aquaculture* 70:345-354.

Sikoki, F.D., and A.O. Ekwu (2000). Comparison of growth and survival of two trains of *Oreochromis niloticus* (Trewavas) fed with methyltestosterone treated diet. *Global Journal of Pure and Applied Sciences* 6:401-406.

Smith, C.E. (1979). The prevention of liver lipid degeneration (ceroidosis) and microcytic anaemia in rainbow trout, *Salmo gairdneri* Richardson, fed rancid diets: A preliminary report. *Journal of Fish Diseases* 2:429-437.

Steffens, W. (1989). *Principles of fish nutrition.* New York: Halsted Press.

Tacon, A.G. (1981). Speculative review of possible careteniod function in fish. *Progressive Fish-Culturist* 43:205-208.

Tacon, A.G. (1985). *Nutritional fish pathology: Morphological signs of nutrient deficiency and toxicity in farmed fish.* ADCP/REP/85/22. Food and Agriculture Organization of the United Nations, Rome, Italy.

Teichert-Coddington, D., B. Manning, J. Eya, and D. Brock (2000). Concentration of 17 α-methyltestosterone in hormone-treated feed: Effects of analytical technique, fabrication, and storage temperature. *Journal of the World Aquaculture Society* 31:42-50.

Teshima, S., A. Kanazawa, and S. Koshio (1987). Effect of feeding rate, fish size and dietary protein and cellulose levels on the growth of *Tilapia nilotica. Memorial of the Faculty of Fisheries Kagoshima University* 35:7-15.

Torrissen, O.J., R.W. Hardy, and K.D. shearer (1989). Pigmentation of salmonids-carotenoid deposition and metabolism. *Reviews in Aquatic Sciences* 1:209-225.

Tsai, H.-J., W.K. Chi, C.-C. Chang, T.-T. Kuo, and C.-F. Chang (1993). Enhancement of tilapia growth by dietary administration of recombinant yeast lysates as a supplement. *Journal of the Fisheries Society of Taiwan* 20:339-345.

Tuan, N.A., J.M. Grizzle, R.T. Lovell, B.M. Manning, and G.E. Rottinghaus (2002). Growth and hepatic lesions of Nile tilapia *(Oreochromis niloticus)* fed diets containing aflatoxin B1. *Aquaculture* 212:311-319.

Varadaraj, K., S. Sindu Kumari, and T.J. Pandian (1994). Comparison of conditions for hormonal sex reversal of Mozambique tilapias. *Progressive Fish-Culturist* 56:81-90.

Viola, S., N. Gur, and G. Zohar (1985). Effects of pelleting temperature, binders and basic grains on water-stability of pellets and on growth of tilapia. *Bamidgeh* 37(1):19-26.

Yong, W.-Y., T. Takeuchi, and T. Watanabe (1989). Relationship between digestible energy contents and optimum energy to protein ration on *Oreochromis niloticus* diets. *Bulletin of the Japanese Society of Scientific Fisheries* 55:869-873.

Zohar, Y. (1989). Endocrinology and fish farming: Aspects in reproduction, growth, and smoltification. *Fish Physiology and Biochemistry* 7:395-405.

Chapter 14

Feed Formulation and Manufacture

Menghe H. Li
Chhorn E. Lim
Carl D. Webster

INTRODUCTION

Tilapia are cultured worldwide and culture methods vary among different regions. However, as tilapia farming continues to expand, production methods are moving toward intensive culture systems where fish are stocked at high densities and are given nutritionally balanced, high quality feeds. Since feed represents the major expense in intensive fish production, development of cost-effective, nutritionally efficient feeds is imperative to successful tilapia farming.

To formulate and manufacture high-quality fish feeds, including tilapia feeds, one should have knowledge of nutrient requirements, nutrient composition, digestibility, and availability of feed ingredients; impact of the manufacturing process on nutrient characteristics of the feed; and the factors affecting physical properties of the feed (i.e., pellet water stability and floatability of extruded feeds). It is essential that the feed be formulated using highly digestible feedstuffs to meet fish's nutrient requirements and be manufactured into a physical form with an appropriate size that is readily consumed by fish. The feed manufacturing process may have positive or negative effects on certain nutrients. The manufacturing conditions, such as high temperature, pressure, and moisture encountered during extrusion, and high temperature experienced during drying, destroy certain nutrients and improve the availability of others. Some vitamins are sensitive to

---

This article is approved for publication as Book Chapter No. BC-10444 of the Mississippi Agricultural and Forestry Experiment Station (MAFES), Mississippi State University.

doi:10.1300/5513_14

destruction; thus, fish feeds are normally overfortified with vitamins to account for losses during feed manufacture and storage. Starch digestibility appears to be improved by the extrusion process. Extrusion and drying processes may also inactivate certain undesirable substances present in feedstuffs (i.e., trypsin inhibitors) and reduce the occurrence of molds and bacteria in the feed pellet. This chapter focuses on formulation and manufacture of nutritionally adequate feeds for tilapia.

FEED INGREDIENTS

Commercial fish feeds comprise a mixture of feedstuffs and vitamin and mineral premixes that provide adequate amounts of essential nutrients and digestible energy. To be qualified as a suitable candidate for a dietary ingredient, several criteria must be met. Dietary ingredients must be highly digestible, be available on a consistent basis, be easily handled in the manufacturing process, be able to withstand the rigors of the manufacturing process, and be economical. This section provides a brief summary of the commonly available feed ingredients in the United States that may be used in tilapia feeds.

Protein Sources

Feedstuffs containing 20 percent crude protein or more are generally considered as protein sources. Protein sources are classified as either animal or plant proteins based on their origin. Animal proteins used in fish feeds come from tissues unsuitable for direct human consumption from meat-packing or rendering plants, milk products, or marine sources. Those commonly used in tilapia feeds include fish meal, meat meal, meat and bone meal, blood meal, and poultry by-product meals. Animal proteins are generally considered to be of a higher nutritional value than plant proteins primarily because of their balanced indispensable amino acids. Fish meal prepared from whole fish appears to be a better protein source than fish meal made of fish processing residues or other animal protein sources.

The primary plant protein sources used in tilapia feeds are oilseed meals, such as soybean meal, cottonseed meal, peanut meal, and canola meal. Compared with animal proteins, most plant proteins (except for soybean meal and perhaps for canola meal), are deficient in lysine and methionine/cystine, the two most limiting essential amino acids in tilapia feeds. Also, certain plant proteins contain antinutritional factors and toxins that may or may not be inactivated during processing. For example, trypsin inhibitors are inactivated by heat, but phytic acid is unaffected by heat. A brief description of various animal and plant protein sources that are commonly used in tilapia feeds is given in the following sections.

Fish Meal

Fish meal is processed by either direct drying or cooking followed by pressing to remove liquid and oil before drying, and grinding of whole, undecomposed fish and/or fish cuttings. Major commercial fish meals are made from the main catch of a single species of small, oily marine fish (e.g., anchovy, herring, and menhaden), from bycatch of mixed species, or from processing residues (e.g., cod, haddock, whiting, and pollock). Differences in the method of manufacturing result in considerable variation in the quality of fish meal derived from the same raw material (i.e., high- and low-temperature fish meals). Fish meals made from the bycatch and fish cuttings are more variable in protein content and nutritional quality than those made from a single species because of the differences in the composition of fish bycatch and trimmings.

Fish meal contains 60 to 70 percent protein (50 to 60 percent for fish meal made from bycatch of mixed species) that is of good quality and highly palatable to fish. Since fish meal is a good source of essential amino acids, particularly lysine and methionine/cystine, it is often used to supplement feeds containing plant proteins. Fish meal is also rich in energy, minerals, and essential fatty acids, particularly the n-3 highly unsaturated fatty acids. The fat content of deoiled fish meal ranges from about 4 to 11 percent. The ash content is highly variable and ranges from 11 percent in herring meal to above 23 percent for meals made from fish trimmings. Fish meal is used sparingly in tilapia feeds because of its limited availability and high cost.

Meat Meal

Meat meal is the rendered product from beef, pork, or lamb tissues and should not contain blood, hair, hoof, horn, hide trimmings, manure, or stomach and rumen contents except in trace amounts unavoidable during processing. Meat meal contains approximately 55 percent crude protein, 7 percent crude fat, and 25 percent ash.

Meat and Bone Meal

Meat and bone meal is similar to meat meal except with added bone. Meat and bone meal contains approximately 45 to 50 percent crude protein and 8.5 percent fat. Its protein quality is inferior to whole fish meal because it contains less lysine and methionine/cystine, and the consistency of the product may vary considerably. Although it is a good source of minerals, high ash content (33 to 37 percent) may limit its use because of a possible mineral imbalance. El-Sayed (1998) reported that a dietary level of 40 percent meat and bone meal could completely replace herring fish meal in the diet (35 percent of diet) for Nile tilapia *Oreochromis niloticus* without affecting their growth, feed efficiency, or protein efficiency ratio. Wu et al. (1999) also reported that meat and bone meal, used at 6 percent of the diet to completely replace fish meal, did not affect the growth of Nile tilapia.

Owing to concerns of bovine spongiform encephalopathy or "mad cow" disease, many countries prohibit the use of meat meal and meat and bone meal in animal feeds, especially feeds for ruminant animals. The European Community has placed a total ban on the use of meat meal and meat and bone meal in diets for all farmed animals. In 2002, the European Community also passed regulations on animal by-products and classified these products into three categories. Category 3 products—materials that are derived from healthy animals—may be allowed in animal feeds (intra-species recycling is still prohibited) after proper treatment in approved processing plants in the future if the total ban on meat meal and meat and bone meal is ever lifted. The U.S. Food and Drug Administration (FDA) prohibits the use of meat meal and meat and bone meal from ruminant animals in feeds for ruminants. Some fish feed manufacturers in the United States have voluntarily discontinued the use of meat meal and meat and bone meal for precautionary measures. Owing to government regulations and public safety concerns, there is uncertainty about the future use of these products in fish feeds.

Blood Meal

Blood meal is prepared from clean, fresh animal blood, excluding hair, stomach contents, and urine, except in unavoidable trace quantities. It contains about 80 percent crude protein and is a good source of lysine and leucine, but is deficient in arginine, isoleucine, and methionine/cystine.

Meat and Bone/Blood Meal Blend

A mixture of meat and bone meal and blood meal is available in the United States for use in fish feeds. The two feedstuffs are mixed to produce the desired nutritional profile, usually mimicking that of fish meal, and provide 60 to 65 percent protein.

Poultry By-Product Meal

Poultry by-product meal is made up of ground, or rendered clean parts of the carcass of slaughtered poultry. It contains heads, feet, underdeveloped eggs, and visceral organs but does not contain feathers or the contents of gizzards and intestines. The product contains approximately 58 percent protein, 14 percent fat, and 16 percent ash. It has adequate amount of methionine and cystine, but is deficient in lysine. El-Sayed (1998) reported that complete replacement of fish meal with poultry by-product meal in Nile tilapia diets (fish meal was the sole protein source) did not affect fish growth, but reduced feed efficiency.

Hydrolyzed Poultry Feather Meal

Hydrolyzed poultry feather meal is prepared by treatment with $Ca(OH)_2$ under the pressure of clean, undecomposed feathers from slaughtered poultry. At least 75 percent of the protein should be digestible as measured by pepsin digestion. It is high in protein (85 percent), but deficient in several essential amino acids, especially lysine. Its use in fish feeds is restricted due to its low in vivo protein digestibility by fish.

Soybean Meal

Dehulled, solvent-extracted soybean meal is prepared by grinding the flakes, after removal of the oil, from dehulled soybeans by solvent extraction. Dehulled, solvent-extracted soybean meal contains about 48 percent protein, and is the predominant protein source used in feeds for warmwater fish species including tilapia. Mechanically extracted and solvent-extracted soybean meals with hulls contain 42 and 44 percent crude protein, respectively. Solvent-extracted soybean meals contain about 1 percent fat whereas mechanically extracted meal has about 4 to 5 percent fat. Soybean meal has the best amino acid profile among all common plant protein sources and is highly palatable and digestible to tilapia. Antinutritional factors, particularly trypsin inhibitors, are inactivated or reduced to insignificant levels by heat that is applied during the extraction process. Depending on the protein content of the diets, soybean meal at levels ranging from 35 to 55 percent have been successfully used in tilapia diets.

Heated, Full-Fat Soybean Meal

Full-fat soybean meal is prepared by grinding heated soybean from which oil has not been removed. The meal contains about 38 percent protein and 18 percent fat. The nutritional value of full-fat soybean meal for tilapia is similar to defatted soybean meal with added soybean oil. Thus, full-fat soybean meal can be used to replace defatted soybean meal in tilapia feeds on an equal protein basis, provided that the total fat level in the finished feed does not exceed the desired level. Oil in full-fat soybean meal contains approximately 54 percent linoleic acid (18:2 n-6) and 7 percent linolenic acid (18:3 n-3): fatty acids reported to be essential for tilapia. Wee and Shu (1989) reported that full-fat soybean boiled at 100°C for 1 h, could completely replace defatted soybean meal in the diet of Nile tilapia without affecting their growth and feed efficiency. However, carcass fat level was much higher in fish that were fed boiled full-fat soybean meal than fish that were fed defatted soybean meal.

Cottonseed Meal

Cottonseed meal is obtained by grinding the cake that is remaining after oil has been solvent-or mechanically extracted from cottonseed. Cottonseed meal generally contains 41 percent protein and 11 to 13 percent fiber. Solvent-extracted cottonseed meal contains less fat than mechanically extracted cottonseed meal (1 to 2 percent versus 4 percent). Cottonseed meal is highly palatable to tilapia. It is high in arginine, but is deficient in lysine. Glanded cottonseed meal contains free gossypol and cyclopropenoic acids that can be toxic to fish if used at high levels. It has been reported that up to 30 to 35 percent cottonseed meal can be used in tilapia feeds without affecting their growth and feed efficiency (Jackson et al. 1982; Mbahinzireki et al. 2001). However, other studies have indicated that lower levels should be used in tilapia feeds (Ofojekwu and Ejike 1984; Robinson et al. 1984). Earlier research suggests that the amount of cottonseed meal that can be used in tilapia feeds depends on tilapia species, available amino acid concentrations, and free gossypol levels. Studies on gossypol toxicity, however, suggest that blue and Nile tilapia can tolerate very high levels of dietary gossypol (Robinson et al. 1984). Thus, the poor performance of tilapia that are fed high levels of cottonseed meal may be attributed to deficiency of other amino acids in addition to lysine, or the presence of other toxic constituents, such as cyclopropenoic acids. Generally, a level of 15 to 20 percent of cottonseed meal can be used in tilapia feeds without affecting fish growth and feed efficiency.

Peanut Meal

Peanut (groundnut) meal is obtained by grinding shelled peanuts after the oil has been removed either mechanically or by solvent extraction.

Solvent-extracted peanut meal contains about 48 percent protein and 1.5 percent fat, whereas the mechanically extracted product contains about 45 percent protein and 5 percent fat. Peanut meal is high in arginine and leucine, but deficient in lysine, methionine, cystine, and threonine; it is highly palatable to fish. Peanut meal is prone to the growth of molds, which produces aflatoxin.

Corn Gluten Meal

Corn gluten is the protein portion of the corn kernel, and corn gluten meal is the dried residue from corn after the removal of the larger part of the starch and germ, and the separation of the bran by wet milling of corn to produce starch and syrup. Corn gluten meal contains 40 to 60 percent crude protein and is a good source of methionine and cystine, but is deficient in lysine and tryptophan. It contains high levels of yellow pigment xanthophylls (200 to 350 $mg \cdot kg^{-1}$). In channel catfish, over 11 mg xanthophylls$\cdot kg^{-1}$ feed imparts yellow color in the flesh, which is undesirable to consumers (Lovell 1989). For this reason, corn gluten meal is not used in catfish grow-out feeds. We are not aware of any research on the effect of xanthophylls on tilapia flesh coloration. Wu et al. (2000) reported that up to 34 percent corn gluten meal could be used in an amino acid-supplemented diet for Nile tilapia without affecting the growth and flavor characteristics of the fillets.

Distillers' Dried Grains with Solubles

Distillers' dried grains with solubles are primarily the insoluble and soluble residues or by-products, after removal of the alcohol by distillation, from the yeast fermentation of cereal grains. The distillers' grains or insoluble residues are comprised mostly of grains (minus the starch) that can be separated from the fermentation mixture by filtration. The liquid (soluble) portion is comprised of microorganisms and small suspended particles, such as emulsified and soluble nutrients. The combination of these two by-products are referred to as "distillers' dried solid" or "distillers' dried grains with solubles." The product contains approximately 30 percent protein and is highly palatable to fish, but is deficient in lysine and tryptophan. Distillers' dried grains with solubles can be used at levels of up to 30 percent as a substitute, on an equal protein basis, for plant protein in tilapia diets.

Sunflower Meal

Sunflower meal is prepared by grinding the residue left over after mechanical or solvent-extraction of the oil from sunflower seeds. Solvent-extracted, dehulled sunflower meal contains about 44 percent protein, 2 percent fat, and 11 percent fiber. Partially dehulled sunflower meal contains about 30 to 35 percent protein and 20 percent fiber. Sunflower meal is deficient in lysine, but methionine and arginine contents are higher than in soybean meal. Its low lysine content and high level of fiber limit its usefulness in fish feeds.

Canola Meal

Canola meal is prepared from a special rapeseed (canola) after solvent-extraction to remove the oil. Canola refers to new varieties of rapeseed specially bred to contain much lower levels of glucosinolates and erucic acid, that may be detrimental to fish growth. Canola meal contains about 38 percent protein, 4 percent fat, and 11 percent fiber. Its amino acid profile is comparable to that of soybean meal and it appears to be sufficient in all essential amino acids for Nile tilapia. A level of up to 30 percent canola meal can be used in the diets of hybrid tilapia *Oreochromis mossambicus* × *O. aureus* without affecting growth and feed efficiency (Higgs et al. 1989).

Energy Sources

Energy feedstuffs are those that contain less than 20 percent crude protein. These include grain and grain by-products, and animal fats or vegetable oils. Energy sources typically used in commercial tilapia feeds include corn grain, corn screenings, corn gluten feed, wheat grain, wheat middlings, rice bran, animal fat, and fish oil.

Corn Products

Corn grain and corn screenings are used interchangeably in commercial tilapia feeds as a relatively inexpensive source of energy. Corn screenings are obtained in the cleaning of corn and include light and broken corn grains. Corn is highly digestible by tilapia, but cooking improves its energy digestibility. However, mixtures of corn and sorghum or wheat bran resulted in better growth than when fed corn alone (Degani and Revach 1991). Both raw and extruded corn meal can be utilized at levels up to 50 percent of the diet by hybrid tilapia, *O. niloticus* × *O. aureus* (Takeuchi et al. 1994).

Corn gluten feed is the part of corn that remains after the extraction of most of the starch, gluten, and germ by the process of wet milling of corn to produce starch and syrup. The difference between corn gluten feed and corn gluten meal is that corn gluten feed contains more bran and less gluten and thus less protein and higher fiber than corn gluten meal. Corn gluten feed is a potential energy source for tilapia feeds. This product typically contains about 18 to 20 percent protein, 2 percent fat, and 10 percent fiber and is usually competitively priced relative to corn and wheat middlings. Unlike high-protein corn gluten meal, corn gluten feed contains xanthophylls at a level similar to that in corn grain and does not cause yellow pigmentation in fish flesh. A level of 20 percent can be used in Nile tilapia diet without affecting growth performance (Wu et al. 1995).

Wheat Products

Wheat milling, which produces flour, also produces by-products such as wheat bran, germ, shorts, and mill run. Wheat is generally more expensive than corn. As a result, ground wheat has been used sparingly (about 3 to 5 percent) in tilapia feeds, primarily for its pellet binding properties. Wheat middlings are fine particles of wheat bran, shorts, germ, and flour recovered from milling of wheat grain. Depending on availability and cost, wheat middlings are used to replace corn or corn screenings in tilapia feeds and can be used at levels up to about 20 percent of the diet. In humid areas, using wheat middlings at a level greater than 25 percent may cause the diet to become sticky, resulting in clumping of feed pellets.

Rice Products

Rice bran is the bran layer and germ of rice grain with hulls or broken rice at levels that are unavoidable in milling rice grain. It has a higher protein (13.5 percent) content than the grain (7 percent). Unextracted rice bran has a high oil content (12 to 17 percent). Rice bran oil is rich in unsaturated fatty acids that undergo rapid oxidation under normal storage conditions. Rancid rice bran has markedly reduced nutritional values. Rancidity can be prevented or minimized by the addition of antioxidants. Rice bran is also high in crude fiber (about 13 percent). Its high content of oil and fiber may limit its use in tilapia feeds. The oil, because of its high commercial value, is sometimes removed by solvent extraction. Defatted rice bran can be used at higher levels in fish feeds.

Animal and Plant Fats and Oils

Animal and plant fats and oils are highly concentrated sources of energy, essential fatty acids, and fat-soluble vitamins. Animal fats used in tilapia feeds include poultry fat and fish oils. Plant oils can be used, but animal fats are generally preferred because they are generally less expensive. Supplemental fat is generally sprayed on the finished feed pellets at a rate of 1 to 1.5 percent to reduce fines and increase dietary fat content.

Vitamin and Mineral Supplements

Commercial tilapia feeds are supplemented with a vitamin premix that provides vitamins in quantities necessary to meet dietary requirements and compensate for losses due to feed processing. Phosphorus and trace mineral supplements are commonly added in tilapia feeds to ensure that mineral needs are met. Due to lack of information on some vitamin and mineral requirements for tilapia, vitamin and trace mineral premixes available for other species, such as those for common carp and channel catfish, are used in tilapia feeds.

Feed Additives

Nonnutritive feed ingredients, such as pellet binders, hormones, antioxidants, enzymes, and attractants are discussed in Chapter 13 "Nonnutrient Components of Fish Diets" of this book. Steam-pelleted (sinking) feeds, in addition to the binding properties of regular feed ingredients, may require binding agents to improve pellet water stability and withstand handling and shipping. Potential binders that can be used in steam-pelleted tilapia feeds with fair to good binding properties and relatively low cost are bentonite, lignin sulfonate, and processed milo. Extruded (floating) fish feeds, however, depend on starches inherent in feedstuffs for pellet binding.

FEED FORMULATION

Feed formulation is a process that involves a combination of several suitable feed ingredients to obtain a mixture that is palatable, pelletable, and meets or exceeds the minimum essential nutrient requirements of the fish at a reasonable cost. To formulate high-quality, low-cost fish feeds, the following factors associated with nutrition and the manufacturing process must be considered.

Nutrient Requirements

Nutrient requirements of tilapia are discussed in Chapter 12 "Nutrient Requirements"; these have not been as well defined as those for common carp, channel catfish, and rainbow trout. However, more information is available now than a few years ago. Nutrient requirements of tilapia were determined primarily with small fish raised under laboratory conditions. Most of the requirements were determined from single experiments, which may or may not be supported by repeated studies. In addition, many factors such as environmental conditions, presence or absence of natural food organisms, management practices, and fish size greatly affect dietary nutrient levels for optimum fish growth and feed efficiency. Thus, one must realize that the levels of nutrients recommended are not necessarily etched in stone, but can only be used as an approximate guide. However, feed formulations based on these nutrient requirement data have worked well for various species of fish. In formulating tilapia feeds, where the requirement for a specific nutrient or nutrients is not known, the requirements for common carp or channel catfish are typically used, since they too are warmwater species and nutrient requirements are assumed to be similar. More precise nutrient requirements for various sizes of fish cultured under different conditions will allow for more refined and cost-effective feed formulations.

In formulating and manufacturing fish feeds, it is essential that the finished feed meet the nutrient requirements of the fish. Depending on the circumstances, feed formulators may not be concerned with all known nutrients since some are more critical than others. Among the major dietary nutrients required, protein is considered the first because it is the most expensive among all macronutrients and cannot be replaced by other nutrients, for instance carbohydrates can be replaced by fats. A protein level in moderate excess of the requirement is generally not harmful, but makes the feed cost higher than necessary. Among the ten essential amino acids required by tilapia, only the levels of lysine and methionine/cystine are considered. Experience has shown that, using a combination of commonly available feed ingredients, if the minimum requirements for lysine and methionine/cystine are satisfied, usually the requirements for the other eight essential amino acids will also be met. The dietary levels of available energy are also of prime importance. Too much energy in the feed may result in excessive fattiness and limit the amount of feed consumption since, like terrestrial animals, tilapia eat to satisfy their energy needs. When feed is deficient in energy, protein and other nutrients are not utilized to their optimum potential for growth. Levels of other nutrients, such as available phosphorus, fat, and essential fatty acids should meet the minimum requirement but should not be used in excess since high dietary phosphorus levels will increase the phosphorus load in the pond water and excessive dietary fat will increase body fat in the fish. Vitamin and mineral requirements are usually not considered in formulating fish feeds since fixed amounts of vitamin and mineral premixes are included in the formulation.

Ingredients

Nutrient content, and physical and chemical properties of each ingredient to be used in the feed should be known. Nutrient concentrations are determined by chemical analysis. This information for commonly used ingredients is available in tables of feed composition (NRC 1993; Dale 2000). The values given in such tables are averages of gross nutrient content, not the bioavailability of the individual nutrients. Nutrient content of individual lots of ingredients may vary depending on many factors such as stage of maturity at harvest, and processing method and conditions.

Chemical analysis can detect the levels of nutrients within ingredients, but does not provide information on how much each particular nutrient is available to fish. Nutrient bioavailability is determined through biological evaluation, particularly digestibility trials. Nutrient bioavailability varies depending on several factors such as sources of ingredients, ingredient processing methods, inclusion levels, interactions with other dietary nutrients, fish size or age, and methods used for its determination. Thus, it is essential that nutrient bioavailability values be used in feed formulation. Limited information, however, is available on nutrient bioavailability of various ingredients for tilapia.

The availability and cost of feed ingredients are important factors that determine the cost of feed. Feed formulators can manipulate the levels of ingredients or interchange one ingredient with another as the availability and price change. However, feed formulators should maintain strict quality control since there are limitations on the use of certain feed ingredients. Some feed ingredients can be incorporated in the diet of a particular species up to a certain level, above which the nutritional value and physical property of the diet may be adversely affected. For instance, cottonseed meal and *Leucaena* meal, due to the presence of free gossypol and memosine, respectively, which are toxic to fish, should not be included in large quantities. The amount of meat and bone meal and full-fat soybean meal, because of their high ash and fat contents, respectively, that can be used in the diet should be limited to prevent possible mineral and energy imbalances. Also, a nutrient in a particular feed ingredient may not have the same nutritional value as the same nutrient in another ingredient. Changes of ingredients or their proportions may also lead to change in physical properties and palatability of the diet. Thus, limits (minimum, maximum, or fixed amounts) are normally imposed on the levels of certain ingredients, regardless of cost.

Other Considerations

The physical form and type of feed (the purpose for which the feed is intended to be used) determine to a large extent the levels of ingredients and nutrient content of the formulation. Under extensive pond culture systems, where natural food is abundant, supplemental feeds (nutritionally incomplete) comprised of locally available, inexpensive feed ingredients can be used to increase fish production. Supplemental feeds usually contain low protein and high energy, and may be deficient in some vitamins and trace minerals. It is assumed that the lacking dietary nutrients will be compensated by natural food organisms available in the culture systems. However, under intensive culture systems, or when fish are grown in environments where natural foods are absent or limited, nutritionally complete feeds are required.

Whether the feed is to be steam-pelleted or extruded will affect its composition. Extrusion processing of floating pellets requires that the formulation contains relatively high amounts of starch-containing ingredients such as corn meal and cereal products/by-products to facilitate gelatinization and expansion. Steam-pelleting for sinking feeds requires the use of binding agents, in addition to the binding property of starch-containing ingredients. Binders can be activated by low heat and moisture to reduce fines and improve the pellet hardness and water stability. High fibrous ingredients must be limited to low levels (< 6 percent) because high levels of fiber reduce pellet quality. Feed processing and storage conditions are also important with regard to the fate of certain nutrients such as vitamins. Considerations must therefore be taken to include in the formulae a margin of safety to compensate for nutrient losses during processing and storage.

LEAST-COST FEED FORMULATION

Least-cost feed formulation, or linear programming, is a mathematical procedure that involves the simultaneous solution of a series of equations by the use of computers. Linear programming has been used widely in animal feed formulations for the past four decades but its use in fish feed formulations has been relatively recent (occurring in the past 10 years or so). Earlier feeds for fish have been based on fixed formulae because of the lack of nutritional information. Presently, fish feeds can be formulated on a least-cost basis. However, the use of least-cost formulation in fish feeds is not as extensive as that in feeds for other farmed animals. The primary constraint limiting the use of least-cost programs for formulating fish feeds is that relatively few feedstuffs are available that can be used in fish feeds.

Many feedstuffs are unsuitable for use in fish feeds because of their poor nutritional content or because of feed manufacturing constraints.

To carry out least-cost feed formulation, the following information is needed:

1. nutrient requirements;
2. nutrient concentrations in feedstuffs;
3. nutrient digestibility or availability from feedstuffs;
4. cost of ingredients;
5. nutritional and nonnutritional restrictions.

The desired or recommended nutrient levels are specified in terms of restrictions such as minimum, maximum, and fixed amounts, a range, or a certain ratio. The same type of constraints may also be imposed on feed ingredients depending on the physical and chemical characteristics of individual ingredients. To obtain effective least-cost formulations, the prices

TABLE 14.1. Example of restrictions for least-cost formulation of extruded 32 percent protein feeds for grow-out of Nile tilapia.

Item	Restriction	Amount[a]
Crude protein (percent)	Minimum	32.0
Crude fiber (percent)	Maximum	7.0
Lipid (percent)	Maximum	6.0
Linoleic series (n-6) fatty acid	Minimum	1.0
Available phosphorus (percent)	Maximum	0.5
Digestible energy ($kcal \cdot g^{-1}$)	Minimum	2.8
Digestible energy ($kcal \cdot g^{-1}$)	Maximum	3.1
Available lysine[b]	Minimum	1.63
Available methionine[b]	Minimum	0.51
Available methionine + cystine[b]	Minimum	1.02
Grain or grain by-products	Minimum	25.0
Cottonseed meal (percent)	Maximum	15.0
Whole fish meal or other high quality animal protein source	Minimum	6.0
Soybean meal	None	Unlimited
Vitamin premix[c]	Fixed	
Trace mineral premix[c]	Fixed	

[a]*Amounts are expressed on an as-fed basis (about 10 percent moisture).* [b] Amounts vary depending on the dietary level of protein. [c]Must meet recommendations.

used should be current. Examples of restrictions placed on nutrients and feed ingredients for least-cost formulation of tilapia feeds are presented in Table 14.1.

FEED MANUFACTURING PROCESSES

Feed manufacture is the process of mixing and processing finely ground ingredients into uniform feed particles (meals, crumbles, and sinking or floating pellets) that are suitable for consumption by various sizes of fish. Fish feeds are unique compared to livestock feeds because fish feeds must be pelleted (except for meal-type feeds for larvae), be water stable, and in most cases be made to float on the water surface (extruded feeds). Depending on the fish size, the manufacturing process may range from a simple reduction of particle size to forming pellets through steam-pelleting or extrusion. Fish feeds in the United States are manufactured in modern feed mills specifically designed for manufacturing fish feeds. Regardless of whether a feed is sinking or floating, the general manufacturing scheme is the same (Figure 14.1). First, whole grains or coarse ingredients are ground (through a hammer mill, pulverizer, or attrition mill) to reduce particle size prior to batching. The dietary ingredients are then batched, weighed, mixed, and reground. After regrinding, mixed ingredients are either steam-pelleted or extruded, then cooled or dried, fat coated, and stored in bins for load-out or bagging. During preparation for load-out or bagging, the feed is screened to remove fines and then loaded into trucks for bulk delivery or bagged. Operations of the various phases of feed manufacture are computer controlled by operators.

Ingredient Receiving and Storage

Feedstuffs and other ingredients are either received at the mill in bulk or in bags (minor ingredients), by rail or by truck. Rail is generally more economical. Feedstuffs are unloaded from the railcars or trucks and transferred to storage houses or bins. Bulk ingredients are commonly screened to remove contamination before being conveyed to storage bins. As feedstuffs are needed they are moved by belt conveyers or screw conveyers to the appropriate section of the feed mill for processing.

Grinding, Batching, and Mixing

Grinding is a major task and an expensive part in feed manufacturing. Generally, the finer the ingredients, the harder and more durable are the

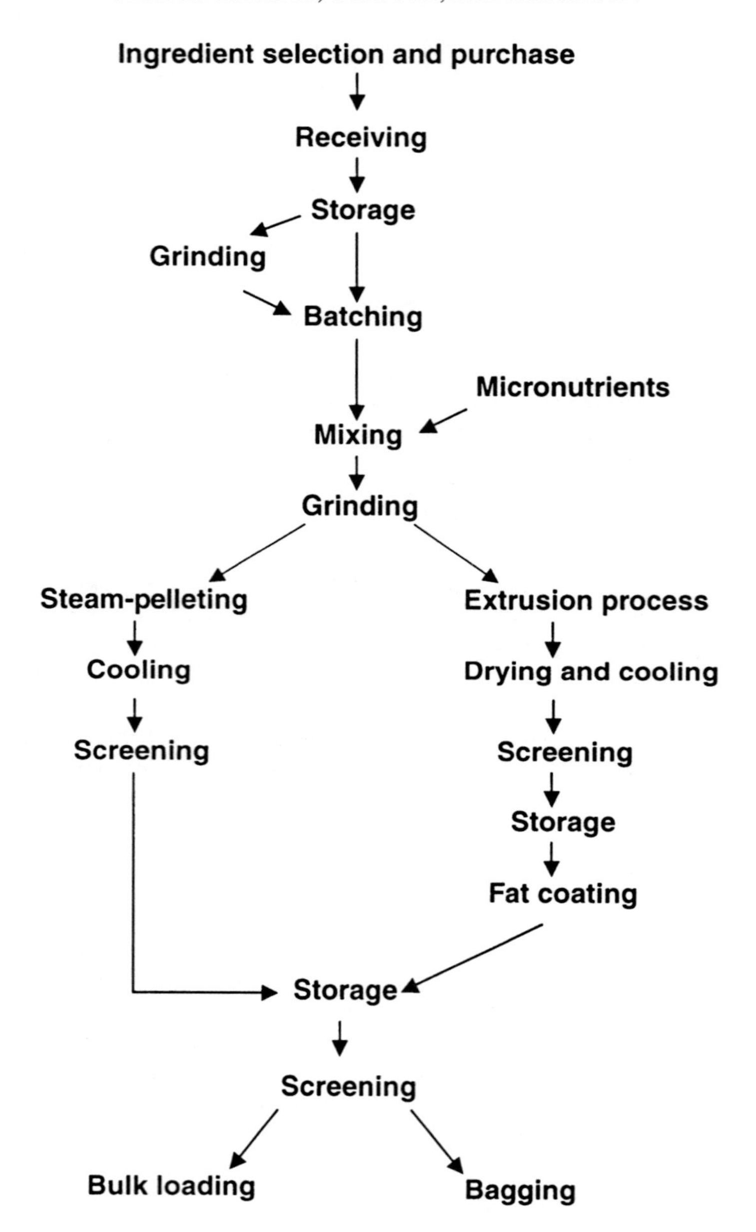

FIGURE 14.1. Block diagram of typical scheme for manufacturing fish feeds. *Source:* Adapted from Robinson et al. (2004).

pellets. The smaller particles tend to fill a larger percentage of void spaces allowing surface-to-surface contact between particles. This results in more compact and durable pellets and reduces the rate of pellet fracture that normally occurs between large pieces of ingredients. Grinding increases the surface area of ingredient particles, thus allowing more space for steam condensation during the conditioning process leading to more water absorption, higher temperature, and better starch gelatinization. Fine grinding also enhances proper and homogeneous mixing, uniform pellet texture, and final product appearance.

Whole grains (corn, wheat, etc.) or coarse ingredients are usually ground through a number 7 screen (2.8 mm) in a hammer mill prior to batching and mixing. During batching, feed ingredients are conveyed to a hopper above the mixer and weighed prior to mixing. After batching, the ingredients are dropped into a mixer and mixed for about 1.5 to 3 min. Mixing for too long may result in particle segregation or stratification. If liquid ingredients are used, they should be added and mixed after dry ingredient mixing. When mixing has been completed, the feed mixture is reground through a smaller screen, a number 4 or 6 (1.6 to 2.4 mm) depending on the type of feed being manufactured, and moved into a hopper above the pellet mill or the extruder.

Steam Pelleting

Commercial tilapia feeds in developing countries are usually produced by steam pelleting because production of steam-pelleted (sinking) feeds uses less sophisticated equipment and is less expensive. Steam-pelleted feeds are manufactured by using moisture, heat, and pressure to form ground feed ingredients into larger homogeneous feed particles. Most pellet mills (Figure 14.2) are equipped with 1 to 3 direct, or indirect, steam conditioning barrels where steam can be injected into the feed mixture or into the conditioner jackets. The jackets maintain high temperature within the conditioning barrel. Steam is generally added to the ground feed mixture to increase the moisture level to 15 to 16 percent and temperature to 70°C to 85°C. The feed mixture remains in the conditioning barrels for 30 to 60 s. Steam helps to partially gelatinize starch and activate any added binder that holds the feed particles together. The preconditioning, or hot, "mash" is then discharged into a rotating die ring, compressed by rollers through die holes into pellets that emerge from the outer edge of the die, and are cut into desirable length by adjustable knives. The uniformity and speed at which the mash is discharged into the die, the rotational speed of the die and rollers, and die thickness and hole diameter are essential for proper compaction,

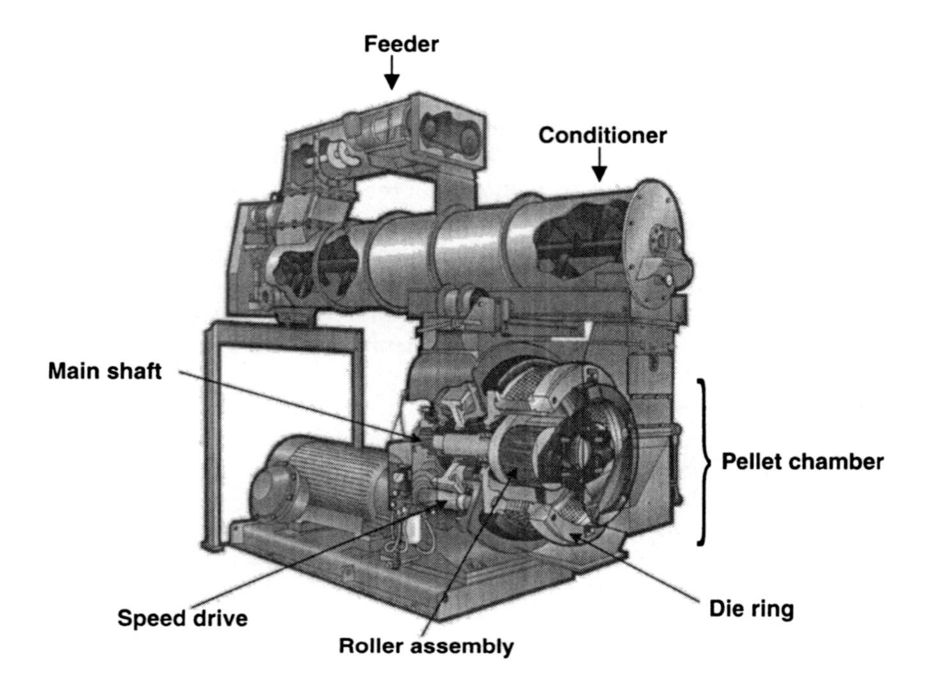

FIGURE 14.2. Pellet mill for manufacturing pelleted (sinking) fish feeds. *Source:* Image courtesy of California Pellet Mill/Pacific private Ltd.

good binding, and high production rate. Die hole diameter is dependent on the size of pellet desired. The most common pellet size for tilapia grow-out is approximately 3 to 4 mm in diameter and 6 to 10 mm in length. The pellets exit the die at about 14 to 15 percent moisture and at a temperature about 5°C above the temperature of the incoming mash. The moist, hot pellets are very fragile and must be immediately cooled and dried in the pellet dryer/cooler. Steam-pelleted feeds are generally less expensive to manufacture than extruded feeds because lower equipment costs and less energy is expended in their manufacture. Also, less destruction of nutrients occurs during steam pelleting as compared to extrusion.

Extrusion Processing

Although the extruded feed is more expensive to produce, it is the future of the fish feed and aquaculture industries, because floating feeds are more water stable and allow the person feeding to observe fish-feeding activities, and avoid over feeding and minimize feed wastes. Extrusion cooking (Figures 14.3A and 14.3B) is a process that involves the plasticizing and cooking of feed ingredients in the extruder barrel by a combination of pressure (shear), moisture, heat, and friction. Fish feed ingredients are a mixture of starchy and proteinaceous materials that are moistened to form a mash. The mash may be preconditioned in a pressurized or an atmospheric conditioning chamber for 3 to 5 min during which moisture is added in the form of steam or hot water to increase the moisture level of the mash to about 25 percent. During this period, the mash is cooked as heat and moisture penetrate the ingredient particles. Preconditioning not only permits proper cooking but also may improve flavor development and feed digestibility, reduce extruder barrel wear, and increase throughput from the extruder.

After preconditioning, the mash is discharged into the extruder assembly (single or twin screw design) consisting of a barrel and screws. The feed mixture moves through the extruder barrel toward the die by a single or twin rotating screws with tapered flights. Temperatures in the extruder barrel generally range from about 120°C to 150°C and are generated from the injection of steam into the feed mixture and friction of the feed moving through the barrel. The superheated mixture is then forced through a die located at the end of the extruder barrel. The die restricts product flow; thus causing development of the necessary pressure and shear. The die is also used to shape the product (extrudate) passing through it. As the product passes through the die, a sudden reduction in pressure results in the vaporization of part of the water from within the pellet causing the formation of air pockets and expansion of pellets. Various die sizes can be used to produce different sizes of feeds. Fingerling feeds can be made as small as 1 to 1.5 mm in diameter. The common pellet size for food fish grow-out is usually about 5 mm in diameter.

Drying and Cooling

Steam-pelleted feeds exit the die at a moisture content of 14 to 15 percent and require cooling and drying. The hot, moist pellets are transferred to the pellet cooler where temperature and moisture content are reduced by evaporative cooling, which is achieved by passing large volumes of high-speed air (generated by a fan) at ambient temperature (unheated) through the pellets. Final temperatures should be equivalent to ambient temperature and moisture content should be about 8 to 10 percent. Dry pellets are discharged into a conveyer that carries the finished feed to storage bins.

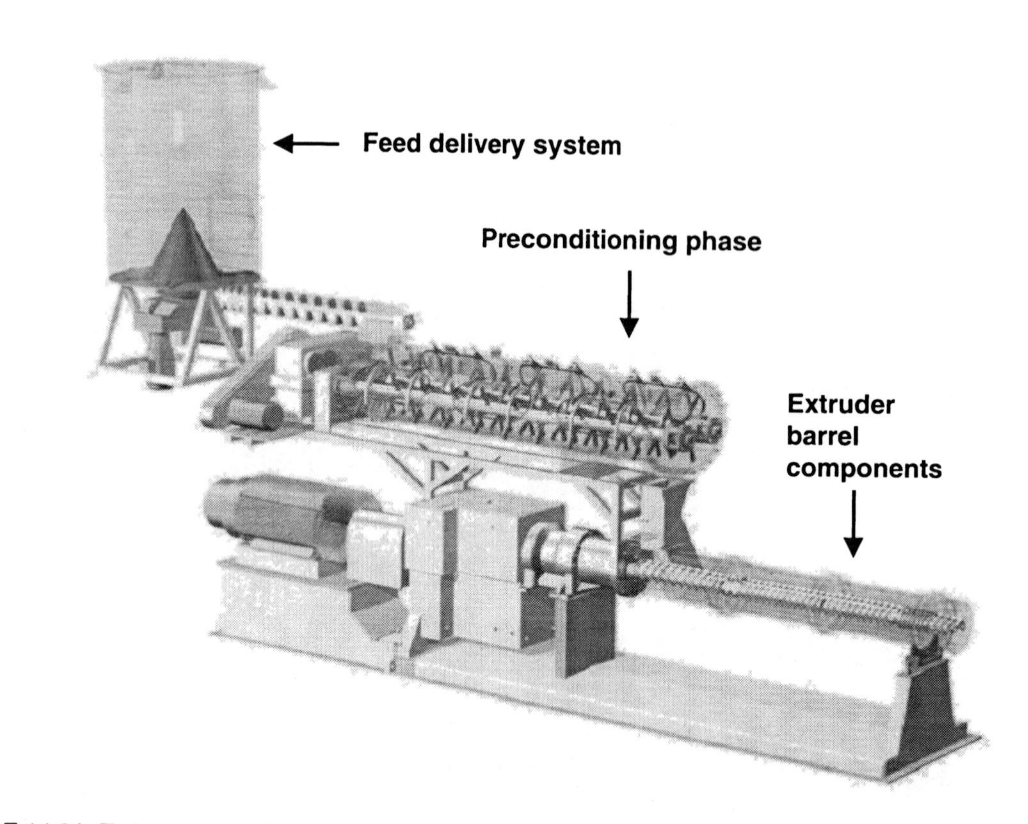

FIGURE 14.3A. Twin screw cooker-extruder for processing expanded (floating) fish feeds. *Source:* Image courtesy of Wenger Manufacturing, Inc.

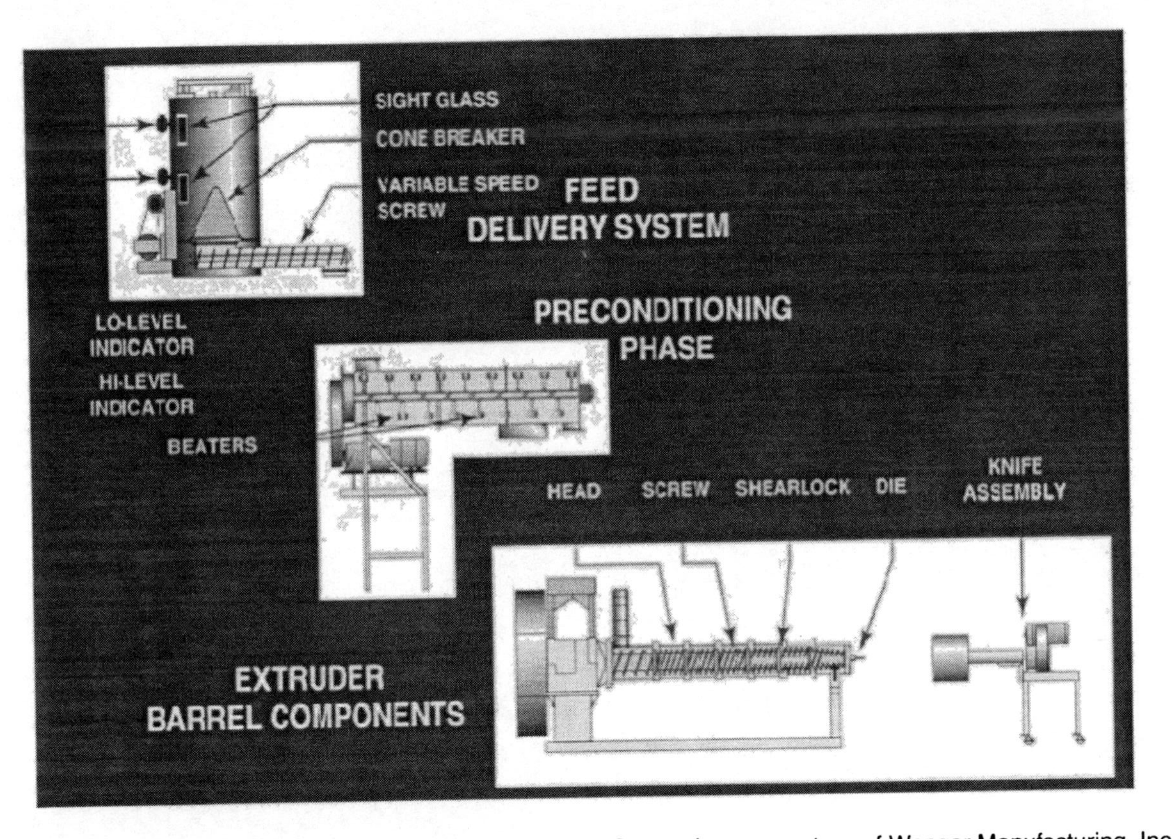

FIGURE 14.3B. Major components of the cooker-extruder. *Source:* Image courtesy of Wenger Manufacturing, Inc.

The moisture content of the pellets leaving the extruder is higher (18 to 21 percent) than that of steam-pelleted feed; thus, extruded pellets must be dried with heat. Extruded feeds lose some moisture by flash evaporation and evaporative cooling (about 2 percent). Extruded feeds should be dried to a moisture content of 8 to 10 percent. At this level of moisture, the "shelf life" of the product is extended. Drying is generally accomplished using a dryer/cooler (Figures 14.4A and 14.4B) which has different temperature zones. For extruded feeds, drying time is around 30 min and temperatures range between 135°C and 150°C.

Screening, Fat Coating, Storage, and Delivery

After drying, pellets are screened to remove fines, which are reclaimed and used as a feed ingredient. Extruded feeds are normally passed through a fat coater that applies a thin layer of fat to the pellet surface to increase dietary fat levels and help reduce fines. After fat coating, the product is then stored in bins for load-out or bagging. Just before load-out or bagging, feeds are screened again to remove fines. Commercial tilapia feeds are delivered to the farm in bulk or bags.

FIGURE 14.4A. Multipass conveyor drier/cooler for drying extruder fish feeds. *Source:* Image courtesy of Aeroglide Corporation.

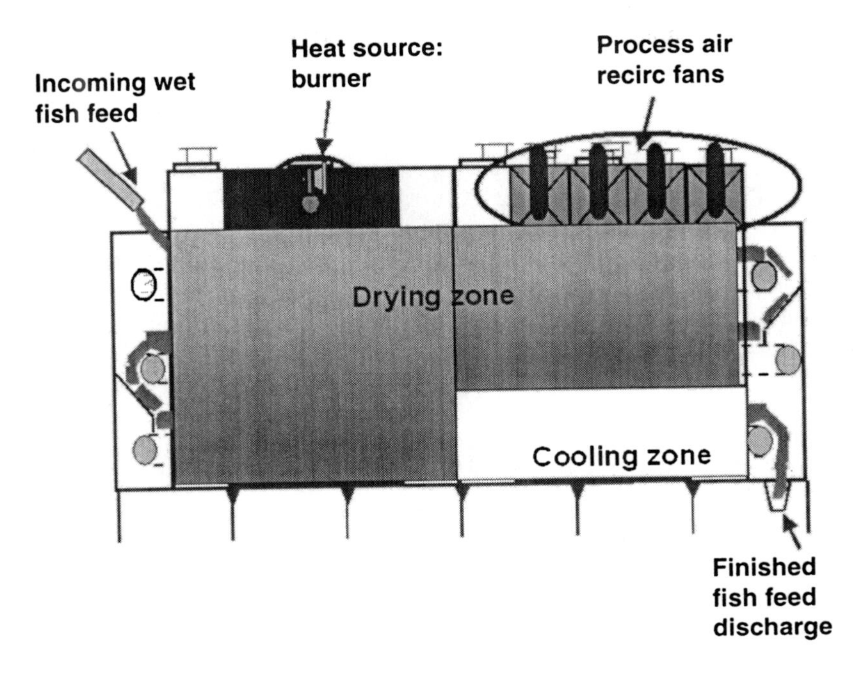

FIGURE 14.4B. Feed-flow diagram through the dryer/cooler. *Source:* Image courtesy of Aeroglide Corporation.

QUALITY CONTROL

Feed mills should have a continuous and comprehensive quality control program in place whereby various quality control measures are appropriately carried out to ensure that high-quality feeds are nutritionally balanced, are cost effective, are free of contaminants, have proper physical properties, and are consistently manufactured. A stringent quality control program must be the responsibility of all personnel involved in feed manufacturing to encompass all aspects of feed production from ingredient purchase to delivery of the finished feed.

Feed Ingredients

Feed quality is dependent on the quality of feed ingredients, balanced nutrient composition, and the manufacturing process. The purchasing agent should have an understanding of feed ingredients and knowledge of the suppliers who can consistently provide ingredients as needed to ensure that high-quality ingredients are available on a timely basis at a reasonable cost. Working with the nutritionist and the production manager, the purchasing agent establishes and uses ingredient specifications to ensure that ingredients meet the standards desired. Ingredients are inspected for color, odor, and texture prior to acceptance. Although subjective, visual and sensory inspection provides useful information on the quality of ingredients prior to use. An in-house test for moisture or toxins may be performed. Each batch of samples is taken for chemical analyses to determine if ingredients meet nutrient specifications. Also, analyses may be conducted to determine the presence of toxins, pesticides, or heavy metals. Since chemical tests lag behind ingredient use, a particular ingredient will be used prior to receiving the analytical results. However, if specifications are not met, a deficiency claim is filed with the supplier. In addition to ensuring quality by inspecting ingredients, ingredient inventories are maintained, which provide information on the amount of an ingredient used over a certain time period. This can be used to check and correct errors in the manufacturing process.

Formulation

Tilapia feed formulations are based on nutrient requirements established for different species of tilapia and data available for other species of similar food and feeding habits when specific nutrient requirements for tilapia are not available. Nutrient requirement data should be updated frequently to ensure that current data are available for formulating least-cost feeds. Nutrient profiles of feedstuffs should be continually updated based on actual assays conducted over a number of years on feedstuffs used and on information supplied by various suppliers of feedstuffs. Feeds are generally formulated to meet nutrient requirements at the least possible cost. A safety margin is used to account for variations in the nutrient content of feed ingredients and expected losses during processing and storage.

Manufacturing

Quality control measures continue during each phase of production to ensure that a feed containing the proper nutrient content with desirable physical characteristics is produced. All equipment used is selected to produce quality products. Equipment is continually inspected, and maintained according to specifications. Since a uniform mix is essential, mixing is checked periodically by assaying for particular vitamins or other micronutrients.

Finished Feed

The finished feed is routinely assayed for moisture, protein, fat, and fiber, and periodically assayed for selected micronutrients to ensure nutritional quality. Each batch of feed is checked for physical characteristics, such as pellet hardness/density, water stability, and floatability. Most states in the United States require certain guaranteed analyses, such as minimum crude protein and fat, and maximum crude fiber and ash for commercial feeds.

TYPE OF FEED

Steam-pelleted and extruded feeds have been discussed in the section "Feed Manufacturing Process." The following section focuses on meal-type feeds, crumbles, moist feeds, and medicated feeds. Other types of feeds such as microdiets (microencapsulated, microbound, and microcoated diets) are not commonly used in tilapia culture since these types of feeds are generally more expensive than meal-type feeds and tilapia fry can readily and quickly consume fine particles of formulated feeds after their yolk sac is absorbed. Therefore, these types of feeds are not discussed in this chapter. Various forms of fish feeds are shown in Figure 14.5.

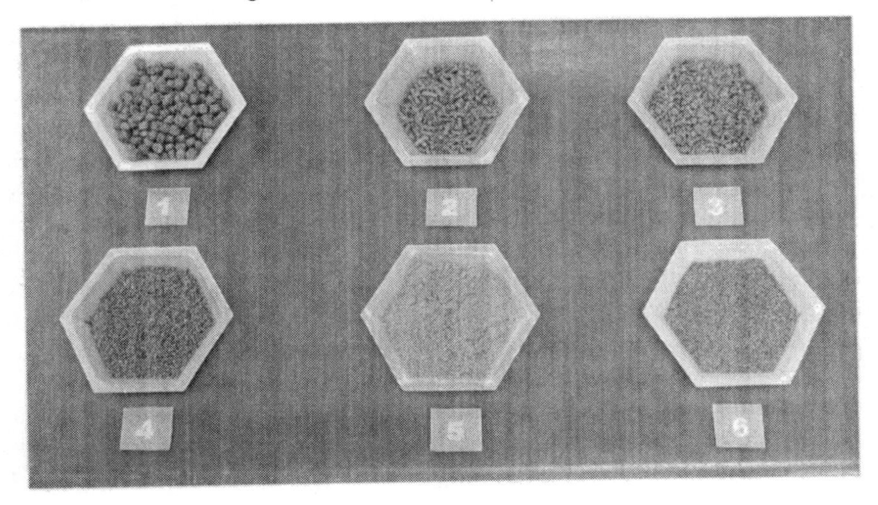

FIGURE 14.5. Various types of tilapia feeds: (1) extruded (floating); (2) steam-pelleted (sinking); (3) large crumbles; (4) small crumbles; (5) fine meal (<0.5 mm); and (6) coarse meal (0.5 to 1.0 mm). *Source:* Image courtesy of California Pellet Mill/Pacific Ltd.

Meals and Crumbles

Feeds of a small particle size (meals or crumbles) are needed for feeding fry and small fingerlings. Meal-type feeds are usually prepared by either reducing the particle size of a steam-pelleted or extruded feed by grinding and screening to the appropriate size or by finely grinding feed ingredients to a particle size of less than 0.5 mm and mixing the ground ingredients according to the proportion given in the formulae. Crumbles are usually prepared by crushing (crumbling) pelleted or extruded feeds and screening for proper size. If meal-type feeds are made from feed pellets and then reground to the proper particle size instead of simply grinding and mixing feed ingredients, water-soluble nutrients are less likely to be lost to the water. Since water-soluble vitamins are easily lost to the water, fry feeds should be overfortified with those vitamins. Spraying fat on the surface of meal or crumble feeds improves water stability and floatability, and reduces nutrient loss to the water.

Moist Feeds

In some developing countries, moist feeds using locally available ingredients are made and used on the farm. Moist feeds are usually less expensive than commercial dry feeds. Equipment to make such feed is simple since only a hammer mill (for grinding dry ingredients if needed) and a food grinder or a pellet machine is needed, and no heating and drying are required. Moist feeds are prepared by adding water and/or fresh tissues (e.g., liver, blood, ground fish, or fish processing waste) to dry ingredients and processing to form moist feed strands by using a food grinder or a pellet machine to produce pellets. The resulting feed strands or pellets generally contain 20 to 40 percent moisture. Addition of a binding agent to the diet improves water stability.

Moist feeds are susceptible to microorganism or oxidation spoilage unless they are used immediately or stored frozen. Fresh fish tissues should be heated to destroy possible pathogens and thiaminase activities. Humectants like propylene glycol and sodium chloride may be added to lower water activity to prohibit bacterium growth. Fungistats like propionic acid or sorbic acid is used to retard mold growth. These diets do not require frozen storage, but must be packaged in hermetically sealed containers and stored at low temperatures. They also need to be overfortified with vitamins because moisture enhances oxidative loss of vitamins. Stabilized vitamin C should be used.

Medicated Feeds

Systemic bacterial infections in cultured fish can be successfully treated with medicated feeds containing antibiotics if the diseased fish are treated at the early stage of the disease outbreak. However, feeding medicated feeds should not be substituted for proper management practices, since medicated feeds are expensive and their efficacy varies depending on the state and severity of the disease. Efforts should be made to identify and eliminate possible stress factors that weaken the fish's immune system and induce infections.

There are only four antibiotics that have been approved by the U.S. Food and Drug Administration (FDA) to be incorporated in feeds to treat specific diseases of certain fish species cultured for human foods. These are Romet 30® (Alpharma, Inc., Animal Health Division, Fort Lee, New Jersey, USA), Terramycin® (Philbro Animal Health, Fort Lee, New Jersey, USA), Aquaflor® (Schering-Plough Animal Health, Summit New Jersey, USA); and sulfamerizine. Sulfamerizine is no longer available in the United States. Presently there are no FDA-approved antibiotics to treat bacterial infections of tilapia in the United States.

Romet 30, a combination of two drugs, sulfadimethoxine and ormetoprim (5:1), is registered by the FDA for control of enteric septicemia of catfish (ESC) and furunculosis in salmonids. Romet-medicated feed is fed at a rate sufficient to deliver 50 mg of the active antibiotics per kilogram fish per day for five days. Commercial Romet-medicated feed for channel catfish typically contains 5 g of the active antibiotics per kilogram feed and is fed at a rate of 1 percent body weight per day. Romet-medicated feeds are not as palatable as regular feeds to channel catfish due to the bitter taste of the ormetoprim component. However, palatability problems can be partially overcome by adding at least 16 percent fish meal to the feed (Robinson et al. 1990). Romet is heat stable and therefore is typically incorporated in a floating feed. Mandatory withdrawal periods are 3 and 42 days for catfish and salmonids, respectively, before fish can be slaughtered and processed.

Terramycin, or oxytetracycline, is a broad-spectrum antibiotic that is registered by FDA to control ESC and pseudomonas disease in catfish, and several bacterial infections in salmonids (ulcer disease, furunculosis, bacterial hemorrhagic septicemia, and pseudomonas disease). Terramycin-medicated feed is fed at a rate sufficient to deliver 2.5 to 3.75 g active oxytetracycline per 45 kg fish per day for 10 days. Commercial Terramycin-medicated feed for channel catfish typically contains 5.5 g active oxytetracycline per kilogram diet and is fed at a rate of 1 to 1.5 percent body weight per day. Terramycin-medicated feeds have been manufactured as sinking pellets because the antibiotic is heat-labile and does not withstand the high temperatures required to make floating pellets. However, a new "cold" extrusion process has been developed to make floating Terramycin-feeds that allow the feeder to observe the fish feed during a bacterial disease episode. A 21-day withdrawal period for catfish and salmonids is required before fish are slaughtered and processed.

Aquaflor, or florfenicol, is also a broad-spectrum antibiotic, registered by the FDA in 2005 to control ESC in channel catfish. The antibiotic can be used only under the professional supervision of a licensed veterinarian through a veterinary feed directive order. Aquaflor-medicated feed is fed at a rate sufficient to deliver 10 mg active florfenicol per kilogram fish per day for 10 days. Commercial Aquaflor-medicated feed for channel catfish typically contains 500 mg active florfenicol per kilogram diet and is fed at a rate of 2 percent body weight per day. Aquaflor is heat stable and therefore is typically incorporated in a floating feed. A 12-day withdrawal period for channel catfish is required before fish are slaughtered and processed.

REFERENCES

Dale, N. (2000). Ingredient analysis tables: 2000. *Feedstuffs Reference Issue* 71:24-25.

Degani, G., and A. Revach (1991). Digestive capabilities of three commensal fish species: Carp, *Cyprinus carpio* L., tilapia, *Oreochromis aureus* × *O. niloticus,* and African catfish, *Clarias gariepinus* (Burchell 1822). *Aquaculture and Fisheries Management* 22:397-403.

El-Sayed, A.F.M. (1998). Total replacement of fish meal with animal protein sources in Nile tilapia, *Oreochromis niloticus. Aquaculture Research* 29:275-280.

Higgs, D.A., B.S. Dosanjh, M. Little, R.J.J. Roy, and J.R. McBride (1989). Potential for including canola products (meal and oil) in diets fro *Oreochromis mossambicus* × *O. aureus* hybrids. In M. Taeda and T. Watanabe (Eds.), *Proceedings of the Third International Symposium on Feeding and Nutrition in Fish* (pp. 301-314). Toba, Japan.

Jackson, A.J., B.S. Capper, and A.J. Matty (1982). Evaluation of some plant proteins in complete diets for the tilapia *Sarotherodon mossambicus. Aquaculture* 27:97-109.

Lovell, T. (1989). *Nutrition and feeding of fish.* New York: Van Nostrand Reinhold.

Mbahinzireki, G.B., K. Dabrowski, K.-J. Lee, D. El-Saidy, and E.R. Wisner (2001). Growth, feed utilization and body composition of tilapia (*Oreochromis* sp.) fed with cottonseed meal-based diets in a recirculating system. *Aquaculture Nutrition* 7:189-200.

NRC (National Research Council) (1993). *Nutritional requirements of fish.* Washington, DC: National Academy Press.

Ofojekwu, P.C., and C. Ejike (1984). Growth response and feed utilization in the tropical cichlid *Oreochromis niloticus* (Linn.) fed on cottonseed-based artificial diets. *Aquaculture* 42:27-36.

Robinson, E.H., J.R. Brent, J.T. Crabtree, and C.S. Tucker (1990). Improved palatability of channel catfish feeds containing Romet-30. *Journal of Aquatic Animal Health* 2:43-48.

Robinson, E.H., B.B. Manning, and M.H. Li (2004). Feeds and feeding practices. In C.S. Tucker and J.A. Hargreaves (Eds.), *Biology and culture of channel* catfish (pp. 324-348). Amsterdam, the Netherlands: Elsevier Science Publishers.

Robinson, E.H., S.D. Rawles, P.W. Oldenberg, and R.R. Stickney (1984). Effect of feeding glandless or glanded cottonseed products and gossypol to *Tilapia aurea. Aquaculture* 38:145-154.

Takeuchi T., M. Hernandez, and T. Watanabe (1994). Nutritive value of gelatinized corn meal as a carbohydrate source to grass carp and hybrid tilapia *Oreochromis niloticus* × *O. aureus. Fisheries Sciences* 60:573-577.

Wee, K.L., and S.W. Shu (1989). The nutritive value of boiled full-fat soybean in pelleted feed for Nile tilapia. *Aquaculture* 81:303-314.

Wu, Y.V., R. Rosati, D.J. Sessa, and P. Brown (1995). Utilization of corn gluten feed by Nile tilapia. *The Progressive Fish-Culturist* 57:305-309.

Wu, Y.V., R. Rosati, K. Warner, and P. Brown (2000). Growth, feed conversion, protein utilization, and sensory evaluation of Nile tilapia fed diets containing corn gluten meal, full-fat soy, and synthetic amino acids. *Journal of Aquatic Food Product Technology* 9:77-88.

Wu, Y.V., K.W. Tudor, P.B. Brown, and R.R. Rosati (1999). Substitution of plant proteins or meat and bone meal for fish meal in diets for Nile tilapia. *North American Journal of Aquaculture* 61:58-63.

Chapter 15

Feeding Practices

Chhorn E. Lim
Carl D. Webster
Menghe H. Li

INTRODUCTION

Aquaculture production of tilapia has expanded rapidly throughout the world over the past decade and is expected to continue to grow in the foreseeable future: as world population grows, demand for quality seafood continues to rise and supply from capture fisheries becomes limited. Parallel to the growth of the industry has been a trend toward intensification of culture practices where fish are stocked at high densities aiming to obtain higher yield per unit area. In contrast to extensive and semi-intensive culture system where fish derive all or most of their nutritional needs from natural pond food organisms, tilapia reared under intensive systems depend largely on compounded feeds. Successful operations of these production systems are dependent, among other factors, on good nutrition. Since feed costs represent the major proportion of the overall production cost in intensive culture operations, the availability of low-cost, nutritionally balanced diets of appropriate water stability, size, shape, texture, odor, and taste, and the use of good feeding practices are the two most important requisites for sustainable and successful fish production. Without adequate intake of suitable quality feeds, fish are unable to grow and reproduce efficiently and remain healthy. Dupree (1984) indicated that good feeding practices are as important to the aquaculturist as the availability of good feeds. Since fish are fed in the water, excess or uneaten feed represents not only a direct economic loss, but also causes environmental degradation that can bring about stress, diseases, slow growth, low survival, and poor harvest. Thus, the use of good feeding management for different sizes and species cultured under diverse environmental conditions and production systems is of critical importance for aquaculturists.

doi:10.1300/5513_15

FEEDS

Natural Foods

In early developmental stages of almost all fishes, including herbivores, the larvae, fry and young juveniles are typically carnivores and feed on zooplankton and small invertebrates, especially crustaceans (Bowen 1982; Beveridge and Baird 2000). These authors reported that the characteristic diet of adult tilapia is plant material and/or detritus of plant origin. Phytoplankton, blue-green and green algae, macrophytes, periphyton, and bacteria are all common constituents of natural diets of all species, although fish of the genus *Tilapia* are macrophyte-feeders, in which the adult feed mainly on filamentous algae and higher aquatic plants whereas fish of the genus *Oreochromis* are microphagous; their feeding regime consists principally of phytoplankton, zooplankton, detritus, and benthic organisms. However, there is an extensive overlapping among the diet composition of tilapia. Tilapia species that feed on macrophytes also ingest the attached algae, bacteria, and detritus. This attached material is an important component of the diet and, on a dry weight basis, may constitute more than 25 percent of the supporting macrophyte consumed by tilapia (Bowen 1982). Epiphyte feeders also frequently ingest some of the supporting macrophyte. Despite the diversity of food resources consumed by adult *Tilapia* spp. and *Oreochromis* spp., they are commonly regarded as opportunistic omnivores with a strong tendency toward herbivory. Their choice of diets is determined by species, size, age, seasonal variations, abundance of food resources, and inter- and intraspecific competition (Philippart and Ruwet 1982; Beveridge and Baird 2000).

Tilapia are often regarded as filter-feeders because they can effectively ingest phytoplankton as small as 5 μ in diameter. However, because tilapia have short and widely spaced gill rakers, they cannot physically filter water through gill rakers as efficiently as silver and bighead carps. The collecting process of the minute food particles is achieved by entrapment among gill rakers and filaments in strands and aggregates of mucus secreted by mucous cells associated with the gill arches (Beveridge and Baird 2000). Food and feeding habit of tilapia are described in Chapter 1, "Biology," of this book.

Artificial Feeds

Artificial or compound feeds are mixtures of feedstuffs, and vitamin and mineral premixes that are formulated to contain desired levels of essential nutrients and energy. Feeds for tilapia may contain adequate levels of all of the required nutrients (complete feeds) or they may be formulated with the notion that certain nutrients essential for growth and other physiological functions will be available in the culture environments (supplemental feeds). Supplemental feeds that may be low in protein, high in energy, and deficient in some vitamins and minerals are commonly used for feeding tilapia stocked in ponds at low to moderate densities. It is assumed that the deficient nutrients will be provided by natural food organisms available in the culture environments. However, due to the extreme variation in the culture practices and the availability of natural food, it is practically impossible to formulate feeds to efficiently supplement the contribution of natural foods (Lim 1989). In intensive culture, such as in cages, raceways, tanks, or heavily stocked ponds, tilapia rely on the prepared feeds as a source of nutrients and energy. Thus, a nutritionally complete feed containing adequate levels of essential nutrients and digestible energy to meet the nutritional needs of tilapia is required. More detail information on least-cost formulation of nutritionally balanced diets and feed manufacturing processes can be found in Chapter 14 of this book.

Normally, feeds of different nutrient contents and physical forms are used at various life stages. Larvae and fry of tilapia have been reared successfully using diets containing 25 to 45 percent crude protein. It is suggested, however, that larval or starter feeds for tilapia reared intensively in hatcheries, such as in tanks or hapas, should contain about 45 to 50 percent crude protein and 8 to 10 percent fat. Since the digestive tract of first-feeding larvae is not fully functional as that of juveniles or adults, larval feeds must be formulated using high-quality and highly digestible ingredients; they must have appropriate particle size (≤0.5 mm), texture, density or buoyancy, taste, and odor so that they are readily consumed. Larval or starter feeds are commonly fed to swim-up fry for approximately 2 to 4 weeks. The final size of the larvae varies depending on the rearing systems, stocking rate, feed quality, feeding management, and environmental conditions. At the end of larval rearing phase, fish are transferred to nursery tanks or ponds and fed larger particle size (0.5 to 1.2 mm) larval feeds or fry feeds formulated to contain approximately 40 to 45 percent crude protein and 8 percent fat. When fish reach an average weight of 5 to 6 g, they are switched to fingerling feeds (crumbles or small pellets with particle sizes of 1.2 to 2 mm) containing 35 to 40 percent crude protein. Fish are reared in nursery systems to an advanced fingerling size (30 to 50 g) prior to stocking in grow-out systems. A variety of feeds formulated for other species such as common carp and channel catfish have been used successfully for grow-out of tilapia (Lim 1989, 1997). Production diets for tilapia reared in ponds usually contain 24 to 28 percent protein, whereas those used in intensive culture systems such in cages, tanks, or raceways contain 28 to 32 percent protein. Most farmers use grow-out feeds for broodstocks. These diets have been used successfully for broodfish held in various seed production facilities such as ponds, tanks, or hapas. In clear water tank systems, Nile tilapia *O. niloticus* broodstock that were fed a complete diet containing 30 percent crude protein sustained high output of seeds over periods of up to 200 days (Little 1989). De Silva and Radampola (1990) reported that, under intensive tank systems, the fecundity and frequency of spawning of Nile tilapia broodstock are optimized at a dietary protein level of 25 to 30 percent. Feeding good-quality feed containing 25 to 30 percent protein for both conditioning and spawning results in accelerated fish growth and high seed production, but limits the reproductive life span of broodstock as the fish grow too quickly to a size that is difficult to manage (Little and Hulata 2000).

FEEDING PRACTICES

Feeding Larval Fish

Newly hatched fry use their yolk sac as a source of nutrient and energy. As soon as the yolk sac is absorbed they begin to swim up and search for food. At this stage a variety of foods is used depending on the species. Species with immature digestive systems at the initiation of feeding are more difficult to feed and usually require live foods as a part of their diet (NRC 1993). Those that, at the time of first feeding, have structurally and functionally differentiated digestive tracts such as in the case of tilapia will accept prepared diets immediately from the onset of feeding (Stickney 1979; NRC 1983, 1993). Fry in larval rearing units should be fed with the starter diets or larval feeds 8 to 10 times daily (see Table 15.1) or continuously. This is usually accomplished by the use of automatic feeders. At this stage, fry are generally fed in excess or at least to satiation to permit larvae to feed on demand. A feeding rate of 30 to 50 percent of body weight daily has been successfully used for larval rearing of tilapia (Table 15.1). Due to the fast growth rate of fry, frequent adjustment of daily feed allowances is required. Overfeeding of fry is not unusually expensive as overfeeding larger fish because the total amount of feed used at this stage is relatively small. However, precaution should be taken to prevent fouling of water by removal of the accumulated waste from the tank bottom and replenishing the tank with clean water. The daily feeding rate (percent of biomass) and frequency gradually decrease as the fish grow. Feed particle size should be increased in proportion to the size of fish. When necessary, feeds should be screened to remove fines or particles that are too small or too large. If the feed particles are too small or too large, they are not efficiently consumed by fish and

TABLE 15.1. Suggested daily feeding rate and frequency for various sizes of tilapia at 28°C.

Fish size	Feeding rate (percent of body weight)	Feeding frequency (No. of times/day)
2 days old to 1 g	30-10	8
1-5 g	10-6	6
5-20 g	6-4	4
20-100 g	4-3	3-4
>100 g	2-3	2-3

Source: Modified after Lim (1989).

more feed is wasted. Also, the amount of nutrient leaching into the water is greater for smaller particle size feed (Lim and Cuzon 1994).

Feeding Grow-Out Fish

Daily Feed Allowance

Daily feed allowance, or ration, is the quantity of feed offered to fish per day. Feed allowance is calculated based on a percentage of body weight (feeding rate) multiplied by total fish weight or total biomass. The latter can be obtained based on previous records of fish growth and survival, by sampling (average fish weight × total number of fish) or estimating weight gain based on a feed conversion ratio [(total feed fed:feed conversion ratio) + (total initial weight of fish)] or feed efficiency ratio [(total feed fed × feed efficiency ratio) + (total initial weight of fish)]. Ideally, feed allowances should be adjusted as frequently as possible. In practice, however, weekly or fortnightly adjustment is sufficient.

Determining the optimum feeding rate is difficult because feed intake is under close metabolic, endocrine, and neural controls that are affected by species, fish size, water temperature, salinity, dissolved oxygen (DO), concentrations of metabolites or pollutants, and photoperiod. The availability of natural food, dietary levels of nutrient and energy, diet palatability or taste, feeding frequency, the amount of feed consumed the previous day, social factors, and fish health also affect feed intake. *Tilapia randalli* consume more feed than *O. niloticus* of comparable size (Ballarin and Haller 1982).

Fish size influences feeding rates. It is generally recognized that younger or smaller fish require more energy for metabolism per unit body weight and have a faster growth rate than larger or adult fish. Thus, smaller fish consume more feed on a percent weight basis than larger fish. The suggested feeding rates for different sizes of tilapia are given in Table 15.1.

Water temperature influences the metabolic rate and energy expenditure, and these have profound effects on feed consumption and growth. Fish have been shown to adjust feed intake to the same level at which the metabolic rate was affected by temperatures (Brett 1979). The optimum temperature for growth of various species of tilapia has been reported to range from 26°C to 32°C. Tilapia are very sensitive to cold. They become generally inactive and feeding stops at about 16°C and death may occur when exposed to temperatures below 12°C for a few days, even though some species may be more tolerant than others (Ross 2000). Fish will consume less feed in cold weather than in warmer weather. Examples of suggested feeding rates at temperatures below the optimum range is given in Table 15.2.

Although tilapia are principally freshwater fish, many species can tolerate and grow well in a wide range of water salinity. Nile tilapia can grow as well in 18 ppt saline water as in freshwater and can adapt to at least 50 ppt (Chervinski 1961). Ross (2000), however, indicated that optimum salinity for growth for Nile and blue tilapia *(O. aureus)* and their hybrids is lower than 18 ppt. Brett (1979) reported that fish regulate their plasma ions such that the osmotic pressure of their body fluids is equivalent to about 10 ppt salinity. Thus, salinity fluctuation beyond the optimum range would affect feed consumption due to changes in metabolic processes such as an increase in energy expenditure associated with active transport of ions to

TABLE 15.2. Suggested feeding rate and feeding frequency for tilapia at different water temperatures.

	Average body weight <100 g		Average body weight >100 g	
Water temperature (°C)	**Feeding rate (percent of NR[a])**	**Feeding frequency (No./day)**	**Feeding rate (percent of NR)**	**Feeding frequency (No./day)**
32 > t > 26	100	4	100	2-3
26 > t > 24	90	3	90	2
24 > t > 22	70	2	60	2
22 > t > 20	50	2	40	2
20 > t > 18	30	1-2	20	1
18 > t > 16	20	1	10	1
16 > t	No feeding		No feeding	

Source: Modified after Luquet (1991).

Note: [a]Normal feeding rate.

maintain the internal body fluid osmotic pressure. Morgan et al. (1997) showed that the physiological changes associated with acclimation of Mozambique tilapia to seawater represent a significant short-term energy loss and may represent as much as 20 percent of total body metabolism after four days in seawater.

Tilapia, compared to most other fish species, can tolerate relatively poor water quality and even survive after a short-term exposure in water with DO of 0.1 ppm or less (Chervinski 1982). This leads many to believe that tilapia can grow and thrive in such conditions. However, low DO and high concentrations of metabolites, such as ammonia and nitrite, negatively affect feed consumption, digestion, nutrient metabolism and growth. Rappaport et al. (1976) observed that the growth of tilapia is reduced at DO levels below 25 percent saturation.

Un-ionized ammonia (NH_3) and nitrite are toxic to fish if present at high concentrations. Nitrite, when enters into the blood through the gill, binds with hemoglobin to form methemoglobin that has very poor oxygen carrying capacity leading to inadequate supply of oxygen to the tissues.

Fish, including tilapia, regulate the level of feed intake such that certain amount of energy and nutrients are taken in to satisfy their needs. Thus, the energy content of the diet determines the amount of feed consumed. There is also evidence that, when fed to satiation with isocaloric diets, increasing dietary protein beyond the requirement level decreases feed consumption.

Palatability of the diets also influence the amount of feed intake. This is particularly important for early stage of larval rearing. In semi-intensive culture systems, natural food can make a significant contribution to the nutrient requirements of fish. Under these conditions, the amount of feed offered should be less than that for fish grown under intensive systems. Feeding frequency also influences the amount of feed consumed by fish. Research at Auburn University showed that pond-grown channel catfish consumed 20 percent more feed when fed twice daily than fed once (Lovell 1989).

Feed intake is also affected by nutritional status and the amount of feed consumed on previous days. Fish may consume less feed following a period of intense feeding but feed intake increases following periods of starvation or feed deprivation. It has also been recognized that the presence of dominant, more aggressive fish (usually male or larger fish) in a population affects the appetite or feeding activity of subordinate fish. Social hierarchy commonly occurs in mixed-sex culture of tilapia (males grow faster than females) or when nonuniform size fish are stocked. High-stocking densities decrease the aggression of the dominant fish.

Like any other animal, diseased fish eat poorly, if at all. Feeding with medicated feeds in which drugs or chemicals are incorporated in the formula before processing or coated on the surface of the feed by oil or fat is probably the only means to treat diseased fish in ponds. Robinson et al. (1994) reported that some catfish producers do not feed fish during the outbreaks of certain diseases whereas some producers reduce feeding to only once every other day.

Since there are many factors that influence feed consumption, feeding as much as the fish can consume in a given time period (satiation feeding) may be a better alternative to feeding a prescribed amount based on percentage of fish biomass. If different sizes of fish are present in the culture system such as for mixed-sex culture, it is recommended that they be fed to apparent satiation. Providing as much feed as the fish will eat gives a better opportunity for the smaller, less aggressive fish to receive their share, and therefore improves feed efficiency and minimizes size heterogeneity. Satiation feeding can be easily done for tilapia that can complete their meal within 15 to 20 min and also when floating feeds are used.

Although it is recommended that fish should be fed as much as they will eat, at high-standing crops of fish it may be impossible to satiate the fish and maintain the water quality at an acceptable level (Robinson et al. 1994). The quantity of feed offered must not exceed the capacity of the culture system to assimilate or oxidize the waste products. It has been suggested that the daily feed allowances for channel catfish in static ponds without aeration should not exceed 35 to 45 $kg \cdot ha^{-1} \cdot day^{-1}$ (Swingle 1968; Shell 1968; NRC 1983). Cole and Boyd (1986) reported that minimal aeration was required at a feeding level of 56 $kg \cdot ha^{-1} \cdot day^{-1}$. They reported, however, that in ponds with a feed allowance of 112 $kg \cdot ha^{-1} \cdot day^{-1}$ and above, aeration was required constantly at night. Overfeeding should be avoided to maintain good feed efficiency, and minimize feed wastage and deterioration of water quality. Clark et al. (1990) obtained maximum growth of Florida red tilapia reared in marine cages at feeding rates near (90 percent) satiation. They observed, however, that feed cost could be significantly reduced with minimal growth reduction by feeding at 70 percent of satiation.

Feeding Method

Feeding methods can be grouped into two categories, hand feeding and the use of mechanical devices (feed blower, automatic feeder, and demand feeder). For farms with small ponds, tanks, raceways, or cages and where labor is inexpensive, hand feeding is commonly practiced. Hand feeding, even though labor intensive, would save some capital expenditure for the purchase of feeding devices. It also has other advantages over other methods of feeding. Hand feeding enables the feeder to better observe fish feeding activities and behaviors, thereby gauging the health of the fish. By observing the fish while they eat, the feeder can regulate the amount of feed fed to prevent overfeeding or minimize wastage. Floating or expanded pellets can be a very helpful management tool in assessing feeding response, especially for the feeder who lacks experience. Floating feed remains on the water surface allowing observation of leftover feed whereas conventional compressed pellet sinks to the bottom. However, the cost of floating feed is higher than that of sinking feed.

Although much can be gained from hand feeding, this method is not feasible in large commercial farms and/or developed countries where labor costs are high. The most common method of feeding large ponds is blowing the feed uniformly onto the water surface along the dikes using feed blowers that are either mounted on or pulled behind vehicles. A funnel-shaped hopper mounted on a scale measures the amount of feed delivered from the hopper. Feed drops from the hopper into the air stream generated by a high-volume blower and is blown onto the surface of the ponds. Robinson et al. (1994) suggested that, on large channel catfish farms, feed should be blown over a large pond area to provide equal feeding opportunities for as many fish as possible. It is desirable to feed all sides of the pond, but this is generally not practiced on large farms where several ponds of fish must be fed in a limited time period. Also, prevailing winds dictate that feed must be distributed along the upwind levee to prevent feed from washing ashore.

Automatic feeders, driven by timeclock (belt feeders) or electrical devices (timed feeders), allow farmers to specify or preset the amount of feed, and the number and time of feeding. This enables the farmers to feed the fish many times each day and at any time during a 24-h period. Belt feeders are often used in larval rearing, whereas electrically driven devices are used in conjunction with fish cultured in tanks, cages, or raceways. A major disadvantage associated with this method of feeding is the potential malfunctioning of the devices and waste of feed because automatic feeders deliver feed according to the schedule set by the operator regardless of the appetite of fish.

Another alternative feeding device is the demand feeder or self-feeder. Demand feeders consist of a feed hopper with a top opening for loading the feed and a tapered conical bottom opening that serves as a movable gate for feed delivery. Attached to the gate is a rod (pendulum) with the tip extending down into the water where it can be activated by the fish. Lateral movements of the rod cause the gate mechanism to open, allowing small amount of feed to discharge from the hopper into the water. As long as the fish continue to trigger the rod, feed will continue to flow out. Thus, the use of demand feeders permit *ad libitum* feeding and do not require determination of daily feed allowances as in other methods of feeding. A major disadvantage of the self-feeder is the unnecessary operation of the gate mechanism by fish that results in feed waste. Comparison of hand feeding and self-feeding for common carp cultured in raceways showed that hand feeding provided 10 percent better feed efficiency than feeding by use of demand feeder (Lim 1985). Clark et al. (1990), however, indicated that self-feeder may prove to be the best feeding method for tilapia grown in off-shore marine cages since it provides good growth and feed conversion with reduced labor.

Whichever method of feeding is employed, it is necessary that fish be fed to meet their nutritional needs and feed be administered in such a way that it can be consumed immediately by fish and all fish receive their share. Fry are relatively weak swimmers and cannot travel a long distance to obtain the feed. In this case, it is essential that feed should be distributed over the pond area as uniformly as possible. Tilapia, like other animals, are creatures of habit; therefore, feeding them at the same time, in the same manner and in the same area each day is important for immediate feed consumption.

Feeding Frequency

Frequency of feeding varies with fish size or life stage. As has been previously indicated swim-up fry should be fed 8 to 10 times daily. Feeding frequency decreases as the fish grows. Three to four times feeding per day is very common for small fingerlings. Feeding frequency decreases to 1 to 3 times daily for grow-out (Tables 15.1 and 15.2) and 1 to 2 times daily for broodstock.

Tilapia have been reported to benefit from multiple daily feedings. Due to continuous feeding behavior and smaller stomach capacity of tilapia they respond better to more frequent feeding than do channel catfish and salmonids. Too frequent feeding, however, is not beneficial and is uneconomical. Nile tilapia fingerlings grew faster when fed four times a day than when fed eight times a day (Kubaryk 1980).

Multiple daily feeding also reduces the exposure time of feed in the water, thus reducing the disintegration of pellet and leaching rate of water-soluble nutrients. More frequent feeding also increases the rate of feed consumption. At water temperature below the optimum level for growth, less frequent feeding is required as compared to feeding at optimum temperature (Table 15.2). Less frequent feeding may also be desirable under certain circumstances such as during disease episodes, extreme high temperature and unfavorable water quality conditions.

Daily Feeding Schedule

The time of day to feed the fish is also of critical importance. It has been reported that tilapia show high feeding activity at dawn and dusk, whereas channel catfish feed more actively during the dark phases of the day. A pond study with channel catfish showed, however, no difference in growth, feed intake, feed efficiency, and survival among fish fed to apparent satiation at 8:30 a.m., 4:00 p.m. or 8:00 p.m. (Robinson and Li 1996). This suggests that fish are able to adapt/adjust their feeding rhythms according to the daily feeding schedule. Lovell (1977, 1989) suggested that the best time of the day for feeding channel catfish is influenced by the water DO and temperature. Low DO depresses feeding activity and reduces feed consumption. Dissolved oxygen content in static ponds without supplemental aeration is normally lowest in early morning, before day break. Thus, fish should not be fed until well after sunrise when the DO has risen to a level for active feeding by fish. Some scientists argue that fish should not be fed until the DO has risen to 75 percent of the saturation level. In large commercial operations, due to large number of pond to be fed, it is most practical to begin feeding in the morning as the DO begins to increase (Robinson et al. 1994). Also, fish should not be fed late at night so that their increased oxygen requirement subsequent to ingesting the feed will not coincide with the period of low DO in the pond. Robinson et al. (1994) reported that peak oxygen demand for channel catfish usually occurs 6 h after feeding.

Good quality feed can result in poor and costly production unless proper feeding practices are used. However, because numerous factors influences fish feeding, feeding practices for tilapia have not been worked out. Even for channel catfish production in which fish have been intensively fed for many years, there is still no one best method of feeding catfish (Robinson and Li 1996). They indicated that no two fish ponds are exactly alike, and, consequently, feeding behavior in individual pond may differ greatly or feeding activity in a particular pond may differ greatly from day to day. It is recommended that feeding practices that have been used successfully for some species such as channel catfish and common carp can be adapted for tilapia by proper consideration of their physiological and nutritional needs, feeding behaviors and habits, culture systems, and environmental conditions.

REFERENCES

Balarin, J.D., and R.D. Haller (1982). The intensive culture of tilapia in tanks, raceways and cages. In J.E. Muir and R.J. Roberts (Eds.), *Recent advances in aquaculture* (pp. 265-256). Boulder, CO: Westview Press.

Beveridge, M.C.M., and Baird, D.J. (2000). Diets, feeding and digestive physiology. In M.C.M. Malcomb, and B.J. Mc.Andrew (Eds.), *Tilapias: Biology and exploitation* (pp. 59-87). Boston: Kluwer Academic Publishers.

Bowen, S.H. (1982). Feeding, digestion and growth—Qualitative considerations. In R.S.V. Pullin, and R.H. Lowe-McConnell (Eds.), *The biology and culture of tilapias* (pp. 141-156). Manila, Philippines: International Center for Living Aquatic Resources Management.

Brett, J.R. (1979). Environmental factors and growth. In W.S Hoar, D.J. Randall, and J.R. Brett (Eds.), *Fish physiology* (Vol. VIII, pp. 121-159). New York: Academic Press, Inc.

Chervinski, J. (1961). Laboratory experiments on the growth of *Tilapia nilotica* in various saline concentrations. *Bamidgeh* 13:8-14.

———. (1982). Environmental physiology of tilapias. In R.S.V. Pullin, and R.H. Lowe-McConnell (Eds.), *The biology and culture of tilapias* (pp. 119-128). Manila, Philippines: International Center for Living Aquatic Resources Management.

Clark, J.H., W.O. Watanabe, and D.H. Ernest (1990). Effect of feeding rate on growth and feed conversion of Florida red tilapia reared in floating marine cages. *Journal of the World Aquaculture Society* 21:16-24.

Cole, B.A., and C.E. Boyd (1986). Feeding rate, water quality and channel catfish production in ponds. *The Progressive Fish-Culturist* 81:25-29.

De Silva, S., and K. Radampola (1990). Effect of dietary protein level on the reproductive performance of *Oreochromis niloticus*. In R. Hirano, and I. Hanyu (Eds.), *Proceedings of the Second Asian Fisheries Forum* (pp. 559-563). Manila, Philippines: Asian Fisheries Society.

Dupree, H.K. (1984). Practical feeding. In E.H. Robinson, and R.T. Lovell (Eds.), *Nutrition and feeding of channel catfish* (Revised, pp. 34-40). Southern Cooperative Series Bulletin No. 296, Auburn University, Auburn, Alabama.

Kubaryk, J.M. (1980). Effect of diet, feeding schedule and sex on food consumption, growth and retention of protein and energy by tilapia. Doctoral dissertation, Auburn University, Auburn, Alabama.

Lim, C. (1985). *Fish nutrition and feeding technology research in Indonesia*. Jakarta, Indonesia: Central Research Institute for Fisheries, Agency for Agricultural Research and Development, Ministry of Agriculture.

———. (1989). Practical feeding—Tilapias. In T. Lovell (Ed.), *Nutrition and feeding of fish* (pp. 163-183). New York: Van Nostrand Reinhold.

———. (1997). Practical feeding of tropical aquatic species. In *Proceedings, International Symposium on Finfish and Crustacean Nutrition, November 8-11, 1995,* Sao Paulo, Brazil, pp. 159-171.

Lim, C., and G. Cuzon (1994). Water stability of shrimp pellet: A review. *Asian Fisheries Science* 7:115-127.

Little, D.C. (1989). An evaluation of strategies for production of Nile tilapia *(Oreochromis niloticus)* fry suitable for hormonal treatment. Doctoral dissertation, University of Sterling, Scotland.

Little, D.C., and G. Hulata (2000). Strategies for tilapia seed production. In M.C.M. Malcomb and B.J. Mc.Andrew (Eds.), *Tilapia: Biology and exploitation* (pp. 267-326). Boston: Kluwer Academic Publishers.

Lovell, R.T. (1977). Feeding practices. In R.R. Stickney, and R.T. Lovell (Eds.), *Nutrition and feeding of catfish* (pp. 50-55). Southern Cooperative Series Bulletin 218, Auburn University, Auburn, Alabama.

———. (1989). *Nutrition and feeding of fish.* New York: Van Nostrand Reinhold.

Luquet, P. (1991). Tilapia, *Oreochromis* spp. In R.P. Wilson (Ed.), *Handbook of nutrient requirement of finfish* (pp. 169-179). Boca Raton, LA: CRC Press.

Morgan, J.D., T. Sakamoto, E.G. Grau, and E.K. Ewama (1997). Physiological and respiratory responses of the Mozambique tilapia *(Oreochromis mossambicus)* to salinity acclimation. *Comparative Biochemistry and Physiology* 117A:391-398.

NRC (National Research Council) (1983). *Nutrient requirements of warmwater fishes and shellfishes.* Washington, DC: National Academy Press.

———. (1993). *Nutrient requirements of fish.* Washington, DC: National Academy Press.

Philippart, J-Cl., and J-Cl. Ruwet (1982). Ecology and distribution of tilapia. In R.S.V. Pullin, and R.H. Lowe-McConnell (Eds.), *The biology and culture of tilapias* (pp. 15-60). Manila, Philippines: International Center for Living Aquatic Resources Management.

Rappaport, A., S. Sarig, and M. Marek (1976). Results of tests of various aeration systems on the oxygen regime in Genosar experimental ponds and growth of fish there in 1975. *Bamidgeh* 28:35-49.

Robinson, E.H., and M.H. Li (1996). A practical guide to nutrition, feeds and feeding of catfish. Bulletin 1041, Mississippi Agricultural and Forestry Experiment Station, Mississippi State, Mississippi.

Robinson, E.H., C.R. Weirich, and M.H. Li (1994). *Feeding catfish.* Bulletin 1019, Mississippi Agricultural and Forestry Experiment Station, Mississippi State, Mississippi.

Ross, L.G. (2000). Environmental physiology and energetics. In M.C.M. Malcomb, and B.J. McAndrew (Eds.), *Tilapia biology and exploitation* (pp. 89-162). Boston: Kluwer Academic Publishers.

Shell, E.W. (1968). Feed and feeding of warm-water fish in North America. In *Proceedings of the World Symposium on Warm-Water Pond Fish Culture* (Vol. 3, pp. 310-325). FAO, United Nations, Rome, Italy.

Stickney, R.R. (1979). *Principles of warmwater aquaculture.* New York: John Wiley & Sons.

Swingle, H.S. (1968). Estimation of standing crops and rates of feeding fish in ponds. In *Proceedings of the World Symposium on Warm-Water Pond Fish Culture* (Vol. 3, pp. 416-423). FAO, United Nations, Rome, Italy.

Chapter 16

Parasites and Diseases

Craig A. Shoemaker
De-Hai Xu
Joyce J. Evans
Phillip H. Klesius

INTRODUCTION

Aquaculture of tilapia has expanded rapidly in the past decade and this trend is expected to continue due to an increased demand for tilapia in both domestic and international markets. Intensive water reuse culture systems have become established in the United States and other parts of the world (Engle 1997). Water reuse aquaculture systems have also expanded to desert areas of Israel and increased usage of these types of systems will continue where, or when, water is limited. Water reuse systems, as well as the intensification of pond culture, have resulted in an increase in the incidence and severity of disease agents present in tilapia. Prior to the development of intensive-culture systems, tilapia were considered to be more resistant to disease agents than many other cultured fish species (Roberts and Sommerville 1982). This chapter reviews the limited information available on parasites, fungi, bacterial pathogens, and viral agents of tilapia and suggests possible health management strategies to prevent and control diseases. The most important parasites include *Ichthyophthirius multifillis* (Ich) and other protozoans (e.g., *Tichodina, Trichodenella, Epistylus, Ichthyobodo*) parasitic crustaceans (*Learnea, Argulus,* and *Ergasilis*), and monogenetic and digenetic trematodes. Fungi also cause problems in cultured tilapia, with most problems occurring in hatcheries when eggs become infected. Bacterial infections caused by both Gram-positive and Gram-negative bacteria are probably responsible for greater monetary losses than the other disease agents at the present time. *Streptococcus iniae* (a Gram-positive bacterium) alone has been suggested to be responsible for more than $150 million in

doi:10.1300/5513_16

losses annually to world aquaculture. Health management strategies to prevent, control, or limit the effects of disease-causing agents in tilapia include

1. quarantine of new fish stocks;
2. cleanliness of culture facilities;
3. maintenance of water quality within normal limits;
4. nutritionally balanced diets and proper feeding strategies;
5. proper stocking rates;
6. chemotherapy or antibiotics; and
7. vaccination.

Vaccination is probably the most cost-effective and suitable approach for preventing infectious agents in tilapia aquaculture.

PROTOZOAN PARASITES

Protozoan parasites mainly cause external infections of tilapia and are commonly seen in different sizes of fish, at many geographic locations, and may be seen in different seasons. Some internal protozoans are present in tilapia, but no clinical disease has been reported. Most protozoans have a simple life cycle and reproduce by binary fission, which does not involve an intermediate host. Usually light loads of protozoans in fish do not cause problems. However, protozoan parasites may cause significant loss of fish, especially for fry and fingerlings, when parasite density is high. External protozoans cause damage to the skin, fins, and gills of fish. When the skin is the primary site of infection, fish show signs of disease including thickening of mucus, depigmentation, rubbing on the side or bottom of tanks/ponds, loss of scales, and lesions on the body surface. The fish with severe infection of the gills show respiratory distress, lethargy, listlessness, and loss of appetite. Many protozoans cause damage to both skin and gills. Protozoans are the major cause of parasitic diseases and are responsible in nearly all losses of cultured fish due to parasites (Rogers 1985; Klesius and Rogers 1995).

Ichthyophthirius multifiliis

Ichthyophthirius multifiliis (Ich or white spot) is one of the most important ciliated protozoan parasites of tilapia (Figure 16.1A,B). Ich is distributed throughout the world (Hoffman 1978) and was first described in wild tilapia. Ich has a direct life cycle and requires a fish host to survive. Infection is characterized by white spots (Figure 16.1A) where the Ich organisms

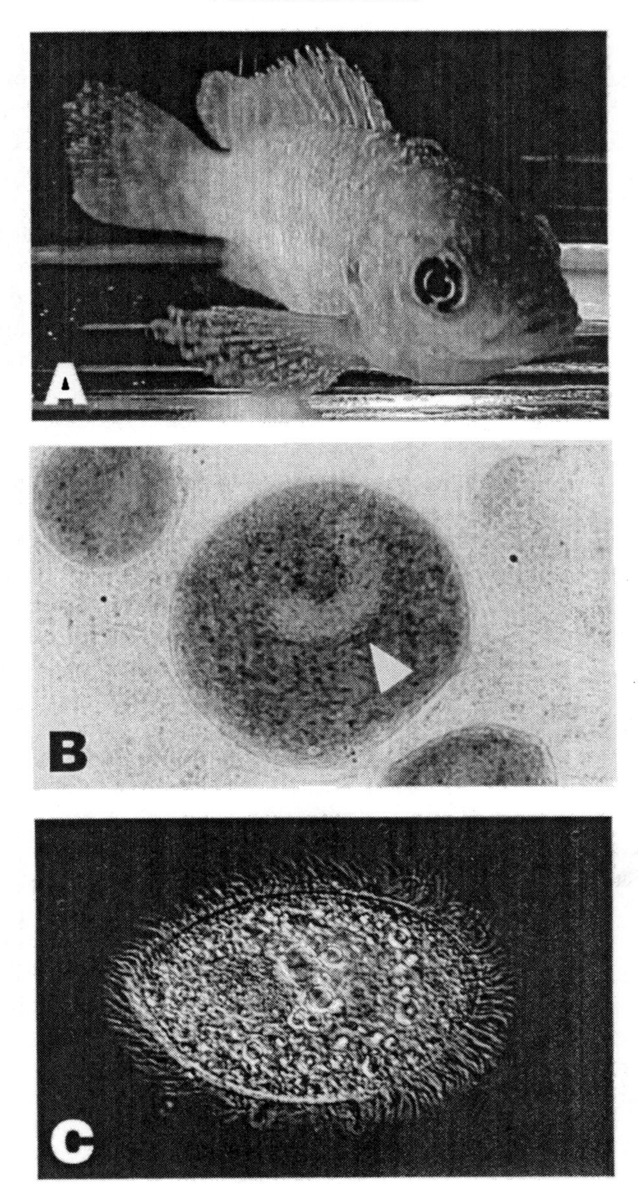

FIGURE 16.1. Photographs of *Ichthyophthirius multifiliis: (A)* tilapia infected with *I. multifiliis* showing the typical white spots; *(B)* an adult Ich, note the "C-shaped" nucleus (arrow); *(C)* an infective theront under phase-contrast microscopy.

live in the epithelium of skin and gills of infected fish (Ewing and Kochan 1992). The infective stage of Ich is the theront, possessing a pyriform or fusiform shape (Figure 16.1C). The theronts swim actively in water and attach to a fish host within 48 h after releasing from cysts (Xu et al. 2000). Once established, the theront becomes a feeding trophont and feeds on fish tissue until it reaches mature size with a "C-shaped" nucleus (Figure 16.1B). The mature trophont detaches from the epithelium and settles to a substrate where a cyst wall is excreted around the newly formed tomont. Repeated asexual division occurs within the settled tomont and results in the production of several hundreds to thousands of theronts. Tomonts then release the infective theronts and upon attachment to a new fish host, the life cycle is completed.

Subasinghe and Sommerville (1992) demonstrated that the most-susceptible life stage of the Mozambique tilapia *(O. mossambicus)* is the fry. Significant mortality can result and parasites were found in the nares, pharynx, gills, and skin of young fry. When conditions are favorable for rapid multiplication of the parasite, epizootics occur. The water temperature is one important factor of Ich development and greatly influences the duration of the Ich life cycle (Nigrelli et al. 1976). The life cycle is completed in a shorter time at warm temperatures than at lower temperatures. Outbreaks of Ich are most likely to occur when fish are stressed. Many factors are known causes of stress in fish: low dissolved oxygen, poor water quality, crowding, spawning activities, improper handling, and poor nutrition.

Ich is also one of the most significant pathogens in water reuse systems because of the density at which the fish are raised and the potential for rapid fish to fish transmission in these systems. Water temperatures of 20°C to 25°C result in high mortality due to Ich in a short time. Plumb (1997) suggested that the optimum temperature for tilapia growth may limit the severity of Ich disease since the production temperature is 5°C to 10°C warmer than the optimum temperature for Ich. There are no FDA-approved chemotherapeutic agents effective against the encysted trophonts in the fish epithelium. Formalin is often used to control Ich infection, but it only kills theronts in the water. Quarantine and monitoring of new stocks for 7 to 10 days at temperatures optimum for Ich (20°C to 25°C) is one possible means of control by preventing the introduction of infected stocks into culture systems.

Other Ciliated Protozoa

Trichodina spp., *Trichodinella* spp., *Chilodonella, Apiosoma, Ambiphyra,* and *Epistylis* all attach to tilapia. *Trichodina* spp. is probably the most reported (Natividad et al. 1986; Plumb 1997). This parasite is saucer shaped and usually causes damage to the skin and/or gills of fish (Figure 16.2A). Signs of disease include respiratory distress, loss of appetite, depigmentation, or loss of scales. *Chilodonella* is a leaf or heart-shaped ciliate and is distinct in having parallel rows of cilia along the body margin. The center of the ventral surface is free from cilia. *Chilodonella* infects skin, fins, and gills of tilapia. A heavy infection can cause detached scales, necrosis of branchial epithelium, and mass mortalities (Paperna and Van As 1983). *Apiosoma* and *Ambiphyra* are ciliates that both have an urn-shaped body with a ring of cilia. The major difference is the shape of macronucleus, with a rounded nucleus seen in *Apiosoma* and a ribbon-shaped nucleus seen in *Ambiphyra.* These protozoans have a direct life cycle and reproduce by binary fission on the skin and gills. *Epistylis* gets attached to fish with its stalk (Figure 16.2B) and commonly forms branched colonies that distinguishes it from *Apiosoma* and *Ambiphyra*. A severe infection with *Epistylis* causes erosion of skin, scales, spines, and bloody lesions, so the common name of the disease is Red Sore Disease. Among these ciliates, *Trichodina* and *Chilodonella* were reported to cause severe problems in cultured tilapia (Roberts and Sommerville 1982). Other parasites may cause problems by damaging the host skin, thus resulting in a portal of entry for other pathogens (i.e., bacteria; Plumb 1997). Most of these parasites may be treated with salt, formalin, or other approved parasiticides, however, it may take multiple treatments to effectively control these ciliates.

Flagellated Protozoa

Ichthyobodo (Costia) necatrix is the most important flagellated parasite of tilapia (Plumb 1997). These small parasites are similar to Ich in that they need a fish host for survival. Transmission of *Ichthyobodo* is by direct fish-to-fish contact. Common signs indicating the presence of this parasite are respiratory distress and flashing. *Ichthyobodo* can cause severe losses in a short time if not treated. Effective treatments include the use of copper sulfate and formalin. Management practices, such as quarantine, to avoid transmission to fish, should be used. *Ichthyobodo*'s life cycle is influenced by water temperature. The parasite favors water temperatures in the 20°C range (Klesius and Rogers 1995). Plumb (1997) suggested that if the temperature at which the tilapia were raised was >25°C, *Ichthyobodo* would not be a severe problem. However, *Ichthyobodo* has been seen as a problem in young tilapia cultured in water reuse systems in the United States. Water temperatures at these facilities were greater than 25°C when the parasites were observed (Shoemaker and Klesius, unpublished data).

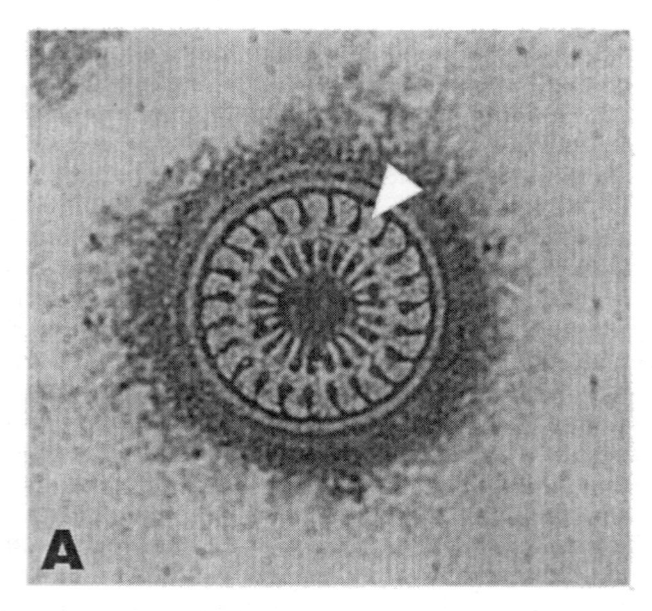

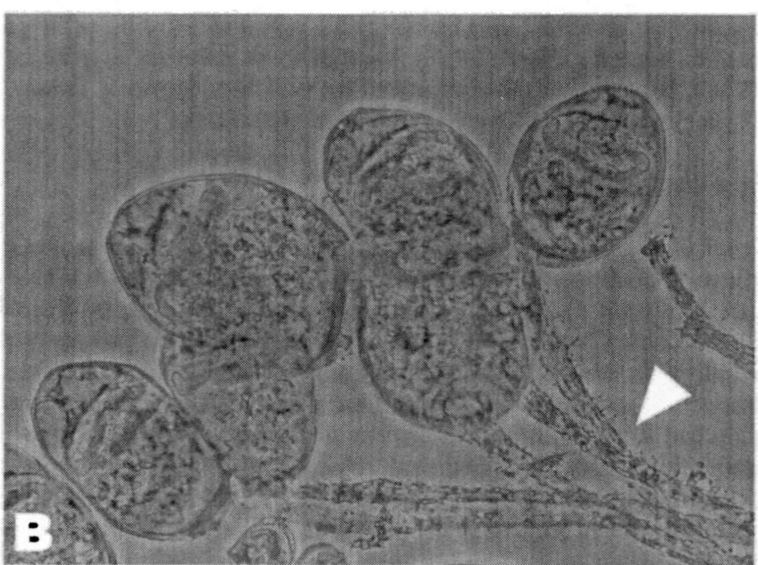

FIGURE 16.2. Ciliated parasites of tilapia: *(A) Trichodina* sp. from the skin of a tilapia, note the ring of denticles (arrow); *(B) Epistylis* from the skin of a tilapia, note the stalk (arrow).

Sporozoans

Sporozoan parasites have been reported from wild tilapia and those cultured in ponds (Okaeme et al. 1988; Gbankoto et al. 2001a,b). Myxosporea are commonly observed when the parasite is in the spore stage. Opaque white cysts containing spores may be seen distributed in gill, skin, eye, spleen, kidney, and ovarian tissue. The infection results in lesions in the infected tissues. The reproduction in female tilapia was affected by infection with Myxosporea in the ovary (Gbankoto et al. 2001a). However, no heavy mortality due to Sporozoan parasites has been reported in tilapia.

METAZOAN PARASITES

Monogenetic Trematodes

Many species of monogenetic trematodes have been reported on the skin and/or gills of tilapia. The most common are *Gyrodactylus* sp. (Figure 16.3) or *Dactylogyrus* sp. The latter is most often found on the gills, whereas, *Gyrodactylus* can be found on either skin or gills. Monogenetic trematodes use haptors at the posterior end to attach to fish. There are large anchors located centrally and marginal hooks at the edge of the haptor. The numbers of large anchors and marginal hooks vary in species of monogenetic trematodes. Anchors and hooks of monogenetic trematodes penetrate into the surface layer of skin, fin, and gills and cause tissue damage. These worms move about the body surface and feed on dermal and gill debris (Post 1987). Monogenetic trematodes have a direct life cycle and must have a fish host to survive. Transmission is by fish-to-fish contact or water. Monogenetic trematodes are usually not a severe problem unless they increase to tremendous numbers. In heavy gill infections, fish become lethargic, swim near the water surface, exhibit signs of partial suffocation, and refuse food (Post 1987). With large numbers of monogenetic trematodes on skin, fish show excessive mucus production, rubbing against sides of holding tanks, and occasional jumping out of water. Sites of parasite attachment on the gills and/or skin of the fish may act as portals of entry for primary or secondary bacterial and fungal pathogens as suggested by Klesius and Rogers (1995) and Plumb (1997). Monogenetic trematodes may become problematic in water reuse and/or cage culture systems because of the high densities at which fish are cultured. A marine species of monogenetic trematode *(Neobenedenia mellani)* was reported to be a problem in tilapia cultured in seawater cages in Hawaii (Kaneko et al. 1988).

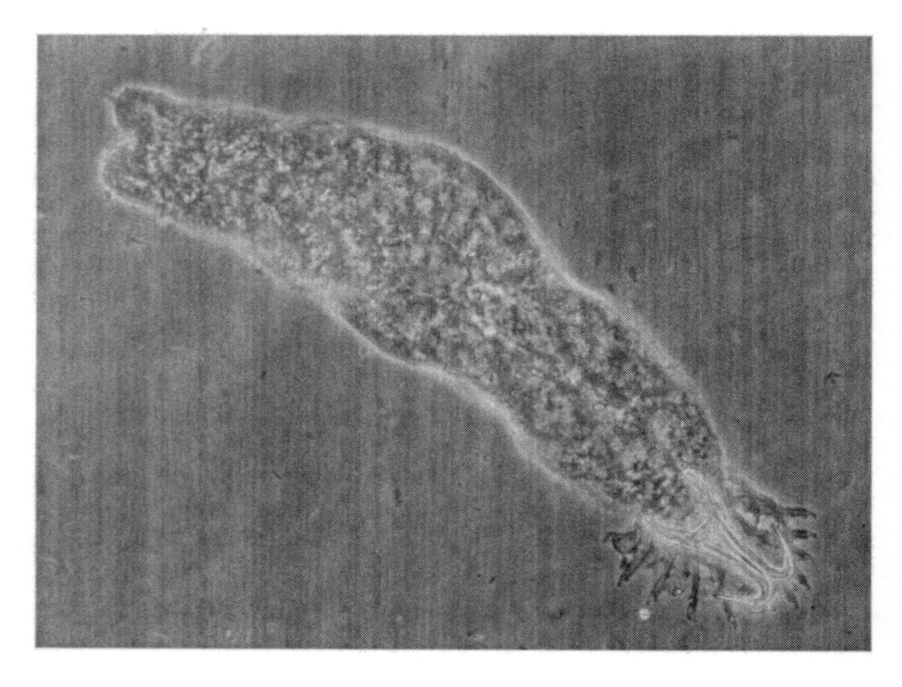

FIGURE 16.3. The monogenetic trematode, *Gyrodactylus* sp. from the skin of an infected tilapia.

Digenetic Trematodes

Roberts and Somerville (1982) reported that a variety of digenean parasites were potentially capable of causing heavy losses in tilapia. Digenean parasites are present in fish either as sexually mature adults or as immature larvae. Adult digenetic trematodes usually inhabit the intestine, gall bladder, and urinary bladder. Typically, they do not cause significant clinical symptoms or problems. The metacercaria is a larval stage of digenetic trematodes in fish and reach adulthood in fish-eating vertebrates. Eggs from adult parasites leave the vertebrate and hatch in water to become a free-swimming miracidia. Miracidia infect aquatic snails and continue to develop into free-swimming cercaria. After penetration into specific target organs of fish, cercaria transform into metacercariae and complete their life cycle. Because the life-cycle of these parasites involve three hosts, they probably will not cause a significant problem in enclosed water reuse systems. However, in open and/or nonclosed outdoor systems, these parasites may be important. Death to small fish or juveniles may occur when the cercariae of digenetic trematodes migrate to target organs for development into metacercaria. A digenetic trematode that deserves mention is the yellow grub (*Clinostomum* sp.). This trematode may be a problem because the aesthetic value of the fish is compromised and consumers will not buy the fish. This particular parasite may also result in human infection because its definitive host is a bird. If raw or undercooked fish flesh containing the metacercaria is consumed, parasitemia may occur in humans. Cooking of fish suspected of being infected with this parasite is recommended.

Tapeworms, Acanthocephalons, and Nematodes

These parasites do not appear to present much of a problem in cultured tilapia (Plumb 1997). However, a significant population of adult worms in the intestine may limit growth and nutrient uptake. Also, in some instances, the migrating larvae may cause problems or may encyst in muscles of fish resulting in a loss of value of the product. These parasitic worms will be more important in pond production.

PARASITIC CRUSTACEA

Anchor worm, *Lernaea,* fish louse, *Argulus,* and *Ergasilus* are parasitic crustacea that may cause problems in tilapia (Plumb 1997). These crustacea are relatively large ectoparasites that can be seen by the naked eye. *Argulus* (fish louse) has two prominent sucking discs, ventrally, and moves on skin and fins (sometimes gills) of fish (Hoffman 1999). *Argulus* punctures the fish skin and injects a toxin via oral sting and feeds on the host's blood (Mitchum 1995). Both parasitic *Lernaea* and *Ergasilus* are the females that possess a pair of distinct egg sacs when mature. Males are free-living and usually die after mating. The female *Lernaea* become adult on fish only after fertilization by the male. The head of the parasite burrows into the skin and develops into an "anchor" to hold the parasite on the fish surface (Rogers 1985). *Ergasilus* commonly attaches to gill filaments and obtains its nourishment from the epithelial and mucous cells of the gills. *Ergasilus* may cause damage of the gills due to the parasite moving on gills and firmly attaching at multiple sites. These parasitic crustacea cause hemorrhage, ulcerated lesions, damage to skin or gills, and may retard fish growth.

VIRAL DISEASES

Two viral diseases have been reported in tilapia. One agent is an Iridovirus that causes lymphocystis disease and was first described in wild tilapia by Paperna (1973). Lymphocystis will not cause mortality in infected tilapia or other fish, however, the aesthetic look of the fish is compromised. Lymphocystis is characterized by hypertrophied cells forming clusters or bumps on the skin and/or fins of infected fish. Plumb (1997) reported that this virus has not been identified in cultured tilapia.

Avtalion and Shlapobersky (1994) reported a second viral disease in tilapia in 1994. Larval stage of three species (*O. aureus, O. niloticus,* and *S. galilaeus*) were highly susceptible to this unidentified virus. The viral particles were contained in brain cell endoplasmic reticulum and were reported to be about 100 nm. Disease signs included whirling in water followed by darkening of the skin and anorexia (Avtalion and Shlapobersky 1994). There are no effective treatments for viral agents of tilapia. The best approach is quarantine of larval fish or avoidance of larval fish, juvenile and/or adult, with a history of viral diseases or problems to avoid introduction into culture facilities. With the increase in the intensification of tilapia production, diagnosticians should test for the possibility of viral agents if bacterial or water quality problems cannot be identified as the cause of mortalities in production systems.

FUNGAL INFECTIONS

El-Sharouny and Badran (1995) recovered 17 fungal species from tilapia. Of those recovered, *Saprolegnia parasitica* and *S. ferax* resulted in mortalities in experimental animals. Plumb (1997) suggested that most problems with fungal species result when water temperatures are near the lethal level for tilapia (12°C or lower). The chance of infection by fungi is increased markedly when fish are injured physically or by parasite and bacterial infection. Fish could not produce protective external mucus in the injured area because goblet cells are destroyed and the epidermis is sloughed (Xu and Rogers 1991). Spores of the fungus usually attach to damaged or unhealthy tissue of fish, germinate, and grow. Fungal hyphae may spread outwards on the skin surface (often looking like cotton or fuzz) or penetrate to deeper tissues to create ulceration and erosion. Roberts and Sommerville (1982) also reported *Branchiomyces* spp. in the gills of affected tilapia. This fungal species may cause high levels of mortality (Plumb 1997). *Branchiomyces* spp. has an affinity for organically rich water and, thus, may become established in water reuse systems. Producers should be aware of this fungal species if conditions may favor the growth of *Branchiomyces* spp. In the 1980s, *Paecilomyces lilacinus* was isolated from tilapia (*O. aureus* and *O. mossambicus*) in Puerto Rico (Bunkley-Williams and Williams 1994; Rand et al. 2000). This fungal species was believed to be responsible for a tilapia wasting disease that is a common and fatal disease of tilapia in eutrophic waters of Puerto Rico. Rand et al. (2000) suggested this fungal problem is also associated with organically rich waters and poor water quality parameters as associated with *Branchiomyces*.

Fungal problems of tilapia may be significant in hatcheries producing fry. If *Saprolegnia* or other fungi become established in incubating eggs, all may be lost (Plumb 1997). The best way to avoid this problem is to remove injured or nonfertilized eggs (Post 1987) to avoid establishment of fungal infection in the healthy eggs.

BACTERIAL DISEASES

Columnaris Disease

Flavobacterium columnare (Flexibacter columnaris) is the cause of columnaris disease of many fish species including tilapia. This bacterium is often difficult to grow on media, and initial identification of the bacteria is from wet mounts of the fish's skin and/or gills. Long slender rods (4 to 10 μ*m)* seen flexing or gliding often forming haystacks in wet mounts is a presumptive identification of this bacterium (Plumb 1999). Signs of columnaris disease include frayed fins, depigmented lesions on skin, and necrotic lesions in gills. Often, the margins of the lesions will have a yellow appearance because of the pigment in the bacteria. Isolated bacterial colonies are yellow in color with rhizoid margins. Amin et al. (1988) described columnaris in cultured Nile tilapia *(O. niloticus)* and were only able to experimentally induce mortality if the gills of fish were damaged, or if the fish were held in water with elevated ammonia. They also suggested that the pathogenicity of the isolates (seven tested) was different. In our laboratory, working with channel catfish *(Ictalurus punctatus)*, we demonstrated that nutrition influenced *F. columnare*-induced mortality (Klesius, et al. 1999a). Fish withheld from feed for a period of 7 to 10 days exhibited significantly higher mortality than fish that were fed, suggesting innate resistance was negatively influenced by poor nutrition.

Current treatments for columnaris disease are often impractical in ponds and can cause problems with biological filters of water reuse systems. Development of efficacious vaccines will probably be the solution. We (unpublished data) showed that channel catfish that survived challenge with

F. columnare were immune upon second challenge. These data suggest that development of a vaccine against *F. columnare* is possible. Work is in progress to develop an effective killed *F. columnare* or modified live *F. columnare* vaccine for use in food fish.

Edwardsiella tarda

Plumb (1997) described *E. tarda* infections as a septicemia in tilapia especially if cultured at high densities and under adverse conditions. In the United States, *E. tarda* has been isolated and has resulted in losses to producers raising tilapia in water reuse systems. *Edwardsiella tarda* was recently isolated from Nile tilapia *(O. niloticus)* in Brazil cultured in an integrated system that used pig manure as a feed source (Muratori et al. 2000). Muratori et al. (2000) cultured *E. tarda* from the skin, gills, and fins of tilapia in this system; however, they did not report it as causing disease in these fish.

Edwardsiella tarda is a member of the enteric bacteria and is a motile Gram-negative rod (1-2 µm). This bacterium is presumptively identified by being cytochrome oxidase negative, and producing an alkaline over acid and gas and H_2S on triple sugar iron agar (Plumb 1999). Plumb (1997) described disease symptoms of tilapia infected with *E. tarda* as having lethargic swimming, enlarged abdomen, exophthalmic, and hemorrhaged eyes. Necrosis and gas-filled pockets may be observed in the muscle in some cases. Treatment for *E. tarda* is by use of medicated feeds; however, these treatments may or may not be effective. Development of an effective vaccine is needed to help prevent this infection in tilapia. *Edwardsiella tarda* has also been reported in domestic livestock, amphibians, and birds and cases have been reported in humans following puncture wounds. Care should be taken when handling fish known to be infected with *E. tarda*. Human infection has not resulted from consuming cooked fish.

Motile Aeromonads and Vibriosis

Disease signs seen in tilapia with *Aeromonas* infections and *Vibrio* infections are similar; these groups will be described together. Bacteria responsible for motile *Aeromonas* septicemia include *Aeromonas hydrophila, A. sobria, A. caviae, Psuedomonas* sp., and other related species (Plumb 1999). These bacterial species are short Gram-negative highly motile rods. *Vibrio* sp. described from tilapia include *V. parahaemolyticus* (Hubert 1989) and *V. mimicus* (Plumb 1997) in freshwater. Undoubtedly, other *Vibrio* sp. may occur in saltwater. All are members of Vibrionaceae and, thus, are cytochrome oxidase positive. A positive cytochrome oxidase reaction and a negative Gram-stain are commonly used to presumptively identify the *Aeromonas* sp. *Vibrio* sp. are Gram-negative comma-shaped or curved short rods. *Vibrio* sp. are sensitive to vibriostat 0/129, whereas, the *Aeromonas* sp. are resistant. Most researchers suggest that motile *Aeromonas* septicemia and vibriosis of tilapia are secondary infections and will occur following stressful conditions (Plumb 1997). We (unpublished data) have isolated *A. hydrophila* and *A. sobria* from tilapia raised at high densities in closed culture systems. The bacterial isolations typically occurred following a decrease in water temperature or after handling and/or transport of fish. Management practices to maintain constant water temperature and reduce handling will be the best solution to vibriosis and motile *Aeromonas* septicemia of tilapia.

Streptococcal Disease

Gram-positive bacteria have been reported to be a significant problem in over 22 species of marine and freshwater fish, including tilapia. These bacteria have a worldwide distribution. Ringo and Gatesoupe (1998) present a review of lactic acid bacteria in fish. They suggested that the pathogenic bacterial species included *Streptococcus, Enterococcus, Lactobacillus, Carnobacterium,* and *Lactococcus* that have been isolated from ascites fluid, kidney, liver, heart, and spleen of sick fish. Streptococcal species are commonly isolated on blood agar (Figure 16.4A) and *Streptococcus iniae* is positive for starch hydrolysis (Figure 16.4B). Pathogenic streptococci of tilapia (Table 16.1) include *Streptococcus iniae* (Klesius, Shoemaker, and Evans 1999; Shoemaker et al. 2001), *group B S. agalactiae* (Vandamme et al. 1997; Evans et al. 2002), nonhemolytic *Streptoccoccus* sp. (Chang and Plumb 1996), and alpha hemolytic *Streptococcus* sp. (Al-Harbi 1994). Sites of bacterial isolation that Ringo and Gatesoupe (1998) failed to mention were the brain and nare of diseased fish. Evans et al. (2000) demonstrated that tilapia and hybrid striped bass (*Morone chrysops* × *M. saxatilis*) could be infected with *S. iniae* following inoculation of the nares, but not by inoculation of the eyes. Evans et al. (2001) also demonstrated dissemination of *S. iniae* throughout the fish following inoculation of the nares. Diseased fish were culture-positive in every organ in which culture was attempted (i.e., nare, brain, eye, blood, anterior kidney, and heart). Typical signs of disease early in the infectious process include slow to no feeding response followed by darkening of skin, erratic swimming (Figure 16.5A), lethargy, body curvature (Figure 16.5B), exophthalmia and death. Most often, exophthalmia,

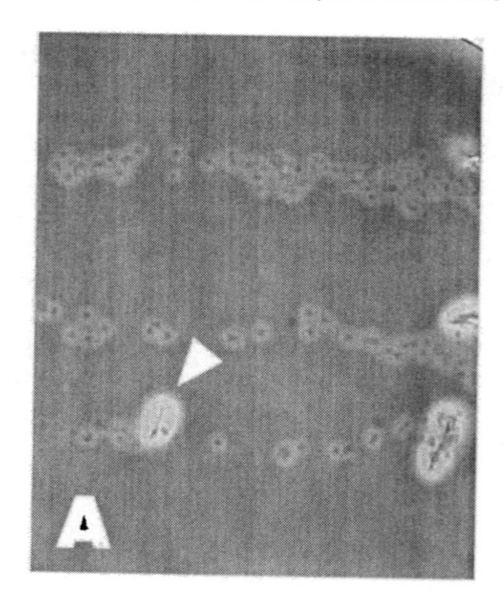

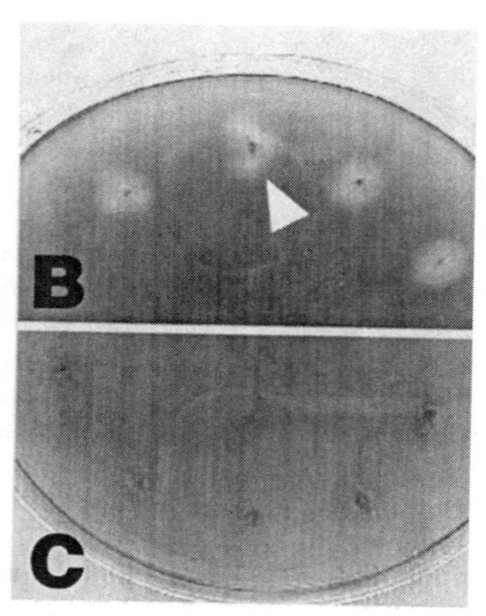

FIGURE 16.4. Characteristics of *Streptococcus iniae* grown on culture media: *(A) Streptococcus iniae* in pure culture on a 5 percent sheep blood agar plate showing *Beta*-hemolysis that is especially evident in the stabbed areas (arrow); *(B)* starch hydrolysis by *Streptococcus iniae* on a starch agar plate (note the zone of clearing, arrow); and *(C)* streptococci negative for starch hydrolysis.

TABLE 16.1. Characteristics of *Streptococcus* sp. in tilapia.

	Bacterium		
Phenotypic Characteristic	**Streptococcus iniae[a]**	**Streptococcus sp.[b]**	**Streptococcus agalactiae[c]**
Gram stain	+ cocci	+ cocci	+ cocci
Motility	Negative	Negative	Negative
Hemolysis	Beta	Non	Beta
Catalase	Negative	Negative	Negative
CAMP test	Positive (variable)	Negative	ND
Starch hydrolysis	Positive	Negative	Negative
Hippurate hydrolysis	Negative	Positive	Positive
Growth at 10°C	Positive	ND	Negative
Growth at 45°C	Negative	ND	Negative
Growth 6.5 percent NaCl	Variable	Negative	Variable
Bile-esculin	Negative	Negative	Negative
Sorbitol	Negative	Negative	Negative
MRS broth	Negative	Negative	ND
VP	Negative	Weak +	Positive
Arginine	Positive (variable)	Positive	Positive
LAP	Positive	Weak +	Positive
PYR	Positive	Negative	Negative
Bacitracin	Resistant	Susceptible	ND
Vancomycin	Susceptible	Susceptible	Susceptible
Lancefield group	None	None	B

Notes: [a]Shoemaker and Klesius (1997). [b]Identified as *Streptococcus* sp. in our lab. Isolated by Dr. John Plumb from diseased tilapia in Mexico. [c]Evans et al. 2002.

and eye opacity are the last signs observed. Eye opacity or cloudiness appears to be more associated with chronic disease.

In 1997-1998, Shoemaker et al. (2001) determined the prevalence of *S. iniae* in tilapia (*Oreochromis* spp), hybrid striped bass, and channel catfish on commercial fish farms in the United States. Overall prevalence in tilapia and hybrid striped bass was found to be 3.81 and 7.23 percent, respectively (Shoemaker et al. 2001). However, prevalence in market-sized fish (both tilapia and hybrid striped bass) was only 1.75 percent (5 of 286 sampled). Earlier reports from Canada (Weinstein et al. 1997) indicated that prevalence in market-sized fish sampled in their study was 81.8 percent.

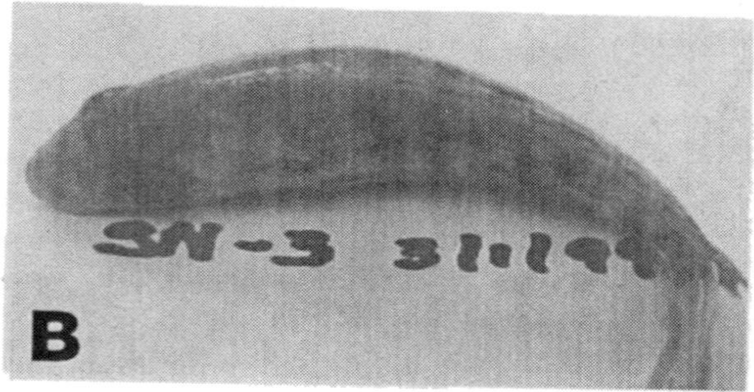

FIGURE 16.5. Characteristics of *Streptococcus iniae* infected tilapia: *(A)* an infected tilapia showing erratic swimming and body curvature; the other tilapia (in background) are exhibiting normal behavior; *(B)* dead tilapia showing the body curvature following inoculation via the nares.

Weinstein et al. (1997) suggested that *S. iniae* was a human pathogen with infections believed to result from handling fresh whole tilapia bought at retail stores. However, no fish implicated in the infections were tested (Weinstein et al. 1997). Results of our prevalence survey did not support the contention that *S. iniae* is a serious public health threat associated with farm-raised fish. *Streptococcus iniae* probably represents a limited risk to older or immunocompromised people who incur puncture wounds while handling and preparing infected fish (Shoemaker et al. 2001).

Artificial infection of tilapia with *S. iniae* has been accomplished by intraperitoneal injection, immersion, cohabitation, oral intubation, and nare inoculation. Shoemaker et al. (2000) demonstrated that immersion infection of tilapia was possible and that high fish density favored *S. iniae* transmission and resulted in a significant increase in mortality. Low density groups (5.6 $g \cdot L^{-1}$) had mortalities of 4 to 5 percent compared to medium (11.2 $g \cdot L^{-1}$) and high density groups (22.4 $g \cdot L^{-1}$), with mortalities ranging from 25 to 40 percent (Shoemaker et al. 2000). Cohabitation had previously been attempted by Perera et al. (1997), however, they were unsuccessful in induction of *S. iniae* mortality. Shoemaker et al. (2000) succeeded in establishing the disease in tilapia by this route if cohabitiation was done at high fish density. Barcellos et al. (1999) documented that high-stocking density and, thus, social interaction, affected the acute stress response in tilapia fingerlings. This acute stress response was seen as an increase in plasma cortisol levels and may be responsible for the increased susceptibility of fish stocked at high density to streptococcal disease that we previously demonstrated. However, we did not measure plasma cortisol levels. Fish to fish transmission of *Streptococcus* sp. is favored in water reuse systems because of the high densities at which fish are cultured and low amount of water exchange. A reduction of stocking density and removal of dead and dying fish may significantly enhance production and economic benefit by limiting *S. iniae* transmission and mortality and, thus, the loss of fish in water reuse systems.

Other environmental factors that have been studied with regard to streptococcal disease in tilapia include temperature, dissolved oxygen, and nitrite level. Bunch and Bejerano (1997) evaluated low oxygen and high nitrite in relation to streptococcal disease in hybrid tilapia (*Oreochromis niloticus* × *O. aureus*). In their study fish were infected with two isolates of *Streptococcus* sp. Their results suggested that low oxygen or high nitrite increased mortality in hybrid tilapia, but no additive effect of low oxygen and high nitrite in combination was observed. Bowser et al. (1998) also suggested low oxygen was responsible for *Streptococcus* sp. mortalities in a water reuse production system. In the laboratory, they (Bowser et al. 1998) were unable to induce streptococcal disease following exposure to low oxygen for 8 h. In their laboratory trials, however, the density of fish was significantly less than at the production facility. Prevention of streptococcal disease at present will rely on vaccination and good health management practices.

CONCLUSIONS

Health management or maintenance (Plumb 1999) will be an important component of intensive aquaculture systems. New fish to be introduced into production facilities should be examined for disease agents as well as use of a quarantine period to allow for latent or unseen disease agents to manifest. Quarantine should limit the likelihood of introduction of new pathogens into a facility. Clean facilities and removal of dead and or dying fish is essential (Kitao 1993). Maintenance of water quality parameters within optimum levels for the overall health of fish should be practiced (Boyd 1990). Nutritionally balanced diets and proper feed management practices should be employed to meet the needs of the tilapia for growth and overall health (Lim 1997). Stocking rates also influence the health of tilapia. The industry standards use high stocking densities (30 to 290 $g{\cdot}L^{-1}$) to increase yield per unit area. However, the economics of this practice should be reexamined because in water reuse systems losses greater than 70 percent have been documented (Perera et al. 1994; Stoffergen et al. 1996). Our recent laboratory results (Shoemaker et al. 2000) using densities $\geq 11\ g{\cdot}L^{-1}$ suggest that a reduction in fish density reduced mortality due to *S. iniae*.

Another method of disease control is chemotherapy and/or use of antibiotics. Antibiotic treatments have not been effective in eliminating streptococcal problems in water reuse systems. The treatments appear to suppress disease signs (Evans, Klesius, and Shoemakers, unpublished data). More fish may survive following antibiotic treatment (Stoffregen et al. 1996); however, because of the ability of these bacteria to survive in macrophages (Zimmerman et al. 1975), reappearance of infection is often seen. Part of the problem with current antibiotic use is that at the time of intervention, the sick fish often do not eat the medicated feed, or treatments are not of long enough duration. In the United States, no antibiotics are presently approved for treating streptococcal infections in tilapia. Extra-label use is permitted following approval of the United States Food and Drug Administration-granted Investigational New Animal Drug (INAD) application, veterinary prescription, or in cases of emergency experimental INAD's are granted. Approval of antibiotics for use on food fish likely will not occur and these drugs will not be the silver bullet against bacterial disease but may be implemented in an overall health management plan. Chemicals used to treat parasitic infections are often expensive and may have a negative effect on environments.

The best approach to limit the effect of disease in tilapia is the use of vaccination. An effective vaccine results in the prevention of disease, rather than treatment. Klesius et al. (1999b, 2000, 2001, 2002) describe the use of vaccines in tilapia against *S. iniae* and the need for development of other vaccines for the other significant diseases of tilapia. With the rapidly expanding knowledge in fish vaccinology, it is likely that in the next 5 to 10 years efficacious vaccines will be developed against the major bacterial pathogens of tilapia. Development of these vaccines will allow for expanded growth of the tilapia industry in water reuse systems.

REFERENCES

Al-Harbi, A.H. (1994). First isolation of *Streptococcus* sp. from hybrid tilapia *(Oreochromis niloticus × O. aureus)* in Saudi Arabia. *Aquaculture* 128:195-201.

Amin, N.E., I.S. Abdallah, M. Faisal, M. Easa, T. Alaway, and S.A. Alyan (1988). Columnaris infection among cultured Nile tilapia *Oreochromis niloticus. Antonie van Leeuwenhoek* 54:509-520.

Avtalion, R.R., and M. Shlapobersky (1994). A whirling viral disease of tilapia larvae. *The Israeli Journal of Aquaculture-Bamidgeh* 46:102-104.

Barcellos, L.J.G., S. Nicolaiewsky, S.M.G. de Souza, and F. Lulhier (1999). The effects of stocking density and social interaction on acute stress response in Nile tilapia *Oreochromis niloticus* (L.) fingerlings. *Aquaculture Research* 30:887-892.

Bowser, P.R., G.A. Wooster, R.G. Getchell, and M.B. Timmons (1998). *Streptococcus iniae* infection of tilapia *Oreochromis niloticus* in a recirculation production facility. *Journal of the World Aquaculture Society* 29:335-339.

Boyd, C.E. (1990). *Water quality in ponds for aquaculture.* Auburn: Alabama Agricultural Experiment Station, Auburn University.

Bunch, E.C., and I. Bejerano (1997). The effect of environmental factors on the susceptibility of hybrid tilapia *(Oreochromis niloticus × O. aureus)* to streptococcosis. *The Israeli Journal of Aquaculture-Bamidgeh* 49:67-76.

Bunkley-Williams, L.B., and E.H. Williams (1994). *Parasites of Puerto Rican freshwater sport fishes.* San Juan: Puerto Rico Department of Natural and Environmental Resources.

Chang, P.H., and J.A. Plumb (1996). Histopathology of experimental *Streptococcus* sp. infection in tilapia, *Oreochromis niloticus* (L.) and channel catfish, *Ictalurus punctatus* (Rafinesque). *Journal of Fish Diseases* 19:235-241.

El-Sharouny, H.M., and R.A.M. Badran (1995). Experimental transmission and pathogenicity of some zoosporic fungi to tilapia fish. *Mycopathologia* 132:95-103.

Engle, C.R. (1997). Marketing tilapias. In B.A. Costa-Pierce and J.E. Rakocy (Eds.), *Tilapia aquaculture in the Americas,* (Volume 1, pp. 244-258). Baton Rouge, LA: World Aquaculture Society.

Evans, J.J., P.H. Klesius, P.M. Glibert, C.A. Shoemaker, M.A. Al Sarawi, J. Landsberg, R. Duremdez, A. Al Marzouk, and S. Al Zenki (2002). Characterization of beta-hemolytic group B *Streptococcus agalactiae* in cultured sea bream, *Sparus auratus* (L.) and wild mullet, *Liza klunzingeri* (Day), in Kuwait. *Journal of Fish Diseases* 25:1-9.

———. (2001). Distribution of *Streptococcus iniae* in hybrid striped bass (*Morone chrysops × Morone saxatilis*) following nare inoculation. *Aquaculture* 194:233-243.

Evans, J.J., C.A. Shoemaker, and P.H. Kleisus (2000). Experimental *Streptococcus iniae* infection of hybrid striped bass (*Morone chrysops × M. saxatilis*) and tilapia *(Oreochromis niloticus)* by nares inoculation. *Aquaculture* 189:197-210.

Ewing, M.S., and K.M. Kochan (1992). Invasion and development strategies of *Ichthyophthirius multifiliis,* a parasitic ciliate of fish. *Parasitology Today* 8:204-208.

Food and Agriculture Organization (FAO) (2001). World review of fisheries and aquaculture, Part 1. Available online at <http://www.fao.org/DOCREP/003/X8002E/x8002e04.htm>.

Gbankoto, A., C. Pampoulie, A. Marques, and G.N. Sakiti (2001a). *Myxobolus dahomeyensis* infection in ovaries of tilapia species from Benin (West Africa). *Journal of Fish Diseases* 44:217-222.

———. (2001b). Occurrence of myxosporean parasites in the gills of two tilapia species from Lake Nokoué (Bénin, West Africa): effect of host size and sex, and seasonal patterns of infection. *Diseases of Aquatic Organisms* 44:217-222.

Hoffman, G.L. (1978). Ciliates of freshwater fishes. In J.P. Krier (Ed.), *Parasitic protozoa: Intestinal flagellates, histomonads, trichomonads, amoeba, opalindas and ciliates* (pp. 583-632). New York: Academic Press Inc.

———. (1999). *Parasites of North American freshwater fishes* (2nd Ed.). Ithaca, NY: Cornell University Press.

Hubert, R.M. (1989). Bacterial diseases in warm water fish. In M. Shilo and J. Sarig (Eds.), *Fish culture in warm water systems: Problems and trends* (pp. 187-194). Boca Raton, FL: CRC Press.

Kaneko, J.J., R. Yamada, J.A. Brock, and R.M. Nakamura (1988). Infection of tilapia, *Oreochromis mossambicus* (Trewavas), by a marine monogenean, *Neobenedenia melli* (MacCallum, 1927) in Kaneohe Bay, Hawaii, USA, and its treatment. *Journal of Fish Diseases* 11:295-300.

Kitao, T. (1993). Streptococcal infections. In V. Inglis, R.J. Roberts, and N.R. Bromage (Eds.), *Bacterial diseases of fish* (pp. 196-210). Oxford, UK: Blackwell Scientific Publications.

Klesius, P.H., C. Lim, and C. Shoemaker (1999). Effect of feed deprivation on innate resistance and antibody response to *Flavobacterium columnare* in channel catfish, *Ictalurus punctatus. Bulletin of the European Association of Fish Pathologist* 19:156-158.

Klesius, P.H., and W. Rogers (1995). Parasitisms of catfish and other farm raised food fish. *Journal of the American Veterinary Medical Association* 207:1473-1478.

Klesius, P.H., C.A. Shoemaker, and J.J. Evans (1999). Efficacy of a killed *Streptococcus iniae* vaccine in tilapia *(Oreochromis niloticus). Bulletin European Association of Fish Pathologists* 19:39-41.

———. (2000). Efficacy of single and combined *Streptococcus iniae* isolate vaccine administered by intraperitoneal and intramuscular routes in tilapia *(Oreochromis niloticus). Aquaculture* 188:237-246.

———. (2002). *Streptococcus iniae* vaccine. US Patent No. 6,379,677.

Klesius, P.H., C.A. Shoemaker, J.J. Evans, and C. Lim (2001). Vaccines: prevention of diseases in aquatic animals. In C. Lim and C.D. Webster (Eds.), *Nutrition and fish health* (pp. 317-335). Binghamton, NY: The Haworth Press, Inc.

Lim, C. (1997). Nutrition and feeding of Tilapias. In *Fourth Symposium on Aquaculture in Central America: focusing on Shrimp and Tilapia* (pp. 94-107). Tegucigalpa, Honduras: Latin America Chapter of the World Aquaculture Society.

Mitchum, D.L. (1995). *Parasites of fishes in Wyoming*. Cheyenne, Wyoming: Wyoming Game and Fish Department.

Muratori, M., A. DeOliveira, L. Ribeiro, R. Leite, A. Costa, and M. DaSilva (2000). *Edwardsiella tarda* isolated in integrated fish farming. *Aquaculture Research* 31:481-483.

Natividad, J.M., M.G. Bondad-Reantaso, and J.R. Arthur (1986). Parasites of Nile tilapia *(Oreochromis niloticus)* in the Phillipines. In J. Macalean, L.D. Zon, and L. Hosillos (Eds.), *First Asian fisheries forum* (pp. 255-259). Manila, Phillipines: Asian Fisheries Society.

Nigrelli, R.F., K.S. Pokorny, and G.D. Ruggieri (1976). Notes on *Ichthyophthirius multifiliis,* a ciliate parasite on freshwater fishes, with some remarks on possible physiological races and species. *Transactions of the American Microscopical Society* 95:607-613.

Okaeme, A.N., A.I. Obiekezie, J. Lehman, E.E. Antai, and C.T. Madu (1988). Parasites and diseases of cultured fish of Lake Kainji area, Nigeria. *Journal of Biology* 32:479-481.

Paperna, J. (1973). Lymphocystis in fish from East African lakes. *Journal of Wildlife Diseases* 9:331-335.

Paperna, I., and J.G. Van As (1983). The pathology of *Chilodonella hexasticha* (Kiernik) infections in cichlid fishes. *Journal of Fish Biology* 23:441-450.

Perera, R.P., S.K. Johnson, M.D. Collins, and D.H. Lewis (1994). *Streptococcus iniae* associated with mortaltiy of *Tilapia nilotica* × *T. aurea* hybrids. *Journal of Aquatic Animal Health* 6:335-340.

Perera, R.P., S.K. Johnson, and D.H. Lewis (1997). Epizootiological aspects of *Streptococcus iniae* affecting tilapia in Texas. *Aquaculture* 152:25-33.

Plumb, J.A. (1997). Infectious diseases of tilapia. In B.A. Costa-Pierce and J.E. Rakocy (Eds.), *Tilapia aquaculture in the Americas* (Volume 1, pp. 212-228). Baton Rouge, LA: World Aquaculture Society.

———. (1999). *Health maintenance and microbial diseases of cultured fishes*. Ames: Iowa State University Press.

Post, G. (1987). *Textbook of fish health*. Neptune City, NJ: T.F.H. Publications, Inc.

Rand, T.G., L. Bunkley-Williams, and E.H. Williams (2000). A hyphomycete fugus, *Paecilomyces lilacinus,* associated with wasting disease in two species of tilapia from Puerto Rico. *Journal of Aquatic Animal Health* 12:149-156.

Ringo, E., and F.-J. Gatesoupe (1998). Lactic acid bacteria in fish: a review. *Aquaculture* 160:177-203.

Roberts, R.J. and C. Sommerville (1982). Diseases of tilapias. In R.S.V. Pullin and R.H. Lowe-McConnel (Eds.), *The biology and culture of tilapias* (pp. 247-263). Manila, Philippines: ICLARM Conference Proceedings 7.

Rogers, W.A. (1985). Protozoan parasites. In J.A. Plumb (Ed.), *Principal diseases of farm raised catfish*. pp. 28-37 Auburn: Southern Cooperative Series Bulletin 225.

Shoemaker, C., and P. Klesius (1997). Streptococcal disease problems and control a review. In K. Fitzsimmons (Ed.), *Tilapia aquaculture* (Vol. 2, pp. 671-682). Ithaca, NY: Northeast Regional Agricultural Engineering Service.

Shoemaker, C.A., J.J. Evans, and P.H. Klesius (2000). Density and dose: factors affecting mortality of *Streptococcus iniae* infected tilapia *(Oreochromis niloticus). Aquaculture* 188:229-235.

Shoemaker, C.A., P.H. Klesius, and J.J. Evans (2001). Prevalence of *Streptococcus iniae* in tilapia, hybrid striped bass and channel catfish from fish farms in the United States. *American Journal of Veterinary Research* 62:174-177.

Stoffregen, D.A., S.C. Backman, R.E. Perham, P.R. Bowser, and J.G. Babish (1996). Initial disease report of *Streptococcus iniae* in hybrid striped (sunshine) bass and successful therapeutic intervention with fluoroquinolone antibacterial enrofloxacin. *Journal of the World Aquaculture Society* 27:420-434.

Subasinghe, R.P., and C. Sommerville (1992). Susceptibility of *Oreochromis mosambicus* (Peters) fry to the ciliate ectoparasite *Ichthyophthirius multifiliis* (Fouquet). In I.M. Shariff, R.P. Subasinghe, and J.R. Arthur (Eds.), *Diseases of Asian aquaculture* (pp. 335-360). Manila, Philippines: Asian Fisheries Society.

Vandamme, P., Devriese, L.A., Pot, B., Keresters, K., and Melin.P. (1997) *Streptococcus difficle* is a non-hemolytic group B, type Ib *Streptococcus. International Journal of Systematic Bacteriology* 47(1):81-85.

Weinstein, M.R., M. Litt, D.A. Kertesz, P. Wyper, D. Rose, M. Coulter, A. McGeer, R. Facklam, C. Ostach, B.M. Willey, A. Borczyk, and D.E. Low (1997). Invasive infections due to a fish pathogen, *Streptococcus iniae. New England Journal of Medicine* 337:589-594.

Xu, D., P.H. Klesius, C.A. Shoemaker, and J.J. Evans (2000). The early development of *Ichthyophthirius multifiliis* in channel catfish tissue *in vitro. Journal of Aquatic Animal Health* 12:290-296.

Xu, D., and W.A. Rogers (1991). Electron microscopy of infection by *Saprolegnia* spp. in channel catfish. *Journal of Aquatic Animal Health* 3:63-69.

Zimmerman, R.A., P.H. Klesius, D.H. Krushak, and T. Mathews (1975). Effect of antibiotic treatment on the immune response following group A streptococcal pharyngitis. *Canadian Journal of Internal Medicine* 39:227-230.

Chapter 17

Streptococcal Vaccinology in Aquaculture

Phillip H. Klesius
Joyce J. Evans
Craig A. Shoemaker
David J. Pasnik

INTRODUCTION

Minimizing the effects of diseases is crucial to prevent morbidity and mortality and to promote optimal growth of farmed tilapia in fresh and marine waters. Since the inception of aquaculture, the control of diseases has been dependent on the use of therapeutics. The resolute demands of consumer, environmental, and governmental groups for wholesome fish and for an environment free of potentially harmful drugs in aquaculture production have increased. In addition, issues related to increased emergence of antibiotic resistant pathogens have made headlines and stimulated serious public concern. In 2004, world tilapia production was 2,007,087 metric tons and worldwide farm value of tilapia was greater than $3,000,000,000 (http://ag.arizona.edu/azaqua/ista/markets.htm). Tilapia ranked second in the world among the species of farmed fish produced, and it is predicted that tilapia will become the most important aquaculture crop in this century.

The continued growth and well-being of tilapia aquaculture requires that the industry meet the challenges of minimizing the effects of disease, providing a wholesome product, and preventing increases in pathogen resistance to antibiotics and chemotherapeutics. The tilapia industry can meet these challenges with more rapid and expanded health management practices that make use of vaccines to increase the survival and optimal growth of farmed tilapia.

We thank Dr. Victor Panangala and Lisa Biggar for their helpful editorial comments and suggestions. We also thank Kristie Butts for her thoughtful help in preparation of the manuscript.

doi:10.1300/5513_17

Vaccination is among the most successful veterinary practices to prevent deaths and to provide safeguards for animal production and biosecurity (Klesius et al. 2000a,b, 2001). In recent years, fish vaccinology has made real progress in both the safety and efficacy of vaccines, especially against bacterial diseases of salmonids (Gudding et al. 1997). Fish vaccines that are currently available for immunization of fish may be found on the <http://www.aphis.usda.gov> website. However, very limited information is available in the literature on tilapia vaccination. One of the earliest studies showed that Nile tilapia *(Oreochromis niloticus)* vaccinated intraperitoneally (IP) with a killed *Aeromonas hydrophila* vaccine were protected against infection (Rungpan et al. 1986). The formalin-killed bacterin provided 53 to 61 percent protection 1 week after immunization and 100 percent protection 2 weeks after immunization. This study demonstrated that vaccination could prove a useful and effective tool for tilapia aquaculture.

Tilapia are widely cultured throughout the world, though multiple factors affect their successful aquaculture production. One of the major issues is streptococcal disease, caused by a variety of *Streptococcus* spp. that produce significant mortalities among farmed tilapia worldwide. Tilapia growers consider streptococcal diseases caused by *S. iniae* and *S. agalactiae* the most serious economic threat to profit loss (Klesius et al. 2000b; Shoemaker et al. 2000, 2001a; Evans et al. 2002). *Streptococcus agalactiae* infection is responsible for severe economic losses in seabream and tilapia production (Evans et al. 2002; Glibert et al. 2002). In addition, seabream, seabass, yellowtail, amberjack, and rainbow trout producers suffer severe economic losses due to *Lactococcus garviae* infection (Kitao 1993; Eldar et al. 1996, 1999; Schmidtke and Carson 1999). Losses to all of these bacterial pathogens are estimated in the millions of dollars annually, worldwide. However, there are currently no vaccines commercially available for use in tilapia. This chapter will focus on the development of vaccines against *Streptococcus* spp. according to the known factors, strategies, and benefits germane to tilapia vaccination (Tables 17.1 and 17.2).

DEVELOPMENT OF STREPTOCOCCAL VACCINES

Results of investigations into the nature of pathogenesis of *S. iniae* (Evans et al. 2000, 2001) and mechanisms of the immune response against *S. iniae* resulted in the development of streptococcal vaccines. Passive immunization experiments and the development of enzyme-linked immunosorbent assay (ELISA) demonstrated that infection resulted in a serum antibody response and that passive immunization with antibody from immune fish

TABLE 17.1. Effects of fish species, vaccine, vaccine type, intraperitoneal (IP), intramuscular (IM), or oral route of vaccination, and size of fish on vaccine efficacy against streptococcal infection.

Fish species	Vaccine against	Vaccine type	Route of vaccination	Size (g)	Relative percentage survival (RPS)
Tilapia[a]	*S. agalactiae*	ECP[j] modified bacterin	IP	5	25
Striped bass[a]	*S. iniae*	ECP-modified bacterin	IP	10	95
Striped bass[a]	*S. iniae*	ECP-modified bacterin	IM	10	87
Striped bass[a]	*S. iniae*	ECP	IP	10	48
Striped bass[a]	*S. iniae*	ECP	IM	10	76
Rainbow trout[b]	*S. iniae*	Bacterin	IP	10	56
Tilapia[c]	*S. iniae*	ECP-modified bacterin	IP	13	100
Tilapia[c]	*S. iniae*	ECP-modified bacterin	Oral	13	63
Tilapia[d]	*S. iniae*	ECP-modified bacterin	IP	18	94
Tilapia[d]	*S. iniae*	ECP-modified bacterin	IM	18	60
Tilapia[e]	*S. iniae*	ECP-modified bacterin	IP	25	95
Tilapia[a]	*S. agalactiae*	ECP-modified bacterin	IP	30	70-80
Tilapia[a]	*S. agalactiae*	ECP-modified bacterin	IP	30	83
Turbot[f]	*Enterococcus* sp.	Toxoid bacterin	IP	45	89-100
Rainbow trout[g]	*S. iniae*	Bacterin	IP	30-50	86-88
Rainbow trout[h]	*Enterococcus* sp.	Bacterin CFA[k]	IP	57	89
Turbot[f]	*Enterococcus* sp.	Toxoid bacterin	Oral feed glucans/IP	45	89-94
Tilapia[e]	*S. iniae*	ECP-modified bacterin	IP	100	84-95
Turbot[f]	*Enterococcus* sp.	Toxoid bacterin	IP	150	67-86
Tilapia[i]	*S. difficile*	Bacterin	IP/booster	150-180	100
Tilapia[i]	*S. difficile*	Protein extract	IP/booster	150-180	92

Notes: [a]Evans, Klesius, and Shoemaker (2004); Evans, Shoemaker, and Klesius (2004); Evans et al. (in press); [b]Nakanishi et al. (2002); [c]Shoemaker et al. (in press); [d]Klesius et al. (2000a,b); [e] Klesius et al. (1999); [f]Toranzo et al. (1995); Romalde et al. (1996); Romalde et al. (1999); [g]Eldar et al. (1997); [h]Akhlaghi et al. (1996); [i]Eldar et al. (1995). [j]ECP is the designation for extracellular product; [k]Complete Freund's adjuvant.

TABLE 17.2. Effects of fish species, vaccine, vaccine type, immersion route of vaccination, and size of fish on vaccine efficacy against streptococcal infection.

Fish species	Vaccine against	Vaccine type	Route of vaccination	Size (g)	Relative percent survival (RPS)
Tilapia[a]	*S. agalactiae*	ECP[e]-modified bacterin	Immersion	5	34
Rainbow trout[b]	*S. iniae*	Bacterin	Immersion	10	0
Rainbow trout[b]	*S. iniae*	Bacterin-diluted	Multiple puncture and immersion	10	50
Rainbow trout[b]	*S. iniae*	Bacterin full strength	Multiple puncture and immersion	10	50
Tilapia[a]	*S. agalactiae*	ECP-modified bacterin	Immersion	30	35
Turbot[c]	*Enterococcus* sp.	Toxoid bacterin	Immersion	45	0
Turbot[c]	*Enterococcus* sp.	Toxoid bacterin/oil	Immersion	45	0
Rainbow trout[d]	*Streptococcus* sp.	Bacterin	Immersion	57	11
Turbot[c]	*Enterococcus* sp.	Toxoid bacterin	Immersion	150	0

Notes: [a]Evans, Kelsius, and Shoemaker (2004); [b]Nakanishi et al. (2002); [c]Toranzo et al. (1995); Romalde et al. (1999); [d]Akhlaghi et al. (1996). [e]ECP is the designation for extracellular product.

resulted in protection against challenge infection (Shelby et al. 2001 2002, 2003, 2004). The fact that *S. iniae* is a pathogen of both fish (12 species in both fresh and marine water) and humans (Lau et al. 2003), the development of a genetically modified live vaccine was considered to be an unacceptable approach for safety reasons. A modified killed *S. iniae* vaccine was developed that incorporated extracellular products (ECP) and the formalin-killed cells together (Klesius et al. 1999, 2000b, 2002). Protective serum antibody responses were produced following vaccination that resulted in immunity in tilapia and hybrid striped bass to *S. iniae.* At the present time, the results of vaccine experiments have shown that the *S. iniae* vaccine is primarily efficacious by parenteral injection. However, the *S. iniae* vaccine was not protective against *S. agalactiae* (Evans et al. 2004a), and therefore, an ECP modified killed vaccine to *S. agalactiae* was developed for use in tilapia. Immersion immunization was accomplished with the *S. agalactiae* vaccine, but the protective effects were lower than in fish administered the vaccine parenterally. Similarly, protection against *S. agalactiae* challenge was significantly correlated with specific antibody responses in Nile tilapia vaccinated with an *S. agalactiae* vaccine (Evans, Klesius, and Shoemaker 2004; Pasnik, Evans, and Kelsius 2005; Pasnik, Evans, Panangala et al. 2005). The reasons are unknown for the difference in results related to routes of immunization with *S. iniae* and *S. agalactiae* vaccines.

Types of Streptococcal Vaccines

Extracellular Product Vaccines

Extracellular products (ECP) that are excreted or secreted from the bacterial cells during fermentation are believed to consist of protective antigens and immunostimulatory substances (such as toxins) that may be highly beneficial to augment the effectiveness and duration of protection of the vaccine. Klesius et al. (1999, 2000a, 2002) developed an ECP-modified killed bacterin from encapsulated *S. iniae* to prevent streptococcosis in Nile tilapia and hybrid striped bass (*Morone chrysops* × *Morone saxatilis*). Previously, Toranzo et al. (1995) and Romalde et al. (1996) developed a toxoid-enriched bacterin that consisted of formalin-killed *Enterococcus* sp. and culture fluid containing a toxid produced after formalin treatment. This toxoid-enriched bacterin appeared to protect farmed turbot *(Scophthalmus maximus)* against *Enterococcus* sp. infection.

The formulation for the *S. iniae* vaccine was partially based on this concept of combining ECP with whole cell antigens. The ECP of *S. iniae* were secreted or excreted during three days of *S. iniae* culture. The cultures were then treated with 10 percent neutral buffered formalin to give a final concentration of 3 percent formalin in culture for 24 h. The killed cells and culture fluid were then separated and the cell-free culture fluid was concentrated 20-fold using a hollow fiber concentrator, which also removed substances at or below 2 kDa. This concentrate was filter-sterilized, combined with formalin-killed cells at a concentration of 4×10^9 cells·mL^{-1}, and tested for sterility by plating on sheep blood agar. The ECP-modified *S. iniae* bacterin was given to 25 or 100 g Nile tilapia by a single IP injection of vaccine doses at 4×10^8 or 8×10^8 per fish, respectively. No booster or adjuvant was used in the immunization protocol. At 30 days postvaccination, the result showed that mortality was reduced by 91.3 percent in these vaccinates over a period of 2 months. In addition, the challenged fish were evaluated for clinical signs of disease, and the vaccinates showed no signs of erratic swimming, hemorrhagic exophthalmia, or ocular opacity as exhibited in the controls. The relative percent survival (RPS; Amend 1981) was 95 in the 25 g fish and 84 to 95 in the 100 g fish. The duration of the vaccine response was at least 6 months (Klesius et al. 2002).

Evans et al. (2004a) developed a modified killed *S. agalactiae* vaccine that was formulated to contain killed whole cells and ECP (3 kDa). This vaccine was found to be highly efficacious in 25 to 30 g Nile tilapia after IP administration with an RPS of 70 to 83. The duration of the vaccine response was at least 6 months (Pasnik et al. 2005a). Additional detailed information is provided to the reader (see "Route of Administration" section) on the efficacy of this *S. agalactiae* vaccine by different routes of administration.

Bacterin and Acellular Vaccines

A number of vaccines were produced to control severe and costly outbreaks of streptococcosis in farmed rainbow trout and tilapia in Israel caused by *S. difficile* and by *S. iniae* (formerly *S. shiloi*), respectively (Eldar et al. 1995). They estimated losses ranging between 30 and 45 percent, respectively, in the Israeli farmed rainbow trout *(Oncorhynchus mykiss)* and tilapia (*O. aureus* × *O. nilotica* hybrids) industries. A killed *S. difficile* bacterin was administered by intraperitoneal injection into 150 to180 g tilapia, and the fish were given an IP booster immunization 4 weeks later. Three weeks postimmunization, the vaccinates and nonvaccinates were challenged by IP injection with 100 LD_{50} of *S. difficile.* They reported no mortality in the vaccinates for a period of 75 days postchallenge, and all of the nonvaccinates had died at 60 days postchallenge. A protein vaccine (acellular) was also prepared by first sonicating live *S. difficile* cells followed by overnight treatment of whole sonicate with 5 volumes of acetone. The resulting precipitate was collected, made into a 50 percent protein suspension in PBS, and then complexed with aluminum hydroxide to give a 4 $mg \cdot mL^{-1}$ protein vaccine. Then 150 to 180 g tilapia were twice immunized with 100 μL delivered intramuscularly (IM) at an interval of 2 weeks. Vaccinates and nonvaccinates were then challenged with *S. difficile*. The reported results showed that 8 percent of vaccinates and 100 percent of nonvaccinates died, resulting in an RPS of 92.

The success of the *S. difficile* bacterin in tilapia prompted the development, through this study, of a formalin-killed *S. iniae* vaccine for Israeli farmed trout (Eldar et al. 1997). A *S. iniae* bacterin was administered at doses of 10^9 and 10^{10} cells per fish. This vaccine was also evaluated at higher vaccine concentrations of 3 × 10^{11} cells than for the *S. difficile* bacterin (10^9 cells). In addition, the bacterin dose of 3 × 10^{11} cells was given either once or twice by IP injection. Furthermore, the trout exposed to the vaccine were smaller (30 to 50 g) than the tilapia previously immunized with *S. difficile* vaccine (150 to 180 g). The results reported indicated that 90 percent protection was achieved in fish receiving a single IP injection of 3×10^{11} cells, and the duration of immunity was 6 months. Fish that were twice immunized with the same dose also had 90 percent protection for 1 month. The fish immunized with 10^9 had only 50 percent protection at 1 month postimmunization, while the fish immunized with 10^{10} cells showed 90 percent protection at 1 month.

PISCINE IMMUNITY

Innate Immunity

Successful vaccination of any animal depends on the stimulation of immune responses, and researchers must understand the available mechanisms of immune protection against diseases. The two principal immune mechanisms are innate and acquired immunity, and these responses may be stimulated by vaccination. The mechanisms of innate immunity include a variety of immune system cells and substances (Shoemaker et al. 2001b). The principal cells of innate immune mechanisms include natural killer cells and phagocytes such as granulocytes (neutrophils) and monocytes (macrophages). Substances important in the innate immune responses include cytokines, lectins, lysozyme, complement components, C-reactive proteins, and ceruloplasmin. Cations, such as calcium, magnesium, and iron are also important mineral components that are required in a variety of innate immune responses. Cytokines are produced by a number of cell types that include lymphocytes, granulocytes, and accessory cells. These reactive molecules function in a variety of pathways that may cause injury to the pathogen, though they may also cause injury to host tissue as well. Innate immune responses are important because they provide rapid, albeit nonspecific responses upon initial exposure to pathogens, and they have been implicated in the evolutionary differences between disease-susceptible and resistant strains of tilapia (Beacham and Evelyn 1992).

Acquired Immunity

Humoral Antibody Responses

Acquired immune mechanisms include humoral (antibody) and cell-mediated responses that are slower than innate responses but are pathogen-specific (Shoemaker et al. 2001b). Fish appear to have only one class of antibody, which is a tetrameric IgM-like immunoglobulin (Ig) produced by bursa-equivalent lymphocytes (B-lymphocytes). The antibody responses are generally important against extracellular pathogens, though IgM responses following primary pathogen exposure generally only peak 21 to 30 days later. However, immune fish will produce secondary IgM responses within 7 days after a second exposure, and successful immunization with a suitable antigen would allow for this rapid secondary antibody response. Thus, vaccinated tilapia that produce antibody responses to the vaccine would be expected to be immune before challenge infection with the pathogen. Immune antibody responses also play an important role in aiding phagocytosis by macrophages or neutrophils through immobilizing or entrapping the pathogen and facilitating its uptake by phagocytes. Complement killing activity for a pathogen can also be activated by antibody-pathogen interactions.

Cellular Immune Responses

Cell-mediated immunity (CMI) is generally important against intracellular pathogens. The thymus-dependent lymphocyte (T-lymphocyte) and cytotoxic lymphocytes are among the principal cells of CMI responses. Attempts to correlate CMI responses and acquired immunity by in vitro assays have commonly produced poor results in fish, and thus CMI in fish is not well-characterized. However, it will be important to fully characterize immunity in fish, because a number of factors influence innate and acquired immunity in fish and many of these factors are associated with the successes or failures of the vaccines. Finally, killed types of vaccines (bacterins) are generally less effective in producing CMI responses and are thus normally less effective against intracellular pathogens. In general, bacterins can produce protective antibody responses and are often more effective against pathogens that remain in the extracellular milieu.

Vaccine-Induced Immunity to* Streptococcus *Sp.

Enzyme-Linked Immunosorbent Assay

Previous research has indicated that immunity to streptococci in humans and animals is dependent upon specific antibody responses, though this was formerly difficult to evaluate in fish. The only method previously available for measuring the antibody response in tilapia to *S. iniae* infection was agglutination, because enzyme-linked immunosorbent assays (ELISA) were not available at that time and an antibody against tilapia Ig had not been produced or characterized. Shelby et al. (2002) developed an indirect ELISA to measure the serum antibody responses in tilapia to *S. iniae*. The ELISA was developed using a goat anti-tilapia Ig, a rabbit anti-goat Ig conjugate, and soluble coating antigens prepared from whole-cell *S. iniae* sonicates. A correlation of $R^2 = 0.85$ was found between the ELISA and agglutination assay using sera from *S. iniae*-infected and uninfected tilapia. The major advantage of the ELISA is that antibody response can be quantitatively measured using an ELISA reader, thus eliminating the subjectivity inherent with visual observation of an agglutination titration endpoint. Additional advantages of the ELISA include a lower serum requirement and a shorter assay time than the agglutination test, and the ELISA also has potential for fish producers to identify natural infections by testing sera from exposed fish.

An ELISA can also be used to monitor vaccine efficacy since production of humoral anti-*S. iniae* antibodies is one parameter by which the quality of vaccines can be measured. Shelby et al. (2003, 2004) initially measured this function in hybrid striped bass that were vaccinated with a *S. iniae* vaccine. The antibody levels in vaccinates were shown to progressively increase 14 days postvaccination and reach a peak at 22 and 28 days postvaccination. The antibody responses were observed to show a further increase when the vaccinated fish were infected at 28 days postvaccination. These vaccinated fish were protected, indicating that anti-*S. iniae* antibody was at least partially responsible for the protective immunity. Complement levels were also significantly increased following vaccination, and therefore both antibody and complement appear to play an important role in mediating the protective immunity against *S. iniae* following vaccination. This conclusion supports previous results showing that passively transferred sera from tilapia immune to *S. iniae* conferred protection to naïve tilapia against *S. iniae* infection (Shelby et al. 2002). Similarly, protection against *S. agalactiae* challenge was significantly correlated with specific antibody reponses in Nile tilapia vaccinated with an *S. agalactiae* vaccine (Evans, Klesius, and Shoemaker 2004; Pasnik, Evans, and Klesius 2005; Pasnik, Evans, Panangala et al. 2005).

Agglutinating Antibody Assay

As stated previously, Eldar et al. (1995) produced an *S. difficile* bacterin and an acellular protein vaccine and inoculated tilapia twice with either of the vaccines. The bacterin had an RPS of 100 and the protein vaccine had an RPS of 92 after challenge with *S. difficile*. Vaccinates produced agglutinating antibodies against the *S. difficile* at low titers (1/60 serum dilution) with both bacterin and protein vaccines. Western blot analysis of these antibodies revealed that sera from bacterin-immunized fish recognized only one antigen (49 kDa), whereas sera from fish immunized with the protein vaccine reacted with two antigens (36 and 49 kDa). Furthermore, the authors indicated that the bacterin offered a broad degree of protection against different isolates of *S. difficile*. These findings suggest that humoral responses against certain components of streptococci confer species-specific, but broad-range protection against *Streptococcus* sp. Eldar et al. (1997) also developed a formalin-killed *S. iniae* vaccine that conferred significant long-term protection at higher doses. Antibody titers against *S. iniae* were produced at the higher vaccine concentrations (1×10^{10} or 3×10^{11} cells), while the lowest dose (1×10^{9} cells) produced no antibody response. In addition, hyper-immune rabbit serum was shown to partially, passively protect trout against *S. iniae* infections, further indicating that antibody against *S. iniae* was in part responsible for protective immunity. The antigens recognized by sera from these vaccinated fish were between 18.5 and 49 kDa, with the most significant response against the latter. These immunogens most likely represent streptococcal capsular antigens. The results of this study were summarized in Bercovier et al. (1997), and the authors concluded that bacterins against *S. difficile* or *S. iniae* were protective in tilipia for at least 4 months. Antibodies were produced after vaccination and appeared to mediate the protective immune responses against cell surface antigens.

Cellular Immunity Assays

However, other researchers have indicated alternate modes of immune protection. In Japan, β-hemolytic streptococcal infections were prevalent in many rainbow trout and Nile tilapia producing fish farms (Sakai et al. 1989), and the *Streptococcus* sp. was presumed to be a biotype of *S. iniae*. A formalin-killed bacterin was prepared using an isolate obtained from moribund rainbow trout. The bacterin was used to immunize two groups of fish by IP injection or by bath immersion. The study evaluated agglutination titers, serum bactericidal activities, phagocytic activities of leukocytes from the head kidney, and clearance of challenge infection from the blood, brain, kidney, and liver. Titers of agglutinating antibodies were only demonstrated in the IP-injected fish, though the titers were low. The sera from IP and immersion-immunized fish did not demonstrate any bactericidal activities against *S. iniae*. The phagocytic activities of kidney leukocytes against *S. iniae* were significantly different between the vaccinated and the control fish. However, no difference was observed between the IP and immersion-immunized fish. The IP and immersion-vaccinated fish cleared the *S. iniae* infection by 72 h postchallenge. No difference was observed in the rate of clearance between the IP and immersion vaccinates. The control fish were observed not to have cleared their infections by 72 h postchallenge. Sakai et al. (1989) concluded that the results of their study indicated that streptococcal cells are not killed by complement and specific antibody and that serum bactericidal activity was not associated with protective immunity in the vaccinates. Further, they indicated that phagocytic activity and protective immunity were associated, but other cellular immunities played a role in protective immunity following vaccination.

Passive Immunization

Though the findings of these tilapia studies may contradict each other, studies of vaccines against streptococcosis in other fish species add more information. Akhlaghi et al. (1996) used passive and active immunizations to prevent streptococcosis in rainbow trout. The results of passive immunization with hyperimmune *Enterococcus* sp. sera from sheep, rabbit, or fish showed that the sheep and rabbit hyperimmune sera given 1 month after immunization with virulent *Enterococcus* sp. provided protection in tilapia. The sheep-derived hyperimmune sera provided the greatest protection, and the hyperimmune fish sera failed to provide protection. Intraperitoneal injection or a 30 s immersion immunization with the *Enterococcus* sp. bacterin emulsified with Freund's complete adjuvant only provided an RPS of 89 and 11, respectively, at 1 month postimmunization. The authors concluded that antistreptococcal antibodies indeed played a role in protection against streptococcosis following vaccination, but cellular immunity may also play a role in the protection.

When Romalde et al. (1999) evaluated their toxiod-enriched *Streptococcus* sp. bacterin, the vaccine exhibited significant, long-term protective effects, and the results showed 100 RPS 4 weeks postimmunization. The duration of protective immunity was 24 months postimmunization with an RPS of 70 when the bacterin was combined with adjuvant. Meanwhile, the bacterin without adjuvant conferred an RPS of 45 at 24 months postvaccination. However, the study showed no association between antibody titers and protective immunity. Nonetheless, the authors observed an increased rate of phagocytosis at 4 days postimmunization, possibly indicating another crucial characteristic of protective antistreptococcal immune responses.

FACTORS AFFECTING VACCINE EFFICACY

Several factors have been evaluated to determine their effects on protective immunity in fish that were administered streptococcal vaccines. Not all factors have been studied nor has the exploration of examined factors been exhaustive or definitive, but the following still give important insights into successful vaccination of tilapia against *Streptococcus* sp.

Bacterial Isolate Selection

The development of a vaccine is based on its components, and the efficacy of streptococcal vaccines appears to be related to the bacterial isolate used. Klesius et al. (2000b) immunized tilapia against *S. iniae* using the ECP-modified bacterin. The results indicated that the IP route provided an RPS of 93 after challenge with a heterologous strain of *S. iniae*. Interestingly, if the fish were challenged with the same isolate of *S. iniae* that was used to prepare the vaccine, the RPS was significant, but decreased below that found in the heterologous challenge. These results confirm that the isolate candidate chosen to prepare the ECP-modified bacterin is central to its apparent efficacy and that differences do exist in the protective antigens between *S. iniae* isolates. The definitive basis for selection of the master seed for licensing and production of *S. iniae* vaccine is currently unknown, but the selection of an encapsulated isolate may be an important consideration. Klesius et al. (2000b) suggest that protective antibody responses are dependent on the antigenic composition of the *S. iniae* used to prepare the vaccine.

A study by Bachrach et al. (2001) reported mass mortalities in rainbow trout that had been previously immunized with the *S. iniae* bacterin (Eldar et al. 1997). Though the *S. iniae* bacterin had been previously reported to give very broad protection against a number of different isolates of *S. iniae* (Eldar et al. 1997), a bacterin failure was reported in these fish apparently due to a new serotype of *S. iniae* in the rainbow trout. Using PCR and serological methods, the new serotype was classified as serotype II, whereas the bacterin in use was prepared from the serotype I of *S. iniae*. Thus new bacterial serotypes may emerge among fish populations and thereby previously efficacious vaccines may be rendered ineffective.

Route of Administration

One of the major factors determining the immunostimulatory and protective effects of a vaccine is its route of administration. Generally, vaccines are given intramuscularly, intraperitoneally, orally, or by immersion. Although all routes have provided some level of protection against streptococcosis (Klesius et al. 1999; Clark et al. 2000; Klesius et al. 2000b; Evans, Shoemaker, and Klesius 2004; Evans et al. in press; Shoemaker et al.

in press), some appear more efficacious than others. Klesius et al. (2000b) also evaluated the IP and IM routes of vaccination with the ECP-modified *S. iniae* bacterin. The results indicated that IP route provided an RPS of 93 versus 59 by the IM route after challenge with a heterologous isolate of *S. iniae*. If the fish were challenged with a homologous isolate of *S. iniae*, the IP route still gave better protection even though the RPS was lower for both routes of vaccination than observed for the heterologous challenge. These results indicate that the IP route of injection is more efficacious than the IM route regardless of the challenge isolate.

Evans et al. (2004a) used an ECP-modified *S. agalactiae* vaccine to determine the effects of immersion and IP vaccination in 5 g tilapia. The RPS among the fish vaccinated by immersion were moderately higher than among the fish inoculated intraperitoneally. Though the RPS differences between the two routes may not be significant, immersion inoculation might be considerably more desirable because of its relative ease of administration to fry. Furthermore, small increases in vaccine efficacy translate to higher numbers of fry surviving to adulthood. This efficacy may not apply to all species and sizes of fish. When Akhlaghi et al. (1996) used IP injection or a 30-s immersion immunization of *Enterococcus* sp. bacterin emulsified in Freund's complete adjuvant, the vaccine provided an RPS of 89 or 11, respectively. In addition, these investigators found that a streptococcal vaccine provided 70 percent protection in rainbow trout following IP injection. Previously, the results of immersion immunization of rainbow trout with a streptococcal bacterin showed that 70 percent protection was produced after a 3-min immersion period (Sakai et al. 1987).

Nakanishi et al. (2002) demonstrated that protection of 10-g rainbow trout immersed in a formalin-killed *S. iniae* vaccine suspension, following skin puncture, matched that obtained by IP injection. Mortality of IP-vaccinated and multiple puncture/immersion-vaccinated fish was 40 percent, whereas those vaccinated by immersion alone and nonvaccinated fish were each 80 percent. *Streptococcus agalactiae* vaccination by both IP injection and immersion without percutaneous puncture in a study by Evans et al. (2004a) had greater efficacy than *S. iniae* vaccination by these routes of administration performed by Nakanishi et al. (2002).

Influence of Size of Tilapia on Efficacy

Vaccination needs to be accomplished at the earliest life stage to be practical and economical for the fish producer. Many investigators believe that successful vaccination of fish is not possible until at least 21 days of age. However, Warr (1997) demonstrated that functional lymphocytes in lymphoid organs are present a few days after hatching in fish. Streptococcal vaccines administered either IP or IM are shown to be efficacious between 18 g to 150 to180 g tilapia (Table 17.1). Yet an ECP-modified *S. agalactiae* vaccine was significantly less efficacious in 5 g tilapia (RPS 25) than in 30 g tilapia (RPS 80) when administrated by IP injection (Evans et al. 2004a). Thus the trend indicates that of the vaccines are efficacious in 18 g or larger size tilapia by IP or IM injection, though there is no information on the efficacy on the vaccines in tilapia between 6 and 18 g.

Despite this, smaller fish may better benefit from inoculation by means other than injection. Evans et al. (2004a) indicated that a higher level of protection was found in 5 g tilapia vaccinated against *S. agalactiae* by immersion (RPS 34) than 5 g tilapia vaccinated intraperitoneally (RPS 25). Clark et al. (2000) found an RPS of approximately 70 to 80 after administration of an *S. iniae* vaccine by either immersion or orally to 1, 2, and 7 g tilapia. Nakanishi et al. (2002) demonstrated that the protection of 10 g rainbow trout immersed in a formalin-killed *S. iniae* vaccine suspension, following skin puncture, had an RPS equivalent to that obtained by IP vaccination. In contrast, neither 45 g turbot nor 57 g rainbow trout were protected by immersion vaccination. These findings indicate that tilapia develop immunocompetence early in life sufficient for successful vaccination.

Interestingly, multiple researchers have also found that streptococcal vaccination itself can enhance fish growth, regardless of vaccine construct. Eldar et al. (1997) found that fish administered a *S. iniae* bacterin gained 20 percent more weight than nonvaccinates over 4 months. The initial weight of nonimmunized fish was 55 to 59 g, whereas the average weight of vaccinates was 410 g versus 340 g among nonvaccinates after 4 months. Similar positive growth effects have been seen among juvenile tilapia administered the ECP-modified *S. agalactiae* vaccine (Klesius and Evans personal communication). The basis for vaccinating small tilapia is strengthened by the fact that fish may grow faster after vaccination, which, per se, may positively enhance the immune system development and decrease the production cycle duration.

Methods of Assessing Vaccine Efficacy

The challenge methods cited in the literature include IP injection, IM injection bath immersion, and cohabitation. Nordmo (1997) concluded that bath immersion and cohabitation aptly mimic natural exposure to a pathogen and ensure that cutaneous antibodies play a role in the immune response. On the other hand, the injection method ensures that all fish receive an equal exposure to the pathogen and lends to standardization of a defined pathogen dose. Thus, challenge infections are often accomplished by IP injection. The RPS method (Amend 1981) is most often used to determine the efficacy of a fish vaccine. However, the RPS may vary significantly between the challenge methods employed (Nordmo and Ramstad 1997). Genetic and stress factors also influence the RPS due to change in the susceptibility of fish to a defined pathogen dose (Simon et al. 1981; Barton et al. 1985; Maule et al. 1987; Evans, Klesius, Shoemaker et al. 2004; Evans, Shoemaker, and Klesius 2004). In addition, the RPS may vary between batches of vaccine and the results of experimental trials may not agree with those obtained in the field. On the other hand, vaccines may produce very good results in the field, but may not produce RPS results equal to or greater than 60 in experimental trials. The statistical aspects of vaccine trials are very important to the design, analysis, and assessment of vaccine effectiveness. There is a need to employ a random block design, determine the observation unit, calculate the sample size, and determine the statistical methods that are best applicable to the design of the trial (Jarp and Tverdal 1997). In many cases, vaccine trials should be conducted as single-blind studies. The life function test (Kaplan-Meier) may offer advantages over the RPS method. Care must be taken of differences if cluster effect is influencing the results of vaccine trial where multiple replicate tanks are employed. Finally, the proper vaccine trial should meet the requirements for federal licensing of fish vaccines. There is a need to develop challenge and assessment methods for fish vaccines that may more closely mimic field conditions.

Nutrition

Only limited research has evaluated the correlations between nutritional components and streptococcal vaccines for tilapia. Whittington et al. (2003) studied the nutritional supplement, β-hydroxy-β-methylbutyrate, which can increase disease resistance and growth in other animals. Fingerling tilapia were fed diets supplemented with 0, 12.5, 25, or 50 $mg{\cdot}kg^{-1}$ β-hydroxy-β-methylbutyrate for 2 weeks prior to vaccination with a killed *S. iniae* vaccine. Although the vaccination protected the fish after *S. iniae* challenge, diets with β-hydroxy-β-methylbutyrate failed to enhance the level of protection. Romalde et al. (1999) fed fish diets containing the immunostimulant, glucan, and vaccinated them with a toxoid-enriched *Enterococcus* sp. bacterin in water or emulsified in a mineral oil adjuvant. Once again, the vaccine itself provided significant protection against bacterial challenge, but the glucan-containing diets did not enhance the protection of the bacterin. Though their vaccine was evaluated in turbot *(Scophthalmus maximus)*, similar effects would likely be seen in tilapia that were administered a streptococcal vaccine. Further studies on nutritional supplements are warranted despite these initial findings, because vaccination efficacy is often limited and supplements may further increase the level of disease resistance.

Vaccine Failure

Roth (1991) and Klesius et al. (2001) provide some of the reasons for the failure of vaccines to provide protection. The most relevant of these are improper handling and administration, host and environmental factors, overwhelming infection, inadequate duration of immunity, and major serological differences between the vaccine and field strains. Highly stressful husbandry practices or environmental factors are harmful to the fish immune system and thus may weaken both the innate and the acquired immunity against the pathogen (Klesius et al. 2003). Evans et al. (2004b) reported that Nile tilapia exposed to low dissolved oxygen concentration resulted in a stress response and increased susceptibility to *S. agalactiae* infection. The stress response was determined by changes in blood glucose concentration using an electronic glucose monitor. Stressed tilapia were found to be susceptible to infection by 95 colony-forming units of *S. agalactiae.*

Endo et al. (2002) found that tilapia that were fed free-choice feed had increased immune responses and reduced stress levels. Plasma cortisol levels were significantly lower in the free-choice fed fish than in scheduled-fed fish. Antibody production, macrophage phagocytosis, and the number of blood circulating lymphocytes of free-choice fed fish significantly exceeded those of scheduled-fed fish. Immunization of stressed fish will most likely not provide an effective response to the vaccination. Finally, poor nutrition is likely to produce a diminished immune response to vaccination because of the low availability of energy required by the immune system to mount an effective response against the pathogen (Klesius et al. 1999).

Antibiotics

Antibiotic treatment may also suppress antibody production and cause vaccine failure. Wise and Johnson (1998) found that catfish fed Romet-medicated rations on a daily basis had lower antibody titers against *Edwardsiella ictaluri* than fish that were fed nonmedicated rations. Siwiki et al. (1989) found that IP injection of oxytetracycline suppressed the immune function in salmonids. Rainbow trout vaccinated intraperitonally simultaneously against *Vibrio anguillarum* and *Aeromonas salmonicida* and then treated with oxytetracycline by immersion were reported to have low antibody titers, which led to vaccine failure (Lunden et al. 1998). However, in another investigation, Tafalla et al. (1999) reported that oxytetracycline did not suppress the immune responses in turbot. Thus, different antibiotics, dosages, routes of administration, and fish species appear to have variable influences on the outcome of the immune responses in fish (Tafalla et al. 1999).

Economic Benefits of Vaccination

Producers are recognizing the long-term benefits of fish vaccination. The benefits include reduced fish disease problems, reduction of costly chemical and antibiotic treatments (Markestad and Graves 1997), and enhanced growth rates in the absence of the pathogen. The application of vaccines may reduce the stocking densities because of greater survival of the vaccinated fish over the nonvaccinated fish. Tilapia losses were estimated to be 45 percent of 8,000 tons per year due to *S. iniae* in Israel (Bercovier et al. 1997). Immunization of rainbow trout with the specific *S. iniae* bacterin was reported to reduce mortality from 50 percent to less than 5 percent, annually, in Israeli farms between 1995 and 1997 (Eldar et al. 1997). The *S. iniae* bacterin vaccine was estimated to return $12 for every dollar spent in Israel (Bercovier et al. 1997). Evans, Klesius, and Shoemaker (2004) and Evans, Shoemaker, and Klesius (2004) reported that immunization of Nile tilapia with an ECP-modified *S. agalactiae* vaccine. Significantly reduced the *S. agalactiae* infection stress response. Furthermore, the IP immunized tilapia had a short-term stress response following vaccination. As mentioned earlier, these ECP-modified *S. iniae* or *S. agalactiae* vaccines may be expected to enhance the growth rate of tilapia in addition to reducing mortalities due to streptococcosis. In other words, vaccination may achieve faster growing tilapia that consume more feed and significantly reduce the production time to market size. Attaining this level of disease prevention would result in significant economic benefits to the tilapia producer worth hundreds of million dollars annually.

Strategic Application of Vaccines

Formulation and application of vaccines needs to include the characteristics of the target disease agent, the type of protective immune responses required, the earliest life stage of the fish for immunization, routes of administration, duration of protective immunity, type of fish culture system, manufacturing considerations, and licensing and cost of vaccine with regard to the species being immunized. Table 17.3 summarizes the ideal parameters for a *S. iniae* vaccine for use in tilapia production. The ideal time to vaccinate

TABLE 17.3. Strategic application of the modified *Streptococcus iniae* vaccine against streptococcosis.

Factors	Application
Size	1 to 5 g wet weight
Cost-effective vaccine	ECP-modified vaccine
Route	Immersion followed by oral fed vaccine
Timing	21 to 30 days postprimary immunization
Duration	9 to 12 months
Benefits	85 to 95 percent reduction in mortality and enhanced growth rate

tilapia is before the fish are stocked into production ponds, tanks, or cages. The size of these fish is generally 0.5 to 1 g, and the most cost-effective route of vaccination is bath immersion. A killed vaccine produced by fermentation would be easier to manufacture and more acceptable to licensing because of both its safety and efficacy in the target species. These vaccines can provide significant efficacy, but their utility may be improved if the primary immersion immunization is followed by oral booster immunization in the production system. The timing between the primary and the booster immunization should be 21 to 30 days or after the fish obtain a size the of 5 to 10 g and are showing good growth rates and feed consumption.

CONCLUSIONS

Vaccination is a valid, environmentally sound approach to producing specific immune responses that can be effective in the prevention of streptococcal diseases in farmed tilapia. The information presented here will enhance our understanding of vaccination against streptococcal disease and enable us to modify and improve our vaccination procedures so that maximum protection can be achieved with cost-saving benefits to the fish producer.

REFERENCES

Akhlaghi, M., B.L. Munday, and R.J. Whittington (1996). Comparison of passive and active immunization of fish against streptococcosis (enterococcosis). *Journal of Fish Diseases* 19:251-258.

Amend, D.F. (1981). Potency testing of fish vaccines. In D.P. Anderson, and W. Hennssen (Eds.), *Fish biologics: Serodiagnostics and vaccines* Volume 49: *Development in biological standardization* (pp. 447-454). Basel, Switzerland: Karger.

Bachrach, G., A. Zlotkin, A. Hurvitz, D.L. Evans, and A. Eldar (2001). Recovery of *Streptococcus iniae* from diseased fish previously vaccinated with a *Streptococcus* vaccine. *Applied and Environmental Microbiology* 67:3756-3758.

Barton, B.A., C.B. Schreck, R.D. Ewing, A.R. Hemmingsen, and R. Patino (1985). Changes in plasma cortisol during stress and smoltification in Coho salmon, *Oncorhyncus kisutch. General and Comparative Endocrinology* 59:468-471.

Beacham, T.D., and T.P.T. Evelyn (1992). Genetic variation in disease resistance and growth to chinook, coho, and chum salmon with respect to vibriosis, furunculosis, and bacterial kidney disease. *Transactions of American Fisheries Society* 121:456-485.

Bercovier, H., C. Ghittino, and A. Eldar (1997). Immunization with bacterial antigens: Infections with streptococci and related organisms. In R. Gudding, A. Lillenhaug, P. Midtlyng, and F. Brown (Eds.), *Fish vaccinology*: Developments in biological standardization (vol. 90, pp. 153-160). Basel, Switzerland: Karger.

Clark, J.S., B. Paller, and P.D. Smith (2002). Prevention of *Steptococcus* in tilapia by vaccination: The Philippine experience, In K. Fitzsimmons and J.C. Filho (Eds.), *Tilapia aquaculture in the 21st century* (pp. 545-551). Rio de Janerio, Brazil.

Eldar, A., C. Ghittino, L. Asanta, E. Bozzetta, M. Goria, M. Prearo, and H. Bercovier (1996). *Entercoccus seriolicida* is a junior synonym of *Lactococcus garvieae* and meningoencephalitis in fish. *Current Microbiology* 32:85-88.

Eldar, A., M. Goria, C.L. Ghittino, A. Zlotkin, and H. Bercovier (1999). Biodiversity of *Lactococcus garvieae* strains isolated from fish in Europe, Asia and Australia. *Applied Environmental Microbiology* 65:1005-1008.

Eldar, A., A. Horovitz, and H. Bercovier (1997). Development and efficacy of a vaccine against *Streptococcus iniae* infection in farmed rainbow trout. *Veterinary Immunology and Immunopathology* 56:175-183.

Eldar, A., O. Shapiro, Y. Bejerano, and H. Bercovier (1995). Vaccination with whole-cell vaccine and bacterial protein extract protects tilapia against *Streptococcus difficile* meningoencephalitis. *Vaccine* 13:867-870.

Endo, M., C. Kumahara, T. Yoshida, and M. Tabata (2002). Reduced stress and increased immune responses in Nile tilapia kept under self-feeding conditions. *Fisheries Science* 68:253-257.

Evans, J.J., P.H. Klesius, P.M. Glibert, C.A. Shoemaker, M.A. Al Sarawi, J. Landsberg, R. Duremdez, A. Al Marzouk, and S. Al Zenki (2002). Characterization of beta-hemolytic group B *Streptococcus agalactiae* in cultured seabream, *Sparus auratus* (L.), and wild mullet, *Liza klunzingeri* (Day), in Kuwait. *Journal of Fish Diseases* 25(9):1-9.

Evans, J.J., P.H. Klesius, and C.A. Shoemaker (2004). Efficacy of *Streptococcus agalactiae* (Group B) vaccine in tilapia *(Oreochromis niloticus)* by intraperitoneal and bath immersion administration. *Vaccines* 22 (27-28):3769-3773.

Evans, J.J., P.H. Klesius, and C.A. Shoemaker (in press). Therapeutic and prophylatic immunization against Streptococcus iniae infecton in hybrid striped bass (*Morone chrysops* × *M. saxatilis*). *Aquaculture Research.*

Evans, J.J., P.H. Klesius, C.A. Shoemaker, and B. Fitzpatrick (2004). *Streptococcus agalactiae*: Vaccination and infection stress in Nile tilapia, *Oreochromis niloticus*. *Journal of Applied Aquaculture*. 16(3/4):105-115.

Evans, J.J., C.A. Shoemaker, and P.H. Klesius (2000). Experimental *Streptococcus iniae* infection of hybrid striped bass (*Morone chrysops* × *Morone saxatilis*) and tilapia (*Oreochromis niloticus*) by nares inoculation. *Aquaculture* 189(3-4):197-210.

———. (2001). Distribution of *Streptococcus iniae* in hybrid striped bass (*Morone chrysops* × *Morone saxatilis*) following nare inoculation. *Aquaculture* 194(3-4): 233-243.

———. (2004). Effects of sublethal dissolved oxygen stress on blood glucose and susceptibility to *Streptococcus agalactiae* in Nile tilapia, *Oreochromis niloticus*. *Journal of Aquatic Animal Health* 15:202-208.

Glibert, P.M., J.H. Landsberg, J.J. Evans, M.A. Al-Sarawi, M. Faraj, M.A. Al-Jarallah, A. Haywood, S. Ibrahem, P.H. Klesius, C. Powell et al. (2002). A fish kill of massive proportion in Kuwait Bay, Arabian Gulf, 2001: The roles of bacterial disease, harmful algae, and eutrophication. *Harmful Algae* 1(2):1-17.

Gudding R., A. Lillehaug, P. Midtlyng, and F. Brown (Eds.) (1997). *Fish vaccinology: Developments in biological standardization* (vol. 90). Basel, Switzerland: Karger.

Jarp, J. and A. Tverdal (1997). Statistical aspects of fish vaccination trials. In R. Gudding, A. Lillehaug, P.J. Midtlyng, and F. Brown, (Eds.), *Fish vaccinology: Development in biological standardization* (Vol. 90). Basel, Switzerland: Karger.

Kitao, T. (1993). Streptococcal infection. In V. Inglis, R.J. Roberts, and N.R. Bromage (Eds.), *Bacterial diseases of fish* (pp. 196-210). Oxford, United Kingdom: Blackwell Scientific Publications.

Klesius, P.H., C.A. Shoemaker, and J.J. Evans (1999). Efficacy of a killed *Streptococcus iniae* vaccine in tilapia (*Oreochromis niloticus*). *Bulletin of the European Association of Fish Pathologists* 19(1):39-41.

———. (2000a). Efficacy of a single and combined *Streptococcus iniae* isolate vaccine administered by intraperitoneal and intramuscular routes in tilapia (*Oreochromis niloticus*). *Aquaculture* 188: 237-246.

———. (2000b). Vaccination: A health management practice for preventing diseases caused by *Streptococcus* in tilapia and other cultured fish. In K. Fitzsimmons, and J.C. Filho (Eds.), *Tilapia aquaculture in the 21st century* (pp. 558-564). Rio de Janeiro, Brazil: Panorama da Aqüicultura.

———. (2002). *Streptococcus iniae* vaccine. U.S. Patent 6,379,677 B1.

———. (2003). The disease continuum model: Bi-directional response between stress and infection linked by neuroimmune change. In C.-S. Lee, and P.J. O'Bryen (Eds.), *Biosecurity in aquaculture production systems: Exclusion of pathogens and other undesirables* (pp. 13-34). Baton Rouge, LA: The World Aquaculture Society.

Klesius, P.H., C.A. Shoemaker, J.J. Evans, and C. Lim (2001). Vaccines: Prevention of diseases in aquatic animals. In C. Lim, and C. Webster (Eds.), *Nutrition and fish health* (pp. 317-335). Binghamton, NY: The Haworth Press, Inc.

Lau, S.K.P., P.C.Y. Woo, H. Tse, K.W. Leung, S.S.Y. Wong, and K.Y. Yuen, (2003). Invasive *Streptococcus iniae* infections outside North America. *Journal of Clinical Microbiology* 41(3):1004-1009.

Lunden, T., S. Miettinen, L.G. Lonnstrom, E.M. Lilius, and G. Bylund (1998). Influence of oxytetracycline and oxolinic acid on the immune response of rainbow trout *(Oncorhynchus mykiss). Fish and Shellfish Immunology* 8:217-230.

Markestad, A., and K. Grave (1997). Reduction of antibacterial drug use in Norwegian fish farming due to vaccination. In R. Gudding, A. Lillehaug, P. Midtylyng, and F. Brown (Eds.), *Fish vaccinology: Developmens in biological standardization* (Vol. 90, pp. 365-369). Basel, Switzerland: Karger.

Maule, A.G., C.B. Schreck, and S. Kaattari (1987). Changes in the immune system of coho salmon *(Oncorhynchus kisutch)* during parr-to-smolt transformation and after implantation of cortisol. *Canadian Journal of Fisheries and Aquatic Science* 44:161-166.

Nakanishi, T., I. Kiryu, and M. Ototake (2002). Development of a new vaccine delivery method for fish: Percutaneous administration by immersion with application of a multiple puncture instrument. *Vaccine* 20: 3764-3769.

Nordmo, R. (1997). Strengths and weaknesses of different challenge methods. In Gudding, R., A. Lillehaug, P.J. Midtlyng, and F. Brown (Eds.), *Fish vaccinology: Developments in biological standardization* (Vol. 90, pp. 303-309). Basel, Switzerland: Karger.

Nordmo R., and A. Ramstad (1997). Comparison of different challenge methods to evaluate the efficacy of furunculosis vaccines in Atlantic salmon (*Salmo salar* L.). *Journal of Fish Diseases* 20:119-126.

Pasnik, D.J., J.J. Evans, and P.H. Klesius (2005). Duration of protective antibodies and correlation with survival in Nile tilapia *(Oreochromis nilotics)* following *Streptococcus agalactiae* vaccination. *Diseases of Aquatic Organisms* 66:129-134.

Pasnik, D.J., J.J. Evans, V.S. Panangala, P.H. Kelsius, R.A. Shelby, and C.A. Shoemaker (2005). Antigenicity of *Streptococcus agalactiae* extracellular products and vaccine efficacy. *Journal of Fish Diseases* 28:205-212.

Romalde, J.L., B. MargariñĐŐ' Òãú A.E. Toranzo (1999). Prevention of streptococcosis in turbot by intraperitoneal vaccination: A review. *Journal of Applied Ichthyology* 15:153-158.

Romalde, J.L., R. Silva, A. Riaza, and A.E. Toranzo (1996). Long-lasting protection against turbot streptococcosis obtained with toxoid-enriched bacterin. *Bulletin of the European Association of Fish Pathologists* 16:169-171.

Roth, J. (1991). The principles of vaccination: The factors behind vaccine efficacy and failure. *Veterinary Medicine* 44:2366-2372.

Rungpan, L., T. Kitao, and Y. Yoshida (1986). Protective efficacy of *Aeromonas hydrophila* vaccine in Nile tilapia. *Veterinary Immunology and Immunopathology* 12:345-360.

Sakai, M., R.S. Atsuta, and M. Kobayashi (1989). Protective immune response in rainbow trout *Oncorhynchus mykiss* vaccinated with beta-haemolytic streptococcal bacterin. *Fish Pathology* 24:169-173.

Sakai, M., R. Kubota, S. Atsuta, and M. Kobayashi (1987). Vaccination of rainbow trout, *Salmo gairdneri,* against beta-haemolytic streptococcal disease. *Nippon Suisan Gakkaishi* 53:1373-1376.

Schmidtke, L.M., and J. Carson (1999). Induction, characterization and pathogenicity in rainbow trout *Oncorhynchus mykiss* (Walbaum) of *Lactococcus garvieae* L-forms. *Veterinary Microbiology* 69:287-300.

Shelby, R.A., C.A. Shoemaker, and P.H. Klesius (2002). Detection of humoral response to *Streptococcus iniae* infection of Nile tilapia, *Oreochromis niloticus,* by a monoclonal antibody-based ELISA. *Journal of Applied Aquaculture* 12(3): 23-31.

———. (2003). Development of an immunoassay to measure the humoral response of hybrid striped bass *Morone chrysops* × *M. saxatilis. Veterinary Immunology and Immunopathology* 91:217-235.

———. (2004). Development of an ELISA to meaeure the humoral immune response of hybrid striped bass *Morone chrysops* × *M. saxatilis* to *Streptococcus iniae. Aquaculture Research* 35:997-1001.

Shelby, R., C.A. Shoemaker, P.H. Klesius, and J.J. Evans (2000). Detection of humoral response to *Streptococcus iniae* infection of tilapia by indirect ELISA. *Journal of Applied Aquaculture.*

———. (2001). Development of an indirect ELISA to detect humoral response to *Streptococcus iniae* infection of Nile tilapia *(Oreochromis niloticus). Journal of Applied Aquaculture* 11(3):35-44.

Shoemaker, C.A., P.H. Klesius, and J.J. Evans (2000). Diseases of tilapia with emphasis on economically important pathogens. In K. Fitsimmons, and J.C. Filho (Eds.), *Tilapia aquaculture in the 21st century,* (pp. 565-572). Rio de Janeiro, Brazil: Panorama da Aqüicultura.

———. (2001). Prevalence of *Streptococcus iniae* in tilapia, hybrid striped bass, and channel catfish from farms in the U.S. *American Journal of Veterinary Research* 62(2):174-177.

Shoemaker, C.A., P.H. Klesius, and C. Lim (2001). Immunity and disease resistance in fish. In C. Lim, and C. Webster (Eds.), *Nutrition and fish health* (pp. 149-162). Binghamton, NY: The Haworth Press, Inc.

Shoemaker, C.A., G.W. Vandenburg, A. Desormeaux, P.H. Klesius, and J.J. Evans (in press). Efficacy of a *Streptococcus iniae* modified bacterin delivered using Oralject™ technology in Nile tilapia *(Oreochromis niloticus). Aquaculture.*

Simon, R.C., C.B. Shill, and H.L. Kincaid (1981). Causes and importance of genetic differences between groups of test fish. In D.P. Anderson, and W. Hennessen (Eds.), *Fish biologics: Serdiagnoostics and vaccines: Developments in biological standardization* (Vol. 90, pp. 267-272). Basel, Switzerland: Karger.

Siwiki, A.K., D.P. Anderson, and O.W. Dixon (1989). Comparisons of nonspecific and specific immunomodulation by oxolinic acid, oxytetracycline and levamisol in salmonids. *Veterinary Immunology and Immunopathology* 23:195-200.

Tafalla, C., B. Novoa, J.M. Alvarez, and A. Figueras (1999). In vivo and in vitro effect of oxytetracycline treatment on the immune response of turbot, *Scophthalmus maximus* (L.). *Journal of Fish Diseases* 22:271-276.

Toranzo, A.E., S. Deresa, J.L. Romalde, J. Lamas, A. Riaza, J. Leiro, and J.L. Barja (1995). Efficacy of intraperitoneal and immersion vaccination against *Enterococcus* spp. infection in turbot. *Aquaculture* 134:17-27.

Warr, G. (1997). The adaptive immune system of fish. In R. Gudding, A. Lillehaug, P. Midtlyng, and F. Brown (Eds.), *Fish vaccinology,* Volume 90: *Developments in biological standardization* (pp. 153-160). Basel, Switzerland: Karger.

Whittington, R., C.A. Shoemaker, C. Lim, and P.H. Klesius (2003). Effects of β-hydroxy-β-methylbutyrate on growth and survival of Nile tilapia, *Oreochromis niloticus,* vaccinated against *Streptococcus iniae. Journal of Applied Aquaculture* 14:25-36.

Wise, D.J., and M.R. Johnson (1998). Effect of feeding frequency and Romet-medicated feed on survival, antibody response, and weight gain of fingerling channel catfish *Ictalurus punctatus* after natural exposure to *Edwardsiella ictaluri. Journal of the World Aquaculture Society* 29:169-175.

Chapter 18

Harvest, Handling, and Processing

Kevin Fitzsimmons

INTRODUCTION

Quality control of tilapia products has been one of the most critical aspects of the success of the industry. Maintaining and improving the quality of the various product forms have been central to the rapid growth of demand for tilapia products in the market. This attention to detail starts while the fish are still growing in their various production systems. Processors and farmers work together to ensure that fish are not contaminated by chemical pollutants or by parasites. Virtually all farms check their water sources on a regular basis to ensure high quality. Many farms now use bird nets or greenhouse covers to keep out birds and other sources of potential contamination. The following are brief descriptions of some of the preharvest and postharvest considerations for growing, harvesting, and processing tilapia.

Off-Flavor

The objectionable taste and smell of some cultured fish fillets, commonly referred to as off-flavor, is one of the most important factors of product quality. Monitoring for off-flavor is a process that begins before harvest and continues throughout processing. Fish from ponds will sometimes accumulate geosmin and/or methylisoborneol (MIB), compounds that are produced by blue-green algae, at levels that impart objectionable tastes and/or odors. Occasionally, even intensive recirculating systems have been known to develop off-flavors. The most common method for determining whether fish is "off-flavor" is to smell and taste a sample. Normally, a whole fish or a freshly cut fillet is cooked in a microwave oven. Most testers will cook the fish or fillet inside a closed brown paper bag to concentrate any odors. (Note: plastic bags are not good as they can emit their own odor when

doi:10.1300/5513_18

heated in a microwave). The odor may be obvious just by smelling the contents of the bag. If there is no obvious odor, the tester will taste some of the fish to detect off-flavor. Some testers have the ability to detect geosmin and MIB at levels of 4 or 5 parts per billion.

Fish are normally sampled a week before a tentative harvest. If detected, the standard method to eliminate off-flavor is to place the fish in clean flowing water, without feeding, for several days. This is normally sufficient to allow for elimination of the offending compounds. Another method is to leave the fish in ponds, with minimal feeding, until the off-flavor subsides. Taste testing is repeated to ensure that the fish is free of off-flavors. Most processors will repeat testing at several points in processing as part of their quality control.

Harvest Techniques

Tilapia harvesting varies considerably depending upon the culture system. Ponds are normally partially drained and then the fish are concentrated in a corner using a seine net. The fish may be collected by hand nets or lifted out with a large scoop net, often suspended from a crane or back-hoe. More sophisticated ponds and raceways may use a harvest box that concentrates fish for removal using nets or baskets. Cage culture typically uses a large bar placed across the top of the cage. One side of the cage will be pulled up and over the bar concentrating the fish in the smaller part of the net. The process continues until the fish are concentrated into one end or corner where they can be lifted out by hand or scoop net. The largest farms may use fish pumps or other mechanical means to remove fish. Many farms use graders to separate harvest size fish and either leave small fish in the production system or remove them to another production unit.

Depuration

Most major farms, and even some processing plants, now incorporate a depuration stage between harvest and processing. This is normally a specially designed pond or tank system designed to clear the fish of off-flavors and eliminate materials from the gastrointestinal system. Purging fish in this manner may lead to a 4 percent loss in weight. This may be a significant additional cost for the grower, but it greatly reduces the chances of off-flavor, the amount of fish waste in the transport water, and the threat of contamination of product with fish waste.

Hauling

Most fish are delivered live to the processing plant to assure the highest quality of the processed product. At larger, fully integrated farms, the processing plant may be on-site and fish may be delivered by flume or other mechanical means. When delivered from a remote farm, fish are delivered in live-haul truck. Crude live-haulers may utilize an open top canvas bag suspended by rails on a stake-bed truck. More sophisticated haulers use specially designed fish hauling boxes equipped with aerators or bottled oxygen. In all cases, it is important to deliver fish to the processing plant alive and with a minimum of physical damage. Some haulers chill the hauling water, but most deliver fish using water at ambient temperatures.

POSTHARVEST HANDLING AND PROCESSING

Processing and food quality requirements vary considerably from country to country. United States, European Union, and International Standardization Organization (ISO) guidelines are continually updated as public health concerns and technology evolve. Hazard Analysis at Critical Control Points (HACCP) and other processing guidelines should be examined carefully before deciding on a particular design and operating plan for a processing plant. Likewise, practices vary from plant to plant regarding how and when products are weighed, how glazes are applied, and how product is labeled. Buyers should inspect and agree on product specifics before purchasing.

Processing Lines

There are two basic designs of tilapia processing plants. The first uses a batch process whereby quantities of fish are acted upon at a station and then the product is bunched in totes or baskets and transferred to the next station. The other basic design is a continuous line with product continuing down the line as portions are removed and the final product gets packaged.

Bleeding/Chilling

Many processors prefer to bleed fish as a preliminary step. Most often this entails hand-cutting the gills of the fish. Some plants also cut the caudal blood vessels in front of the tail. The intention is to quickly remove much of the blood from the fish, which improves the quality and appearance of the final fillet. Fish are typically placed in vats of water to bleed. The vats may be at ambient temperature, which will encourage rapid bleeding, or in chilled or iced water, which will begin the chilling process but slow the bleeding. Some plants will bleed in ambient water and then add ice to chill in the same vat. Some processors prefer to put newly arrived fish directly into an ice slurry to immediately kill the fish and rapidly chill the carcass. This is more common for fish that are frozen whole or gutted. It can be counterproductive to chill fish before bleeding is completed, but for some processing plants this has to be a requirement in their guidelines.

Scale Removal

Some plants use hand labor to remove scales from the carcass whereas others use mechanical equipment. The most common equipment is a rotating drum with slotted surfaces that tumble the fish to remove scales. Mechanical scrappers were used for a short time, but none appear to be in use currently. The drum scalers are not used at some plants that deal primarily with hand-filleted products.

Deheading

Removal of the head from the carcass is increasingly becoming the standard method at processing plants. This operation can be accomplished using either a food grade band saw, rotating knives mounted in a mechanical deheader, or in some cases using a large hand knife or cleaver. Most plants use either a curved cut or a "v-shaped" cut in order to recover the flesh behind the head. A few plants still directly remove the fillet from the carcass, leaving the head intact on the skeleton. This was common in plants with an abundance of low-cost labor, but even these plants are increasingly moving toward more automation and recovery of the head as a marketable by-product.

Evisceration

Removal of the viscera is another common procedure. Typically an incision is made from the anus up to the pectoral fins by hand or machine. Some machines may make an incision from where the head has been removed down to the anus. The viscera may be removed by hand, with the help of a high-pressure water jet, or by a suction device. A good depuration system will minimize the amount of undigested feed and fecal material. Again, there are some plants that do not eviscerate as the fillet is taken directly from the carcass.

Fillet

In recent years, the percentage of tilapia being filleted has rapidly increased. There are several automated fillet machines that take the entire fish, make several cuts and leave finished fillets. These will be described in more detail under "Multifunction Machines." Hand-filleting is still the most common method of filleting tilapia. There are several methods of hand-filleting. Variations depend on whether the cutter is right- or left-handed, which side of the fish is being cut, and whether the head has already been removed. The type of knife used also varies considerably. Some prefer to use a heavy long shank knife, while others prefer a thin knife, which allows the cutter to easily feel the bones. Some filleters prefer to cut through the rib cage and then remove the ribs as a separate operation; others leave the rib cage intact and cut the fillets from around the bones, leaving the skeleton intact. Most processing plants use a bonus system to reward especially skilled filleters. Typically, the bonus is based on the number of fillets that a cutter can recover per time period (hour, shift, pay-period).

Skinning

Automatic or mechanical skinners are ubiquitous in the industry. A skin-on fillet is hand-fed to the skinner, which has rotating rollers that grab the skin and pull it down while a knife blade set on the aperture cuts the fillet from the skin. The depth of the cut can be adjusted to leave more or less of the flesh on the fillet. A deeper cut, leaving more of the darker flesh on the skin, has become more popular with buyers in recent years. A deeper skinning will typically decrease the fillet weight by 5 percent. New skinners that freeze the skin to a roller and then use a movable blade to cut off the fillet are being tested and may eventually replace the current models if consumers and processors appreciate the results. The new skinners leave a smoother cut, which has been requested by some buyers.

Trimming

The next step is to remove pin bones and trim off the outer edges of the fillet. Normally, several small pin bones are left in the fillet when it is cut off the skeleton. Typically a "v-cut" is made to remove these bones. An accomplished trimmer can do this removal with a minimum of wastage. The loose supportive tissue along the top of the fillet is often removed, as are thin pieces along the belly portion. These tissues often come off during handling and cooking so the buyers prefer to have them removed during processing.

Some plants will also rub the fillet against a roughened plastic surface as a final step to remove any remaining subdermal facia.

Ozone and Chlorine Baths

Most plants run their trimmed fillets through a water bath at this stage. In the past, some plants used a mild chlorine solution in the water to reduce the bacteria and lengthen the shelf life. Most plants have now replaced chlorine with ozone gas that is bubbled into the tank. Ozone does not have the disinfection by-products that chlorine does, nor does it leave any disagreeable taste that can be discerned by some consumers. Most plants use an on-site ozone generation system that supplies the small amounts of ozone needed to effectively disinfect fillets. Studies conducted at the University of Arizona demonstrated that bacterial counts could be lowered by several degrees of magnitude and shelf life could be extended by several days when fillets were rinsed with ozonated water compared to untreated fillets.

Carbon Monoxide and Liquid Smoke

Carbon monoxide (CO) gas and liquid smoke have been used in some countries to maintain the appearance of the red meat on the fillet. It appears that the gas is absorbed by the flesh and reacts with myoglobin in the muscle tissue. By binding CO to the myoglobin, fillets maintain a fresh, bright red color in the myomeres for extended periods. This gas is applied by placing fillets on a tray, which is then placed into a large plastic bag. The bag is inflated with gas, tied off, and fillets are left in the bag to absorb CO for 5 to 10 min. An alternate method is to place the trays of fillets into a large cabinet that is filled with the gas. Several countries do not allow the treatment of fish fillets with CO and will not accept imports that have been treated with CO. The United States is reviewing the practice and may restrict its use, or may impose labeling of the procedure on the packaging. Many of the major buyers of fillet products in the United States will not accept fillets that have been treated with CO.

Freezing

Rapid freezing of the fillet, or whole fish, is critical to maintain the product quality. Fillets are normally placed on large trays on the top of a conveyer that passes through a tunnel freezer. Often the fish are given a quick dip in water, or hand-sprayed with water, to form a glaze over the fillets.

This avoids freezer burn (and adds weight). Whole or gutted fish may go through a tunnel freezer or a blast freezer.

Packaging

Marketing of whole or gutted tilapia in international trade necessitated that the products frequently be transported in large containers holding hundreds of individually quick frozen (IQF) fish. These were repackaged by placing onto individual styrofoam trays with plastic wrap for retail sales. Today, with more sophisticated processing in the tilapia producing countries, virtually any style of packaging is available. Many fillets, and even whole fish, are now packed into individual bags that are heat-sealed or vacuum-packed. The bags are normally put into a 5 or 10 pound cardboard or plastic box. These boxes may be placed into an insulated master pack. Fresh fillets are normally packaged in 5 or 10 pound plastic packs that can be resealed and these are preferred by the restaurant trade.

The fillets are normally graded by size. Most common grades are 3 oz and under, 3 to 5 oz, 4 to 6 oz, 5 to 7 oz, 6 to 8 oz, and over 7 oz . Many plants have automatic sorting machines that separate fillets by weight. In developing countries hand sorting is common and highly accurate, with scales used only for checking. The variety of fillet forms continues to grow with size variations, skinning variations, and various treatments available.

Multifunction Machines

There are several automated fillet machines that are capable of accepting a whole fish at one end, and discharging finished fillets at the other. Many processors feel that these machines are still not cost-effective, primarily because they do not recover as much as through hand-filleting and cannot compete with the low labor costs in most of the major tilapia producing countries. Additional innovations and increasing labor costs should eventually close the cost gap. There are also several machines that conduct one or more processing functions, and these are being used in numerous processing plants around the world, even in countries where labor costs are low, as the move toward more automated processing continues.

By-Products

Skins have become the most valuable by-product from processed tilapia. There are three primary markets. First, skins have been used to make a variety of leather goods. In Brazil, several companies have extensive product lines, including clothing and accessories made from tilapia leather. The second market is as a snack food: descaled skins can be cut into thin strips and deep-fried. These are especially popular in Thailand and the Philippines. A third market for tilapia skins is as a pharmaceutical product. European companies are substituting material from tilapia skins for mammalian products for gelatin used to make time-released medicines.

Other by-products are the trimmings and heads. Heads are used for soups in some countries. Postocular and throat muscles can be recovered and used for ceviche and other preparations using small amounts of meat. Recovery of flesh through deboning of pin bone cuts and skeletons can provide a base for fish sticks or other highly processed forms. Carcasses, heads, and trimmings can be used for animal feeds, especially hogs.

PROCESSING FOR INTERNATIONAL MARKETS

Taiwan was the first country to produce and export significant quantities of tilapia, starting in the late 1980s. Most of the exports were of whole, or gutted, frozen fish sent to the United States. Jamaica was the second major exporter on the world market, sending fresh and frozen fillets to the United States and Europe. Just a year or two later, Indonesia began processing cage-cultured tilapia from reservoirs and exporting frozen fillets. In 1990, Colombia and Costa Rica began processing fish grown in raceways and semi-intensive ponds, and exporting fresh fillets to the United States. After a series of major disease outbreaks, several large shrimp farmers in Ecuador switched to tilapia production.

Using existing production, processing, and marketing channels, Ecuadorian farms have a significant share of the fresh-fillet market in the United States. Using technology and investment from Taiwan, several southern provinces in the People's Republic of China have become major producers and exporters. Large quantities of frozen fillets are now exported to the United States and Europe. Production in Zimbabwe is based on cage operations in Lake Kariba. Fillets from the processing plant are marketed in Europe. Brazil and Thailand are major producers that have sophisticated processing plants and minor exporters, who are beginning to have an impact on the international markets for processed tilapia. Mexico and the Philippines each has major producers who expect to develop international quality processing plants and products, in the near future.

The market for fresh fillets has grown to be the most valuable sector (Figures 18.1 and 18.2). Although sales of whole, frozen fish have stagnated and frozen fillets have grown steadily, sales of fresh fillets have increased dramatically. In virtually every region of the United States, fresh fillets are

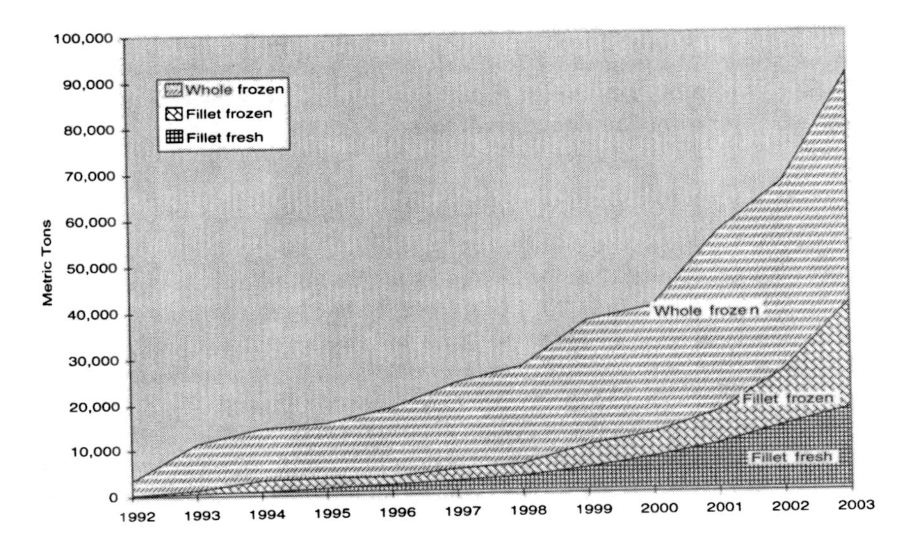

FIGURE 18.1. Volume of tilapia products imported to the United States.

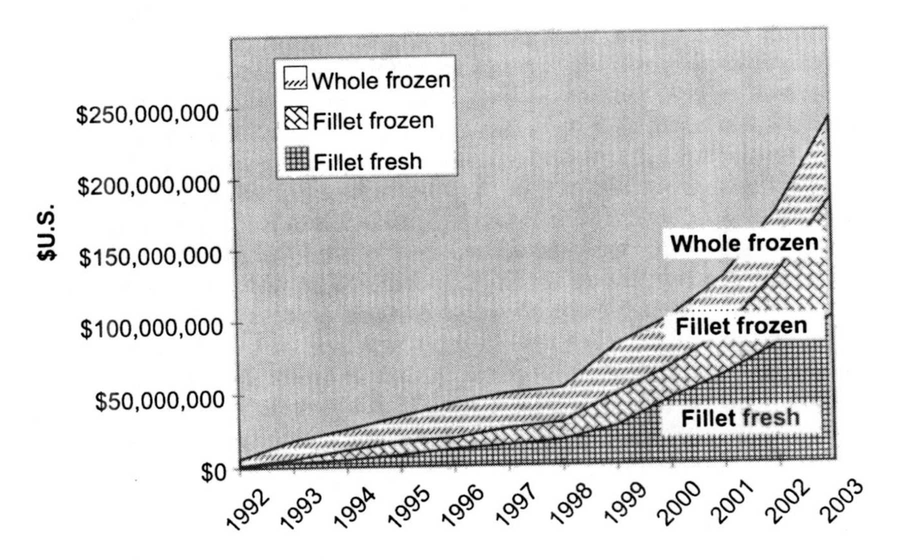

FIGURE 18.2. Relative and total value of U.S. tilapia imports.

available in grocery stores and restaurants. In 2001, tilapia made the top ten list of commonly consumed seafood in the United States and in 2002, tilapia moved up to ninth most popular on the list. Increased sales in 2005 may well move tilapia into the sixth most popular seafood slot.

PROCESSING IN THE UNITED STATES

The market for tilapia in the United States has grown at an incredible rate for the past 20 years. In 2005, 127 million kg of tilapia products representing 281 million kg of global production was exported to the United States. In 1980, the primary form of tilapia available on the market was live fish, grown on a handful of farms scattered across the United States. Most of these fish were live-hauled to Asian and other ethnic markets. Today, these are still the primary markets for over 9 million kg of tilapia grown in the United States. Several pioneers who saw the potential for processed tilapia have operated processing facilities supplying U.S. customers. However, none of these operations has been able to compete with non-U.S. farms and processors, and no one produces significant quantities of fillets in the United States at the present time.

Fish Breeders of Idaho was one of the earliest producers of tilapia in the United States. The owner, Leo Ray, added tilapia to his catfish raceway farm, in Buhl, Idaho, around 1980. Tilapia consumed uneaten food and other waste materials, and improved water quality before it left the raceways and entered the Snake River. Tilapia were an ideal choice and grew quickly, while reducing the solids load of his effluent. His initial marketing plan for tilapia was to include free samples of tilapia packed with his regular shipments of processed catfish. As customers tried the fish, orders for the new product followed. As his stocks increased, he added tilapia fillets to the product list, along with the gutted and headed/gutted fish that he first supplied. The processing facility also evolved as well, adding automated equipment to replace the predominately hand processing. Ray's company continues to provide a full list of tilapia products, from whole, iced fish to fillets. However, the majority of his fish are delivered live. The Simplot organization, also in Idaho, built a large tilapia farm in the late 1980s. This farm was focused completely at the fillet market. In the two or three years of operation, Simplot provided a very high-quality fillet product that was well received and brought a premium above the price that had been expected in the business plan. Simplot used a highly automated processing line but still had an experienced crew that checked and trimmed each fillet before packaging. Unfortunately, production costs at the farm were higher than expected and the geothermal water resources were insufficient to support the size of the farm. However, Simplot did provide an excellent market opening by introducing many chefs and consumers to high-quality tilapia fillets and a professional marketing program.

In 1986, Solar Aquafarms began producing tilapia in southern California. This farm, like the Simplot farm, used a recirculation system, but of a different design. The processing plant at Solar Aquafarms used a depuration tank to guard against off-flavor from their unique "Organic Detrital Algal Soup" production system. However, this system was not always successful, and off-flavor product occasionally made its way to consumers. Chiquita Brands purchased Solar Aquafarm's facility in the early 1990s and began an active marketing program. Unfortunately, the fillet products had difficulty competing against the rising tide of low-cost tilapia coming into the California market from Taiwan. The occasional problem with off-flavor was never completely solved and Solar Aquafarms eventually went back to live product and the processing plant was closed.

In the late 1980s and the early 1990s, four small processors tried to sell tilapia fillets in Arizona. Two of these farmer/processors used food-grade band saws and automatic skinners, along with hand-filleting; the other two relied exclusively on hand-filleting. All of the farms produced quality fillets and rarely had any off-flavor problems. None of these operations spent much money on marketing or promotion. Deliveries were made directly to restaurants and grocery stores. However, the availability was inconsistent and the farms were undercapitalized and unable to increase the product supply. All four farms eventually reverted back to live sales. Three farms have since closed and the other has changed hands three times.

In the mid- to late 1990s, Integrated Food Technologies (IFT), in eastern Pennsylvania, began growing and processing tilapia, yellow perch, and striped bass. IFT used an indoor raceway recirculating system, coupled with a constructed wetland for solids treatment. The processing system was mostly automated that produced a high-quality fillet. Unfortunately, IFT made almost all of the business mistakes possible: production costs were higher than anticipated; production species were changed several times; more money was spent finding investors than on marketing products; and customers were not provided a consistent product.

The Southern States Coop tilapia program was planned as an integral component of the larger Southern States farmers cooperative that already sells tilapia feed to growers in the southeastern United States. The plan to sell processed tilapia never developed as advertised and all of the growers have put their fish into the live trade.

So far, no U.S. processor has been able to compete with the imports. Domestic production and processing could be successful, but only under certain circumstances. A U.S. processor would need to start with a lucrative niche market and then expand. A strong marketing program, which is based on service and quality, would be needed. Consistent supply and high-quality products are critical factors. Time is needed to develop customer relationships and then these must be cultivated. However, competition from offshore producers is likely to be fierce. The tilapia products being imported are continuing to improve as international growers upgrade their production technologies, purchase state-of-the-art processing lines, and transportation becomes faster and more efficient. Several processors are also moving into value-added products. U.S. producers and processors would have to work hard to get and retain their customers. The more likely scenario is that U.S. producers will continue to supply the live fish markets, and international products will dominate the processed markets.

Chapter 19

Marketing and Economics

Carole R. Engle

INTRODUCTION

The economic history of the development of the tilapia industry worldwide is a fascinating study of a fish enterprise that has been managed successfully on nearly every scale of business. This by no means implies that all attempts to raise tilapia have been successful, but rather that examples of successful tilapia enterprises can be found over a wide range of sizes, scales, and business organization.

In 2002, tilapia were being grown and sold in 81 different countries, on every major continent, and in tropical, subtropical, and temperate climates. Tilapia are produced by near-subsistence farmers as a savings account for hard times; caught and consumed by subsistence fishermen; raised and sold to local village markets and upscale domestic markets; exported to high-end sales outlets in the United States, Japan, and Europe; and raised by hobby farmers in the United States and Europe. Tilapia are positioned, often in the same countries, as low-priced products for the poor; as ethnic products; and as gourmet, luxury, upscale products for white tablecloth restaurants. Tilapia are raised in virtually all conceivable types of production systems and in both fresh and saltwater. Regardless of an individual's particular perspective, the tilapia are undoubtedly the most ubiquitous, the most successful, and the most adaptable aquaculture species in the world.

This chapter explores the development of tilapia markets, from the early markets that were developed for wild-caught tilapia, home consumption of the farmed tilapia, sales to local markets and upscale domestic markets, and export marketing. Issues and challenges related to these various targeted markets are discussed. The costs of producing tilapia on a variety of scales of production and in different production systems are also examined.

doi:10.1300/5513_19

TILAPIA MARKETS AND MARKETING

Tilapia Supply Around the World

Tilapia culture has expanded worldwide at an average annual rate of 14.2 percent since 1984 (FAO 1998; Figure 19.1). Global tilapia production through aquaculture reached 800,000 metric tonnes (MT) in 1996 (Food and Agriculture Organization of the United Nations (1998). Cultured tilapia production has increased to constitute 57 percent of total tilapia production (wild-caught and cultured) worldwide.

The Asian continent, where tilapia were introduced in the late 1940s, leads the world in tilapia production and includes four of the top five tilapia-producing countries (Figure 19.2). China is the world's largest producer of tilapia with a production of 500,198 MT in 1997 (FAO 1998). The Philippines, Indonesia, and Thailand are the third, fourth, and fifth largest world producers of tilapia.

The rest of the world produces less than 20 percent of the total aquaculture production of tilapia (FAO 1998). Egypt is the second largest producer, worldwide, of tilapia and generates 58 percent of the total African production. Mexico is the leading producer in North America with 81,490 MT in 1994 (World Bank 1997). Brazil leads Latin America in tilapia production, although much of this production is from tilapia caught by fee-fishermen in reservoirs that receive little management. In Europe, Israel is the only country with any noticeable amount of tilapia production.

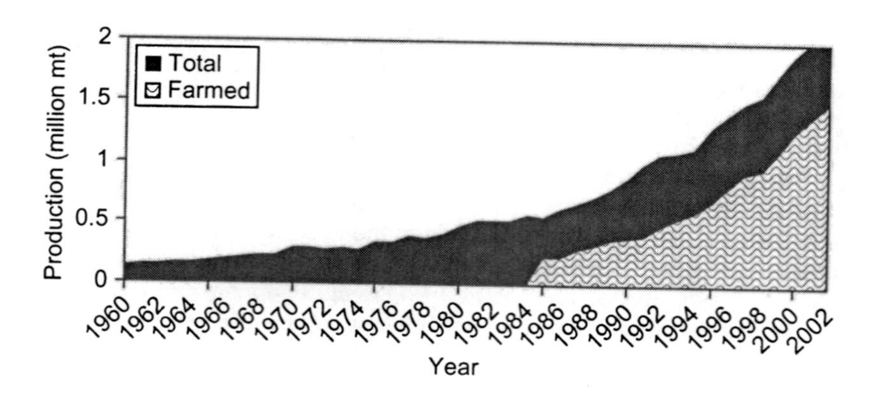

FIGURE 19.1. Global supply of tilapia over time. *Source:* Data from FAO database.

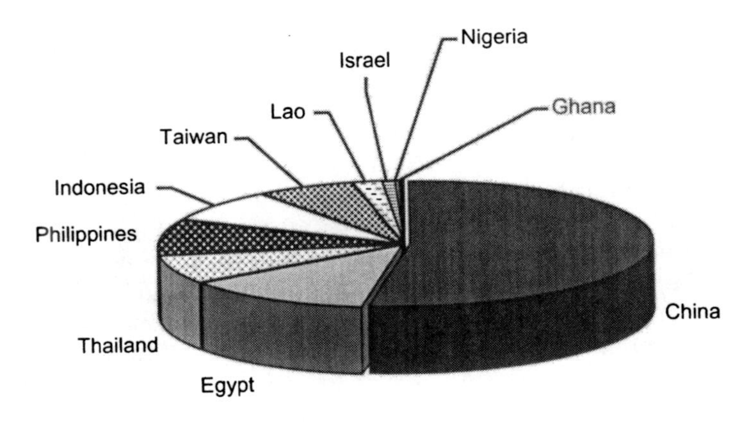

FIGURE 19.2. Top tilapia-producing countries worldwide, 2002. *Source:* Data from FAO database.

As a tropical fish, the distribution of tilapia production would be expected to be concentrated in areas with tropical climates. This is largely the case, and tilapia production has grown rapidly in tropical areas. Nevertheless, tilapia production has also expanded into areas with temperate climates. The United States alone produced over 9,000 MT of tilapia in 2000 (American Tilapia Association 2001). Much of the U.S. production is in indoor or greenhouse facilities. Given the well-developed air freight systems in most regions of the world, the tropical areas would be expected to continue to have a competitive advantage in tilapia production for many years to come.

Development of Early Markets for Wild-Caught Tilapia

It is common knowledge that tilapia are native to Africa. They were distributed throughout the world in the 1950s by international development agencies with the intention of providing a solution for world hunger. This extensive effort to distribute tilapia worldwide resulted in two distinct consequences: (1) tilapia, through government stocking programs and perhaps escapement from production facilities, established wild populations in most tropical and subtropical areas and (2) tilapia quickly developed a reputation as an inferior good, or as a product that is primarily consumed by people of the poorer classes.

The first consequence of the development of wild-spawning populations of tilapia was the establishment of fisheries for tilapia in many countries.

Historically, there have been significant commercial catches of tilapia in African countries such as Egypt, Uganda, Tanzania, Kenya, Madagascar, Malawi, Sudan, Mali, and Ethiopia. However, today, there are also significant commercial catches in China, Thailand, the Philippines, Mexico, Brazil, Nicaragua, and even in the United States.

Tilapia have often been first introduced as wild-caught fish into fish markets. Wild-caught tilapia in many countries are small, low-priced fish sold in open-air markets where many people of the lower economic strata purchase their food. In Ghana, for example, 50 g tilapia are sold for \$0.50 kg^{-1}, deep-fried, and are a staple food for many. Early government stocking programs in natural lakes and man-made impoundments in Mexico and other Latin American countries were directed toward creating a supply of low-priced fish for the rural poor (Fitzsimmons 2000).

In Florida, tilapia were thought to have been released into the wild by aquarium owners who tired of them. Tilapia quickly established spawning populations in subtropical lakes in the state. Throughout the 1990s, millions of kilograms of "wild fresh" tilapia were caught and sold in Florida. They are sold in the poorest areas of major cities. Fishermen have been observed to sell 0.9 kg wild-caught tilapia for \$0.25 kg^{-1}. Competition from these wild-caught fish has made it more difficult for a tilapia foodfish industry to develop in Florida.

Growing tilapia fisheries have led to the export of wild-caught tilapia. The tilapia fishery in Lake Nicaragua provides a substantial volume of wild-caught tilapia for neighboring countries of Honduras and El Salvador (Neira et al. 2003). In 1997, Nicaragua exported 77 MT primarily of wild-caught tilapia. Increasing quantities of wild-caught tilapia are imported into Jamaica and sold at lower prices (Hanley 2000).

Wild-caught tilapia have also been exported as fillets. For example, in the early 1980s, wild tilapia fillets from large hydroelectric dams in Ceara, Brazil, were exported to the United States. These fillets were sold for low prices of \$3.08 to \$4.18 kg^{-1} for a time, but problems with the quality of the product prevented development of a stable market.

Home Consumption of Farmed Tilapia

The early attempts to culture tilapia in many countries around the world focused on raising tilapia as a source of animal protein in rural areas characterized by animal protein deficiencies. Throughout Latin America, Africa, and Asia, tilapia continue to be raised to supplement household supplies of animal protein (Edwards et al. 1991; Engle and Skladany 1992; Engle 1997a). Early producers of tilapia in Mexico used tilapia to supply the household with animal protein, and even today the majority of the Mexican tilapia grown on farms is consumed by growers and fishermen in their local communities (World Bank 1997; Fitzsimmons 2000). In Africa, Engle et al. (1993) demonstrated that fish culture in Rwanda had potential to provide high-quality animal protein affordably for home consumption. Another example is Jamaica where subsistence farmers raise tilapia as an extra crop produced in the farmers' spare time (Chakalall and Noriega-Curtis 1988).

In countries that began to experience declines in wild-caught tilapia, local market niches for farm-raised tilapia developed. These were often based on pond-bank sales of tilapia. The decline in wild catches of fish in Thailand led to increases in fish production in multipurpose ponds that were constructed to store water for domestic use and for agriculture (Little 1995). Tilapia raised in ponds often are raised with dual objectives of producing a cash crop and also as a source of food for the family (Engle 1997c).

Brazil consumes its production domestically, but through fee-fishing. Most farmed tilapia in Brazil are sold live to operators of fee-fishing ("pesque e pague") ponds (Martin et al. 1995). Fish prices paid by fee-fishing operators are higher than those paid by food-fish consumers and processors. Farmers can sell whole tilapia directly to consumers for U.S. \$0.90-1.40 kg^{-1} and to processors for U.S. \$0.75-1.10 kg^{-1}, but receive \$1.20 to \$1.90 kg^{-1} from fee-fishing businesses. Nevertheless, tilapia is viewed as a low-priced, poor-quality fish in the southern areas of Brazil where greater volumes of fish are sold (Lovshin 2000). Consumers in these larger fish market areas prefer purchasing either a whole-dressed fish, or a 5 to 7 ounce fillet from supermarkets or luxury fish shops.

Local Markets for Farmed Tilapia

As farm-raised tilapia production increased, farmers began to sell tilapia to local markets. Tilapia production for supply into local markets has become common in many countries. Sometimes these marketing efforts were organized by a cooperative of growers. Tilapia sold in local markets are primarily sold fresh on ice.

In Mexico, for example, surplus whole fish are sold in local markets in addition to home consumption, but tilapia fillets are also sold in grocery stores (World Bank 1997). Both large (600 g) and small (250 g) tilapia are available in many urban retail markets. Beginning in the early 2000s, a market for live fish (held in tanks in grocery stores), has developed in Guadalajara and Mexico City. Some farms operate small roadside restaurants and deep-fry tilapia in hot oil. Given the nature of the domestic market in Mexico, it has exported little to the United States. Some processing capacity has enabled businessmen to develop upscale markets for value-added tilapia products in Mexico.

Middlemen purchase tilapia from small-scale farmers for sale in local markets in Jamaica where live tilapia (preferred over frozen or fresh-chilled products) are sold from live-haul tanks in retail markets (Hanley 1991; Engle 1997b). Tilapia have rapidly become the leading freshwater species raised on Venezuelan farms and are marketed primarily through local markets (Chiappe 1998).

In China and the Philippines, two of the largest tilapia-producing nations, most of the production is sold locally (Fitzsimmons and Posadas 1997). Tilapia farms are also widespread in Thailand, and tilapia has become one of the most common fish sold in markets and restaurants. Similarly, Indonesia has many small farms that produce tilapia for local markets. Israel is another country that has developed a large domestic tilapia market and exports little.

Overfishing at a time when farmed tilapia were becoming abundant resulted in the development of a sizeable domestic market for tilapia in Colombia (Popma and Rodriguez 2000). This led to a decrease in exports to the United States and increase in sales in the domestic market. Colombian consumers prefer whole fish to fillets. The fillet market is only 3 percent of the total tilapia market in Colombia. Whole fish, weighing 135 to 150 g, are sold to factory and institutional restaurants; fish weighing 200 to 350 g are marketed through restaurants and supermarkets; and tilapia weighing over 400 g are sold to specialized seafood restaurants. The combined demand for whole tilapia and the availability of supply resulted in per capita consumption of tilapia reaching 8 kg in 1998 (Alceste 2001a). This volume of consumption has attracted tilapia imports from Venezuela and Ecuador (Jory et al. 1999). Ecuador exports whole tilapia on ice to Colombia at reduced processing and transportation costs as compared to selling tilapia in the U.S. market.

Development of Upscale Domestic Markets for Tilapia

Over time, opportunities to sell a larger, higher-quality tilapia to supermarkets and restaurants became available. As enterprising individuals moved beyond production for supplemental income to a more business-oriented approach, more attention was paid to cultivating higher-valued market outlets.

Nevertheless, the move to develop higher-end supermarket and restaurant markets for tilapia has proved difficult in many countries. This is due to consumer attitudes of tilapia as a poor-quality, low-priced fish for poor people. These attitudes are derived from the intentions of the early tilapia introductions and the low-priced wild-caught tilapia prevalent in many markets. Negative consumer perceptions like these can constitute a major obstacle to the development of domestic markets.

For example, Honduran markets have abundant supplies of wild-caught tilapia imported from Lake Nicaragua (Monestime et al. 2003; Funez et al. 2003a; Engle and Neira 2003). Viewed as poor people's food, supermarkets and restaurants are hesitant to test tilapia in their establishments. In Nicaragua, these same consumer perceptions are exacerbated by fears that freshwater fish may have been caught in Lake Managua and are feared to be contaminated (Neira et al. 2003; Engle and Neira 2003).

Even in the United States, the problems of poor quality, often off-flavor, wild-caught tilapia that were sold at a low price were documented in the 1970s (Lauenstein 1978). Identical in appearance to most farm-raised tilapia, poor quality tilapia on the market can have a negative effect on market development.

Declining fisheries in Puerto Rico (Alston et al. 2000; Collazo and Calderon 1988; Caraballo and Sadovy 1990) and growing per capita consumption of seafood (Tucker and Jory 1991) have created opportunities for aquaculture. However, there is a cultural bias against freshwater fish and fish with the silver-black appearance (most common varieties of tilapia). This cultural bias has likely limited the market demand and commercial production in some areas (Sandifer 1991; Chakalall and Noriega-Curtis 1992). Red tilapia hybrids grown in saltwater could produce a more broadly accepted product with good commercial potential (Watanabe et al. 1989; Sandifer 1991; Watanabe et al. 1991; Chakalall and Noriega-Curtis 1992; Head et al.1996). Market tests in Puerto Rico appeared to demonstrate good consumer acceptance of large tilapia (> 454 g), particularly if differentiated from wild-caught, freshwater tilapia. However, in rural markets, saltwater-cultured red tilapia were priced as a low value species ($0.90 kg^{-1}$) (Watanabe et al. 1997), similar to that for freshwater, river-caught black tilapia. In contrast, in urban markets in Haiti, saltwater-cultured red tilapia was treated as a premium-quality marine fish. Retailers consistently paid $2.20 kg^{-1}$ for cleaned, whole red tilapia, the going price for snapper and grouper.

In the United States, tilapia production has increased from 2,700 MT in 1990 (USDC 1999) to 7,648 MT in 1997 (FAO 1998). Most of the tilapia have been marketed live to ethnic markets in metropolitan areas where they fetch a premium price. The U.S. tilapia market, as elsewhere in the world, is highly segmented among live fish, whole-frozen fish, frozen fillets, and fresh fillets (Fitzsimmons 2001). Live tilapia are marketed through ethnic markets for live and fresh whole fish (Redmayne 1993). However, growth in the live market has slowed in recent years (Fitzsimmons 2001). Part of the new market demand for tilapia was created by immigrants who provided niche markets for tilapia in the United States and Canada (Fitzsimmons and Posadas 1997).

Export Marketing of Tilapia

As tilapia businesses grew in many tropical areas, export markets were developed. Japan, the United States, and Europe comprise the largest seafood markets in the world. Japan and the United States represent major export targets for tilapia growers whereas European markets for tilapia are restricted to consumers of certain ethnic markets. The Japanese market is particularly attractive because of the high prices for seafood and Japanese consumers' recognition of tilapia as a high-quality fish. Japanese buyers prefer live, whole, and fresh tilapia (Homziak and Posadas 1992; Fitzsimmons and Posadas 1997). The preference for red-colored fish flesh led to experimentation with carbon monoxide treatments of fillets that impart a red color to the fat line. This process generated a great deal of controversy and led to bans of CO-treated tilapia in Japan. The preferred color is either pink or red because it is sold as an inexpensive substitute for red snapper.

High-quality fillets are used as sashimi (thinly sliced raw fish). The largest and highest quality fish are marketed to Japan for prices of up to \$10 kg^{-1} (Fitzsimmons and Posadas 1997). In the Tsukiji market, the estimated sale volume is around 1 MT per week and most of this demand is supplied by imports from Taiwan.

Although the United States represented an attractive and rapidly growing market for tilapia through the 1990s and into the 2000s, this market has been a relatively recent development. However, the potential consumer acceptance of tilapia was documented as early as the 1960s (Avault and Shell 1966; Shell 1966). Later studies further documented the potential market demand for tilapia (Crawford et al. 1978; Engle 1978; Galbreath and Barnes 1981). In 2001, tilapia imports into the United States reached 56.4 million kg, an increase of 39 percent over the previous year (Harvey 2002). Quantities of tilapia imported in 2001 were 38.7 million kg of frozen whole tilapia (69 percent), 10.3 million kg of fresh fillets (18 percent), and 7.4 million kg of frozen fillets (13 percent) of total tilapia imports. Nevertheless, in terms of value, fresh and frozen tilapia fillets accounted for 70 percent of all tilapia imports. The equivalent weight of tilapia estimated to be required to supply these imports was 95.5 million kg of live fish in 2002.

Engle (1997b) indicated that the majority of companies handling tilapia were small distributors and wholesalers with total annual sales of less than \$50 million. Homziak and Posadas (1992) showed that tilapia products were primarily handled by larger seafood marketing companies to expand

and diversify their product lines. This apparent discrepancy may be due to the fact that the early imports of tilapia were primarily whole-dressed product from Taiwan that were being imported by a single large importing firm. Much of the increases in imports of tilapia fillets were what was documented by Engle (1997b) as imports by smaller importing companies.

Frozen whole fish come mostly from Taiwan (Engle 1997b; Harvey 2002); Fitzsimmons and Posadas 1997). Taiwan accounted for 71 percent of frozen whole tilapia imported into the United States in 2001 (Harvey 2002). Most of these tilapia are small fish that are sold in ethnic markets. Taiwan has monopolized this low-value frozen whole tilapia market in the United States for a number of years.

The People's Republic of China has emerged as one of the most rapidly growing exporters of tilapia to the United States. Over half of China's exports to the United States were of whole, frozen tilapia, and China appeared to have gained a substantial portion of this market from Taiwan by the year 2004, Taiwan regained its market share in 2005 (Figure 19.3).

On the high end, fresh and frozen fillets are mostly imported and sold through importers, wholesalers, and distributors to retail grocers and restaurants (Stutzman 1995). Frozen tilapia fillets have demonstrated some growth in imports since 1994, with most frozen fillets coming from Indonesia (Harvey 2002). Taiwan historically dominated the frozen tilapia supply to the United States, but has lost market share over the last several years. Both Indonesia and China exported a greater value of frozen tilapia fillets to the United States in 2001 than did Taiwan (Figure 19.4).

Fresh tilapia fillets have demonstrated the most rapid growth of any tilapia product form. Imports of fresh fillets come primarily from Central and South America, regions close enough for rapid transport to the United States (Figure 19.5). Costa Rica has been producing tilapia for export to the United States since the early 1980s when a small tilapia canning operation was started. It became a major exporter to the United States in the 1990s by producing and marketing fresh fillets (Fitzsimmons and Posadas 1997). In 1999, Costa Rica was the world leader in supplying fresh tilapia fillets to markets in the United States. Of its tilapia exports, 99.8 percent were fresh fillets.

However, Ecuador quickly became one of the leading exporters of fresh fillets to the United States. Ecuador surpassed even Costa Rica, in the year 2000, in the value of fresh tilapia fillets exported to the United States. By the end of 2001, Ecuador had gained a market share of 52 percent in spite of a 21 percent increase in fresh fillet imports from Costa Rica. Viral epidemics on shrimp farms in Ecuador resulted in the adaptation of many shrimp ponds to tilapia production. Larger fish are exported to the United States whereas smaller fish are exported as a whole product on ice to Colombia. For U.S. destinations, Ecuadorian producers use mostly East Coast ports,

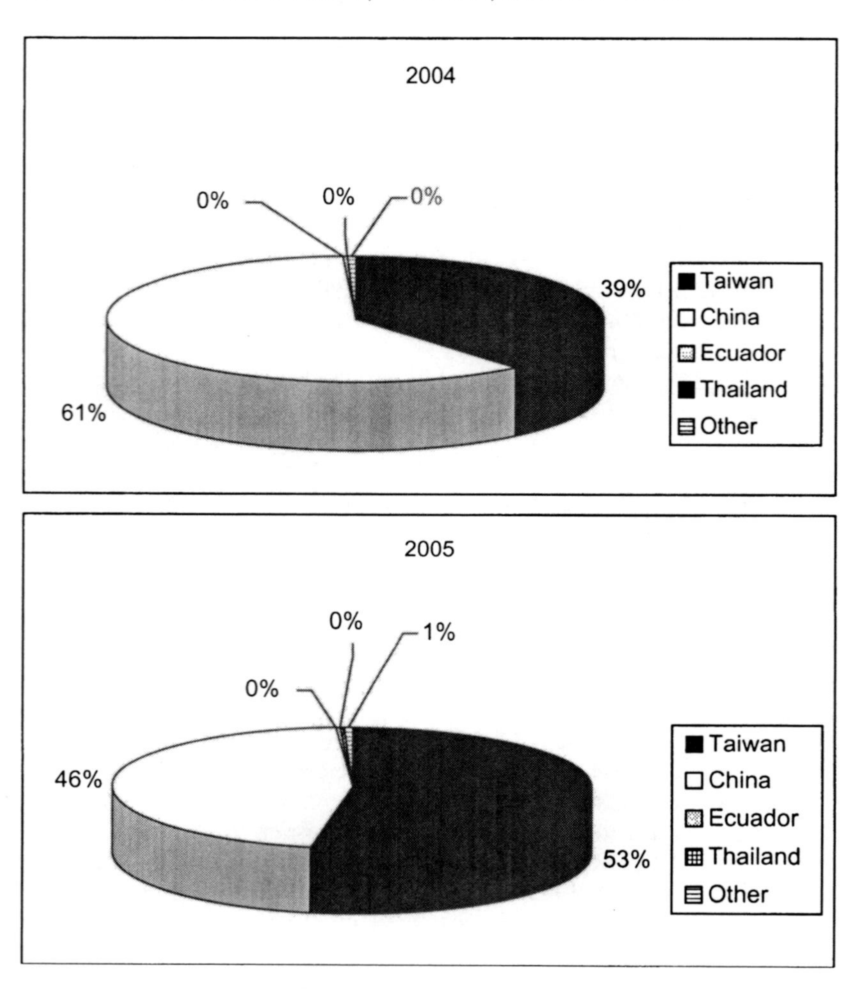

FIGURE 19.3. Market share by country of frozen whole tilapia, 2004-2005. *Source:* Data from USDA Aquaculture and Situation Outlook Reports.

routing exports through Florida (Alceste 2001b). Exports consist of "standard high-quality" frozen fillets (Alceste 2001b). These are fillets from fish grown under good conditions, purged in clean water, well trimmed (with bloodline removed), graded, and blast frozen. Nearly all (99.7 percent) of the fresh fillets are sold as 114 to 171 g (4 to 6 oz fillets) to the United States.

There has been a marked increase in fresh fillet imports from Honduras and China as well. Tilapia production grew rapidly in Honduras through the 1990s

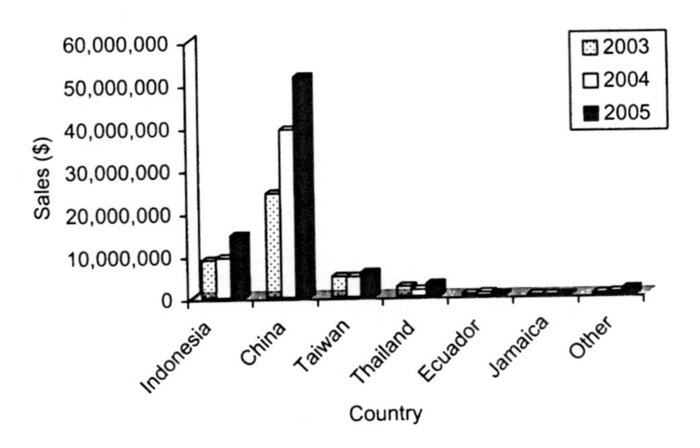

FIGURE 19.4. Market share ($) of frozen tilapia fillets, 2003-2005. *Source:* Data from USDA Aquaculture and Situation Outlook Reports.

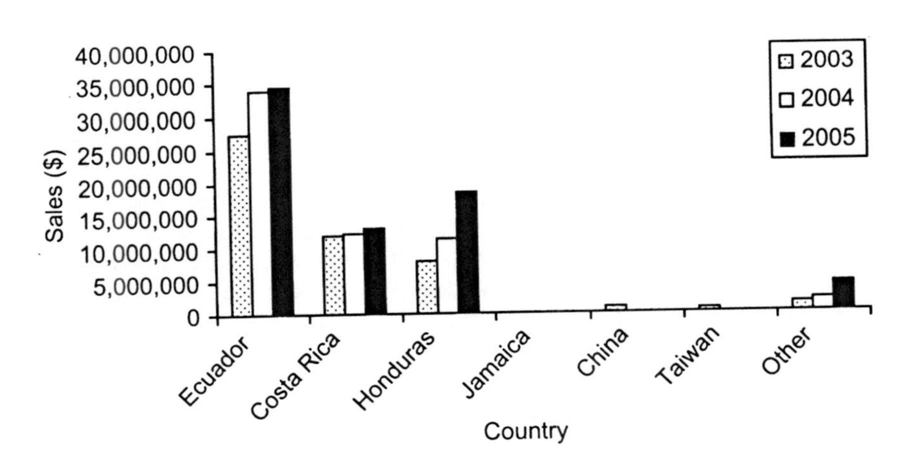

FIGURE 19.5. Market share ($) of fresh tilapia fillets, 2003-2005. *Source:* Data from USDA Aquaculture and Situation Outlook Reports.

into the 2000s as a result of the expansion of markets in the United States. Honduras has attracted several large tilapia agribusinesses and has grown to be the third largest source of fresh tilapia fillets for the United States.

Improved consistency and increased consumer awareness of tilapia explain much of the increase in tilapia fillet sales (Seafood Business 2002).

Most tilapia fillets are now deep-skinned. Deep-skinning removes the fat layer that has a tendency to turn brown upon exposure to air, in spite of the reduction in yield from 35 to 30 percent. Some Chinese and Taiwanese producers market tilapia fillets that have been treated with carbon monoxide to develop a cherry-red fat layer. FDA requires this product to be labeled. Japanese bans on treated tilapia fillets led to declining sales in Japan and increased interest in the U.S. market.

COSTS OF PRODUCING TILAPIA

Tilapia are probably raised by more individuals worldwide than any other cultured species. Given that many individuals who grow tilapia are very small-scale, near-subsistence farmers or hobbyist farmers makes it very difficult to obtain accurate statistics on the numbers of tilapia growers. Nevertheless, the literature documents that tilapia are cultured on every conceivable scale of production, from very small ponds used for subsistence-level production to very intensive, large-scale, sophisticated production systems managed by agribusiness corporations. Moreover, tilapia are raised in many varied types of production systems, including earthen ponds, integrated with confinement animal operations, floating cages in both freshwater and marine environments, raceways, indoor tanks in greenhouses, and indoor recirculating tank systems with and without hydroponics (Costa-Pierce and Rakocy 1997).

Costs of production clearly vary with each specific farming operation, and also vary according to local, regional, and national economic factors over time. This section of the chapter will discusses production costs presented in the scientific literature by tracing the key cost factors and changes in these costs as both the scale of production and the intensity of production increase.

Subsistence Production of Tilapia

Tilapia production was, for many years, promoted as an alternative means to combat world hunger. The hardiness, ability to feed low on the food chain, and fecundity of the tilapia seemed to point to a solution to the problem of world hunger. Tilapia clearly have not saved the world from hunger; world hunger is a far more complex issue than can be resolved with a small fish pond. Furthermore, research has shown that, in reality, there are few viable examples of tilapia being raised in a truly subsistence capacity. Athough there are likely multiple interrelated factors that explain why this is the case, economics can shed some light on this question.

Athough it is true that tilapia can be produced with low levels of inputs, what constitutes a "low" level of input is relative. In reality, aquaculture production requires relatively higher levels of resources than many other types of food production, particularly the crops considered as staple food items. In many areas, land and fertilizers are very precious. In order to truly survive, families must first meet caloric requirements. Scarce land resources may dictate that what land and sources of fertilizer are available be devoted to production of staple crops to meet caloric requirements. In these situations, meeting protein requirements is something of a luxury.

Subsistence farmers typically are confronted with a shortage of resources that can be used for food production. Fish production in some areas is used as a type of "savings account" (Smith and Peterson 1982), whereas land and fertilizers have other competing uses in staple crop field production. Inorganic fertilizers are often the cheapest source of nutrients; however, true subsistence farmers often have no cash with which to purchase inorganic fertilizers.

Pond construction, regardless of whether heavy equipment or family labor is used, represents a very high investment. This level of investment, in turn, requires a higher return than what is often possible from subsistence operations that have very low levels of resource availability.

Traditional economic analyses treat fish ponds as a capital investment because, in most countries, ponds are constructed with heavy equipment and financed with either debt or equity capital. However, in subsistence economies, ponds may be constructed by hand, not by hired labor or by machinery. Thus, the investment is a labor, not a capital investment. Nevertheless, even if the investment is in labor, it still represents a significant investment.

Rwanda presents an example that is as close to true subsistence as may be possible. The Rwandan agricultural economy remains predominantly a subsistence economy with few opportunities for cash income. The potential for fish culture to supplement family income and to provide high-quality animal protein for home consumption appears to be great. In a study of subsistence aquaculture in Rwanda, labor was the primary resource utilized in fish farming (Engle et al. 1993). In Rwanda, the primary input was compost made from vegetative matter. Given the general lack of cash money, most farmers cut grass or used weeds pulled from other crops to add to a compost pile located in the pond itself. Thus, the primary variable resource used to add nutrients to ponds was the operator's and family's labor.

However, even at this near-subsistence level, fish were still considered primarily as a cash crop and secondarily as a source of food for the family (Engle 1997c). Cash returns from fish farming were competitive with daily wage rates even though realistic employment opportunities were scarce.

Income from fish farming represented a significant proportion of total cash income in addition to its food value for the household.

In northeast Brazil, Greenfield et al. (1974) showed that, if land costs were subsidized by the government, tilapia hybrid culture was feasible on a small scale (1 ha). This study further highlighted that land scarcity can be a serious impediment to true subsistence production of tilapia. This example provides additional evidence of the costs associated with fish production. For farmers, a decision to convert crop-cultivating land to pond production of fish is normally a major one. Once a pond has been dug it cannot easily be reconverted to crop cultivation.

Near-Subsistence Production of Tilapia

There are more examples of viable tilapia production at what should be described more as "near-subsistence" production of tilapia. As Engle et al. (1997) showed in Rwanda, most farmers who raise tilapia on a very small scale, even in primarily subsistence economies, do so to generate cash. Tilapia production in many of these situations is based on a slightly higher level of intensity than that of the nearly exclusive use of green manures found in Rwanda. Most tilapia production at a near-subsistence level involves the use of organic fertilizers or possibly the combination of organic and inorganic fertilization. The use of organic fertilizer implies an overall organization and scale of production large enough to support some type of livestock as well as the ability to collect manure from the livestock. Manure collection requires additional labor and, if inorganic fertilizers are used, some amount of cash is needed for purchases.

Teichert-Coddington et al. (1992) conducted studies to identify a baseline of chemical and biological data on tilapia production with minimal nutrient inputs in Panama and in Honduras. The economic analysis demonstrated that a sole input of phosphorus resulted in an economic loss. Additional nutrient input in the form of higher rates of fertilization and/or feeds would be necessary to make tilapia culture profitable.

Near-subsistence farmers may have dual farm objectives of maximizing net farm income while meeting food security concerns of the family. Setboonsarng and Edwards (1998) showed that tilapia production can be integrated effectively into small-scale farming systems in a manner that meets both food security and income-generating goals. Little (1995) showed that, in Thailand, the best mix of production activities was to produce rice, cassava, sugarcane, vegetables, and fish produced with inorganic fertilization. Fish production generated 23 percent of total income, and, overall, used less than half of the capital available to a typical farmer in northeast Thailand.

Land and labor constrained resources especially during rice transplanting season. Fish production with inorganic fertilization fit into this particular farming system better because it used fewer of the scarce labor resources during this time period than did other production activities. However, cash resources would be required to purchase inorganic fertilizer.

Setboonsarng and Edwards (1998) showed that fish production could contribute significantly to improving the incomes of resource-poor farming systems in northeast Thailand with improved fertilization techniques. Neither capital nor labor resource requirements were excessive with the improved fertilization technologies and farmers continued to have adequate resources to produce staple crops for home consumption while raising some fish to generate income.

Integrating fish production directly with livestock production represents an increased level of organization, capital, and operating capital resources than collecting manure or, even, purchasing inorganic fertilizer. Integrated fish/livestock production systems were widely recommended in the late 1970s as an efficient use and reuse of limited resources. However, several different economic studies demonstrated that the level of resource utilization is, in fact, relatively intensive. The fish component may not require any other resources, but confinement livestock production requires a complete, balanced ration for the livestock. This requirement often constitutes a significant amount of cash resources for limited-resource farmers. Well-studied examples are available from Thailand, Panama, Guatemala, and Honduras.

Duck/fish integrated systems were shown to require relatively high capital inputs in Thailand and the feasibility of such an integrated system on farms depended on the availability of capital (Little 1995). Much of the capital required was for the purchase of feed to grow ducks. However, this more intensive use of resources can result in higher levels of financial returns.

Engle and Skladany (1992) showed that net returns increased by 85 percent and 98 percent for chicken manure collected for use on fish ponds and for that applied directly from integrated systems, respectively. Returns on average investment on ponds and equipment increased by 60 to 70 percent whereas returns on average total investment (including land) increased by 5 percent by adopting chicken manure fertilization technologies.

Hatch and Engle (1987) demonstrated that fish produced with either duck, chicken, or hog manure were economically feasible alternatives in small-scale, semisubsistence production systems in Panama. However, this study also pointed out the capital requirements of confinement livestock systems. Hatch and Hanson (1992) found that production costs of low-input fish production were lower than costs of producing poultry and were similar to those of corn and mixed vegetable crops. With sales of 40 percent of the fish crop, farm income increased by 17 percent.

The use of feed in tilapia production can increase fish yields, but requires higher levels of cash resources. For example, in Thailand, Edwards et al. (1991) found that using rice bran as feed for fish had much higher operating capital requirements than fertilizing the pond using manure because rice bran would have to be purchased from the rice mill. On the other hand, labor requirements were lower for feeding fish with rice bran than by fertilizing with manures. Thus, capital can be substituted for labor to intensify production as long as capital resources are available.

Green et al. (1994) evaluated 41 semi-intensive tilapia pond management systems in Honduras. Of these, only 16 systems had positive returns to land and management in 1992. The most profitable treatment included both organic fertilization plus supplemental feed.

However, commercial production systems that target a local market could be designed on a small scale. Engle (1987) showed in Panama that polyculture production designed to supply the local market could be more profitable than monoculture systems given the level of resource availability, export market infrastructure, and risk. Similarly, Hanley (1991) showed that tilapia production on a small scale that targeted the local market could be more profitable than other types of crop production.

Commercial, Small-Scale Production of Tilapia

Tilapia businesses have evolved in many countries beyond near-subsistence production to small-scale commercial production. Adequate levels of capital resources allow farmers to look for profit-maximizing levels of production as their primary objective instead of focusing on family subsistence with only supplemental sales.

Green et al. (1989), for example, evaluated tilapia yields in ponds fertilized with chicken litter, fresh cow manure, or chemical fertilizer. Results showed that the long-term profitability of small-scale commercial tilapia production was dependent on high chicken litter input. Recommendations were made that chicken litter should be applied weekly at 250 $kg \cdot ha^{-1}$, but subsequent research demonstrated that chicken litter supplemented with moderate inputs of nitrogen would increase primary productivity and fish growth (Teichert-Coddington and Green 2000). However, at high chicken litter application rates, early morning dissolved oxygen levels in the pond were often zero and tilapia survived by gulping oxygen-rich surface water (Teichert-Coddington and Green 1993a). Growth rate decreases when dissolved oxygen chronically drops below 10 to 20 percent of saturation even though tilapia survived (Teichert-Coddington and Green 1993b).

As tilapia production increased in Honduras, intensified production systems were implemented to be able to supply the developing markets (Teichert-Coddington and Green 2000). These systems rely on feed as a major component of the production system. Feed-based systems frequently are more profitable than systems that rely on fertilization alone, but are only possible if adequate capital resources are available with which to purchase feed.

Studies with various feeding and fertilization rates showed that heavy fertilization was most profitable when sufficient fish biomass was present to graze on augmented natural productivity (Teichert-Coddington and Green 2000). Positive returns to land and management in fed systems were obtained only at stocking rates of 2 m^{-2}. These production systems require a high-quality feed that would result in a market price that would require the fish to be sold in export markets.

Teichert-Coddington and Green (1993b) showed that aeration increased tilapia yield and individual final size. Thus, those firms that target U.S. fillet markets need to be large and managed intensively to be competitive.

Head et al. (1996) developed an enterprise budget analysis of a proposed commercial scale (20 ha) saltwater pond culture operation for Florida red tilapia in northern Puerto Rico. Breakeven price was \$3.86 kg^{-1} and suggested that the proposed operation was marginally feasible.

Hatchery Costs

Commercial production of tilapia requires cost-effective hatchery production of tilapia fingerlings. Watanabe et al. (1997) developed an enterprise budget analysis of a tilapia hatchery in the Bahamas. In this analysis, breakeven prices ranged from \$0.06 to \$0.16 per fry at an annual production level of 1 million fry. Cost savings occur at larger production levels due to moderate increases in feed, packaging, and shipping costs as production is increased.

Engle (1986) estimated the costs of producing tilapia fingerlings at \$0.017 per fingerling for pond-raised *O. niloticus* and \$0.071 per fingerling and \$0.143 per fingerling for hapa-produced and pond-produced hybrids, respectively. Bailey and Rakocy (1996) estimated the variable costs of producing Florida mixed-sex red tilapia fingerlings (hybrid *Oreochromis urolepis hornorum* × *O. mossambicus*). The variable costs decreased to \$0.16 per fingerling with sex-reversal technologies. Head et al. (1996) estimated a similar level of costs of sex-reversed fry (\$0.11 per fry).

Commercial Cage Production of Tilapia

Cages have been promoted as a low-investment option for raising fish like tilapia. The total investment required for the construction of some types of cages is less than that required for constructing ponds. Yet cage culture entails other costs and risks that must be taken into consideration.

Watanabe et al. (1997) judged that saltwater cage production of Florida red tilapia was not economically viable at the artisanal scale in Haiti. Moreover, the high financial risk associated with saltwater cage production on a commercial scale was due to the difficult access to inputs and to the unstable and undeveloped market conditions in Haiti.

Colombia, Brazil, and Honduras all have cage production of tilapia (Alceste 2001b). Mixed-sex tilapia can be raised in cages because cages disrupt tilapia breeding cycles. However, tilapia raised in cages are at a greater risk from losses due to poaching, are totally dependent on nutritionally complete feeds, and cannot cope with poor water quality in ponds.

In Colombia, cages are placed in reservoirs constructed behind hydroelectric dams, such as Hidroprado, Betania (Colombia), and Yojoa in Honduras. In Hidroprado in Colombia, there are an estimated 2,300 cages of roughly 2.25 m^3 each, whereas there are fewer but larger cages (50 to 90 m^3) in Betania.

On a smaller scale, up to 100 2.25 m^3 cages can be placed in 1 ha reservoirs that have been constructed primarily for irrigating sugar cane (Alceste 2001a). These cages produce up to 120 $kg \cdot m^{-3}$ of fish at costs of approximately \$1.20 to \$1.30 kg^{-1} (roundweight). At retail prices of \$1.50 to 1.75 kg^{-1}, cage culture in Colombia appears to be profitable.

Cages in Brazil are typically 6 to 18 m^3 in size and are placed in the many reservoirs in Brazil to raise tilapia for stocking reservoirs for fee-fishing. Cage farmers receive \$0.40 to \$1.10 kg^{-1} for tilapia with production costs above \$0.80 kg^{-1}.

Commercial Production of Tilapia for Export

There are several models of successful tilapia production to supply export markets in the United States. Honduras, the third largest source of fresh tilapia fillets to the United States, has used an intensive, monoculture, phased production system to produce export-quality tilapia. Green and Engle (2000) developed enterprise budgets for two scenarios for production in Honduras of export-quality tilapia for sale in U.S. markets: (1) producing tilapia on a 24-ha farm with the primary goal of exporting fillets to the United States and (2) smaller scale production of tilapia on a 6-ha farm to supply local markets in Honduras. Results indicated a margin of profit to operate in both the short and the long term. Export fillet prices could drop by 19 percent and the business would still be profitable in the long term. Short-term prices could drop by as much as 31 percent and the business could continue to operate for a period of time.

For domestic production, the break-even prices to cover variable costs alone and break-even prices to cover total costs were less than the market price, indicating that this type of production system remains profitable even with market price fluctuations of 21 percent in the short term and 7 percent in the long term. Smaller farms utilized less intensive production practices and targeted a market-size of 250 g, instead of the 300 to 400 g tilapia sold domestically from the larger farms. Budgets for smaller farms indicated that growing a 250 g fish for domestic sale with a less intensive management strategy than that employed by producers exporting tilapia is profitable.

A number of Ecuadorian shrimp farmers have converted shrimp ponds into tilapia production. Other than pond renovation costs to divide shrimp ponds into smaller sizes, no additional expense has been incurred for water supply for the polyculture systems. However, extruded feeds were necessary for tilapia and adds some additional costs for feed. Many of the major Ecuadorian producers have integrated vertically; this vertical integration may be one of the keys to Ecuador's success in tilapia production and export (Alceste 2001b).

The lack of well-developed market channels for a new species like tilapia has required farms to create their own. Many export market-oriented farms have constructed their own processing units. Figure 19.6 presents investment costs for a 62 ha integrated business in Central America. Pond construction is the largest cost, followed by land costs, pond production, equipment, and the water supply. The processing plant, including the purchase of two vehicles for marketing, resulted in an additional 43 percent investment cost.

Feeds and fertilizers constitute the single largest operating expense (Figure 19.7). This is followed by management, transportation costs, processing materials, and interest on operating capital. Labor, utilities, and fuel constitute relatively low costs in the production of tilapia for export.

Intensive Indoor Production of Tilapia in the United States

The most intensive level of tilapia production is in indoor recirculating systems (RAS). Although feed remains an important cost in indoor production, other costs assume additional importance. Management and the capital investment, or annual fixed costs become much more significant than

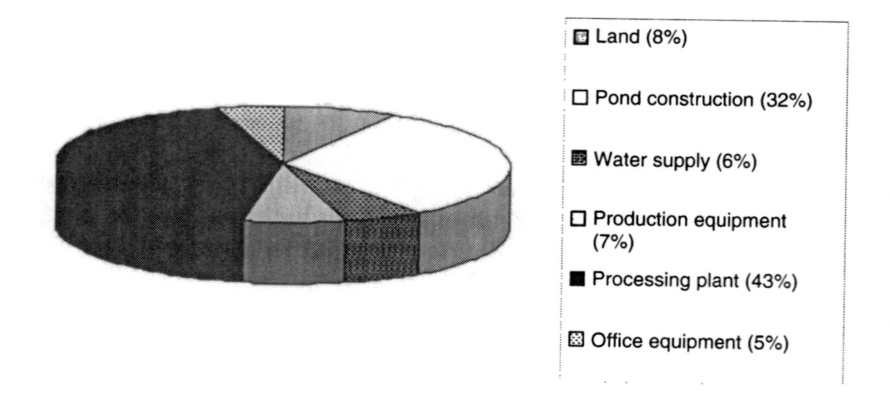

FIGURE 19.6. Investment costs for a 62-ha integrated tilapia pond-processing business. *Source:* Engle, unpublished data.

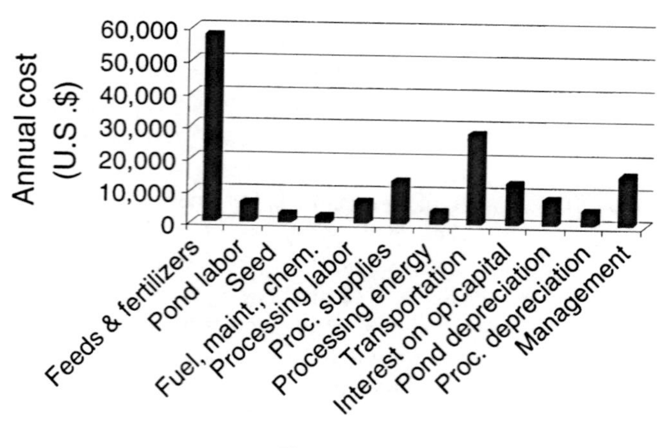

FIGURE 19.7. Annual cost components, variable and fixed, for a 62-ha integrated pond-processing business. *Source:* Engle, unpublished data.

other types of production costs in RAS as compared to pond-production systems.

Production cost estimates of tilapia in indoor systems in the United States are higher than those of tilapia production in other systems. O'Rourke (1996) published total costs of more than \$3.50 kg^{-1} in indoor recirculating systems. Lasordo and Westerman (1994) reported an initial investment requirement of \$4.34 kg^{-1} annual capacity for a bioeconomic model of an intermediate scale recirculating facility in North Carolina with an overall production cost of roughly \$2.79 kg^{-1}. When they examined the economic impact of a 10 percent increase in system capacity, total production cost estimates were reduced by 4.1 percent. Watanabe et al. (1997) provided cost estimates that equate to roughly \$2.58 kg^{-1} for whole Florida red tilapia produced in a 20-ha saltwater pond facility in Puerto Rico and \$2.33 kg^{-1} for whole fish from a 6,240 m^3 flow-through tank facility.

Wade et al. (1996) showed that indoor recirculating systems have significant economies of scale. This results in a strong incentive for indoor systems to expand production to operate at efficient (low) costs. Yet large systems require processing capability that need very large production volumes to meet the economies of scale in processing (Engle 1997a). In the absence of a tilapia processing sector, tilapia growers have relied on live and niche markets where they can capture higher prices. A significant challenge for the tilapia industry is this paradox in which the existing market can handle only small quantities at a time, but economies of scale require that efficient costs exist only at large sizes of businesses.

According to Lutz (2000), the lowest production costs occur in tropical ponds (\$1.63 kg^{-1}) and raceways (\$1.64 kg^{-1}), followed by ponds in the southern United States (\$2.27 kg^{-1}) and subtropical greenhouse systems (\$2.46 kg^{-1}). The highest costs of production were those in temperate indoor tanks (\$3.57 kg^{-1}). The costs in temperate indoor tanks were more than double the costs of tilapia production in ponds and raceways in the tropics. The more intensive levels of management of indoor systems result in higher costs that require a high-end, often niche market price to be profitable.

The production costs of tilapia raised in ponds and raceways in the tropics have transportation costs added before entering the market place in the United States. Air freight costs for seafood cargo from Central America and many Caribbean basin locations to Miami approximate \$2.20 kg^{-1}. An additional \$0.81 to \$0.88 kg^{-1} is needed to transport cargo from Miami to New Orleans. Wholesale prices for fresh and frozen imported tilapia fillets as low as \$6.95 and \$6.17 kg^{-1} suggest that production costs for tropical pond/raceway facilities may be lower than that estimated here.

These higher costs of production in indoor systems in the United States have raised questions as to whether or not U.S. growers can compete with imported fillets from the tropics. Clancy et al. (1994) indicated fresh fillet processing costs of \$0.21 to \$0.24 kg^{-1} for whole channel catfish *(Ictalurus punctatus)* in the United States. Costs for frozen fillet production increased to \$0.35 to \$0.42 kg^{-1} (of live weight). If processing costs for whole tilapia were similar, with a raw material factor of 2.94, costs of \$0.62 to \$0.71 kg^{-1}

of fresh fillet or \$1.03 to \$1.23 kg^{-1} of frozen fillets would be incurred. In the southern United States, wholesale prices for frozen imported tilapia fillets often are \$2.00 kg^{-1} less than those for fresh fillets. Posadas (2000) found that marketing margins in large wholesale markets are not sufficiently large for locally produced tilapia to be processed into fillets.

REFERENCES

Alceste, C.C. (2001a). Red tilapia cage farming in Colombia and Brazil. *Aquaculture Magazine* (September/October):82-86.

———. (2001b). Tilapia farming industry in Ecuador. *Aquaculture Magazine* (May/June):77-83.

Alston, D.E., S. Wiscovich, F. Richardson, and M. Nicolás (2000). *Tilapia culture in Puerto Rico and the Dominican Republic.* In B.A. Costa-Pierce and J.E. Rakocy (Eds.), *Tilapia aquaculture in the Americas* (Vol. 2, pp. 215-244). Baton Rouge, LA: The World Aquaculture Society.

American Tilapia Association (2001). *2000 outlook report.* University of Arizona.

Avault, J.W., and E.W. Shell (1966). Preliminary studies with the hybrid tilapia *Tilapia nilotica* × *Tilapia mossambica*. FAO World Symposium on Warm Water Pond Fish Culture. FR:IV/E-14.

Bailey, D.S., and J.E. Rakocy (1996). Management practices and costs for small-scale production of Florida red tilapia fingerlings. Book of Abstracts. Aquaculture America 1996. Annual Meeting of U.S. Chapter of World Aquaculture Society, Arlington, Texas.

Caraballo, D.M., and Y. Sadovy (1990). Overview of Puerto Rico's small scale fisheries statistics, 1988-1989. Corporation for the Development and Administration of the Marine, Lacustrine and Fluvial Resources of Puerto Rico. *Fisheries Research Laboratory Technical Report* 1(4):1-17.

Chakalall, B., and P. Noriega-Curtis (1988). Tilapia farming in Jamaica. Proceedings of the Forty-First Annual Gulf and Caribbean Fisheries Institute, Charleston, South Carolina.

———. (1992). Tilapia farming in Jamaica. *Gulf and Caribbean Fisheries Institute* 41:545-569.

Chiappe, D. (1998). Aquaculture expands its options. *Conapri Venezuela Now,* Year 3, N. 8, August 3.

Clancy, C.M., T.H. Spreen, D.J. Zimet, and S.O. Olowolayemo (1994). Analyzing production feasibility and market potential for Florida aquaculture catfish products. *Journal of the World Aquaculture Society* 25:250-260.

Collazo, J., and J.A. Calderon (1988). Status of fisheries in Puerto Rico, 1979-1982. Corporation for the Development and Administration of Marine, Lacustrine and Fluvial Resources of Puerto Rico. *Fisheries Research Laboratory Technical Report* 1(2):1-30.

Costa-Pierce, B.A., and J.E. Rakocy (1997). *Tilapia aquaculture in the Americas,* Volume 1. Baton Rouge, LA: The World Aquaculture Society.

Crawford, K.W., D.R. Dunseth, C.R. Engle, M.L. Hopkins, E.W. McCoy, and R.O. Smitherman (1978). Marketing tilapia and Chinese carps. In R.O. Smitherman, W.L. Shelton, and J.H. Grover (Eds.), *Symposium on culture of exotic fishes* (pp. 240-257). Atlanta, GA: Fish Culture Section, American Fisheries Society.

Edwards, P., H. Demaine, and S. Komolmarl (1991). Toward the improvement of fish culture by small-scale farmers in Northeast Thailand. *Journal of Asian Farm System Association* 1:287-302.

Engle, C.R. (1978). Preliminary market tests of several exotic fish species. Master's thesis, Auburn University, Alabama.

———. (1986). Costos de produccion de semilla de entrega en Panama. *Revista Latinoamericana de Acuicultura* 30:43-57.

———. (1987). Análisis Económico de la Produccion Comercial de Tilapia, colossoma y *Macrobrachium rosenbergii* en Mono- y Policultivo en Panama. *Revista Latinoamericana de Acuicultura* 33:7-25.

———. (1997a). Economics of tilapia aquaculture. In B.A. Costa-Pierce, and J.E. Rakocy (Eds.), *Tilapia aquaculture in the Americas* (Vol. 1, pp. 229-243). Baton Rouge, LA: The World Aquaculture Society.

———. (1997b). Marketing tilapias. In B.A. Costa-Pierce, and J.E. Rakocy (Eds.), *Tilapia aquaculture in the Americas* (Vol. 1, pp. 244-258). Baton Rouge, LA: The World Aquaculture Society.

———. (1997c). Optimal resource allocation by fish farmers in Rwanda. *Journal of Applied Aquaculture* 7(1):1-17.

———. (2003). Potential for supermarket outlets for tilapia in Nicaragua. PD/A CRSP Research Report 03-190. Corvallis, OR: Oregon State University.

Engle, C.R., R. Balakrishnan, T.R. Hanson, and J.J. Molnar (1997). Economic considerations. In H.S. Egna and C.E. Boyd (Eds.), *Dynamics of pond aquaculture* (pp. 377-396). New York: CRC Press LLC.

Engle, C.R., M. Brewster, and F. Hitayezu (1993). An economic analysis of fish production in a subsistence agricultural economy: The case of Rwanda. *Journal of Aquaculture in the Tropics* 8:151-165.

Engle, C.R., and I. Neira (2003). Potential for open-air fish market outlets for tilapia in Nicaragua. PD/A CRSP Research Report 03-194. Corvallis, OR: Oregon State University.

Engle, C.R., and M. Skladany (1992). *The economic benefit of chicken manure utilization in fish production in Thailand.* Pond Dynamics/Aquaculture CRSP Research Report 92-45. Corvallis: Office of International Research and Development, Oregon State University.

Fitzsimmons, K. (2000). Tilapia aquaculture in Mexico. In B.A. Costa-Pierce and J.E. Rakocy (Eds.), *Tilapia aquaculture in the Americas* (Vol. 2, pp. 171-183). Baton Rouge, LA: The World Aquaculture Society.

———. (2001). Tilapia markets in the Americas, 2001 and beyond. Memoria: Sesiones de Tilapia, 6to. Simposio Centroamericano de Acuacultura, Asociación Nacional de Acuicultores de Honduras, Tegucigalpa, Honduras.

Fitzsimmons, K., and B.C. Posadas (1997). Consumer demand for tilapia products in the U.S. and the effects on local markets in exporting countries. In K. Fitzsimmons (Ed.), *Tilapia aquaculture: Proceedings from the Fourth*

International Symposium on Tilapia in Aquaculture (Vol. 2, pp. 613-632). Ithaca, NY: Northeast Regional Agricultural Engineering Service.

Food and Agriculture Organization of the United Nations (FAO) (1998). *Aquaculture Production Statistics.* Rome, Italy: FAO.

Fúnez, O., I. Neira, and C. Engle (2003a). *Open-air fish market outlets for tilapia in Honduras: An overview of survey results.* PD/A CRSP Research Report 03-193. Corvallis, OR: Oregon State University.

———. (2003b). *Supermarket outlets for tilapia in Honduras: An overview of survey results.* PD/A CRSP Research Report 03-189. Corvallis OR: Oregon State University.

Galbreath, P.F., and T.A. Barnes (1981). Consumer preference for color and size of tilapia sold in supermarkets. *Proceedings Catfish Farmers of America Research Workshop* 3:47-48.

Green, B.W., and C.R. Engle (2000). Commercial tilapia aquaculture in Honduras. In B.A. Costa-Pierce and J.E. Rakocy (Eds.), *Tilapia aquaculture in the Americas* (Vol. 1, pp. 151-170). Baton Rouge, LA: The World Aquaculture Society.

Green, B.W., R.P. Phelps, and H.R. Alvarenga (1989). The effect of manures and chemical fertilizers on the production of *Oreochromis niloticus* in earthen ponds. *Aquaculture* 76:37-42.

Green, B.W., D.R. Teichert-Coddington, and T.R. Hanson (1994). *Development of semi-intensive aquaculture technologies in Honduras: Summary of freshwater aquacultural research conducted from 1983 to 1992.* Research and Development Series Number 39. International Center for Aquaculture and Aquatic Environments, Auburn University, Alabama.

Greenfield, J.E., E.R. Lira, and J.W. Jensen (1974). *Economic evaluation of tilapia hybrid culture in northeast Brazil.* FAO/CARPAS Symposium on Aquaculture in Latin America. Food and Agriculture Organization of the United Nations, Rome, Italy.

Hanley, F. (1991). Freshwater tilapia culture in Jamaica. *World Aquaculture* 22(1): 42-48.

———. (2000). Tilapia aquaculture in Jamaica. In B.A. Costa-Pierce and J.E. Rakocy (Eds.), *Tilapia aquaculture in the Americas* (Vol. 2, pp. 204-214). Baton Rouge, LA: The World Aquaculture Society.

Harvey, D. (2002). *Aquaculture outlook.* Electronic outlook report. Washington, DC: Economic Research Service, U.S. Department of Agriculture.

Hatch, U., and C. Engle (1987). Economic analysis of aquaculture as a component of integrated agro-aquaculture systems: Some evidence from Panama. *Journal of Aquaculture in the Tropics* 2:93-105.

Hatch, L.U., and T.R. Hanson (1992). *Economic viability of farm diversification through tropical freshwater aquaculture in less developed countries.* San Francisco, CA: Westview Press.

Head, W.D., A. Zerbi, and W.O. Watanabe (1996). Economic evaluation of commercial-scale, saltwater pond production of Florida red tilapia in Puerto Rico. *Journal of the World Aquaculture Society* 27:275-289.

Homziak, J., and B.C. Posadas (1992). A preliminary survey of tilapia markets in North America. *Proceedings of the Annual Gulf and Caribbean Fisheries Institute* 42:83-102.

Jory, D., T. Cabrera, B. Polanco, J. Millán, J. Rosas, E. García, R. Sanchéz, M. Useche, R. Agudo, and C. Alceste (1999). Aquaculture in Venezuela: Culture status and perspectives. *World Aquaculture* 30(3):20-26.

Lasordo, T.M., and P. Westerman (1994). An analysis of biological, economic and engineering factors affecting the costs of fish production in recirculating aquaculture systems. *Journal of the World Aquaculture Society* 25:193-203.

Lauenstein, P. (1978). Intensive culture of tilapia with geothermally heated water. In R.O. Smitherman, W.L. Shelton, and J.H. Grover (Eds.), *Symposium on Culture of Exotic Fishes, Fish Culture Section* (pp. 82-85). Atlanta, GA: American Fisheries Society.

Little, D. (1995). The development of small-scale poultry-fish integration in northeast Thailand: Potential and constraints. In J. Symoens and J. Micha (Eds.), *The management of integrated fresh water agro-piscicultural eco-systems in tropical areas: Proceedings of seminar* (pp. 25-276). Brussels, Belgium.

Lovshin, L. (2000). Tilapia culture in Brazil. In B.A. Costa-Pierce and J.E. Rakocy (Eds.), *Tilapia aquaculture in the Americas* (Vol. 2, pp. 133-140). Baton Rouge, LA: The World Aquaculture Society.

Lutz, C.G. (2000). Production economics and potential competitive dynamics of commercial tilapia culture in the Americas. In B.A. Costa-Pierce and J.E. Rakocy (Eds.), *Tilapia aquaculture in the Americas* (Vol. 2, pp. 119-132). Baton Rouge, LA: The World Aquaculture Society.

Martin, N.B., J.S. Scorvo Filho, E.G. Sanches, P.F.C. Novato, and L.S. Ayrosa (1995). Custos e retornos na piscicultura em São Paulo. *Informação Econômicas* 25:9-47.

Monestime, D., I. Neira, O. Fúnez, and C.R. Engle (2003). *The potential market for tilapia in Honduras: Results of a survey of restaurants.* PD/A CRSP Research Report 03-191. Corvallis, OR: Oregon State University.

Neira, I., C.R. Engle, and K. Quagrainie (2003). Potential restaurant markets for farm-raised tilapia in Nicaragua. *Aquaculture Economics and Management* 7(3/4):1-17.

O'Rourke, P.D. (1996). The economics of recirculating aquaculture systems. In G.S. Libey and M.B. Timmons (Eds.), *Successes and failures in commercial recirculating aquaculture* (pp. 61-78). Ithaca, NY: Northeast Regional Agricultural Engineering Service.

Popma, T.J., and F. Rodriquez B. (2000). Tilapia aquaculture in Colombia. In B.A. Costa-Pierce and J.E. Rakocy (Eds.), *Tilapia aquaculture in the Americas* (Vol. 2, pp. 141-150). Baton Rouge, LA: The World Aquaculture Society.

Posadas, B.C. (2000). Tilapia marketing in the northern Gulf of Mexico region. In B.A. Costa-Pierce and J.E. Rakocy (Eds.), *Tilapia aquaculture in the Americas* (Vol. 2, pp. 91-99). Baton Rouge, LA: The World Aquaculture Society.

Redmayne, P. (1993). *Tilapia update. Seafood Leader.* Seattle, WA: Waterfront Press Company.

Sandifer, P.A. (1991). Species with aquaculture potential for the Caribbean. In J.A. Hargreaves, and D.E. Alston (Eds.), *Status and potential of aquaculture in the Caribbean* (pp. 30-60). Baton Rouge, LA: The World Aquaculture Society.

Seafood Business (2002). Buyer's guide: Tilapia. *Seafood Business* September:46.

Setboonsarng, S. and P. Edwards (1998). An assessment of alternative strategies for the integration of pond aquaculture into the small-scale farming system of North-east Thailand. *Aquaculture Economics & Management* 2(3):151-162.

Shell, E.W. (1966). Monosex culture of male *Tilapia nilotica* (Linn.) in ponds stocked at three rates. FAO World Symposium on Warm Water Fish Culture. FR:V/E-5.

Smith, L.J., and S. Peterson (1982). *Aquaculture development in less developed countries: Social, economic, and political problems*. Boulder, CO: Westview Press.

Stutzman, C. (1995). Annual tilapia situation and outlook report. In G.J. Gallagher (Ed.), *Aquaculture magazine* (pp. 6-11). Asheville, NC.

Teichert-Coddington, D.R., and B.W. Green (1993a). Influence of daylight and incubation interval on water column respiration in tropical fish ponds. *Hydrobiologia* 250:159-165.

———. (1993b). Yield improvement through maintenance of minimal oxygen concentration in experimental grow-out ponds in Honduras. *Aquaculture* 118:63-71.

———. (2000). Experimental and commercial culture of tilapia in Honduras. In B.A. Costa-Pierce and J.E. Rakocy (Eds.), *Tilapia aquaculture in the Americas* (Vol. 1, pp. 142-162). Baton Rouge, LA: The World Aquaculture Society.

Teichert-Coddington, D.R., Green B.W., and R.P. Phelps (1992). Influence of site and season on water quality and tilapia production in Panama and Honduras. *Aquaculture* 105:297-314.

Tucker, J.W., Jr., and D.E. Jory (1991). Marine fish culture in the Caribbean region. *World Aquaculture* 22(1):10-27.

U.S. Department of Commerce (1999). *U.S. imports of tilapia by product form.* Washington, DC: Bureau of Census, U.S. Department of Commerce.

Wade, E.M., S.T. Summerfelt, and J.A. Hankins (1996). Economies of scale in recycle systems. In *Successes and failures in commercial recirculating aquaculture*. Conference Proceedings NRAES-98. Ithaca, NY: Northeast Regional Agricultural Engineering Service.

Watanabe, W.O., B.L. Olla, R.I. Wicklund, and W.D. Head (1997). Saltwater culture of the Florida red tilapia and other saline-tolerant tilapias: A review. In B.A. Costa-Pierce and J.E. Rakocy (Eds.), *Tilapia aquaculture in the Americas* (Vol. 1, pp. 55-141). Baton Rouge, LA: The World Aquaculture Society.

Watanabe, W.O., S.J. Smith, R.I. Wicklund, and B.L. Olla (1991). Evaluation of methods for prevention and control of monogenetic trematode *(Neobenedenia melleni)* parasitosis of Florida red tilapia reared in seawater pools. *Journal of the World Aquaculture Society* 22:63A (abstract).

Watanabe, W.O., R.I. Wicklund, B.L. Olla, D.H. Ernst, and L.J. Ellingson (1989). Potential for saltwater tilapia culture in the Caribbean. *Gulf and Caribbean Fisheries Institute* 39:435-445; *Journal of the World Aquaculture Society* 20:223-229.

World Bank (1997). *Mexico aquaculture development project.* Report 16476-ME. Washington, DC: World Bank.

Index

Note to the reader: Page numbers followed by the letter "f" indicate figures; those followed by the letter "t" indicate tables.

doi:10.1300/5513_20

LaVergne, TN USA
14 September 2010

196891LV00009B/1/P

9 781560 223184